THE MOLECULAR NUTRITION OF AMINO ACIDS AND PROTEINS

THE MOLECULAR NUTRITION OF AMINO ACIDS AND PROTEINS

A Volume in the Molecular Nutrition Series

Edited by

Dominique Dardevet
Institut National de la Recherche Agronomique (INRA), Ceyrat, France

AMSTERDAM • BOSTON • HEIDELBERG • LONDON
NEW YORK • OXFORD • PARIS • SAN DIEGO
SAN FRANCISCO • SINGAPORE • SYDNEY • TOKYO

Academic Press is an imprint of Elsevier

Academic Press is an imprint of Elsevier
125 London Wall, London EC2Y 5AS, UK
525 B Street, Suite 1800, San Diego, CA 92101-4495, USA
50 Hampshire Street, 5th Floor, Cambridge, MA 02139, USA
The Boulevard, Langford Lane, Kidlington, Oxford OX5 1GB, UK

British Library Cataloguing-in-Publication Data
A catalogue record for this book is available from the British Library.

Library of Congress Cataloging-in-Publication Data
A catalog record for this book is available from the Library of Congress.

ISBN: 978-0-12-802167-5

For Information on all Academic Press publications
visit our website at http://www.elsevier.com/

Publisher: Nikki Levy
Acquisition Editor: Megan Ball
Editorial Project Manager: Karen Miller
Production Project Manager: Caroline Johnson
Designer: Victoria Pearson

Typeset by MPS Limited, Chennai, India

Contents

I
GENERAL AND INTRODUCTORY ASPECTS

II
CELLULAR ASPECTS OF PROTEIN AND AMINO ACIDS METABOLISM IN ANABOLIC AND CATABOLIC SITUATIONS

13. Amino Acids, Protein, and the Gastrointestinal Tract

M.J. BRUINS, K.V.K. KOELFAT AND P.B. SOETERS

14. Regulation of Macroautophagy by Nutrients and Metabolites

S. LORIN, S. PATTINGRE, A.J. MEIJER AND P. CODOGNO

III

CELLULAR AND MOLECULAR ACTIONS OF AMINO ACIDS IN NON PROTEIN METABOLISM

15. Dietary Protein and Colonic Microbiota: Molecular Aspects

G. BOUDRY, I. LE HUËROU-LURON AND C. MICHEL

16. Control of Food Intake by Dietary Amino Acids and Proteins: Molecular and Cellular Aspects

G. FROMENTIN, N. DARCEL, C. CHAUMONTET, P. EVEN, D. TOMÉ AND C. GAUDICHON

17. Dietary Protein and Hepatic Glucose Production

C. GAUDICHON, D. AZZOUT-MARNICHE AND D. TOMÉ

18. Impact of Dietary Proteins on Energy Balance, Insulin Sensitivity and Glucose Homeostasis: From Proteins to Peptides to Amino Acids

G. CHEVRIER, P. MITCHELL, M.-S. BEAUDOIN AND A. MARETTE

19. Sulfur Amino Acids Metabolism From Protein Synthesis to Glutathione

G. COURTNEY-MARTIN AND P.B. PENCHARZ

IV

DIETARY AMINO ACID AND PROTEIN ON GENE EXPRESSION

20. Adaptation to Amino Acid Availability: Role of GCN2 in the Regulation of Physiological Functions and in Pathological Disorders

J. AVEROUS, C. JOUSSE, A.-C. MAURIN, A. BRUHAT AND P. FAFOURNOUX

21. Amino Acid-Related Diseases

I. KNERR

22. Genes in Skeletal Muscle Remodeling and Impact of Feeding: Molecular and Cellular Aspects

Y.-W. CHEN, M.D. BARBERIO AND M.J. HUBAL

23. Brain Amino Acid Sensing: The Use of a Rodent Model of Protein-Malnutrition, Lysine Deficiency

K. TORII AND T. TSURUGIZAWA

List of Contributors

J.M. Argilés Cancer Research Group, Departament de Bioquímica i Biologia Molecular, Facultat de Biologia, Universitat de Barcelona, Barcelona, Spain; Institut de Biomedicina de la Universitat de Barcelona, Barcelona, Spain

P.J. Atherton MRC-ARUK Centre for Musculoskeletal Ageing Research, School of Medicine, University of Nottingham, Nottingham, United Kingdom

D. Attaix Clermont Université, Université d'Auvergne, Unité de Nutrition Humaine, Clermont-Ferrand, France; INRA, UMR 1019, UNH, CRNH Auvergne, Saint Genès Champanelle, France

J. Averous Unité de Nutrition Humaine, UMR 1019, INRA, Université d'Auvergne, Centre INRA de Clermont-Ferrand-Theix, Saint Genès Champanelle, France

D. Azzout-Marniche UMR Physiologie de la Nutrition et du Comportement Alimentaire, AgroParisTech, INRA, Université Paris Saclay, Paris, France

M.D. Barberio Center for Genetic Medicine Research, Children's National Healthy System, Washington DC, USA

E. Barreiro Pulmonology Department, Muscle and Lung Cancer Research Group, IMIM-Hospital del Mar, Parc de Salut Mar, Health and Experimental Sciences Department (CEXS), Universitat Pompeu Fabra (UPF), Barcelona Biomedical Research Park (PRBB), Barcelona, Spain; Centro de Investigación en Red de Enfermedades Respiratorias (CIBERES), Instituto de Salud Carlos III (ISCIII), Barcelona, Spain

M.-S. Beaudoin Department of Medicine, Faculty of Medicine, Cardiology Axis of the Québec Heart and Lung Institute, Québec, QC, Canada; Institute of Nutrition and Functional Foods, Laval University, Québec, QC, Canada

D. Béchet Clermont Université, Université d'Auvergne, Unité de Nutrition Humaine, Clermont-Ferrand, France; INRA, UMR 1019, UNH, CRNH Auvergne, Saint Genès Champanelle, France

Y. Boirie Clermont Université, Université d'Auvergne, Unité de Nutrition Humaine, Clermont-Ferrand, France; INRA, UMR 1019, UNH, CRNH Auvergne, Clermont-Ferrand, France; CHU Clermont-Ferrand, service de Nutrition Clinique, Clermont-Ferrand, France

G. Boudry INRA UR1341 ADNC, St-Gilles, France

R. Boutrou INRA, UMR 1253, Science et Technologie du lait et de l'œuf, Rennes, France

A. Bruhat Unité de Nutrition Humaine, UMR 1019, INRA, Université d'Auvergne, Centre INRA de Clermont-Ferrand-Theix, Saint Genès Champanelle, France

M.J. Bruins The Hague, The Netherlands

S. Busquets Cancer Research Group, Departament de Bioquímica i Biologia Molecular, Facultat de Biologia, Universitat de Barcelona, Barcelona, Spain; Institut de Biomedicina de la Universitat de Barcelona, Barcelona, Spain

J.W. Carbone School of Health Sciences, Eastern Michigan University, Ypsilanti, MI, United States

C. Chaumontet UMR Physiologie de la Nutrition et du Comportement Alimentaire, AgroParisTech, INRA, Université Paris-Saclay, Paris, France

Y.-W. Chen Department of Integrative Systems Biology, George Washington University, Washington DC, USA; Center for Genetic Medicine Research, Children's National Healthy System, Washington DC, USA

G. Chevrier Department of Medicine, Faculty of Medicine, Cardiology Axis of the Québec Heart and Lung Institute, Québec, QC, Canada; Institute of Nutrition and Functional Foods, Laval University, Québec, QC, Canada

P. Codogno INEM, Institut Necker Enfants-Malades, Paris, France; INSERM U1151-CNRS UMR 8253, Paris, France; Université Paris Descartes, Paris, France

L. Combaret Clermont Université, Université d'Auvergne, Unité de Nutrition Humaine, Clermont-Ferrand, France; INRA, UMR 1019, UNH, CRNH Auvergne, Saint Genès Champanelle, France

G. Courtney-Martin Faculty of Kinesiology & Physical Education, Department of Clinical Dietetics, University of Toronto, The Hospital for Sick Children, Toronto, ON, Canada

N. Darcel UMR Physiologie de la Nutrition et du Comportement Alimentaire, AgroParisTech, INRA, Université Paris-Saclay, Paris, France

E.L. Dillon Department of Internal Medicine, Division of Endocrinology and Metabolism, The University of Texas Medical Branch, Galveston, TX, United States

C. Domingues-Faria Clermont Université, Université d'Auvergne, Unité de Nutrition Humaine, Clermont-Ferrand, France; INRA, UMR 1019, UNH, CRNH Auvergne, Clermont-Ferrand, France

P. Even UMR Physiologie de la Nutrition et du Comportement Alimentaire, AgroParisTech, INRA, Université Paris-Saclay, Paris, France

P. Fafournoux Unité de Nutrition Humaine, UMR 1019, INRA, Université d'Auvergne, Centre INRA de Clermont-Ferrand-Theix, Saint Genès Champanelle, France

G. Fromentin UMR Physiologie de la Nutrition et du Comportement Alimentaire, AgroParisTech, INRA, Université Paris-Saclay, Paris, France

C. Gaudichon UMR Physiologie de la Nutrition et du Comportement Alimentaire, AgroParisTech, INRA, Université Paris-Saclay, Paris, France

J. Gea Pulmonology Department, Muscle and Lung Cancer Research Group, IMIM-Hospital del Mar, Parc de Salut Mar, Health and Experimental Sciences Department (CEXS), Universitat Pompeu Fabra (UPF), Barcelona Biomedical Research Park (PRBB), Barcelona, Spain; Centro de Investigación en Red de Enfermedades Respiratorias (CIBERES), Instituto de Salud Carlos III (ISCIII), Barcelona, Spain

C. Guillet Clermont Université, Université d'Auvergne, Unité de Nutrition Humaine, Clermont-Ferrand, France; INRA, UMR 1019, UNH, CRNH Auvergne, Clermont-Ferrand, France

M.J. Hubal Center for Genetic Medicine Research, Children's National Healthy System, Washington DC, USA; Department of Exercise and Nutrition Sciences, George Washington University, Washington DC, USA

C. Jousse Unité de Nutrition Humaine, UMR 1019, INRA, Université d'Auvergne, Centre INRA de Clermont-Ferrand-Theix, Saint Genès Champanelle, France

I. Knerr National Centre for Inherited Metabolic Disorders, Temple Street Children's University Hospital, Dublin, Ireland

K.V.K. Koelfat Maastricht University Medical Center, Maastricht, The Netherlands

I. Le Huërou-Luron INRA UR1341 ADNC, St-Gilles, France

F.J. López-Soriano Cancer Research Group, Departament de Bioquímica i Biologia Molecular, Facultat de Biologia, Universitat de Barcelona, Barcelona, Spain; Institut de Biomedicina de la Universitat de Barcelona, Barcelona, Spain

S. Lorin Faculté de Pharmacie, Université Paris-Saclay, Châtenay-Malabry, France; INSERM UMR-S-1193, Châtenay-Malabry, France

A. Marette Department of Medicine, Faculty of Medicine, Cardiology Axis of the Québec Heart and Lung Institute, Québec, QC, Canada; Institute of Nutrition and Functional Foods, Laval University, Québec, QC, Canada

L.M. Margolis Military Nutrition Division, US Army Research Institute of Environmental Medicine, Natick, MA, United States

F. Mariotti UMR Physiologie de la Nutrition et du Comportement Alimentaire, AgroParisTech, INRA, Université Paris-Saclay, Paris, France

A.-C. Maurin Unité de Nutrition Humaine, UMR 1019, INRA, Université d'Auvergne, Centre INRA de Clermont-Ferrand-Theix, Saint Genès Champanelle, France

C. McGlory Exercise Metabolism Research Group, Department of Kinesiology, McMaster University, Hamilton, ON, Canada

A.J. Meijer Department of Medical Biochemistry, Academic Medical Center, University of Amsterdam, Amsterdam, The Netherlands

C. Michel INRA UMR1280 PhAN, Nantes, France

P. Mitchell Department of Medicine, Faculty of Medicine, Cardiology Axis of the Québec Heart and Lung Institute, Québec, QC, Canada; Institute of Nutrition and Functional Foods, Laval University, Québec, QC, Canada

S.M. Pasiakos Military Nutrition Division, US Army Research Institute of Environmental Medicine, Natick, MA, United States

S. Pattingre IRCM, Institut de Recherche en Cancérologie de Montpellier, Montpellier, France; INSERM, U1194, Montpellier, France; Université de Montpellier, Montpellier, France; Institut régional du Cancer de Montpellier, Montpellier, France

P.B. Pencharz Department of Paediatrics and Nutritional Sciences (Emeritus), Senior Scientist Research Institute, University of Toronto, The Hospital for Sick Children, Toronto, ON, Canada

S.M. Phillips Exercise Metabolism Research Group, Department of Kinesiology, McMaster University, Hamilton, ON, Canada

C. Polge Clermont Université, Université d'Auvergne, Unité de Nutrition Humaine, Clermont-Ferrand, France; INRA, UMR 1019, UNH, CRNH Auvergne, Saint Genès Champanelle, France

D. Rémond INRA, UMR 1019-Unité de Nutrition Humaine, St Genès-Champanelle, France

I. Savary-Auzeloux INRA, UMR 1019-Unité de Nutrition Humaine, St Genès-Champanelle, France

K. Smith MRC-ARUK Centre for Musculoskeletal Ageing Research, School of Medicine, University of Nottingham, Nottingham, United Kingdom

P.B. Soeters Maastricht University Medical Center, Maastricht, The Netherlands

D. Taillandier Clermont Université, Université d'Auvergne, Unité de Nutrition Humaine, Clermont-Ferrand, France; INRA, UMR 1019, UNH, CRNH Auvergne, Saint Genès Champanelle, France

P.M. Taylor Division of Cell Signalling & Immunology, School of Life Sciences, University of Dundee, Sir James Black Centre, Dundee, United Kingdom

D. Tomé UMR Physiologie de la Nutrition et du Comportement Alimentaire, AgroParisTech, INRA, Université Paris-Saclay, Paris, France

K. Torii Torii Nutrient-Stasis Institute, Inc., Tokyo, Japan

T. Tsurugizawa Neurospin, Commissariat à l'Energie Atomique et aux Energies Alternatives, Gif-sur-Yvette, France

S. Walrand Clermont Université, Université d'Auvergne, Unité de Nutrition Humaine, Clermont-Ferrand, France; INRA, UMR 1019, UNH, CRNH Auvergne, Clermont-Ferrand, France

P.J.M. Weijs Department of Nutrition and Dietetics, Internal Medicine, VU University Medical Center; Department of Intensive Care Medicine, VU University Medical Center; Department of Nutrition and Dietetics, School of Sports and Nutrition, Amsterdam University of Applied Sciences, Amsterdam, The Netherlands

D.J. Wilkinson MRC-ARUK Centre for Musculoskeletal Ageing Research, School of Medicine, University of Nottingham, Nottingham, United Kingdom

Preface

In this series on *Molecular Nutrition*, the editors of each book aim to disseminate important material pertaining to molecular nutrition in its broadest sense. The coverage ranges from molecular aspects to whole organs, and the impact of nutrition or malnutrition on individuals and whole communities. It includes concepts, policy, preclinical studies, and clinical investigations relating to molecular nutrition. The subject areas include molecular mechanisms, polymorphisms, SNPs, genomic-wide analysis, genotypes, gene expression, genetic modifications, and many other aspects. Information given in the *Molecular Nutrition* series relates to national, international, and global issues.

A major feature of the series that sets it apart from other texts is the initiative to bridge the transintellectual divide so that it is suitable for novices and experts alike. It embraces traditional and nontraditional formats of nutritional sciences in different ways. Each book in the series has both overviews and detailed and focused chapters.

Molecular Nutrition is designed for nutritionists, dieticians, educationalists, health experts, epidemiologists, and health-related professionals such as chemists. It is also suitable for students, graduates, postgraduates, researchers, lecturers, teachers, and professors. Contributors are national or international experts, many of whom are from world-renowned institutions or universities. It is intended to be an authoritative text covering nutrition at the molecular level.

V.R. Preedy
Series Editor

SECTION I

GENERAL AND INTRODUCTORY ASPECTS

C H A P T E R

1

Bioactive Peptides Derived From Food Proteins

D. Rémond[1], I. Savary-Auzeloux[1] and R. Boutrou[2]

[1]INRA, UMR 1019-Unité de Nutrition Humaine, St Genès-Champanelle, France

[2]INRA, UMR 1253, Science et Technologie du lait et de l'œuf, Rennes, France

The value of dietary proteins is classically assessed using amino acid composition and protein digestibility (Leser, 2013). However, other parameters, such as their digestion rate (Dangin et al., 2002) or their potential to release bioactive peptides during digestion (Kitts and Weiler, 2003), would be of interest to fully describe dietary proteins value. The term bioactive peptide was mentioned for the first time by Mellander and Isaksson in 1950 (Mellander, 1950) who observed that casein phosphorylated peptides were favoring calcium binding in bones of children suffering from rachitis. In 1979, Zioudrou et al. (1979) showed an opioid effect of peptides derived from gluten hydrolysis. Since then, a large spectrum of studies has been devoted to bioactive peptides (also called functional peptides) and their potential beneficial effect on human health and metabolism, with effects on digestive, immune, cardiovascular, and nervous systems. Many bioactive peptides have been discovered in foods from both animal or plant origin. Actually the largest part of the investigation has been carried out on milk proteins (Nagpal et al., 2011; Boutrou et al., 2015). Bioactive peptides generally correspond to molecules with fewer than 20 amino acids (down to two), but several bigger molecules, such as caseinomacropeptide, have been equally identified as bioactive peptides. Inactive within their precursor proteins, bioactive peptides have to be released by proteolysis in order to become functional. Any food protein source can provide bioactive peptides. Apart from milk and milk products, bioactive peptides have also been isolated from hydrolysates of proteins from egg, fish, cereals, and legumes. These peptides can be produced directly in the food by the action of endogenous proteases in various food technological processing, such as milk fermentation, or meat ripening and cooking, but also can be already present in the ingested food (eg, glutathione, carnosine, or peptides produced during food processing implying fermentations). They can also be generated in vitro by the use of exogenous proteases. In this last case, the peptides should be resistant as much as possible to intestinal digestion to be able to trigger a biological effect. However, most bioactive peptides are formed during digestion in the body.

In this chapter, we present the main biological activities attributed to peptides derived from food proteins, the mechanisms by which they are produced in the digestive tract, and potentially absorbed across its wall/barrier.

1.1. PHYSIOLOGICAL EFFECTS OF FOOD-DERIVED PEPTIDES

1. Impact on the digestive tract

   Once released in the digestive tract, peptides derived from food proteins can act on digestive processes (secretions and transit) or modulate nutrients absorption (Shimizu, 2004).

   a. Regulation of digestion

      The potential involvement of food-derived peptides on the regulation of digestive processes can be explained partially and indirectly via the secretion of a gut hormone, cholecystokinin (CCK), known to stimulate biliary and pancreatic secretion, and inhibit gastric secretion of enzymes. Furthermore, this hormone increases intestinal motility, inhibits gastric emptying, and is considered as a strong anorexigenic

The Molecular Nutrition of Amino Acids and Proteins.
DOI: http://dx.doi.org/10.1016/B978-0-12-802167-5.00001-3

gut hormone. Casein, ovalbumin, soya, meat, and gluten enzymatic hydrolysates have been shown to stimulate CCK secretion in perfused rat intestine (Cuber et al., 1990), isolated intestinal cells (Nishi et al., 2001), or tumorous intestinal cells (Nemoz-Gaillard et al., 1998), showing a direct action of some compounds issued from these hydrolysates. Some of the corresponding bioactive peptides have been identified. For instance, the caseinomacropeptide (obtained through hydrolysis of κ-casein by gastric proteinases) or the derived peptides were shown to stimulate CCK (Yvon et al., 1994) and pancreatic secretions (Pedersen et al., 2000) and to inhibit gastric acid secretion (Yvon et al., 1994). Furthermore, CCK antagonists have also been shown to inhibit the satietogenic effect of CCK induced by a casein meal (Froetschel et al., 2001). In soy hydrolysate, the 51–63 fragment of β-conglycinin, presenting a high affinity for intestinal brush border cells, has also been shown to induce an increase of CCK secretion and hence indirectly impact on appetite control (Nishi et al., 2003). Again, this latter effect is blunted by administration of a CCK antagonist (Nishi et al., 2003). A similar effect was reported for the tripeptide RIY that is released from the rapeseed napin.

Food-derived peptides could also modulate the gastric emptying rate and intestinal food transit via an activation of the opioid receptors that are present in the intestine. Indeed it was shown in rats that β-casomorphins (obtained from α_{S1}- and β-casein) slow down gastric emptying, this effect being blunted by treatment with naloxone, an opioid antagonist (Daniel et al., 1990).

In addition, some food-derived peptides could also interact with intestinal barrier function whose role is to selectively allow the absorption of nutrients and ions while preventing the influx of microorganisms from the intestinal lumen (Martinez-Augustin et al., 2014). For example, the β-casein fragment (94–123) evidenced in yogurts is able to specifically stimulate MUC2 production, a crucial factor of intestinal protection (Plaisancie et al., 2013, 2015).

b. Modulation of nutrients uptake

This mainly concerns the capacity of some peptides, such as caseinophosphopeptides (CPPs), to favor the uptake of micronutrients, such as minerals. CPPs are obtained from casein by trypsin or chymotrypsin hydrolysis (Sato et al., 1991). They have been detected in the human stomach and duodenum after milk ingestion (Chabance et al., 1998). Although primary sequences of these CPPs greatly differ, they all share a phosphorylated seryl-cluster (SpSpSpEE) (Silva and Malcata, 2005) where 30% of the phosphate ions from milk are bound. These sites, negatively charged, are one of the sites of minerals binding (Meisel, 1998), especially for calcium. This latter property was first demonstrated in the 1950s by Mellander and Isaksson who showed that casein phosphorylated peptides (via their ability to fix milk calcium; Sato et al., 1986) had a beneficial effect on calcium uptake by bones of rachitic children. Phosphorylation and mineral binding prevent CPPs from intestinal peptidases hydrolysis until they reach epithelial cells, where minerals are released by phosphatase activity (Boutrou et al., 2010). However, subsequent calcium absorption was not improved when associated with CPPs (Teucher et al., 2006). Other ions such as iron, zinc, copper, and magnesium can also bind to CPPs (FitzGerald, 1998). The type of bound cation deeply modifies the intestinal enzyme action; for example the coordination of bound copper to CPP inhibits the action of both phosphatase and peptidases (Boutrou et al., 2010).

Egg yolks represent another source of phosphopeptides (phosvitin) with calcium-binding capacity (Choi et al., 2005). And, aside from phosphopeptides, some calcium-binding peptides have been evidenced in whey and wheat proteins hydrolysates (Zhao et al., 2014; Liu et al., 2013).

2. Immunomodulation

The immunomodulatory activities (proliferation, activity, antibody synthesis, and cytokines production/regulation) of peptides issued from milk and soy proteins have mainly been described in vitro, on lymphocytes and macrophages (Singh et al., 2014; Chakrabarti et al., 2014). Peptides derived from milk β-and α-casein as well as α-lactalbumin, have been proven efficient to stimulate lymphocytes proliferation in vitro (Kayser and Meisel, 1996; Coste et al., 1992) and to increase the resistance of mice to *Klebsiella pneumonia* infection (Fiat et al., 1993). Caseinomacropeptide from κ-casein presents similar properties on proliferation and phagocytic activities in human macrophage-like cells (Li and Mine, 2004). The underlying mechanisms responsible for these immunomodulatory activities are not known. The μ opioid receptors, that are present in lymphocytes, could be involved in the stimulation of the immunoreactivity (Kayser and Meisel, 1996).

3. Antimicrobial effect

Antimicrobial peptides have been identified mainly from milk protein hydrolysates (Walther and Sieber, 2011; Clare et al., 2003). More precisely, lactoferricins (derived from lactoferrin) (Wakabayashi et al., 2003) and casein fragments were proven efficient to exhibit bactericidal activity (Lahov and Regelson, 1996). Bactericidal activity of lactoferricidins results from a direct interaction of the peptide (sequences 17–41 and 20–30) with the bacterial membrane, by increasing its permeability. Their action covers a relatively wide spectrum of microbes (gram ± bacteria, some yeasts and mushrooms) (Tomita et al., 1994). Caseinomacropeptide has also been shown to inhibit the binding of actinomyces and streptococci to enterocytes (Neeser et al., 1988). Although less studied, peptides from other food-proteins seem to present antimicrobial properties: pepsin hydrolysates from bovine hemoglobin (Nedjar-Arroume et al., 2006), hydrolysates from sarcoplasmic proteins (Jang et al., 2008), or peptides issued from barley and soybean (McClean et al., 2014).

4. Impact on the cardiovascular system

a. Antithrombotic effect

During blood clotting, fibrinogen binding to its platelet receptor induces platelets aggregation. Analogies between peptide sequences from κ-casein and from the C-terminal peptide of the γ chain of fibrinogen lead to a competition between casein peptides and fibrinogen for platelet receptors, causing the antithrombotic property of peptides issued from κ-casein (Jolles et al., 1986). This is also true for a lactotrasferrin peptide, whose antithrombotic effect has been demonstrated in vivo (Drouet et al., 1990).

b. Antihypertensive effect

Antihypertensive peptides act by inhibiting the angiotensin-converting enzyme (ACE), a key step in the cascade of events involved in the regulation of blood pressure. The first inhibitors of ACE have been identified in snake venom (Ondetti et al., 1971). The capacity of peptides to bind to ACE and inhibit its activity lies in their C-terminal tripeptide sequence, often rich in proline, branched chain, aromatic, and basic amino acids (FitzGerald and Meisel, 2000). Various peptides, from 2 to 10 amino acids residues, presenting these characteristics have been identified. Many of them come from hydrolysis of milk proteins, such as casein α_{S1} (Maruyama et al., 1987) and β (Maruyama et al., 1985), as well as muscle proteins (Vercruysse et al., 2005). The antihypertensive activity of these peptides has been demonstrated in vivo on hypertensive rats with a reduced systolic blood pressure and a lower ACE activity (Masuda et al., 1996; Nakamura et al., 1996) and in humans (Seppo et al., 2003). Peptides presenting similar properties have also been isolated from various food proteins (nonexhaustive list): fish (Yokoyama et al., 1992), egg (ovalbumin) (Fujita et al., 1995), and several vegetable proteins like soya (Yang et al., 2004), rapeseed (Marczak et al., 2003), or pea (Pedroche et al., 2002).

5. Impact on the nervous system

Because some food-derived peptides can present similar opioid activities as the enkephalins and endorphins released by brain and pituitary gland, they have been called exorphins (Zioudrou et al., 1979). They have been detected in hydrolysates from wheat gluten, casein α (Zioudrou et al., 1979), casein β (Brantl et al., 1979), and lactalbumin (Yoshikawa et al., 1986). Usually, food-derived opioid peptides present the following N-terminal sequence: YXF or YX_1X_2F. The tyrosine residue in the N-terminal position and the presence of another aromatic amino acid in the 3rd or 4th position favor the interaction of the peptide with μ receptors at the brain level. The absence of this sequence leads to no biological effect (Chang et al., 1981). Antiopioid effects also exist among the food-derived peptides; they derive from casein κ and are called casoxins (Chiba et al., 1989).

Some food-derived peptides could have anxiolytic activity. Indeed, it was shown that by binding to a benzodiazepine receptor, a α-casein fragment decreased anxiousness and improved sleep quality in animals subject to a slight chronic stress (Guesdon et al., 2006; Miclo et al., 2001).

6. Antiproliferative activity

Some peptides from animal or vegetable origins have been proven efficient in preventing initiation, promotion, or progression of cancer both in vivo and in vitro (de Mejia and Dia, 2010). It was, for instance, shown that a pentapeptide isolated from rice possesses cancer growth inhibitory properties on colon, breast, lung, and liver cancer cells (Kannan et al., 2010).

7. Anti-inflammatory and antioxidant activity

Food-derived peptides having anti-inflammatory activity have been evidenced in different animal- or plant-derived foods. In vitro approaches showed that this effect is mediated by an inhibition of the NF-κB signaling (Majumder et al., 2013), or the c-Jun N-terminal kinase pathway (Aihara et al., 2009). For instance, the bioactive peptide lactoferricin, released from bovine lactoferrin through hydrolysis, demonstrated an anti-inflammatory effect on human cartilage and synovial cells (Yan et al., 2013). In vivo, casein hydrolysates were shown to decrease inflammation in animal models of arthritis (Hatori et al., 2008), corn gluten hydrolysates decreased inflammation in animal models of inflammatory bowel disease (Mochizuki et al., 2010), and fish protein hydrolysate reduced inflammatory markers in high fat-fed mice (Bjorndal et al., 2013). In vivo evidence of such an effect in humans are lacking, however a meta-analysis of the literature suggests that dairy products, in particular fermented products, have anti-inflammatory properties in humans, in particular in subjects with metabolic disorders, which would match with the presence of bioactive peptides in these products (Bordoni et al., 2015).

On the basis of chemical assays, many peptides feature antioxidant properties. However, evidence of in vivo effects is scarce. Nevertheless, long-term consumption of egg white hydrolyzed with pepsin was shown to improve the plasma antioxidant capacity, and decrease the malondialdehyde levels in the aortic tissues of hypertensive rats (Manso et al., 2008).

8. Glycemia management

Theoretically a large number of food-derived peptides could help to regulate glycemia through their inhibitory effect on α-glucosidase enzyme, or dipeptidyl peptidase-IV (Patil et al., 2015). However, in vivo evidence of such an effect is currently lacking. A study in humans, showed a better effectiveness of whey protein hydrolysate in postprandial glycemia regulation compared to intact whey consumption (Goudarzi and Madadlou, 2013). Although indirect this observation supports a potential effect of peptides.

1.2. IN VIVO EVIDENCE OF FOOD-DERIVED PEPTIDE EFFECTS

The biological activities of food-derived peptides have been highlighted with various approaches (in vitro, in vivo) depending on the targeted activity, and the nature of the tested substance (hydrolysates, specific fragments). It is noticeable that it is often difficult to know which dose of peptide, and even more which amount of food, is necessary in order to observe an in vivo effect.

The best known activity is probably the antihypertensive one, for which an IC_{50} (concentration necessary to achieve 50% inhibition) can be measured in vivo, for example, in hypertensive rats. This parameter which largely varies among peptides (from 3 to 2349 μM) allows at least the comparison of the potential activity of different peptides. Lactotripeptides derived from casein digestion have been shown to have very low IC50, and the antihypertensive effect of a daily consumption of 150 g of fermented milk observed in humans was attributed to these peptides (Seppo et al., 2003). This study argued in favor of an action of food-derived peptides on physiological parameters, with food consumption compatible with a balanced diet. However, a recent meta-analysis of all clinical trials, in which lactotripepetides were tested, highlighted an inconsistency of the antihypertensive effect of these peptides in humans (Fekete et al., 2013).

Concerning the anxiolytic effect of a α_{S1}-casein hydrolysate, it was demonstrated in rats by intraperitoneal injection (0.4 mg/kg) that the peptide 91–101 (named α-casozepine) has an anxiolytic effect (Miclo et al., 2001). The daily intake of 15 mg/kg of a tryptic hydrolysate of α_{S1}-casein, which provided a maximum of 0.7 mg/kg of α-casozepine (but also other opioid peptides), was shown to improve sleep quality in rats subjected to chronic stress (Guesdon et al., 2006). In humans, ingestion of 1200 mg of a trypsic hydrolysate of α_{S1}-casein mitigated the effects of stress on blood pressure and plasma cortisol (Messaoudi et al., 2005). This dose of hydrolysate corresponded to about 60 mg of α-casozepine (but possibly also to other peptides), ie, to a consumption of about 120 g of milk.

1.3. BIOACTIVE PEPTIDES RELEASED DURING DIGESTION

Dietary protein degradation starts in the stomach where the secretion of hydrochloric acid by the parietal cells, stimulated by gastrin, causes their denaturation, which favors the exposure of peptide bonds to gastric proteases. Pepsins secreted by the gastric mucosa as a pepsinogen, are activated by the acidity of the stomach. They fragment the protein into polypeptides of varying sizes. They preferentially hydrolyze peptide bonds located within the polypeptide chain involving aromatic amino acids (phenylalanine, tyrosine, or tryptophan) or leucine, in a way that peptides released by gastric digestion often contain an aromatic amino acid in the N-terminal position (Bauchart et al., 2007). Many peptides derived from the degradation of caseins have been identified in the gastric contents of humans after ingestion of milk or yogurt (Chabance et al., 1998). For instance, caseinomacropeptide is released in the stomach from κ-casein, and its presence has been identified in the gastric chyme of humans after ingestion of dairy products. It was shown that the structure of the dairy matrices has little influence on the nature of the released peptides, which relies on the mechanism of proteolysis itself (cleavage sites), but significantly affects their amount in the stomach effluents and the kinetics of their appearance (Barbe et al., 2014). Similarly, after meat or fish consumption a large number of peptides deriving from actin and myosin (the main muscle proteins) have been identified in the stomach effluent (Bauchart et al., 2007). Interestingly, none of the peptides identified in the ready to eat meat were still present in the chyme flowing out the stomach, which well illustrates the intensity of pepsin activity. Approximately 20% of identified peptides were reproducibly observed in stomach effluent, showing, as for dairy products, that the occurrence of peptides at the entry of the small intestine is not only a matter of chance and that we can also expect some reproducibility in the biological effect of these peptides. Moreover, it was particularly interesting to note that six peptide sequences among the 18 reproducibly identified in duodenal contents after trout flesh intake were exactly the same as those derived from beef fragments of actin (96–106, 171–178, 24–33), of myosin heavy chain (835–842), creatine kinase (195–204), and GA3PDH (232–241). It thus seems that some peptides are generated consistently during gastric hydrolysis, regardless of the original muscle and its mode of preparation.

Protein digestion then proceeds in the intestinal lumen by the action of five proteolytic enzymes synthesized and secreted by the pancreatic acinar cells as inactive zymogens: trypsinogen, chymotrypsinogen, proelastase, and the procarboxypeptidases A and B. In slightly alkaline medium (pH 7.6–8.2), trypsinogen is activated to trypsin by the enterokinase, an enzyme of the intestinal mucosa. Trypsin, in turn, activates chymotrypsinogen, proelastase, and the procarboxypeptidases in chymotrypsin, elastase, and carboxypeptidases, respectively. Trypsin is the most abundant enzyme, representing 20% of pancreatic proteins. This endopeptidase cleaves the peptide bonds after hydrophilic amino acids, particularly lysine and arginine. Chymotrypsin preferentially acts after aromatic amino acid (phenylalanine, tyrosine), tryptophan, leucine, or methionine. The action of elastase is at the level of neutral amino acids (alanine, glycine, and serine). Carboxypeptidase A cleaves preferably at an aromatic or aliphatic amino acid and carboxypeptidase B at C-terminal basic amino acids. The action of these enzymes is completed by peptidases associated with the brush border membrane of the intestine. Many aminopeptidases are present at this level, including aminopeptidase N and A which release the neutral amino acids and anionic amino acids in the N-terminal position, respectively. Aminopeptidase P and W hydrolyze N-terminal X-Pro and X-Trp bonds, respectively. Dipeptidyl aminopeptidase IV releases dipeptide from fragments having proline or alanine in the penultimate position of the N-terminal extremity. In addition to these aminopeptidases, the intestinal brush border also contains endopeptidases and carboxypeptidases. The endopeptidases 24.11 and 24.18, which have similar activity to chymotrypsin, cleave peptide bonds at a hydrophobic or aromatic amino acid. Carboxypeptidase P releases the amino acid in the C-terminal position when proline, alanine, or glycine is in the penultimate position. Carboxypeptidase M releases C-terminal lysine and arginine. Finally, dipeptidyl carboxypeptidase hydrolyzes Pro-X, Phe-X, and Leu-X at the C-terminal position.

The activity of all these enzymes is considerable and it rapidly completes the action of the gastric proteases. Thus, peptide nitrogen that flows into the proximal jejunum, within 2 h after a milk or yogurt intake, was reported to account for about two-thirds of the dietary nitrogen intake (Gaudichon et al., 1995). A wide number of bioactive peptides have been identified in the jejunal content of humans after casein or milk whey proteins ingestion (Boutrou et al., 2013). Most of the casein-derived peptides were from β-casein, and a few derived from whey proteins. The most frequent activities for these peptides were antihypertensive and opioid-like activities. CPPs (mineral absorption enhancers) have been also identified in the jejunum of mini pigs after dairy products ingestion. Their presence in the digestive effluent at the distal ileum suggests a high resistance to gastrointestinal digestion (Meisel et al., 2003). In vivo studies with other sources of dietary proteins are scarce; meat proteins digestion in the small intestine was shown to reproducibly release actin, myosin, and creatine kinase fragments, in which antihypertensive sequences have been identified (Bauchart et al., 2007).

1.4. PEPTIDE BIOAVAILABILITY

Until the 1970s, it was generally accepted that the dietary α-amino nitrogen is exclusively absorbed from the small intestine in the form of free amino acids, after hydrolysis of proteins and peptides in the digestive lumen. It is now known that a considerable amount of amino acids cross the brush border of the enterocytes in the form of di- and tripeptides, via a specific transporter, the H^+-coupled PEPT1 transporter, which is located at the apical membrane of mature enterocytes all along the small intestine, but whose occurrence decreases from the duodenum to the ileum. Once they are inside, the enterocyte peptides are extensively hydrolyzed by cytosolic peptidases, before being released into the bloodstream in the form of free amino acids. However, peptides that are resistant to intracellular hydrolysis can be transported intact across the basolateral membrane of enterocytes and reach the bloodstream. Since the discovery of PEPT1 carrier, other peptide carriers have been highlighted in the intestinal epithelium, such as OATP and PHT1. In humans, OATP-B was clearly localized to the apical membrane of the enterocytes (Kobayashi et al., 2003), it could transport peptides with a mass greater than 450 Da (Hagenbuch and Meier, 2004). Similarly, the peptide/histidine transporter hPHT1 has been evidenced in epithelium of the different sections of the small intestine (Bhardwaj et al., 2006). The role of these last two carriers in dietary peptide absorption is however still unclear. The occurrence of a peptide carrier at the basolateral membrane of the enterocyte, allowing passage of the peptide from the enterocyte to blood vessels, has also been suggested. This carrier seems to have lower substrate affinity, but similar substrate specificity, than PEPT1 (Terada et al., 1999; Irie et al., 2004). Cooperation with PEPT1 would thus allow the transfer of di- and tripeptides across the epithelium. Absorption of peptides of more than 4 amino acids seems also possible by transcytosis (Shimizu et al., 1997) or by the paracellular pathway (Pappenheimer et al., 1994). Passive diffusion across the phospholipid bilayer of apical and basolateral membranes of the enterocytes is limited due to the hydrophilicity of most peptides.

It is generally considered that very few peptides are absorbed intact through the intestinal epithelium, and that the absorption of peptides contributes little to the absorption of amino acids from dietary proteins. However, experimental data to support this claim are lacking. We have seen that several mechanisms may allow the crossing of the epithelium by peptides of varying size, and studies in adult animals (sheep) suggested that intestinal absorption of low molecular weight peptides (<3000 Da) may account for a quarter of the amino acid absorption (Remond et al., 2000, 2003). Furthermore it was shown that up to 5% of the lactotripeptide VPP present in a casein hydrolysate can cross intact the gut epithelium of pigs after intragastric dosing (Ten Have et al., 2015). In humans, the possibility of food-derived peptide absorption through the epithelium has been little studied, but has been demonstrated for some peptides: the proline- and hydroxyproline-rich peptides after ingestion of gelatin (Prockop et al., 1962), carnosine after a meat meal (Park et al., 2005), peptides from the CMP and casein fragments detected in plasma after ingestion of dairy products (Chabance et al., 1998), and the lactotripeptide Ileu-Pro-Pro after ingestion of a yogurt beverage (Foltz et al., 2007). However there is currently little evidence that dietary bioactive peptides longer than tripeptides can cross the gut wall intact and be present in plasma in physiological relevant concentrations (Miner-Williams et al., 2014).

1.5. CONCLUSION

All dietary proteins are potential sources of bioactive peptides, with a large range of beneficial effects on health. However, although technical progresses, especially in mass spectrometry (Sanchez-Rivera et al., 2014), has allowed significant breakthroughs in the identification of peptides issued from in vivo protein digestion, some links in the chain between protein ingestion and the physiological effect of the derived peptides are still lacking. Peptides released from protein are rapidly cut into smaller fragments in the gut, and the true quantification of the peptides at each step of the degradation would be useful in order to explore a potential activity at the gut level (digestion, nutrient absorption, gut barrier). For peptides having peripheral effects (cardiovascular or nervous system), the major uncertainty is on their ability to cross the gut epithelium and to present a sufficiently long half-life in the plasma to be able to trigger a physiologic response. Clearly, clinical evidence supporting the health effects of food-derived bioactive peptides is currently too weak to translate this promising area of research into a solid criterion of the description of the nutritional quality of a food protein (Nongonierma and FitzGerald, 2015).

References

Aihara, K., Ishii, H., Yoshida, M., 2009. Casein-derived tripeptide, val-pro-pro (VPP), modulates monocyte adhesion to vascular endothelium. J. Atheroscler. Thromb. 16, 594–603.

Barbe, F., Le Feunteun, S., Remond, D., Menard, O., Jardin, J., et al., 2014. Tracking the in vivo release of bioactive peptides in the gut during digestion: mass spectrometry peptidomic characterization of effluents collected in the gut of dairy matrix fed mini-pigs. Food Res. Int. 63, 147–156.

Bauchart, C., Morzel, M., Chambon, C., Mirand, P.P., Reynes, C., et al., 2007. Peptides reproducibly released by in vivo digestion of beef meat and trout flesh in pigs. Br. J. Nutr. 98, 1187–1195.

Bhardwaj, R.K., Herrera-Ruiz, D., Eltoukhy, N., Saad, M., Knipp, G.T., 2006. The functional evaluation of human peptide/histidine transporter 1 (hPHT1) in transiently transfected COS-7 cells. Eur. J. Pharm. Sci. 27, 533–542.

Bjorndal, B., Berge, C., Ramsvik, M.S., Svardal, A., Bohov, P., et al., 2013. A fish protein hydrolysate alters fatty acid composition in liver and adipose tissue and increases plasma carnitine levels in a mouse model of chronic inflammation. Lipids Health Dis. 12, 143.

Bordoni, A., Danesi, F., Dardevet, D., Dupont, D., Fernandez, A.S., et al., 2015. Dairy products and inflammation: a review of the clinical evidence. Crit. Rev. Food Sci. Nutr. 19. Available from: http://dx.doi.org/10.1080/10408398.2014.967385.
Boutrou, R., Coirre, E., Jardin, J., Leonil, J., 2010. Phosphorylation and coordination bond of mineral inhibit the hydrolysis of the beta-casein (1–25) peptide by intestinal brush-border membrane enzymes. J. Agric. Food Chem. 58, 7955–7961.
Boutrou, R., Gaudichon, C., Dupont, D., Jardin, J., Airinei, G., et al., 2013. Sequential release of milk protein-derived bioactive peptides in the jejunum in healthy humans. Am. J. Clin. Nutr. 97, 1314–1323.
Boutrou, R., Henry, G., Sanchez-Rivera, L., 2015. On the trail of milk bioactive peptides in human and animal intestinal tracts during digestion: a review. Dairy Sci. Technol. 1–15.
Brantl, V., Teschemacher, H., Henschen, A., Lottspeich, F., 1979. Novel opioid peptides derived from casein (beta-casomorphins). I. Isolation from bovine casein peptone. Hoppe Seylers Z. Physiol. Chem. 360, 1211–1216.
Chabance, B., Marteau, P., Rambaud, J.C., Migliore-Samour, D., Boynard, M., et al., 1998. Casein peptide release and passage to the blood in humans during digestion of milk or yogurt. Biochimie 80, 155–165.
Chakrabarti, S., Jahandideh, F., Wu, J.P., 2014. Food-derived bioactive peptides on inflammation and oxidative stress. Biomed. Res. Int. 2014. Available from: http://dx.doi.org/10.1155/2014/608979.
Chang, K.J., Lillian, A., Hazum, E., Cuatrecasas, P., Chang, J.K., 1981. Morphiceptin (NH4-tyr-pro-phe-pro-COHN2): a potent and specific agonist for morphine (mu) receptors. Science 212, 75–77.
Chiba, H., Tani, F., Yoshikawa, M., 1989. Opioid antagonist peptides derived from kappa-casein. J. Dairy Res. 56, 363–366.
Choi, I., Jung, C., Choi, H., Kim, C., Ha, H., 2005. Effectiveness of phosvitin peptides on enhancing bioavailability of calcium and its accumulation in bones. Food Chem. 93, 577–583.
Clare, D.A., Catignani, G.L., Swaisgood, H.E., 2003. Biodefense properties of milk: the role of antimicrobial proteins and peptides. Curr. Pharm. Des. 9, 1239–1255.
Coste, M., Rochet, V., Leonil, J., Molle, D., Bouhallab, S., et al., 1992. Identification of C-terminal peptides of bovine beta-casein that enhance proliferation of rat lymphocytes. Immunol. Lett. 33, 41–46.
Cuber, J.C., Bernard, G., Fushiki, T., Bernard, C., Yamanishi, R., et al., 1990. Luminal CCK-releasing factors in the isolated vascularly perfused rat duodenojejunum. Am. J. Physiol. 259, G191–G197.
Dangin, M., Boirie, Y., Guillet, C., Beaufrere, B., 2002. Influence of the protein digestion rate on protein turnover in young and elderly subjects. J. Nutr. 132, 3228S–3233S.
Daniel, H., Vohwinkel, M., Rehner, G., 1990. Effect of casein and beta-casomorphins on gastrointestinal motility in rats. J. Nutr. 120, 252–257.
de Mejia, E.G., Dia, V.P., 2010. The role of nutraceutical proteins and peptides in apoptosis, angiogenesis, and metastasis of cancer cells. Cancer Metastasis Rev. 29, 511–528.
Drouet, L., Bal dit Sollier, C., Cisse, M., Pignaud, G., Mazoyer, E., et al., 1990. The antithrombotic effect of KRDS, a lactotransferrin peptide, compared with RGDS. Nouv. Rev. Fr. Hematol. 32, 59–62.
Fekete, A.A., Givens, D.I., Lovegrove, J.A., 2013. The impact of milk proteins and peptides on blood pressure and vascular function: a review of evidence from human intervention studies. Nutr. Res. Rev. 26, 177–190.
Fiat, A.M., Migliore-Samour, D., Jolles, P., Drouet, L., Bal dit Sollier, C., et al., 1993. Biologically active peptides from milk proteins with emphasis on two examples concerning antithrombotic and immunomodulating activities. J. Dairy Sci. 76, 301–310.
FitzGerald, R.J., 1998. Potential uses of caseinophosphopeptides. Int. Dairy J. 8, 451–457.
FitzGerald, R.J., Meisel, H., 2000. Milk protein-derived peptide inhibitors of angiotensin-I-converting enzyme. Br. J. Nutr. 84 (Suppl. 1), S33–S37.
Foltz, M., Meynen, E.E., Bianco, V., van Platerink, C., Koning, T.M., et al., 2007. Angiotensin converting enzyme inhibitory peptides from a lactotripeptide-enriched milk beverage are absorbed intact into the circulation. J. Nutr. 137, 953–958.
Froetschel, M.A., Azain, M.J., Edwards, G.L., Barb, C.R., Amos, H.E., 2001. Opioid and cholecystokinin antagonists alleviate gastric inhibition of food intake by premeal loads of casein in meal-fed rats. J. Nutr. 131, 3270–3276.
Fujita, H., Sasaki, R., Yoshikawa, M., 1995. Potentiation of the antihypertensive activity of orally administered ovokinin, a vasorelaxing peptide derived from ovalbumin, by emulsification in egg phosphatidylcholine. Biosci. Biotechnol. Biochem. 59, 2344–2345.
Gaudichon, C., Mahe, S., Roos, N., Benamouzig, R., Luengo, C., et al., 1995. Exogenous and endogenous nitrogen flow rates and level of protein hydrolysis in the human jejunum after [15N]milk and [15N]yoghurt ingestion. Br. J. Nutr. 74, 251–260.
Goudarzi, M., Madadlou, A., 2013. Influence of whey protein and its hydrolysate on prehypertension and postprandial hyperglycaemia in adult men. Int. Dairy J. 33, 62–66.
Guesdon, B., Messaoudi, M., Lefranc-Millot, C., Fromentin, G., Tome, D., et al., 2006. A tryptic hydrolysate from bovine milk alphaS1-casein improves sleep in rats subjected to chronic mild stress. Peptides 27, 1476–1482.
Hagenbuch, B., Meier, P.J., 2004. Organic anion transporting polypeptides of the OATP/SLC21 family: phylogenetic classification as OATP/SLCO superfamily, new nomenclature and molecular/functional properties. Pflugers Arch. 447, 653–665.
Hatori, M., Ohki, K., Hirano, S., Yang, X.P., Kuboki, H., et al., 2008. Effects of a casein hydrolysate prepared from *Aspergillus oryzae* protease on adjuvant arthritis in rats. Biosci. Biotechnol. Biochem. 72, 1983–1991.
Irie, M., Terada, T., Okuda, M., Inui, K., 2004. Efflux properties of basolateral peptide transporter in human intestinal cell line Caco-2. Pflugers Arch. 449, 186–194.
Jang, A., Jo, C., Kang, K.S., Lee, M., 2008. Antimicrobial and human cancer cell cytotoxic effect of synthetic angiotensin-converting enzyme (ACE) inhibitory peptides. Food Chem. 107, 327–336.
Jolles, P., Levy-Toledano, S., Fiat, A.M., Soria, C., Gillessen, D., et al., 1986. Analogy between fibrinogen and casein. Effect of an undecapeptide isolated from kappa-casein on platelet function. Eur. J. Biochem. 158, 379–382.
Kannan, A., Hettiarachchy, N.S., Lay, J.O., Liyanage, R., 2010. Human cancer cell proliferation inhibition by a pentapeptide isolated and characterized from rice bran. Peptides 31, 1629–1634.
Kayser, H., Meisel, H., 1996. Stimulation of human peripheral blood lymphocytes by bioactive peptides derived from bovine milk proteins. FEBS Lett. 383, 18–20.

Kitts, D.D., Weiler, K., 2003. Bioactive proteins and peptides from food sources. Applications of bioprocesses used in isolation and recovery. Curr. Pharm. Des. 9, 1309–1323.

Kobayashi, D., Nozawa, T., Imai, K., Nezu, J., Tsuji, A., et al., 2003. Involvement of human organic anion transporting polypeptide OATP-B (SLC21A9) in pH-dependent transport across intestinal apical membrane. J. Pharmacol. Exp. Ther. 306, 703–708.

Lahov, E., Regelson, W., 1996. Antibacterial and immunostimulating casein-derived substances from milk: casecidin, isracidin peptides. Food Chem. Toxicol. 34, 131–145.

Leser, S., 2013. The 2013 FAO report on dietary protein quality evaluation in human nutrition: recommendations and implications. Nutr. Bull. 38, 421–428.

Li, E.W., Mine, Y., 2004. Immunoenhancing effects of bovine glycomacropeptide and its derivatives on the proliferative response and phagocytic activities of human macrophagelike cells, U937. J. Agric. Food Chem. 52, 2704–2708.

Liu, F.R., Wang, L., Wang, R., Chen, Z.X., 2013. Calcium-binding capacity of wheat germ protein hydrolysate and characterization of peptide-calcium complex. J. Agric. Food Chem. 61, 7537–7544.

Majumder, K., Chakrabarti, S., Davidge, S.T., Wu, J., 2013. Structure and activity study of egg protein ovotransferrin derived peptides (IRW and IQW) on endothelial inflammatory response and oxidative stress. J. Agric. Food Chem. 61, 2120–2129.

Manso, M.A., Miguel, M., Even, J., Hernandez, R., Aleixandre, A., et al., 2008. Effect of the long-term intake of an egg white hydrolysate on the oxidative status and blood lipid profile of spontaneously hypertensive rats. Food Chem. 109, 361–367.

Marczak, E.D., Usui, H., Fujita, H., Yang, Y., Yokoo, M., et al., 2003. New antihypertensive peptides isolated from rapeseed. Peptides 24, 791–798.

Martinez-Augustin, O., Rivero-Gutierrez, B., Mascaraque, C., de Medina, F.S., 2014. Food derived bioactive peptides and intestinal barrier function. Int. J. Mol. Sci. 15, 22857–22873.

Maruyama, S., Nakagomi, K., Tomizuka, N., Suzuki, H., 1985. Angiotensin I-converting enzyme-inhibitor derived from an enzymatic hydrolysate of casein. 2. Isolation and Bradykinin-potentiating activity on the uterus and the ileum of rats. Agric. Biol. Chem. 49, 1405–1409.

Maruyama, S., Mitachi, H., Tanaka, H., Tomizuka, N., Suzuki, H., 1987. Angiotensin I-converting enzyme-inhibitors derived from an enzymatic hydrolysate of casein. 4. Studies on the active-site and antihypertensive activity of angiotensin I-converting enzyme-inhibitors derived from casein. Agric. Biol. Chem. 51, 1581–1586.

Masuda, O., Nakamura, Y., Takano, T., 1996. Antihypertensive peptides are present in aorta after oral administration of sour milk containing these peptides to spontaneously hypertensive rats. J. Nutr. 126, 3063–3068.

McClean, S., Beggs, L.B., Welch, R.W., 2014. Antimicrobial activity of antihypertensive food-derived peptides and selected alanine analogues. Food Chem. 146, 443–447.

Meisel, H., 1998. Overview on milk protein-derived peptides. Int. Dairy J. 8, 363–373.

Meisel, H., Bernard, H., Fairweather-Tait, S., FitzGerald, R.J., Hartmann, R., et al., 2003. Detection of caseinophosphopeptides in the distal ileostomy fluid of human subjects. Br. J. Nutr. 89, 351–359.

Mellander, O., 1950. The physiological importance of the casein phosphopeptide calcium salts. II. Peroral calcium dosage of infants. Acta Soc. Med. Ups 55, 247–255.

Messaoudi, M., Lefranc-Millot, C., Desor, D., Demagny, B., Bourdon, L., 2005. Effects of a tryptic hydrolysate from bovine milk alphaS1-casein on hemodynamic responses in healthy human volunteers facing successive mental and physical stress situations. Eur. J. Nutr. 44, 128–132.

Miclo, L., Perrin, E., Driou, A., Papadopoulos, V., Boujrad, N., et al., 2001. Characterization of alpha-casozepine, a tryptic peptide from bovine alpha(s1)-casein with benzodiazepine-like activity. FASEB J. 15, 1780–1782.

Miner-Williams, W.M., Stevens, B.R., Moughan, P.J., 2014. Are intact peptides absorbed from the healthy gut in the adult human? Nutr. Res. Rev. 27, 308–329.

Mochizuki, M., Shigemura, H., Hasegawa, N., 2010. Anti-inflammatory effect of enzymatic hydrolysate of corn gluten in an experimental model of colitis. J. Pharm. Pharmacol. 62, 389–392.

Nagpal, R., Behare, P., Rana, R., Kumar, A., Kumar, M., et al., 2011. Bioactive peptides derived from milk proteins and their health beneficial potentials: an update. Food Funct. 2, 18–27.

Nakamura, Y., Masuda, O., Takano, T., 1996. Decrease of tissue angiotensin I-converting enzyme activity upon feeding sour milk in spontaneously hypertensive rats. Biosci. Biotechnol. Biochem. 60, 488–489.

Nedjar-Arroume, N., Dubois-Delval, V., Miloudi, K., Daoud, R., Krier, F., et al., 2006. Isolation and characterization of four antibacterial peptides from bovine hemoglobin. Peptides 27, 2082–2089.

Neeser, J.R., Chambaz, A., Del Vedovo, S., Prigent, M.J., Guggenheim, B., 1988. Specific and nonspecific inhibition of adhesion of oral actinomyces and streptococci to erythrocytes and polystyrene by caseinoglycopeptide derivatives. Infect. Immun. 56, 3201–3208.

Nemoz-Gaillard, E., Bernard, C., Abello, J., Cordier-Bussat, M., Chayvialle, J.A., et al., 1998. Regulation of cholecystokinin secretion by peptones and peptidomimetic antibiotics in STC-1 cells. Endocrinology 139, 932–938.

Nishi, T., Hara, H., Hira, T., Tomita, F., 2001. Dietary protein peptic hydrolysates stimulate cholecystokinin release via direct sensing by rat intestinal mucosal cells. Exp. Biol. Med. (Maywood) 226, 1031–1036.

Nishi, T., Hara, H., Asano, K., Tomita, F., 2003. The soybean beta-conglycinin beta 51–63 fragment suppresses appetite by stimulating cholecystokinin release in rats. J. Nutr. 133, 2537–2542.

Nongonierma, A.B., FitzGerald, R.J., 2015. The scientific evidence for the role of milk protein-derived bioactive peptides in humans: a review. J. Funct. Foods 17, 640–656.

Ondetti, M.A., Williams, N.J., Sabo, E.F., Pluscec, J., Weaver, E.R., et al., 1971. Angiotensin-converting enzyme inhibitors from the venom of *Bothrops jararaca*. Isolation, elucidation of structure, and synthesis. Biochemistry 10, 4033–4039.

Pappenheimer, J.R., Dahl, C.E., Karnovsky, M.L., Maggio, J.E., 1994. Intestinal absorption and excretion of octapeptides composed of D amino acids. Proc. Natl. Acad. Sci. U.S.A. 91, 1942–1945.

Park, Y.J., Volpe, S.L., Decker, E.A., 2005. Quantitation of carnosine in humans plasma after dietary consumption of beef. J. Agric. Food Chem. 53, 4736–4739.

Patil, P., Mandal, S., Tomar, S.K., Anand, S., 2015. Food protein-derived bioactive peptides in management of type 2 diabetes. Eur. J. Nutr. 54, 863–880.
Pedersen, N.L., Nagain-Domaine, C., Mahe, S., Chariot, J., Roze, C., et al., 2000. Caseinomacropeptide specifically stimulates exocrine pancreatic secretion in the anesthetized rat. Peptides 21, 1527–1535.
Pedroche, J., Yust, M.M., Giron-Calle, J., Alaiz, M., Millan, F., et al., 2002. Utilisation of chickpea protein isolates for production of peptides with angiotensin I-converting enzyme (ACE)-inhibitory activity. J. Sci. Food Agric. 82, 960–965.
Plaisancie, P., Claustre, J., Estienne, M., Henry, G., Boutrou, R., et al., 2013. A novel bioactive peptide from yoghurts modulates expression of the gel-forming MUC2 mucin as well as population of goblet cells and Paneth cells along the small intestine. J. Nutr. Biochem. 24, 213–221.
Plaisancie, P., Boutrou, R., Estienne, M., Henry, G., Jardin, J., et al., 2015. beta-Casein (94–123)-derived peptides differently modulate production of mucins in intestinal goblet cells. J. Dairy Res. 82, 36–46.
Prockop, D.J., Keiser, H.R., Sjoerdsma, A., 1962. Gastrointestinal absorption and renal excretion of hydroxyproline peptides. Lancet 2, 527–528.
Remond, D., Bernard, L., Poncet, C., 2000. Free and peptide amino acid net flux across the rumen and the mesenteric- and portal-drained viscera of sheep. J. Anim. Sci. 78, 1960–1972.
Remond, D., Bernard, L., Chauveau, B., Noziere, P., Poncet, C., 2003. Digestion and nutrient net fluxes across the rumen, and the mesenteric- and portal-drained viscera in sheep fed with fresh forage twice daily: net balance and dynamic aspects. Br. J. Nutr. 89, 649–666.
Sanchez-Rivera, L., Martinez-Maqueda, D., Cruz-Huerta, E., Miralles, B., Recio, I., 2014. Peptidomics for discovery, bioavailability and monitoring of dairy bioactive peptides. Food Res. Int. 63, 170–181.
Sato, R., Noguchi, T., Naito, H., 1986. Casein phosphopeptide (CPP) enhances calcium absorption from the ligated segment of rat small intestine. J. Nutr. Sci. Vitaminol. (Tokyo) 32, 67–76.
Sato, R., Shindo, M., Gunshin, H., Noguchi, T., Naito, H., 1991. Characterization of phosphopeptide derived from bovine beta-casein: an inhibitor to intra-intestinal precipitation of calcium phosphate. Biochim. Biophys. Acta 1077, 413–415.
Seppo, L., Jauhiainen, T., Poussa, T., Korpela, R., 2003. A fermented milk high in bioactive peptides has a blood pressure-lowering effect in hypertensive subjects. Am. J. Clin. Nutr. 77, 326–330.
Shimizu, M., 2004. Food-derived peptides and intestinal functions. Biofactors 21, 43–47.
Shimizu, M., Tsunogai, M., Arai, S., 1997. Transepithelial transport of oligopeptides in the human intestinal cell, Caco-2. Peptides 18, 681–687.
Silva, S.V., Malcata, F.X., 2005. Caseins as source of bioactive peptides. Int. Dairy J. 15, 1–15.
Singh, B.P., Vij, S., Hati, S., 2014. Functional significance of bioactive peptides derived from soybean. Peptides 54, 171–179.
Ten Have, G.A., van der Pijl, P.C., Kies, A.K., Deutz, N.E., 2015. Enhanced lacto-tri-peptide bio-availability by co-ingestion of macronutrients. PLoS One 10, e0130638.
Terada, T., Sawada, K., Saito, H., Hashimoto, Y., Inui, K., 1999. Functional characteristics of basolateral peptide transporter in the human intestinal cell line Caco-2. Am. J. Physiol. 276, G1435–1441.
Teucher, B., Majsak-Newman, G., Dainty, J.R., McDonagh, D., FitzGerald, R.J., et al., 2006. Calcium absorption is not increased by caseinophosphopeptides. Am. J. Clin. Nutr. 84, 162–166.
Tomita, M., Takase, M., Bellamy, W., Shimamura, S., 1994. A review: the active peptide of lactoferrin. Acta Paediatr. Jpn 36, 585–591.
Vercruysse, L., Van Camp, J., Smagghe, G., 2005. ACE inhibitory peptides derived from enzymatic hydrolysates of animal muscle protein: a review. J. Agric. Food Chem. 53, 8106–8115.
Wakabayashi, H., Takase, M., Tomita, M., 2003. Lactoferricin derived from milk protein lactoferrin. Curr. Pharm. Des. 9, 1277–1287.
Walther, B., Sieber, R., 2011. Bioactive proteins and peptides in foods. Int. J. Vitam. Nutr. Res. 81, 181–192.
Yan, D.Y., Chen, D., Shen, J., Xiao, G.Z., Van Wijnen, A.J., et al., 2013. Bovine lactoferricin is anti-inflammatory and anti-catabolic in human articular cartilage and synovium. J. Cell Physiol. 228, 447–456.
Yang, H.Y., Yang, S.C., Chen, J.R., Tzeng, Y.H., Han, B.C., 2004. Soyabean protein hydrolysate prevents the development of hypertension in spontaneously hypertensive rats. Br. J. Nutr. 92, 507–512.
Yokoyama, K., Chiba, H., Yoshikawa, M., 1992. Peptide inhibitors for angiotensin I-converting enzyme from thermolysin digest of dried bonito. Biosci. Biotechnol. Biochem. 56, 1541–1545.
Yoshikawa, M., Tani, F., Yoshimura, T., Chiba, H., 1986. Opioid-peptides from milk-proteins. Agric. Biol. Chem. 50, 2419–2421.
Yvon, M., Beucher, S., Guilloteau, P., Le Huerou-Luron, I., Corring, T., 1994. Effects of caseinomacropeptide (CMP) on digestion regulation. Reprod. Nutr. Dev. 34, 527–537.
Zhao, L., Huang, Q., Huang, S., Lin, J., Wang, S., et al., 2014. Novel peptide with a specific calcium-binding capacity from whey protein hydrolysate and the possible chelating mode. J. Agric. Food Chem. 62, 10274–10282.
Zioudrou, C., Streaty, R.A., Klee, W.A., 1979. Opioid peptides derived from food proteins. The exorphins. J. Biol. Chem. 254, 2446–2449.

CHAPTER

2

Protein Intake Throughout Life and Current Dietary Recommendations

F. Mariotti

UMR Physiologie de la Nutrition et du Comportement Alimentaire, AgroParisTech, INRA, Université Paris-Saclay, Paris, France

2.1. INTRODUCTION

Protein nutrition is a much more complex issue than might be thought at first glance, and this complexity has major implications for the evaluation of dietary requirements. The term "protein" covers both all amino acids taken together, which are used to make protein, and a series of specific amino acids, with specific metabolism and physiological properties. Considered together, all amino acids include an alpha-amino nitrogen moiety; the nitrogen is not synthesized by the body and is thus indispensable. Twenty amino acids are the most abundant in the body and in our diet because they are used for protein synthesis. However, all amino acids have specific metabolism and properties, and it has long been established that some cannot be synthesized de novo in quantities that are commensurable with metabolic demands for protein synthesis. Basically, for nitrogen and indispensable amino acids, the final criteria used to evaluate requirements have always been based on the utilization of amino acids for body protein turnover. There is thus a large body of data concerning estimates of nitrogen and amino acid requirements for the general adult population; these data have been used to draw up current reference values, although there are certain limitations for theoretical and practical reasons. The same criteria have also been used to define the requirements in other situations, such as throughout the life cycle.

As we will be discussing below, and as dealt with by other contributors to this book, the background metabolism of amino acids in protein synthesis is indeed extremely intricate, and the same applies to factors that impact the utilization of amino acids for body protein turnover in the context of health and disease prevention. Furthermore, amino acid metabolism interplays with other metabolisms and numerous specific tissue functions. In particular, some amino acids (eg, leucine, arginine, cysteine), which are present at varying amounts in dietary proteins, are linked to key cellular signaling processes (eg, mechanistic target of rapamycin cell signaling pathway, nitrergic signaling, redox signaling, etc.). This has provided the foundations for a great deal of current research on the possible relationship between protein intake, amino acids, and health-related parameters, which could ultimately be used as alternative criteria for the determination of protein recommendations. However, our understanding of protein and amino acid requirements is also based on data obtained using simpler criteria, and as we shall discuss, most guidelines are still being built on this old, but solid, body of data. We will also present and discuss the protein intakes of different populations, by comparison with the recommendations, so as to further identify current issues regarding protein nutrition.

The Molecular Nutrition of Amino Acids and Proteins.
DOI: http://dx.doi.org/10.1016/B978-0-12-802167-5.00002-5

2.2. CURRENT ESTIMATES FOR PROTEIN AND AMINO ACID REQUIREMENTS THROUGHOUT LIFE

For more than a century, criteria to define protein requirements have been based on the utilization of protein to renew body protein and balance nitrogenous losses (Sherman, 1920). From the series of nitrogen balance studies conducted in adult humans, a large set of data has been developed to estimate the minimum amount of protein nitrogen that can balance such losses (Fig. 2.1), and then derive a total protein requirement according to this basic, simple criterion. In line with earlier estimates (FAO/WHO/UNU, 1985), reviews and meta-analysis of these data found that the average requirement was 0.66 g/kg body weight per day (Li et al., 2014; Rand et al., 2003; WHO/FAO/UNU, 2007). The Dietary Reference Intake, to use the US term, also referred to as the Population Reference Intake in Europe (EFSA Panel on Dietetic Products Nutrition and Allergies, 2010), or in other words the intake that covers virtually all (~97.5%) the requirements of the population, was estimated at 0.83 g/kg per day of a mixture of proteins with adequate value. Although there were some differences related to gender, these were not ultimately considered as being significant or strong enough to be retained in the recommendations. Differences in protein metabolism between men and women are largely attributable to differences in body composition, except at critical periods of hormonal changes (puberty and menopause; Markofski and Volpi, 2011). A population reference intake of ~0.8 g/kg for the general adult population has now been endorsed by virtually all countries and organizations (AFSSA (French Food Safety Agency), 2007; EFSA Panel on Dietetic Products Nutrition and Allergies, 2012).

Although the reference value for adults was derived from a large set of experimental data, those regarding other populations or conditions are much less robust. In cases where little or no experimental data are available, such as infants, young children, pregnant and lactating women, a factorial approach has been adopted. This combines estimates for standard maintenance requirements based on classical nitrogen balance data applied to the specific population reference for body weight, with an additional component to account for the specific requirements of a population due to protein deposition during growth (in children or pregnant women) or extra protein demand (during lactation). On this basis, for instance, the population reference intakes for children aged 1, 2, 3, and 8 years are 1.14, 0.97, 0.90, and 0.92 g/kg per day, respectively (WHO/FAO/UNU, 2007). The reference value varies along with the growth component, which rapidly decreases during the first years. When setting these values, the references were taken from normal development and the expected normal energy requirement, that is, for children with an appropriate body composition and a moderate level of activity (WHO/FAO/UNU, 2007). Likewise, the population reference intakes for pregnant women have been derived by adding to the standard (maintenance) value extra components of 0.7, 9.6, and 31.2 g protein for the first, second, and third trimesters, respectively, when the efficiency of protein deposition is taken as 42%, as in the FAO/WHO report (WHO/FAO/UNU, 2007), while they are 1, 9, and 28 g protein for the first, second, and third trimesters, respectively, when considering that the efficiency of protein deposition is 47%, as according to the EFSA report (EFSA Panel on Dietetic Products Nutrition and Allergies, 2012). Using different background estimates and hypotheses, the French agency published values that were not markedly different (with values of +14.7 and +27.3 during

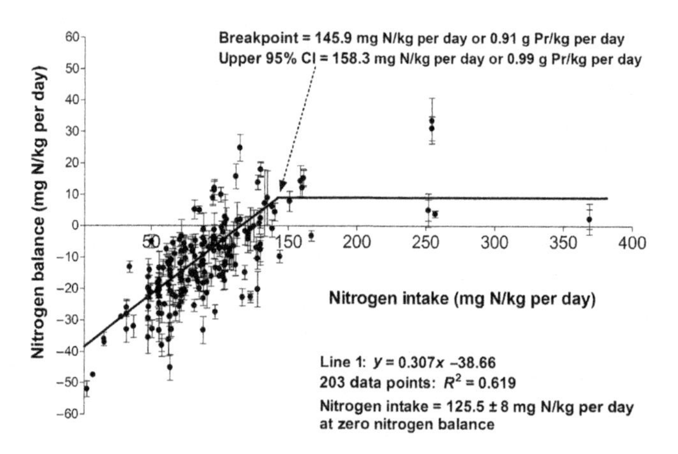

FIGURE 2.1 Evaluation of relationship between various nitrogen intakes and the mean nitrogen balances from 28 nitrogen balance studies using a biphase linear regression to identify the mean nitrogen requirement as a breakpoint. *From Humayun et al. (2007). Reprint with permission.*

the last two trimesters of pregnancy; AFSSA (French Food Safety Agency), 2007). The extra component increases during pregnancy, as the specific metabolic demand rises to sustain the growth in protein mass. These values were set while considering an average weight gain considered to be normal.

Whether protein requirement increases with age has long been a subject of debate (Millward and Roberts, 1996; Morais et al., 2006). Based on an analysis of the less numerous good nitrogen balance studies in older people, some authors have considered that their nitrogen utilization is lower, thus justifying a higher reference value. Likewise, some studies have reported negative balances or altered protein status in older people consuming the reference intake for adults, indicating that this value may not be appropriate as the reference in this older population (eg, Pannemans et al., 1997). In the famous meta-analysis by Rand and collaborators, the lower efficiency of utilization in older people was confirmed and estimated at 31% in individuals aged over 55 years, compared to 48% in younger individuals (Rand et al., 2003). Taken together, these data would argue in favor of setting the Population Reference Intake (PRI) for older people at a level of around 0.9–1.0 g/kg per day (AFSSA (French Food Safety Agency), 2007). However, it is accepted that data from nitrogen balance studies are scarce in older people and may have been biased by confounding factors, such as the low energy intake in nitrogen balance studies. Because the evidence remains limited, the FAO/WHO and the EFSA have chosen not to endorse a higher estimate for protein requirements in older people, whereas the French agency has proposed setting the PRI at ~1 g/kg (AFSSA (French Food Safety Agency), 2007; EFSA Panel on Dietetic Products Nutrition and Allergies, 2012; WHO/FAO/UNU, 2007).

2.3. THEORETICAL AND PRACTICAL LIMITATIONS AND UNCERTAINTIES

Although the nitrogen balance method is considered as robust, and has produced a large set of estimates that are still the most useful when estimating requirements, the meta-regression between nitrogen intake and balance yields estimates that are imprecise. This imprecision originates from the modeling of the relationship between nitrogen intake and balance in meta-regression analyses, where the use of simple linear regression has been criticized. Other higher (biphasic) models have reached PRI estimates of 0.99 g/kg per day (Humayun et al., 2007), see Fig. 2.1. Imprecision also originates from the intrinsic and methodological factors that affect nitrogen balance data. Imprecisions regarding nitrogen intakes and nitrogen losses (which are also considered as underestimated) are well-known and may explain in part the findings of positive nitrogen balances (Fig. 2.1), which is not realistic in the long-term in adults. Furthermore, nitrogen balance data are known to be markedly influenced by the energy balance. Lastly, and more importantly, there has been criticism of the fact that these balance studies were mostly performed in the short term (less than 2 weeks), which would not account for the adaptation of metabolism. Adaptive phenomena are a critical factor in such studies because they have probably led to an underestimation of the efficiency of utilization, which will have directly overestimated the intercept, that is, the estimated requirement. There has been considerable controversy regarding the extent to which this adaptation is not captured by multilevel nitrogen balance studies, and the resulting overestimation of protein requirements (Millward and Jackson, 2004; Pillai et al., 2010), which indeed dates back to the early 20th century (Sherman, 1920).

Further to the discussion about uncertainties regarding the existence of specific requirements in older people, due to the paucity of nitrogen balance data, a few recent studies which used the oxidation of an indicator amino acid in response to graded protein intakes, challenged the current estimates for requirements and proposed that the population reference intake might in fact be as high as ~1.2–1.3 g/kg per day (Rafii et al., 2015; Tang et al., 2014). This method is elegant and easily applicable to vulnerable groups, but it has been criticized on practical and theoretical grounds, merely because it is a short-term method (Fukagawa, 2014; Millward, 2014; Millward and Jackson, 2012). Furthermore, the estimates in older people are finally quite similar to those obtained using the same technique in younger adults (population reference intake: 1.2 g/kg per day; Humayun et al., 2007), which might indeed be taken as evidence for no marked increase in requirement with advancing age. According to most authors, the different estimates that are higher or lower than those currently prevailing in older people are plagued by uncertainties, and a consensus may be out of reach (Fukagawa, 2014; Marini, 2015). This therefore shows the need for other approaches, involving the use of other criteria, a point we will be addressing below (Volpi et al., 2013).

It should also be noted that in specific populations such as children and pregnant women, the additional components in the factorial method remain indirect and highly approximate, involving assumptions for the efficiency of deposition that have not been confirmed under the specific conditions of these populations and are rather gross estimates derived from data in the general population. If the metabolism adapts to the high demand under

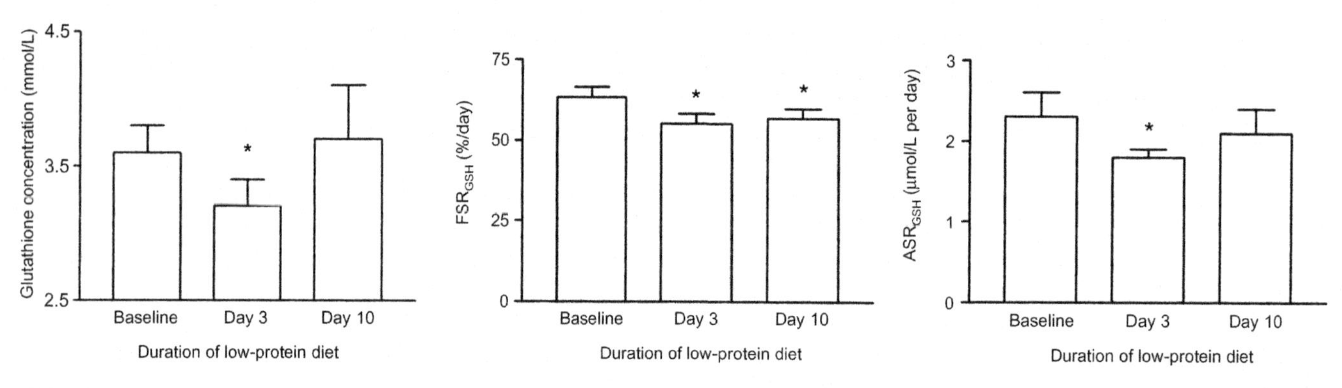

FIGURE 2.2 Mean (±SEM) erythrocyte glutathione concentrations and mean (± SEM) fractional synthesis rates and absolute synthesis rates of erythrocyte glutathione (FSRGSH and ASRGSH, respectively) in 12 healthy adults (6 men and 6 women) during consumption of their habitual amount of dietary protein at baseline and on days 3 and 10 of consumption of a diet that provided the safe amount of protein. *Significantly different from baseline, $p < 0.05$ (repeated-measures ANOVA followed by post hoc analysis with Bonferroni correction for multiple comparisons). *From Jackson et al. (2004). Reprint with permission.*

these conditions, leading to an improvement in the efficiency of protein utilization, the factorial method would result in an overestimation of requirements. By contrast, recent data obtained by measuring the oxidation of an indicator amino acid in response to graded amino acid levels have argued that the protein requirement may be much higher than that currently proposed during pregnancy (estimated average requirement of 1.22 and 1.52 g/kg per day in early and late gestation, respectively, compared to a current estimate of 0.88 g/kg per day; Stephens et al., 2015). Likewise, similarly higher estimates have been reported in children (Elango et al., 2011). However, once again, this method has been the subject of criticism (Fukagawa, 2014; Hoffer, 2012).

The supply of nitrogen to maintain body nitrogen pools is considered to be a basic, minimum criterion to estimate requirements. Even under this apparently simple theoretical approach, questions are raised concerning evaluation of the consequences of metabolic adaptation and accommodation to enable the final homeostasis of body nitrogen, for example, changes to protein fluxes and reductions in lean mass (Millward and Roberts, 1996). In the general population, adaptive/accommodative phenomena may be considered as acceptable, on condition that they do not adversely impact health. However, there is almost no evidence to confirm this, apart from that of a purely theoretical type. During adaptation to the protein reference intake, healthy adults have changes in glutathione kinetics (Fig. 2.2) and the turnover of some specific protein, suggesting a functional cost (Afolabi et al., 2004; Jackson et al., 2004). In more specific populations such as older people at risk of developing sarcopenia, accommodation to a marginal protein intake may secure the nitrogen balance but the associated metabolic cost may have implications for the optimal maintenance of muscle function during aging (Campbell et al., 2002). Likewise, it has been shown that dietary proteins (eg, milk and soy proteins) with varying amino acid compositions that succeed in meeting the requirements for maintenance and growth in rodents will indeed leave a footprint, as identified in the natural isotopic abundance in tissues, which shows that the utilization of these proteins in response to metabolic demand is not allowed by the same arrangements in the underlying metabolism (Poupin et al., 2011, 2014). The consequences for health of these underlying metabolic changes remain unknown. Finally, all these different considerations show that metabolic data alone are not sufficient to determine an optimal level toward the lower end of the range of intakes that the body can adapt to or accommodate. A more detailed characterization of the accommodative metabolic processes involved, and an assessment of their physiological and pathophysiological impacts, are necessary.

At a broader scale, the maintenance of body nitrogen is indeed considered to be a minimum criterion for determining requirements because a very large number of functions and health-related parameters may be influenced by protein intake. This means that the application of other criteria would result in higher protein reference intakes than those defined at present, and they would still remain far below the upper level of intake, despite the scarcity of data. This is the rationale for the utilization of the wording "safe level of intake" by the FAO/WHO/UNU, although this does not differ markedly from the standard usage and conception of the "Population Reference Intake," "Recommended Dietary Allowance," or "Apport Nutritionnel conseillé" in Europe, USA, and France. At a practical level, this means that the recommendation is not to reduce the protein intake to values close to the PRI. From a scientific point of view, further studies are necessary to consider criteria other than the minimal criterion that is nitrogen balance.

Amino acid requirements are also based on quite basic criteria. The requirement for an individual amino acid no longer depends on the amount required to achieve the overall nitrogen balance, but on the minimum quantity that balances the oxidative loss of (the carbon skeleton of) this amino acid, or limits the oxidative loss of another indispensable proteinogenic amino acid, determined using various tracer-based methods and protocols. However, the criteria relates to the utilization of amino acids in their quantitatively major utilization pathway, that is, protein synthesis. The requirements for individual amino acids can be estimated in absolute amounts (ie, mg amino acid per kg body weight per day), but because amino acids are consumed as protein in the diet, these values have been used to determine the amino acid composition of protein intake, which, when consumed in a quantity sufficient to meet nitrogen requirements, will also meet those of individual amino acids (WHO/FAO/UNU, 2007; Young and Borgonha, 2000). This amino acid profile is used as a reference pattern to assess the nutritional quality of dietary proteins. Several reference patterns are available for children in specific age groups, calculated using the amino acid and protein requirements of each group. By contrast, the reference pattern for newborns (0–6 months) is taken directly as the amino acid profile found in human milk, although these figures may overestimate actual requirement (WHO/FAO/UNU, 2007). In older people, insufficient data are available to consider differences in individual amino acid requirements and hence different amino acid reference patterns (Pillai and Kurpad, 2012). Indeed, the debate concerning potentially higher individual amino acid requirements is similar to that about a possibly high overall protein requirement, inasmuch as it relates to potential differences in the efficiency of utilization of amino acids. One reason for a higher indispensable amino acid requirement in older people may indeed be their higher first pass splanchnic extraction (Boirie et al., 1997; Morais et al., 2006; Volpi et al., 1999), which limits the efficiency of utilization for retention, although other authors have argued that metabolic demand is lower in older individuals, which may result in a similar apparent amino acid requirement (WHO/FAO/UNU, 2007).

2.4. EVIDENCE FOR DEFINING REQUIREMENTS BASED ON MEALS RATHER THAN AN AVERAGE DAILY INTAKE IN OLDER PEOPLE

The uncertainties concerning protein and amino acid requirements in older people clearly indicate that the traditional approach to the overall daily nitrogen and amino acid balance remains limited. Amino acid balance methods (eg, leucine) do not withstand alternating fasted and fed states, but study metabolism in the artificial steady fed state and fasted state, whereas differences in metabolism throughout life, and particularly during aging, may in fact stem from an altered dynamic of changes in protein metabolism as impacted by the intake of a meal. The specificity of protein metabolism, compared to that of other energy nutrient, is that there is no inactive form of protein that can be used to store dietary protein in the postprandial state, so that the precise regulation of postprandial metabolism is critical to protein homeostasis. Our current understanding of dietary protein and amino acids in the context of aging is that older people are resistant to postprandial anabolic stimulation by dietary protein, and that this resistance can be overcome by supplying daily protein in the form of protein-rich meals (Paddon-Jones and Leidy, 2014; Rodriguez, 2014). A higher level of postprandial anabolism has been evidenced in older people (but not younger adults) following a single large protein meal versus several smaller ones (Arnal et al., 1999, 2000; Mamerow et al., 2014). There is now consensus that a protein-rich meal in this context contains more than 30 g protein, which is considered to be the amount necessary to pass the "anabolic threshold" and optimize postprandial anabolism (Paddon-Jones and Leidy, 2014; Paddon-Jones and Rasmussen, 2009). Similarly, proteins that are absorbed and delivered rapidly elicit a better postprandial amino acid balance than those which are absorbed slowly, in the older people, while the reverse holds true in younger adults (Beasley et al., 2013; Dangin et al., 2003; Fouillet et al., 2009). This argues in favor of an age-related decrease in the ability of the available amino acids to stimulate anabolism, lending further credence to the "anabolic threshold" paradigm (Dardevet et al., 2012). Indispensable amino acids, and particularly branched-chain amino acids, are considered to be key in eliciting this anabolic response in the postprandial state, so that a threshold (at 3 g) for peak anabolism has also been proposed for meal leucine (Gryson et al., 2014), which triggers a signal for anabolic utilization of the bulk of amino acids (Dardevet et al., 2002; Magne et al., 2012). In line with this, at a relatively low dose (20 g), whey protein (a leucine-rich, "fast" protein) causes a greater increase in postprandial anabolism in older people than casein (slow and lower in leucine) and casein hydrolysate (fast, but lower in leucine) (Pennings et al., 2011). Of note, long-term benefit of leucine-rich protein and/or high protein diets in older people may also proceed from benefits in the limitation of muscle proteolysis (Mosoni et al., 2014). Beyond the specific case of leucine, there is a need to define the optimum amino acid profile that maximizes postprandial anabolism in older individuals, and could thus be used to refine amino acid requirements and the amino acid template using more precise metabolic criteria. However, achieving this goal is still a long way off.

The timing and conditions under which this anabolic resistance appears during aging remain uncertain. However, it has been suggested that resistance may start long before the classically considered age of 70 years, and be accentuated by the appearance of a catabolic stressor such as inactivity or low-grade inflammation (Balage et al., 2010; Breen and Churchward-Venne, 2012; Glover et al., 2008; Paddon-Jones and Leidy, 2014; Rieu et al., 2009). This difference in the features of protein and amino acid metabolism with aging can be explained by changes to the molecular signaling of amino acids in the body, as described and discussed throughout this book.

Lastly, the relationship between protein intake and protein metabolism in older people needs to be studied while bearing in mind the different factors that may impact their protein metabolism. Of particular importance in this respect are energy intake and physical activity, levels of which largely impact nitrogen balance and muscle protein metabolism (Carbone et al., 2012). The (low-grade) inflammatory status of older people may also modify protein requirements for anabolism and muscle strength (Balage et al., 2010; Bartali et al., 2012; Buffiere et al., 2015; Guadagni and Biolo, 2009; Rieu et al., 2009). Indeed, there is no general consensus regarding whether the anabolic resistance of muscle protein synthesis rates is truly an intrinsic characteristic of aging muscle or the self-induced product of a sedentary lifestyle (Knuiman and Kramer, 2012). The higher protein requirement estimates produced by studies in older people might be explained by their lower energy intake and reduced physical activity. In other words, it is not certain that healthy and active older people whose energy intake matches the energy expenditure corresponding to their physical activity, do indeed have lower overall protein requirements. Likewise, some authors have suggested that an increase in protein intake in this age group will only be beneficial when associated with an increase in physical activity (Bauer et al., 2013; EFSA Panel on Dietetic Products Nutrition and Allergies, 2012; Paddon-Jones and Rasmussen, 2009).

2.5. TOWARD OTHER CRITERIA TO DEFINE REQUIREMENTS, USING HEALTH-RELATED PARAMETERS?

Even refined metabolic criteria, such as those based on postprandial effects on protein and some amino acids in older people, are very limited when defining requirements where health is the central reference criterion. Indeed, using metabolic criteria faces two obstacles. The first, as we have already discussed, is that protein and amino acid intakes that are close to the minimum amount required not to disrupt basic metabolic function (such as nitrogen homeostasis or protein turnover) may go along with accommodative phenomena, which are difficult to characterize and for which little information is available regarding their possible adverse impacts on health. The second problem is more directly related to the absence in pure metabolic studies of a marker that would be interpretable in terms of health. Such data, will tend to be obtained from observational studies relating to protein intake and physical function, disease risk factors and disease incidence, and during interventional trials that have studied the relationship between protein intake and disease risk factors.

There is a large body of such data, which have been intensively reviewed by different institutions and agencies but have been considered as inconclusive (AFSSA (French Food Safety Agency), 2007; EFSA Panel on Dietetic Products Nutrition and Allergies, 2012; FNB/IOM, 2005; WHO/FAO/UNU, 2007). The first criterion to have been largely considered, because it is the most directly related to metabolic criteria, is muscle mass and function. However, there is little evidence that a protein intake above the requirement defined using metabolic criteria (such as nitrogen balance) can increase muscle mass and improve function in adults and even in individuals engaged in exercising programs, in children, or in older people, and this criterion has tended to be used in the context of older individuals when discussing the idea that their requirements may be higher than those of adults, or that the pattern of protein intake over a day should be considered as critical. A recent review of the literature concluded that the evidence for a higher PRI for protein in the elderly population remains limited, ranging from *suggestive* to *inconclusive* (Pedersen and Cederholm, 2014).

Other health criteria that have been considered include body weight and body composition. There is a large body of data which suggests that, under conditions of energy restriction, high protein diets are effective for losing weight, limiting a decrease in lean mass, and with benefits that persist after the weight loss program (Clifton et al., 2014). However, in this energy restriction context, high protein diets are only high in protein on a relative basis, and indeed such diets supply a normal protein intake when considered quantitatively. Furthermore, high protein diets are necessarily also low-carbohydrate diets so it remains difficult to wholly ascribe their effects to the protein component alone. In the longer term, high protein diets have not been shown to perform better than other types of diet (Sacks et al., 2009), as concluded by a recent systematic review and meta-analysis of high protein diets as a variant of low-carbohydrate diets for weight loss (Naude et al., 2014).

Lastly, based on the results of studies that have controlled energy intake, it is now considered that the level of protein per se in the diet does not relate to weight loss (Halkjaer et al., 2011). The benefit of high protein diets may therefore be related more to greater compliance with energy restriction in ad libitum programs, which could be in part could be related to changes in appetite regulation through the use of high protein foods in low energy meals (Clifton, 2009; Leidy et al., 2007; Martens and Westerterp-Plantenga, 2014). More importantly, there is a paucity of data resulting from investigations of the relationship between protein intake and the maintenance of body weight and composition in a normal energy balance situation. In rodents, high protein diets have been shown to limit the development of diet-induced obesity (Petzke et al., 2014) but data in humans are lacking.

What is true for protein and body weight or composition is even more true for individual amino acids. The type of protein (casein versus whey) or its distribution throughout the day (pulse or spread) was reported not to impact changes in body composition during a short-term weight loss program (Adechian et al., 2012). There are only very limited, preliminary data from rodent studies, and observational data from human studies, that suggest a relationship between the intake of certain amino acids, body weight and body composition. For instance, animal data have shown that arginine supplementation can impact body composition (Jobgen et al., 2009; Tan et al., 2009; Wu et al., 2012). Observational data in humans have reported inverse associations between the intake of branched-chain amino acids and being overweight or obese (Qin et al., 2011). However, the concentrations of branched-chain amino acids are elevated in obese subjects with insulin resistance and/or metabolic syndrome (Newgard et al., 2009), and they are associated with cardiovascular risk factors (Yang et al., 2014) and predictive of diabetes (Wang et al., 2011). Although a higher plasma concentration of branched-chain amino acids is the result of a complex change in their metabolism (Lynch and Adams, 2014; She et al., 2013), supplementation with branched-chain amino acids has also been reported to contribute to the development of insulin resistance (Balage et al., 2011) in particular in the context of high-fat feeding (Newgard, 2012), although these findings were controversial, because completely opposite results were found with leucine alone in mice (Macotela et al., 2011; Zhang et al., 2007). What these examples show is that the amino acid requirements were estimated from the quantitative requirement for protein turnover, while emerging science has shown that the intake of certain amino acids, including those not considered to be "indispensable" (such as arginine) or "conditionally indispensable" (such as cysteine) may impact signaling in many important pathways and have a profound effect on key functions for long-term health. Likewise, dietary proteins which differ in their amino acid profiles, and the supplementation of meals or the diet with certain amino acids, may have a differential impact on redox status, insulin sensitivity and vascular homeostasis (Borucki et al., 2009; Jones et al., 2011; Magne et al., 2009; Mariotti et al., 2008). This opens up a very important area of research to define the requirements of individual amino acids based on health-related criteria.

Likewise, many studies in the literature have further examined the relationship between protein and amino acid intake and health-related parameters, including bone health, insulin sensitivity and the risk of disease. Unfortunately, this body of evidence remains small, and using these criteria is not currently helping to resolve the controversy regarding a possibly higher protein requirement when considered in terms of the amount required to obtain improvements in body composition. Finally, and as recently concluded by a systematic literature review by Pedersen and colleagues, although the evidence is assessed as *probable* regarding the estimated requirement based on nitrogen balance studies, it is considered as *suggestive* to *inconclusive* for protein intake and mortality and morbidity (Pedersen and Cederholm, 2014).

As far as dietary reference values are concerned, this chapter would be incomplete without briefly considering the issue of the upper level value for protein intake. This issue has been studied for a long time. From a metabolic point of view, few data have identified a set threshold for an adverse impact of protein intake on nitrogen metabolism. Based on a study of urea synthesis with different protein intakes, it was estimated that maximum urea synthesis was reached with an average of 3.5 g/kg per day, so that, accounting for typical intraindividual variability, levels below 2.2 g/kg per day for an entire population would never saturate urea synthesis (AFSSA (French Food Safety Agency), 2007). The data were obtained in subjects who had not been adapted to the protein level. The values were proposed initially to qualify intake levels but were not considered as tolerable upper level intake levels, because of the overall lack of data and characterization of their impact. At the physiological and pathophysiological levels, there have long been concerns that high levels of protein intake might adversely impact renal function and thereby may contribute to initiating renal dysfunction or hastening the progression of renal disease. Indeed, in healthy adults and older people, data are scarce and little conclusive, at least when it comes to characterizing the physiological impact (such as changes in glomerular filtration rates) in terms of risk (Walrand et al., 2008). The current recommendations regarding limitations on protein intake are restricted to older people with severe kidney disease (Bauer et al., 2013). When considering other health-related criteria and other populations, there are few data to identify and characterize the risk of excessive protein intake.

2.6. CURRENT DIETARY INTAKE OF PROTEIN AND AMINO ACIDS

In developing countries, protein-energy malnutrition remains a central issue, but interventional programs for the prevention and treatment of malnutrition mostly target a large set of macro- and micronutrients to improve nutritional status (Desjeux, 2006) and focus specifically on critical populations at their most vulnerable stages, that is, children, adolescents, and pregnant women (Jacob and Nair, 2012). It is particularly important that epidemiological and animal studies in these populations have documented that protein malnutrition during pregnancy and lactation result in a change to so-called fetal programming, attended by long-term health risks which include a risk of obesity, metabolic dysregulation, and abnormal neurobehavioral development (Belluscio et al., 2014; Levin, 2009; Michaelsen and Greer, 2014; Seki et al., 2012).

In western countries, protein intake has increased markedly during the past century, in line with the increase in the consumption of animal products, and notably meat in countries with the highest levels of income (WHO/FAO, 2003). Furthermore, as far as we can trace it, the increase in the contribution of animal products to total energy intake may be a central feature in the nutritional transition that is affecting the whole world. For instance, total protein intake in Spain rose from 79 g in 1961 to 106 g in 2009, with the proportion of animal proteins increasing from 33% to 61%, according to food balance sheets (F. Mariotti, from FAO, 2012). In most industrialized countries, the protein intake is around 100 g/day, that is, 1.3–1.4 g/kg per day and ~16% total energy intake (Dubuisson et al., 2010; Elmadfa, 2009; Fulgoni, 2008). However, as a function of country or a specific region, or gender, total protein intake varies little, at between 13% and 18% of overall energy intake (Elmadfa, 2009; Halkjaer et al., 2009).

Therefore, for the general adult population in western countries, the average protein intake (~1.3 g/day) is about twice the estimated average requirement (0.66 g/kg per day). Accordingly, when comparing protein intake in the whole population with a theoretical distribution of requirements, it has been concluded that virtually everyone in the general population consumes more than the requirements (AFSSA (French Food Safety Agency), 2007). Even subpopulations with lower protein intakes, such as nonstrict vegetarians and even most vegans, have total protein intakes that clearly cover their requirements, because the contribution of total protein to energy remains reasonably high (Clarys et al., 2014; Halkjaer et al., 2009). Likewise, although pregnancy increases protein requirements, protein intake by pregnant women is considered to largely cover their requirements.

The protein intake in children in industrialized countries is high. For instance, from the European collection of survey results (Elmadfa, 2009) it can be calculated that the average intake of protein in children aged 4–6 years is 56 g/day. The values differ according to country (with averages ranging from 49 to 69 g in Europe) and there are quite considerable interindividual variations, which result in 32 g/day as the lowest estimate in the 5th percentile across European countries for this age group (EFSA Panel on Dietetic Products Nutrition and Allergies, 2012). The contrast between this level of intake and protein requirements is striking, since the PRI is about 15 g/day. Accordingly, the issue with such levels of protein intake may in fact concern the risk of them being excessive. However, and as discussed above, a tolerable upper level of intake has not yet been set. In its absence, and especially in children, if the value defined by the French Food Agency is applied, most of them, and particularly the youngest age groups, have "high" or "very high" intakes, the latter being in the majority (ie, exceeding 3.5 g/kg per day).

In older people, protein intake remains an important issue. The contribution of protein to total energy intake in older people is similar to that in adults (~16% of energy across European countries) but because older people have a lower energy intake, their protein intake is usually slightly lower (the averages in male Europeans being 86 g/day in those aged over 65 years vs 96 g/day in people aged 19–64 years). When compared to the population reference value of 0.83 g/kg per day in adults, or even with higher estimates of protein requirements, such as the 1.0 g/kg per day proposed by the French Food Agency, once again virtually all older people have intakes that exceed this requirement (AFSSA (French Food Safety Agency), 2007). It is however necessary to look at these findings more closely. Indeed, the estimated prevalence concerns 3–5% of the older population (>65 year) in France, who are usually aged ~70 year. The intakes of even older people (~80 year) have been little studied (Volkert et al., 2004) but they are expected to be slightly lower than those of the less older counterparts, leading to an insufficient intake by a considerable proportion of the population (Berner et al., 2013). To this increase in nutritional risk with age should be added the fact that although protein intake varies little, it may be considerably lower in some regions. For instance, it is 86 g/day on average in Europe but ~70 g in Austria and Greece (Elmadfa, 2009). Lastly, protein intake has been shown to be lower in institutionalized older people, as illustrated by a recent comparison of different Dutch populations, which reported a protein intake of 0.8 g/kg per day in institutionalized elderly compared to 1.1 g/kg per day among those of a similar age living at home (Tieland et al., 2012). Therefore, if specific populations of older people with lower protein and energy intakes are considered, bearing in mind the possibility that protein requirements may be higher in this population than in adults (with a population reference intake ~1 g/kg per day), then protein intake may be insufficient in many of the most vulnerable older age groups. If higher estimates of protein requirement in older people (such as >1.2 g/kg per day) are to be endorsed (Bauer et al., 2013), then most of them would be considered as having an insufficient protein intake. This shows how critical it is to define the optimal intake, and thus choose the best criteria to determine protein requirements.

We have also mentioned that as well as overall daily values, protein and amino acid requirements should be discussed at the meal level in older people. Accordingly, the distribution of protein intake throughout the day will also impact protein status in this population. Although indirect, the data available suggest that most meals consumed by older people include less than the 30 g protein that is taken as their postprandial anabolic threshold. Indeed, as reasoned by Volpi and collaborators from the US national survey data, only dinner is on average likely to contain 30 g protein (~31 g protein), while other meals will not (Volpi et al., 2013). That only one meal a day (either dinner or lunch, depending on the country and population) contains protein in quantities clearly above the threshold has been evidenced in other populations worldwide (Berner et al., 2013; Valenzuela et al., 2013). Protein intake may be more evenly distributed throughout the day in the frail elderly population than in healthy adults (Bollwein et al., 2013).

This chapter does not discuss amino acid intakes relative to the amino acid requirement or the derived amino acid pattern of protein. In western countries, the general population consumes a wide variety of proteins, and as we have just mentioned, the total protein intake is much higher than that required. Therefore, even among populations whose diet contains markedly different protein intakes from different protein sources, such as vegetarians, there should be no risk of a marginal intake of amino acids. One exception may concern the lysine intake, in some subpopulations in countries such as India and the UK, but this observation has been taken as evidence that the lysine requirement may have been overestimated and should in fact be chosen from the lower range of estimates, in order to account for possible adaptive phenomena that probably operate to match intake to metabolic demand (Millward and Jackson, 2004; Wiseman, 2004). These observations also highlight the fact that individual amino acid requirements should be considered at both the meal level (ie, taking account of their effects on the dynamic of postprandial metabolism; Fouillet et al., 2009; Mariotti et al., 2001; Millward et al., 2002), and using criteria that go beyond the protein balance and could be used to identify the impact of specific amino acids on regulatory metabolic and physiological pathways (Magne et al., 2009; Mariotti et al., 2013). It is necessary to directly investigate the impact of changing the intake levels of some specific amino acids within the natural nutritional range on the metabolic and physiological effect of meals. Such investigations should address protein intake in terms of the nutritional value of the dietary protein consumed, under a broader consideration of nutritional quality, that is, beyond the nitrogen balance (Millward et al., 2008).

2.7. CONCLUSION AND PERSPECTIVES

We close this chapter by admitting that there remain major limitations to our understanding of protein requirements, even when studied using simple criteria such as the nitrogen balance in specific populations corresponding to the different stages in life. This can be ascribed to a lack of direct data on specific populations, such as infants and pregnant women, but also to shortcomings in identifying the adaptive or accommodative phenomena that probably operate under low protein and amino acid intakes and their possible impacts on long-term health. Advancing beyond basic criteria related to growth or the nitrogen balance has been advocated for nearly two decades and has stimulated research in the field, but the data remain fragmented and very scarce. In some specific populations, such as older people, a body of evidence has been built to refine the framework of protein requirements; this has been made possible by focusing on postprandial metabolism and on the metabolic and physiological criteria related to sarcopenia. By contrast, the body of evidence concerning the general adult population remains evanescent, which may be related to difficulties in characterizing the specific relationship between protein and amino acid intakes and endpoints that will be wholly adequate to describe numerous health-related parameters and disease risks—a classic conundrum in nutrition. Such research is necessary to analyze the value of protein to our diets—and our meals—and to rationalize the usefulness of different protein sources as a function of their characteristics, and particularly their amino acid contents. A wide-ranging analysis of the impact of protein nutrition on health must also take account of the association of protein with other nutrients in foodstuffs, so we need to better understand the consequences of changes to total protein intake and/or intake from different protein sources on the global nutritional adequacy of diets and their relevance to dietary guidelines (Camilleri et al., 2013; Estaquio et al., 2009). Although this will further increase the complexity of this research area, a more global evaluation is also required in order to transform specific protein-related recommendations into optimum and pragmatic dietary guidelines for the population.

References

Adechian, S., Balage, M., Remond, D., Migne, C., Quignard-Boulange, A., Marset-Baglieri, A., et al., 2012. Protein feeding pattern, casein feeding, or milk-soluble protein feeding did not change the evolution of body composition during a short-term weight loss program. Am. J. Physiol. Endocrinol. Metab. 303, E973–982.

Afolabi, P.R., Jahoor, F., Gibson, N.R., Jackson, A.A., 2004. Response of hepatic proteins to the lowering of habitual dietary protein to the recommended safe level of intake. Am. J. Physiol. Endocrinol. Metab. 287, E327–330.

AFSSA (French Food Safety Agency), 2007. Report "Protein Intake: Dietary Intake, Quality, Requirements and Recommendations".

Arnal, M.A., Mosoni, L., Boirie, Y., Houlier, M.L., Morin, L., Verdier, E., et al., 1999. Protein pulse feeding improves protein retention in elderly women. Am. J. Clin. Nutr. 69, 1202–1208.

Arnal, M.A., Mosoni, L., Boirie, Y., Houlier, M.L., Morin, L., Verdier, E., et al., 2000. Protein feeding pattern does not affect protein retention in young women. J. Nutr. 130, 1700–1704.

Balage, M., Averous, J., Remond, D., Bos, C., Pujos-Guillot, E., Papet, I., et al., 2010. Presence of low-grade inflammation impaired postprandial stimulation of muscle protein synthesis in old rats. J. Nutr. Biochem. 21, 325–331.

Balage, M., Dupont, J., Mothe-Satney, I., Tesseraud, S., Mosoni, L., Dardevet, D., 2011. Leucine supplementation in rats induced a delay in muscle IR/PI3K signaling pathway associated with overall impaired glucose tolerance. J. Nutr. Biochem. 22, 219–226.

Bartali, B., Frongillo, E.A., Stipanuk, M.H., Bandinelli, S., Salvini, S., Palli, D., et al., 2012. Protein intake and muscle strength in older persons: does inflammation matter? J. Am. Geriatr. Soc. 60, 480–484.

Bauer, J., Biolo, G., Cederholm, T., Cesari, M., Cruz-Jentoft, A.J., Morley, J.E., et al., 2013. Evidence-based recommendations for optimal dietary protein intake in older people: a position paper from the PROT-AGE Study Group. J. Am. Med. Dir. Assoc. 14, 542–559.

Beasley, J.M., Shikany, J.M., Thomson, C.A., 2013. The role of dietary protein intake in the prevention of sarcopenia of aging. Nutr. Clin. Pract. 28, 684–690.

Belluscio, L.M., Berardino, B.G., Ferroni, N.M., Ceruti, J.M., Canepa, E.T., 2014. Early protein malnutrition negatively impacts physical growth and neurological reflexes and evokes anxiety and depressive-like behaviors. Physiol. Behav. 129, 237–254.

Berner, L.A., Becker, G., Wise, M., Doi, J., 2013. Characterization of dietary protein among older adults in the United States: amount, animal sources, and meal patterns. J. Acad. Nutr. Diet. 113, 809–815.

Boirie, Y., Gachon, P., Beaufrere, B., 1997. Splanchnic and whole-body leucine kinetics in young and elderly men. Am. J. Clin. Nutr. 65, 489–495.

Bollwein, J., Diekmann, R., Kaiser, M.J., Bauer, J.M., Uter, W., Sieber, C.C., et al., 2013. Distribution but not amount of protein intake is associated with frailty: a cross-sectional investigation in the region of Nurnberg. Nutr. J. 12, 109.

Borucki, K., Aronica, S., Starke, I., Luley, C., Westphal, S., 2009. Addition of 2.5 g L-arginine in a fatty meal prevents the lipemia-induced endothelial dysfunction in healthy volunteers. Atherosclerosis 205, 251–254.

Breen, L., Churchward-Venne, T.A., 2012. Leucine: a nutrient 'trigger' for muscle anabolism, but what more? J. Physiol. 590, 2065–2066.

Buffiere, C., Mariotti, F., Savary-Auzeloux, I., Migne, C., Meunier, N., Hercberg, S., et al., 2015. Slight chronic elevation of C reactive protein is associated with lower aerobic fitness but does not impair meal-induced stimulation of muscle protein metabolism in healthy old men. J. Physiol. 593, 1259–1272.

Camilleri, G.M., Verger, E.O., Huneau, J.F., Carpentier, F., Dubuisson, C., Mariotti, F., 2013. Plant and animal protein intakes are differently associated with nutrient adequacy of the diet of French adults. J. Nutr. 143, 1466–1473.

Campbell, W.W., Trappe, T.A., Jozsi, A.C., Kruskall, L.J., Wolfe, R.R., Evans, W.J., 2002. Dietary protein adequacy and lower body versus whole body resistive training in older humans. J. Physiol. 542, 631–642.

Carbone, J.W., McClung, J.P., Pasiakos, S.M., 2012. Skeletal muscle responses to negative energy balance: effects of dietary protein. Adv. Nutr. 3, 119–126.

Clarys, P., Deliens, T., Huybrechts, I., Deriemaeker, P., Vanaelst, B., De Keyzer, W., et al., 2014. Comparison of nutritional quality of the vegan, vegetarian, semi-vegetarian, pesco-vegetarian and omnivorous diet. Nutrients 6, 1318–1332.

Clifton, P., 2009. High protein diets and weight control. Nutr. Metab. Cardiovasc. Dis. 19, 379–382.

Clifton, P.M., Condo, D., Keogh, J.B., 2014. Long term weight maintenance after advice to consume low carbohydrate, higher protein diets—a systematic review and meta analysis. Nutr. Metab. Cardiovasc. Dis. 24, 224–235.

Dangin, M., Guillet, C., Garcia-Rodenas, C., Gachon, P., Bouteloup-Demange, C., Reiffers-Magnani, K., et al., 2003. The rate of protein digestion affects protein gain differently during aging in humans. J. Physiol. 549, 635–644.

Dardevet, D., Sornet, C., Bayle, G., Prugnaud, J., Pouyet, C., Grizard, J., 2002. Postprandial stimulation of muscle protein synthesis in old rats can be restored by a leucine-supplemented meal. J. Nutr. 132, 95–100.

Dardevet, D., Remond, D., Peyron, M.A., Papet, I., Savary-Auzeloux, I., Mosoni, L., 2012. Muscle wasting and resistance of muscle anabolism: the "anabolic threshold concept" for adapted nutritional strategies during sarcopenia. ScientificWorldJournal 2012, 269531.

Desjeux, J.F., 2006. Recent issues in energy-protein malnutrition in children. Nestle Nutr. Workshop Ser. Pediatr. Programme 58, 177–184, (discussion 184–178).

Dubuisson, C., Lioret, S., Touvier, M., Dufour, A., Calamassi-Tran, G., Volatier, J.L., et al., 2010. Trends in food and nutritional intakes of French adults from 1999 to 2007: results from the INCA surveys. Br. J. Nutr. 103, 1035–1048.

EFSA Panel on Dietetic Products Nutrition and Allergies, 2010. Scientific opinion on principles for deriving and applying dietary reference values. EFSA J. 8, 1458 [1430 p].

EFSA Panel on Dietetic Products Nutrition and Allergies, 2012. Scientific opinion on dietary reference values for protein. EFSA J. 10, 2257 [2266 p].

Elango, R., Humayun, M.A., Ball, R.O., Pencharz, P.B., 2011. Protein requirement of healthy school-age children determined by the indicator amino acid oxidation method. Am. J. Clin. Nutr. 94, 1545–1552.

Elmadfa, I., 2009. European nutrition and health report 2009. Forum Nutr. Karger. 62, 1–405.

Estaquio, C., Kesse-Guyot, E., Deschamps, V., Bertrais, S., Dauchet, L., Galan, P., et al., 2009. Adherence to the French Programme National Nutrition Sante Guideline Score is associated with better nutrient intake and nutritional status. J. Am. Diet. Assoc. 109, 1031–1041.

FAO, 2012. FAOSTAT Statistics Division. Food balance sheets. <http://faostat3.fao.org/download/FB/*/E>.

FAO/WHO/UNU, 1985. Energy and protein requirements. Report of a Joint FAO/WHO/UNU Expert Consultation. World Health Organization, WHO Technical Report Series, No 724.

FNB/IOM, 2005. Dietary Reference Intakes for Energy, Carbohydrate, Fiber, Fat, Fatty Acids, Cholesterol, Protein, and Amino Acids (Macronutrients). The National Academies Press, Washington, D.C.

Fouillet, H., Juillet, B., Gaudichon, C., Mariotti, F., Tome, D., Bos, C., 2009. Absorption kinetics are a key factor regulating postprandial protein metabolism in response to qualitative and quantitative variations in protein intake. Am. J. Physiol. Regul. Integr. Comp. Physiol. 297, R1691–1705.

Fukagawa, N.K., 2014. Protein requirements: methodologic controversy amid a call for change. Am. J. Clin. Nutr. 99, 761–762.

Fulgoni 3rd, V.L., 2008. Current protein intake in America: analysis of the National Health and Nutrition Examination Survey, 2003–2004. Am. J. Clin. Nutr. 87 (5), 1554S–1557S.

Glover, E.I., Phillips, S.M., Oates, B.R., Tang, J.E., Tarnopolsky, M.A., Selby, A., et al., 2008. Immobilization induces anabolic resistance in human myofibrillar protein synthesis with low and high dose amino acid infusion. J. Physiol. 586, 6049–6061.

Gryson, C., Walrand, S., Giraudet, C., Rousset, P., Migne, C., Bonhomme, C., et al., 2014. "Fast proteins" with a unique essential amino acid content as an optimal nutrition in the elderly: growing evidence. Clin. Nutr. 33, 642–648.

Guadagni, M., Biolo, G., 2009. Effects of inflammation and/or inactivity on the need for dietary protein. Curr. Opin. Clin. Nutr. Metab. Care 12, 617–622.

Halkjaer, J., Olsen, A., Bjerregaard, L.J., Deharveng, G., Tjonneland, A., Welch, A.A., et al., 2009. Intake of total, animal and plant proteins, and their food sources in 10 countries in the European prospective investigation into cancer and nutrition. Eur. J. Clin. Nutr. 63 (Suppl. 4), S16–36.

Halkjaer, J., Olsen, A., Overvad, K., Jakobsen, M.U., Boeing, H., Buijsse, B., et al., 2011. Intake of total, animal and plant protein and subsequent changes in weight or waist circumference in European men and women: the Diogenes project. Int. J. Obes. (Lond) 35, 1104–1113.

Hoffer, L.J., 2012. Protein requirement of school-age children. Am. J. Clin. Nutr. 95, 777 (author reply 777–778).

Humayun, M.A., Elango, R., Ball, R.O., Pencharz, P.B., 2007. Reevaluation of the protein requirement in young men with the indicator amino acid oxidation technique. Am. J. Clin. Nutr. 86, 995–1002.

Jackson, A.A., Gibson, N.R., Lu, Y., Jahoor, F., 2004. Synthesis of erythrocyte glutathione in healthy adults consuming the safe amount of dietary protein. Am. J. Clin. Nutr. 80, 101–107.

Jacob, J.A., Nair, M.K., 2012. Protein and micronutrient supplementation in complementing pubertal growth. Indian J. Pediatr. 79 (Suppl. 1), S84–91.

Jobgen, W., Meininger, C.J., Jobgen, S.C., Li, P., Lee, M.J., Smith, S.B., et al., 2009. Dietary L-arginine supplementation reduces white fat gain and enhances skeletal muscle and brown fat masses in diet-induced obese rats. J. Nutr. 139, 230–237.

Jones, D.P., Park, Y., Gletsu-Miller, N., Liang, Y., Yu, T., Accardi, C.J., et al., 2011. Dietary sulfur amino acid effects on fasting plasma cysteine/cystine redox potential in humans. Nutrition 27, 199–205.

Knuiman, P., Kramer, I.F., 2012. Contributions to the understanding of the anabolic properties of different dietary proteins. J. Physiol. 590, 2839–2840.

Leidy, H.J., Carnell, N.S., Mattes, R.D., Campbell, W.W., 2007. Higher protein intake preserves lean mass and satiety with weight loss in preobese and obese women. Obesity (Silver Spring) 15, 421–429.

Levin, B.E., 2009. Synergy of nature and nurture in the development of childhood obesity. Int. J. Obes. (Lond) 33 (Suppl. 1), S53–56.

Li, M., Sun, F., Piao, J.H., Yang, X.G., 2014. Protein requirements in healthy adults: a meta-analysis of nitrogen balance studies. Biomed. Environ. Sci. 27, 606–613.

Lynch, C.J., Adams, S.H., 2014. Branched-chain amino acids in metabolic signalling and insulin resistance. Nat. Rev. Endocrinol. 10, 723–736.

Macotela, Y., Emanuelli, B., Bang, A.M., Espinoza, D.O., Boucher, J., Beebe, K., et al., 2011. Dietary leucine—an environmental modifier of insulin resistance acting on multiple levels of metabolism. PLoS One 6, e21187.

Magne, H., Savary-Auzeloux, I., Migne, C., Peyron, M.A., Combaret, L., Remond, D., et al., 2012. Contrarily to whey and high protein diets, dietary free leucine supplementation cannot reverse the lack of recovery of muscle mass after prolonged immobilization during ageing. J. Physiol. 590, 2035–2049.

Magne, J., Huneau, J.F., Tsikas, D., Delemasure, S., Rochette, L., Tome, D., et al., 2009. Rapeseed protein in a high-fat mixed meal alleviates postprandial systemic and vascular oxidative stress and prevents vascular endothelial dysfunction in healthy rats. J. Nutr. 139, 1660–1666.

Mamerow, M.M., Mettler, J.A., English, K.L., Casperson, S.L., Arentson-Lantz, E., Sheffield-Moore, M., et al., 2014. Dietary protein distribution positively influences 24-h muscle protein synthesis in healthy adults. J. Nutr. 144, 876–880.

Marini, J.C., 2015. Protein requirements: are we ready for new recommendations? J. Nutr. 145, 5–6.

Mariotti, F., Pueyo, M.E., Tome, D., Berot, S., Benamouzig, R., Mahe, S., 2001. The influence of the albumin fraction on the bioavailability and postprandial utilization of pea protein given selectively to humans. J. Nutr. 131, 1706–1713.

Mariotti, F., Hermier, D., Sarrat, C., Magne, J., Fenart, E., Evrard, J., et al., 2008. Rapeseed protein inhibits the initiation of insulin resistance by a high-saturated fat, high-sucrose diet in rats. Br. J. Nutr. 100, 984–991.

Mariotti, F., Petzke, K.J., Bonnet, D., Szezepanski, I., Bos, C., Huneau, J.F., et al., 2013. Kinetics of the utilization of dietary arginine for nitric oxide and urea synthesis: insight into the arginine-nitric oxide metabolic system in humans. Am. J. Clin. Nutr. 97, 972–979.

Markofski, M.M., Volpi, E., 2011. Protein metabolism in women and men: similarities and disparities. Curr. Opin. Clin. Nutr. Metab. Care 14, 93–97.

Martens, E.A., Westerterp-Plantenga, M.S., 2014. Protein diets, body weight loss and weight maintenance. Curr. Opin. Clin. Nutr. Metab. Care 17, 75–79.

Michaelsen, K.F., Greer, F.R., 2014. Protein needs early in life and long-term health. Am. J. Clin. Nutr. 99, 718S–722S.

Millward, D.J., 2014. Protein requirements and aging. Am. J. Clin. Nutr. 100, 1210–1212.

Millward, D.J., Jackson, A.A., 2004. Protein/energy ratios of current diets in developed and developing countries compared with a safe protein/energy ratio: implications for recommended protein and amino acid intakes. Public Health Nutr. 7, 387–405.

Millward, D.J., Jackson, A.A., 2012. Protein requirements and the indicator amino acid oxidation method. Am. J. Clin. Nutr. 95, 1498–1501 (author reply 1501–1492).

Millward, D.J., Roberts, S.B., 1996. Protein requirements of older individuals. Nutr. Res. Rev. 9, 67–87.

Millward, D.J., Fereday, A., Gibson, N.R., Cox, M.C., Pacy, P.J., 2002. Efficiency of utilization of wheat and milk protein in healthy adults and apparent lysine requirements determined by a single-meal [1-13C]leucine balance protocol. Am. J. Clin. Nutr. 76, 1326–1334.

Millward, D.J., Layman, D.K., Tome, D., Schaafsma, G., 2008. Protein quality assessment: impact of expanding understanding of protein and amino acid needs for optimal health. Am. J. Clin. Nutr. 87, 1576S–1581S.

Morais, J.A., Chevalier, S., Gougeon, R., 2006. Protein turnover and requirements in the healthy and frail elderly. J. Nutr. Health Aging 10, 272–283.

Mosoni, L., Gatineau, E., Gatellier, P., Migne, C., Savary-Auzeloux, I., Remond, D., et al., 2014. High whey protein intake delayed the loss of lean body mass in healthy old rats, whereas protein type and polyphenol/antioxidant supplementation had no effects. PLoS One 9, e109098.

Naude, C.E., Schoonees, A., Senekal, M., Young, T., Garner, P., Volmink, J., 2014. Low carbohydrate versus isoenergetic balanced diets for reducing weight and cardiovascular risk: a systematic review and meta-analysis. PLoS One 9, e100652.

Newgard, C.B., 2012. Interplay between lipids and branched-chain amino acids in development of insulin resistance. Cell Metab. 15, 606–614.

Newgard, C.B., An, J., Bain, J.R., Muehlbauer, M.J., Stevens, R.D., Lien, L.F., et al., 2009. A branched-chain amino acid-related metabolic signature that differentiates obese and lean humans and contributes to insulin resistance. Cell Metab. 9, 311–326.

Paddon-Jones, D., Leidy, H., 2014. Dietary protein and muscle in older persons. Curr. Opin. Clin. Nutr. Metab. Care 17, 5–11.

Paddon-Jones, D., Rasmussen, B.B., 2009. Dietary protein recommendations and the prevention of sarcopenia. Curr. Opin. Clin. Nutr. Metab. Care 12, 86–90.

Pannemans, D.L., Wagenmakers, A.J., Westerterp, K.R., Schaafsma, G., Halliday, D., 1997. The effect of an increase of protein intake on whole-body protein turnover in elderly women is tracer dependent. J. Nutr. 127, 1788–1794.

Pedersen, A.N., Cederholm, T., 2014. Health effects of protein intake in healthy elderly populations: a systematic literature review. Food Nutr. Res. 58, 23364. Available from: http://dx.doi.org/10.3402/fnr.v58.23364.

Pennings, B., Boirie, Y., Senden, J.M., Gijsen, A.P., Kuipers, H., van Loon, L.J., 2011. Whey protein stimulates postprandial muscle protein accretion more effectively than do casein and casein hydrolysate in older men. Am. J. Clin. Nutr. 93, 997–1005.

Petzke, K.J., Freudenberg, A., Klaus, S., 2014. Beyond the role of dietary protein and amino acids in the prevention of diet-induced obesity. Int. J. Mol. Sci. 15, 1374–1391.

Pillai, R.R., Kurpad, A.V., 2012. Amino acid requirements in children and the elderly population. Br. J. Nutr. 108 (Suppl. 2), S44–49.

Pillai, R.R., Elango, R., Muthayya, S., Ball, R.O., Kurpad, A.V., Pencharz, P.B., 2010. Lysine requirement of healthy, school-aged Indian children determined by the indicator amino acid oxidation technique. J. Nutr. 140, 54–59.

Poupin, N., Bos, C., Mariotti, F., Huneau, J.F., Tome, D., Fouillet, H., 2011. The nature of the dietary protein impacts the tissue-to-diet 15N discrimination factors in laboratory rats. PLoS One 6, e28046.

Poupin, N., Mariotti, F., Huneau, J.F., Hermier, D., Fouillet, H., 2014. Natural isotopic signatures of variations in body nitrogen fluxes: a - compartmental model analysis. PLoS Comput. Biol. 10, e1003865.

Qin, L.Q., Xun, P., Bujnowski, D., Daviglus, M.L., Van Horn, L., Stamler, J., et al., 2011. Higher branched-chain amino acid intake is associated with a lower prevalence of being overweight or obese in middle-aged East Asian and Western adults. J. Nutr. 141, 249–254.

Rafii, M., Chapman, K., Owens, J., Elango, R., Campbell, W.W., Ball, R.O., et al., 2015. Dietary protein requirement of female adults >65 years determined by the indicator amino Acid oxidation technique is higher than current recommendations. J. Nutr. 145, 18–24.

Rand, W.M., Pellett, P.L., Young, V.R., 2003. Meta-analysis of nitrogen balance studies for estimating protein requirements in healthy adults. Am. J. Clin. Nutr. 77, 109–127.

Rieu, I., Magne, H., Savary-Auzeloux, I., Averous, J., Bos, C., Peyron, M.A., et al., 2009. Reduction of low grade inflammation restores blunting of postprandial muscle anabolism and limits sarcopenia in old rats. J. Physiol. 587, 5483–5492.

Rodriguez, N.R., 2014. Protein-centric meals for optimal protein utilization: can it be that simple? J. Nutr. 144, 797–798.

Sacks, F.M., Bray, G.A., Carey, V.J., Smith, S.R., Ryan, D.H., Anton, S.D., et al., 2009. Comparison of weight-loss diets with different compositions of fat, protein, and carbohydrates. N. Engl. J. Med. 360, 859–873.

Seki, Y., Williams, L., Vuguin, P.M., Charron, M.J., 2012. Minireview: epigenetic programming of diabetes and obesity: animal models. Endocrinology 153, 1031–1038.
She, P., Olson, K.C., Kadota, Y., Inukai, A., Shimomura, Y., Hoppel, C.L., et al., 2013. Leucine and protein metabolism in obese Zucker rats. PLoS One 8, e59443.
Sherman, H.C., 1920. The protein requirement of maintenance in man. Proc. Natl. Acad. Sci. USA 6, 38–40.
Stephens, T.V., Payne, M., Ball, R.O., Pencharz, P.B., Elango, R., 2015. Protein requirements of healthy pregnant women during early and late gestation are higher than current recommendations. J. Nutr. 145, 73–78.
Tan, B., Yin, Y., Liu, Z., Li, X., Xu, H., Kong, X., et al., 2009. Dietary L-arginine supplementation increases muscle gain and reduces body fat mass in growing-finishing pigs. Amino Acids 37, 169–175.
Tang, M., McCabe, G.P., Elango, R., Pencharz, P.B., Ball, R.O., Campbell, W.W., 2014. Assessment of protein requirement in octogenarian women with use of the indicator amino acid oxidation technique. Am. J. Clin. Nutr. 99, 891–898.
Tieland, M., Borgonjen-Van den Berg, K.J., van Loon, L.J., de Groot, L.C., 2012. Dietary protein intake in community-dwelling, frail, and institutionalized elderly people: scope for improvement. Eur. J. Nutr. 51, 173–179.
Valenzuela, R.E., Ponce, J.A., Morales-Figueroa, G.G., Muro, K.A., Carreon, V.R., Aleman-Mateo, H., 2013. Insufficient amounts and inadequate distribution of dietary protein intake in apparently healthy older adults in a developing country: implications for dietary strategies to prevent sarcopenia. Clin. Interv. Aging 8, 1143–1148.
Volkert, D., Kreuel, K., Heseker, H., Stehle, P., 2004. Energy and nutrient intake of young-old, old-old and very-old elderly in Germany. Eur. J. Clin. Nutr. 58, 1190–1200.
Volpi, E., Mittendorfer, B., Wolf, S.E., Wolfe, R.R., 1999. Oral amino acids stimulate muscle protein anabolism in the elderly despite higher first-pass splanchnic extraction. Am. J. Physiol. 277, E513–520.
Volpi, E., Campbell, W.W., Dwyer, J.T., Johnson, M.A., Jensen, G.L., Morley, J.E., et al., 2013. Is the optimal level of protein intake for older adults greater than the recommended dietary allowance? J. Gerontol. A Biol. Sci. Med. Sci. 68, 677–681.
Walrand, S., Short, K.R., Bigelow, M.L., Sweatt, A.J., Hutson, S.M., Nair, K.S., 2008. Functional impact of high protein intake on healthy elderly people. Am. J. Physiol. Endocrinol. Metab. 295, E921–928.
Wang, T.J., Larson, M.G., Vasan, R.S., Cheng, S., Rhee, E.P., McCabe, E., et al., 2011. Metabolite profiles and the risk of developing diabetes. Nat. Med. 17, 448–453.
WHO/FAO, 2003. Diet, nutrition and the prevention of chronic diseases. World Health Organ Tech Rep Ser, 1–160, back cover.
WHO/FAO/UNU, 2007. Protein and amino acid requirements in human nutrition. World Health Organ Tech Rep Ser, 1–265, back cover.
Wiseman, M., 2004. The feast of the assumptions. Public Health Nutr. 7, 385.
Wu, Z., Satterfield, M.C., Bazer, F.W., Wu, G., 2012. Regulation of brown adipose tissue development and white fat reduction by L-arginine. Curr. Opin. Clin. Nutr. Metab. Care 15, 529–538.
Yang, R., Dong, J., Zhao, H., Li, H., Guo, H., Wang, S., et al., 2014. Association of branched-chain amino acids with carotid intima-media thickness and coronary artery disease risk factors. PLoS One 9, e99598.
Young, V.R., Borgonha, S., 2000. Nitrogen and amino acid requirements: the Massachusetts Institute of Technology amino acid requirement pattern. J. Nutr. 130, 1841S–1849S.
Zhang, Y., Guo, K., LeBlanc, R.E., Loh, D., Schwartz, G.J., Yu, Y.H., 2007. Increasing dietary leucine intake reduces diet-induced obesity and improves glucose and cholesterol metabolism in mice via multimechanisms. Diabetes 56, 1647–1654.

CHAPTER

3

Cellular Mechanisms of Protein Degradation Among Tissues

L. Combaret[1,2], D. Taillandier[1,2], C. Polge[1,2], D. Béchet[1,2] and D. Attaix[1,2]

[1]Clermont Université, Université d'Auvergne, Unité de Nutrition Humaine, Clermont-Ferrand, France
[2]INRA, UMR 1019, UNH, CRNH Auvergne, Saint Genès Champanelle, France

3.1. INTRODUCTION

Proteolysis or intracellular protein degradation has key roles in mammalian cells. First this process is involved in the immune response and in the elimination of invasive pathogens. Second, proteolysis rapidly eliminates abnormal or defective proteins, preventing a deleterious accumulation of such proteins. Third, protein breakdown provides the body with free amino acids when dietary protein and/or energy requirements are not met. These amino acids can be used as either an energy source or for the synthesis of proteins essential for survival. Fourth, proteolysis can quickly alter functional protein levels resulting in a fine-tuning of cell metabolism in response to any challenge. For example, it has become clear over recent decades that proteolysis plays a key role in both cell division and proliferation or death by apoptosis and is involved in the regulation of intercellular and intracellular protein trafficking. Detailing all these roles is out of scope of the present review. We focus here on the tissue-specific features of protein breakdown, which are still poorly understood.

3.2. PROTEOLYTIC SYSTEMS

At least five major proteolytic systems (lysosomal, Ca^{2+}-dependent, caspase-dependent, ubiquitin-proteasome-dependent, and metalloproteinases) operate in the body. Although ubiquitous, the relative importance of each pathway varies in a given tissue or organ depending on intrinsic and extrinsic factors (ie, health status, genetics, exercise, dietary habits...).

3.2.1 Ca^{2+}-Dependent Proteolysis

This pathway is composed of cysteine proteases named calpains. They are ubiquitous (μ and m-calpains) or tissue-specific enzymes, and are involved in limited proteolytic events. Ubiquitous calpain activities play a role in a large number of physiological and pathological processes, for example, cell motility by remodeling cytoskeletal anchorage complexes, control of cell cycle, or apoptosis. In skeletal muscle, calpains are involved in regenerative processes (for a review see Dargelos et al., 2008). In addition, in muscular dystrophies characterized by an increased efflux of calcium, calpain expression and activity increased concomitantly with enhanced proteolysis (Alderton and Steinhardt, 2000; Combaret et al., 1996). Mutations in the *capn3* gene coding for the skeletal muscle-specific isoform of calpain, calpain-3, result in LGMD2A and other calpainopathies (see Ono et al., 2016 for a recent review), and partial inhibition of calpain-3 leads to disorganization of the sarcomeres (Poussard et al., 1996). Like ubiquitous calpains, calpain-3 cleaves many cytoskeletal proteins and is involved in cytoskeleton regulation, and adaptive

The Molecular Nutrition of Amino Acids and Proteins.
DOI: http://dx.doi.org/10.1016/B978-0-12-802167-5.00003-7

responses to exercise or regeneration after muscle wasting. A putative role of calpains might be the initial cleavage of several myofibrillar proteins, making them accessible for further degradation by the ubiquitin–proteasome pathway (see Section 3.2.3). Calpain activities increased in several tissues (red cells, nervous cells) in aging. Although little is known on the regulation of this pathway in sarcopenic muscle, calcium homeostasis is modified in skeletal muscle, so that resting calcium concentrations increased (Fulle et al., 2004; Fraysse et al., 2006). The resulting enhanced calpain activity may account for myofibrillar degradation.

3.2.2 Caspases

Caspases are proteases with a well-defined role in apoptosis (see Jin and El-Deiry, 2005 for a complete description of apoptotic pathways and of the regulation of caspases). They are involved in limited proteolysis of substrates. The role of caspases in muscle proteolysis will be described below as they may participate in the disorganization of the myofibrillar structure of skeletal muscles (see Section 3.3.3). Increased evidence indicates that caspases play multiple functions outside apoptosis (eg, in inflammation, necroptosis, immunity, tissue differentiation. . .). For a recent review, see Shalini et al. (2015).

3.2.3 The Ubiquitin-Proteasome System

Basically, there are two main steps in this pathway: (1) the covalent attachment of a polyubiquitin degradation signal to the substrate by ubiquitination enzymes; and (2) the specific recognition of the polyubiquitin chain and the subsequent breakdown into peptides of the targeted protein by the 26S proteasome.

3.2.3.1 Ubiquitination

Covalent modification of proteins by ubiquitin (Ub) is highly sophisticated and polyvalent. The attachment of Ub to a substrate can be monomeric, attached in chains using any of the seven internal lysine residues of Ub or even combined with other Ub-like modifiers (Ravid and Hochstrasser, 2008; Kravtsova-Ivantsiv and Ciechanover, 2012; Ciechanover and Stanhill, 2014). The type of Ub chains built onto proteins is associated with known functions such as targeting the substrate to proteasome-dependent proteolysis (Lys48, Lys11), NFκB activation, DNA repair or targeting to lysosomes (Lys63), and unknown functions (Lys6, Lys27, Lys29, Lys33) (for review see Ye and Rape, 2009; Polge et al., 2013). The whole process is highly specific and tightly regulated in response to catabolic stimuli to avoid unwanted degradation of proteins. The first steps of the ubiquitin-proteasome system (UPS) are dedicated to substrate recognition and thus represent a crucial point for controlling the substrate fate. This is also a potential entry for developing therapeutic strategies. Ubiquitination of substrates involves several hundreds of enzymes distributed in three classes that act in cascade (Polge et al., 2013).

Ub is first activated by a single E1 (Ub-activating enzyme) that transfers high energy Ub to one of the 35 E2s (Ub-conjugating enzymes) in humans (van Wijk and Timmers, 2010). The E2s transfer Ub on target proteins in conjunction with the third class of enzymes, named E3 ligases (>600, Metzger et al., 2012). An E2 is able to cooperate with different E3s and vice versa, which enables the specific targeting of virtually any cellular protein. E3s recognize the target protein to be degraded and thus bring specificity to the ubiquitination machinery but most E3s lack enzymatic activity. Therefore, each E2–E3 couple is functionally more relevant. Proteins carrying Ub chains linked through Lys48 are bona fide substrates for the 26S proteasome. The latter recognizes these Ub chains as a degradation signal, trims the Ub moieties, and degrades the target protein into small peptides.

3.2.3.2 Proteasome Degradation

The eukaryotic 26S proteasome is constituted of a proteolytic chamber referred to as the 20S core particle (CP) and a regulatory particle (RP) that contains ATPases.

The CP consists of four axially stacked hetero-heptameric rings. The outer rings consist of seven different α-subunits (α1–α7). The inner rings contain seven distinct β-subunits (β1–β7). The β1-, β2-, and β5-subunits contain the proteolytic active sites, that is, chymotrypsin-, trypsin-, or caspase-like activities that cleave preferentially after particular amino acid residues. In mammals, three additional β-subunits (ie, β1i, β2i, β5i) are induced by specific stimuli, namely interferon-γ production. These inducible subunits replace the canonical β1-, β2-, and β5-subunits and modulate CP proteolytic activity. This results in the generation of peptides that can be loaded onto the class I major histocompatibility complex for immune presentation to killer T cells (for review see Kniepert and Groettrup, 2014).

The RP is responsible for the gating of the CP α-rings and for the binding, deubiquitination, unfolding, and translocation of substrates into the proteolytic chamber of the CP. The RP contains at least 19 subunits and is composed of two subcomplexes, the lid and the base. The base consists of 10 subunits: six ATPases (S4, S6, S6′, S7, S8, and S10b) that form a hexameric ring, and four RP non-ATPase subunits (S1, S2, S5a, and ADRM1). The lid consists of nine different subunits, S3, S9, S10a, S11, S12, S13, S14, S15, and p55 (for review see Tomko and Hochstrasser, 2013).

3.2.4 Autophagy

Lysosomes are a major component of the degradative machinery in mammalian cells. They are membrane-bounded vesicles containing high concentrations of various acid hydrolases, which typically present an acidic lumen (pH 4–5) and a high density (Kirschke and Barrett, 1985). Lysosomal hydrolases contain proteases, glycosidases, lipases, nucleases, and phosphatases. Lysosomes therefore act as intracellular compartments dedicated to the degradation of a variety of macromolecules. Should they escape from lysosomes, acid hydrolases can be devastating for cellular or extracellular constituents. Therefore, accurate synthesis, processing, and sorting of lysosomal hydrolases to endosomes/lysosomes, not only determine the capacity for lysosomal proteolysis, but are also vital for cellular homeostasis. The lumen of lysosomes topographically corresponds to the extracellular milieu. Lysosomal hydrolases are therefore implicated in the degradation of extracellular constituents, which may reach lysosomes by endocytosis, pinocytosis, or phagocytosis. Endocytosis and secretion pathways also deliver cell membranes and vesicles to endosomes/lysosomes, and hence lysosomes play a central role in the turnover of membrane lipids and transmembrane proteins. Lysosomes are further implicated in the turnover of cytoplasmic soluble constituents, and in the breakdown of cellular organelles including mitochondria, peroxisomes, and even nuclei (Roberts, 2005). In contrast to the other proteolytic systems (proteasomes, calpains) involved in the degradation of intracellular proteins, lysosomal hydrolases are physically isolated from cytoplasmic constituents by the lysosomal membrane. Various mechanisms of autophagy are then essential to deliver cytoplasmic substrates inside lysosomes. Delivery of substrates, together with lysosomal hydrolytic capacity, will specify the role of lysosomes in overall intracellular proteolysis.

Schematically, lysosomal-dependent degradation of cytoplasmic constituents (autophagy) involves the initial sequestration of protein substrates into the vacuolar system and their subsequent hydrolysis by lysosomal hydrolases. Different pathways may be used to deliver intracellular protein substrates to lysosomes, including macroautophagy, named autophagy in the next sections (for a detailed description of autophagy and of its regulation by nutrients and metabolites, see chapter: Regulation of Macroautophagy by Nutrients and Metabolites by Lorin et al.).

3.2.5 Metalloproteinases

These enzymes are involved in the degradation of the extracellular matrix (ECM), but also regulate ECM assembly, structure, and quantity, and are key participants in diverse immune and inflammatory processes. For a review of metalloproteinases and their role, see Tallant et al. (2010) and Gaffney et al. (2015). The role of these proteinases will not be described in this chapter.

3.3. SKELETAL MUSCLE PROTEOLYSIS

3.3.1 UPS: The Main Player for Myofibrillar Protein Degradation

3.3.1.1 Role of the E1 Enzyme

E1 has low expression in skeletal muscle and its mRNA level is not regulated in catabolic states (Lecker et al., 1999). This is not surprising because (1) E1 is an extremely active enzyme capable of charging excess amounts of E2s with ubiquitin (K_m values for E2s of ~100 pM) and (2) E1 is a common element in all pathways of ubiquitin conjugation. Thus, any E1 impairment affects the whole downstream ubiquitination cascade.

3.3.1.2 Role of E2 Enzymes

E2 enzymes determine the type of chain built on the substrate and thus whether the ubiquitination of the target protein is dedicated to degradation or to other fates (signaling, modulation of activity, etc.). Thus, E2s are central players in the ubiquitination machinery but the exact role of E2s in the development of skeletal muscle atrophy is still an open question. Indeed, our knowledge on the role of E2s during skeletal muscle atrophy relies almost exclusively on descriptive observations (mRNA levels) or on in vitro ubiquitination assays. The former are not really informative about mechanisms and specific features of E2s may bias the latter. Thirty-five E2s (plus 2 putative) are described in the human genome and have been grouped into four different classes (van Wijk and Timmers, 2010).

Class I E2s are the most studied. UBE2B/14-kDa E2 is abundant in skeletal muscle (Wing and Banville, 1994). UBE2B mRNA levels are tightly linked to muscle wasting whatever the catabolic stimuli is, suggesting a major role for UBE2B in a ubiquitous atrophying program (for a recent review see Polge et al., 2015a). In addition, UBE2B mRNA levels are also downregulated by anabolic stimuli (IGF-1, insulin, reloading) (Taillandier et al., 2003; Wing and Banville, 1994; Wing and Bedard, 1996). However, depressing UBE2B had only a limited impact on muscle protein ubiquitination during fasting suggesting compensatory mechanisms (Adegoke et al., 2002). Few studies confirmed a role for UBE2B at the protein level, but most antibodies cross-react with the isoform UBE2A. The latter is suspected to compensate for the loss of function of UBE2B (Adegoke et al., 2002). Expression at the mRNA levels may thus not be sufficient for proving that UBE2B is important for muscle homeostasis, but in rats submitted to unweighting atrophy, increased UBE2B mRNA levels correlated with efficient translation (Taillandier et al., 1996). In fasting, UBE2B protein levels were not modified while mRNA levels were elevated (Adegoke et al., 2002), possibly because UBE2B turnover increased in atrophying muscles. However, UBE2B interacts with several E3s and seems implicated in myofibrillar protein loss in catabolic C2C12 myotubes in the soluble protein fraction (Polge et al., 2015b).

The UBE2D family of E2s (also belonging to Class I) exhibits ubiquitination activity in vitro with a large number of E3s towards various substrates, including the major contractile proteins (actin, myosin heavy chain, troponin) along with the MuRF1 E3 enzyme in vitro (see below). However, UBE2D (1) is not upregulated in any catabolic situation, (2) exhibits low specificity toward E3s and substrates in vitro, and (3) lacks specificity for Ub chain linkage in vitro. Altogether, these observations do not support a major role of UBE2D in the muscle atrophying program (for a recent review see Polge et al., 2015a).

There are few studies addressing the role of other E2 enzymes in skeletal muscle. Among these E2s, UBE2G1, UBE2G2, UBEL3, UBEO, UBE2J1 were regulated in skeletal muscle at the mRNA levels upon catabolic stimuli (chronic renal failure, diabetes, fasting, cancer, and disuse).

Altogether, studies on the role of E2s in muscle are lacking and the paucity of available data does not enable the emergence of a clear picture of their precise role in muscle wasting.

3.3.1.3 *Role of E3 Enzymes*

Different E3 ligases have been implicated in muscle atrophy and/or development. A single report has described a very Large E3 (E3L), which was involved in the in vitro breakdown of ubiquitinated actin, troponin T, and MyoD (Gonen et al., 1996). In catabolic muscles, several groups (eg, Lecker et al., 1999) have reported increases in mRNA levels for E3α1, the ubiquitous N-end rule RING (Really Interesting New Genes) finger ligase that functions with the UBE2B/14-kDa E2. However, such changes were not associated with altered protein levels of E3α1. Furthermore, E3α1 has presumably little significant physiological role in atrophying muscles. First, E3α1 is involved in the ubiquitination of soluble muscle proteins, not of myofibrillar proteins (Lecker et al., 1999). Second, mice lacking the E3α1 gene are viable and fertile, and only exhibited smaller skeletal muscles than control animals (Kwon et al., 2001).

There are several muscle-specific RING finger E3s that include MuRF-1, -2, -3 (Muscle RING Finger proteins-1–3; Centner et al., 2001), SMRZ (Striated Muscle RING Zinc finger; Dai and Liew, 2001), ANAPC11 (ANAphase Promoting Complex; Chan et al., 2001), and MAFbx/Atrogin-1 (Bodine et al., 2001; Gomes et al., 2001).

Multiple studies showed that MAFbx/Atrogin-1 and MuRF1 expression increased by at least 6–10 times in several catabolic conditions including muscle disuse, hindlimb suspension, denervation, and glucocorticoid- or interleukin-1-induced muscle atrophy (Bodine et al., 2001) as well as in fasting, cancer cachexia, diabetes, and renal failure (Gomes et al., 2001). Moreover, knockout mice for either E3 were partially resistant to muscle wasting (Bodine et al., 2001). MAFbx/Atrogin-1 is overexpressed in nearly any catabolic situation (Bodine and Baehr, 2014). However, studies in different laboratories reported no correlation between the expression of MAFbx/Atrogin-1 and rates of protein breakdown both in rat muscles (Krawiec et al., 2005; Fareed et al., 2006) and in C2C12 myotubes (Dehoux et al., 2007). Attaix and Baracos (2010) pointed out such discrepancies that also prevailed in human studies (Murton et al., 2008; Murton and Greenhaff, 2010). Well characterized MAFbx/Atrogin-1 substrates include the MyoD transcription factor (Tintignac et al., 2005) and the elongation factor eIF3f

(Lagirand-Cantaloube et al., 2008). Proteolysis of MyoD or eIF3f is expected to influence muscle differentiation and protein synthesis. Overall, the exact role of this E3 could be much more complex. Indeed, a recent study has shown that MAFbx/atrogin-1 mediated the interplay between the UPS and the autophagy/lysosome system with beneficial effects in cardiomyocytes (Zaglia et al., 2014).

MuRF1 is upregulated in nearly any catabolic situation by different transcription factors that include FoxOs (see Sandri, 2013; Bodine and Baehr, 2014 for recent reviews). Interestingly, MuRF1 (and perhaps the MuRF3 isoform) targets the major myofibrillar proteins for subsequent degradation by the 26S proteasome (Kedar et al., 2004; Clarke et al., 2007; Fielitz et al., 2007; Polge et al., 2011). However, a yet unanswered question is the identity of the E2(s) that work(s) in pair(s) with MuRF1. Indeed, MuRF1 belongs to the RING finger E3 ligase family. These E3s are the most numerous but do not possess any catalytic activity. They rely on E2 enzymes for conjugating Ub to the proteins to be degraded.

Other RING E3 ligases like TRAF6 may also have a role in the atrophying program. They also require specific E2s for properly targeting substrates for degradation (Kudryashova et al., 2005; Hishiya et al., 2006). Furthermore, recent studies have identified other FoxO-dependent E3s called MUSA1 and SMART (Specific of Muscle Atrophy and Regulated by Transcription) (Milan et al., 2015) that also seem important for muscle atrophy.

3.3.1.4 Role of the Proteasome

The demonstration that only proteasome inhibitors (lactacystin, MG132) suppress the enhanced rates of overall proteolysis in atrophying muscles provided strong support for a major role of the proteasome in the breakdown of myofibrillar proteins. Proteasomes are tightly associated with myofibrils in mature skeletal muscle (Bassaglia et al., 2005). Studies with artificial substrates have shown that the chymotrypsin-like peptidase activity increases in some muscle wasting conditions, but is unchanged in diabetes (reviewed in Attaix et al., 2005). Discrepancies between rates of overall muscle proteolysis and some specific proteasome activities may have several explanations. First, the 20S proteasome population comprises at least six distinct subtypes in skeletal muscle, including constitutive proteasomes, immunoproteasomes, and their intermediate forms. Thus the properties of a 20S proteasome population isolated from muscle represent the average properties of the whole set of proteasomes subtypes. Secondly, the hydrolysis of artificial substrates may not reflect the in vivo situation with endogenous substrates. However, and by contrast, both chymotrypsin- and trypsin-like peptidase activities were reduced when skeletal muscle proteasome-dependent proteolysis was impaired by chemotherapy (Tilignac et al., 2002; Attaix et al., 2005).

Numerous groups have reported that enhanced ATP- and/or proteasome-dependent rates of muscle proteolysis correlate with elevated mRNA levels for the catalytic and noncatalytic subunits of the 20S proteasome (Attaix et al., 2005; Jagoe and Goldberg, 2001; Combaret et al., 2002, 2004; Price, 2003). However, gene array experiments have shown that a small number of subunits are actually overexpressed in different muscle wasting conditions (Lecker et al., 2004). There is very limited information about the protein levels of 20S proteasome subunits in catabolic states. Increased protein abundance of one 20S proteasome subunit correlates with enhanced mRNA levels for other subunits in cancer cachexia (reviewed in Attaix et al., 2005). Conversely, when proteasome-dependent proteolysis was inhibited by chemotherapy to below basal levels, mRNA levels for 20S proteasome subunits correlated with reduced protein levels of the two subunits (Tilignac et al., 2002). The overexpressed α-4 subunit entered active translation in the atrophying unweighted soleus muscle (Taillandier et al., 2003), and an increase in transcribed proteasome subunit mRNA was observed in acidosis (Price, 2003). Glucocorticoids (Attaix et al., 2005; Jagoe and Goldberg, 2001; Price, 2003) and TNF-α (tumor necrosis factor-α) (Attaix et al., 2005; Combaret et al., 2003) upregulate mRNA levels for 20S proteasome subunits. Glucocorticoids induce proteasome α-2 subunit transcription in L6 muscle cells by opposing the suppression of its transcription by NF-κB, whereas the glucocorticoid-dependent increased transcription of ubiquitin involves Sp1 and MEK1 (Price, 2003). Thus the increased coordinated transcription of several genes in the UPS results from the activation of alternative signaling pathways.

Some, but not all, mRNA levels for ATPase and non-ATPase subunits of the 19S complex are also upregulated in muscle wasting. However, this upregulation clearly depends on a given catabolic state (Attaix et al., 2005; Combaret et al., 2004). Furthermore, the mRNA levels and protein contents of the individual 19S subunits are regulated independently, and do not systematically correlate with rates of proteolysis (Combaret et al., 2003; Tilignac et al., 2002). The selective increased expression of some 20S or 19S proteasome subunits strongly suggests that these subunits may be rate-limiting in the assembly of the mature complex (Lecker et al., 2004). Furthermore, these findings also suggest that, in muscle, in contrast to findings in yeast, different transcription factors or coregulators affect the expression of subgroups of proteasome subunits.

3.3.2 Autophagy-Lysosome System in Skeletal Muscle

3.3.2.1 *Role of Cathepsins*

High levels of expression of cathepsins are found in tissues with high rates of protein turnover (kidney, spleen, liver, and placenta). By contrast, low concentrations of cathepsins prevail in slowly turning-over skeletal muscles. Slow-twitch oxidative muscles exhibit higher levels of cathepsins than fast-twitch glycolytic muscles. The most abundant cathepsins in skeletal muscle are 3 cysteine proteases (cathepsin B, H, and L) and the aspartic protease cathepsin D. Cathepsins are involved in muscle development (proliferation and fusion of myoblasts, generation of secondary myotubes, formation and alignment of myofibers, and organization into bundle; see Bechet et al., 2005 for a review). A coordinated stimulation of the lysosomal process with the ubiquitin-dependent proteasome pathway (Baracos et al., 1995; Wing and Goldberg, 1993), with Ca^{2+}-dependent calpains (Combaret et al., 1996), or with both (Mansoor et al., 1996; Taillandier et al., 1996; Voisin et al., 1996) prevails in different models of muscle wasting. Amongst all endopeptidases implicated in muscular proteolysis, cathepsin L has been identified as a reliable marker of muscle atrophy (Deval et al., 2001). mRNA levels for cathepsin L increased by several fold in different catabolic conditions (ie, dexamethasone treatment, disuse atrophy, cancer, IL-6 overexpression, fasting, etc.). Compared to other muscle lysosomal enzymes, cathepsin L is induced early in catabolic states and its expression strongly correlates with increased protein breakdown. Increased mRNA levels for other cathepsins (ie, cathepsin B and D) also occur in atrophying muscles. However, this increase is not systematically observed in all models of muscle wasting, and when any is less pronounced than those of cathepsin L (Bechet et al., 2005).

3.3.2.2 *Autophagy: A Crucial Pathway for Muscle Mass Maintenance*

Autophagy is the major proteolytic pathway implicated in the amino acid-dependent regulation of proteolysis in myotubes (Mordier et al., 2000), with a key role for the phosphatidylinositol 3-kinase activity of PI3KIII-Beclin1/Atg6 complex in the mediation of autophagy induction (Tassa et al., 2003). A very clear induction of autophagy in response to starvation has been revealed using transgenic mice expressing LC3 fused to green fluorescent protein (Mizushima et al., 2004). This response is muscle-fiber type specific, as autophagy appears rapidly and intensively in the fast-twitch extensor digitorum longus muscle, but is moderate and slow to develop in the slow-twitch soleus muscle. It is noteworthy that skeletal muscle generates only small autophagosomes, even in starved mice, whereas hepatocytes produce large autophagosomes (Mizushima et al., 2004). This observation may explain why little attention has been previously paid to autophagy in skeletal muscle. The regulation of autophagy is different in skeletal muscle compared to other tissues, such as liver. Indeed, activation of autophagy upon starvation is rapid and transient in liver, while skeletal muscle exhibited a sustained induction (Mizushima et al., 2004). This suggests that the autophagosome formation may be controlled by distinct signaling pathways during short or long periods of induced autophagy (eg, Runx1, Jumpy, Akt, P38, mTOR . . .; for a review see Sandri, 2010).

However, autophagy is now recognized as a key pathway in the control of muscle mass and function. Two autophagy genes (LC3 and Gabarap) are upregulated atrogenes, which encode proteins that are degraded when autophagosomes fuse with lysosomes. Autophagy is required for muscle atrophy induced by overexpression of the transcription factor FoxO3 (Mammucari et al., 2007). Oxidative stress that prevailed in several situations of muscle wasting (eg, disuse) has been also reported to increase autophagy (Dobrowolny et al., 2008). During aging, skeletal muscle exhibits an alteration of mitochondrial function and an activation of autophagy. Maintenance of mitochondrial biogenesis in skeletal muscles from aged animals ameliorates loss of muscle mass and prevents the increase of autophagy (Wenz et al., 2009). Thus, autophagy plays a role in acute and chronic situations of muscle wasting. If excessive autophagy is detrimental to muscle mass, suppression of autophagy (ie, using knockout mice for the critical Atg7 gene to block autophagy specifically in skeletal muscle) is not beneficial and results in atrophy, weakness, and several features of myopathy (Masiero et al., 2009). This was associated with accumulation of protein aggregates, appearance of abnormal mitochondria, and induction of oxidative stress and activation of the unfolded protein response. The inhibition of autophagy leads to abnormalities in motor neuron synapses that ultimately lead to the denervation of skeletal muscle, causing a decrease in force generation (Carnio et al., 2014).

3.3.3 Functional Cooperation of Proteolytic Systems for Myofibrillar Protein Degradation

The 26S proteasome degrades proteins only into peptides. Except when presented on MHC class I molecules, these peptides must undergo further hydrolysis into free amino acids (Attaix et al., 2001). Studies showed that the extralysosomal peptidase tripeptidyl-peptidase II (TPP II) degrades peptides generated by the proteasome (Hasselgren et al., 2002). TPP II expression, protein content and activity increased in septic muscles. In addition, the glucocorticoid receptor antagonist RU 38486 blunted these adaptations, indicating that glucocorticoids participate in the upregulation of TPP II (Wray et al., 2002). Conversely, other proteases may act upstream of the proteasome. Specific interactions between the myofibrillar proteins appear to protect them from Ub-dependent degradation, and the rate-limiting step in their degradation is probably their dissociation from the myofibril (Solomon and Goldberg, 1996). Calpains play key roles in the disassembly of sarcomeric proteins and in Z-band disintegration, resulting in the release of myofilaments (Williams et al., 1999). These data suggest that calpains are acting upstream of the proteasome (Hasselgren et al., 2002). In addition, the expression of several proteolytic genes (including cathepsin L and several components of the UPS) was downregulated in mice knocked out for the muscle-specific calpain-3 (Combaret et al., 2003). In any case, it seems important to elucidate proteolytic mechanisms both upstream and downstream of the proteasome that result in the complete degradation of muscle proteins.

3.4. PROTEOLYSIS IN VISCERA

3.4.1 Liver and Autophagy: For Regulation of Energy Metabolism

It was demonstrated several decades ago that the bulk of liver proteolysis is mainly regulated by the lysosomal pathway in the liver (see Mortimore and Pösö, 1987). More recent studies have shown that liver autophagy makes a large contribution to the maintenance of cell homeostasis and health and that amino acids released by autophagic degradation can be metabolized via gluconeogenesis for the maintenance of blood glucose (Ezaki et al., 2011; Ueno et al., 2012). Moreover, an alternative pathway of lipid metabolism through autophagy, called lipophagy, has been first described in hepatocytes. In this process of macroautophagy cells break down triglycerides and cholesterol stored in lipid droplets (for reviews see Liu and Czaja, 2013; Madrigal-Matute and Cuervo, 2016). Lipid breakdown through lipophagy leads to the release into the cytoplasm of degradation products such as free fatty acids that sustain rates of mitochondrial β-oxidation for the generation of ATP to maintain cellular energy homeostasis.

Other lines of evidence also suggest that lysosomal proteolysis plays a key role in the regulation of energy metabolism in other tissues. Pancreatic β-cell autophagy is altered during diabetes (Kaniuk et al., 2007; Masini et al., 2009) and is involved in the maintenance of β-cell mass, structure, and function. Hypothalamic inhibition of autophagy increased energy consumption and reduced energy expenditure, leading to impaired adipose lipolysis. An intriguing hypothesis would suggest that defective autophagy might cause hypothalamic inflammation and dysfunction, leading to obesity, systemic insulin resistance, and probably β-cell dysfunction. Mice with defective autophagy exhibit a default in basal and glucose-stimulated insulin release and consequently develop hyperglycemia due to insufficient insulin action in target tissues such as liver, adipose tissue, and skeletal muscle (Ebato et al., 2008; Jung et al., 2008) (for further details see Kim and Lee, 2014).

Autophagy also increased in adipose tissue from obese subjects (Kovsan et al., 2011; Jansen et al., 2012). This may reflect a compensatory role of autophagy to mitigate obesity-induced inflammation and to prevent aggravation of obesity-induced insulin resistance. Indeed, inhibition of autophagy leads to an increase of inflammatory markers (ie, IL-6 and IL-1β) in adipose tissue in correlation to the degree of obesity or adiposity (Jansen et al., 2012). However, further investigations are clearly required to address this assumption.

3.4.2 A Major Role of Autophagy in Small Intestine

The gut is highly heterogeneous and has one of the highest rates of protein turnover in the body, up to 100%/day in rodents. Proteolysis has been poorly investigated in the small intestine due to the lack of a suitable technique to quantify this process. This is particularly unfortunate since the protein mass of the small intestine is mainly controlled by proteolysis. For example in adult rats fasted for 5 days the fractional rate of protein synthesis only decreased by 26% while the small intestinal protein mass was depressed by 47% (Samuels et al., 1996). This paper also reported increased mRNA levels for cathepsins and components of the UPS in the small intestine of fasted rats. Further experiments confirmed that the latter adaptations reflected increased cathepsin and proteasome activities (Combaret et al., unpublished data). Immunofluorescence labeling of various markers of autophagy (Atg 16, LAMP1) demonstrated intense labeling of the fasted mucosa, while the labeling mainly prevailed in the serosa of fed animals. Furthermore the administration of glucagon-like peptide-2 (GLP-2, a very potent intestinal trophic factor) blunted the wasting of the small intestine of fasted rats by suppressing increased cathepsin activities, but not proteasome activities (Combaret et al., unpublished data). Altogether these data strongly suggest that small intestinal proteolysis is mainly lysosomal.

3.4.2.1 *For Amino Acids Supply to Peripheral Tissues*

Since the small intestine contributes like the liver to a significant percentage of whole body protein synthesis and, by inference, of proteolysis (approximately 10%), the rapid wasting of the viscera in response to nutrient deprivation provides the body with large amounts of free amino acids that can be used both for providing energy and for protein synthesis in peripheral tissues like skeletal muscle.

In related unpublished experiments we have shown that there is a competition between the gut and skeletal muscle for the utilization of amino acids. We did observe in previous experiments that muscle recovery is delayed in skeletal muscle compared with intestinal recovery pending totally unrelated catabolic episodes, that is, following refeeding after starvation (Kee et al., 2003) or in chemotherapy treated cancer mice (Samuels et al., 2000; Tilignac et al., 2002). This prompted us to hypothesize that priority must be given to the gut. To demonstrate this concept we limited intestinal wasting by GLP-2 in fasted and refed rats. Muscle recovery was small and incomplete in the untreated fasted and refed rats. By contrast, muscle recovery was immediate and almost total following 24 h of refeeding in the GLP-2 treated rats (Combaret et al., unpublished data). These data demonstrate that there is a cross-talk between the small intestine and peripheral tissues like skeletal muscle for the utilization of amino acids that we called the global protein turnover concept.

3.4.2.2 *For Regulation of the Epithelial Barrier*

Microbial sensing through pattern recognition receptors (PRRs) drives complementary functions in Intestinal Epithelial Cells (IECs). Basal PRR activation maintains barrier function and commensal composition, but aberrant PRR signaling may be a central contributor to the pathophysiology of inflammatory bowel diseases. Dysregulation of PRR pathways influences other processes implicated in intestinal homeostasis, such as autophagy. For example, the PRR NOD2 protein stimulates autophagy by directly interacting with the autophagy gene ATG16L1, which allows the recruitment of ATG16L1 to sites of bacterial entry. Conversely, mutant forms of NOD2 and ATG16L1 showed reduced autophagy resulting in impaired antigen presentation and bacterial killing. Thus, IECs employ autophagy to contain and eliminate invading bacteria, and deregulation of autophagy is linked to susceptibility to inflammatory intestinal diseases (Maloy and Powrie, 2011; Elson and Alexander, 2015).

Besides autophagy, the UPS can also play a role in inflammatory bowel pathogenesis. Indeed, infection of IECs with adherent-invasive *Escherichia coli* (AIEC) modulated the UPS turnover by reducing polyubiquitin conjugate accumulation, increasing 26S proteasome activities and decreasing protein levels of the NF-κB regulator CYLD, resulting in NF-κB activation. This activity was very important for the pathogenicity of AIEC since decreased CYLD resulted in increased ability of AIEC LF82 to replicate intracellularly (Cleynen et al., 2014).

3.5. CONCLUDING REMARKS

Except in part in skeletal muscle and liver, the complexity of the role of the major pathways of proteolysis, namely the UPS and autophagy, is still poorly understood. It is now becoming clear that these pathways not only play a role in endogenous proteolysis per se, but are also implicated in the control of key functions as cell division for the UPS and of lipid and/or carbohydrate metabolism for autophagy. The field of autophagy is rapidly expanding, although the signaling pathways that turn on the different processes of autophagy have only started to be elucidated. Future studies will certainly point out tissue-specific differences, but also give us a better picture of metabolism.

Acknowledgments

Studies in the laboratory of the authors are supported by grants from the Institut National de la Recherche Agronomique, the Association Française contre les Myopathies, and the Société Française de Nutrition.

References

Adegoke, O.A.J., Bedard, N., Roest, H.P., Wing, S.S., 2002. Ubiquitin-conjugating enzyme E2(14k)/HR6B is dispensable for increased protein catabolism in muscle of fasted mice. Am. J. Physiol. Endocrinol. Metab. 283, E482–E489.

Alderton, J.M., Steinhardt, R.A., 2000. How calcium influx through calcium leak channels is responsible for the elevated levels of calcium-dependent proteolysis in dystrophic myotubes. Trends Cardiovasc. Med. 10, 268–272.

Attaix, D., Baracos, V.E., 2010. MAFbx/Atrogin-1 expression is a poor index of muscle proteolysis. Curr. Opin. Clin. Nutr. Metab. Care 13, 223–224.
Attaix, D., Combaret, L., Pouch, M.N., Taillandier, D., 2001. Regulation of proteolysis. Curr. Opin. Clin. Nutr. Metab. Care 4, 45–49.
Attaix, D., Ventadour, S., Codran, A., Béchet, D., Taillandier, D., Combaret, L., 2005. The ubiquitin-proteasome system and skeletal muscle wasting. Essays Biochem. 41, 173–186.
Baracos, V.E., DeVivo, C., Hoyle, D.H., Goldberg, A.L., 1995. Activation of the ATP-ubiquitin-proteasome pathway in skeletal muscle of cachectic rats bearing a hepatoma. Am. J. Physiol. Endocrinol. Metab. 268, E996–E1006.
Bassaglia, Y., Cebrian, J., Covan, S., Garcia, M., Foucrier, J., 2005. Proteasomes are tightly associated to myofibrils in mature skeletal muscle. Exp. Cell Res. 302, 221–232.
Bechet, D., Tassa, A., Taillandier, D., Combaret, L., Attaix, D., 2005. Lysosomal proteolysis in skeletal muscle. Int. J. Biochem. Cell Biol. 37, 2098–2114.
Bodine, S.C., Baehr, L.M., 2014. Skeletal muscle atrophy and the E3 ubiquitin ligases MuRF1 and MAFbx/atrogin-1. Am. J. Physiol. Endocrinol. Metab. 307, E469–E484.
Bodine, S.C., Latres, E., Baumhueter, S., Lai, V.K.M., Nunez, L., Clarke, B.A., et al., 2001. Identification of ubiquitin ligases required for skeletal muscle atrophy. Science 294, 1704–1708.
Carnio, S., LoVerso, F., Baraibar, M.A., Longa, E., Khan, M.M., Maffei, M., et al., 2014. Autophagy impairment in muscle induces neuromuscular junction degeneration and precocious aging. Cell Rep. 8, 1509–1521.
Centner, T., Yano, J., Kimura, E., McElhinny, A.S., Pelin, K., Witt, C.C., et al., 2001. Identification of muscle specific ring finger proteins as potential regulators of the titin kinase domain. J. Mol. Biol. 306, 717–726.
Chan, A.H., Lee, S.M., Chim, S.S., Kok, L.D., Waye, M.M., Lee, C.Y., et al., 2001. Molecular cloning and characterization of a RING-H2 finger protein, ANAPC11, the human homolog of yeast Apc11p. J. Cell Biochem. 83, 249–258.
Ciechanover, A., Stanhill, A., 2014. The complexity of recognition of ubiquitinated substrates by the 26S proteasome. Biochim. Biophys. Acta 1843, 86–96.
Clarke, B.A., Drujan, D., Willis, M.S., Murphy, L.O., Corpina, R.A., Burova, E., et al., 2007. The E3 ligase MuRF1 degrades myosin heavy chain protein in dexamethasone-treated skeletal muscle. Cell Metab. 6, 376–385.
Cleynen, I., Vazeille, E., Artieda, M., Verspaget, H.W., Szczypiorska, M., Bringer, M.A., et al., 2014. Genetic and microbial factors modulating the ubiquitin proteasome system in inflammatory bowel disease. Gut 63, 1265–1274.
Combaret, L., Taillandier, D., Voisin, L., Samuels, S.E., Boespflug-Tanguy, O., Attaix, D., 1996. No alteration in gene expression of components of the ubiquitin-proteasome proteolytic pathway in dystrophin-deficient muscles. FEBS Lett. 393, 292–296.
Combaret, L., Tilignac, T., Claustre, A., Voisin, L., Taillandier, D., Obled, C., et al., 2002. Torbafylline (HWA 448) inhibits enhanced skeletal muscle ubiquitin-proteasome-dependent proteolysis in cancer and septic rats. Biochem. J. 361, 185–192.
Combaret, L., Bechet, D., Claustre, A., Taillandier, D., Richard, I., Attaix, D., 2003. Down-regulation of genes in the lysosomal and ubiquitin-proteasome proteolytic pathways in calpain-3-deficient muscle. Int. J. Biochem. Cell Biol. 35, 676–684.
Combaret, L., Taillandier, D., Dardevet, D., Béchet, D., Rallière, C., Claustre, A., et al., 2004. Glucocorticoids regulate mRNA levels for subunits of the 19 S regulatory complex of the 26 S proteasome in fast-twitch skeletal muscles. Biochem. J. 378, 239–246.
Dai, K.S., Liew, C.C., 2001. A novel human striated muscle RING zinc finger protein, SMRZ, interacts with SMT3b via its RING domain. J. Biol. Chem. 276, 23992–23999.
Dargelos, E., Poussard, S., Brule, C., Daury, L., Cottin, P., 2008. Calcium-dependent proteolytic system and muscle dysfunctions: a possible role of calpains in sarcopenia. Biochimie 90, 359–368.
Dehoux, M., Gobier, C., Lause, P., Bertrand, L., Ketelslegers, J.M., Thissen, J.P., 2007. IGF-I does not prevent myotube atrophy caused by proinflammatory cytokines despite activation of Akt/Foxo and GSK-3beta pathways and inhibition of atrogin-1 mRNA. Am. J. Physiol. Endocrinol. Metab. 292, E145–E150.
Deval, C., Mordier, S., Obled, C., Béchet, D., Combaret, L., Attaix, D., et al., 2001. Identification of cathepsin L as a differentially expressed message associated with skeletal muscle wasting. Biochem. J. 360, 143–150.
Dobrowolny, G., Aucello, M., Rizzuto, E., Beccafico, S., Mammucari, C., Boncompagni, S., et al., 2008. Skeletal muscle is a primary target of SOD1G93A-mediated toxicity. Cell Metab. 8, 425–436.
Ebato, C., Uchida, T., Arakawa, M., Komatsu, M., Ueno, T., Komiya, K., et al., 2008. Autophagy is important in islet homeostasis and compensatory increase of beta cell mass in response to high-fat diet. Cell Metab. 8, 325–332.
Elson, C.O., Alexander, K.L., 2015. Host-microbiota interactions in the intestine. Dig. Dis. 33, 131–136.
Ezaki, J., Matsumoto, N., Takeda-Ezaki, M., Komatsu, M., Takahashi, K., Hiraoka, Y., et al., 2011. Liver autophagy contributes to the maintenance of blood glucose and amino acid levels. Autophagy 7, 727–736.
Fareed, M.U., Evenson, A.R., Wei, W., Menconi, M., Poylin, V., Petkova, V., et al., 2006. Treatment of rats with calpain inhibitors prevents sepsis-induced muscle proteolysis independent of atrogin-1/MAFbx and MURF1 expression. Am. J. Physiol. Reg. Integr. Comp. Physiol. 290, R1589–R1597.
Fielitz, J., Kim, M.S., Shelton, J.M., Latif, S., Spencer, J.A., Glass, D.J., et al., 2007. Myosin accumulation and striated muscle myopathy result from the loss of muscle RING finger 1 and 3. J. Clin. Invest. 117, 2486–2495.
Fraysse, B., Desaphy, J.F., Rolland, J.F., Pierno, S., Liantonio, A., Giannuzzi, V., et al., 2006. Fiber type-related changes in rat skeletal muscle calcium homeostasis during aging and restoration by growth hormone. Neurobiol. Dis. 21, 372–380.
Fulle, S., Protasi, F., Di Tano, G., Pietrangelo, T., Beltramin, A., Boncompagni, S., et al., 2004. The contribution of reactive oxygen species to sarcopenia and muscle ageing. Exp. Gerontol. 39, 17–24.
Gaffney, J., Solomonov, I., Zehorai, E., Sagi, I., 2015. Multilevel regulation of matrix metalloproteinases in tissue homeostasis indicates their molecular specificity in vivo. Matrix Biol. 44–46, 191–199.
Gomes, M.D., Lecker, S.H., Jagoe, R.T., Navon, A., Goldberg, A.L., 2001. Atrogin-1, a muscle-specific F-box protein highly expressed during muscle atrophy. Proc. Natl. Acad. Sci. U.S.A. 98, 14440–14445.

Gonen, H., Stancovski, I., Shkedy, D., Hadari, T., Bercovich, B., Bengal, E., et al., 1996. Isolation, characterization, and partial purification of a novel ubiquitin-protein ligase, E3—targeting of protein substrates via multiple and distinct recognition signals and conjugating enzymes. J. Biol. Chem. 271, 302–310.

Hasselgren, P.O., Wray, C., Mammen, J., 2002. Molecular regulation of muscle cachexia: it may be more than the proteasome. Biochem. Biophys. Res. Commun. 290, 1–10.

Hishiya, A., Iemura, S., Natsume, T., Takayama, S., Ikeda, K., Watanabe, K., 2006. A novel ubiquitin-binding protein ZNF216 functioning in muscle atrophy. EMBO J. 25, 554–564.

Jagoe, R.T., Goldberg, A.L., 2001. What do we really know about the ubiquitin-proteasome pathway in muscle atrophy?. Curr. Opin. Clin. Nutr. Metab. Care 4, 183–190.

Jansen, H.J., van Essen, P., Koenen, T., Joosten, L.A., Netea, M.G., Tack, C.J., et al., 2012. Autophagy activity is up-regulated in adipose tissue of obese individuals and modulates proinflammatory cytokine expression. Endocrinology 153, 5866–5874.

Jin, Z., El-Deiry, W.S., 2005. Overview of cell death signaling pathways. Cancer Biol. Ther. 4, 139–163.

Jung, H.S., Chung, K.W., Won Kim, J., Kim, J., Komatsu, M., Tanaka, K., et al., 2008. Loss of autophagy diminishes pancreatic beta cell mass and function with resultant hyperglycemia. Cell Metab. 8, 318–324.

Kaniuk, N.A., Kiraly, M., Bates, H., Vranic, M., Volchuk, A., Brumell, J.H., 2007. Ubiquitinated-protein aggregates form in pancreatic beta-cells during diabetes-induced oxidative stress and are regulated by autophagy. Diabetes 56, 930–939.

Kedar, V., McDonough, H., Arya, R., Li, H.H., Rockman, H.A., Patterson, C., 2004. Muscle-specific RING finger 1 is a bona fide ubiquitin ligase that degrades cardiac troponin I. Proc. Natl. Acad. Sci. U.S.A. 101, 18135–18140.

Kee, A.J., Combaret, L., Tilignac, T., Souweine, B., Aurousseau, E., Dalle, M., et al., 2003. Ubiquitin-proteasome-dependent muscle proteolysis responds slowly to insulin release and refeeding in starved rats. J. Physiol. 546, 765–776.

Kim, K.H., Lee, M.S., 2014. Autophagy as a crosstalk mediator of metabolic organs in regulation of energy metabolism. Rev. Endocr. Metab. Disord. 15, 11–20.

Kirschke, H., Barrett, A.J., 1985. Cathepsin L-a lysosomal cysteine proteinase. Prog. Clin. Biol. Res. 180, 61–69.

Kniepert, A., Groettrup, M., 2014. The unique functions of tissue-specific proteasomes. Trends Biochem. Sci. 39, 17–24.

Kovsan, J., Bluher, M., Tarnovscki, T., Kloting, N., Kirshtein, B., Madar, L., et al., 2011. Altered autophagy in human adipose tissues in obesity. J. Clin. Endocrinol. Metab. 96, E268–E277.

Kravtsova-Ivantsiv, Y., Ciechanover, A., 2012. Non-canonical ubiquitin-based signals for proteasomal degradation. J. Cell Sci. 125, 539–548.

Krawiec, B.J., Frost, R.A., Vary, T.C., Jefferson, L.S., Lang, C.H., 2005. Hindlimb casting decreases muscle mass in part by proteasome-dependent proteolysis but independent of protein synthesis. Am. J. Physiol. Endocrinol. Metab. 289, E969–E980.

Kudryashova, E., Kudryashov, D., Kramerova, I., Spencer, M.J., 2005. Trim32 is a ubiquitin ligase mutated in limb girdle muscular dystrophy type 2H that binds to skeletal muscle myosin and ubiquitinates actin. J. Mol. Biol. 354, 413–424.

Kwon, Y.T., Xia, Z., Davydov, I.V., Lecker, S.H., Varshavsky, A., 2001. Construction and analysis of mouse strains lacking the ubiquitin ligase UBR1 (E3alpha) of the N-end rule pathway. Mol. Cell. Biol. 21, 8007–8021.

Lagirand-Cantaloube, J., Offner, N., Csibi, A., Leibovitch, M.P., Batonnet-Pichon, S., Tintignac, L.A., et al., 2008. The initiation factor eIF3-f is a major target for atrogin1/MAFbx function in skeletal muscle atrophy. EMBO J. 27, 1266–1276.

Lecker, S.H., Solomon, V., Mitch, W.E., Goldberg, A.L., 1999. Muscle protein breakdown and the critical role of the ubiquitin-proteasome pathway in normal and disease states. J. Nutr. 129, 227S–237S.

Lecker, S.H., Jagoe, R.T., Gilbert, A., Gomes, M., Baracos, V., Bailey, J., et al., 2004. Multiple types of skeletal muscle atrophy involve a common program of changes in gene expression. FASEB J. 18, 39–51.

Liu, K., Czaja, M.J., 2013. Regulation of lipid stores and metabolism by lipophagy. Cell Death Differ. 20, 3–11.

Madrigal-Matute, J., Cuervo, A.M., 2016. Regulation of liver metabolism by autophagy. Gastroenterology 150, 328–339.

Maloy, K.J., Powrie, F., 2005. Fueling regulation: IL-2 keeps CD4+ Treg cells fit. Nat. Immunol. 6 (11), 1071–1072.

Mammucari, C., Milan, G., Romanello, V., Masiero, E., Rudolf, R., Del Piccolo, P., et al., 2007. Foxo3 controls autophagy in skeletal muscle in vivo. Cell Metab. 6, 458–471.

Mansoor, O., Beaufrère, B., Boirie, Y., Rallière, C., Taillandier, D., Aurousseau, E., et al., 1996. Increased mRNA levels for components of the lysosomal, Ca^{2+}-activated, and ATP-ubiquitin-dependent proteolytic pathways in skeletal muscle from head trauma patients. Proc. Natl. Acad. Sci. U.S.A. 93, 2714–2718.

Masiero, E., Agatea, L., Mammucari, C., Blaauw, B., Loro, E., Komatsu, M., et al., 2009. Autophagy is required to maintain muscle mass. Cell Metab. 10, 507–515.

Masini, M., Bugliani, M., Lupi, R., del Guerra, S., Boggi, U., Filipponi, F., et al., 2009. Autophagy in human type 2 diabetes pancreatic beta cells. Diabetologia 52, 1083–1086.

Metzger, M.B., Hristova, V.A., Weissman, A.M., 2012. HECT and RING finger families of E3 ubiquitin ligases at a glance. J. Cell Sci. 125, 531–537.

Milan, G., Romanello, V., Pescatore, F., Armani, A., Paik, J.H., Frasson, L., et al., 2015. Regulation of autophagy and the ubiquitin-proteasome system by the FoxO transcriptional network during muscle atrophy. Nat. Commun. 6, 6670.

Mizushima, N., Yamamoto, A., Matsui, M., Yoshimori, T., Ohsumi, Y., 2004. In vivo analysis of autophagy in response to nutrient starvation using transgenic mice expressing a fluorescent autophagosome marker. Mol. Biol. Cell 15, 1101–1111.

Mordier, S., Deval, C., Béchet, D., Tassa, A., Ferrara, M., 2000. Leucine limitation induces autophagy and activation of lysosome-dependent proteolysis in C2C12 myotubes through a mammalian target of rapamycin-independent signaling pathway. J. Biol. Chem. 275, 29900–29906.

Mortimore, G.E., Pösö, A.R., 1987. Intracellular protein catabolism and its control during nutrient deprivation and supply. Annu. Rev. Nutr. 7, 539–564.

Murton, A.J., Constantin, D., Greenhaff, P.L., 2008. The involvement of the ubiquitin proteasome system in human skeletal muscle remodelling and atrophy. Biochim. Biophys. Acta 1782, 730–743.

Murton, A.J., Greenhaff, P.L., 2010. Physiological control of muscle mass in humans during resistance exercise, disuse and rehabilitation. Curr. Opin. Clin. Nutr. Metab. Care 13, 249–254.
Ono, Y., Ojima, K., Sjinkai-Ouchi, F., Sorimachi, H., 2016. An eccentric calpain, CAPN3/p94/calpain-3. Biochimie 122, 169–187.
Polge, C., Heng, A.E., Jarzaguet, M., Ventadour, S., Claustre, A., Combaret, L., et al., 2011. Muscle actin is polyubiquitinylated in vitro and in vivo and targeted for breakdown by the E3 ligase MuRF1. FASEB J. 25, 3790–3802.
Polge, C., Heng, A.E., Combaret, L., Bechet, D., Taillandier, D., Attaix, D., 2013. Recent progress in elucidating signalling proteolytic pathways in muscle wasting: potential clinical implications. Nutr. Metab. Cardiovasc. Dis. 23, S1–S5.
Polge, C., Attaix, D., Taillandier, D., 2015a. Role of E2-Ub-conjugating enzymes during skeletal muscle atrophy. Front. Physiol. 6, 59.
Polge, C., Leulmi, R., Jarzaguet, M., Claustre, A., Combaret, L., Béchet, D., et al., 2015b. UBE2B is implicated in myofibrillar protein loss in catabolic C2C12 myotubes. J. Cachexia Sarcopenia Muscle (in press).
Poussard, S., Duvert, M., Balcerzak, D., Ramassamy, S., Brustis, J.J., Cottin, P., et al., 1996. Evidence for implication of muscle-specific calpain (p94) in myofibrillar integrity. Cell Growth Differ. 7, 1461–1469.
Price, S.R., 2003. Increased transcription of ubiquitin-proteasome system components: molecular responses associated with muscle atrophy. Int. J. Biochem. Cell Biol. 35, 617–628.
Ravid, T., Hochstrasser, M., 2008. Diversity of degradation signals in the ubiquitin-proteasome system. Nat. Rev. Mol. Cell Biol. 9, 679–690.
Roberts, R., 2005. Lysosomal cysteine proteases: structure, function and inhibition of cathepsins. Drug News Perspect. 18, 605–614.
Samuels, S.E., Taillandier, D., Aurousseau, E., Cherel, Y., Le Maho, Y., Arnal, M., et al., 1996. Gastrointestinal tract protein synthesis and mRNA levels for proteolytic systems in adult fasted rats. Am. J. Physiol. 271, E232–E238.
Samuels, S.E., Knowles, A.L., Tilignac, T., Debiton, E., Madelmont, J.C., Attaix, D., 2000. Protein metabolism in the small intestine during cancer cachexia and chemotherapy in mice. Cancer Res. 60, 4968–4974.
Sandri, M., 2010. Autophagy in skeletal muscle. FEBS Lett. 584, 1411–1416.
Sandri, M., 2013. Protein breakdown in muscle wasting: role of autophagy-lysosome and ubiquitin-proteasome. Int. J. Biochem. Cell Biol. 45, 2121–2129.
Shalini, S., Dorstyn, L., Dawar, S., Kumar, S., 2015. Old, new and emerging functions of caspases. Cell Death Differ. 22, 526–539.
Solomon, V., Goldberg, A.L., 1996. Importance of the ATP-ubiquitin-proteasome pathway in the degradation of soluble and myofibrillar proteins in rabbit muscle extracts. J. Biol. Chem. 271, 26690–26697.
Taillandier, D., Aurousseau, E., Meynial-Denis, D., Bechet, D., Ferrara, M., Cottin, P., et al., 1996. Coordinate activation of lysosomal, Ca^{2+}-activated and ATP-ubiquitin-dependent proteinases in the unweighted rat soleus muscle. Biochem. J. 316, 65–72.
Taillandier, D., Aurousseau, E., Combaret, L., Guezennec, C.Y., Attaix, D., 2003. Regulation of proteolysis during reloading of the unweighted soleus muscle. Int. J. Biochem. Cell Biol. 35, 665–675.
Tallant, C., Marrero, A., Gomis-Ruth, F.X., 2010. Matrix metalloproteinases: fold and function of their catalytic domains. Biochim. Biophys. Acta 1803, 20–28.
Tassa, A., Roux, M.P., Attaix, D., Béchet, D.M., 2003. Class III phosphoinositide 3-kinase-Beclin1 complex mediates the amino acid-dependent regulation of autophagy in C2C12 myotubes. Biochem. J. 376, 577–586.
Tilignac, T., Temparis, S., Combaret, L., Taillandier, D., Pouch, M.N., Cervek, M., et al., 2002. Chemotherapy inhibits skeletal muscle ubiquitin-proteasome-dependent proteolysis. Cancer Res. 62, 2771–2777.
Tintignac, L.A., Lagirand, J., Batonnet, S., Sirri, V., Leibovitch, M.P., Leibovitch, S.A., 2005. Degradation of myod mediated by the SCF (Mafbx) ubiquitin ligase. J. Biol. Chem. 280, 2847–2856.
Tomko Jr., R.J., Hochstrasser, M., 2013. Molecular architecture and assembly of the eukaryotic proteasome. Annu. Rev. Biochem. 82, 415–445.
Ueno, T., Ezaki, J., Kominami, E., 2012. Metabolic contribution of hepatic autophagic proteolysis: old wine in new bottles. Biochim. Biophys. Acta 1824, 51–58.
van Wijk, S.J., Timmers, H.T., 2010. The family of ubiquitin-conjugating enzymes (E2s): deciding between life and death of proteins. FASEB J. 24, 981–993.
Voisin, L., Breuille, D., Combaret, L., Pouyet, C., Taillandier, D., Aurousseau, E., et al., 1996. Muscle wasting in a rat model of long-lasting sepsis results from the activation of lysosomal, Ca^{2+}-activated, and ubiquitin-proteasome proteolytic pathways. J. Clin. Invest. 97, 1610–1617.
Wenz, T., Rossi, S.G., Rotundo, R.L., Spiegelman, B.M., Moraes, C.T., 2009. Increased muscle PGC-1alpha expression protects from sarcopenia and metabolic disease during aging. Proc. Natl. Acad. Sci. U.S.A. 106, 20405–20410.
Williams, A.B., Decourten-Myers, G.M., Fischer, J.E., Luo, G., Sun, X., Hasselgren, P.O., 1999. Sepsis stimulates release of myofilaments in skeletal muscle by a calcium-dependent mechanism. FASEB J. 13, 1435–1443.
Wing, S.S., Banville, D., 1994. 14-kDa ubiquitin-conjugating enzyme: structure of the rat gene and regulation upon fasting and by insulin. Am. J. Physiol. 267, E39–E48.
Wing, S.S., Bedard, N., 1996. Insulin-like growth factor I stimulates degradation of an mRNA transcript encoding the 14 kDa ubiquitin conjugating enzyme. Biochem. J. 319, 455–461.
Wing, S.S., Goldberg, A.L., 1993. Glucocorticoids activate the ATP-ubiquitin-dependent proteolytic system in skeletal muscle during fasting. Am. J. Physiol. 264, E668–E676.
Wray, C.J., Tomkinson, B., Robb, B.W., Hasselgren, P.O., 2002. Tripeptidyl-peptidase II expression and activity are increased in skeletal muscle during sepsis. Biochem. Biophys. Res. Commun. 296, 41–47.
Ye, Y., Rape, M., 2009. Building ubiquitin chains: E2 enzymes at work. Nat. Rev. Mol. Cell Biol. 10, 755–764.
Zaglia, T., Milan, G., Ruhs, A., Franzoso, M., Bertaggia, E., Pianca, N., et al., 2014. Atrogin-1 deficiency promotes cardiomyopathy and premature death via impaired autophagy. J. Clin. Invest. 124, 2410–2424.

CHAPTER

4

Cellular and Molecular Mechanisms of Protein Synthesis Among Tissues

J.W. Carbone[1], *L.M. Margolis*[2] *and S.M. Pasiakos*[2]

[1]School of Health Sciences, Eastern Michigan University, Ypsilanti, MI, United States [2]Military Nutrition Division, US Army Research Institute of Environmental Medicine, Natick, MA, United States

4.1. INTRODUCTION

Protein synthesis is a core component of cellular regulation and, as such, has profound effects at tissue, organ, and whole body levels. Protein synthesis can be stimulated or inhibited in response to both endogenous and exogenous stimuli, allowing for adaptation and appropriate reactions to various stressors (Fig. 4.1). Newly synthesized proteins function as enzymes, transporters, hormones, and antibodies, and can assist in the critical functions of fluid and electrolyte balance, acid-base balance, energy production, immune defense, and cellular growth, maintenance, and repair. As such, protein synthesis is highly regulated and the consequences of impaired protein synthesis can increase the likelihood of disease development, progression, and even death.

The objective of this chapter is to review the cellular and molecular mechanisms which regulate protein synthesis, with an emphasis on muscle protein synthesis. In particular, this chapter will:

- Provide an overview of mRNA translation (Section 4.1.1) and the intracellular pathways (Section 4.1.2) that modulate these processes.
- Introduce common endogenous and exogenous factors that modulate molecular and cellular control of protein synthesis (Section 4.1.3).
- Provide a detailed description of the intracellular regulation of muscle hypertrophy, myogenesis, and muscle regeneration (Sections 4.2 and 4.3).
- Illustrate the importance of well-regulated protein synthesis by highlighting the consequences of impairment (Section 4.4).

4.1.1 Molecular Basics of Protein Synthesis

Protein synthesis is the process of translating mRNA to functional peptides. Translation begins with the formation of the 80S initiation complex, composed of mRNA, charged methionyl-tRNA, and both the 40S and 60S ribosomal subunits (Fig. 4.2). This cascade of events begins with the binding of eukaryotic initiation factor 2 (eIF2) to methionyl-tRNA which is then moved into the P-site of the 40S ribosomal subunit, establishing the 43S preinitiation complex. Through an exergonic reaction involving guanosine triphosphate (GTP) cleavage, eIF2 is released. The 43S preinitiation complex subsequently binds to mRNA with assistance from eIF4F, composed of multiple initiation factors that unwind secondary structures upstream of the translation start codon, bind to the 5′-methyl mRNA cap and 3′-polyadenylated tail, and facilitate linkage with the 60S ribosomal subunit to form the 80S ribosome. The anticodon of methionyl-tRNA (3′-UAC-5′) is aligned with its complementary mRNA codon (5′-AUG-3′) and the ribosome then moves along the mRNA, with specific amino acid-charged tRNA moving into the A-site of the ribosome based on the subsequent mRNA codon sequences. Peptide bonds form between

The Molecular Nutrition of Amino Acids and Proteins.
DOI: http://dx.doi.org/10.1016/B978-0-12-802167-5.00004-9

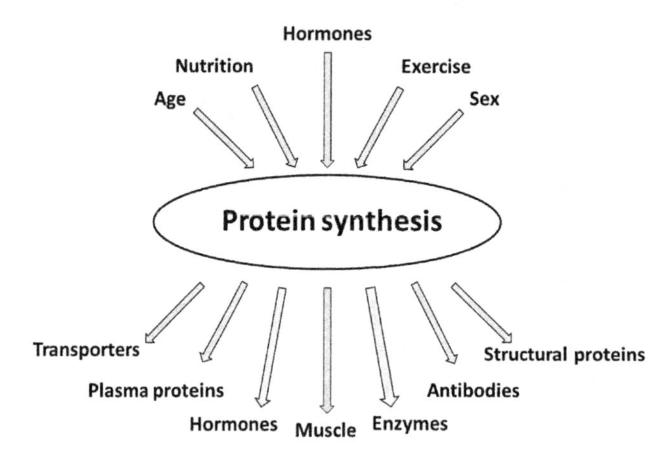

FIGURE 4.1 Endogenous and exogenous modulators of protein synthesis and downstream products.

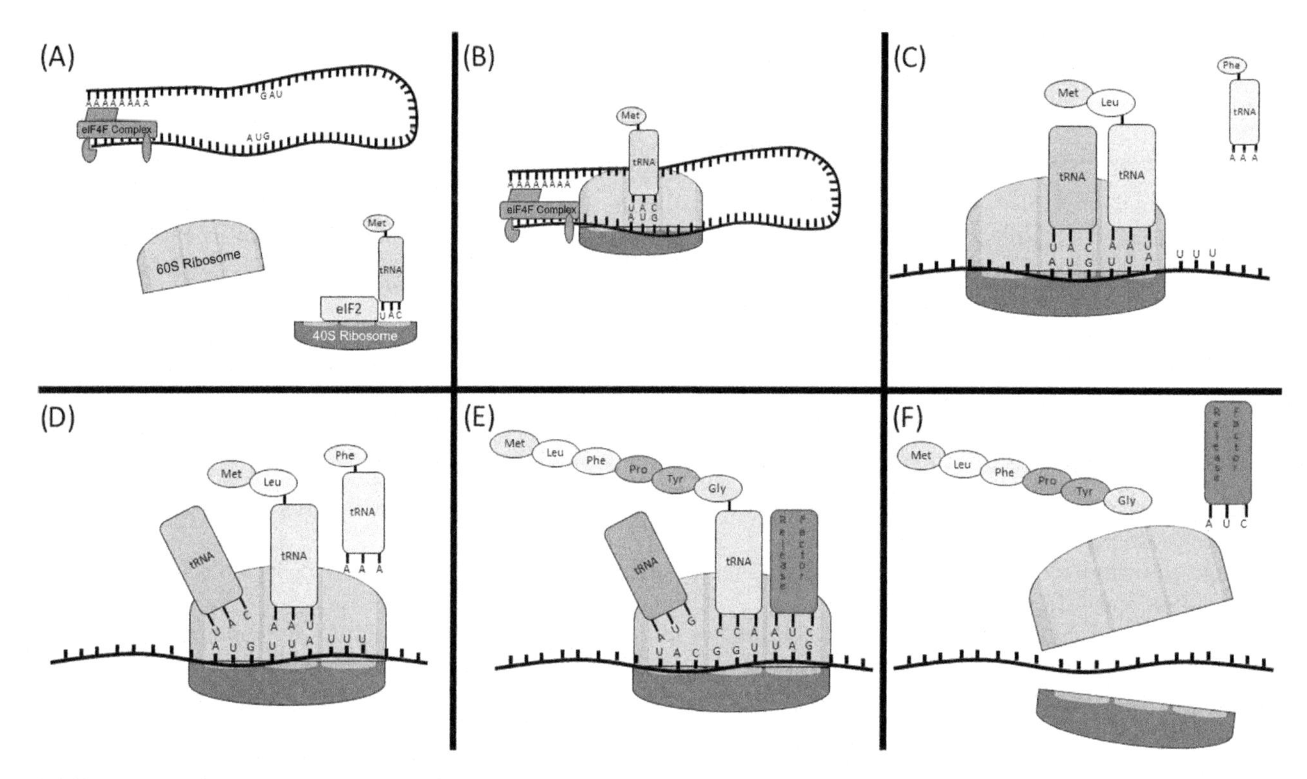

FIGURE 4.2 The translation of mRNA to polypeptides. (A) Formation of the 43S preinitiation complex and eIF4F binding of mRNA; (B) Creation of the 80S ribosome and translation initiation; (C) Peptide elongation; (D) Continued elongation and release of uncharged tRNA; (E) Commencement of translation termination as the ribosome encounters a nonsense/stop codon; (F) Translation termination and dissociation of the 80S ribosome (factors not drawn to scale).

adjacent amino acids and the P-site tRNA is uncharged and released via the E-site as the ribosome moves toward the 3′ end of the mRNA. Peptide elongation continues in this manner until the ribosome encounters an mRNA nonsense/stop codon (UAA, UAG, UGA) and an uncharged release factor moves into the A-site of the ribosome, facilitating the release of the newly synthesized polypeptide and dissociation of the 80S ribosome. Following any posttranslational modifications—including addition of functional groups (eg, methylation), structural changes (eg, disulfide bonds), and chemical modification (eg, deamidation)—protein synthesis results in a myriad of products capable of regulating intra- and extracellular processes.

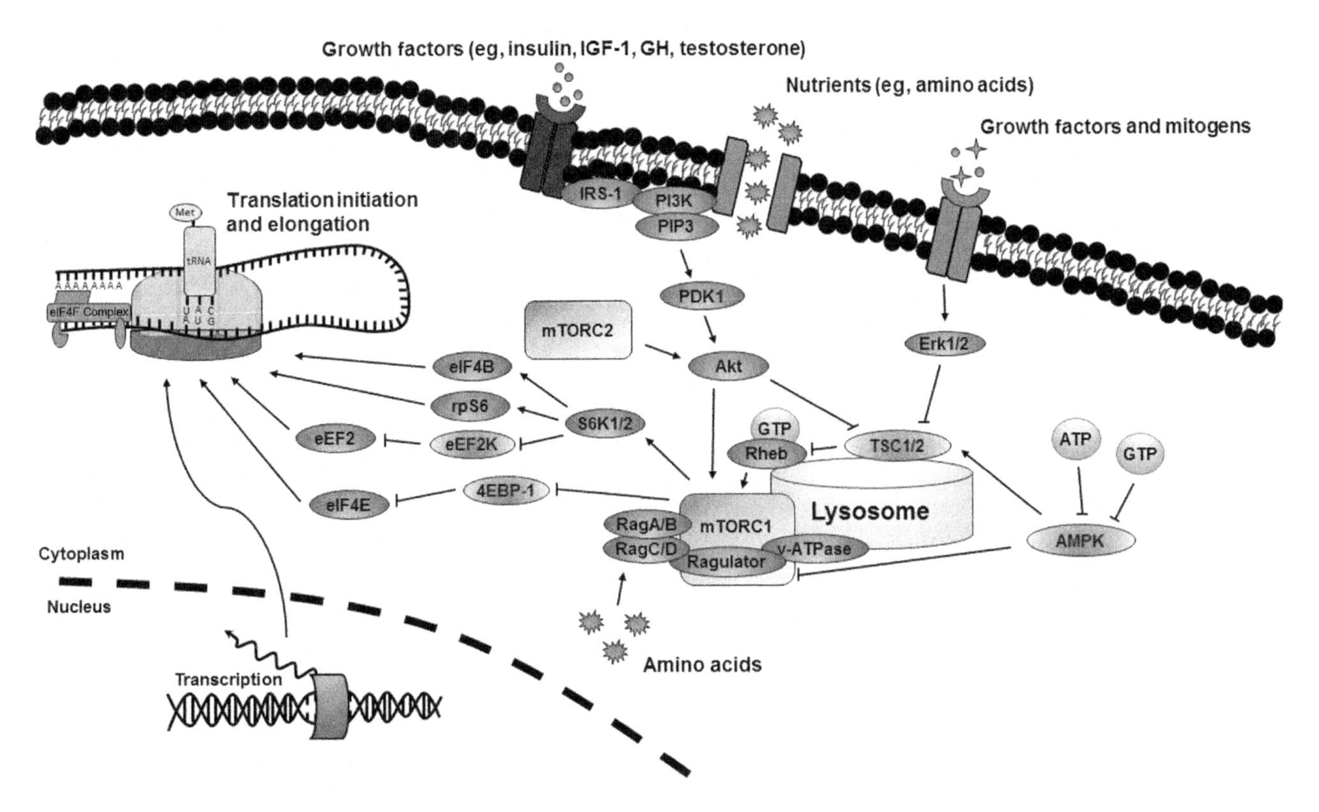

FIGURE 4.3 Extracellular mediators and the coordinated intracellular responses regulating protein synthesis. Green factors positively influence rates of protein synthesis while red factors work to inhibit protein synthesis. For simplicity of illustration, the ribosome is shown independent of the endoplasmic reticulum (factors not drawn to scale).

4.1.2 Introduction of the Intracellular Regulation of Protein Synthesis

Protein synthesis is regulated by complex intracellular stimulating and inhibiting factors, all competing for the end effect of increasing or decreasing mRNA translation initiation and elongation (Fig. 4.3). The mechanistic (ie, mammalian) target of rapamycin (mTOR) is, perhaps, the most critical and well-characterized intracellular regulator of protein synthesis (discussed in greater detail in Section 4.2). mTOR exists in two regulatory complexes, mTOR complex 1 (mTORC1) and mTOR complex 2 (mTORC2), with mTORC1 being the primary regulator of protein synthesis (Tavares et al., 2015). Protein synthesis is energy demanding and, as such, mTORC1 signaling is downregulated in response to diminished intracellular energy stores (ie, ↓ ATP:AMP) (Bolster et al., 2002; Jiang et al., 2008a,b; Tyagi et al., 2015).

Changes in mTORC1 activity can have widespread influence, particularly on glucose homeostasis and insulin sensitivity, aging, adipocyte metabolism, and energy balance (Magnuson et al., 2012). Downstream targets of mTORC1, notably the ribosomal proteins S6 kinase 1 (S6K1) and S6 kinase 2 (S6K2), also regulate apoptosis, metabolism, transcription and mRNA splicing, inflammation, and cytoskeletal organization (Tavares et al., 2015). S6K1 and S6K2 act to phosphorylate the ribosomal protein S6, a component of the 40S ribosome. As such, dysregulation of mTORC1-S6K signaling can have detrimental organism-level effects.

4.1.3 Endogenous and Exogenous Regulators of Protein Synthesis

Hormones (ie, growth factors), nutrition, and exercise are key determinants of protein synthesis. Hormones modulate protein synthesis via their respective membrane receptors and signaling cascades. For example, insulin binds to the insulin receptor and stimulates protein synthesis by activating class I phosphoinositide 3-kinase (PI3K). Activated PI3K subsequently phosphorylates Akt (protein kinase B) (Cesena et al., 2007; Kimball et al., 2002; White et al., 2013), which in turn stimulates mTORC1 signaling.

Energy and protein intakes are the primary nutritional effectors of protein synthesis. Protein synthesis is downregulated to spare endogenous energy stores during periods of fasting and/or sustained energy deficit.

Dietary protein provides the essential amino acids necessary to support the synthesis of new functional proteins. The amount of protein consumed in the diet directly affects whole-body and muscle protein synthetic rates (Kim et al., 2015), although there appears to be a plateau effect whereby consuming more dietary protein fails to further stimulate protein synthesis (Moore et al., 2009). Amino acids, particularly the branched-chain amino acid leucine, directly activate mTORC1 (Pasiakos and McClung, 2011). Human studies have demonstrated stimulation of the mTORC1 pathway, and resulting increased muscle protein synthesis, in response to increased leucine consumption (Drummond and Rasmussen, 2008; Hoffer et al., 1997; Nair et al., 1992; Pasiakos et al., 2011; Pasiakos and McClung, 2011). In neonatal pigs, supplementing a low protein diet with leucine significantly increases muscle protein synthesis compared to a low protein diet without additional leucine (Murgas Torrazza et al., 2010). Furthermore, phosphorylation of mTOR and S6K1 (p70 isoform) are similar between pigs consuming a leucine-supplemented low protein diet or a high protein diet without additional leucine, illustrating the importance of leucine in the stimulation and regulation of protein synthesis.

Exercise also regulates protein synthesis (Pasiakos, 2012). It is well accepted that the mechanical strain induced by resistive and endurance-type exercise stimulates protein synthesis via the mTORC1 pathway (Cuthbertson et al., 2006; Mascher et al., 2007; Sakamoto et al., 2004). Training status (ie, untrained vs trained), exercise mode, total volume, and load placed on the muscle, length of time the muscle is under tension, and the velocity of contractile forces during exercise modulate the magnitude and duration of increases in muscle protein synthesis (Burd et al., 2010a,b, 2012; Mitchell et al., 2012; Sheffield-Moore et al., 2004; Shepstone et al., 2005; Tipton et al., 1996). The repetitive protein synthetic stimulus of exercise training may lead to protein accretion (ie, hypertrophy), especially when the exercise stimulus is combined with the anabolic effects of exogenous protein and/or amino acid consumption.

4.2. CELLULAR AND MOLECULAR REGULATION OF HYPERTROPHY

Hypertrophy occurs when the synthesis of new proteins exceeds the rate at which endogenous proteins are degraded, resulting in positive protein balance. In healthy individuals, protein synthesis is the primary determinant of hypertrophy, as sustained periods of positive protein balance resulting from increased protein synthesis (with or without reductions in protein breakdown) can result in an increase in muscle fiber size and muscle protein accretion. As previously discussed, the intracellular regulation of protein synthesis is anything but simplistic, involving a highly regulated, integrated system of intracellular signaling proteins that modulate gene transcription, translation, and posttranslational modifications (Carbone et al., 2012; Drummond et al., 2009). The activation or inhibition of the mTORC1 pathway is the central intracellular mechanism controlling protein synthesis (Fig. 4.3).

This complex consists of five proteins, each with a functional role in the regulation of protein synthesis, including the mTOR kinase itself, the regulatory-associated protein of mTOR (RAPTOR), 40 kDa Pro-rich Akt substrate (PRAS40), DEP domain-containing mTOR-interacting protein (DEPTOR), and the mammalian lethal with SEC13 protein 8 (mLST8; also known as GβL) (Fig. 4.4) (Laplante and Sabatini, 2009). The upstream mechanisms, sensitive to endogenous (eg, insulin, growth hormone, insulin-like growth factor-1 (IGF-1)) and exogenous (eg, exercise and amino acid availability) stimuli, that modulate the components of mTORC1 and, ultimately, the activity of the complex have been studied extensively (Drummond et al., 2009; Proud, 2007).

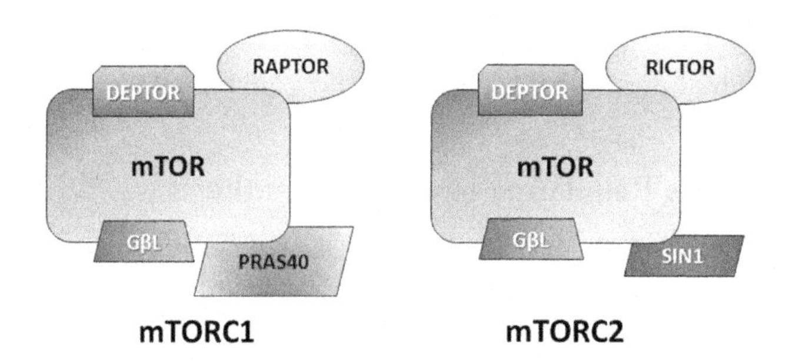

FIGURE 4.4 mTOR and the additional regulatory components required to form the mTORC1 and mTORC2 complexes (not to scale). mTORC1 is composed of mTOR, DEPTOR, GβL (mLST8), RAPTOR, and the complex inhibitor PRAS40. mTORC2 is composed of mTOR, DEPTOR, GβL (mLST8), RICTOR, and SIN1.

Akt, extracellular-signal-regulated kinase 1/2 (ERK 1/2), and p90 ribosomal S6 kinase 1 (p90SK1) are examples of upstream, positive regulators of mTORC1 (Fig. 4.3) (Inoki et al., 2002; Laplante and Sabatini, 2009; Ma and Blenis, 2009; Potter et al., 2002; Roux et al., 2004). The AMP-activated protein kinase (AMPK) is an example of an upstream, negative regulator of mTORC1 (Laplante and Sabatini, 2009). The GTPase-activating proteins (GAP) tuberous sclerosis (TSC) 1 and 2 appear to be the convergence point by which these upstream signaling proteins exert their effects to stimulate or inhibit mTORC1. TSC 1 and 2 transmit energy-mediated signals to activate or suppress mTORC1 based on the intracellular ratio of GTP:GDP and its interaction with the Ras homolog enriched with brain (Rheb) (Inoki et al., 2002; Long et al., 2005). Growth factor-mediated phosphorylation of TSC by Akt, ERK1/2, or p90S6K causes a guanosine diphosphate (GDP) to GTP exchange that results in Rheb being bound to GTP, thus activating mTORC1, whereas the opposite occurs in response to intracellular energy depletion, such that AMPK phosphorylation of TSC results in a diminished Rheb-GTP:Rheb-GDP ratio and downregulation of mTORC1 activity (Inoki et al., 2002; Kimball, 2014).

Although the interaction of mTORC1 and Rheb-GTP is necessary for upstream activation of the mTOR kinase, amino acid starvation overrides this stimulation and mTOR signaling is suppressed (Nobukuni et al., 2005; Smith et al., 2005). This discovery suggested that, unlike growth factors and energy status, amino acids modulate mTORC1 activity via a distinct mechanism capable of independently stimulating translation initiation and protein synthesis. Intracellular translocation of mTORC1 to the lysosomal membrane appears to be the mechanism by which amino acids modulate mTORC1 activity (Sancak et al., 2008; Sancak and Sabatini, 2009). Amino acid-dependent translocation of mTORC1 is mediated by the Ras-related small guanosine triphosphatases (Rag GTPases) that form a heterodimer that binds to the mTORC1 protein Raptor (Sancak et al., 2010). Following translocation to the lysosomal membrane, the Rag heterodimer binds with Ragulator, a complex consisting of p18, p14, and MP1, proteins that serve as scaffolding to tether the Rag GTPase-mTORC1 complex to the lysosome (Bar-Peled et al., 2012).

This interaction of the Rag GTPase complex with Ragulator and subsequent activation of mTORC1 is mediated by vacuolar H^+-ATPase (v-ATPase), a protein found at the lysosomal membrane that transmits an amino acid-induced signal to activate the guanine nucleotide exchange factors of Ragulator. The translocation and binding of mTORC1 to the lysosome increases the proximity of the mTOR kinase to Rheb (also located at the lysosome) and promotes signal transduction and stimulation of mTOR and, ultimately, protein synthesis (Sancak et al., 2010).

Activation of the mTOR kinase triggers the downstream signaling and phosphorylation and activation of the translational regulator proteins S6K1 (p70S6K1) and eukaryotic translation initiation factor 4E (eIF4E) (Bar-Peled et al., 2012). As mentioned previously, phosphorylated p70S6K1, in turn, phosphorylates ribosomal protein S6 (rpS6) and upregulates mRNA translation initiation (Baar and Esser, 1999; Wang et al., 2001). Furthermore, p70S6K1 activates eukaryotic elongation factor 2 (eEF2) through phosphorylation and deactivation of eEF2 kinase (eEF2K), an inhibitor of eEF2. Upregulated eEF2 increases mRNA translation elongation efficiency, and ultimately increases rates of protein synthesis. Activated mTORC1 also phosphorylates the eIF 4E-binding protein 1 (4E-BP1), which when dephosphorylated binds to its downstream target eIF4E, inhibiting the formation of the multisubunit complex, eIF4F, critical for cap-dependent mRNA translation (Beugnet et al., 2003). The eIF4F multisubunit complex is complete when eIF4E and eIF4G bind with eIF4A, recruiting the 40S ribosomal subunit to form the preinitiation complex required for translation initiation (Kimball, 2014). It is important to recognize that while mTORC1 signaling is often depicted as a linear, causal regulator of muscle hypertrophy via upstream and downstream signaling proteins, there are several complex branch points of this signal transduction pathway that also contribute to the modulation of protein synthesis and hypertrophy (Fig. 4.3) (Bolster et al., 2004).

4.3. MYOGENESIS: THE DEVELOPMENT AND REGENERATION OF MUSCLE

Myogenesis is the fundamental synthesis and development of new muscle fibers (ie, building muscle) and the regeneration of existing muscle that occurs throughout life, reliant on heavily coordinated synthesis of multiple types of muscle protein (Bentzinger et al., 2012; Devlin and Emerson, 1978). Proliferation, the initial phase of myogenesis, occurs in embryonic cells, as founder stem cells (ie, nonspecified cells capable of self-renewal and becoming tissue/organ-specific cells) differentiate into myocytes and progress during fetal development to become mature myofibers (Fig. 4.5) (Ivanova et al., 2002; Tajbakhsh, 2009). Postnatal juvenile

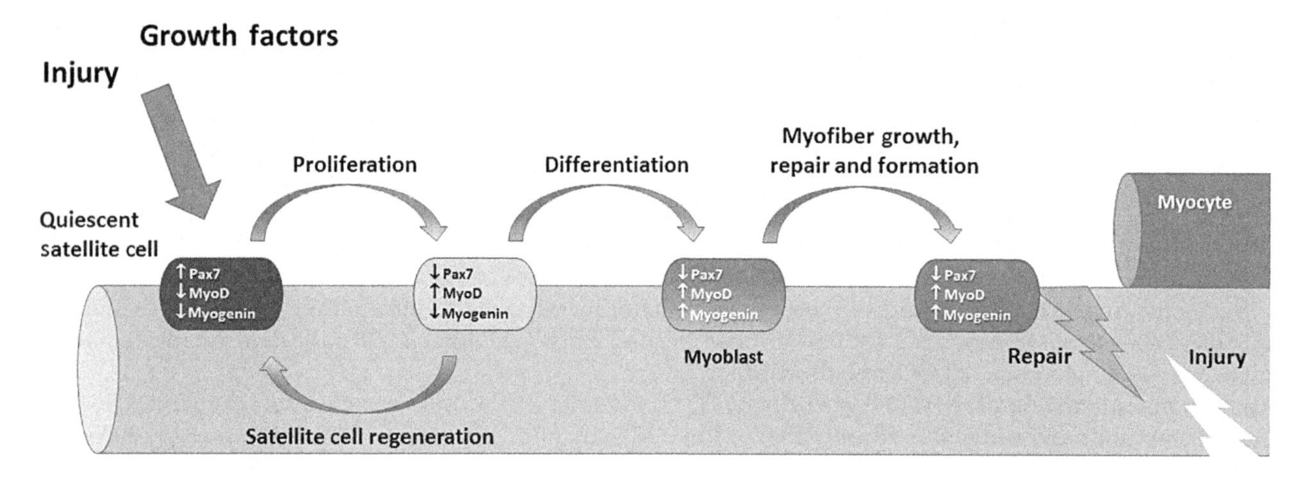

FIGURE 4.5 The role of muscle regulatory factors in satellite cell response to injury and growth factors. The satellite cell may leave the quiescent state, proliferating and differentiating into active myoblasts, capable of repairing myofiber damage and contributing to hypertrophy. The satellite cell lineage is maintained via divergence from myoblast formation and restoration of the quiescent state (not to scale).

skeletal muscle stem cells (ie, satellite cells) continue to proliferate, relocating between the basal lamina of the myofiber and the sarcolemma which envelops it (Schultz and McCormick, 1994). These juvenile satellite cells eventually serve a functional role in muscle growth, as they transition to become adult, mitotically quiescent satellite cells, the primary source of myoblasts required for muscle repair and regeneration (Relaix and Zammit, 2012; Yin et al., 2013).

The activation of satellite cells occurs in response to several of the same exogenous and endogenous stimuli that upregulate mTORC1 signaling. For example, satellite cells are activated in response to muscle damage, which can occur following vigorous exercise (Fig. 4.5). Muscle injury causes satellite cells to mobilize and migrate to the site of tissue damage (Sanes, 2003). Satellite cells can also be activated by hypertrophic stimulants and an increase in circulating growth factors, including IGF-1, vascular endothelial growth factor (VEGF), and fibroblast growth factor (FGF) (Ceafalan et al., 2014).

Once activated, several regulatory factors are necessary to assist satellite cells during myogenesis (Tajbakhsh, 2009). These include myogenic regulator factors (MRFs), notably the paired-box (Pax) transcription factors Pax3 and Pax7, which are expressed in adult satellite cells and have both distinct and overlapping functions in MRF regulation (Relaix et al., 2006). Pax3 and Pax7 act to modulate myogenic progenitor cells and regulate their differentiation, as well as the expression of downstream transcription factors, leading to a net increase in myogenic transcription initiation and increased protein synthesis. They are also necessary for cell specificity during development, particularly Pax3, which is required for embryonic myogenesis (Messina et al., 2009).

The exact mechanisms for how Pax3 and Pax7 regulate satellite cell biogenesis, survival, and self-renewal remain elusive. It appears that they serve as the master regulators of myogenesis and regeneration, orchestrating the involvement of other important MRFs to control the commitment of satellite cells to myoblast formation and differentiation, and the fusion of myocytes to myofibers; these important MRFs include MyoD, myogenic factor 5 (Myf5), Myf6, and myogenin (Khan et al., 1999; Olguin and Pisconti, 2012; Olguin et al., 2007; Rudnicki and Jaenisch, 1995; Seale et al., 2004).

The earliest markers of myogenic cells are Myf5 and MyoD, which are involved in the commitment of proliferating satellite cells to myogenesis and regeneration (Rudnicki et al., 1993). These two MRFs appear to have somewhat redundant functionality, as knocking out either of these genes still allows for myogenesis and regeneration, albeit at limited capacity (Rudnicki et al., 1993). If both genes are silenced, however, myocyte formation does not occur.

MyoD has been shown to regulate cell cycling and the ability of the satellite cell to move from G1 to the S phase of mitosis (Zhang et al., 2010). Following mitotic division, satellite cell progeny undergo divergent fates (Fig. 4.5). Decreased Pax7 expression and associated increases in MyoD expression drive differentiation, leading to increased expression of myogenin. This cascade results in increased protein synthesis, and associated muscle growth and repair, via terminal differentiation, whereby the myocytes are either fused with an existing myofiber or join with other myocytes to form a new myofiber (Olguin and Pisconti, 2012). Alternatively, satellite cell progeny can maintain Pax7 expression and suppress MyoD and myogenin (Zammit et al., 2004). This pathway allows for continuance of the myogenic lineage, renewing the quiescent satellite cell pool. Maintenance of satellite cells provides for future myogenesis in response to exercise or muscle injury.

4.4. APPLIED IMPLICATIONS OF PROTEIN SYNTHESIS IN VIVO

The highly regulated processes of protein synthesis are critical for the maintenance of human health and life. Protein synthesis can be positively impacted by exercise, both aerobic and resistive, and nutrition, meaning sufficient energy and amino acid intakes to meet metabolic demand. In the case of muscle protein, disuse and/or denervation can lead to atrophy, with rates of proteolysis exceeding dramatically reduced protein synthesis. Certain disease states—including sepsis, cancer, and uremia—are also associated with diminished protein synthesis, upregulated proteolysis, and overall degradation of muscle protein.

Similarly, dietary energy and protein deficiencies can deprive the body of the building blocks needed to generate new proteins. Decreased whole body protein synthesis can have serious consequences on health and longevity. Diminished hepatic protein synthesis can lead to impaired plasma protein concentrations which may manifest as edema and ascites. In the case of plasma transporters, movement of hydrophobic compounds through the aqueous blood becomes diminished if transporter protein synthesis is impaired. This can severely impact the transport of fat-soluble vitamins and plasma lipids, as well as interfere with correct dosing of hydrophobic pharmaceuticals. Similarly, decreased antibody and immune factor production increases susceptibility to infection. Diminished enzyme and protein hormone synthesis can have system-wide implications and, depending on the severity and duration of these limitations, can drive increased mortality. It is these deficiencies of protein synthesis, and their associated outcomes, that illustrate the critical importance of proper regulation and maintenance of protein synthesis.

4.5. CONCLUSIONS AND SUMMARY OF KEY POINTS

As described within this chapter:

- protein synthesis is a highly regulated process essential for life;
- polypeptides are created, based on an mRNA code, through the process of translation, requiring multiple initiation and elongation factors working in concert with the ribosome and amino acid-charged tRNA;
- mTORC1 is the primary regulator of protein synthesis, sensitive to multiple upstream effectors and working through various downstream factors to modulate ribosomal activity;
- satellite cells serve a key role in regulating muscle protein synthesis in order to maintain, repair, and augment myofibers and myocytes;
- impairment of protein synthesis can result in severe physiological manifestations that may increase the risk of morbidity and mortality.

DISCLOSURES

The opinions or assertions contained herein are the private views of the authors and are not to be construed as official or as reflecting the views of the US Army or Department of Defense. Any citations of commercial organizations and trade names in this report do not constitute an official Department of the Army endorsement of approval of the products or services of these organizations. The authors have no potential conflicts of interest to report.

References

Baar, K., Esser, K., 1999. Phosphorylation of p70(S6k) correlates with increased skeletal muscle mass following resistance exercise. Am. J. Physiol. 276, C120–127.

Bar-Peled, L., Schweitzer, L.D., Zoncu, R., Sabatini, D.M., 2012. Ragulator is a GEF for the rag GTPases that signal amino acid levels to mTORC1. Cell 150, 1196–1208.

Bentzinger, C.F., Wang, Y.X., Rudnicki, M.A., 2012. Building muscle: molecular regulation of myogenesis. Cold Spring Harb. Perspect. Biol. 4.

Beugnet, A., Tee, A.R., Taylor, P.M., Proud, C.G., 2003. Regulation of targets of mTOR (mammalian target of rapamycin) signalling by intracellular amino acid availability. Biochem. J. 372, 555–566.

Bolster, D.R., Crozier, S.J., Kimball, S.R., Jefferson, L.S., 2002. AMP-activated protein kinase suppresses protein synthesis in rat skeletal muscle through down-regulated mammalian target of rapamycin (mTOR) signaling. J. Biol. Chem. 277, 23977–23980.

Bolster, D.R., Jefferson, L.S., Kimball, S.R., 2004. Regulation of protein synthesis associated with skeletal muscle hypertrophy by insulin-, amino acid- and exercise-induced signalling. Proc. Nutr. Soc. 63, 351–356.

Burd, N.A., Holwerda, A.M., Selby, K.C., West, D.W., Staples, A.W., Cain, N.E., et al., 2010a. Resistance exercise volume affects myofibrillar protein synthesis and anabolic signalling molecule phosphorylation in young men. J. Physiol. 588, 3119–3130.

Burd, N.A., West, D.W., Staples, A.W., Atherton, P.J., Baker, J.M., Moore, D.R., et al., 2010b. Low-load high volume resistance exercise stimulates muscle protein synthesis more than high-load low volume resistance exercise in young men. PLoS One 5, e12033.

Burd, N.A., Andrews, R.J., West, D.W., Little, J.P., Cochran, A.J., Hector, A.J., et al., 2012. Muscle time under tension during resistance exercise stimulates differential muscle protein sub-fractional synthetic responses in men. J. Physiol. 590, 351–362.

Carbone, J.W., McClung, J.P., Pasiakos, S.M., 2012. Skeletal muscle responses to negative energy balance: effects of dietary protein. Adv. Nutr. 3, 119–126.

Ceafalan, L.C., Popescu, B.O., Hinescu, M.E., 2014. Cellular players in skeletal muscle regeneration. Biomed. Res. Int. 2014, 957014.

Cesena, T.I., Cui, T.X., Piwien-Pilipuk, G., Kaplani, J., Calinescu, A.A., Huo, J.S., et al., 2007. Multiple mechanisms of growth hormone-regulated gene transcription. Mol. Genet. Metab. 90, 126–133.

Cuthbertson, D.J., Babraj, J., Smith, K., Wilkes, E., Fedele, M.J., Esser, K., et al., 2006. Anabolic signaling and protein synthesis in human skeletal muscle after dynamic shortening or lengthening exercise. Am. J. Physiol. Endocrinol. Metab. 290, E731–738.

Devlin, R.B., Emerson Jr., C.P., 1978. Coordinate regulation of contractile protein synthesis during myoblast differentiation. Cell 13, 599–611.

Drummond, M.J., Rasmussen, B.B., 2008. Leucine-enriched nutrients and the regulation of mammalian target of rapamycin signalling and human skeletal muscle protein synthesis. Curr. Opin. Clin. Nutr. Metab. Care 11, 222–226.

Drummond, M.J., Dreyer, H.C., Fry, C.S., Glynn, E.L., Rasmussen, B.B., 2009. Nutritional and contractile regulation of human skeletal muscle protein synthesis and mTORC1 signaling. J. Appl. Physiol. 106, 1374–1384.

Hoffer, L.J., Taveroff, A., Robitaille, L., Hamadeh, M.J., Mamer, O.A., 1997. Effects of leucine on whole body leucine, valine, and threonine metabolism in humans. Am. J. Physiol. 272, E1037–1042.

Inoki, K., Li, Y., Zhu, T., Wu, J., Guan, K.L., 2002. TSC2 is phosphorylated and inhibited by Akt and suppresses mTOR signalling. Nat. Cell Biol. 4, 648–657.

Ivanova, N.B., Dimos, J.T., Schaniel, C., Hackney, J.A., Moore, K.A., Lemischka, I.R., 2002. A stem cell molecular signature. Science 298, 601–604.

Jiang, W., Zhu, Z., Thompson, H.J., 2008a. Dietary energy restriction modulates the activity of AMP-activated protein kinase, Akt, and mammalian target of rapamycin in mammary carcinomas, mammary gland, and liver. Cancer Res. 68, 5492–5499.

Jiang, W., Zhu, Z., Thompson, H.J., 2008b. Modulation of the activities of AMP-activated protein kinase, protein kinase B, and mammalian target of rapamycin by limiting energy availability with 2-deoxyglucose. Mol. Carcinog. 47, 616–628.

Khan, J., Bittner, M.L., Saal, L.H., Teichmann, U., Azorsa, D.O., Gooden, G.C., et al., 1999. cDNA microarrays detect activation of a myogenic transcription program by the PAX3-FKHR fusion oncogene. Proc. Natl. Acad. Sci. U.S.A. 96, 13264–13269.

Kim, I.Y., Schutzler, S., Schrader, A., Spencer, H., Kortebein, P., Deutz, N.E., et al., 2015. Quantity of dietary protein intake, but not pattern of intake, affects net protein balance primarily through differences in protein synthesis in older adults. Am. J. Physiol. Endocrinol. Metab. 308, E21–28.

Kimball, S.R., 2014. Integration of signals generated by nutrients, hormones, and exercise in skeletal muscle. Am. J. Clin. Nutr. 99, 237S–242S.

Kimball, S.R., Farrell, P.A., Jefferson, L.S., 2002. Invited review: role of insulin in translational control of protein synthesis in skeletal muscle by amino acids or exercise. J. Appl. Physiol. (1985) 93, 1168–1180.

Laplante, M., Sabatini, D.M., 2009. mTOR signaling at a glance. J. Cell Sci. 122, 3589–3594.

Long, X., Lin, Y., Ortiz-Vega, S., Yonezawa, K., Avruch, J., 2005. Rheb binds and regulates the mTOR kinase. Curr. Biol. 15, 702–713.

Ma, X.M., Blenis, J., 2009. Molecular mechanisms of mTOR-mediated translational control. Nat. Rev. Mol. Cell Biol. 10, 307–318.

Magnuson, B., Ekim, B., Fingar, D.C., 2012. Regulation and function of ribosomal protein S6 kinase (S6K) within mTOR signalling networks. Biochem. J. 441, 1–21.

Mascher, H., Andersson, H., Nilsson, P.A., Ekblom, B., Blomstrand, E., 2007. Changes in signalling pathways regulating protein synthesis in human muscle in the recovery period after endurance exercise. Acta Physiol. (Oxf) 191, 67–75.

Messina, G., Sirabella, D., Monteverde, S., Galvez, B.G., Tonlorenzi, R., Schnapp, E., et al., 2009. Skeletal muscle differentiation of embryonic mesoangioblasts requires pax3 activity. Stem Cells 27, 157–164.

Mitchell, C.J., Churchward-Venne, T.A., West, D.D., Burd, N.A., Breen, L., Baker, S.K., et al., 2012. Resistance exercise load does not determine training-mediated hypertrophic gains in young men. J. Appl. Physiol. 113, 71–77.

Moore, D.R., Robinson, M.J., Fry, J.L., Tang, J.E., Glover, E.I., Wilkinson, S.B., et al., 2009. Ingested protein dose response of muscle and albumin protein synthesis after resistance exercise in young men. Am. J. Clin. Nutr. 89, 161–168.

Murgas Torrazza, R., Suryawan, A., Gazzaneo, M.C., Orellana, R.A., Frank, J.W., Nguyen, H.V., et al., 2010. Leucine supplementation of a low-protein meal increases skeletal muscle and visceral tissue protein synthesis in neonatal pigs by stimulating mTOR-dependent translation initiation. J. Nutr. 140, 2145–2152.

Nair, K.S., Schwartz, R.G., Welle, S., 1992. Leucine as a regulator of whole body and skeletal muscle protein metabolism in humans. Am. J. Physiol. 263, E928–934.

Nobukuni, T., Joaquin, M., Roccio, M., Dann, S.G., Kim, S.Y., Gulati, P., et al., 2005. Amino acids mediate mTOR/raptor signaling through activation of class 3 phosphatidylinositol 3OH-kinase. Proc. Natl. Acad. Sci. U.S.A. 102, 14238–14243.

Olguin, H.C., Pisconti, A., 2012. Marking the tempo for myogenesis: Pax7 and the regulation of muscle stem cell fate decisions. J. Cell Mol. Med. 16, 1013–1025.

Olguin, H.C., Yang, Z., Tapscott, S.J., Olwin, B.B., 2007. Reciprocal inhibition between Pax7 and muscle regulatory factors modulates myogenic cell fate determination. J. Cell Biol. 177, 769–779.

Pasiakos, S.M., 2012. Exercise and amino acid anabolic cell signaling and the regulation of skeletal muscle mass. Nutrients 4, 740–758.

Pasiakos, S.M., McClung, J.P., 2011. Supplemental dietary leucine and the skeletal muscle anabolic response to essential amino acids. Nutr. Rev. 69, 550–557.

Pasiakos, S.M., McClung, H.L., McClung, J.P., Margolis, L.M., Andersen, N.E., Cloutier, G.J., et al., 2011. Leucine-enriched essential amino acid supplementation during moderate steady state exercise enhances postexercise muscle protein synthesis. Am. J. Clin. Nutr. 94, 809–818.

Potter, C.J., Pedraza, L.G., Xu, T., 2002. Akt regulates growth by directly phosphorylating Tsc2. Nat. Cell Biol. 4, 658–665.

Proud, C.G., 2007. Amino acids and mTOR signalling in anabolic function. Biochem. Soc. Trans. 35, 1187–1190.

Relaix, F., Zammit, P.S., 2012. Satellite cells are essential for skeletal muscle regeneration: the cell on the edge returns centre stage. Development 139, 2845–2856.

Relaix, F., Montarras, D., Zaffran, S., Gayraud-Morel, B., Rocancourt, D., Tajbakhsh, S., et al., 2006. Pax3 and Pax7 have distinct and overlapping functions in adult muscle progenitor cells. J. Cell Biol. 172, 91–102.

Roux, P.P., Ballif, B.A., Anjum, R., Gygi, S.P., Blenis, J., 2004. Tumor-promoting phorbol esters and activated Ras inactivate the tuberous sclerosis tumor suppressor complex via p90 ribosomal S6 kinase. Proc. Natl. Acad. Sci. U.S.A. 101, 13489–13494.

Rudnicki, M.A., Jaenisch, R., 1995. The MyoD family of transcription factors and skeletal myogenesis. Bioessays 17, 203–209.

Rudnicki, M.A., Schnegelsberg, P.N., Stead, R.H., Braun, T., Arnold, H.H., Jaenisch, R., 1993. MyoD or Myf-5 is required for the formation of skeletal muscle. Cell 75, 1351–1359.

Sakamoto, K., Arnolds, D.E., Ekberg, I., Thorell, A., Goodyear, L.J., 2004. Exercise regulates Akt and glycogen synthase kinase-3 activities in human skeletal muscle. Biochem. Biophys. Res. Commun. 319, 419–425.

Sancak, Y., Sabatini, D.M., 2009. Rag proteins regulate amino-acid-induced mTORC1 signalling. Biochem. Soc. Trans. 37, 289–290.

Sancak, Y., Peterson, T.R., Shaul, Y.D., Lindquist, R.A., Thoreen, C.C., Bar-Peled, L., et al., 2008. The Rag GTPases bind raptor and mediate amino acid signaling to mTORC1. Science 320, 1496–1501.

Sancak, Y., Bar-Peled, L., Zoncu, R., Markhard, A.L., Nada, S., Sabatini, D.M., 2010. Ragulator-Rag complex targets mTORC1 to the lysosomal surface and is necessary for its activation by amino acids. Cell 141, 290–303.

Sanes, J.R., 2003. The basement membrane/basal lamina of skeletal muscle. J. Biol. Chem. 278, 12601–12604.

Schultz, E., McCormick, K.M., 1994. Skeletal muscle satellite cells. Rev. Physiol. Biochem. Pharmacol. 123, 213–257.

Seale, P., Ishibashi, J., Scime, A., Rudnicki, M.A., 2004. Pax7 is necessary and sufficient for the myogenic specification of CD45 + :Sca1 + stem cells from injured muscle. PLoS Biol. 2, E130.

Sheffield-Moore, M., Yeckel, C.W., Volpi, E., Wolf, S.E., Morio, B., Chinkes, D.L., et al., 2004. Postexercise protein metabolism in older and younger men following moderate-intensity aerobic exercise. Am. J. Physiol. Endocrinol. Metab. 287, E513–522.

Shepstone, T.N., Tang, J.E., Dallaire, S., Schuenke, M.D., Staron, R.S., Phillips, S.M., 2005. Short-term high- vs. low-velocity isokinetic lengthening training results in greater hypertrophy of the elbow flexors in young men. J. Appl. Physiol. 98, 1768–1776.

Smith, E.M., Finn, S.G., Tee, A.R., Browne, G.J., Proud, C.G., 2005. The tuberous sclerosis protein TSC2 is not required for the regulation of the mammalian target of rapamycin by amino acids and certain cellular stresses. J. Biol. Chem. 280, 18717–18727.

Tajbakhsh, S., 2009. Skeletal muscle stem cells in developmental versus regenerative myogenesis. J. Intern. Med. 266, 372–389.

Tavares, M.R., Pavan, I.C., Amaral, C.L., Meneguello, L., Luchessi, A.D., Simabuco, F.M., 2015. The S6Kprotein family in health and disease. Life Sci. 131, 1–10.

Tipton, K.D., Ferrando, A.A., Williams, B.D., Wolfe, R.R., 1996. Muscle protein metabolism in female swimmers after a combination of resistance and endurance exercise. J. Appl. Physiol. (1985) 81, 2034–2038.

Tyagi, R., Shahani, N., Gorgen, L., Ferretti, M., Pryor, W., Chen, P.Y., et al., 2015. Rheb inhibits protein synthesis by activating the PERK-eIF2alpha signaling cascade. Cell Rep. S2211-1247, 00027-0003.

Wang, X., Li, W., Williams, M., Terada, N., Alessi, D.R., Proud, C.G., 2001. Regulation of elongation factor 2 kinase by p90(RSK1) and p70 S6 kinase. EMBO J. 20, 4370–4379.

White, J.P., Gao, S., Puppa, M.J., Sato, S., Welle, S.L., Carson, J.A., 2013. Testosterone regulation of Akt/mTORC1/FoxO3a signaling in skeletal muscle. Mol. Cell Endocrinol. 365, 174–186.

Yin, H., Price, F., Rudnicki, M.A., 2013. Satellite cells and the muscle stem cell niche. Physiol. Rev. 93, 23–67.

Zammit, P.S., Golding, J.P., Nagata, Y., Hudon, V., Partridge, T.A., Beauchamp, J.R., 2004. Muscle satellite cells adopt divergent fates: a mechanism for self-renewal? J. Cell Biol. 166, 347–357.

Zhang, K., Sha, J., Harter, M.L., 2010. Activation of Cdc6 by MyoD is associated with the expansion of quiescent myogenic satellite cells. J. Cell Biol. 188, 39–48.

CHAPTER

5

Role of Amino Acid Transporters in Protein Metabolism

P. M. Taylor

Division of Cell Signalling & Immunology, School of Life Sciences, University of Dundee, Sir James Black Centre, Dundee, United Kingdom

5.1. AMINO ACID TRANSPORTERS: STRUCTURE AND MOLECULAR FUNCTION

Proteolipid cell membranes act as selective barriers to polar solutes such as amino acids (AAs) and sugars (Stein, 1986). As AAs do not readily diffuse across lipid membranes, membrane-spanning protein "pores" are needed to help move (or "transport") AA between the cytosol and extracellular fluid and also between membrane-bound compartments of cells, for example, cytosol and endosomal lumen (see Broer, 2008; Forrest et al., 2011 for review). AA transporter proteins form such pores and act as "gatekeepers" to permit selective AA transport across intracellular and plasma membranes. There are six major families of AA transporters in the Solute Carrier (*SLC*) gene superfamily (*SLC1,6,7,36,38,43* families), an "orphan" *SLC16* (monocarboxylate family) transporter, which transports aromatic AA, and three *SLC* families (*SLC17,25,32*), which include AA transporters expressed only on intracellular membranes (see Tables 5.1 and 5.2; note that a subset of transporters localize to both endosomal and plasma membranes). The protein products of these transporter genes are characterized by having a primary protein structure with multiple (typically 10–12) stretches of hydrophobic AA (Fig. 5.1A). Within the membrane, these hydrophobic regions are arranged into a series of transmembrane helices (TMH) which typically fold themselves into two (or more) distinct opposed or intertwined domains organized around a central "pore" translocation pathway (Fig. 5.1B). A relatively small number of underlying protein folds have been recognized across seemingly disparate solute transporter gene families (see Forrest et al., 2011 for review). The *SLC3* gene family, although classed as AA transporters, are single TMH glycoproteins acting as regulatory subunits for the glycoprotein-associated amino acid transporter/heteromeric amino acid transporter subfamily of *SLC7* transporters (see Table 5.1). There are also a group of 7-TMH AA transporter proteins from the LCT (lysosomal cystine transporter) gene family expressed at the lysosomal membrane (Jézégou et al., 2012; Zhai et al., 2001) (see Table 5.2).

Transporters (or "carriers") function essentially in similar ways to enzymes, except that they catalyze a vectorial movement of substrate across the membrane rather than a chemical reaction converting substrate to product. Both types of catalytic mechanism are describable using principles of energetics in terms of the free-energy of different conformational states and the energy barriers between them (see Forrest et al., 2011 for review). A generally accepted "alternate access" model of transport involves a catalytic cycle in which sequential conformational changes "switch" the transporter protein between states in which the substrate binding site (or "binding center") is open to one or other side of the membrane (see Fig. 5.1C). This conformational switch is achieved by movement of the opposing transmembrane domains of the protein, which swivel or "rock" about the binding center (the "rocker-switch" mechanism). For most AA transporters, this does not require a significant movement of the actual binding center, which remains located near the center of the membrane (in cross-section). Secondary "gates" may regulate accessibility of the bound substrate to free solution

The Molecular Nutrition of Amino Acids and Proteins.
DOI: http://dx.doi.org/10.1016/B978-0-12-802167-5.00005-0

TABLE 5.1 Amino acid transporters of plasma membranes

AA transporter family	Human gene (HGNC)[a]	Common acronym	Transport mechanism designation (see Fig. 5.1)	AA substrates	Transport system(s)
SLC1 (2.A.23)					
Excitatory AA transporters	*SLC1A1-3, A6-7*	EAAT1-5	S5	AAA	X^{-}_{AG}
	SLC1A4, A5	ASCT1, 2	A2[b]	SNAA	ASC
SLC6 (2.A.22)					
Neurotransmitter transporters	*SLC6A1, A11, A13*	GAT1-3	S4	GABA	GABA
	SLC6A5, A9	GLYT1,2	S4	Gly	Gly
	SLC6A6	TauT	S4	Tau	Tau
	SLC6A7	PROT	S4	Pro	Pro
	SLC6A12	BGT1	S4	Betaine, GABA	
	SLC6A14	$ATB^{0,+}$	S4	NAA, CAA	$B^{0,+}$
	SLC6A15, A18, A19	B^0AT1-3 (XT2)	S1	NAA	B^0
	SLC6A17	XT1	S4	NAA	
	SLC6A20	IMINO (XT3)	S4	Pro, Sarcosine	Pro/β
SLC7 (2.A.3)					
(i) Cationic AA transporters	*SLC7A1-A4*	CAT1-4	U	CAA	y^+
(ii) Glycoprotein-associated AA transporters (gpaAT/HAT)[c]	*SLC7A5, A8*	LAT1,2	A1	LNAA, TH	L (L1)
	SLC7A6, A7	y^+LAT1,2	A2[d]	NAA, CAA	y^+L
	SLC7A9	$b^{0,+}$AT	A1	NAA, CAA, Cystine	$b^{0,+}$
	SLC7A10	ascT	A1	SNAA	asc
	SLC7A11	xCT	A1	Cystine, Glu	X^{-}_{C}
	SLC7A13	AGT1	A1	AAA	
SLC16 (2.A.1)					
Monocarboxylate transporters	*SLC16A10*	TAT1	U	Aromatic AA, TH	T
SLC36 (2.A.18)					
Proton-coupled AA transporters	*SLC36A1-A4*	PAT1-4 (LYAAT)	S2	Pro, Gly, Ala, GABA	PAT
SLC38 (2.A.18)					
Small neutral AA transporters	*SLC38A1,A2,A4*	SNAT1,2,4	S1	SNAA	A
	SLC38A3, A5	SNAT3,5	S3	Gln, Asn, His	N
SLC43 (2.A.1)					
Large neutral AA transporters	*SLC43A1,A2*	LAT3,4	U	LNAA	L (L2)

[a]*Only AA transporters with human orthologues are included.*
[b]*Na^+-dependent NAA antiport, exact mechanism uncertain.*
[c]*The gpaATs form heteromeric transporters with an "accessory" SLC3* (8.A.9) *subunit, either SLC3A1 (rBAT/NBAT) or SLC3A2 (F42hc/CD98).*
[d]*Primarily Na^+-NAA/CAA antiport.*
First column shows transporter families as classified by Human Genome Organisation (HGNC), see http://www.bioparadigms.org, and (in parentheses) by the International Union of Biochemistry and Molecular Biology (IUBMB), see http://www.tcdb.org. See also, for example, Broer and Palacin (2011); Hediger et al. (2013); Schiöth et al. (2013); for original sources and further details on substrate ranges and tissue expression.
AAA, anionic AA; (L/S)NAA, (large/small) neutral AA; CAA, cationic AA; TH, thyroid hormone.

TABLE 5.2 Amino acid transporters of intracellular membranes

Gene family	HGNC	Acronym	AA substrates	Mechanism (see Fig. 5.1)	Principal localization
SLC17 (2.A.1)					
Vesicular glutamate transporters	*SLC17A6-A8*	VGLUTs	Glu	A3[a]	Endosome
SLC25 (2.A.29)					
Mitochondrial transporters	*SLC25A2, A15*	ORNT2, 1	CAA, citrulline	A1[b]	Mitochondria
	SLC25A 12, A13	AGC1,2	AAA	A1[b]	Mitochondria
	SLC25A 18,A22	GC2, 1	Glu	S2	Mitochondria
	SLC25A29	ORNT3	Om, acylcarnitine	?	Mitochondria
SLC32 (2.A.18)					
Vesicular inhibitory AA transporters	*SLC32A1*	VGAT	GABA, Gly	A3[a]	Endosome
SLC36 (2.A.18)					
Proton-coupled AA transporters	*SLC36A1*	PAT1 (LYAAT1)	Pro, Gly, GABA	S2	Lysosome
	SLC36A4	PAT4 (LYAAT2)	Pro, Trp	S2	Lysosome
SLC38 (2.A.18)					
Small neutral AA transporters	*SLC38A7*	SNAT7	NAA, CAA	S1/S3?	Lysosome
	SLC38A9	SNAT9	Gln, Arg	S1/S3?	Lysosome
LCT (2.A.43)					
Lysosomal cystine transporters	*PQLC2*	LAAT1	CAA	S2?	Lysosome
	PQLC4	CTNS (cystinosin)	Cystine	S2	Lysosome

[a] *Cl^- dependent, exact mechanism uncertain.*
[b] *H^+ dependent.*
First column shows transporter families as classified by HGNC and (in parentheses) IUBMB. See, for example, Hediger et al. (2013); Schiöth et al. (2013); Jézégou et al. (2012) for more detailed information and original sources.
AAA, anionic AA; NAA, neutral AA; CAA, cationic AA.

independent of substrate translocation (these gates may be as small as a single AA residue in the transporter protein structure), effectively locking substrate in place at the binding center during certain stages of the transport cycle (the "gated-pore" mechanism). The gated, substrate-bound states are of low energy and the reorientation of the unloaded transporter across the membrane bilayer is generally identified as the rate-limiting energy barrier (see, eg, Forrest et al., 2011; Fotiadis et al., 2013 for detailed reviews of transport mechanisms). AA transport may be coupled to movements of other substrates including Na^+, H^+, K^+ and/or Cl^- (typically Na^+-AA symport, *aka* cotransport) as well as movement of other AA by antiport (Fig. 5.1D). Binding of multiple substrates to a transporter may follow a specific order.

Functionally, the rate of AA transport increases in proportion (determined by K_m) to the *cis* AA concentration until the transporters approach saturation (see Fig. 5.2), at which point transport rate can only be further increased by altering certain aspects of the driving force(s) or increasing the number of transporters (V_{max}). For AA transporters with cosubstrates (symport), *cis* concentration of all substrates will influence AA transport rate. Equally, for AA antiport the *trans* concentration of the counter-substrate will influence *cis* to *trans* AA transport rate.

The binding centers of most AA transporters encoded by mammalian genomes recognize a range of structurally-similar AA as cargo for transport (see Tables 5.1 and 5.2), typically either large neutral AA (LNAA), small neutral AA (SNAA), cationic AA (CAA), or anionic AA (AAA). Of the eight dietary essential (*aka* indispensable) AAs, Lys is a CAA whilst the other seven (Leu, Ile, Val, Phe, Trp, Met, Thr) are considered as LNAA in the current context. Most AA transporters are stereoselective for the proteinogenic L-forms of AA, although the *SLC36* transporter family and SLC1A4/A5 are characterized by a relatively-low stereoselectivity and have particular importance for transport of small, biologically-active D-AA such as D-Ser (Edwards et al., 2011).

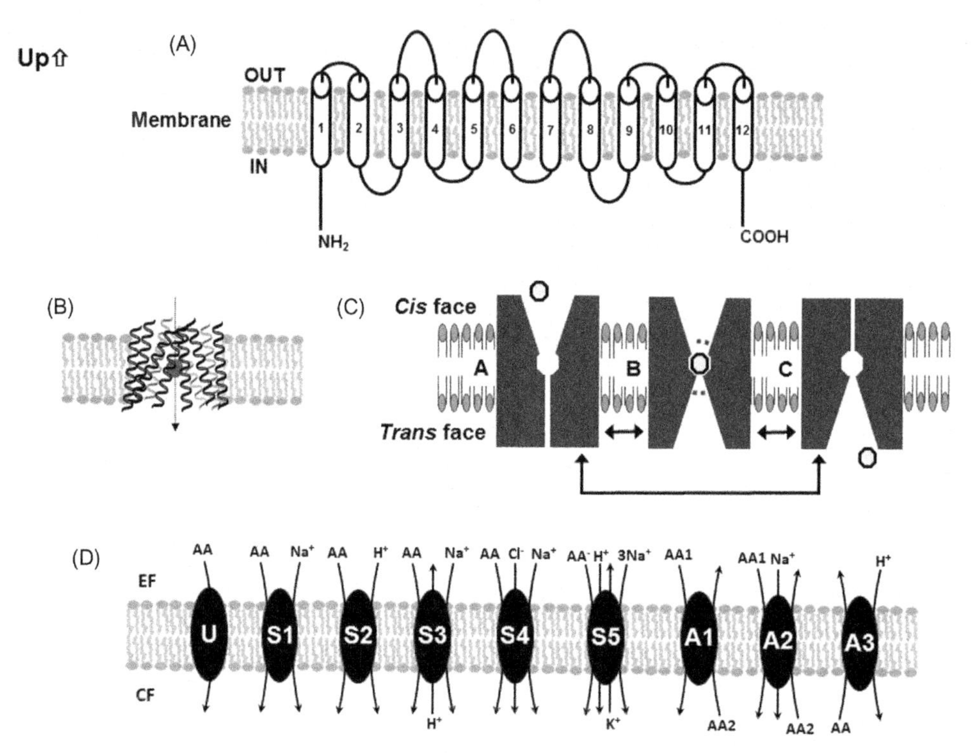

FIGURE 5.1 **Prototypic AA transporter structure and mechanism.** (A) The transporter polypeptide typically forms 10 or 12 transmembrane helixes (depending on fold type) within the membrane. IN, OUT shown in relation to the cytosol. Outward facing regions may be glycosylated. (B) The helices fold within the membrane to form a central hydrophilic translocation pathway (dotted arrow) with a localized binding center (gray circle). For clarity, peptide regions peripheral to the membrane are not shown. (C) The Alternate Access transport mechanism. AA transport is proposed to proceed stepwise from *cis* to *trans* compartments through a sequence of states: (1) AA adsorbs or "binds" to the binding center of the free transporter open to the *cis* compartment (State A); (2) Conformational changes induced by AA binding produce intermediate transition states, which may include gated states (gates indicated by dashed lines), such as a fully-occluded state (State B). These result in bound AA becoming exposed at the *trans* membrane surface through the central translocation pathway; (3) AA is released into the *trans* compartment (State C); and (4) the free transporter reorientates to the *cis*-facing conformation. These steps are reversible. Only three states of uniport are shown here for simplicity. AA transport may be coupled to symport/antiport of ions and/or antiport of another AA. Symport proceeds similarly to uniport when all substrates are bound or released. For antiport, the final reorientation step (4) only occurs after the antiported substrate has bound to the free transporter in the *trans*-facing conformation. (D) Major transport-cycle stoichiometries for mammalian AA transporters, directionality shown in relation to cytosolic fluid (CF) and extracellular/endosomal fluid (EF). U, Uniport; S, cation—AA Symport (primary Na^+-coupling except for S2); A, Antiport (primary AA antiport except for A3). Transport cycles generating a net electrical charge (rheogenic mechanisms, eg, S1, S2, S5, A2 for neutral amino acids) are influenced by both chemical gradient of substrates and electrical gradient (membrane potential): for Na^+-AA symport, this leads potentially to a 100-times accumulation of substrate AA per Na^+ transported (Stein, 1986). Electroneutral transport cycles are only influenced by chemical gradient of substrate.

The classification of AA transporters based on genetic similarity now supersedes an earlier "Transport System" classification based on substrate selectivity and transport mechanism (see Table 5.1), although the latter remains useful as a functional descriptor (Broer, 2008; Hediger et al., 2013; Taylor, 2014).

5.2. AA TRANSPORTERS AND CELLULAR FUNCTION

5.2.1 Cellular Nutrient Supply

Alongside being the basic structural units for protein synthesis, many AA also function as important metabolic intermediates and/or as substrates for biosynthesis of low-molecular weight molecules of physiological importance (see, eg, Manjarin et al., 2014; Wu, 2013 for review). Sources of cytosolic AA for these processes include (1) the extracellular fluid, (2) proteolytic endosomes such as lysosomes, and (3) cytosolic biosynthetic/proteolytic pathways (see Fig. 5.3). Sources (1) and (2) both require the AA to be transported into the cytosol. AA transporter expression is tissue-specific and many cell types express several AA transporters with overlapping substrate selectivities, such that transport of a specific AA across a cell membrane frequently involves the integrated

Up⇧

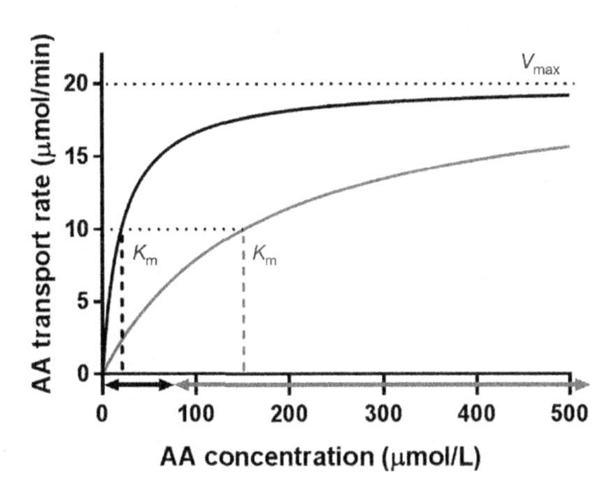

FIGURE 5.2 **A simple AA transport versus concentration relationship.** Transport rate increases from zero to a maximum plateau value (V_{max}) as substrate concentration increases, tracing a single rectangular hyperbola. Values shown on *y*- and *x*-axes are nominal. V_{max} represents the transport rate when the system has reached full capacity and K_m is the substrate concentration at which transport rate is half-maximal (0.5 V_{max}). The graph shows two representative curves for different AA transporters with similar V_{max} but different K_m values (20 and 165 μmol/L for black and gray curves respectively). The arrows on the *x*-axis indicate the ranges of AA concentration over which each transporter is the more sensitive to changes in AA concentration.

activity of uniporters, symporters, and antiporters operating in parallel (see Fig. 5.3). The functional interactions between AA for transport across cell membranes in vivo, which include *cis*-competition and (for antiporters) *trans*-effects, are influenced by (and may impact on) the nutritional and physiological status of the body (Christensen, 1990; Taylor, 2014 for review). Many cell types express both high-K_m and low-K_m transporters for particular AA types (*cf.* Fig. 5.2). Such dual-transporter systems act to "fine-tune" sensing of nutrient depletion through integration of inputs relating to internal and external nutrient availability and may facilitate both preparation for and recovery from cellular starvation (Levy et al., 2011).

In a typical mesodermal cell, CAA and LNAA are taken up by uniport or antiport mechanisms whereas SNAA and AAA are transported by concentrative Na^+-coupled symport (see Fig. 5.3). This results in SNAA such as glutamine and alanine becoming highly-concentrated in cell types such as skeletal muscle, where they may function at least partly as labile nitrogen stores (Brosnan, 2003; He et al., 2010). In contrast, LNAA do not accumulate to any great extent and indeed tend to equilibrate between intracellular and extracellular fluids. Facilitative transport (or "facilitated diffusion") by AA uniporters such as SLC7A1-A4 and SLC43A1-A2 provides a route for CAA and LNAA respectively to move down a transmembrane (electro)chemical potential gradient where simple diffusion across the lipid bilayer is energetically unfavorable. A sequential relationship between primary, secondary, and tertiary active transport systems (*P, S, T*; see Fig. 5.3) contributes substantially to transport of LNAA across cell membranes; this includes symport (cotransport) and antiport (exchange) mechanisms operating in series downstream of the Na^+ pump. This relationship hinges on the ability of a subset of AA (notably glutamine) to be accepted as substrates by both secondary and tertiary active AA transporters. The best-studied of these AA transporters are SLC38A2, SLC1A5 (both secondary active transporters mediating Na^+-glutamine symport), and SLC7A5 (an LNAA antiporter mediating tertiary active transport when acting downstream of either symporter) (Baird et al., 2009; Nicklin et al., 2009; Sinclair et al., 2013). In epithelial cells, broad-scope Na^+-AA symporters, including SLC6A19 (Broer et al., 2011) and SLC6A14 (Van Winkle et al., 2006), enable all types of AA to be transported directly into cells.

AA transport from extracellular fluid appears not to be a limiting source of AA for the cytosolic pool used in cellular mRNA translation (protein synthesis) in quiescent or terminally differentiated cells, but may become a more important factor in rapidly growing and proliferating cells as well as those with high rates of protein secretion. Notably, upregulation of plasma membrane AA transport is a characteristic feature of cell activation and transformation (often driven by growth factor signaling; McCracken and Edinger, 2013), thereby increasing intracellular availability of AA for protein synthesis and/or cell metabolism. Furthermore, rapid growth/proliferation of mammalian cells is inhibited by genetic or functional inactivation of specific AA transporters (eg, Heublein et al., 2010; Nicklin et al., 2009; Sinclair et al., 2013; Usui et al., 2006). Lysosomal AA transport mechanisms are poorly studied but represent a key step in the recycling of AA between processes of

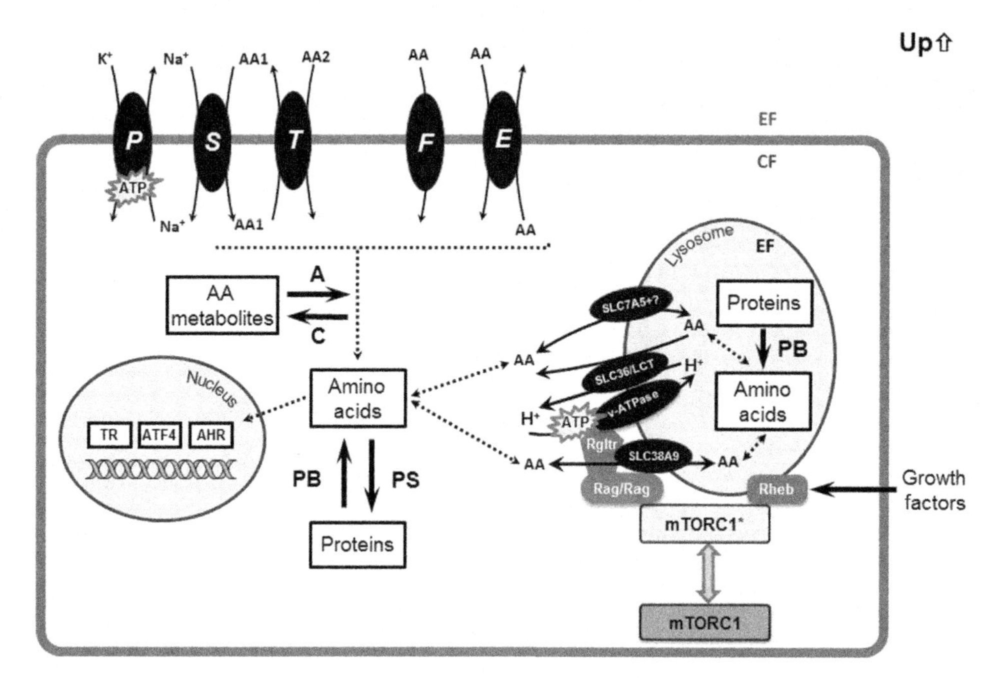

FIGURE 5.3 **AA transporters, cell metabolism, and nutrient signaling.** CF, cytosolic fluid; EF, extracellular/endosomal fluid. AA may enter (or exit) cells down chemical gradients generated by metabolic processes (eg, protein synthesis (PS)/breakdown (PB), AA biosynthesis (A), AA catabolism (C)) either through facilitative transport (*F*, a uniport mechanism) or exchange (*E*, an antiport mechanism). Charged AA movements will also depend on electrical gradient. AA may be accumulated in the CF from the extracellular fluid against an (electro)chemical gradient, downstream of primary active transport (*P*, the Na^+-K^+ ATPase pump at the plasma membrane), by secondary active transport (*S*, Na^+-AA symport mechanism using the Na^+ electrochemical gradient generated by *P*) or by tertiary active transport (*T*, AA antiport mechanism using AA1 gradient generated by *S* to drive accumulation of AA2). AA transporters on endosomal membranes (eg, the H^+-AA transporter SLC36A1 and the lysosomal cystine transporters (LCT)) more typically utilize an H^+ gradient generated by the v-ATPase H^+ pump to drive AA movement from endosomal fluid into the CF. AA delivered to the CF by AA transporters may directly modulate transcription factor activity in the cell nucleus, for example, the SLC7A5 substrates T_3 and T_4 are ligands for Thyroid Receptors (TR) and kynurenine is a ligand for Aryl Hydrocarbon Receptors (AHR). AA transporter activity contributes to maintenance of free AA concentrations in CF: fluctuations in cytosolic AA concentrations indirectly modulate abundance of transcription factors such as ATF4 (which binds to AARE on AA-responsive genes). AA may also move from CF into the lysosomal lumen through transporters such as SLC7A5 and SLC38A9 (the exact nature of substrate coupling for SLC38A9 is not clear at present). Fluctuations in intralysosomal AA concentrations appear to be sensed by SLC38A9 as part of the mechanism for recruitment and subsequent activation (denoted *) of the mTORC1 signaling complex at the lysosomal membrane. The Rag–Ragulator (Rgltr) protein complexes act as scaffold/signaling elements. Rheb activates mTORC1 on the lysosome and is a focal point for integration of upstream AA and growth factor signals. AA metabolites such as amines (eg, dopamine, tryptamine) and NO (a product of Arg metabolism) may also act as intra- or extracellular signaling molecules.

protein synthesis and breakdown and may be of particular importance for maintaining protein mass of cells at steady-state (eg, Liu et al., 2012).

AA transport at the cell surface may also be increased for scavenging purposes during periods of AA deprivation. This process, known as "adaptive regulation," is exemplified by SLC38A2 and involves transcriptional upregulation of transporter gene expression via translational upregulation of activating transcription factor 4 (ATF4) (a transcription factor which binds AA response elements (AARE) in target genes such as *SLC38A2*, activating transcription) (Palii et al., 2006), maintenance of *SLC38A2* mRNA translation through an internal ribosome entry site (Gaccioli et al., 2006) and increased stability (reduced degradation) of SLC38A2 transporter proteins (eg, Hyde et al., 2007).

5.2.2 Nutrient Sensing

Nutrient-responsive signaling pathways involved in control of cell growth, proliferation, and metabolic rate utilize a variety of enzymes, receptor and transporter proteins as nutrient sensors (Duan et al., 2015; Kimball, 2014; Taylor, 2014 for review). The major AA sensing-signaling pathways in mammalian cells are the GCN (general control nonderepressible) and mTORC1 (mammalian/mechanistic target of rapamycin complex 1) pathways. The GCN pathway is activated when one or more AA are scarce and primarily senses cytosolic AA availability at the level of AA "charging" on tRNA (Gallinetti et al., 2013; Sonenberg and Hinnebusch, 2009); its activation inhibits global protein synthesis. The mTORC1 pathway is activated when certain AA (eg, leucine, glutamine, arginine) are abundant and includes sensors which monitor AA availability in both cytosol and subcellular organelles such as lysosomes (Duan et al., 2015; Kim et al., 2013); mTORC1 activation promotes net protein accretion by simultaneously stimulating protein synthesis and inhibiting autophagic protein breakdown (Kimball, 2014 for review). Several putative cytosolic AA sensors linked to mTORC1 activation have been reported, including some capable of directly binding AA such as leucine (eg, leucyl-tRNA synthetase) (Han et al., 2012; Kim et al., 2013), alongside membrane-bound AA sensors (Duan et al., 2015 for review). AA transporters are now recognized to have important roles in both sensor and effector arms of the GCN and mTORC1 pathways (see Kim et al., 2013; Kriel et al., 2011; Taylor, 2014 for review). AA transporters may act directly as the initiating sensor for a signaling pathway (see Section 5.2.2.1) or serve as a conduit for delivery of extracellular AA to intracellular AA sensors (see Section 5.2.2.2). They may also generate indirect nutrient-related signals related to effects of transported substrates on intracellular pH and volume (Hundal and Taylor, 2009 for review).

5.2.2.1 AA Transporters as AA Sensors

AA transporter activity reflects substrate quantity (or "availability") as functions of both binding-site occupancy and transport-cycle rate (over a particular range of substrate concentration; see Fig. 5.2) and the intrinsic link between substrate binding and protein conformational change (see Fig. 5.1) affords them considerable potential as nutrient "sensors" (although outwith their range of sensitivity they can only qualitatively sense presence or absence of substrate for a signaling mechanism unless the gain of the transduction mechanism is modulated). Furthermore, AA transporters are uniquely positioned to access and "sample" both cytosolic and extracytosolic (extracellular/endosomal) AA pools. Mammalian AA transporters such as SLC38A2/A9 and SLC36A1/A4 may also act as AA-sensing receptors (ie, AA substrate binding to the transporter protein induces an intracellular nutrient signal independent of AA transport) and thus act as multifunctional "AA transceptors" on either endosomal or plasma membranes (Ögmundsdóttir et al., 2012; Pinilla et al., 2011; Rebsamen et al., 2015; Wang et al., 2015). The AA-dependent recruitment of the mTORC1 signaling complex to lysosomal membranes (a key stage in mTORC1 activation) is associated with AA accumulation into lysosomes (Zoncu et al., 2011). A lysosomal-anchored "nutrisome" protein complex appears to act as a sensor of intralysosomal AA levels upstream of mTORC1 activation. Both SLC38A9 and SLC36A1 are included as part of the "nutrisome" (Ögmundsdóttir et al., 2012; Rebsamen et al., 2015; Wang et al., 2015) alongside the v-ATPase which helps develop the acidic lumen of the lysosome (pH 5) by actively pumping H^+ from the cytosol (see Fig. 5.3). Detection of influx, efflux, and/or accumulation of AA into the lysosomal lumen by the nutrisome (a so-called "inside-out" method of AA sensing; Rebsamen et al., 2015; Zoncu et al., 2011), causes activation of the guanine exchange factor function of the Ragulator complex which in turn promotes the guanosine triphosphate charging of RagA/B subunits necessary for mTORC1 engagement and activation (see Fig. 5.3; see Duan et al., 2015 for review). SLC38A9 accepts both glutamine and arginine as substrates (Rebsamen et al., 2015; Wang et al., 2015) and an AA-dependent conformational change of SLC38A9 involving its cytosolic N-terminus (a domain binding directly to Ragulator) may directly initiate an AA signal through this transceptor (Wang et al., 2015). Transceptor signaling is likely to be initiated during substrate-occluded states of the transport cycle (see Fig. 5.1; see Kriel et al., 2011; Taylor, 2014 for review). SLC38A7 is another *SLC38*-family transporter localized to lysosomal membranes (Chapel et al., 2013) and it appears to accept a broad range of neutral and CAA as substrates (Hägglund et al., 2011). The *SLC38*-family AA transporters are highly pH-sensitive and a subset operate as Na^+-dependent H^+-AA antiporters (see Schiöth et al., 2013 for recent review), a mechanism which might favor glutamine and arginine accumulation into lysosomes by both SLC38A7 and A9. The requirement for lysosomal localization of mTORC1 for its activation may differ between leucine and other AA (Averous et al., 2014). Nevertheless, lysosomal AA transport mechanisms resembling the L and T Systems for LNAA in plasma membranes have been characterized functionally (Andersson et al., 1990; Pisoni and Thoene, 1991; Stewart et al., 1989) which could mediate net lysosomal uptake or release of essential LNAA depending upon cellular AA status. Indeed very recent findings (Milkereit et al., 2015) indicate that SLC7A5 (as the functional SLC7A5—SLC3A2 System L heterodimer) appears to be recruited constitutively to lysosomes by the lysosomal-associated transmembrane protein 4b, where it promotes mTORC1 activation via the v-ATPase nutrisome complex. As an AA antiporter, SLC7A5 would require a sufficiency of exchangeable AA in the lysosomal lumen (produced by proteolysis and/or uptake through other AA transporters) in order for it to accumulate leucine into this compartment. SLC36A1 may exert a negative influence on lysosomal mTORC1 signaling by mediating H^+-dependent efflux of SNAA from the lysosomal lumen into the cytosol (Zoncu et al., 2011). PQLC2 is a lysosomal exporter of CAAs (Jézégou et al., 2012; Liu et al., 2012).

The "inside-out" lysosomal AA sensing mechanism implies sensing of AA availability in the lysosomal lumen, which for a membrane-bound transceptor such as SLC38A9 is spatially equivalent to the extracellular space. Another *SLC38* family member, SLC38A2, functions as a transceptor at the plasma membrane linking extracellular AA availability to cell function by a mechanism also involving its cytosolic N-terminal protein domain (Hyde et al., 2007; Pinilla et al., 2011). Characteristic properties of cell-surface nutrient transceptors in lower eukaryotic organisms include (1) substrate transport/binding promotes internalization and degradation of the transceptor protein when substrate is abundant and (2) induction of transceptor gene expression in cells starved of their substrates (Conrad et al., 2014; Kriel et al., 2011). SLC38A2 displays both these transceptor properties (Hyde et al., 2007) and its upregulation as an AA scavenger during periods of AA deficiency is an important effector arm of the GCN pathway in mammalian cells.

5.2.2.2 AA Transporters Upstream of Intracellular AA Sensors

AA transporters at the plasma membrane deliver extracellular AAs to the intracellular sensor molecules associated with mTORC1 and GCN pathway regulation. Changes in cell-surface AA transporter expression or extracellular AA concentration (the latter within relative limits highlighted in Fig. 5.2) will alter these rates of AA delivery, which may influence nutrient-dependent signaling. SLC7A5 appears to be of particular importance for leucine-dependent activation of mTORC1 and may provide a direct, two-stage route for leucine movement between extracellular and lysosomal compartments (Milkereit et al., 2015; see Fig. 5.3). The level of functional cell-surface SLC7A5 expression in fibroblasts correlates directly with the effectiveness of leucine-induced mTORC1 activation in these cells (Schriever et al., 2012). Induction of functional *SLC7A5* gene expression is an initiating factor for mTORC1 pathway activation by the proliferation factor HIF2α in vivo (Elorza et al., 2012) and for mTORC1-dependent T-lymphocyte activation (see Section 5.5); induction of the functionally similar SLC7A8 has also been linked to cell proliferation in vivo (Kurayama et al., 2011). Leucine transport by SLC7A5/A8 requires obligatory AA antiport, typically leucine-glutamine antiport. The ability of glutamine transporters such as SLC38A2 and SLC1A5 to develop or maintain a suitable large intracellular glutamine pool for antiport may become limiting for leucine-dependent mTORC1 activation downstream of SLC7A5 (Baird et al., 2009; Nakaya et al., 2014; Nicklin et al., 2009). SLC7A5/SLC3A2 and SLC1A5 have been reported (Xu and Hemler, 2005) to colocalize with CD147 as part of a cell-surface protein "supercomplex" linked to activation of cell metabolism and proliferation.

Delivery of extracellular arginine via the SLC7A1 and A3 transporters may also influence mTORC1 pathway activation (Huang et al., 2007).

5.2.3 Cell-Cell Communication

The *SLC1* high-affinity glutamate transporters have key functions in the control of excitatory neurotransmission within the central nervous system (CNS) (Kanai et al., 2013), principally in the clearance of neurotransmitter AA from the synaptic cleft, whilst the *SLC6* family transporters for gamma-aminobutyric acid (GABA) and glycine have similar roles in inhibitory neurotransmission (Pramod et al., 2013). The neurotransmitter AA are concentrated within neuronal synaptic vesicles by vesicular AA transporters (*SLC17/32* transporter families; Table 5.2) which utilize the endosomal pH gradient (see Fig. 5.3) as a driving force. The SNAA transporter SLC7A10 (which is unusual in that it can operate by either antiport or uniport mechanisms) acts in the CNS to regulation synaptic concentration of NMDA (glutamate) receptor coagonists D-serine and glycine (Xie et al., 2005).

Certain AA transporters have specific roles in transport of biologically-relevant AA derivatives involved in intercellular signaling, for example, SLC7A5 accepts thyroid hormones and L-DOPA as substrates (Taylor and Ritchie, 2007). SLC7A2 and SLC7A7 have opposing roles in the regulation of intracellular levels of the NO precursor arginine (Ogier de Baulny et al., 2012; Yeramian et al., 2006).

5.3. AA TRANSPORTERS IN WHOLE-BODY NUTRITION

5.3.1 Absorption of AA and Peptides

The recommended daily allowance for dietary protein intake by an adult person is 0.8 g/kg per day, irrespective of overall energy intake (Layman et al., 2015). Dietary protein of high quality includes sufficient amounts of all essential AA required for a particular stage of life (Wu, 2013). This exogenous protein (along with a similar

TABLE 5.3 Peptide transporters

Gene family	HUGO	Acronym	Substrates	Mechanism (see Fig. 5.1)
SLC15 (2.A.17)				
Peptide transporters	*SLC15A1, A2*	PEPT1, 2	Tri- and di-peptides, β-lactams	S2
	SLC15A3, A4	PHT2, 1	Tri- and di-peptides, Histidine	S2

First column shows transporter families as classified by HGNC and (in parentheses) IUBMB. See, for example, Hediger et al. (2013); Gilbert et al. (2008) for more detailed information and original sources.

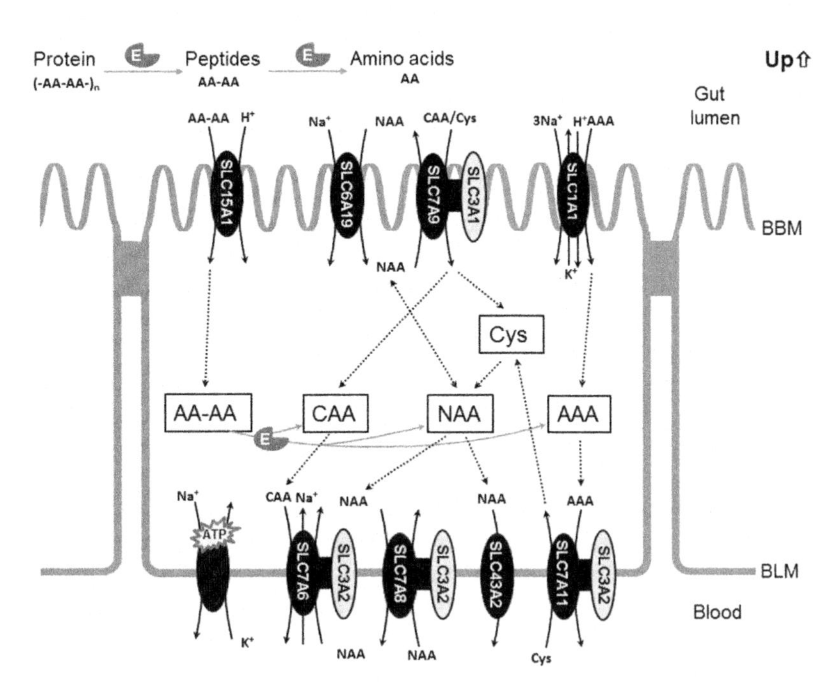

FIGURE 5.4 Intestinal digestion and absorption of amino acids and peptides. Dietary proteins are digested by hydrolysis (proteolytic enzymes indicated E in Figure) to absorbable peptides (shown as AA-AA) and amino acids (AA). Peptides (di/tripeptides) are absorbed into intestinal epithelial cells across the brush-border membrane (BBM) by H^+ coupled transport using the H^+ gradient produced by the acid microclimate exterior to the BBM. Most of these peptides are hydrolyzed to AA within the epithelial cells. The major BBM transport system for neutral AA is the Na^+ coupled transporter SLC6A19 (B^0AT1). The major transporters for cationic and anionic AA at the BBM are SLC7A9 ($b^{0,+}$AT) and SLC1A1 (EAAT3) respectively. Additional BBM transporters for neutral AA (eg, SLC1A5, SLC36A1, SLC6A20) are not shown on Figure. AA antiporters (notably SLC7A6 and SLC7A8) predominate at the basolateral membrane (BLM), enabling physiologically-useful AA exchanges which, in general, are osmotically-neutral. Facilitative transport of neutral AA at the BLM appears to be mediated by SLC43A2 and (not shown) SLC16A10. Note (1) targeting of *SLC7* antiporters to either BBM or BLM by SLC3A1 or SLC3A2 subunits respectively, (2) cycling of NAA at both membranes as part of the mechanism to absorb CAA by AA antiport, and (3) cellular uptake of cystine (Cys; metabolized to cysteine inside cells) by AA antiporters at both membranes.

quantity of endogenous protein released into the digestive tract) is hydrolyzed in the gut lumen by digestive proteases to absorbable constituents which include di- and tripeptides as well as AAs (see, eg, Goodman, 2010 for review). The products of protein digestion must then be moved across both the brush-border membrane and basolateral membrane of the intestinal epithelium (BBM, BLM respectively; see Fig. 5.4) in order to be absorbed by the body. Digestive enzymes include endopeptidases (eg, pepsin, trypsin) and exopeptidases (eg, carboxypeptidase A) secreted by the gastrointestinal tract, as well as numerous BBM-bound and cytosolic enzymes (including aminopeptidases). Active AA and peptide absorption is driven by Na^+ and H^+ electrochemical gradients established by the Na^+/K^+ ATPase pump at the epithelial BLM (see Fig. 5.4) and Na^+/H^+ antiporters at the BBM. Peptides are absorbed by H^+-coupled symport (Table 5.3, Fig. 5.4) and are largely hydrolyzed to AA in the epithelial cells, rather than being absorbed intact into the bloodstream (Gilbert et al., 2008; Nakashima et al., 2011). Neutral and AAA are taken up into intestinal epithelial cells by Na^+-coupled symport (principally by SLC6A19 and SLC1A1 respectively), whereas CAA are transported largely by AA antiport (SLC7A9). All AA pass from the epithelial cells to the bloodstream by AA antiport and/or facilitative transport (uniport) systems at the BLM. Notably, glutamine/essential NAA antiport at the BLM via SLC7A8 has the dual benefit of providing glutamine as an intestinal fuel and completing essential NAA absorption. SLC7A6 mediates electroneutral AA exchanges (balancing Na^+ and CAA positive charges) which tend to favor efflux of CAA from cells. The BLM transport step is typically rate-limiting for AA absorption.

The relative importance of peptide and free AA absorption at the BBM depends upon the digestibility of the ingested protein (Gilbert et al., 2008; Nakashima et al., 2011). Peptide transport facilitates absorption of high-protein loads in the intestine (Nassl et al., 2011), although it does not appear to be nutritionally-essential (Hu et al., 2008). The SLC6A19 AA transporter forms functional "digestive complexes" with specific peptides (eg, aminopeptidase N) in the BBM which improve its absorptive efficiency (Fairweather et al., 2012). SLC6A19 provides an important AA supply for epithelial mTORC1 signaling and SLC6A19 knockout mice exhibit a phenotype of apparent epithelial cell starvation (Broer et al., 2011). Inherited disorders of epithelial AA absorption include AA transport defects at both BBM; for example, cystinuria (*SLC7A9/SLC3A1* defect), Hartnup disorder (*SLC6A19* defect) and BLM; lysinuric protein intolerance (*SLC7A6* defect) (see Broer and Palacin, 2011 for review). These disorders of AA absorption rarely result in any specific AA deficiency because of the overlapping substrate selectivities of intestinal AA transporters.

The absorptive capacity for protein (as AA and peptides) is regulated in relation to both quantity and quality of dietary protein and is generally maintained at a higher level than the normal dietary supply (thus providing a "safety factor" for efficient absorption of high-protein meals) (Diamond and Hammond, 1992). Intestinal AA transporter expression may be induced by raised levels of dietary protein or free AA mixtures, but also by protein intakes below those required for essential AA balance to ensure supply of these nutrients (Buddington et al., 1991). Several endocrine factors may be involved in this type of regulation. Epidermal growth factor (EGF) facilitates intestinal growth and adaptation and signaling downstream of EGF may activate functional expression of SLC6A19 (Bhavsar et al., 2011). The carboxypeptidase ACE2 (a peptidase interacting with SLC6A19 at the intestinal BBM; Fairweather et al., 2012) may also be involved in regulation of its functional capacity for AA transport (Vuille-Dit-Bille et al., 2015). Leptin upregulates expression and transport capacity of the peptide transporter SLC15A1 in the intestine when secreted into the gut lumen (Buyse et al., 2001), although abundance of AA transporters (SLC6A19, SLC1A5) in the BBM of intestinal cells is downregulated by leptin (Ducroc et al., 2010; Fanjul et al., 2012). To add further complexity, rapid regulation of intestinal AA transport activity in vivo may involve local neural circuits (Mourad et al., 2009). SLC38A2 and the SLC15A1 peptide transporter may form part of a nutrient-sensing system in the gastrointestinal tract which causes gut hormone release upon activation (Young et al., 2010; Zietek and Daniel, 2015), indeed an oral AA load has been shown to exert an incretin effect mediated by glucose-dependent insulinotropic polypeptide (Lindgren et al., 2015). External to the gut itself, SLC6A15 forms part of an AA sensor in the hypothalamus involved in regulation of food intake (Drgonova et al., 2013; Hagglund et al., 2013). Similarly, SLC38A2 is a secondary component of an AA-sensing system in the anterior piriform cortex of the brain which allows animals to make rapid food selections on the basis of essential AA quality (Gietzen and Aja, 2012).

5.3.2 Interorgan Nitrogen Flow

AA transporter capacity and competition between different AA for transport at the blood–tissue interface are important factors in the determination and regulation of AA flows between mammalian tissues (see, eg, Brosnan, 2003; Christensen, 1990; van de Poll et al., 2004 for review).

In the fed state, the predominant AA flows are from the intestine to other tissues and AA in excess of those required for protein synthesis or particular metabolic purposes are rapidly catabolized, typically as oxidative fuels. Excess nitrogen is excreted as urea: the urea-cycle intermediates arginine and ornithine are provided to hepatocytes (which unusually do not express SLC7A1) at least partly by SLC7A2 (Closs et al., 2006). Blood AA concentrations increase during the absorptive phase, indicating that the rate of intestinal AA absorption exceeds the capacity for tissue AA uptake. The increase in essential AA availability after protein or AA ingestion upregulates expression of AA transporters (eg, SLC38A2, SLC7A5) in skeletal muscle 2–3 h postmeal (Drummond et al., 2010), an adaptive response to help increase intracellular delivery of AA (and thereby enhance the mTORC1 growth signal from nutrients and insulin) during this anabolic phase of the dietary cycle (see Kimball, 2014; Layman et al., 2015 for review). The strong pH-dependence of SLC38A2 makes it very sensitive to inhibition by plasma acidification and this feature contributes to the protein-catabolic effects of metabolic acidosis (Evans et al., 2008).

In the postabsorptive and fasted states, the dominant AA flows are between muscle, liver, and kidney and these assist in conserving body nitrogen stores by channeling protein and AA catabolism to production of glutamine and alanine rather than urea (Brosnan, 2003; van de Poll et al., 2004). Within the liver, perivenous hepatocytes take up glutamate through SLC1A2 for synthesis of glutamine, which is released into the bloodstream by the Na^+-H^+-antiporting SLC38A3 transporter operating unusually in reverse mode (under circumstances where the combined glutamine and H^+ gradients apparently exceed the inwardly-directed Na^+ gradient) (Baird et al., 2004; Broer et al., 2002). Expression of SLC1A1 in kidney, muscle, and lung provides glutamate for glutamine synthesis in these tissues. This glutamine can be exchanged for essential NAA through plasma membrane AA antiporters (eg, SLC7A5 and A8) in tissues such as skeletal muscle, providing essential NAA to sustain muscle protein synthesis whilst simultaneously increasing glutamine availability in the blood for other tissues to use either as a fuel (eg, lymphocytes, intestinal epithelium) or as an aid to maintaining optimum acid–base and nitrogen stasis (eg, kidney) (Baird et al., 2004; He et al., 2010; Karinch et al., 2007). SLC38A3 is upregulated in the kidney during circumstances of metabolic acidosis to increase glutamine uptake from the blood as a source of NH_4^+ for urinary acid excretion (Moret et al., 2007). The essential NAA uniporter SLC43A1 is upregulated by starvation in both liver and skeletal muscle (Fukuhara et al., 2007). This would tend to favor AA efflux from these tissues into the bloodstream and might participate in the supply of essential NAA to organs such as the brain during prolonged starvation. Alanine is an important carbon source for hepatic gluconeogenesis in fasted states and SLC38A2 is upregulated (under the influence of glucagon) in periportal hepatocytes to help maintain alanine supply from the bloodstream under such circumstances (Varoqui and Erickson, 2002).

5.4. AA TRANSPORTERS IN MAMMALIAN EMBRYONIC DEVELOPMENT AND GROWTH

Several AA transporters are essential for mammalian embryonic development (eg, *SLC7A5*; Poncet et al., 2014) or perinatal survival (eg, *SLC7A1*; Perkins et al., 1997; *SLC43A2*; Guetg et al., 2015). This may reflect either a particular importance for AA supply over these periods or a specific function in processes such as signaling of growth or early tissue differentiation.

AA activation of mTORC1 is an important aspect of blastocyst activation during early stages of development (Gonzalez et al., 2012). Upregulation of SLC6A14 (a broad-scope Na^+/Cl^--coupled AA transporter) at the blastocyst stage enhances uptake of AA which helps promote blastocyst activation and trophoblast outgrowth (Van Winkle et al., 2006). Placental AA transporters supply the growing fetus with AA from the maternal bloodstream (see, eg, Dilworth and Sibley, 2013 for review). Placental growth is also modulated by mTORC1 which regulates the expression and activity of key AA transporters (eg, SLC7A5) at the placental surface (Roos et al., 2009). Intrauterine growth restriction is associated with decreased activity of placental AA transporters in conjunction with reduced placental mTORC1 activity (Roos et al., 2007). During normal pregnancy, expression of AA transporters including the essential NAA uniporters SLC16A10 and SLC43A1 correlates positively with birth-size of infants (Cleal et al., 2011). In contrast, an inverse correlation between SLC38A2 expression and placental weight is observed; that is, smaller placentas appear to upregulate SNAA uptake by this particular route (Coan et al., 2008). This may help to maintain fetal growth rate at least partly independent of placental size.

The high growth rate of newborn and infant tissues (particularly skeletal muscle) is maintained in part by AA-dependent stimulation of the mTORC1 pathway which promotes net protein synthesis (see Suryawan and Davis, 2011 for review). Maternal milk provision for growing infants requires extremely high rates of protein turnover in lactating mammary glands. AA such as lysine, methionine, and tryptophan may become limiting for milk production, at least in livestock diets (see Manjarin et al., 2014 for review). AA transporters including SLC7A1, SLC7A5, SLC6A14, and SLC38A2 are upregulated correspondingly in lactating mammary tissue, correlated with the action of hormones such as prolactin and oestradiol (Aleman et al., 2009; Manjarin et al., 2011; Velazquez-Villegas et al., 2015).

5.5. AA TRANSPORTERS AND THE IMMUNE RESPONSE

AA transporters have functional importance for cells of both the innate and adaptive immune systems. SLC7A2 is induced by activation in macrophages, providing arginine as a substrate for NO or polyamine production depending upon type of activation (Yeramian et al., 2006). SLC7A11 (which mediates electroneutral glutamate/anionic cystine exchange) is also upregulated in macrophages to support synthesis of the antioxidant glutathione (Conrad and Sato, 2012). T-lymphocytes undergo periods of rapid cell growth, proliferation and protein (eg, cytokine) secretion during an immune response, following activation through the T-cell receptor. During these periods of markedly increased protein synthesis, intracellular availability of both LNAA (eg, leucine for protein synthesis) and SNAA (eg, glutamine or glycine for cell metabolism) may become increasingly dependent on the AA transport capacity at the cell surface. T-lymphocyte activation is associated with a high fold-induction of *SLC7A5* which increases high-affinity (low μM K_m) LNAA supply (Sinclair et al., 2013). Removal of a block to mRNA elongation may stimulate induction of *SLC7A5* (and *SLC3A2*) gene transcripts during T-cell activation (Nii et al., 2001). Other AA transporters (notably SLC7A1 and SLC1A5) are also upregulated to a lesser extent during activation of T-cells (Nakaya et al., 2014; Sinclair et al., 2013). Sustained uptake of leucine (and other LNAA) through SLC7A5 in activated T-lymphocytes is required for mTORC1 activation, induction of the c-Myc mitogen, and upregulation of energy-supplying metabolic pathways (Sinclair et al., 2013). The Trp metabolite kynurenine (Kyn) is an SLC7A5 substrate and Trp/Kyn exchanges have been implicated in immunoregulatory mechanisms and immune-escape strategies, partly linked to activation of the Aryl Hydrocarbon Receptor (AHR) by Kyn (Kaper et al., 2007; Ramsay and Cantrell, 2015).

5.6. AA AND PEPTIDE TRANSPORTERS AS THERAPEUTIC TARGETS

Several AA and peptide transporters are of pharmaceutical interest as drug targets (for specific diseases or as cell-growth/proliferation inhibitors) and/or as drug-delivery systems (see, eg, Hediger et al., 2013 for recent review). Glutamate, GABA and glycine transporters of the *SLC1* and *SLC6* transporter families are important pharmacological targets for neuromodulatory therapies (see Kanai et al., 2013; Pramod et al., 2013 for review). The SLC7A5 transporter is recognized as a potential immunosuppressive (Usui et al., 2006) and antitumor (Oda et al., 2010) target and high-affinity (nM K_m) specific inhibitors of SLC7A5 are under development (eg, Oda et al., 2010; Usui et al., 2006). AA transporter-specific positron emission tomography tracers are now in use for tissue and tumor imaging in vivo (eg, Wiriyasermkul et al., 2012).

Essential NAA (especially Leu) are required for full activation of mTORC1 signaling downstream of growth factors, such as insulin, and both *SLC6A19*-null mice (Jiang et al., 2015) and muscle-specific *SLC7A5*-null mice (Poncet et al., 2014) show evidence of reduced baseline insulin-sensitivity, highlighting important links between AA transporter function and endocrine control of metabolism. Leucine may also stimulate increases in tissue oxidative capacity (Sun and Zemel, 2009; Vaughan et al., 2013). Dietary leucine is therefore under evaluation as an adjunct treatment for insulin resistance related to obesity (eg, Adeva et al., 2012; Macotela et al., 2011), although there is controversy over the possible long-term metabolic consequences of resultant AA imbalances in vivo (Layman et al., 2015 for review).

Acknowledgment

Sources of Funding: PMT acknowledges research funding from The Wellcome Trust, Biotechnology and Biological Sciences Research Council UK, Diabetes UK, National Institutes of Health USA, and the Rural and Environment Science and Analytical Services Division (RERAD) of The Scottish Government.

References

Adeva, M.M., Calvino, J., Souto, G., Donapetry, C., 2012. Insulin resistance and the metabolism of branched-chain amino acids in humans. Amino Acids 43, 171–181.

Aleman, G., Lopez, A., Ordaz, G., Torres, N., Tovar, A.R., 2009. Changes in messenger RNA abundance of amino acid transporters in rat mammary gland during pregnancy, lactation, and weaning. Metabolism 58, 594–601.

Andersson, H.C., Kohn, L.D., Bernardini, I., Blom, H.J., Tietze, F., Gahl, W.A., 1990. Characterization of lysosomal monoiodotyrosine transport in rat thyroid cells. Evidence for transport by system h. J. Biol. Chem. 265, 10950–10954.

Averous, J., Lambert-Langlais, S., Carraro, V., Gourbeyre, O., Parry, L., B'Chir, W., et al., 2014. Requirement for lysosomal localization of mTOR for its activation differs between leucine and other amino acids. Cell. Signal. 26, 1918–1927.

Baird, F.E., Beattie, K.J., Hyde, A.R., Ganapathy, V., Rennie, M.J., Taylor, P.M., 2004. Bidirectional substrate fluxes through the system N (SNAT5) glutamine transporter may determine net glutamine flux in rat liver. J. Physiol. 559, 367–381.

Baird, F.E., Bett, K.J., MacLean, C., Tee, A.R., Hundal, H.S., Taylor, P.M., 2009. Tertiary active transport of amino acids reconstituted by coexpression of system A and L transporters in *Xenopus oocytes*. Am. J. Physiol. Endocrinol. Metab. 297, E822–829.

Bhavsar, S.K., Hosseinzadeh, Z., Merches, K., Gu, S., Bröer, S., Lang, F., 2011. Stimulation of the amino acid transporter *SLC6A19* by JAK2. Biochem. Biophys. Res. Commun. 414, 456–461.

Broer, S., 2008. Amino acid transport across mammalian intestinal and renal epithelia. Physiol. Rev. 88, 249–286.

Broer, S., Palacin, M., 2011. The role of amino acid transporters in inherited and acquired diseases. Biochem. J. 436, 193–211.

Broer, A., Albers, A., Setiawan, I., Edwards, R.H., Chaudhry, F.A., Lang, F., et al., 2002. Regulation of the glutamine transporter SN1 by extracellular pH and intracellular sodium ions. J. Physiol. 539, 3–14.

Broer, A., Juelich, T., Vanslambrouck, J.M., Tietze, N., Solomon, P.S., Holst, J., et al., 2011. Impaired nutrient signaling and body weight control in a Na^+ neutral amino acid cotransporter (*Slc6a19*)-deficient mouse. J. Biol. Chem. 286, 26638–26651.

Brosnan, J.T., 2003. Interorgan amino acid transport and its regulation. J. Nutr. 133, 2068S–2072S.

Buddington, R.K., Chen, J.W., Diamond, J.M., 1991. Dietary regulation of intestinal brush-border sugar and amino acid transport in carnivores. Am. J. Physiol. 261, R793–801.

Buyse, M., Berlioz, F., Guilmeau, S., Tsocas, A., Voisin, T., Péranzi, G., et al., 2001. PepT1-mediated epithelial transport of dipeptides and cephalexin is enhanced by luminal leptin in the small intestine. J. Clin. Invest. 108, 1483–1494.

Chapel, A., Kieffer-Jaquinod, S., Sagné, C., Verdon, Q., Ivaldi, C., Mellal, M., et al., 2013. An extended proteome map of the lysosomal membrane reveals novel potential transporters. Mol. Cell. Proteomics 12, 1572–1588.

Christensen, H.N., 1990. Role of amino acid transport and countertransport in nutrition and metabolism. Physiol. Rev. 70, 43–77.

Cleal, J.K., Glazier, J.D., Ntani, G., Crozier, S.R., Day, P.E., Harvey, N.C., et al., 2011. Facilitated transporters mediate net efflux of amino acids to the fetus across the basal membrane of the placental syncytiotrophoblast. J. Physiol. 589, 987–997.

Closs, E.I., Boissel, J.P., Habermeier, A., Rotmann, A., 2006. Structure and function of cationic amino acid transporters (CATs). J. Membr. Biol. 213, 67–77.

Coan, P.M., Angiolini, E., Sandovici, I., Burton, G.J., Constância, M., Fowden, A.L., 2008. Adaptations in placental nutrient transfer capacity to meet fetal growth demands depend on placental size in mice. J. Physiol. 586, 4567–4576.

Conrad, M., Sato, H., 2012. The oxidative stress-inducible cystine/glutamate antiporter, system x (c) (-): cystine supplier and beyond. Amino Acids 42, 231–246.

Conrad, M., Schothorst, J., Kankipati, H.N., Van Zeebroeck, G., Rubio-Texeira, M., Thevelein, J.M., 2014. Nutrient sensing and signaling in the yeast Saccharomyces cerevisiae. FEMS Microbiol. Rev. 38, 254–299.

Diamond, J., Hammond, K., 1992. The matches, achieved by natural selection, between biological capacities and their natural loads. Experientia 48, 551–557.

Dilworth, M.R., Sibley, C.P., 2013. Review: transport across the placenta of mice and women. Placenta 34 (Suppl.), S34–S39.

Drgonova, J., Jacobsson, J.A., Han, J.C., Yanovski, J.A., Fredriksson, R., Marcus, C., et al., 2013. Involvement of the neutral amino acid transporter *SLC6A15* and leucine in obesity-related phenotypes. PLoS One 8, e68245.

Drummond, M.J., Glynn, E.L., Fry, C.S., Timmerman, K.L., Volpi, E., Rasmussen, B.B., 2010. An increase in essential amino acid availability upregulates amino acid transporter expression in human skeletal muscle. Am. J. Physiol. Endocrinol. Metab. 298, E1011–1018.

Duan, Y., Li, F., Tan, K., Liu, H., Li, Y., Liu, Y., et al., 2015. Key mediators of intracellular amino acids signaling to mTORC1 activation. Amino Acids 47, 857–867.

Ducroc, R., Sakar, Y., Fanjul, C., Barber, A., Bado, A., Lostao, M.P., 2010. Luminal leptin inhibits l-glutamine transport in rat small intestine: involvement of ASCT2 and B0AT1. Am. J. Physiol. Gastrointest. Liver Physiol. 299, G179–G185.

Edwards, N., Anderson, C.M.H., Gatfield, K.M., Jevons, M.P., Ganapathy, V., Thwaites, D.T., 2011. Amino acid derivatives are substrates or non-transported inhibitors of the amino acid transporter PAT2 (slc36a2). Biochim. Biophys. Acta 1808, 260–270.

Elorza, A., Soro-Arnáiz, I., Meléndez-Rodríguez, F., Rodríguez-Vaello, V., Marsboom, G., de Cárcer, G., et al., 2012. HIF2α Acts as an mTORC1 Activator through the Amino Acid Carrier *SLC7A5*. Mol. Cell. 48, 681–691.

Evans, K., Nasim, Z., Brown, J., Clapp, E., Amin, A., Yang, B., et al., 2008. Inhibition of SNAT2 by metabolic acidosis enhances proteolysis in skeletal muscle. J. Am. Soc. Nephrol. 19, 2119–2129.

Fairweather, S.J., Bröer, A., O'Mara, M.L., Bröer, S., 2012. Intestinal peptidases form functional complexes with the neutral amino acid transporter B0AT1. Biochem. J. 446, 135–148.

Fanjul, C., Barrenetxe, J., Inigo, C., Sakar, Y., Ducroc, R., Barber, A., et al., 2012. Leptin regulates sugar and amino acids transport in the human intestinal cell line Caco-2. Acta Physiol. 205, 82–91.

Forrest, L.R., Krämer, R., Ziegler, C., 2011. The structural basis of secondary active transport mechanisms. Biochim. Biophys. Acta 1807, 167–188.

Fotiadis, D., Kanai, Y., Palacin, M., 2013. The *SLC3* and *SLC7* families of amino acid transporters. Mol. Aspects Med. 34, 139–158.

Fukuhara, D., Kanai, Y., Chairoungdua, A., Babu, E., Bessho, F., Kawano, T., et al., 2007. Protein characterization of Na + -independent system L amino acid transporter 3 in mice: a potential role in supply of branched-chain amino acids under nutrient starvation. Am. J. Pathol. 170, 888–898.

Gaccioli, F., Huang, C.C., Wang, C., Bevilacqua, E., Franchi-Gazzola, R., Gazzola, G.C., et al., 2006. Amino acid starvation induces the SNAT2 neutral amino acid transporter by a mechanism that involves eukaryotic initiation factor 2alpha phosphorylation and cap-independent translation. J. Biol. Chem. 281, 17929–17940.

Gallinetti, J., Harputlugil, E., Mitchell, J.R., 2013. Amino acid sensing in dietary-restriction-mediated longevity: roles of signal-transducing kinases GCN2 and TOR. Biochem. J. 449, 1–10.

Gietzen, D.W., Aja, S.M., 2012. The brain's response to an essential amino acid-deficient diet and the circuitous route to a better meal. Mol. Neurobiol. 46, 332–348.

Gilbert, E.R., Wong, E.A., Webb, K.E., 2008. Peptide absorption and utilization: implications for animal nutrition and health. J. Anim. Sci. 86, 2135–2155.

Gonzalez, I.M., Martin, P.M., Burdsal, C., Sloan, J.L., Mager, S., Harris, T., et al., 2012. Leucine and arginine regulate trophoblast motility through mTOR-dependent and independent pathways in the preimplantation mouse embryo. Dev. Biol. 361, 286–300.

Goodman, B.E., 2010. Insights into digestion and absorption of major nutrients in humans. Adv. Physiol. Educ. 34, 44–53.

Guetg, A., Mariotta, L., Bock, L., Herzog, B., Fingerhut, R., Camargo, S.M., et al., 2015. Essential amino acid transporter Lat4 (*Slc43a2*) is required for mouse development. J. Physiol. 593, 1273–1289.

Hägglund, M.G.A., Sreedharan, S., Nilsson, V.C.O., Shaik, J.H.A., Almkvist, I.M., Bäcklin, S., et al., 2011. Identification of SLC38A7 (SNAT7) protein as a glutamine transporter expressed in neurons. J. Biol. Chem. 286, 20500–20511.

Hagglund, M.G., Roshanbin, S., Lofqvist, E., Hellsten, S.V., Nilsson, V.C., Todkar, A., et al., 2013. B(0)AT2 (SLC6A15) is localized to neurons and astrocytes, and is involved in mediating the effect of leucine in the brain. PLoS One 8, e58651.

Han, J.M., Jeong, S.J., Park, M.C., Kim, G., Kwon, N.H., Kim, H.K., et al., 2012. Leucyl-tRNA synthetase is an intracellular leucine sensor for the mTORC1-signaling pathway. Cell 149, 410–424.

He, Y., Hakvoort, T.B., Kohler, S.E., Vermeulen, J.L., de Waart, D.R., de Theije, C., et al., 2010. Glutamine synthetase in muscle is required for glutamine production during fasting and extrahepatic ammonia detoxification. J. Biol. Chem. 285, 9516–9524.

Hediger, M.A., Clémençon, B., Burrier, R.E., Bruford, E.A., 2013. The ABCs of membrane transporters in health and disease (*SLC* series): introduction. Mol. Aspects Med. 34, 95–107.

Heublein, S., Kazi, S., Ogmundsdottir, M.H., Attwood, E.V., Kala, S., Boyd, C.A., et al., 2010. Proton-assisted amino-acid transporters are conserved regulators of proliferation and amino-acid-dependent mTORC1 activation. Oncogene 29, 4068–4079.

Hu, Y., Smith, D.E., Ma, K., Jappar, D., Thomas, W., Hillgren, K.M., 2008. Targeted disruption of peptide transporter Pept1 gene in mice significantly reduces dipeptide absorption in intestine. Mol. Pharm. 5, 1122–1130.

Huang, Y., Kang, B.N., Tian, J., Liu, Y., Luo, H.R., Hester, L., et al., 2007. The cationic amino acid transporters CAT1 and CAT3 mediate NMDA receptor activation-dependent changes in elaboration of neuronal processes via the mammalian target of rapamycin mTOR pathway. J. Neurosci. 27, 449–458.

Hundal, H.S., Taylor, P.M., 2009. Amino acid transceptors: gate keepers of nutrient exchange and regulators of nutrient signaling. Am. J. Physiol. Endocrinol. Metab. 296, E603–E613.

Hyde, R., Cwiklinski, E.L., MacAulay, K., Taylor, P.M., Hundal, H.S., 2007. Distinct sensor pathways in the hierarchical control of SNAT2, a putative amino acid transceptor, by amino acid availability. J. Biol. Chem. 282, 19788–19798.

Jézégou, A., Llinares, E., Anne, C., Kieffer-Jaquinod, S., O'Regan, S., Aupetit, J., et al., 2012. Heptahelical protein PQLC2 is a lysosomal cationic amino acid exporter underlying the action of cysteamine in cystinosis therapy. Proc. Natl. Acad. Sci. 109, E3434–E3443.

Jiang, Y., Rose, A.J., Sijmonsma, T.P., Bröer, A., Pfenninger, A., Herzig, S., et al., 2015. Mice lacking neutral amino acid transporter B0AT1 (*Slc6a19*) have elevated levels of FGF21 and GLP-1 and improved glycaemic control. Mol. Metab. 4, 406–417.

Kanai, Y., Clémençon, B., Simonin, A., Leuenberger, M., Lochner, M., Weisstanner, M., et al., 2013. The *SLC1* high-affinity glutamate and neutral amino acid transporter family. Mol. Aspects Med. 34, 108–120.

Kaper, T., Looger, L.L., Takanaga, H., Platten, M., Steinman, L., Frommer, W.B., 2007. Nanosensor detection of an immunoregulatory tryptophan Influx/Kynurenine Efflux cycle. PLoS Biol. 5, e257.

Karinch, A.M., Lin, C.M., Meng, Q., Pan, M., Souba, W.W., 2007. Glucocorticoids have a role in renal cortical expression of the SNAT3 glutamine transporter during chronic metabolic acidosis. Am. J. Physiol. Renal. Physiol. 292, F448–455.

Kim, S., Buel, G., Blenis, J., 2013. Nutrient regulation of the mTOR complex 1 signaling pathway. Mol. Cells 35, 463–473.

Kimball, S.R., 2014. Integration of signals generated by nutrients, hormones, and exercise in skeletal muscle. Am. J. Clin. Nutr. 99, 237S–242S.

Kriel, J., Haesendonckx, S., Rubio-Texeira, M., Van Zeebroeck, G., Thevelein, J.M., 2011. From transporter to transceptor: signaling from transporters provokes re-evaluation of complex trafficking and regulatory controls. Bioessays 33, 870–879.

Kurayama, R., Ito, N., Nishibori, Y., Fukuhara, D., Akimoto, Y., Higashihara, E., et al., 2011. Role of amino acid transporter LAT2 in the activation of mTORC1 pathway and the pathogenesis of crescentic glomerulonephritis. Lab. Invest. 91, 992–1006.

Layman, D.K., Anthony, T.G., Rasmussen, B.B., Adams, S.H., Lynch, C.J., Brinkworth, G.D., et al., 2015. Defining meal requirements for protein to optimize metabolic roles of amino acids. Am. J. Clin. Nutr. 101, 1330S–1338S.

Levy, S., Kafri, M., Carmi, M., Barkai, N., 2011. The competitive advantage of a dual-transporter system. Science 334, 1408–1412.

Lindgren, O., Pacini, G., Tura, A., Holst, J.J., Deacon, C.F., Ahrén, B., 2015. Incretin effect after oral amino acid ingestion in humans. J. Clin. Endocrinol. Metab. 100, 1172–1176.

Liu, B., Du, H., Rutkowski, R., Gartner, A., Wang, X., 2012. LAAT-1 is the lysosomal Lysine/Arginine transporter that maintains amino acid homeostasis. Science 337, 351–354.

Macotela, Y., Emanuelli, B., Bang, A.M., Espinoza, D.O., Boucher, J., Beebe, K., et al., 2011. Dietary leucine—an environmental modifier of insulin resistance acting on multiple levels of metabolism. PLoS One 6, e21187.

Manjarin, R., Steibel, J.P., Zamora, V., Am-In, N., Kirkwood, R.N., Ernst, C.W., et al., 2011. Transcript abundance of amino acid transporters, beta-casein, and alpha-lactalbumin in mammary tissue of periparturient, lactating, and postweaned sows. J. Dairy Sci. 94, 3467–3476.

Manjarin, R., Bequette, B.J., Wu, G., Trottier, N.L., 2014. Linking our understanding of mammary gland metabolism to amino acid nutrition. Amino Acids 46, 2447–2462.

McCracken, A.N., Edinger, A.L., 2013. Nutrient transporters: the Achilles' heel of anabolism. Trends Endocrinol. Metab. 24, 200–208.

Milkereit, R., Persaud, A., Vanoaica, L., Guetg, A., Verrey, F., Rotin, D., 2015. LAPTM4b recruits the LAT1-4F2hc Leu transporter to lysosomes and promotes mTORC1 activation. Nat. Commun. 6, 7250. Available from: http://dx.doi.org/10.1038/ncomms8250.

Moret, C., Dave, M.H., Schulz, N., Jiang, J.X., Verrey, F., Wagner, C.A., 2007. Regulation of renal amino acid transporters during metabolic acidosis. Am. J. Physiol. Renal. Physiol. 292, F555–566.

Mourad, F.H., Barada, K.A., Khoury, C., Hamdi, T., Saadé, N.E., Nassar, C.F., 2009. Amino acids in the rat intestinal lumen regulate their own absorption from a distant intestinal site. Am. J. Physiol. Gastrointest. Liver Physiol. 297, G292–G298.

Nakashima, E.M.N., Kudo, A., Iwaihara, Y., Tanaka, M., Matsumoto, K., Matsui, T., 2011. Application of 13C stable isotope labeling liquid chromatography–multiple reaction monitoring–tandem mass spectrometry method for determining intact absorption of bioactive dipeptides in rats. Anal. Biochem. 414, 109–116.

Nakaya, M., Xiao, Y., Zhou, X., Chang, J.-H., Chang, M., Cheng, X., et al., 2014. Inflammatory T cell responses rely on amino acid transporter ASCT2 facilitation of glutamine uptake and mTORC1 kinase activation. Immunity 40, 692–705.

Nassl, A.M., Rubio-Aliaga, I., Fenselau, H., Marth, M.K., Kottra, G., Daniel, H., 2011. Amino acid absorption and homeostasis in mice lacking the intestinal peptide transporter PEPT1. Am. J. Physiol. Gastrointest. Liver Physiol. 301, G128–137.

Nicklin, P., Bergman, P., Zhang, B., Triantafellow, E., Wang, H., Nyfeler, B., et al., 2009. Bidirectional transport of amino acids regulates mTOR and autophagy. Cell 136, 521–534.

Nii, T., Segawa, H., Taketani, Y., Tani, Y., Ohkido, M., Kishida, S., et al., 2001. Molecular events involved in up-regulating human Na^+-independent neutral amino acid transporter LAT1 during T-cell activation. Biochem. J. 358, 693–704.

Oda, K., Hosoda, N., Endo, H., Saito, K., Tsujihara, K., Yamamura, M., et al., 2010. L-type amino acid transporter 1 inhibitors inhibit tumor cell growth. Cancer Sci. 101, 173–179.

Ogier de Baulny, H., Schiff, M., Dionisi-Vici, C., 2012. Lysinuric protein intolerance (LPI): a multi organ disease by far more complex than a classic urea cycle disorder. Mol. Genet. Metab. 106, 12–17.

Ögmundsdóttir, M.H., Heublein, S., Kazi, S., Reynolds, B., Visvalingam, S.M., Shaw, M.K., et al., 2012. Proton-assisted amino acid transporter PAT1 complexes with Rag GTPases and activates TORC1 on late endosomal and lysosomal membranes. PLoS One 7, e36616.

Palii, S.S., Thiaville, M.M., Pan, Y.-X., Zhong, C., Kilberg, M.S., 2006. Characterization of the amino acid response element within the human sodium-coupled neutral amino acid transporter 2 (SNAT2) System A transporter gene. Biochem. J. 395, 517–527.

Perkins, C.P., Mar, V., Shutter, J.R., del Castillo, J., Danilenko, D.M., Medlock, E.S., et al., 1997. Anemia and perinatal death result from loss of the murine ecotropic retrovirus receptor mCAT-1. Genes Dev. 11, 914–925.

Pinilla, J., Aledo, J.C., Cwiklinski, E., Hyde, R., Taylor, P.M., Hundal, H.S., 2011. SNAT2 transceptor signalling via mTOR: a role in cell growth and proliferation? Front. Biosci. 3, 1289–1299.

Pisoni, R.L., Thoene, J.G., 1991. The transport systems of mammalian lysosomes. Biochim. Biophys. Acta 1071, 351–373.

Poncet, N., Mitchell, F.E., Ibrahim, A.F., McGuire, V.A., English, G., Arthur, J.S., et al., 2014. The catalytic subunit of the system L1 amino acid transporter (*slc7a5*) facilitates nutrient signalling in mouse skeletal muscle. PLoS One 9, e89547.

Pramod, A.B., Foster, J., Carvelli, L., Henry, L.K., 2013. *SLC6* transporters: structure, function, regulation, disease association and therapeutics. Mol. Aspects Med. 34, 197–219.

Ramsay, G., Cantrell, D., 2015. Environmental and metabolic sensors that control T cell biology. Front. Immunol. 6, 99.

Rebsamen, M., Pochini, L., Stasyk, T., de Araujo, M.E., Galluccio, M., Kandasamy, R.K., et al., 2015. SLC38A9 is a component of the lysosomal amino acid sensing machinery that controls mTORC1. Nature 519, 477–481.

Roos, S., Jansson, N., Palmberg, I., Säljö, K., Powell, T.L., Jansson, T., 2007. Mammalian target of rapamycin in the human placenta regulates leucine transport and is down-regulated in restricted fetal growth. J. Physiol. 582, 449–459.

Roos, S., Kanai, Y., Prasad, P.D., Powell, T.L., Jansson, T., 2009. Regulation of placental amino acid transporter activity by mammalian target of rapamycin. Am. J. Physiol. Cell Physiol. 296, C142–C150.

Schiöth, H.B., Roshanbin, S., Hägglund, M.G.A., Fredriksson, R., 2013. Evolutionary origin of amino acid transporter families *SLC32, SLC36* and *SLC38* and physiological, pathological and therapeutic aspects. Mol. Aspects Med. 34, 571–585.

Schriever, S.C., Deutsch, M.J., Adamski, J., Roscher, A.A., Ensenauer, R., 2012. Cellular signaling of amino acids towards mTORC1 activation in impaired human leucine catabolism. J. Nutr. Biochem. 24 (5), 824–831.

Sinclair, L.V., Rolf, J., Emslie, E., Shi, Y.-B., Taylor, P.M., Cantrell, D.A., 2013. Control of amino-acid transport by antigen receptors coordinates the metabolic reprogramming essential for T cell differentiation. Nat. Immunol. 14, 500–508.

Sonenberg, N., Hinnebusch, A.G., 2009. Regulation of translation initiation in eukaryotes: mechanisms and biological targets. Cell 136, 731–745.

Stein, W.D., 1986. Transport and Diffusion Across Cell Membranes. Academic Press, London.

Stewart, B.H., Collarini, E.J., Pisoni, R.L., Christensen, H.N., 1989. Separate and shared lysosomal transport of branched and aromatic dipolar amino acids. Biochim. Biophys. Acta 987, 145–153.

Sun, X., Zemel, M.B., 2009. Leucine modulation of mitochondrial mass and oxygen consumption in skeletal muscle cells and adipocytes. Nutr. Metab. (Lond) 6, 26.

Suryawan, A., Davis, T.A., 2011. Regulation of protein synthesis by amino acids in muscle of neonates. Front. Biosci. (Landmark Ed) 16, 1445–1460.

Taylor, P.M., 2014. Role of amino acid transporters in amino acid sensing. Am. J. Clin. Nutr. 99, 223S–230S.

Taylor, P.M., Ritchie, J.W., 2007. Tissue uptake of thyroid hormone by amino acid transporters. Best. Pract. Res. Clin. Endocrinol. Metab. 21, 237–251.

Usui, T., Nagumo, Y., Watanabe, A., Kubota, T., Komatsu, K., Kobayashi, J., et al., 2006. Brasilicardin A, a natural immunosuppressant, targets amino Acid transport system L. Chem. Biol. 13, 1153–1160.

van de Poll, M.C., Soeters, P.B., Deutz, N.E., Fearon, K.C., Dejong, C.H., 2004. Renal metabolism of amino acids: its role in interorgan amino acid exchange. Am. J. Clin. Nutr. 79, 185–197.

Van Winkle, L.J., Tesch, J.K., Shah, A., Campione, A.L., 2006. System B0, + amino acid transport regulates the penetration stage of blastocyst implantation with possible long-term developmental consequences through adulthood. Hum. Reprod. Update 12, 145–157.

Varoqui, H., Erickson, J.D., 2002. Selective up-regulation of system a transporter mRNA in diabetic liver. Biochem. Biophys. Res. Commun. 290, 903–908.

Vaughan, R.A., Garcia-Smith, R., Gannon, N.P., Bisoffi, M., Trujillo, K.A., Conn, C.A., 2013. Leucine treatment enhances oxidative capacity through complete carbohydrate oxidation and increased mitochondrial density in skeletal muscle cells. Amino Acids 45, 901–911.

Velazquez-Villegas, L.A., Lopez-Barradas, A.M., Torres, N., Hernandez-Pando, R., Leon-Contreras, J.C., Granados, O., et al., 2015. Prolactin and the dietary protein/carbohydrate ratio regulate the expression of SNAT2 amino acid transporter in the mammary gland during lactation. Biochim. Biophys. Acta 1848, 1157–1164.

Vuille-Dit-Bille, R.N., Camargo, S.M., Emmenegger, L., Sasse, T., Kummer, E., Jando, J., et al., 2015. Human intestine luminal ACE2 and amino acid transporter expression increased by ACE-inhibitors. Amino Acids 47, 693–705.

Wang, S., Tsun, Z.Y., Wolfson, R.L., Shen, K., Wyant, G.A., Plovanich, M.E., et al., 2015. Metabolism. Lysosomal amino acid transporter SLC38A9 signals arginine sufficiency to mTORC1. Science 347, 188–194.

Wiriyasermkul, P., Nagamori, S., Tominaga, H., Oriuchi, N., Kaira, K., Nakao, H., et al., 2012. Transport of 3-fluoro-L-alpha-methyl-tyrosine by tumor-upregulated L-type amino acid transporter 1: a cause of the tumor uptake in PET. J. Nucl. Med. 53, 1253–1261.
Wu, G., 2013. Functional amino acids in nutrition and health. Amino Acids 45, 407–411.
Xie, X., Dumas, T., Tang, L., Brennan, T., Reeder, T., Thomas, W., et al., 2005. Lack of the alanine-serine-cysteine transporter 1 causes tremors, seizures, and early postnatal death in mice. Brain Res. 1052, 212–221.
Xu, D., Hemler, M.E., 2005. Metabolic activation-related CD147-CD98 complex. Mol. Cell. Proteomics 4, 1061–1071.
Yeramian, A., Martin, L., Arpa, L., Bertran, J., Soler, C., McLeod, C., et al., 2006. Macrophages require distinct arginine catabolism and transport systems for proliferation and for activation. Eur. J. Immunol. 36, 1516–1526.
Young, S.H., Rey, O., Sternini, C., Rozengurt, E., 2010. Amino acid sensing by enteroendocrine STC-1 cells: role of the Na + -coupled neutral amino acid transporter 2. Am. J. Physiol. Cell Physiol. 298, C1401–1413.
Zhai, Y., Heijne, W.H.M., Smith, D.W., Saier Jr., M.H., 2001. Homologues of archaeal rhodopsins in plants, animals and fungi: structural and functional predications for a putative fungal chaperone protein. Biochim. Biophys. Acta 1511, 206–223.
Zietek, T., Daniel, H., 2015. Intestinal nutrient sensing and blood glucose control. Curr. Opin. Clin. Nutr. Metab. Care 18, 381–388.
Zoncu, R., Bar-Peled, L., Efeyan, A., Wang, S., Sancak, Y., Sabatini, D.M., 2011. mTORC1 senses lysosomal amino acids through an inside-out mechanism that requires the vacuolar H(+)-ATPase. Science 334, 678–683.

SECTION II

CELLULAR ASPECTS OF PROTEIN AND AMINO ACIDS METABOLISM IN ANABOLIC AND CATABOLIC SITUATIONS

CHAPTER

6

Amino Acids and Exercise: Molecular and Cellular Aspects

C. McGlory and S.M. Phillips

Exercise Metabolism Research Group, Department of Kinesiology, McMaster University, Hamilton, ON, Canada

6.1. INTRODUCTION

Skeletal muscle is a critical and often unappreciated organ that not only supports human locomotion but its mass is also a robust predictor for all-cause mortality (Kallman et al., 1990; Metter et al., 2002). Furthermore, given that the human body is 45% skeletal muscle by mass, sustaining skeletal muscle mass throughout life is essential for promoting both athletic performance and longevity. Despite its importance, several of the physiological mechanisms that regulate the size of human muscle mass have only recently been elucidated, and many are still unknown. What is known is that the interaction between amino acid feeding-induced changes in rates of muscle protein synthesis (MPS) and rates of muscle protein breakdown (MPB) ultimately dictate net muscle protein balance (NPB) (Phillips et al., 1997). However, the composition, dose, timing, and daily distribution of protein intake (and subsequent effect on aminoacidemia) throughout the day that induces optimal MPS remains a topic of intense research and debate. Moreover, how these variables influence the cellular and molecular regulators that mediate amino acid-induced increases in MPS, particularly those involved in translation initiation and elongation, remain largely unknown. The aim of this chapter is to provide a critical evaluation of our current understanding of how amino acid ingestion, mainly in the form of intact protein sources, following loading (resistance exercise) influences MPS at both the cellular and molecular level. For the purposes of concision and relevance to the human model, data reported primarily derived from human studies will be cited but, where appropriate, work from other experimental models will be introduced to substantiate points of discussion.

6.2. REGULATION OF THE SIZE OF HUMAN MUSCLE MASS

The size of human skeletal muscle mass is dependent upon the coordinated interaction between changes in rates of MPS and MPB (Phillips et al., 1997). The basal and fasted state rates of MPB are known to exceed those of MPS and thus, skeletal muscle net protein balance (NPB = MPS minus MPB) is negative. Over two decades ago, using stable isotopic infusions, it was demonstrated that a mixed macronutrient meal was capable of increasing rates of MPS above rates of MPB and that the amino acids contained within the meal were primarily responsible for this increase (Bennet et al., 1990; Rennie et al., 1982). However, after approximately 2–3 h rates of MPS are known to decline to postabsorptive levels until the consumption of the next amino acid-containing meal (Areta et al., 2013; Atherton et al., 2010). Interestingly, performing exercise, particularly resistance exercise, prior to consuming amino acids has been shown to potentiate rates of MPS (Phillips et al., 2012; Witard et al., 2009). It is this potentiation by resistance exercise of feeding-induced increases in MPS that is responsible for the hypertrophic phenotype observed with resistance exercise training and protein feeding over time (Fig. 6.1).

The Molecular Nutrition of Amino Acids and Proteins.
DOI: http://dx.doi.org/10.1016/B978-0-12-802167-5.00006-2

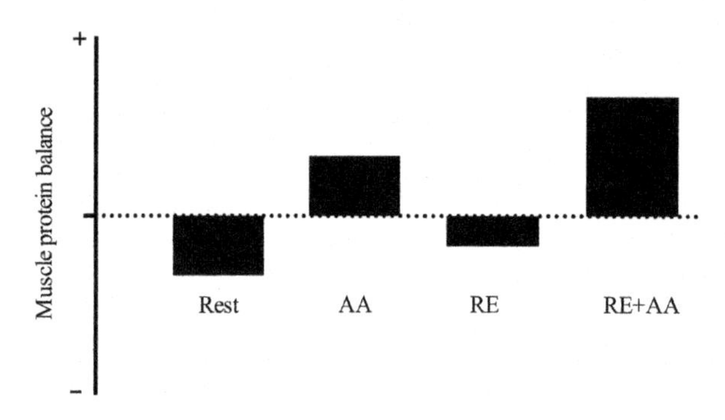

FIGURE 6.1 Changes in net skeletal muscle protein balance at rest, in response to amino acid (AA) consumption, resistance exercise (RE), and when AA ingestion is combined with RE. *Source: Figure redrawn from Phillips (2004) based on original data from Biolo et al. (1995).*

TABLE 6.1 Examples from the literature of rates of muscle protein synthesis captured 4 h following either resistance exercise or endurance exercise

Fractional synthetic rate (%/h)	Rest	RE 4 h	AE 4 h
Mixed	0.04–0.06	0.10–0.16	0.10–0.12
Myofibrillar	0.02–0.05	0.05–0.11	0.03–0.06
Mitochondrial	0.05–0.10	0.10–0.15	0.12–0.15
Sarcoplasmic	0.04–0.06	0.08–0.10	0.08–0.10

Resistance exercise (RE) values obtained from Burd et al. (2010), and aerobic exercise (AE) values obtained from Wilkinson et al. (2008). All mitochondrial protein turnover rates were obtained from Wilkinson et al. (2008).

Importantly, not all amino acids exert the same stimulatory effect of MPS in skeletal muscle either at rest or following exercise, and identifying the relevant amino acids, optimal amino acid/protein dose and/or composition of amino acids to promote gains in muscle size and function is of great scientific interest.

6.3. EXERCISE MODE

While resistance exercise will remain the focus of this chapter it is important to recognize that even aerobic exercise can stimulate MPS (Carraro et al., 1990; Harber et al., 2010) and even result in protein accretion (Harber et al., 2009, 2012). However, the fraction of muscle proteins that are predominantly turned over in response to exercise is specific to the intensity and duration of the exercise bout, as well as the training status of the individual (Burd et al., 2010; Wilkinson et al., 2008). For example, in an untrained state, resistance exercise stimulates an increase in both myofibrillar and mitochondrial MPS; however, in the trained state, resistance exercise only stimulates an increase in myofibrillar MPS (Table 6.1). Endurance exercise on the other hand stimulates an increase in mitochondrial MPS in both the trained and untrained state, but does not result in a stimulation of myofibrillar MPS (Wilkinson et al., 2008). These differences are important considerations when evaluating the efficacy of a given exercise or nutritional stimulus to alter protein turnover as the assessment of mixed MPS can mask any potential protein fraction-specific differences in rates of synthesis (Kim et al., 2005). It is also known that high intensity intermittent exercise training (HIIT), a stimulus that could be considered a "hybrid" of both resistance and endurance exercise, has the capacity to stimulate both myofibrillar and mitochondrial MPS (Bell et al., 2015; Scalzo et al., 2014). Although ingestion of protein (and the subsequent aminoacidemia) is thought to influence the adaptive response to all modes of exercise, the effect is most profound following resistance exercise (Cermak et al., 2012, 2013). As a result, the continuing theme of this chapter will be how protein/amino acid feeding alters the protein synthetic response to resistance exercise with some discussion relating to the role of endurance and HIIT exercise.

6.4. PROTEIN TYPE

A major independent variable that drives MPS is the digestion rate and subsequent aminoacidemia of ingested proteins. This factor can largely be influenced by the quality of intact proteins as defined by various scoring systems such as the Digestible Indispensible Amino Acid Score (for an expanded review see van Vliet et al., 2015). The most widely studied categories of dietary isolated protein are whey, casein and soy, all of which have differing digestion and absorption kinetics as well as amino acid composition. In this regard, isolated whey and soy proteins, since both are acid-soluble, are relatively rapidly digested resulting in a heightened but transient hyperaminoacidemia (Devries and Phillips, 2015). Soy protein however, contains a greater proportion of nonessential amino acids than whey protein (Mahe et al., 1996). Casein in its micellar form (as it exists in milk) is acid insoluble and coagulates in the stomach, which slows transit time into the intestine and thus results in an aminoacidemia that is smaller in amplitude but greater in duration than whey (Boirie et al., 1997).

To date, the influence of protein type on exercise-induced increases in MPS has been understudied. One study has shown that postprandial rates of MPS are greater following the consumption of whey and soy as compared with casein (Tang et al., 2009). Additionally, whey protein consumption has been shown to be superior to both soy and casein for the purposes of stimulating MPS following resistance exercise (Burd et al., 2012; Tang et al., 2009). The superiority of whey to stimulate rates of MPS following resistance exercise is attributed to the high essential amino acid, specifically leucine (Churchward-Venne et al., 2014a) content, the influence of which will be discussed later in this chapter. Another hypothesis is that the rapid increase in blood amino acid concentrations associated with whey protein drives MPS. Indeed, in one study where whey protein was consumed either as a 25 g bolus immediately following resistance exercise or as 10 individual 2.5 g "pulse" drinks separated by 20 min, it was shown that the 25 g bolus of whey induced greater rates of postexercise MPS during a 5 h recovery period (West et al., 2011). When this experiment was repeated, but this time the protein beverages were provided prior to the bout of resistance exercise, there was no difference in postexercise rates of MPS (Burke et al., 2012). However, consuming a single bolus of whey protein following a bout of rigorous exercise as opposed to before is likely a far more practical and attractive strategy to enhance rates of MPS and obviate gastrointestinal stress. Nevertheless, what these data allude to is that the timing, dose, and subsequent hyperaminoacidemia following protein ingestion may be a critical factor modulating the MPS response to resistance exercise.

6.5. DOSE RESPONSE OF MPS TO PROTEIN INGESTION FOLLOWING RESISTANCE EXERCISE

The first study to examine the dose response of MPS to increasing amounts of protein following resistance exercise was conducted by Moore and colleagues who demonstrated that rates of MPS following bilateral lower limb exercises were saturated with the consumption of 20 g of egg protein (Moore et al., 2009). It was also shown that ingestion of 40 g of egg protein resulted in no further enhancement of mixed MPS, but there was a sharp rise in whole-body leucine oxidation indicating oxidative disposal of at least leucine if not other amino acids. Further work on a dose response of MPS with protein ingestion has used a unilateral model of resistance exercise (Witard et al., 2014). These authors reported that rates of both exercise-induced and postprandial myofibrillar MPS were maximally stimulated with 20 g of whey protein (Witard et al., 2014). A unique aspect of this study was that the participants were fed a standard preexercise breakfast meal thus enhancing the practical applicability of the findings to those who consumed food prior to engaging in resistance exercise. Taken together (Moore et al., 2009; Witard et al., 2014), despite the differing sources of protein (egg vs whey), what these data clearly demonstrate is that consuming 20 g of high quality intact protein results in maximal stimulation of both postprandial (Witard et al., 2014) and postresistance exercise (Moore et al., 2009; Witard et al., 2014) rates of MPS in healthy young men.

6.6. TIMING AND DISTRIBUTION

The findings (Moore et al., 2009; Witard et al., 2014) that consumption of protein doses greater than 20 g do not further enhance rates of MPS both at rest and following resistance exercise has lead practitioners of resistance exercise to suggest that frequent small meals may present a feasible strategy to stimulate MPS.

Nonetheless, with prolonged aminoacidemia, lasting approximately 2 h, the rates of MPS are known to decline to postabsorptive values (Atherton et al., 2010) and this effect is apparent whether the amino acids are infused (Bohe et al., 2001) or orally ingested as protein (Atherton et al., 2010). This phenomenon has been coined the "muscle full" effect and has been proposed to be related to the saturation of the translational apparatus (Atherton et al., 2010). However, the characterization of the muscle full effect in these studies was observed in the rested state only. This is an important consideration given that resistance exercise protracts the MPS response to protein feeding (Burd et al., 2011; Churchward-Venne et al., 2012). In fact, as West et al. (West et al., 2011) showed, in a postexercise condition MPS continued to be stimulated despite aminoacidemia having returned back to baseline or despite continued aminoacidemia with oral ingestion. Thus, identification of the optimal pattern and timing of repeated protein feedings to stimulate MPS following a bout of resistance exercise has yet to be determined. Given that consuming 20 g of whey protein is optimal for maximizing postresistance exercise rates of MPS (Moore et al., 2015; Witard et al., 2014), taken together with the knowledge that MPS declines to postabsorptive levels 2 h after feeding at least in the fed state (Atherton et al., 2010), it is physiologically logical to assume that the consumption of 20 g of whey protein every 2–3 h following resistance exercise would be an effective means to enhance daily rates of MPS. This thesis was examined experimentally in a study where an equal amount of protein was distributed either as 8×10 g every 1.5 h, 4×20 g every 3 h, or 2×40 g every 6 h during a 12-h postresistance exercise recovery period (Areta et al., 2013). This approach resulted in differential profiles of postexercise aminoacidemia with the thesis that consumption of 4×20 g of protein 3h would maximize aggregate rates of MPS, which is exactly what this study showed. As such, these data confirm other reports (Moore et al., 2009; Witard et al., 2014) that the consumption of 20 g of high-quality protein postresistance exercise is sufficient to maximize MPS.

While it appears that the strategy to maximize the stimulation of MPS is consumption of 20 g of protein every 3 h (Areta et al., 2013) there are some important caveats to our understanding. Firstly, protein was administered as a liquid bolus and in the absence of other macronutrients. We remain ignorant as to the effect that coingestion of other macronutrients, particularly in the form of a mixed meal, would have on exercise-induced increases in MPS. Moreover, many of these studies employed either a unilateral or bilateral lower limb resistance exercise protocol which are not indicative of the whole-body resistance exercise regimens currently adopted by many physically active individuals, particularly athletes. The relevance of this point is that during whole-body resistance exercise there would be activation of a greater amount of skeletal muscle that would create a greater demand for amino acids to support muscle remodeling. Another important point is that the protein doses in these studies were administered as an absolute value (ie, 20 g) and were not adjusted for body mass. Recent retrospective analysis has showed that MPS is maximally saturated in young men following the consumption of 0.24 g/kg of high-quality protein (with a 90% confidence interval indicating this dose could be as high 0.30 g/kg) at rest (Moore et al., 2015) (Fig. 6.2), but whether this is true with resistance exercise is unknown. Thus, future work addressing these important questions may provide more information that will enable the refinement of current per meal recommendations for protein intake, especially for the athletic population.

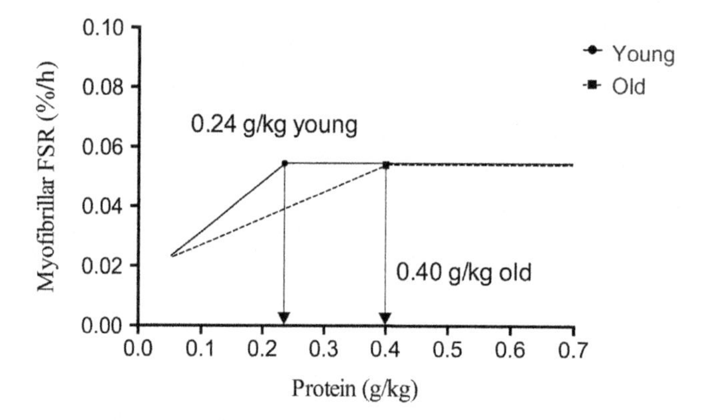

FIGURE 6.2 Biphase linear regression analyses of muscle protein synthesis in response to protein intake expressed as kilogram per body mass in young and older adults. *Source: Redrawn from Moore et al. (2015).*

6.7. THE INFLUENCE OF THE AGING PROCESS

Aging results in a progressive decline in muscle mass and function, collectively termed, sarcopenia. Sarcopenia commences in the fourth decade of life and proceeds on a population basis at a rate of 8% per decade until 70 years of age when the average loss is estimated to occur at 15% per decade (Mitchell et al., 2012). The etiology of sarcopenia is a topic of intense discussion and research. However, an important mechanism is likely related to the inability of skeletal muscle to mount an adequate MPS response to both resistance exercise (Kumar et al., 2009) and protein feeding (Katsanos et al., 2006; for extended discussion see Phillips, 2015). Given resistance exercise (loading) is one of the most potent nonpharmacological methods to enhance muscle mass and strength (Churchward-Venne et al., 2015), this "anabolic resistance" to resistance exercise and protein feeding represents mechanisms that would impair growth of skeletal muscle in aging persons. For instance, whereas 0.24 g/kg/body mass of whey protein is sufficient to maximize MPS following resistance exercise in young adults (Moore et al., 2015; Witard et al., 2014), older adults require 0.40 g/kg/body mass to elicit the same effect (Yang et al., 2012). The physiological mechanisms underlying the reduced sensitivity of older adult skeletal muscle to resistance exercise and protein feeding remain unknown. It has been suggested that older individuals have a reduced skeletal muscle translational capacity compared with their younger counterparts (Chaillou et al., 2014; Kirby et al., 2015) and thus lack the ability to mount a "youthful" response of MPS with the same aminoacidemia; however, this does not explain why consuming a greater amount of protein restores protein synthetic rates (Phillips, 2015). Another theory is that the impairment of insulin-induced increases in arterial blood flow and skeletal muscle microvascular perfusion associated with advancing age compromises the delivery of key amino acids to the translational machinery following exercise. In fact, it has been shown that a prior bout of aerobic exercise enhances microvascular perfusion, blood flow, and subsequent NPB in response to protein–carbohydrate coingestion (Timmerman et al., 2012). Additionally, it is well established that resistance exercise sensitizes skeletal muscle to the anabolic effects of amino acid ingestion. Thus, a combination of resistance exercise and aerobic training could be one means to improve amino acid utilization in skeletal muscle. In this regard, it has been shown that (HIIT) training stimulates increases in both myofibrillar and sarcoplasmic MPS. But whether chronic HIIT improves muscle mass and function to a greater degree than either resistance exercise training or endurance exercise training in the elderly is currently unknown.

Although exercise provides one effective means to counteract sarcopenia, there is some evidence that amino acids such as arginine may be important in determining the anabolic response to a meal. Arginine is an amino acid precursor to nitric oxide production (Stuehr, 2004). In response to feeding there is an increase in nitric oxide-mediated vasodilation that is impaired in older adults (Taddei et al., 2001). One suggestion is that arginine supplementation be used in conjunction with exercise in an attempt to reverse age-related impairments in vasodilation and improve amino acid delivery to skeletal muscle. Though, when ingested a significant proportion of arginine is first-pass cleared resulting in poor bioavailability (Wu et al., 2000a,b). Citrulline on the other hand, is a nonprotein amino acid that serves as the endogenous precursor for arginine that is not metabolized in the intestine or taken up by the liver (Curis et al., 2005). There is also evidence that citrulline of itself can target protein synthetic pathways (Cynober et al., 2010) and thus, could be an effective candidate to support exercise and feeding-induced increases in MPS. However, one study in elderly men who coingested 10 g of citrulline with either 45 g or 15 g of whey, showed no effect of citrulline on either blood flow, microvascular circulation, or postresistance exercise rates of MPS (Churchward-Venne et al., 2014b). Another study also failed to identify an impact of supplementation of 5 g of citrulline on either postprandial, postabsorbtive, or postresistance exercise rates of MPS before and after two weeks of muscle disuse (Devries and Phillips, 2015). The evidence regarding the impact of citrulline on muscle anabolism is, however, equivocal. In agreement with the aforementioned studies, one report has shown little benefit of citrulline on rates of MPS and MPB in the postabsorbtive state (Thibault et al., 2011), whereas others provide evidence that citrulline can improve nitrogen balance (Rouge et al., 2007). Moreover, in one study a high dose (11–24 g) of citrulline supplemented over a period of 8 h did increase rates of MPS compared to nonessential amino acids during a low protein diet (Jourdan et al., 2015), but this was conducted in healthy young adults. Thus, the relevance to this study for the elderly is unknown. Reconciling the differences in the literature is difficult due to differences in the dose, population, as well as exercise stimulus. Clearly more work is now needed to identify the efficacy of citrulline to enhance muscle anabolism, particularly following exercise training in any population.

6.8. THE ROLE OF THE ESSENTIAL AND BRANCHED-CHAIN AMINO ACIDS

Research has shown that the stimulatory impact of protein on MPS following resistance exercise is primarily due to the essential amino acids (Tipton et al., 1999a) with little-to-no role for the nonessential amino acids (Tipton et al., 1999b). This may explain why whey protein exerts such a potent impact on muscle anabolism following resistance exercise (Burd et al., 2012; Tang et al., 2009). In particular, the branched chain amino acid leucine has been shown to be a trigger for muscle anabolism (Averous et al., 2014; Breen and Churchward-Venne, 2012) playing a key role in the activation of the mechanistic/mammalian target of rapamycin complex 1 (mTORC1) and MPS (Jewell et al., 2013). However, while many studies have shown leucine to be stimulatory for MPS, others have failed to replicate the impact of leucine when administered in free form alongside essential amino acids (Glynn et al., 2010). It therefore appears that when other essential amino acids are provided in sufficient quantities, the addition of leucine to protein offers no further benefit. Indeed, the addition of leucine to a lower dose of whey protein (6.5 g), shown to be submaximal for stimulation of MPS, has been shown to be as effective at stimulating MPS at rest compared with a mixture of essential amino acids without leucine (Churchward-Venne et al., 2012).

As leucine is a branched chain amino acid there is a strong belief, especially amongst the sports nutrition and strength and conditioning community, that supplementation with branched chain amino acids may also be anabolic toward skeletal muscle. Indeed, branched chain amino acid supplementation has previously been shown to increase rates of MPS (Shimomura et al., 2006) as well as increase isometric grip strength (Candeloro et al., 1995). Moreover, intense exercise has been shown to result in a reduction in intracellular branched chain amino acid concentration, and thus supplementation with branched chain amino acids may serve to restore endogenous concentrations (Shimomura et al., 2006). Although in one of the most comprehensive studies in this area in which branched chain amino acids were supplemented for 8 weeks, supplementation failed to alter body composition, muscle strength, or muscle endurance (Spillane et al., 2012). Other placebo-controlled trials also have shown no effect of branched chain amino acids supplementation on body composition during resistance exercise training (Ispoglou et al., 2011). In fact, branched chain amino acids are known to compete for the same intestinal- and sarcolemmal-located amino acid transporters and may in fact be antagonistic in their actions with regards to uptake by the muscle (Churchward-Venne et al., 2014a).

6.9. THE MECHANISTIC TARGET OF RAPAMYCIN COMPLEX 1 (mTORC1)

The molecular regulation of MPS is highly integrated and complex. What many years of research have shown is that increases in MPS occur via two main biological processes, enhanced translational efficiency and/or translational capacity. Translational efficiency refers to an increase in rate of protein synthesis per unit of mRNA, presumably through an increased number of ribosomes bound to a single mRNA. Translational capacity refers to an increase in the overall content of ribosomal RNA as well as cellular mRNA and an increased abundance of the 40S and 60S ribosomal subunits. Both of these processes can be regulated by a protein kinase called mTORC1, which is an important signaling protein hub that serves to integrate the MPS response with various stimuli. Such stimuli would include resistance exercise and amino acid ingestion (Apró and Blomstrand, 2010; Areta et al., 2013; Churchward-Venne et al., 2014a; Guertin and Sabatini, 2007; McGlory et al., 2014). It is important to note that there are two mTOR complexes, mTORC1 and mechanistic target of rapamycin complex 2 (mTORC2). Both mTORC1 and mTORC2 impart different biological effects in muscle with mTORC1 being sensitive to the inhibitory effects of the bacterial macrolide rapamycin whereas mTORC2 is not. mTORC1 is the most studied of the complexes and will therefore be the focus of the continuing discussion.

Composed of multiple subunits, the interaction of which dictate its activity, mTORC1 receives and integrates multiple signals. These subunits include the catalytic subunit, mechanistic target of rapamycin (mTOR), regulatory-associated protein of mTOR (Raptor), Mammalian lethal with SEC13 protein 8 (mLST8), DEP domain-containing mTOR-interacting protein (Deptor), and Proline-rich Akt substrate of 40 kDa (PRAS40) (Foster and Fingar, 2010). For mTORC1 assembly to occur it requires interaction with the small guanosine triphosphatase (GTPase) Ras homolog enriched in brain (Rheb). Only when Rheb is in a guanosine triphosphate (GTP)-bound state, is it able to exert a stimulatory impact on mTORC1. In contrast, when in the guanosine diphosphate (GDP)-bound state, Rheb is unable to activate mTORC1, a process that is controlled by the upstream GTPase activating protein (GAP) tuberous sclerosis 2 (TSC2) (Aspuria and Tamanoi, 2004). In this

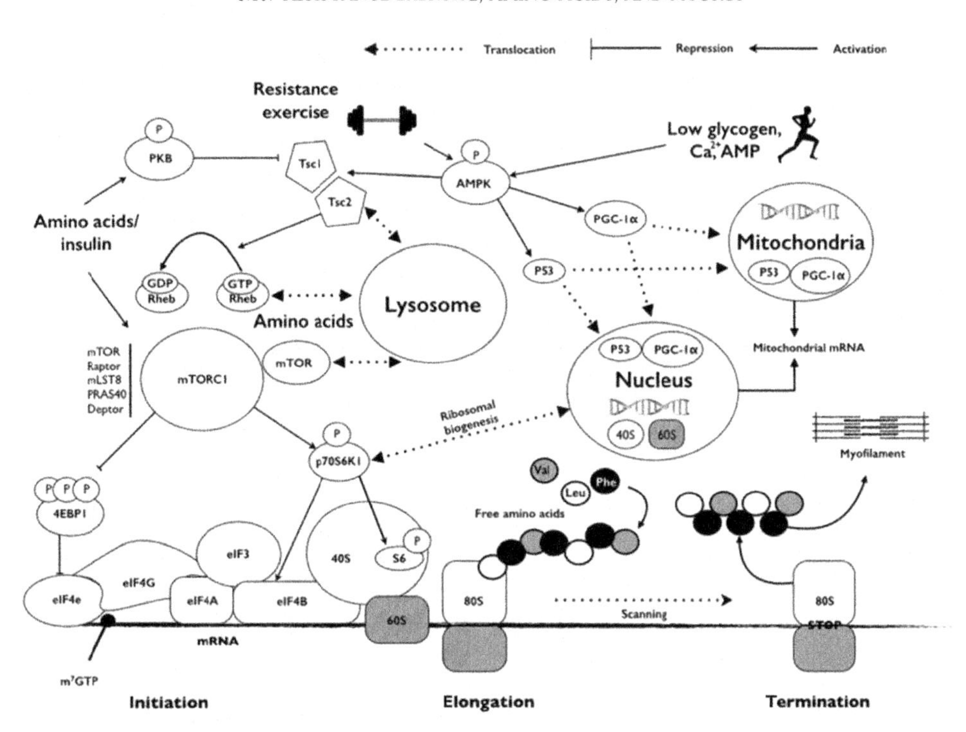

FIGURE 6.3 A schematic overview of the some of the cellular signaling responses regulating muscle protein synthesis and mitochondrial biogenesis in response to exercise. In response to both resistance exercise and endurance exercise there is a stimulation of AMPK that precipitates a series of posttranslational modifications, the intensity of which are highly related to the nature and magnitude of exercise bout. Prolonged, high frequency contractions stimulate an increase in AMPK activity that promotes mitochondrial protein synthesis and biogenesis via PCG-alpha and p53 translocation to the nucleus and mitochondria. Resistance exercise also stimulates AMPK activity and perhaps TSC2 movement away from the lysosomal surface. Following resistance exercise, amino acids promote mTOR and Rheb translocation to the lysosomal membrane as well as promoting PKB activity. The colocalization of Rheb and mTOR to the membrane and/or suppression of TSC2 by PKB promote mTORC1-dependent increases in translation initiation via 4E-BP1 and p70S6K1.

regard, when active, TSC2 activates Rheb by serving to drive its GTPase activity, and increasing its GDP/GTP-bound state, thus inhibiting mTORC1 activation (Inoki et al., 2003). In addition, the nutrient sensitive protein, protein kinase B (PKB) phosphorylates TSC2, preventing its ability to activate Rheb (Inoki et al., 2002). PKB also targets PRAS40, further driving mTORC1 complex assembly (Vander Haar et al., 2007). However, the negative influence of TSC2 on Rheb is enhanced by adenosine monophosphate activated protein kinase (AMPK), which also can inhibit mTORC1 via Raptor (Gwinn et al., 2008). TSC2-Rheb interactions therefore serve as a critical point of mTORC1 activity.

Following its activation, mTORC1 interacts with multiple downstream protein targets. These effectors include the ribosomal protein of 70 kDa S6 kinase 1 (p70S6K1) and 4E-binding protein-1 (4E-BP1) (Dickinson et al., 2011; Gingras et al., 1999; Pearson et al., 1995). Of particular note, p70S6K1 and 4E-BP1 are key signaling molecules that regulate the initiation of protein translation. The phosphorylation status of p70S6K1 and 4E-BP1 are often used as readouts of mTORC1 activity. Indeed, following stimulation mTORC1 phosphorylates p70S6K1 on Thr389 to upregulate translation initiation (Ma et al., 2008; Richardson et al., 2004). Phosphorylated p70S6K1 also serves to enhance translation elongation via eukaryotic elongation factor-2 kinase (eEF2K) (Wang et al., 2001) as well as up regulating ribosomal biogenesis (Chaillou et al., 2014; Hannan et al., 2003). In addition, 4E-BP1 is targeted by mTORC1 via phosphorylation on Thr37/46 reducing its affinity for eukaryotic initiation factor 4E (eIF4E), enabling eIF4E to interact with eukaryotic initiation factor 4G to commence translation initiation (Fig. 6.3).

6.10. RESISTANCE EXERCISE, AMINO ACIDS, AND MTORC1

There are now a significant number of studies that have characterized the impact of amino acid feeding on both MPS and mTORC1 activity. There are even studies to show that specific amino acids such as leucine, arginine and glycine can directly modulate mTORC1 activity (Jewell and Guan, 2013). Resistance exercise even in the absence of protein feeding can result in the cellular uptake of amino acids (Biolo et al., 1995). One of the first studies to show a relationship between resistance exercise and mTORC1 signaling was conducted by Baar and Esser (1999) who demonstrated that the phosphorylation of p70S6K1 6 h following resistance exercise strongly correlated with the degree of muscle hypertrophy in rodents. Additional studies in humans also identified a correlation between resistance exercise training-induced gains in muscle mass and the degree of p70S6K1 phosphorylation (Terzis et al., 2008, 2010). But while many studies have shown correlations between skeletal muscle hypertrophy/MPS and signaling molecule phosphorylation, others have not (Areta et al., 2013; Mitchell et al., 2013). Potential reasons for the lack of congruence between exercise-induced signal phosphorylation and muscle hypertrophy/MPS include differences in the timing of muscle biopsies and or methods of analysis (immunoblotting vs direct measure of kinase activity). In reality, it also is likely due to differences in exercise intensity, as well as redundancy in signaling pathways supporting MPS following stimulation, and likely intersubject variability (Crozier et al., 2005). While informative, these studies (Areta et al., 2013; Terzis et al., 2008, 2010) provide only an association between human muscle hypertrophy and mTORC1 signaling rather than a direct cause and effect. Nonetheless, the publication of a paper in which the mTORC1 inhibitor, rapamycin, was injected into humans prior to a bout of resistance exercise (Drummond et al., 2009) provided unequivocal evidence of the importance of this signaling pathway in humans. In this seminal study the authors were able to show that treatment with ~12 mg of rapamycin reduced resistance exercise-induced increases in MPS as well as $p70S6K1^{Thr389}$ phosphorylation 1 h postexercise. Furthermore, when the authors repeated this experiment to test whether or not rapamycin had any impact on amino acid-induced increases in MPS and mTORC1 activity they found very similar results (Dickinson et al., 2011). Indeed, in response to oral essential amino acid consumption rapamycin completely blocked feeding-induced increases in MPS as well as attenuating $p70S6K1^{Thr389}$ phosphorylation. It should be acknowledged that in both studies rapamycin failed to inhibit 4E-BP1 phosphorylation suggesting that there are other, as yet undefined, mechanisms that act in concert with mTORC1 signaling to regulate MPS. Support for this contention arises from studies which show that rapamycin has no impact on either postabsorptive rates of MPS in humans (Dickinson et al., 2012) or endurance exercise-induced rates of mitochondrial and myofibrillar MPS in rodents (Philp et al., 2015). Although, in the latter study, there was no increase in $mTOR^{Ser2448}$ or $AMPK^{Thr172}$ phosphorylation in the control group either, suggesting that the exercise intensity was not sufficient enough to maximize rates of either myofibrillar or mitochondrial MPS. Nevertheless, the molecular network regulating changes in skeletal muscle morphology in response to endurance exercise is known to be somewhat distinct from that regulating resistance exercise-induced adaptations (for a comprehensive review see Egan and Zierath, 2013).

In addition to phosphorylation, the translocation of signaling proteins may play a key role in the adaptive response to exercise and nutrition. For example, in response to aerobic/endurance exercise the peroxisome proliferator-activated receptor-gamma coactivator (PGC-1) alpha and tumor suppressor protein p53 translocate to the nucleus and mitochondria to promote mitochondrial biogenesis (Safdar et al., 2011), an effect that may be modulated by carbohydrate availability (Bartlett et al., 2013, 2015). Similarly, others using cell models have shown that in response to amino acid provision, mTOR localizes to the Rheb positive lysosomal membrane via rags and the regulator complex (Sancak et al., 2008; Sancak and Sabatini, 2009) (for extensive review see Jewell et al., 2013). In response to amino acid withdrawal, mTOR disassociates from the lysosomal membrane where it is unable to interact with its coactivators to form the fully functional mTORC1 protein complex (Long et al., 2005). Interestingly, resistance exercise is also proposed to alter the trafficking of intracellular signaling molecules. One study in rodents has shown that lengthening contractions resulted not only in movement of mTOR to the lysosome but also disassociation of TSC2 from the lysosomal membrane (Jacobs et al., 2013). This dual effect of amino acid feeding and resistance exercise on mTOR and TSC2 trafficking could in some way be responsible for the potentiation of MPS observed when resistance exercise is performed prior to consuming amino acids. However, it is important to note that these studies did not assess rates of MPS nor were they conducted in human models of exercise thus more work is needed to experimentally corroborate these findings in humans with exercise.

6.11. FUTURE DIRECTIONS

To date, a significant amount of work has been conducted that has provided critical information for the field of exercise physiology and nutrition. With the introduction of the percutaneous muscle biopsy technique (Bergstrom, 1975), together with the application of isotopic tracer methodology in the form of isotopically-labeled amino acids (Rennie et al., 1982), it has been possible to directly track rates of MPS in response to various nutritional and exercise interventions. However, due to the invasive nature of the stable isotope tracers the majority of these studies were limited to the laboratory setting and over a $\sim$12 h period. Refinements in mass spectrometry have now enabled the use of deuterium as a tracer that can be orally consumed and does not require intravenous administration, and thus a laboratory (Wilkinson et al., 2015). The significance of this advance is that the rates of MPS that are measured are indicative of a free-living setting including longer-term periods of fasting, feeding, and sleep. This method also allows participants to engage in everyday tasks while also being under the constraints of the experimental paradigms which would enhance the practical applicability of any findings. Contemporary studies using this method have yielded interesting results including characterization of the adaptations of skeletal muscle in the initial stages of resistance training (Brook et al., 2015), as well as the response of the protein synthetic responses of different muscle fractions to different exercise modes (Bell et al., 2015). However, a common criticism of many studies that assess MPS is that they fail to concomitantly measure MPB and thus are unable to provide a complete picture of muscle protein turnover. There has been, however, one paper that has detailed a method to measure changes in MPB with deuterium that warrants further investigation (Holm et al., 2013). By concurrently measuring both MPS and MPB over a period of days it will be possible to gauge the relative contribution of both MPS and MPB to any given changes in muscle size. Such advances in the measurement of muscle protein turnover, have been accompanied by developments in methods to assess changes in the activity (McGlory et al., 2014) and localization (Jacobs et al., 2013) of protein and protein-kinases that are responsible for regulating MPS at the molecular level. When married together with methods that enable the direct determination of MPS and MPB, these new developments will provide greater insight as to how periods of exercise and feeding as well as inactivity, impact skeletal muscle morphology.

6.12. CONCLUSION

Skeletal muscle is a critical organ, the loss of which is associated with numerous clinical pathologies. Currently, pharmaceutical interventions to mimic the pleiotropic health benefits of exercise, particularly resistance exercise, are nonexistent. Thus, exercise and nutrition remain the key tools to promote muscle mass and improve human health on a population basis. However, there is a significant amount of information that has yet to be discovered, particularly with respect to the molecular processes by which exercise and amino acid ingestion confer an anabolic influence toward skeletal muscle. With the application of deuterium to directly track muscle protein turnover alongside methods to examine changes in the cellular location of anabolic signaling proteins, it is hoped that future work will provide exciting new data for the field.

References

Apró, W., Blomstrand, E., 2010. Influence of supplementation with branched-chain amino acids in combination with resistance exercise on p70S6 kinase phosphorylation in resting and exercising human skeletal muscle. Acta Physiol. (Oxf) 200 (3), 237–248.

Areta, J.L., Burke, L.M., Ross, M.L., Camera, D.M., West, D.W., Broad, E.M., et al., 2013. Timing and distribution of protein ingestion during prolonged recovery from resistance exercise alters myofibrillar protein synthesis. J. Physiol. 591, 2319–2331.

Aspuria, P.J., Tamanoi, F., 2004. The Rheb family of GTP-binding proteins. Cell. Signal. 16, 1105–1112.

Atherton, P.J., Etheridge, T., Watt, P.W., Wilkinson, D., Selby, A., Rankin, D., et al., 2010. Muscle full effect after oral protein: time-dependent concordance and discordance between human muscle protein synthesis and mTORC1 signaling. Am. J. Clin. Nutr. 92, 1080–1088.

Averous, J., Lambert-Langlais, S., Carraro, V., Gourbeyre, O., Parry, L., B'Chir, W., et al., 2014. Requirement for lysosomal localization of mTOR for its activation differs between leucine and other amino acids. Cell. Signal. 26, 1918–1927.

Baar, K., Esser, K., 1999. Phosphorylation of p70(S6k) correlates with increased skeletal muscle mass following resistance exercise. Am. J. Physiol. 276, C120–127.

Bartlett, J.D., Louhelainen, J., Iqbal, Z., Cochran, A.J., Gibala, M.J., Gregson, W., et al., 2013. Reduced carbohydrate availability enhances exercise-induced p53 signaling in human skeletal muscle: implications for mitochondrial biogenesis. Am. J. Physiol. Regul. Integr. Comp. Physiol. 304, R450–458.

Bartlett, J.D., Hawley, J.A., Morton, J.P., 2015. Carbohydrate availability and exercise training adaptation: too much of a good thing? Eur. J. Sport Sci. 15, 3–12.

Bell, K.E., Seguin, C., Parise, G., Baker, S.K., Phillips, S.M., 2015. Day-to-Day changes in muscle protein synthesis in recovery from resistance, aerobic, and high-intensity interval exercise in older men. J. Gerontol. A Biol. Sci. Med. Sci. 70, 1024–1029.

Bennet, W.M., Connacher, A.A., Scrimgeour, C.M., Rennie, M.J., 1990. The effect of amino acid infusion on leg protein turnover assessed by L-[15N]phenylalanine and L-[1-13C]leucine exchange. Eur. J. Clin. Invest. 20, 41–50.

Bergstrom, J., 1975. Percutaneous needle biopsy of skeletal muscle in physiological and clinical research. Scand. J. Clin. Lab. Invest. 35, 609–616.

Biolo, G., Maggi, S.P., Williams, B.D., Tipton, K.D., Wolfe, R.R., 1995. Increased rates of muscle protein turnover and amino acid transport after resistance exercise in humans. Am. J. Physiol. 268, E514–520.

Bohe, J., Low, J.F., Wolfe, R.R., Rennie, M.J., 2001. Latency and duration of stimulation of human muscle protein synthesis during continuous infusion of amino acids. J. Physiol. 532, 575–579.

Boirie, Y., Dangin, M., Gachon, P., Vasson, M.P., Maubois, J.L., Beaufrere, B., 1997. Slow and fast dietary proteins differently modulate postprandial protein accretion. Proc. Natl. Acad. Sci. U.S.A. 94, 14930–14935.

Breen, L., Churchward-Venne, T.A., 2012. Leucine: a nutrient 'trigger' for muscle anabolism, but what more? J. Physiol. 590, 2065–2066.

Brook, M.S., Wilkinson, D.J., Mitchell, W.K., Lund, J.N., Szewczyk, N.J., Greenhaff, P.L., et al., 2015. Skeletal muscle hypertrophy adaptations predominate in the early stages of resistance exercise training, matching deuterium oxide-derived measures of muscle protein synthesis and mechanistic target of rapamycin complex 1 signaling. FASEB J. 29, 4485–4496.

Burd, N.A., West, D.W., Staples, A.W., Atherton, P.J., Baker, J.M., Moore, D.R., et al., 2010. Low-load high volume resistance exercise stimulates muscle protein synthesis more than high-load low volume resistance exercise in young men. PLoS One 5, e12033.

Burd, N.A., West, D.W., Moore, D.R., Atherton, P.J., Staples, A.W., Prior, T., et al., 2011. Enhanced amino acid sensitivity of myofibrillar protein synthesis persists for up to 24 h after resistance exercise in young men. J. Nutr. 141, 568–573.

Burd, N.A., Yang, Y., Moore, D.R., Tang, J.E., Tarnopolsky, M.A., Phillips, S.M., 2012. Greater stimulation of myofibrillar protein synthesis with ingestion of whey protein isolate v. micellar casein at rest and after resistance exercise in elderly men. Br. J. Nutr. 108, 958–962.

Burke, L.M., Hawley, J.A., Ross, M.L., Moore, D.R., Phillips, S.M., Slater, G.R., et al., 2012. Preexercise aminoacidemia and muscle protein synthesis after resistance exercise. Med. Sci. Sports Exerc. 44, 1968–1977.

Candeloro, N., Bertini, I., Melchiorri, G., De Lorenzo, A., 1995. Effects of prolonged administration of branched-chain amino acids on body composition and physical fitness. Minerva Endocrinol. 20, 217–223.

Carraro, F., Stuart, C.A., Hartl, W.H., Rosenblatt, J., Wolfe, R.R., 1990. Effect of exercise and recovery on muscle protein synthesis in human subjects. Am. J. Physiol. 259, E470–476.

Cermak, N.M., Res, P.T., de Groot, L.C., Saris, W.H., van Loon, L.J., 2012. Protein supplementation augments the adaptive response of skeletal muscle to resistance-type exercise training: a meta-analysis. Am. J. Clin. Nutr. 96, 1454–1464.

Cermak, N.M., de Groot, L.C., van Loon, L.J., 2013. Perspective: protein supplementation during prolonged resistance type exercise training augments skeletal muscle mass and strength gains. J. Am. Med. Dir. Assoc. 14, 71–72.

Chaillou, T., Kirby, T.J., McCarthy, J.J., 2014. Ribosome biogenesis: emerging evidence for a central role in the regulation of skeletal muscle mass. J. Cell. Physiol. 229, 1584–1594.

Churchward-Venne, T.A., Burd, N.A., Mitchell, C.J., West, D.W.D., Philp, A., Marcotte, G.R., et al., 2012. Supplementation of a suboptimal protein dose with leucine or essential amino acids: effects on myofibrillar protein synthesis at rest and following resistance exercise in men. J. Physiol. 590, 2751–2765.

Churchward-Venne, T.A., Breen, L., Di Donato, D.M., Hector, A.J., Mitchell, C.J., Moore, D.R., et al., 2014a. Leucine supplementation of a low-protein mixed macronutrient beverage enhances myofibrillar protein synthesis in young men: a double-blind, randomized trial. Am. J. Clin. Nutr. 99, 276–286.

Churchward-Venne, T.A., Cotie, L.M., MacDonald, M.J., Mitchell, C.J., Prior, T., Baker, S.K., et al., 2014b. Citrulline does not enhance blood flow, microvascular circulation, or myofibrillar protein synthesis in elderly men at rest or following exercise. Am. J. Physiol. Endocrinol. Metab. 307, E71–83.

Churchward-Venne, T.A., Tieland, M., Verdijk, L.B., Leenders, M., Dirks, M.L., de Groot, L.C., et al., 2015. There are no nonresponders to resistance-type exercise training in older men and women. J. Am. Med. Dir. Assoc. 16, 400–411.

Crozier, S.J., Kimball, S.R., Emmert, S.W., Anthony, J.C., Jefferson, L.S., 2005. Oral leucine administration stimulates protein synthesis in rat skeletal muscle. J. Nutr. 135, 376–382.

Curis, E., Nicolis, I., Moinard, C., Osowska, S., Zerrouk, N., Benazeth, S., et al., 2005. Almost all about citrulline in mammals. Amino Acids 29, 177–205.

Cynober, L., Moinard, C., De Bandt, J.P., 2010. The 2009 ESPEN Sir David Cuthbertson. Citrulline: a new major signaling molecule or just another player in the pharmaconutrition game? Clin. Nutr. 29, 545–551.

Devries, M.C., Phillips, S.M., 2015. Supplemental protein in support of muscle mass and health: advantage whey. J. Food Sci. 80 (Suppl. 1), A8–A15.

Dickinson, J.M., Fry, C.S., Drummond, M.J., Gundermann, D.M., Walker, D.K., Glynn, E.L., et al., 2011. Mammalian target of rapamycin complex 1 activation is required for the stimulation of human skeletal muscle protein synthesis by essential amino acids. J. Nutr. 141, 856–862.

Dickinson, J.M., Drummond, M.J., Fry, C.S., Gundermann, D.M., Walker, D.K., Volpi, E., et al., 2012. Rapamycin administration does not impair basal protein metabolism in human skeletal muscle. FASEB J. 26, supplement 1075.3.

Drummond, M.J., Fry, C.S., Glynn, E.L., Dreyer, H.C., Dhanani, S., Timmerman, K.L., et al., 2009. Rapamycin administration in humans blocks the contraction-induced increase in skeletal muscle protein synthesis. J. Physiol. 587, 1535–1546.

Egan, B., Zierath, J.R., 2013. Exercise metabolism and the molecular regulation of skeletal muscle adaptation. Cell Metab. 17, 162–184.

Foster, K.G., Fingar, D.C., 2010. Mammalian target of rapamycin (mTOR): conducting the cellular signaling symphony. J. Biol. Chem. 285, 14071–14077.

Gingras, A.C., Gygi, S.P., Raught, B., Polakiewicz, R.D., Abraham, R.T., Hoekstra, M.F., et al., 1999. Regulation of 4E-BP1 phosphorylation: a novel two-step mechanism. Genes Dev. 13, 1422–1437.

Glynn, E.L., Fry, C.S., Drummond, M.J., Timmerman, K.L., Dhanani, S., Volpi, E., et al., 2010. Excess leucine intake enhances muscle anabolic signaling but not net protein anabolism in young men and women. J. Nutr. 140, 1970–1976.

Guertin, D.A., Sabatini, D.M., 2007. Defining the role of mTOR in cancer. Cancer Cell 12, 9–22.

Gwinn, D.M., Shackelford, D.B., Egan, D.F., Mihaylova, M.M., Mery, A., Vasquez, D.S., et al., 2008. AMPK phosphorylation of raptor mediates a metabolic checkpoint. Mol. Cell 30, 214–226.

Hannan, K.M., Brandenburger, Y., Jenkins, A., Sharkey, K., Cavanaugh, A., Rothblum, L., et al., 2003. mTOR-dependent regulation of ribosomal gene transcription requires S6K1 and is mediated by phosphorylation of the carboxy-terminal activation domain of the nucleolar transcription factor UBF. Mol. Cell. Biol. 23, 8862–8877.

Harber, M.P., Konopka, A.R., Douglass, M.D., Minchev, K., Kaminsky, L.A., Trappe, T.A., et al., 2009. Aerobic exercise training improves whole muscle and single myofiber size and function in older women. Am. J. Physiol. Regul. Integr. Comp. Physiol. 297, R1452–1459.

Harber, M.P., Konopka, A.R., Jemiolo, B., Trappe, S.W., Trappe, T.A., Reidy, P.T., 2010. Muscle protein synthesis and gene expression during recovery from aerobic exercise in the fasted and fed states. Am. J. Physiol. Regul. Integr. Comp. Physiol. 299, R1254–1262.

Harber, M.P., Konopka, A.R., Undem, M.K., Hinkley, J.M., Minchev, K., Kaminsky, L.A., et al., 2012. Aerobic exercise training induces skeletal muscle hypertrophy and age-dependent adaptations in myofiber function in young and older men. J. Appl. Physiol. (1985) 113, 1495–1504.

Holm, L., O'Rourke, B., Ebenstein, D., Toth, M.J., Bechshoeft, R., Holstein-Rathlou, N.H., et al., 2013. Determination of steady-state protein breakdown rate in vivo by the disappearance of protein-bound tracer-labeled amino acids: a method applicable in humans. Am. J. Physiol. Endocrinol. Metab. 304, E895–907.

Inoki, K., Li, Y., Zhu, T., Wu, J., Guan, K.L., 2002. TSC2 is phosphorylated and inhibited by Akt and suppresses mTOR signalling. Nat. Cell Biol. 4, 648–657.

Inoki, K., Li, Y., Xu, T., Guan, K.L., 2003. Rheb GTPase is a direct target of TSC2 GAP activity and regulates mTOR signaling. Genes Dev. 17, 1829–1834.

Ispoglou, T., King, R.F., Polman, R.C., Zanker, C., 2011. Daily L-leucine supplementation in novice trainees during a 12-week weight training program. Int. J. Sports Physiol. Perform. 6, 38–50.

Jacobs, B.L., You, J.S., Frey, J.W., Goodman, C.A., Gundermann, D.M., Hornberger, T.A., 2013. Eccentric contractions increase the phosphorylation of tuberous sclerosis complex-2 (TSC2) and alter the targeting of TSC2 and the mechanistic target of rapamycin to the lysosome. J. Physiol. 591, 4611–4620.

Jewell, J.L., Guan, K.L., 2013. Nutrient signaling to mTOR and cell growth. Trends Biochem. Sci. 38, 233–242.

Jewell, J.L., Russell, R.C., Guan, K.L., 2013. Amino acid signalling upstream of mTOR. Nat. Rev. Mol. Cell Biol. 14, 133–139.

Jourdan, M., Nair, K.S., Carter, R.E., Schimke, J., Ford, G.C., Marc, J., et al., 2015. Citrulline stimulates muscle protein synthesis in the post-absorptive state in healthy people fed a low-protein diet—a pilot study. Clin. Nutr. 34, 449–456.

Kallman, D.A., Plato, C.C., Tobin, J.D., 1990. The role of muscle loss in the age-related decline of grip strength: cross-sectional and longitudinal perspectives. J. Gerontol. 45, M82–88.

Katsanos, C.S., Kobayashi, H., Sheffield-Moore, M., Aarsland, A., Wolfe, R.R., 2006. A high proportion of leucine is required for optimal stimulation of the rate of muscle protein synthesis by essential amino acids in the elderly. Am. J. Physiol. Endocrinol. Metab. 291, E381–387.

Kim, P.L., Staron, R.S., Phillips, S.M., 2005. Fasted-state skeletal muscle protein synthesis after resistance exercise is altered with training. J. Physiol. 568, 283–290.

Kirby, T.J., Lee, J.D., England, J.H., Chaillou, T., Esser, K.A., McCarthy, J.J., 2015. Blunted hypertrophic response in aged skeletal muscle is associated with decreased ribosome biogenesis. J. Appl. Physiol. (1985). Available from: http://dx.doi.org/10.1152/japplphysiol.00296.2015.

Kumar, V., Selby, A., Rankin, D., Patel, R., Atherton, P., Hildebrandt, W., et al., 2009. Age-related differences in the dose-response relationship of muscle protein synthesis to resistance exercise in young and old men. J. Physiol. 587, 211–217.

Long, X., Ortiz-Vega, S., Lin, Y., Avruch, J., 2005. Rheb binding to mammalian target of rapamycin (mTOR) is regulated by amino acid sufficiency. J. Biol. Chem. 280, 23433–23436.

Ma, X.M., Yoon, S.O., Richardson, C.J., Julich, K., Blenis, J., 2008. SKAR links pre-mRNA splicing to mTOR/S6K1-mediated enhanced translation efficiency of spliced mRNAs. Cell 133, 303–313.

Mahe, S., Roos, N., Benamouzig, R., Davin, L., Luengo, C., Gagnon, L., et al., 1996. Gastrojejunal kinetics and the digestion of [15N]beta-lactoglobulin and casein in humans: the influence of the nature and quantity of the protein. Am. J. Clin. Nutr. 63, 546–552.

McGlory, C., White, A., Treins, C., Drust, B., Close, G.L., Maclaren, D.P., et al., 2014. Application of the [gamma-32P] ATP kinase assay to study anabolic signaling in human skeletal muscle. J. Appl. Physiol. (1985) 116, 504–513.

Metter, E.J., Talbot, L.A., Schrager, M., Conwit, R., 2002. Skeletal muscle strength as a predictor of all-cause mortality in healthy men. J. Gerontol. A Biol. Sci. Med. Sci. 57, B359–365.

Mitchell, C.J., Churchward-Venne, T.A., Bellamy, L., Parise, G., Baker, S.K., Phillips, S.M., 2013. Muscular and systemic correlates of resistance training-induced muscle hypertrophy. PLoS One 8, e78636.

Mitchell, W.K., Williams, J., Atherton, P., Larvin, M., Lund, J., Narici, M., 2012. Sarcopenia, dynapenia, and the impact of advancing age on human skeletal muscle size and strength; a quantitative review. Front. Physiol. 3, 260.

Moore, D.R., Robinson, M.J., Fry, J.L., Tang, J.E., Glover, E.I., Wilkinson, S.B., et al., 2009. Ingested protein dose response of muscle and albumin protein synthesis after resistance exercise in young men. Am. J. Clin. Nutr. 89, 161–168.

Moore, D.R., Churchward-Venne, T.A., Witard, O., Breen, L., Burd, N.A., Tipton, K.D., et al., 2015. Protein ingestion to stimulate myofibrillar protein synthesis requires greater relative protein intakes in healthy older versus younger men. J. Gerontol. A Biol. Sci. Med. Sci. 70, 57–62.

Pearson, R.B., Dennis, P.B., Han, J.W., Williamson, N.A., Kozma, S.C., Wettenhall, R.E., et al., 1995. The principal target of rapamycin-induced p70s6k inactivation is a novel phosphorylation site within a conserved hydrophobic domain. EMBO J. 14, 5279–5287.

Phillips, B.E., Hill, D.S., Atherton, P.J., 2012. Regulation of muscle protein synthesis in humans. Curr. Opin. Clin. Nutr. Metab. Care 15, 58–63.

Phillips, S.M., 2004. Protein requirements and supplementation in strength sports. Nutrition 20, 689–695.

Phillips, S.M., 2015. Nutritional supplements in support of resistance exercise to counter age-related sarcopenia. Adv. Nutr. 6, 452–460.

Phillips, S.M., Tipton, K.D., Aarsland, A., Wolf, S.E., Wolfe, R.R., 1997. Mixed muscle protein synthesis and breakdown after resistance exercise in humans. Am. J. Physiol. 273, E99–107.

Philp, A., Schenk, S., Perez-Schindler, J., Hamilton, D.L., Breen, L., Laverone, E., et al., 2015. Rapamycin does not prevent increases in myofibrillar or mitochondrial protein synthesis following endurance exercise. J. Physiol. 593, 4275–4284.

Rennie, M.J., Edwards, R.H., Halliday, D., Matthews, D.E., Wolman, S.L., Millward, D.J., 1982. Muscle protein synthesis measured by stable isotope techniques in man: the effects of feeding and fasting. Clin. Sci. (London, England: 1979) 63, 519–523.

Richardson, C.J., Broenstrup, M., Fingar, D.C., Julich, K., Ballif, B.A., Gygi, S., et al., 2004. SKAR is a specific target of S6 kinase 1 in cell growth control. Curr. Biol. 14, 1540–1549.

Rouge, C., Des Robert, C., Robins, A., Le Bacquer, O., Volteau, C., De La Cochetiere, M.F., et al., 2007. Manipulation of citrulline availability in humans. Am. J. Physiol. Gastrointest. Liver Physiol. 293, G1061–1067.

Safdar, A., Little, J.P., Stokl, A.J., Hettinga, B.P., Akhtar, M., Tarnopolsky, M.A., 2011. Exercise increases mitochondrial PGC-1alpha content and promotes nuclear-mitochondrial cross-talk to coordinate mitochondrial biogenesis. J. Biol. Chem. 286, 10605–10617.

Sancak, Y., Sabatini, D.M., 2009. Rag proteins regulate amino-acid-induced mTORC1 signalling. Biochem. Soc. Trans. 37, 289–290.

Sancak, Y., Peterson, T.R., Shaul, Y.D., Lindquist, R.A., Thoreen, C.C., Bar-Peled, L., et al., 2008. The Rag GTPases bind raptor and mediate amino acid signaling to mTORC1. Science 320, 1496–1501.

Scalzo, R.L., Peltonen, G.L., Binns, S.E., Shankaran, M., Giordano, G.R., Hartley, D.A., et al., 2014. Greater muscle protein synthesis and mitochondrial biogenesis in males compared with females during sprint interval training. FASEB J. 28, 2705–2714.

Shimomura, Y., Yamamoto, Y., Bajotto, G., Sato, J., Murakami, T., Shimomura, N., et al., 2006. Nutraceutical effects of branched-chain amino acids on skeletal muscle. J. Nutr. 136, 529S–532S.

Spillane, M., Emerson, C., Willoughby, D.S., 2012. The effects of 8 weeks of heavy resistance training and branched-chain amino acid supplementation on body composition and muscle performance. Nutr. Health 21, 263–273.

Stuehr, D.J., 2004. Enzymes of the L-arginine to nitric oxide pathway. J. Nutr. 134, 2748S–2751S (discussion 2765S–2767S).

Taddei, S., Virdis, A., Ghiadoni, L., Salvetti, G., Bernini, G., Magagna, A., et al., 2001. Age-related reduction of NO availability and oxidative stress in humans. Hypertension 38, 274–279.

Tang, J.E., Moore, D.R., Kujbida, G.W., Tarnopolsky, M.A., Phillips, S.M., 2009. Ingestion of whey hydrolysate, casein, or soy protein isolate: effects on mixed muscle protein synthesis at rest and following resistance exercise in young men. J. Appl. Physiol. (1985) 107, 987–992.

Terzis, G., Georgiadis, G., Stratakos, G., Vogiatzis, I., Kavouras, S., Manta, P., et al., 2008. Resistance exercise-induced increase in muscle mass correlates with p70S6 kinase phosphorylation in human subjects. Eur. J. Appl. Physiol. 102, 145–152.

Terzis, G., Spengos, K., Mascher, H., Georgiadis, G., Manta, P., Blomstrand, E., 2010. The degree of p70 S6k and S6 phosphorylation in human skeletal muscle in response to resistance exercise depends on the training volume. Eur. J. Appl. Physiol. 110, 835–843.

Thibault, R., Flet, L., Vavasseur, F., Lemerle, M., Ferchaud-Roucher, V., Picot, D., et al., 2011. Oral citrulline does not affect whole body protein metabolism in healthy human volunteers: results of a prospective, randomized, double-blind, cross-over study. Clin. Nutr. 30, 807–811.

Timmerman, K.L., Dhanani, S., Glynn, E.L., Fry, C.S., Drummond, M.J., Jennings, K., et al., 2012. A moderate acute increase in physical activity enhances nutritive flow and the muscle protein anabolic response to mixed nutrient intake in older adults. Am. J. Clin. Nutr. 95, 1403–1412.

Tipton, K.D., Ferrando, A.A., Phillips, S.M., Doyle Jr., D., Wolfe, R.R., 1999a. Postexercise net protein synthesis in human muscle from orally administered amino acids. Am. J. Physiol. 276, E628–634.

Tipton, K.D., Gurkin, B.E., Matin, S., Wolfe, R.R., 1999b. Nonessential amino acids are not necessary to stimulate net muscle protein synthesis in healthy volunteers. J. Nutr. Biochem. 10, 89–95.

Vander Haar, E., Lee, S.I., Bandhakavi, S., Griffin, T.J., Kim, D.H., 2007. Insulin signalling to mTOR mediated by the Akt/PKB substrate PRAS40. Nat. Cell Biol. 9, 316–323.

van Vliet, S., Burd, N.A., van Loon, L.J., 2015. The skeletal muscle anabolic response to plant- versus animal-based protein consumption. J. Nutr. 145, 1981–1991.

Wang, X., Li, W., Williams, M., Terada, N., Alessi, D.R., Proud, C.G., 2001. Regulation of elongation factor 2 kinase by p90(RSK1) and p70 S6 kinase. EMBO J. 20, 4370–4379.

West, D.W., Burd, N.A., Coffey, V.G., Baker, S.K., Burke, L.M., Hawley, J.A., et al., 2011. Rapid aminoacidemia enhances myofibrillar protein synthesis and anabolic intramuscular signaling responses after resistance exercise. Am. J. Clin. Nutr. 94, 795–803.

Wilkinson, D.J., Cegielski, J., Phillips, B.E., Boereboom, C., Lund, J.N., Atherton, P.J., et al., 2015. Internal comparison between deuterium oxide (D2O) and L-[ring-13C6] phenylalanine for acute measurement of muscle protein synthesis in humans. Physiol. Rep. 3.

Wilkinson, S.B., Phillips, S.M., Atherton, P.J., Patel, R., Yarasheski, K.E., Tarnopolsky, M.A., et al., 2008. Differential effects of resistance and endurance exercise in the fed state on signalling molecule phosphorylation and protein synthesis in human muscle. J. Physiol. 586, 3701–3717.

Witard, O.C., Tieland, M., Beelen, M., Tipton, K.D., van Loon, L.J., Koopman, R., 2009. Resistance exercise increases postprandial muscle protein synthesis in humans. Med. Sci. Sports Exerc. 41, 144–154.

Witard, O.C., Jackman, S.R., Breen, L., Smith, K., Selby, A., Tipton, K.D., 2014. Myofibrillar muscle protein synthesis rates subsequent to a meal in response to increasing doses of whey protein at rest and after resistance exercise. Am. J. Clin. Nutr. 99, 86–95.

Wu, F., Cholewa, B., Mattson, D.L., 2000. Characterization of L-arginine transporters in rat renal inner medullary collecting duct. Am. J. Physiol. Regul. Integr. Comp. Physiol. 278, R1506–1512.

Wu, G., Meininger, C.J., Knabe, D.A., Bazer, F.W., Rhoads, J.M., 2000. Arginine nutrition in development, health and disease. Curr. Opin. Clin. Nutr. Metab. Care 3, 59–66.

Yang, Y., Breen, L., Burd, N.A., Hector, A.J., Churchward-Venne, T.A., Josse, A.R., et al., 2012. Resistance exercise enhances myofibrillar protein synthesis with graded intakes of whey protein in older men. Br. J. Nutr. 108, 1780–1788.

CHAPTER

7

Protein Metabolism in the Elderly: Molecular and Cellular Aspects

E.L. Dillon

Department of Internal Medicine, Division of Endocrinology and Metabolism, The University of Texas Medical Branch, Galveston, TX, United States

7.1. AGING AND SARCOPENIA

Aging is associated with a gradual loss in skeletal muscle mass and function, collectively identified as sarcopenia of aging (Morley et al., 2014). The incidence of sarcopenia increases with age, affecting about 5% of adults at age 65 and as many as half of people aged 80 and older (Baumgartner et al., 1998; Morley, 2008, 2012). This process of muscle loss takes place slowly over years, when net rates of skeletal muscle protein synthesis are not capable of keeping up with net rates of protein degradation, resulting in a loss of protein mass and skeletal muscle strength (Katsanos et al., 2005; Morley, 2012). However, loss of muscle strength may occur at rates that exceed the loss of mass, contributing to decreased muscle quality with age (Goodpaster et al., 2006; Mitchell et al., 2012). Because of the gradual progression of muscle atrophy, generally starting around middle-age, the onset of sarcopenia is difficult to identify in the individual. Along current existing definitions, sarcopenia is diagnosed in older adults after both muscle mass and muscle function have declined below that of young adult demographic representatives (Chen et al., 2014; Cruz-Jentoft et al., 2010; Dam et al., 2014; Fielding et al., 2011; Morley, 2008; Morley et al., 2011; Muscaritoli et al., 2010; Thomas et al., 2000). The definition of sarcopenia is therefore susceptible to demographic differences between—or changes within—healthy young adult populations as well as to the methodologies used to measure muscle mass and function.

Aging is a continuous process, and defining succinct cutoffs based on chronological age are somewhat arbitrary. While western society generally defines old age by pension age ($\sim$60–65 years), there are no international standards (Roebuck, 1979; WHO). Likewise within aging research, age groups are not clearly defined and the use of the terms like old (or elderly) and young can be somewhat arbitrarily presented in the literature, with young adults often falling somewhere between ages $\sim$18 and 35, older adults represented by ages starting around 60, and middle-age represented in-between (Eskelinen et al., 2015; Straight et al., 2015). Reports on age-related differences can therefore add to loss of clarity in the literature when different cutoffs for inclusion or exclusion of certain demographics are used between studies. Despite the challenges in defining sarcopenia and old age, aging related loss in muscle mass and function is widely recognized as a major public health concern.

7.2. PROTEIN METABOLISM IN THE AGING BODY

Much of our current understanding regarding the changes that occur in protein and amino acid metabolism during aging is derived from research on skeletal muscle. It is important, however, to understand that the effects of aging on protein and amino acid metabolism are not isolated to this tissue alone, although there is no clear

The Molecular Nutrition of Amino Acids and Proteins.
DOI: http://dx.doi.org/10.1016/B978-0-12-802167-5.00007-4

consensus on changes that occur at the whole body level (Dorrens and Rennie, 2003; Morais et al., 2000). While decreases in whole-body protein synthesis have been reported in older rats when compared to adult rats (Jourdan et al., 2011), age-related changes in human protein metabolism are generally defined at the organ level such as skeletal muscle, splanchnic tissues, vascular system, skin, and brain.

Skeletal muscle is the largest organ by mass and provides a large pool of amino acids utilized by other organs (Wolfe, 2006). Aging is generally associated with shifts in body composition. Losses of muscle and bone mass are often observed while fat mass tends to stay unchanged or even increases. Thus, while overall body weight may appear stable, this development of sarcopenic obesity can contribute to significant loss of physical and metabolic function (Santilli et al., 2015). The shifts in body composition with age have been associated with increased risk for dyslipidemia, cardiovascular disease, and metabolic syndrome (Kim and Choi, 2014). Skeletal muscle protein synthesis is very responsive to anabolic stimulation through nutritional and mechanical factors. Although aging results in reduced anabolic efficiency through the mammalian target of rapamycin (mTOR) pathway in response to nutrition, the ingestion of amino acids acutely stimulates muscle protein synthesis similarly in younger and older adults (Paddon-Jones et al., 2004b) and chronic amino acid supplementation stimulates hypertrophy (Borsheim et al., 2008; Dillon et al., 2009), and improves muscle function in older adults (Tieland et al., 2012).

The rate of renewal and repair of skeletal muscle cells declines with age. Muscle hypertrophy can occur independent of satellite cell involvement (Jackson et al., 2012). However, generation of new skeletal muscle requires the activation of precursor satellite cells and proliferation of myogenic daughter cells leading to differentiation and formation of multinucleated muscle fibers (Garcia-Prat et al., 2013; Motohashi and Asakura, 2014). Satellite cells are quiescent, muscle-specific stem cells located under the basal lamina of muscle fibers. While satellite cells from older individuals show no sign of impairment per se in vitro (Hawke and Garry, 2001; Renault et al., 2002), aging results in impaired satellite cell activation and proliferation (Verdijk et al., 2014). A number of factors may contribute to decreased satellite cell activity with age, including the dysregulation of Notch, Wnt, human growth factor (HGF), fibroblast growth factor (FGF), transforming growth factor (TGF-beta), calcineurin, Ras/Mitogen-activated protein kinases/extracellular signal-regulated kinases (Ras/MAPK/ERK), myostatin, follistatin, and nitric oxide (NO) production (Arthur and Cooley, 2012; Conboy and Rando, 2002; Friday et al., 2003; Keren et al., 2006; Pisconti et al., 2006; Suetta et al., 2013). Satellite cell proliferation is controlled through Notch signaling leading to formation of myogenic precursor cells and myoblasts. Notch both induces proliferation and prevents differentiation (Buas et al., 2009; Kitzmann et al., 2006) and the expression of Pax7 during proliferation inhibits MyoD required for cell differentiation (Olguin et al., 2007). Notch signaling may be impaired with aging, possibly implicating increased TGF-beta (and Smads) as molecules competing with Notch signaling pathways. Myoblast differentiation and fusion into myotubes ensues upon deactivation of Notch signaling and induction of the Wnt signaling pathway. MAPK and calcineurin upregulate myoblast differentiation involving the activation of MyoD and myocyte enhancer factor-2 (MEF2) (Friday et al., 2003; Keren et al., 2006). Wnt inhibits phosphorylation of glycogen synthase kinase 3 beta (GSK3beta) and beta-catenin. Inhibition of GSK3 by Wnt promotes protein synthesis by deactivating TSC2 inhibition on mTOR Complex 1 (Inoki et al., 2006). Translocation of dephosphorylated beta-catenin activates transcription factors in the nucleus and results in increased activation of myogenic regulatory factors such as Myf5, and MyoD (Cossu and Borello, 1999). The Wnt pathway is upregulated in aging muscle, and dysregulations in the timing of Wnt signaling events may lead to increased fibrogenic tissue as myogenic precursor cells are diverted into different lineages (Brack et al., 2007). Baseline mRNA expression of several myogenic factors is increased in older compared to younger women (Raue et al., 2006). It is unclear whether these differences indicate compensatory mechanisms to maintain muscle turnover with age.

Splanchnic organs extract and utilize dietary amino acids before they become available to peripheral tissues. Aging has been associated with increased first-pass extraction of amino acids by splanchnic tissues (Boirie et al., 1997; Volpi et al., 1999), although this is not confirmed in all studies (Verhoeven et al., 2009). Increases in the extraction of oral amino acids by the gut and liver could potentially reduce amino acid availability for skeletal muscle protein synthesis (Boirie et al., 1997). This may contribute to decreased protein balance if amino acid availability for systemic utilization is limited due to suboptimal nutritional intake (Jonker et al., 2012; Jourdan et al., 2011). Enterocytes lining the small intestine are continuously sloughed off and replenished, and amino acid nutrition is of great importance for the maintenance of intestinal integrity (Hartl and Alpers, 2010). Despite increased first-pass splanchnic amino acid sequestration, skeletal muscle protein synthesis can be induced similarly in younger and older adults (Volpi et al., 1999). As long as adequate dietary protein or essential amino acids are provided, digestion, absorption, and the subsequent acute anabolic response in muscle protein synthetis is similar in healthy adults regardless of age (Koopman et al., 2009b), despite age-associated increases in splanchnic extraction and utilization of amino acids (Moreau et al., 2013; Volpi et al., 1999). Although alterations in amino acid availability may be a consequence of aging, there is a general consensus that age-related changes in anabolic sensitivity to amino acids at the cellular level are predominant factors responsible for subpar anabolic responses under conditions of low amino acid availability (Breen and Phillips, 2011; Cuthbertson et al., 2005; Dardevet et al., 2000). Furthermore, even under conditions where systemic amino acid availability is high and rates of protein synthesis are similar between older and younger adults, the protein synthetic response is less efficient in older adults compared to younger adults (Durham et al., 2010; Kullman et al., 2013).

Cardiovascular health declines with age and aging is a risk factor for the development of endothelial dysfunction, cardiovascular disease, insulin resistance, and hypertension. Vascular function is in part regulated through the vasodilator effects of nitric oxide (NO). Constitutive endothelial NO synthase (eNOS) produces citrulline and NO from arginine and O_2, and synthesis of NO is regulated by a number of factors including insulin and amino acids (Wu and Meininger, 2002). Aging is associated with reduced bioavailability of NO, due to either reduced eNOS expression or due to decreased eNOS activity. Insulin is a major activator of eNOS through Akt-dependent mechanisms and reduced insulin sensitivity can be an important factor in the age-related decreases in eNOS activity (Du et al., 2006; Montagnani et al., 2001). While the exact contributing mechanisms behind reduced NO availability with age remain to be elucidated, increased oxidative stress appears to play a prominent role (Taddei et al., 2001) and dysregulation in vessel growth, maintenance, and function are believed to be related to impairments in oxidative stress responses (Oellerich and Potente, 2012). In addition to the role of oxidative stress, the accumulation of advanced glycation end-products (AGE) may play a role in age-related cardiovascular dysfunction and AGE have been shown to interfere with endothelial derived NO signaling (Bucala et al., 1991) and inhibit the PI3K/Akt/mTOR pathway in cardiomyocytes (Hou et al., 2014). Insulin-mediated vasodilation in skeletal muscle is NO dependent (Steinberg et al., 1994) and changes in NO signaling may be central to age-related metabolic impairments during exercise. Dysregulations in NO synthesis can contribute to increased oxidative stress during exercise. Uncoupling of eNOS in arterioles in muscle from sedentary older rats results in impaired NO production and an upregulation in the production of O_2^- in response to in vitro stimulated flow when compared to young muscle (Sindler et al., 2009). Despite dysregulation of NO production, the efficiency of skeletal muscle anabolism following the induction of microvascular blood flow with a nitric oxide (NO) donor such as sodium nitroprusside in the presence of amino acids is similar between young and old adults, supporting the notion that impairments following exercise may involve altered endothelial function and impaired NO synthesis opposed to reduced sensitivity to NO signaling (Dillon et al., 2011). Similarly, sodium nitroprusside infusion improves the anabolic response to insulin in absence of exercise in skeletal muscle of older subjects (Timmerman et al., 2010). While these studies suggest that the aging vasculature and muscle remains responsive to the actions of NO, age-related impairments in endogenous NO production may nevertheless be both caused by (Landmesser et al., 2003) and be a contributing factor to (Sindler et al., 2009) increased oxidative stress, thus further contributing to the impaired anabolic responses to changes in blood flow during exercise. Whether age-related changes in insulin sensitivity, sarcolemmal integrity, oxidative stress, endothelial function, and/or responsiveness to exercise play significant roles in the decreased anabolic sensitivity to amino acids remains to be determined.

Skin is the largest organ by surface area and displays a high rate of cellular turnover as dermal layers are continuously lost and replenished. Integrity of the cornified envelope is extremely important as it provides a flexible, water-proof barrier against the hostile and oxidative external environment (Vermeij and Backendorf, 2013). Aging is associated with a gradual thinning of this cornified envelope due to decreased turnover of keratinocytes in the epidermis. Differentiation of keratinocytes is calcium-dependent and the assembly and cross-linking of the various structural proteins, including involucrin, loricrin, and small proline-rich proteins, are catalyzed by epidermal transglutaminases (Hitomi, 2005; Marshall et al., 2001). Expression of several structural proteins is altered with age. Expression of loricrin, the most abundant structural protein in the cornified envelope, decreases with age (Rinnerthaler et al., 2013). In contrast, expression of small proline-rich proteins—which have shown to be protective against, yet sensitive to, external factors including UV light, cigarette smoke, and oxidative stress (Vermeij et al., 2011, 2012)—appear to be increased with age (Rinnerthaler et al., 2013). Thus, both endogenous changes in metabolism as well as exogenous stressors are contributors to changes observed in aging skin.

Brain disorders such as Alzheimer's disease, the major form of age-related dementia, are associated with the accumulation of protein fragments due to incomplete degradation of amyloid-beta-protein precursors (Renziehausen et al., 2014). The dysregulation in proteosomal degradation of proteins is understood to be related to impairments in oxidative stress responses (Cardinale et al., 2014) The accumulation of protein posttranslational modifications such as AGE are likely contributing factors to the development of dementia, and age-related declines in cognitive function are common complications in patients with metabolic dysfunctions that contribute to hyperglycemia (Hishikawa et al., 2014; Toth et al., 2007). Furthermore, dysregulation of mTOR, the key signaling pathway for nutrient sensing, protein synthesis, and regulation of autophagy in skeletal muscle (Yang and Klionsky, 2009), plays an equally crucial role in cell fate in the aging brain (Perluigi et al., 2015).

7.3. AGE-RELATED CHANGES IN NUTRIENT SENSITIVITY

Maintenance of protein balance during aging is affected by many factors including changes in habitual activity and nutrition (Abbatecola et al., 2011; Dillon et al., 2010; Horstman et al., 2012; Kimball et al., 2002; Walker et al., 2011; Wall et al., 2012). Protein loss occurs due to a net imbalance between rates at which existing protein is broken down and rates at which new protein is formed. Fasting rates of protein degradation in skeletal muscle do not appear to be profoundly altered with age, nor are rates of protein degradation following meal ingestion. There is less consensus regarding fasting rates of skeletal muscle protein synthesis with aging. Basal rates of protein synthesis may (Guillet et al., 2004; Rooyackers et al., 1996; Welle et al., 1993; Yarasheski et al., 1993) or may not (Cuthbertson et al., 2005; Dillon et al., 2011; Katsanos et al., 2005, 2006; Moore et al., 2014; Paddon-Jones et al., 2004b; Symons et al., 2009b; Volpi et al., 1999, 2000, 2001) decrease with age, although it is unclear whether some of the reported age-related impairments may be secondary to factors such as habitual activities including suboptimal dietary habits. For instance, some studies have shown beneficial effects of dietary amino acid supplementation on basal muscle protein synthesis rates in older adults (Casperson et al., 2012; Dillon et al., 2009) while other studies show no benefits (Verhoeven et al., 2009; Walrand et al., 2008; Yarasheski et al., 2011). Most experts do agree, however, that the gradual loss of skeletal muscle protein with age is predominantly due to a decline in protein synthesis in response to acute anabolic stimuli and less due to changes in basal protein turnover. There appears to be a shift in nutrient sensitivity with aging that affects the acute anabolic response to meals as well as the duration of positive net balance of protein synthesis following each meal. Thus, while healthy older individuals are still capable of reaching periods of net positive protein balance, the loss in efficiency means that reaching peak positive balance requires higher protein intake, and a return to negative protein balance is reached sooner following food ingestion than in younger adults. Regardless of age, amino acid concentrations must reach some minimum threshold to exert a robust anabolic response in skeletal muscle protein synthesis (Breen and Phillips, 2011; Dardevet et al., 2000; Katsanos et al., 2005, 2006; Kobayashi et al., 2003; Moore et al., 2014; Volpi et al., 2000). This threshold is increased in aged muscle, requiring higher concentrations of amino acids like leucine to elicit maximum anabolic responses comparable to those in muscle from younger individuals (Dardevet et al., 2000; Fig. 7.1). Anabolic responses similar to those observed in young adults can thus be reached in older adults when adequate acute loads of high quality sources of amino acids above a minimum threshold are provided (Cuthbertson et al., 2005; El-Kadi et al., 2012; Gazzaneo et al., 2011; Katsanos et al., 2006; Moore et al., 2014;

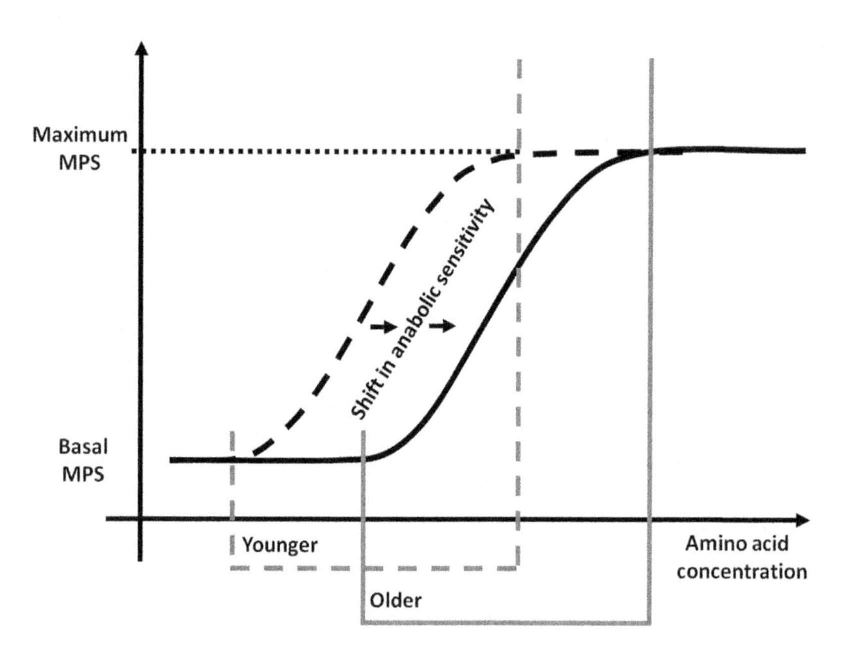

FIGURE 7.1 Anabolic sensitivity to amino acids shifts with aging. Basal and maximally inducible rates of muscle protein synthesis (MPS) remain similar between younger and older individuals. However, the required concentration of amino acids necessary to invoke anabolic responses shifts to the right with increased age. Exceeding acute amino acid intake beyond the range necessary to reach a maximum anabolic response does not further increase MPS in younger and older adults respectively.

Pennings et al., 2012; Rieu et al., 2006; Symons et al., 2007, 2009b; Yang et al., 2012). The general minimum recommendation for adults is to consume at least 0.8 g of protein per kg of body weight per day in order to stay in neutral protein balance. However, this recommendation may not be sufficient to protect all older individuals with anabolic resistance if care is not taken to ensure that the frequency at which this minimum anabolic threshold is reached is adequate to overcome net rates of protein degradation (Paddon-Jones and Rasmussen, 2009). In accordance with the phenomenon of anabolic resistance with aging, the current international consensus is that older adults should consume between 1.0 and 1.2 g/kg/d (recommendations from the ESPEN Expert Group; Bauer et al., 2013; Deutz et al., 2014). This recommendation could be further elevated to 1.2–1.5 g/kg/d for individuals at increased risk for malnutrition or illness. However, the profoundly higher recommendations make compliance more challenging in a demographic that may already have difficulties adhering to the lower range of the recommended intake due to habitually lower dietary intakes.

Several approaches to facilitate and promote optimum amino acid intake and anabolic responses exist. Protein and amino acid supplementation may aid in increasing lean body mass in many older adults (Dillon et al., 2009; Solerte et al., 2008; Volek et al., 2013). Not all studies show chronic benefits of amino acid supplementation on body composition or function in older adults (Balage and Dardevet, 2010; Leenders et al., 2011; Xu et al., 2014) and the benefits of supplementation appear to be highly dependent on additional factors including the formulation of the supplements and habitual dietary intake. Adherence to the minimum recommendations for dietary protein intake may be sufficient for many individuals, with no added benefit of increasing protein intake. However, the minimum recommendations are general and do not necessarily take into consideration how individual dietary habits can influence and fine-tune the anabolic responses to meals. In addition to supplementation strategies aimed at increasing protein intake up to or above the recommendations, altering protein sources or supplementing meals with specific amino acids, such as leucine, changes the overall amino acid quality and possibly the anabolic efficiency of the meals (Luiking et al., 2014). Such approaches can be implemented without increasing overall protein intake. Studies indicate that whey protein has a greater acute anabolic effect on muscle protein synthesis than isocaloric ingestion of casein or soy, possibly by facilitating improved leucine availability due to higher leucine content and faster digestibility (Devries and Phillips, 2015; Gryson et al., 2013; Pennings et al., 2011; Phillips et al., 2009; Yang et al., 2012). Although rates of protein digestion and absorption do not tend to change with age (Koopman et al., 2009b), fast-digesting proteins or protein hydrolysates are beneficial for improved absorption and utilization of amino acids and muscle protein synthesis in older adults (Gryson et al., 2013; Koopman et al., 2009a). Supplementation of dietary protein with essential amino acids may increase the acute anabolic effects of meals by directly stimulating muscle protein synthesis without significantly altering dietary volume. Among the essential amino acids, leucine supplementation is often considered the gold standard, and this amino acid has shown to elicit particularly strong anabolic effects with both acute (Katsanos et al., 2006; Rieu et al., 2006) and chronic benefits (Casperson et al., 2012) on muscle protein synthesis in older adults. While the anabolic effects of leucine remains dose-dependent, it becomes less efficient with age (Crozier et al., 2005).

The distribution of amino acid delivery across daily intake is an important determinant of the efficiency at which protein balance is maintained. In general, higher acute doses of amino acids yield more robust anabolic responses in older individuals (Moore et al., 2014; Pennings et al., 2012). Ingestion of 40 g amino acids divided in small boluses and consumed across a 3-hour period results in blunted anabolic responses in older adults compared to younger adults (Volpi et al., 2000). However, the acute ingestion of 30–35 g protein induces skeletal muscle protein synthesis similarly in older and younger adults (Koopman et al., 2009b; Symons et al., 2009b). Similarly, acute ingestion of 6.7 g of essential amino acids containing 26% leucine stimulates protein synthesis in younger adults but not in older adults. However, no age-related impairment is observed when the concentration of leucine in the same quantity of essential amino acids is increased to 41% (Katsanos et al., 2006). There are both minimum and maximum threshold effects on protein synthesis in response to acute amino acids nutrition (Dardevet et al., 2013; Mitchell et al., 2015). As mentioned, providing adequate amino acid intake acutely is important to ensure positive anabolic stimulation is reached and this may be especially important in populations at risk for muscle loss. Pulse-feeding, a strategy where a significant amount of daily amino acids are provided at a single meal (or bolus) has shown to benefit amino acid availability and lean body mass gain in healthy older adults (Arnal et al., 1999) as well as older undernourished at-risk hospital patients (Bouillanne et al., 2012, 2013). However, in addition to a minimum amino acid concentration threshold that must be met to induce robust protein synthetic responses sufficient to overcome basal rates of protein degradation, there is a maximum threshold beyond which further addition of amino acids no longer increases protein synthesis (Symons et al., 2009b). For instance, there is no further dose response in muscle protein synthesis in older or younger adults when acute intake is increased above ~30 g protein (~10 g essential amino acids), suggesting that at this dose, protein synthesis is induced to the same maximum regardless

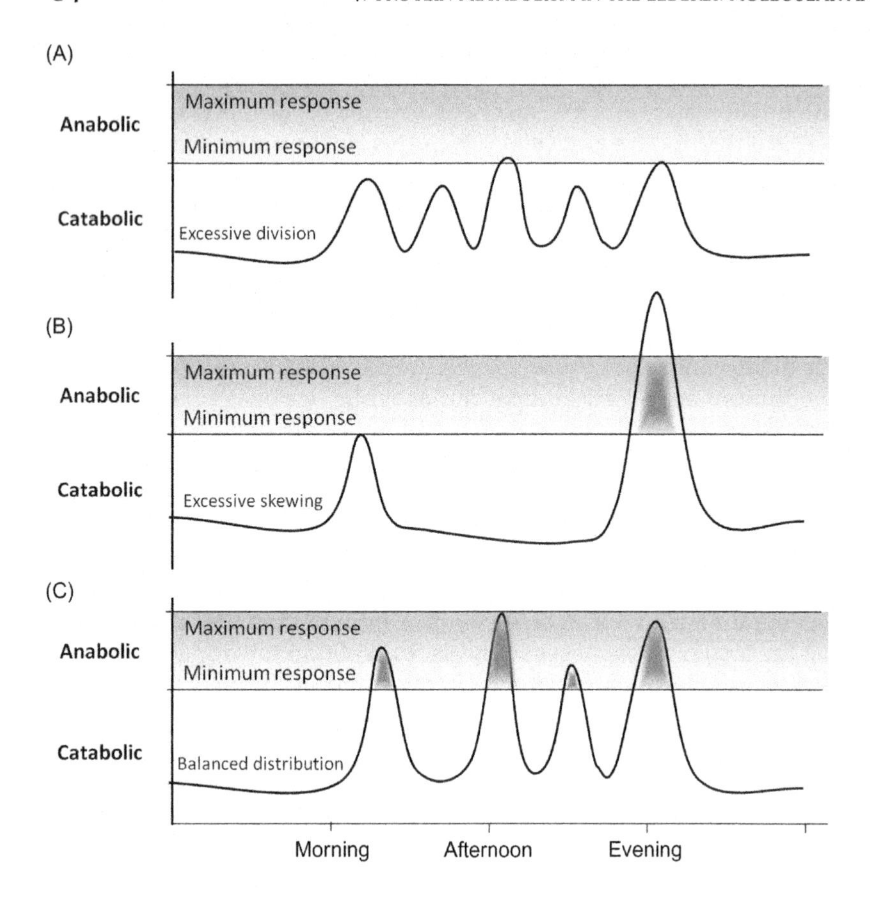

FIGURE 7.2 Amino acid distribution to optimize net protein balance. The distribution of daily amino acid intake across the day can influence protein balance. (A) Excessive division of dietary amino acids over multiple meals across the day can lead to suboptimal acute anabolic periods if these fail to reach the threshold. (B) Excessive skewing of dietary amino acid intake toward few meals can lead to suboptimal acute anabolic periods if the maximum threshold is exceeded at some meals yet the minimum threshold is missed at other meals. (C) Balanced distribution of dietary intake can contribute to multiple optimal acute anabolic periods.

of age (Symons et al., 2009b). Therefore, an optimum target concentration of amino acids exists at each meal that most efficiently stimulates protein anabolism. Careful allocation of dietary amino acids across daily meals may be beneficial and can ensure that maximum numbers of anabolic stimuli are reached over the day (Mamerow et al., 2014; Fig. 7.2). In other words, ingesting the entire daily recommendation for protein in one meal (while technically adhering to the daily dietary recommendations) may provide a single anabolic response of equal potency and duration that could be attained multiple times per day with smaller acute loads. Similarly, excessive division of the daily recommendation into small loads that fail to induce acute anabolic periods of positive net balance may also result in less than efficient utilization for protein synthesis, and possibly net protein loss across the day.

7.4. REGULATION OF mTOR SIGNALING IN AGING

The mammalian (or mechanistic) target of rapamycin (mTOR) has emerged as the major regulatory pathway of protein synthesis. This pathway provides an anabolic response mechanism following nutrient ingestion regardless of age, although its sensitivity and efficiency declines with advancing age (Cuthbertson et al., 2005; Dardevet et al., 2000; O'Connor et al., 2003; Suryawan et al., 2008). While details of the age-related dysregulation of this pathway are still elusive, several potential mechanisms may contribute to this loss in efficiency. A number of possible amino acid sensors have been suggested that may be dysregulated with age and could be involved in the decline in nutrient sensitivity, including amino acid transporters (Rebsamen et al., 2015), vacuolar ATPase (Zoncu et al., 2011), Leucyl-tRNA synthetase (LRS) (Han et al., 2012), p62 (Duran et al., 2011), MAP4K3 (Findlay et al., 2007), and Vps34 (Byfield et al., 2005; Gran and Cameron-Smith, 2011; Nobukuni et al., 2005).

Amino acids predominantly induce protein synthesis through the mTOR Complex 1 (mTORC1) pathway (Dodd and Tee, 2012), although mTOR-independent mechanisms exist (Coeffier et al., 2013; Haegens et al., 2012). The regulation of protein synthesis through mTORC1 is tightly regulated by nutrient availability, especially by leucine in adults (Atherton et al., 2009) but also by other amino acids (Gonzalez et al., 2011; Wang et al., 2012; Xi et al., 2011; Yao et al., 2008) during earlier mammalian development such as the conditionally essential amino acids glutamine and arginine in infants (Wu, 1998; Wu et al., 2004). Both acute energy status as well as amino acid availability affects protein synthesis through separate regulatory actions on this complex. Activation of mTORC1 requires the interaction between mTOR and several regulatory proteins, including Rheb, Rag, raptor, GβL, and LRS, leading to downstream signaling through S6K1 and 4E-BP1 to initiate protein translation (Han et al., 2012; Kimball and Jefferson, 2006; Roccio et al., 2006). In the presence of insulin, amino acids-such as leucine-induce phosphorylation of 4E-BP1, dissociation of 4E-BP1-eIF4E, phosphorylation of eIF4G, association of eIF4G-eIF4E, activation (dephosphorylation) of eEF2, phosphorylation of the p70 S6 kinase (S6K1), and promotion of translation initiation and elongation (Anthony et al., 2002; Balage et al., 2001; Browne and Proud, 2004; Campbell et al., 1999; Hafen, 2004; Nagasawa et al., 2002; Proud, 2004a,b). Amino acid-induced phosphorylation of S6K1 is blunted in older adults when compared to responses in young adults and this is associated with a diminished protein synthetic response (Guillet et al., 2004).

Activation of mTOR by amino acids requires release of inhibition. Amino acid-induced stimulation of the mTOR pathway and translation initiation is permitted upon inhibition of TSC2 through the insulin/PI3K/Akt signaling pathway (Balage et al., 2001; Kimball and Jefferson, 2006). TSC2 in complex with TSC1, functions as a GTPase activating protein and results in reduction of Rheb-bound GTP to GDP and the dissociation of Rheb from the mTOR complex. Rheb (a Ras-like GTPase) and several Rag GTPase activators provide both positive and negative regulation on mTORC1 (Avruch et al., 2009; Groenewoud and Zwartkruis, 2013; Jewell et al., 2015; Martin et al., 2014; Tyagi et al., 2015; Yoon and Chen, 2013). The interaction of Rheb with the mTOR complex requires the presence of amino acids, and low amino acid availability blocks GTP loading of Rheb (Long et al., 2005; Roccio et al., 2006). Upregulation of Rheb expression following resistive exercise and essential amino acid feeding is blunted in older adults and dysregulation in Rheb signaling may contribute to age-related impairments in amino acid induced signaling through mTOR (Drummond et al., 2009).

The inhibition of mTOR through TSC2 is an important mechanism that promotes energy conservation and autophagy under conditions of low energy availability. Under nutrient depleted conditions when ATP, insulin, and amino acid availability is low, full stimulation of the mTOR pathway is prevented. Under these circumstances, AMP activated kinase (AMPK) activity is promoted, leading to energy conservation and ATP production by glycolysis through the TCA cycle. AMPK plays an important role in stress responses and protection during energy deprivation. Nitrositive stress downregulates mTORC1 through the AMPK/TSC2 signaling pathway and promotes autophagy-mediated cell death (Tripathi et al., 2013). Consequently, AMPK and other stress response factors, such as REDD1, phosphorylate and activate TSC2 leading to mTOR inhibition and reduced cell growth (Inoki et al., 2003; Sofer et al., 2005). Following food intake, leading to increased ATP production, insulin/Akt signaling, and amino acid availability, the inhibitory effect of AMPK is released and activation of mTORC1 is favored. These signaling events result in a transient increase in protein anabolism before synthesis rates return to basal levels (Mitchell et al., 2015). Regulation of this "muscle full" effect is not well understood but may involve insulin-mediated accumulation of REDD1 (Frost et al., 2009; Regazzetti et al., 2010), increased degradation of newly synthesized peptides (Atherton et al., 2010), and/or other yet unknown mechanisms. Aging is associated with decreased AMPK activity and this may contribute to a dysregulation of energy metabolism and a decline in resilience against cell stress and regulation of autophagy (Salminen and Kaarniranta, 2011).

Insulin and anabolic sensitivity decrease with aging. The mTOR signaling pathway requires the presence of insulin for its full activation by amino acids (Dennis et al., 2011; Prod'homme et al., 2005). The involvement of insulin/Akt signaling in mTOR activation is mainly permissive and is required to release inhibition of the mTORC1 complex, allowing for amino acid-induced activation of mTOR. While insulin sensitivity and anabolic sensitivity to nutrients declines with age, basal levels of mTOR and S6K1 may be hyperphosphorylated with increased age although rates of protein synthesis are similar to those in younger adults (Markofski et al., 2015). It is unclear whether this is a compensatory mechanism to maintain basal rates of protein synthesis similar to those in young and whether this further contributes to anabolic resistance. Age-associated declines in insulin sensitivity thus not only affects glucose homeostasis and energy metabolism, but directly affects amino acid metabolism. While the addition of glucose to the ingestion of amino acids results in an increased response in insulin release and muscle protein synthesis among young adults, this additive anabolic effect is much smaller in older muscle, despite otherwise similar response in insulin-mediated glucose disposal (Volpi et al., 2000). Insulin is a strong anabolic factor and regulator of mTOR mediated muscle protein metabolism (Prod'homme et al., 2004) and a robust insulin response during feeding is especially important when dietary leucine content is below the upper threshold necessary to reach the maximum anabolic effect in older adults (Chevalier et al., 2011; Katsanos et al., 2008). The decrease in nutrient sensing with aging affects acute nutritional requirements, as higher amino acid concentrations are necessary for full activation of mTOR signaling, and comparatively lower rates of protein synthesis occur under suboptimum amino acid concentrations in older individuals when compared to younger adults.

7.5. THE ROLE OF PHYSICAL ACTIVITY DURING AGING

Exercise induces strong anabolic responses in skeletal muscle, resulting in increased muscle protein synthesis. Habitual levels of activity tend to decline with aging and further contribute to the decline in physical health and function. Active lifestyles have shown to bring significant health benefits during aging, including maintenance of physical function, body composition, aerobic capacity, endocrine and metabolic functions, cognitive function, and reduced AGE accumulation and mechanical stress (Bergouignan et al., 2011; Churchward-Venne et al., 2015; Couppe et al., 2014; Hayes et al., 2013; Straight et al., 2015; Tsai et al., 2015; Wroblewski et al., 2011).

The molecular response to exercise is blunted in older adults. Exercise induces two key anabolic pathways regulating protein synthesis, the mTOR pathway and the MAPK (ERK1/2) pathway (Moore et al., 2010), which are upregulated for up to 24 hours following a single bout of exercise in younger adults (Fry et al., 2011). Aging blunts the immediate phosphorylating events required for the activation of both of these pathways, and the effects are of much shorter duration (up to 6 hours) than those in young, resulting in a much smaller anabolic response in older individuals (Fry et al., 2011). In the fed state, or following an amino acid load, exercise induces immediate anabolic events resulting in increased protein synthesis through the mTOR pathway in young and old (Drummond et al., 2008). However, similar to exercise in the fasted state, the time course of muscle protein synthesis is altered with aging (ie, delayed onset and shorter duration of muscle protein synthesis). Despite blunted nutrient sensitivity in older skeletal muscle, the insulin and amino acid mediated mTOR pathway remains to be the principal anabolic pathway (the MAPK/ERK pathway being the other pathway) during and following exercise in both young and old. While MAPK signaling is activated in response to exercise and amino acids in younger muscle, this pathway is not as strongly induced in older muscle. The combined blunting of the mTOR and MAPK pathways result in shorter periods of positive net anabolic balance following bouts of exercise in the fed state in older muscle (Drummond et al., 2008). In addition to exercise studies, bed rest studies have been instrumental in elucidating the importance of physical activity on health and function and the physiological changes that take place share some striking resemblances to the aging process. Both short-term (ie, several days; Coker et al., 2014; Drummond et al., 2012; Kortebein et al., 2007, 2008; Mulder et al., 2014; Rittweger et al., 2015) and long-term (ie, weeks to months; Haider et al., 2014; Hoff et al., 2014; Koryak, 2014; Paddon-Jones et al., 2004a, 2006; Symons et al., 2009a) bed rest studies have been conducted showing the detrimental effects of inactivity and unloading on systemic physiology including the musculoskeletal, neuromuscular, immune, and cardiovascular systems. Older adults are especially vulnerable to muscle loss during prolonged inactivity (Drummond et al., 2012; Kortebein et al., 2007, 2008) and even short-term bed rest induces cytokine-mediated responses, possibly aggravating inflammatory and catabolic signaling processes contributing to muscle atrophy (Drummond et al., 2013). Furthermore, age-related impairments contribute to reduced recovery of skeletal muscle mass and strength between periods of immobilization, and may accelerate the downward spiral of sarcopenia (English and Paddon-Jones, 2009; Magne et al., 2013).

Physical activity induces blood flow responses leading to increased amino acid delivery (Dreyer et al., 2010; Goodman et al., 2011). Macrovascular blood flow and the recruitment of capillary beds within the muscle further facilitates the delivery and utilization of amino acids for muscle protein synthesis (Biolo et al., 1995; Vincent et al., 2006). While insulin-mediated blood flow and muscle protein synthesis are impaired with age (Fujita et al., 2009; Meneilly et al., 1995), aging does not necessarily reduce the anabolic response to increased activity, and muscle protein turnover increases acutely in older and younger adults during moderate-intensity exercise (Sheffield-Moore et al., 2004). Aerobic exercise acutely improves endothelial function, insulin-induced vasodilation, and protein anabolism, through reduction of the vasoconstrictor endothelin-1 and increased Akt/mTOR mediated skeletal muscle protein synthesis in older subjects (Fujita et al., 2007). However, although acute anabolic responses can be maximized in older adults to levels similar to younger adults, older skeletal muscle still has a reduced efficiency in anabolic responses to amino acids and their utilization following exercise (Durham et al., 2010; Kullman et al., 2013).

Regulation of nitric oxide (NO)-mediated vasodilation declines with age. Skeletal muscle of older adults remains responsive to both the direct anabolic effects on protein synthesis in response to muscle contractions as well as the effects that are brought on by the cardiovascular responses, such as increased systemic blood flow, microvascular blood flow, and muscle perfusion. Maintenance of insulin sensitivity is intricately related to the benefits of exercise, as exercise has rapid effects on improving insulin sensitivity and insulin signaling plays an important role in both energy metabolism during exercise, as well as regulation of nitric oxide (NO)-mediated vasodilation (Fujita et al., 2007; Manrique et al., 2014; Timmerman et al., 2012). The regulation of blood flow is important during stimulation of skeletal muscle anabolism, however, upregulation of blood flow above the rate which is attained during normal feeding responses does not further induce protein anabolism in young adults (Phillips et al., 2013). Older adults retain the physiological response to NO and skeletal muscle protein synthesis increases to the same extent as that in muscle in younger adults (Dillon et al., 2011; Timmerman et al., 2010). However, the anabolic effects of exercise are blunted in older adults compared to young even under conditions where amino acid availability is not a limiting factor (Durham et al., 2010), and it is unknown whether this is in part due to age-related changes that preclude NO signaling to be fully activated in older muscle during exercise.

7.6. AGING AND CHANGES IN ENDOCRINE FUNCTION

Hormone production and regulation changes throughout life, and aging has been associated with dysregulation of the hypothalamic–pituitary–adrenal (HPA) axis (Aguilera, 2010; Born et al., 1995; Veldhuis, 2013). Increased HPA stress and alterations in sex hormone balances are associated with declines in physical and cognitive function in frail older adults (Breuer et al., 2001). Increased physical dependence in frail older males is associated with a decline in production of gonadal hormones. Androgens are produced by the adrenal gland in concentrations that are an order of magnitude lower than produced by the gonads in males. In frail older females there appears to be no association between physical dependence and androgen concentrations (Breuer et al., 2001). In contrast to testosterone, concentrations of the adrenal-derived sex hormone estrogen are associated with increased loss of dependence in frail older women. However, this association may be more closely related to systemic stress responses in frail adults and the regulation of adrenal hormones in general. Cortisol, one of the major stress-response hormones produced by the adrenal gland in males and females, is a potent catabolic hormone to skeletal muscle protein. Thus, while the mechanisms for the contrasting influences of stress and sex hormones on physical and cognitive function in males and females is not yet clear, glucocorticoids and testosterone appear to have potent, yet opposite, concentration-dependent regulatory effects on skeletal muscle metabolism (Urban et al., 2014b).

Androgens promote protein synthesis and reduce protein degradation. Androgens such as testosterone promote the accretion of muscle mass (Bhasin, 2003; Bhasin et al., 1997, 2006a,b, 2001; Huang et al., 2013; Sheffield-Moore, 2000; Sheffield-Moore et al., 1999, 2000, 2006, 2011; Sinha-Hikim et al., 2006; Sinha et al., 2014; Smith et al., 2013). Testosterone regulates a number of anabolic and catabolic signaling factors including insulin-like growth factor–1 (IGF-1), forkhead box O (FOXO1), peroxisome proliferator-activated receptor-gamma coactivator 1 alpha (PGC-1α), and p38 MAPK (Qin et al., 2010; Sculthorpe et al., 2011). Testosterone stimulates hypertrophy via the androgen receptor in myonuclei and satellite cells (Kadi, 2008). Both testosterone and IGF-1 are myogenic, and testosterone activates MAPK/ERK and mTOR signaling cascades (Wu et al., 2010). In addition, testosterone suppresses skeletal muscle catabolism (Ferrando et al., 2002; Sheffield-Moore et al., 2011; Urban et al., 2014a) and its metabolic precursor, dehydroepiandrosterone (DHEA), has anti-inflammatory properties in peripheral tissues (Iwasaki et al., 2004). However, a clear understanding behind the mechanism through which androgens regulate skeletal muscle protein catabolism is still elusive but may involve the repression of atrogin-1, muscle RING-finger protein-1 (MuRF-1) (Pires-Oliveira et al., 2009), and Nf-kB inducing kinase (Urban et al., 2014a). Thus, changes in endocrine regulation, including decreased sex hormone production, contributes toward loss of protection against catabolic signals and increased susceptibility for muscle wasting.

7.7. MOLECULAR DYSREGULATION OF PROTEIN METABOLISM DURING AGING

At the cellular and molecular level, changes in protein and amino acid metabolism during aging are the result of—and contribute to—a number of underlying processes including increased telomere shortening, increased oxidative damage to DNA and proteins, and accumulation of posttranslational modified proteins (Bode-Boger et al., 2005; Li et al., 2015; Mason and Perdigones, 2013; Salahuddin et al., 2014; Winter et al., 2013; Wood et al., 2014; Fig. 7.3). It is not well understood how the efficiency of the basic mechanisms driving even the most ubiquitous cellular processes are precisely affected during aging, but this process appears to involve an increase in posttranslational modifications (ie, methylation, acetylation, carbonylation, nitration, glycosylation, glycation, etc.) of proteins involved in structural, mechanical, and metabolic functions. The molecular changes that occur during aging alter how efficiently these pathways function in the production of new protein in response to anabolic

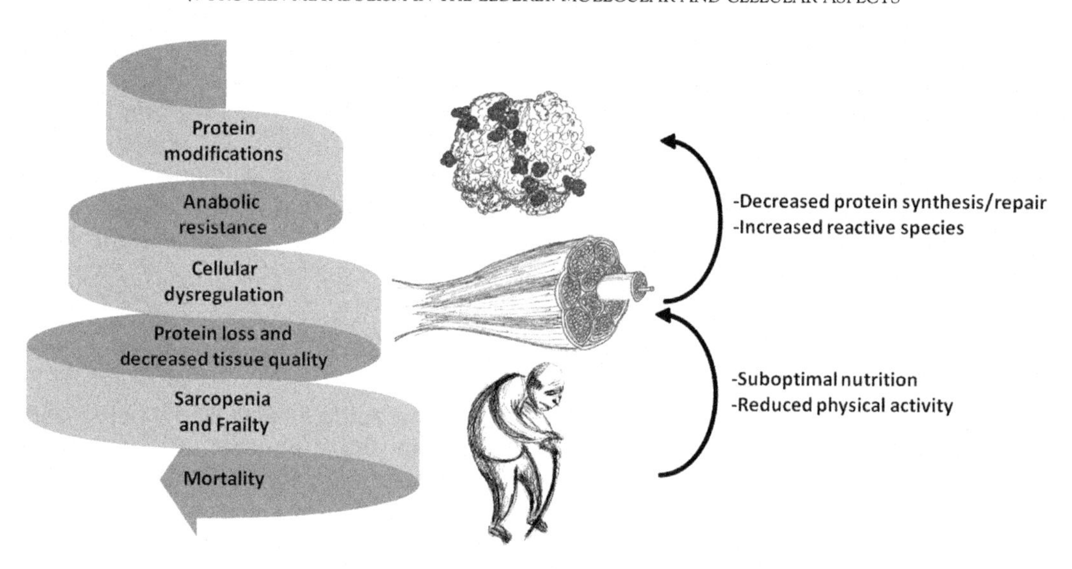

FIGURE 7.3 **Downward spiral of protein metabolism in aging.** The accumulation of altered proteins contributes to the development of anabolic resistance and cellular dysfunction. Cellular dysfunction leads to loss in tissue quality, and to the development of sarcopenia, frailty, and ultimately death. Loss of physical function, suboptimal nutrition, and metabolic dysfunction with aging aggravates the aging process and contributes further to the accumulation of reactive species and protein damage.

stimuli. The process of aging appears to be tipped toward a gradual vicious cycle and a slow but steady accumulation of proteins with altered functional efficiencies as even the housekeeping genes and proteins involved in removal and degradation of modified proteins are increasingly affected by such modifications (Bailey et al., 2011; Mazzola and Sirover, 2001; Uchiki et al., 2011). Oxidative damage to proteins results in posttranslational modifications and possibly altered efficiency or function. These nonenzymatic protein modifications can be either reversible or irreversible depending on the type of adducts that are formed. Damaged proteins are continuously removed through autophagosomal (Rubinsztein et al., 2011) and proteosomal processes (Pickering and Davies, 2012; Korolchuk et al., 2010). However, the efficiency of these processes may be affected with aging and it is thought that a gradual accumulation of damaged proteins contributes to many age- and disease-related physiological changes. Various types of posttranslational modifications are known, including nonenzymatic and irreversible processes, such as protein carbonylation due to reactive oxygen and nitrogen species (ROS and RNS, collectively addressed as RS) and formation of advanced glycation end-products (AGE). The accumulation of protein modified by RS and AGE through nonenzymatic processes are of particular interest in the development of age-related alterations in protein metabolism and cellular signaling (Ott et al., 2014).

Reactive Species (ROS/RNS) contributed to protein damage and mitochondrial dysfunction. Oxidative stress is an unavoidable and ubiquitous challenge for, as well as regulator of, cellular processes (Ji, 2007). While redundant protective mechanisms are in place to repair and remove damaged proteins, aging is associated with a gradual shift toward increased oxidative damage and loss of antioxidant protection. Along with reductions in physical activity, aging is associated with decreased mitochondrial function (Johnson et al., 2013), and dysregulation of mitochondrial function is an important contributing factor to increased formation of RS and loss of protection against oxidative stress with aging (Hepple, 2014).

Advanced glycation end-products (AGE) are posttranslational modifications that occur through nonenzymatic (Maillard) reactions between reducing sugars and proteins, resulting in irreversible additions to proteins. Circulating levels of glucose directly contribute to the increased appearance of such modifications and measures of glycated hemoglobin (hemoglobin A1c) is a commonly used biomarker for long-term elevations in circulating blood glucose (Lyons and Basu, 2012). Elevations in glycated hemoglobin and glycated albumin are observed in several pathologies, including diabetes, nephropathy, retinopathy, and cardiovascular disease (Nathan et al., 2013). Protein modifications may alter the functional efficiency of proteins and often add to protein stability, reducing the rate of degradation by the proteasome and thus further promoting the accumulation of modified proteins (Nowotny et al., 2014). There are many overlaps between the AGE-induced metabolic changes that occur during aging and those in age-associated pathologies such as diabetes (increased blood glucose), muscle stiffness (connective tissue stiffening), vision changes (modified retinal proteins), cardiovascular disease (arterial stiffening), neurodegenerative diseases such as Parkinson's, and Alzheimer's (amyloid formation) (Salahuddin et al.,

2014; Wood et al., 2014). Despite the commonalities between these pathologies and healthy aging, the direct role of AGE in the onset or progression of any of these etiologies is not completely understood at the present time. Enzymatic processes resulting in the formation of collagen cross-links in skeletal muscle do not appear to change with age. However, the nonenzymatic formation of AGE and collagen cross-links increases in connective tissue of muscle from healthy sedentary elderly individuals (Ott et al., 2014; Wood et al., 2014). The accumulation of cross-linked collagen and extracellular matrix hydroxyproline content in muscle contributes to increased muscle stiffness with aging (Wood et al., 2014). The currently most effective strategies to reduce AGE accumulation with age are preventative and include adherence to lifestyles that promote regular exercise and limiting dietary intake of foods containing high AGE content (Seals, 2014; Van Puyvelde et al., 2014; Yoshikawa et al., 2009).

It remains unclear how increased abundance of protein posttranslational modifications with aging directly contributes to changes in anabolic sensitivity to nutrients. While evidence is still needed to show the direct links between specific protein modifications and altered protein metabolism during the human aging process, several of these modifications have been demonstrated as capable of affecting proteins related to mTOR pathway signaling. AGE binding and signaling through its receptor (RAGE) results in the inhibition of the PI3K/Akt/mTOR pathway and induces autophagy (Hou et al., 2014). Oxidative stress plays a particularly interesting role in reduction of nutrient sensitivity with aging as aminoacyl-tRNA synthetases, which are rate limiting enzymes in protein synthesis, can be susceptible to oxidative damage (Takahashi and Goto, 1990) and can be redirected away from their functions in protein synsthesis toward processes required for protection against the stress (Wei et al., 2014). Conversely, antioxidant supplementation has been shown to correct age-related declines in acute leucine activation of protein synthesis in rats, although the mechanism of action is unclear and this benefit was not associated with changes in oxidative damage to p70 S6K (Marzani et al., 2008). Despite our increasing understanding of the importance of posttranslational protein modifications and the metabolic implications during aging, much remains unknown regarding how the individual molecular alterations that occur affect the mechanisms involved in amino acid and protein metabolism.

References

Abbatecola, A.M., Paolisso, G., Fattoretti, P., Evans, W.J., Fiore, V., Dicioccio, L., et al., 2011. Discovering pathways of sarcopenia in older adults: a role for insulin resistance on mitochondria dysfunction. J. Nutr. Health Aging 15, 890–895.

Aguilera, G., 2010. HPA axis responsiveness to stress: implications for healthy aging. Exp. Gerontol. 46, 90–95.

Anthony, J.C., Lang, C.H., Crozier, S.J., Anthony, T.G., MacLean, D.A., Kimball, S.R., et al., 2002. Contribution of insulin to the translational control of protein synthesis in skeletal muscle by leucine. Am. J. Physiol. Endocrinol. Metab. 282, E1092–E1101.

Arnal, M.A., Mosoni, L., Boirie, Y., Houlier, M.L., Morin, L., Verdier, E., et al., 1999. Protein pulse feeding improves protein retention in elderly women. Am. J. Clin. Nutr. 69, 1202–1208.

Arthur, S.T., Cooley, I.D., 2012. The effect of physiological stimuli on sarcopenia; impact of Notch and Wnt signaling on impaired aged skeletal muscle repair. Int. J. Biol. Sci. 8, 731–760.

Atherton, P.J., Smith, K., Etheridge, T., Rankin, D., Rennie, M.J., 2009. Distinct anabolic signalling responses to amino acids in C2C12 skeletal muscle cells. Amino Acids 38, 1533–1539.

Atherton, P.J., Etheridge, T., Watt, P.W., Wilkinson, D., Selby, A., Rankin, D., et al., 2010. Muscle full effect after oral protein: time-dependent concordance and discordance between human muscle protein synthesis and mTORC1 signaling. Am. J. Clin. Nutr. 92, 1080–1088.

Avruch, J., Long, X., Lin, Y., Ortiz-Vega, S., Rapley, J., Papageorgiou, A., et al., 2009. Activation of mTORC1 in two steps: Rheb-GTP activation of catalytic function and increased binding of substrates to raptor. Biochem. Soc. Trans. 37, 223–226.

Bailey, C.E., Hammers, D.W., Deford, J.H., Dimayuga, V.L., Amaning, J.K., Farrar, R., et al., 2011. Ishemia-reperfusion enhances GAPDH nitration in aging skeletal muscle. Aging (Albany NY) 3, 1003–1017.

Balage, M., Dardevet, D., 2010. Long-term effects of leucine supplementation on body composition. Curr. Opin. Clin. Nutr. Metab. Care 13, 265–270.

Balage, M., Sinaud, S., Prod'homme, M., Dardevet, D., Vary, T.C., Kimball, S.R., et al., 2001. Amino acids and insulin are both required to regulate assembly of the eIF4E. eIF4G complex in rat skeletal muscle. Am. J. Physiol. Endocrinol. Metab. 281, E565–E574.

Bauer, J., Biolo, G., Cederholm, T., Cesari, M., Cruz-Jentoft, A.J., Morley, J.E., et al., 2013. Evidence-based recommendations for optimal dietary protein intake in older people: a position paper from the PROT-AGE Study Group. J. Am. Med. Dir. Assoc. 14, 542–559.

Baumgartner, R.N., Koehler, K.M., Gallagher, D., Romero, L., Heymsfield, S.B., Ross, R.R., et al., 1998. Epidemiology of sarcopenia among the elderly in New Mexico. Am. J. Epidemiol. 147, 755–763.

Bergouignan, A., Rudwill, F., Simon, C., Blanc, S., 2011. Physical inactivity as the culprit of metabolic inflexibility: evidence from bed-rest studies. J. Appl. Physiol. (1985) 111, 1201–1210.

Bhasin, S., 2003. Testosterone supplementation for aging-associated sarcopenia. J. Gerontol. A Biol. Sci. Med. Sci. 58, 1002–1008.

Bhasin, S., Storer, T.W., Berman, N., Yarasheski, K.E., Clevenger, B., Phillips, J., et al., 1997. Testosterone replacement increases fat-free mass and muscle size in hypogonadal men. J. Clin. Endocrinol. Metab. 82, 407–413.

Bhasin, S., Woodhouse, L., Storer, T.W., 2001. Proof of the effect of testosterone on skeletal muscle. J. Endocrinol. 170, 27–38.

Bhasin, S., Calof, O.M., Storer, T.W., Lee, M.L., Mazer, N.A., Jasuja, R., et al., 2006a. Drug insight: testosterone and selective androgen receptor modulators as anabolic therapies for chronic illness and aging. Nat. Clin. Pract. Endocrinol. Metab. 2, 146–159.

Bhasin, S., Cunningham, G.R., Hayes, F.J., Matsumoto, A.M., Snyder, P.J., Swerdloff, R.S., et al., 2006b. Testosterone therapy in adult men with androgen deficiency syndromes: an endocrine society clinical practice guideline. J. Clin. Endocrinol. Metab. 91, 1995–2010.

Biolo, G., Maggi, S.P., Williams, B.D., Tipton, K.D., Wolfe, R.R., 1995. Increased rates of muscle protein turnover and amino acid transport after resistance exercise in humans. Am. J. Physiol. 268, E514–E520.

Bode-Boger, S.M., Scalera, F., Martens-Lobenhoffer, J., 2005. Asymmetric dimethylarginine (ADMA) accelerates cell senescence. Vasc. Med. 10 (Suppl. 1), S65–S71.

Boirie, Y., Gachon, P., Beaufrere, B., 1997. Splanchnic and whole-body leucine kinetics in young and elderly men. Am. J. Clin. Nutr. 65, 489–495.

Born, J., Ditschuneit, I., Schreiber, M., Dodt, C., Fehm, H.L., 1995. Effects of age and gender on pituitary-adrenocortical responsiveness in humans. Eur. J. Endocrinol. 132, 705–711.

Borsheim, E., Bui, Q.U., Tissier, S., Kobayashi, H., Ferrando, A.A., Wolfe, R.R., 2008. Effect of amino acid supplementation on muscle mass, strength and physical function in elderly. Clin. Nutr. 27, 189–195.

Bouillanne, O., Curis, E., Hamon-Vilcot, B., Nicolis, I., Chretien, P., Schauer, N., et al., 2012. Impact of protein pulse feeding on lean mass in malnourished and at-risk hospitalized elderly patients: a randomized controlled trial. Clin. Nutr. 32, 186–192.

Bouillanne, O., Neveux, N., Nicolis, I., Curis, E., Cynober, L., Aussel, C., 2013. Long-lasting improved amino acid bioavailability associated with protein pulse feeding in hospitalized elderly patients: a randomized controlled trial. Nutrition 30, 544–550.

Brack, A.S., Conboy, M.J., Roy, S., Lee, M., Kuo, C.J., Keller, C., et al., 2007. Increased Wnt signaling during aging alters muscle stem cell fate and increases fibrosis. Science 317, 807–810.

Breen, L., Phillips, S.M., 2011. Skeletal muscle protein metabolism in the elderly: interventions to counteract the 'anabolic resistance' of ageing. Nutr. Metab. (Lond) 8, 68.

Breuer, B., Trungold, S., Martucci, C., Wallenstein, S., Likourezos, A., Libow, L.S., et al., 2001. Relationships of sex hormone levels to dependence in activities of daily living in the frail elderly. Maturitas 39, 147–159.

Browne, G.J., Proud, C.G., 2004. A novel mTOR-regulated phosphorylation site in elongation factor 2 kinase modulates the activity of the kinase and its binding to calmodulin. Mol. Cell. Biol. 24, 2986–2997.

Buas, M.F., Kabak, S., Kadesch, T., 2009. The Notch effector Hey1 associates with myogenic target genes to repress myogenesis. J. Biol. Chem. 285, 1249–1258.

Bucala, R., Tracey, K.J., Cerami, A., 1991. Advanced glycosylation products quench nitric oxide and mediate defective endothelium-dependent vasodilatation in experimental diabetes. J. Clin. Invest. 87, 432–438.

Byfield, M.P., Murray, J.T., Backer, J.M., 2005. hVps34 is a nutrient-regulated lipid kinase required for activation of p70 S6 kinase. J. Biol. Chem. 280, 33076–33082.

Campbell, L.E., Wang, X., Proud, C.G., 1999. Nutrients differentially regulate multiple translation factors and their control by insulin. Biochem. J. 344 (Pt 2), 433–441.

Cardinale, A., de Stefano, M.C., Mollinari, C., Racaniello, M., Garaci, E., Merlo, D., 2014. Biochemical characterization of sirtuin 6 in the brain and its involvement in oxidative stress response. Neurochem. Res. 40, 59–69.

Casperson, S.L., Sheffield-Moore, M., Hewlings, S.J., Paddon-Jones, D., 2012. Leucine supplementation chronically improves muscle protein synthesis in older adults consuming the RDA for protein. Clin. Nutr. 31, 512–519.

Chen, L.K., Liu, L.K., Woo, J., Assantachai, P., Auyeung, T.W., Bahyah, K.S., et al., 2014. Sarcopenia in Asia: consensus report of the Asian Working Group for Sarcopenia. J. Am. Med. Dir. Assoc. 15, 95–101.

Chevalier, S., Goulet, E.D., Burgos, S.A., Wykes, L.J., Morais, J.A., 2011. Protein anabolic responses to a fed steady state in healthy aging. J. Gerontol. A Biol. Sci. Med. Sci. 66, 681–688.

Churchward-Venne, T.A., Tieland, M., Verdijk, L.B., Leenders, M., Dirks, M.L., de Groot, L.C., et al., 2015. There are no nonresponders to resistance-type exercise training in older men and women. J. Am. Med. Dir. Assoc. 16 (5), 400–411.

Coeffier, M., Claeyssens, S., Bole-Feysot, C., Guerin, C., Maurer, B., Lecleire, S., et al., 2013. Enteral delivery of proteins stimulates protein synthesis in human duodenal mucosa in the fed state through a mammalian target of rapamycin-independent pathway. Am. J. Clin. Nutr. 97, 286–294.

Coker, R.H., Hays, N.P., Williams, R.H., Wolfe, R.R., Evans, W.J., 2014. Bed rest promotes reductions in walking speed, functional parameters, and aerobic fitness in older, healthy adults. J. Gerontol. A Biol. Sci. Med. Sci. 70, 91–96.

Conboy, I.M., Rando, T.A., 2002. The regulation of Notch signaling controls satellite cell activation and cell fate determination in postnatal myogenesis. Dev. Cell 3, 397–409.

Cossu, G., Borello, U., 1999. Wnt signaling and the activation of myogenesis in mammals. EMBO J. 18, 6867–6872.

Couppe, C., Svensson, R.B., Grosset, J.F., Kovanen, V., Nielsen, R.H., Olsen, M.R., et al., 2014. Life-long endurance running is associated with reduced glycation and mechanical stress in connective tissue. Age (Dordr) 36, 9665.

Crozier, S.J., Kimball, S.R., Emmert, S.W., Anthony, J.C., Jefferson, L.S., 2005. Oral leucine administration stimulates protein synthesis in rat skeletal muscle. J. Nutr. 135, 376–382.

Cruz-Jentoft, A.J., Baeyens, J.P., Bauer, J.M., Boirie, Y., Cederholm, T., Landi, F., et al., 2010. Sarcopenia: European consensus on definition and diagnosis: report of the European working group on sarcopenia in older people. Age Ageing 39, 412–423.

Cuthbertson, D., Smith, K., Babraj, J., Leese, G., Waddell, T., Atherton, P., et al., 2005. Anabolic signaling deficits underlie amino acid resistance of wasting, aging muscle. FASEB J. 19, 422–424.

Dam, T.T., Peters, K.W., Fragala, M., Cawthon, P.M., Harris, T.B., McLean, R., et al., 2014. An evidence-based comparison of operational criteria for the presence of sarcopenia. J. Gerontol. A Biol. Sci. Med. Sci. 69, 584–590.

Dardevet, D., Sornet, C., Balage, M., Grizard, J., 2000. Stimulation of in vitro rat muscle protein synthesis by leucine decreases with age. J. Nutr. 130, 2630–2635.

Dardevet, D., Remond, D., Peyron, M.A., Papet, I., Savary-Auzeloux, I., Mosoni, L., 2013. Muscle wasting and resistance of muscle anabolism: the "anabolic threshold concept" for adapted nutritional strategies during sarcopenia. ScientificWorldJournal 2012, 269531.

Dennis, M.D., Baum, J.I., Kimball, S.R., Jefferson, L.S., 2011. Mechanisms involved in the coordinate regulation of mTORC1 by insulin and amino acids. J. Biol. Chem. 286, 8287–8296.

Deutz, N.E., Bauer, J.M., Barazzoni, R., Biolo, G., Boirie, Y., Bosy-Westphal, A., et al., 2014. Protein intake and exercise for optimal muscle function with aging: recommendations from the ESPEN Expert Group. Clin. Nutr. 33, 929–936.

Devries, M.C., Phillips, S.M., 2015. Supplemental protein in support of muscle mass and health: advantage whey. J. Food. Sci. 80 (Suppl. 1), A8–A15.

Dillon, E.L., Sheffield-Moore, M., Paddon-Jones, D., Gilkison, C., Sanford, A.P., Casperson, S.L., et al., 2009. Amino acid supplementation increases lean body mass, basal muscle protein synthesis, and insulin-like growth factor-I expression in older women. J. Clin. Endocrinol. Metab. 94, 1630–1637.

Dillon, E.L., Durham, W.J., Urban, R.J., Sheffield-Moore, M., 2010. Hormone treatment and muscle anabolism during aging: androgens. Clin. Nutr. 29, 697–700.

Dillon, E.L., Casperson, S.L., Durham, W.J., Randolph, K.M., Urban, R.J., Volpi, E., et al., 2011. Muscle protein metabolism responds similarly to exogenous amino acids in healthy younger and older adults during NO-induced hyperemia. Am. J. Physiol. Regul. Integr. Comp. Physiol. 301, R1408–R1417.

Dodd, K.M., Tee, A.R., 2012. Leucine and mTORC1: a complex relationship. Am. J. Physiol. Endocrinol. Metab. 302, E1329–E1342.

Dorrens, J., Rennie, M.J., 2003. Effects of ageing and human whole body and muscle protein turnover. Scand. J. Med. Sci. Sports 13, 26–33.

Dreyer, H.C., Fujita, S., Glynn, E.L., Drummond, M.J., Volpi, E., Rasmussen, B.B., 2010. Resistance exercise increases leg muscle protein synthesis and mTOR signalling independent of sex. Acta Physiol. (Oxf) 199, 71–81.

Drummond, M.J., Dreyer, H.C., Pennings, B., Fry, C.S., Dhanani, S., Dillon, E.L., et al., 2008. Skeletal muscle protein anabolic response to resistance exercise and essential amino acids is delayed with aging. J. Appl. Physiol. (1985) 104, 1452–1461.

Drummond, M.J., Miyazaki, M., Dreyer, H.C., Pennings, B., Dhanani, S., Volpi, E., et al., 2009. Expression of growth-related genes in young and older human skeletal muscle following an acute stimulation of protein synthesis. J. Appl. Physiol. 106, 1403–1411.

Drummond, M.J., Dickinson, J.M., Fry, C.S., Walker, D.K., Gundermann, D.M., Reidy, P.T., et al., 2012. Bed rest impairs skeletal muscle amino acid transporter expression, mTORC1 signaling, and protein synthesis in response to essential amino acids in older adults. Am. J. Physiol. Endocrinol. Metab. 302, E1113–E1122.

Drummond, M.J., Timmerman, K.L., Markofski, M.M., Walker, D.K., Dickinson, J.M., Jamaluddin, M., et al., 2013. Short-term bed rest increases TLR4 and IL-6 expression in skeletal muscle of older adults. Am. J. Physiol. Regul. Integr. Comp. Physiol. 305, R216–R223.

Du, X., Edelstein, D., Obici, S., Higham, N., Zou, M.H., Brownlee, M., 2006. Insulin resistance reduces arterial prostacyclin synthase and eNOS activities by increasing endothelial fatty acid oxidation. J. Clin. Invest. 116, 1071–1080.

Duran, A., Amanchy, R., Linares, J.F., Joshi, J., Abu-Baker, S., Porollo, A., et al., 2011. p62 is a key regulator of nutrient sensing in the mTORC1 pathway. Mol. Cell 44, 134–146.

Durham, W.J., Casperson, S.L., Dillon, E.L., Keske, M.A., Paddon-Jones, D., Sanford, A.P., et al., 2010. Age-related anabolic resistance after endurance-type exercise in healthy humans. FASEB J. 24, 4117–4127.

El-Kadi, S.W., Suryawan, A., Gazzaneo, M.C., Srivastava, N., Orellana, R.A., Nguyen, H.V., et al., 2012. Anabolic signaling and protein deposition are enhanced by intermittent compared with continuous feeding in skeletal muscle of neonates. Am. J. Physiol. Endocrinol. Metab. 302, E674–E686.

English, K.L., Paddon-Jones, D., 2009. Protecting muscle mass and function in older adults during bed rest. Curr. Opin. Clin. Nutr. Metab. Care 13, 34–39.

Eskelinen, J.J., Heinonen, I., Loyttyniemi, E., Saunavaara, V., Kirjavainen, A., Virtanen, K.A., et al., 2015. Muscle-specific glucose and free fatty acid uptake after sprint interval and moderate intensity training in healthy middle-aged men. J. Appl. Physiol. (1985) 118 (9), 1172–1180. Available from: http://dx.doi.org/10.1152/japplphysiol.01122.2014.

Ferrando, A.A., Sheffield-Moore, M., Yeckel, C.W., Gilkison, C., Jiang, J., Achacosa, A., et al., 2002. Testosterone administration to older men improves muscle function: molecular and physiological mechanisms. Am. J. Physiol. Endocrinol. Metab. 282, E601–E607.

Fielding, R.A., Vellas, B., Evans, W.J., Bhasin, S., Morley, J.E., Newman, A.B., et al., 2011. Sarcopenia: an undiagnosed condition in older adults. Current consensus definition: prevalence, etiology, and consequences. International working group on sarcopenia. J. Am. Med. Dir. Assoc. 12, 249–256.

Findlay, G.M., Yan, L., Procter, J., Mieulet, V., Lamb, R.F., 2007. A MAP4 kinase related to Ste20 is a nutrient-sensitive regulator of mTOR signalling. Biochem. J. 403, 13–20.

Friday, B.B., Mitchell, P.O., Kegley, K.M., Pavlath, G.K., 2003. Calcineurin initiates skeletal muscle differentiation by activating MEF2 and MyoD. Differentiation 71, 217–227.

Frost, R.A., Huber, D., Pruznak, A., Lang, C.H., 2009. Regulation of REDD1 by insulin-like growth factor-I in skeletal muscle and myotubes. J. Cell. Biochem. 108, 1192–1202.

Fry, C.S., Drummond, M.J., Glynn, E.L., Dickinson, J.M., Gundermann, D.M., Timmerman, K.L., et al., 2011. Aging impairs contraction-induced human skeletal muscle mTORC1 signaling and protein synthesis. Skelet. Muscle 1, 11.

Fujita, S., Rasmussen, B.B., Cadenas, J.G., Drummond, M.J., Glynn, E.L., Sattler, F.R., et al., 2007. Aerobic exercise overcomes the age-related insulin resistance of muscle protein metabolism by improving endothelial function and Akt/mammalian target of rapamycin signaling. Diabetes 56, 1615–1622.

Fujita, S., Glynn, E.L., Timmerman, K.L., Rasmussen, B.B., Volpi, E., 2009. Supraphysiological hyperinsulinaemia is necessary to stimulate skeletal muscle protein anabolism in older adults: evidence of a true age-related insulin resistance of muscle protein metabolism. Diabetologia 52, 1889–1898.

Garcia-Prat, L., Sousa-Victor, P., Munoz-Canoves, P., 2013. Functional dysregulation of stem cells during aging: a focus on skeletal muscle stem cells. FEBS J. 280, 4051–4062.

Gazzaneo, M.C., Suryawan, A., Orellana, R.A., Torrazza, R.M., El-Kadi, S.W., Wilson, F.A., et al., 2011. Intermittent bolus feeding has a greater stimulatory effect on protein synthesis in skeletal muscle than continuous feeding in neonatal pigs. J. Nutr. 141, 2152–2158.

Gonzalez, I.M., Martin, P.M., Burdsal, C., Sloan, J.L., Mager, S., Harris, T., et al., 2011. Leucine and arginine regulate trophoblast motility through mTOR-dependent and independent pathways in the preimplantation mouse embryo. Dev. Biol. 361, 286–300.

Goodman, C.A., Frey, J.W., Mabrey, D.M., Jacobs, B.L., Lincoln, H.C., You, J.S., et al., 2011. The role of skeletal muscle mTOR in the regulation of mechanical load-induced growth. J. Physiol. 589, 5485–5501.

Goodpaster, B.H., Park, S.W., Harris, T.B., Kritchevsky, S.B., Nevitt, M., Schwartz, A.V., et al., 2006. The loss of skeletal muscle strength, mass, and quality in older adults: the health, aging and body composition study. J. Gerontol. A Biol. Sci. Med. Sci. 61, 1059–1064.

Gran, P., Cameron-Smith, D., 2011. The actions of exogenous leucine on mTOR signalling and amino acid transporters in human myotubes. BMC Physiol. 11, 10.

Groenewoud, M.J., Zwartkruis, F.J., 2013. Rheb and Rags come together at the lysosome to activate mTORC1. Biochem. Soc. Trans. 41, 951–955.

Gryson, C., Walrand, S., Giraudet, C., Rousset, P., Migne, C., Bonhomme, C., et al., 2013. "Fast proteins" with a unique essential amino acid content as an optimal nutrition in the elderly: growing evidence. Clin. Nutr. 33, 642–648.

Guillet, C., Prod'homme, M., Balage, M., Gachon, P., Giraudet, C., Morin, L., et al., 2004. Impaired anabolic response of muscle protein synthesis is associated with S6K1 dysregulation in elderly humans. FASEB J. 18, 1586–1587.

Haegens, A., Schols, A.M., van Essen, A.L., van Loon, L.J., Langen, R.C., 2012. Leucine induces myofibrillar protein accretion in cultured skeletal muscle through mTOR dependent and independent control of myosin heavy chain mRNA levels. Mol. Nutr. Food Res. 56, 741–752.

Hafen, E., 2004. Interplay between growth factor and nutrient signaling: lessons from Drosophila TOR. Curr. Top. Microbiol. Immunol. 279, 153–167.

Haider, T., Gunga, H.C., Matteucci-Gothe, R., Sottara, E., Griesmacher, A., Belavy, D.L., et al., 2014. Effects of long-term head-down-tilt bed rest and different training regimes on the coagulation system of healthy men. Physiol. Rep. 1, e00135.

Han, J.M., Jeong, S.J., Park, M.C., Kim, G., Kwon, N.H., Kim, H.K., et al., 2012. Leucyl-tRNA synthetase is an intracellular leucine sensor for the mTORC1-signaling pathway. Cell 149, 410–424.

Hartl, W.H., Alpers, D.H., 2010. The trophic effects of substrate, insulin, and the route of administration on protein synthesis and the preservation of small bowel mucosal mass in large mammals. Clin. Nutr. 30, 20–27.

Hawke, T.J., Garry, D.J., 2001. Myogenic satellite cells: physiology to molecular biology. J. Appl. Physiol. (1985) 91, 534–551.

Hayes, L.D., Grace, F.M., Sculthorpe, N., Herbert, P., Ratcliffe, J.W., Kilduff, L.P., et al., 2013. The effects of a formal exercise training programme on salivary hormone concentrations and body composition in previously sedentary aging men. Springerplus 2, 18.

Hepple, R.T., 2014. Mitochondrial involvement and impact in aging skeletal muscle. Front. Aging Neurosci. 6, 211.

Hishikawa, N., Yamashita, T., Deguchi, K., Wada, J., Shikata, K., Makino, H., et al., 2014. Cognitive and affective functions in diabetic patients associated with diabetes-related factors, white matter abnormality and aging. Eur. J. Neurol. 22, 313–321.

Hitomi, K., 2005. Transglutaminases in skin epidermis. Eur. J. Dermatol. 15, 313–319.

Hoff, P., Belavy, D.L., Huscher, D., Lang, A., Hahne, M., Kuhlmey, A.K., et al., 2014. Effects of 60-day bed rest with and without exercise on cellular and humoral immunological parameters. Cell. Mol. Immunol.

Horstman, A.M., Dillon, E.L., Urban, R.J., Sheffield-Moore, M., 2012. The role of androgens and estrogens on healthy aging and longevity. J. Gerontol. A Biol. Sci. Med. Sci. 67 (11), 1140–1152.

Hou, X., Hu, Z., Xu, H., Xu, J., Zhang, S., Zhong, Y., et al., 2014. Advanced glycation endproducts trigger autophagy in cadiomyocyte via RAGE/PI3K/AKT/mTOR pathway. Cardiovasc. Diabetol. 13, 78.

Huang, G., Basaria, S., Travison, T.G., Ho, M.H., Davda, M., Mazer, N.A., et al., 2013. Testosterone dose-response relationships in hysterectomized women with or without oophorectomy: effects on sexual function, body composition, muscle performance and physical function in a randomized trial. Menopause 21, 612–623.

Inoki, K., Zhu, T., Guan, K.L., 2003. TSC2 mediates cellular energy response to control cell growth and survival. Cell 115, 577–590.

Inoki, K., Ouyang, H., Zhu, T., Lindvall, C., Wang, Y., Zhang, X., et al., 2006. TSC2 integrates Wnt and energy signals via a coordinated phosphorylation by AMPK and GSK3 to regulate cell growth. Cell 126, 955–968.

Iwasaki, Y., Asai, M., Yoshida, M., Nigawara, T., Kambayashi, M., Nakashima, N., 2004. Dehydroepiandrosterone-sulfate inhibits nuclear factor-kappaB-dependent transcription in hepatocytes, possibly through antioxidant effect. J. Clin. Endocrinol. Metab. 89, 3449–3454.

Jackson, J.R., Mula, J., Kirby, T.J., Fry, C.S., Lee, J.D., Ubele, M.F., et al., 2012. Satellite cell depletion does not inhibit adult skeletal muscle regrowth following unloading-induced atrophy. Am. J. Physiol. Cell Physiol. 303, C854–C861.

Jewell, J.L., Kim, Y.C., Russell, R.C., Yu, F.X., Park, H.W., Plouffe, S.W., et al., 2015. Metabolism. Differential regulation of mTORC1 by leucine and glutamine. Science 347, 194–198.

Ji, L.L., 2007. Antioxidant signaling in skeletal muscle: a brief review. Exp. Gerontol. 42, 582–593.

Johnson, M.L., Robinson, M.M., Nair, K.S., 2013. Skeletal muscle aging and the mitochondrion. Trends. Endocrinol. Metab. 24, 247–256.

Jonker, R., Engelen, M.P., Deutz, N.E., 2012. Role of specific dietary amino acids in clinical conditions. Br. J. Nutr. 108 (Suppl. 2), S139–S148.

Jourdan, M., Deutz, N.E., Cynober, L., Aussel, C., 2011. Features, causes and consequences of splanchnic sequestration of amino acid in old rats. PLoS One 6, e27002.

Kadi, F., 2008. Cellular and molecular mechanisms responsible for the action of testosterone on human skeletal muscle. A basis for illegal performance enhancement. Br. J. Pharmacol. 154, 522–528.

Katsanos, C.S., Kobayashi, H., Sheffield-Moore, M., Aarsland, A., Wolfe, R.R., 2005. Aging is associated with diminished accretion of muscle proteins after the ingestion of a small bolus of essential amino acids. Am. J. Clin. Nutr. 82, 1065–1073.

Katsanos, C.S., Kobayashi, H., Sheffield-Moore, M., Aarsland, A., Wolfe, R.R., 2006. A high proportion of leucine is required for optimal stimulation of the rate of muscle protein synthesis by essential amino acids in the elderly. Am. J. Physiol. Endocrinol. Metab. 291, E381–E387.

Katsanos, C.S., Chinkes, D.L., Paddon-Jones, D., Zhang, X.J., Aarsland, A., Wolfe, R.R., 2008. Whey protein ingestion in elderly persons results in greater muscle protein accrual than ingestion of its constituent essential amino acid content. Nutr. Res. 28, 651–658.

Keren, A., Tamir, Y., Bengal, E., 2006. The p38 MAPK signaling pathway: a major regulator of skeletal muscle development. Mol. Cell. Endocrinol. 252, 224–230.

Kim, T.N., Choi, K.M., 2014. The implications of sarcopenia and sarcopenic obesity on cardiometabolic disease. J. Cell. Biochem. 116 (7), 1171–1178.

Kimball, S.R., Jefferson, L.S., 2006. Signaling pathways and molecular mechanisms through which branched-chain amino acids mediate translational control of protein synthesis. J. Nutr. 136, 227S–231S.

Kimball, S.R., Farrell, P.A., Jefferson, L.S., 2002. Invited review: role of insulin in translational control of protein synthesis in skeletal muscle by amino acids or exercise. J. Appl. Physiol. 93, 1168–1180.

Kitzmann, M., Bonnieu, A., Duret, C., Vernus, B., Barro, M., Laoudj-Chenivesse, D., et al., 2006. Inhibition of Notch signaling induces myotube hypertrophy by recruiting a subpopulation of reserve cells. J. Cell. Physiol. 208, 538–548.

Kobayashi, H., Borsheim, E., Anthony, T.G., Traber, D.L., Badalamenti, J., Kimball, S.R., et al., 2003. Reduced amino acid availability inhibits muscle protein synthesis and decreases activity of initiation factor eIF2B. Am. J. Physiol. Endocrinol. Metab. 284, E488–E498.

Koopman, R., Crombach, N., Gijsen, A.P., Walrand, S., Fauquant, J., Kies, A.K., et al., 2009a. Ingestion of a protein hydrolysate is accompanied by an accelerated in vivo digestion and absorption rate when compared with its intact protein. Am. J. Clin. Nutr. 90, 106–115.

Koopman, R., Walrand, S., Beelen, M., Gijsen, A.P., Kies, A.K., Boirie, Y., et al., 2009b. Dietary protein digestion and absorption rates and the subsequent postprandial muscle protein synthetic response do not differ between young and elderly men. J. Nutr. 139, 1707–1713.

Korolchuk, V.I., Menzies, F.M., Rubinsztein, D.C., 2010. Mechanisms of cross-talk between the ubiquitin-proteasome and autophagy-lysosome systems. FEBS Lett. 584, 1393–1398.

Kortebein, P., Ferrando, A., Lombeida, J., Wolfe, R., Evans, W.J., 2007. Effect of 10 days of bed rest on skeletal muscle in healthy older adults. J. Am. Med. Assoc. 297, 1772–1774.

Kortebein, P., Symons, T.B., Ferrando, A., Paddon-Jones, D., Ronsen, O., Protas, E., et al., 2008. Functional impact of 10 days of bed rest in healthy older adults. J. Gerontol. A Biol. Sci. Med. Sci. 63, 1076–1081.

Koryak, Y.A., 2014. Influence of simulated microgravity on mechanical properties in the human triceps surae muscle in vivo. I: effect of 120 days of bed-rest without physical training on human muscle musculo-tendinous stiffness and contractile properties in young women. Eur. J. Appl. Physiol. 114, 1025–1036.

Kullman, E.L., Campbell, W.W., Krishnan, R.K., Yarasheski, K.E., Evans, W.J., Kirwan, J.P., 2013. Age attenuates leucine oxidation after eccentric exercise. Int. J. Sports Med. 34, 695–699.

Landmesser, U., Dikalov, S., Price, S.R., McCann, L., Fukai, T., Holland, S.M., et al., 2003. Oxidation of tetrahydrobiopterin leads to uncoupling of endothelial cell nitric oxide synthase in hypertension. J. Clin. Invest. 111, 1201–1209.

Leenders, M., Verdijk, L.B., van der Hoeven, L., van Kranenburg, J., Hartgens, F., Wodzig, W.K., et al., 2011. Prolonged leucine supplementation does not augment muscle mass or affect glycemic control in elderly type 2 diabetic men. J. Nutr. 141, 1070–1076.

Li, M., Ogilvie, H., Ochala, J., Artemenko, K., Iwamoto, H., Yagi, N., et al., 2015. Aberrant post-translational modifications compromise human myosin motor function in old age. Aging Cell 14, 228–235.

Long, X., Ortiz-Vega, S., Lin, Y., Avruch, J., 2005. Rheb binding to mammalian target of rapamycin (mTOR) is regulated by amino acid sufficiency. J. Biol. Chem. 280, 23433–23436.

Luiking, Y.C., Deutz, N.E., Memelink, R.G., Verlaan, S., Wolfe, R.R., 2014. Postprandial muscle protein synthesis is higher after a high whey protein, leucine-enriched supplement than after a dairy-like product in healthy older people: a randomized controlled trial. Nutr. J. 13, 9.

Lyons, T.J., Basu, A., 2012. Biomarkers in diabetes: hemoglobin A1c, vascular and tissue markers. Transl. Res. 159, 303–312.

Magne, H., Savary-Auzeloux, I., Remond, D., Dardevet, D., 2013. Nutritional strategies to counteract muscle atrophy caused by disuse and to improve recovery. Nutr. Res. Rev. 26, 149–165.

Mamerow, M.M., Mettler, J.A., English, K.L., Casperson, S.L., Arentson-Lantz, E., Sheffield-Moore, M., et al., 2014. Dietary protein distribution positively influences 24-h muscle protein synthesis in healthy adults. J. Nutr. 144, 876–880.

Manrique, C., Lastra, G., Sowers, J.R., 2014. New insights into insulin action and resistance in the vasculature. Ann. N.Y. Acad. Sci. 1311, 138–150.

Markofski, M.M., Dickinson, J.M., Drummond, M.J., Fry, C.S., Fujita, S., Gundermann, D.M., et al., 2015. Effect of age on basal muscle protein synthesis and mTORC1 signaling in a large cohort of young and older men and women. Exp. Gerontol. 65C, 1–7.

Marshall, D., Hardman, M.J., Nield, K.M., Byrne, C., 2001. Differentially expressed late constituents of the epidermal cornified envelope. Proc. Natl. Acad. Sci. U.S.A. 98, 13031–13036.

Martin, T.D., Chen, X.W., Kaplan, R.E., Saltiel, A.R., Walker, C.L., Reiner, D.J., et al., 2014. Ral and Rheb GTPase activating proteins integrate mTOR and GTPase signaling in aging, autophagy, and tumor cell invasion. Mol. Cell 53, 209–220.

Marzani, B., Balage, M., Venien, A., Astruc, T., Papet, I., Dardevet, D., et al., 2008. Antioxidant supplementation restores defective leucine stimulation of protein synthesis in skeletal muscle from old rats. J. Nutr. 138, 2205–2211.

Mason, P.J., Perdigones, N., 2013. Telomere biology and translational research. Transl. Res. 162, 333–342.

Mazzola, J.L., Sirover, M.A., 2001. Reduction of glyceraldehyde-3-phosphate dehydrogenase activity in Alzheimer's disease and in Huntington's disease fibroblasts. J. Neurochem. 76, 442–449.

Meneilly, G.S., Elliot, T., Bryer-Ash, M., Floras, J.S., 1995. Insulin-mediated increase in blood flow is impaired in the elderly. J. Clin. Endocrinol. Metab. 80, 1899–1903.

Mitchell, W.K., Williams, J., Atherton, P., Larvin, M., Lund, J., Narici, M., 2012. Sarcopenia, dynapenia, and the impact of advancing age on human skeletal muscle size and strength; a quantitative review. Front. Physiol. 3, 260.

Mitchell, W.K., Phillips, B.E., Williams, J.P., Rankin, D., Lund, J.N., Smith, K., et al., 2015. A dose- rather than delivery profile-dependent mechanism regulates the "muscle-full" effect in response to oral essential amino acid intake in young men. J. Nutr. 145, 207–214.

Montagnani, M., Chen, H., Barr, V.A., Quon, M.J., 2001. Insulin-stimulated activation of eNOS is independent of Ca^{2+} but requires phosphorylation by Akt at Ser(1179). J. Biol. Chem. 276, 30392–30398.

Moore, D.R., Atherton, P.J., Rennie, M.J., Tarnopolsky, M.A., Phillips, S.M., 2010. Resistance exercise enhances mTOR and MAPK signalling in human muscle over that seen at rest after bolus protein ingestion. Acta Physiol. (Oxf) 201, 365–372.

Moore, D.R., Churchward-Venne, T.A., Witard, O., Breen, L., Burd, N.A., Tipton, K.D., et al., 2014. Protein ingestion to stimulate myofibrillar protein synthesis requires greater relative protein intakes in healthy older versus younger men. J. Gerontol. A Biol. Sci. Med. Sci. 70, 57–62.

Morais, J.A., Ross, R., Gougeon, R., Pencharz, P.B., Jones, P.J., Marliss, E.B., 2000. Distribution of protein turnover changes with age in humans as assessed by whole-body magnetic resonance image analysis to quantify tissue volumes. J. Nutr. 130, 784–791.

Moreau, K., Walrand, S., Boirie, Y., 2013. Protein redistribution from skeletal muscle to splanchnic tissue on fasting and refeeding in young and older healthy individuals. J. Am. Med. Dir. Assoc. 14, 696–704.

Morley, J.E., 2008. Sarcopenia: diagnosis and treatment. J. Nutr. Health Aging 12, 452–456.

Morley, J.E., 2012. Sarcopenia in the elderly. Fam. Pract. 29 (Suppl. 1), i44–i48.

Morley, J.E., Abbatecola, A.M., Argiles, J.M., Baracos, V., Bauer, J., Bhasin, S., et al., 2011. Sarcopenia with limited mobility: an international consensus. J. Am. Med. Dir. Assoc. 12, 403–409.

Morley, J.E., Anker, S.D., von Haehling, S., 2014. Prevalence, incidence, and clinical impact of sarcopenia: facts, numbers, and epidemiology-update 2014. J. Cachexia Sarcopenia Muscle 5, 253–259.

Motohashi, N., Asakura, A., 2014. Muscle satellite cell heterogeneity and self-renewal. Front. Cell Dev. Biol. 2, 1.

Mulder, E., Clement, G., Linnarsson, D., Paloski, W.H., Wuyts, F.P., Zange, J., et al., 2014. Musculoskeletal effects of 5 days of bed rest with and without locomotion replacement training. Eur. J. Appl. Physiol. 115 (4), 727–738.

Muscaritoli, M., Anker, S.D., Argiles, J., Aversa, Z., Bauer, J.M., Biolo, G., et al., 2010. Consensus definition of sarcopenia, cachexia and pre-cachexia: joint document elaborated by Special Interest Groups (SIG) "cachexia-anorexia in chronic wasting diseases" and "nutrition in geriatrics". Clin. Nutr. 29, 154–159.

Nagasawa, T., Kido, T., Yoshizawa, F., Ito, Y., Nishizawa, N., 2002. Rapid suppression of protein degradation in skeletal muscle after oral feeding of leucine in rats. J. Nutr. Biochem. 13, 121–127.

Nathan, D.M., McGee, P., Steffes, M.W., Lachin, J.M., 2013. Relationship of glycated albumin to blood glucose and HbA1c values and to retinopathy, nephropathy, and cardiovascular outcomes in the DCCT/EDIC study. Diabetes 63, 282–290.

Nobukuni, T., Joaquin, M., Roccio, M., Dann, S.G., Kim, S.Y., Gulati, P., et al., 2005. Amino acids mediate mTOR/raptor signaling through activation of class 3 phosphatidylinositol 3OH-kinase. Proc. Natl. Acad. Sci. U.S.A. 102, 14238–14243.

Nowotny, K., Jung, T., Grune, T., Hohn, A., 2014. Accumulation of modified proteins and aggregate formation in aging. Exp. Gerontol. 57, 122–131.

O'Connor, P.M., Kimball, S.R., Suryawan, A., Bush, J.A., Nguyen, H.V., Jefferson, L.S., et al., 2003. Regulation of translation initiation by insulin and amino acids in skeletal muscle of neonatal pigs. Am. J. Physiol. Endocrinol. Metab. 285, E40–E53.

Oellerich, M.F., Potente, M., 2012. FOXOs and sirtuins in vascular growth, maintenance, and aging. Circ. Res. 110, 1238–1251.

Olguin, H.C., Yang, Z., Tapscott, S.J., Olwin, B.B., 2007. Reciprocal inhibition between Pax7 and muscle regulatory factors modulates myogenic cell fate determination. J. Cell Biol. 177, 769–779.

Ott, C., Jacobs, K., Haucke, E., Navarrete Santos, A., Grune, T., Simm, A., 2014. Role of advanced glycation end products in cellular signaling. Redox Biol. 2, 411–429.

Paddon-Jones, D., Rasmussen, B.B., 2009. Dietary protein recommendations and the prevention of sarcopenia. Curr. Opin. Clin. Nutr. Metab. Care 12, 86–90.

Paddon-Jones, D., Sheffield-Moore, M., Urban, R.J., Sanford, A.P., Aarsland, A., Wolfe, R.R., et al., 2004a. Essential amino acid and carbohydrate supplementation ameliorates muscle protein loss in humans during 28 days bedrest. J. Clin. Endocrinol. Metab. 89, 4351–4358.

Paddon-Jones, D., Sheffield-Moore, M., Zhang, X.J., Volpi, E., Wolf, S.E., Aarsland, A., et al., 2004b. Amino acid ingestion improves muscle protein synthesis in the young and elderly. Am. J. Physiol. Endocrinol. Metab. 286, E321–E328.

Paddon-Jones, D., Sheffield-Moore, M., Cree, M.G., Hewlings, S.J., Aarsland, A., Wolfe, R.R., et al., 2006. Atrophy and impaired muscle protein synthesis during prolonged inactivity and stress. J. Clin. Endocrinol. Metab. 91, 4836–4841.

Pennings, B., Boirie, Y., Senden, J.M., Gijsen, A.P., Kuipers, H., van Loon, L.J., 2011. Whey protein stimulates postprandial muscle protein accretion more effectively than do casein and casein hydrolysate in older men. Am. J. Clin. Nutr. 93, 997–1005.

Pennings, B., Groen, B., de Lange, A., Gijsen, A.P., Zorenc, A.H., Senden, J.M., et al., 2012. Amino acid absorption and subsequent muscle protein accretion following graded intakes of whey protein in elderly men. Am. J. Physiol. Endocrinol. Metab. 302, E992–E999.

Perluigi, M., Di Domenico, F., Butterfield, D.A., 2015. mTOR signaling in aging and neurodegeneration: at the crossroad between metabolism dysfunction and impairment of autophagy. Neurobiol. Dis. 84, 39–49.

Phillips, S.M., Tang, J.E., Moore, D.R., 2009. The role of milk- and soy-based protein in support of muscle protein synthesis and muscle protein accretion in young and elderly persons. J. Am. Coll. Nutr. 28, 343–354.

Phillips, B.E., Atherton, P.J., Varadhan, K., Wilkinson, D.J., Limb, M., Selby, A.L., et al., 2013. Pharmacological enhancement of leg and muscle microvascular blood flow does not augment anabolic responses in skeletal muscle of young men under fed conditions. Am. J. Physiol. Endocrinol. Metab. 306, E168–E176.

Pickering, A.M., Davies, K.J., 2012. Degradation of damaged proteins: the main function of the 20S proteasome. Prog. Mol. Biol. Transl. Sci. 109, 227–248.

Pires-Oliveira, M., Maragno, A.L., Parreiras-e-Silva, L.T., Chiavegatti, T., Gomes, M.D., Godinho, R.O., 2009. Testosterone represses ubiquitin ligases atrogin-1 and Murf-1 expression in an androgen-sensitive rat skeletal muscle in vivo. J. Appl. Physiol. (1985) 108, 266–273.

Pisconti, A., Brunelli, S., Di Padova, M., De Palma, C., Deponti, D., Baesso, S., et al., 2006. Follistatin induction by nitric oxide through cyclic GMP: a tightly regulated signaling pathway that controls myoblast fusion. J. Cell Biol. 172, 233–244.

Prod'homme, M., Rieu, I., Balage, M., Dardevet, D., Grizard, J., 2004. Insulin and amino acids both strongly participate to the regulation of protein metabolism. Curr. Opin. Clin. Nutr. Metab. Care 7, 71–77.

Prod'homme, M., Balage, M., Debras, E., Farges, M.C., Kimball, S., Jefferson, L., et al., 2005. Differential effects of insulin and dietary amino acids on muscle protein synthesis in adult and old rats. J. Physiol. 563, 235–248.

Proud, C.G., 2004a. mTOR-mediated regulation of translation factors by amino acids. Biochem. Biophys. Res. Commun. 313, 429–436.

Proud, C.G., 2004b. Role of mTOR signalling in the control of translation initiation and elongation by nutrients. Curr. Top. Microbiol. Immunol. 279, 215–244.

Qin, W., Pan, J., Wu, Y., Bauman, W.A., Cardozo, C., 2010. Protection against dexamethasone-induced muscle atrophy is related to modulation by testosterone of FOXO1 and PGC-1alpha. Biochem. Biophys. Res. Commun. 403, 473–478.

Raue, U., Slivka, D., Jemiolo, B., Hollon, C., Trappe, S., 2006. Myogenic gene expression at rest and after a bout of resistance exercise in young (18–30 yr) and old (80–89 yr) women. J. Appl. Physiol. (1985) 101, 53–59.

Rebsamen, M., Pochini, L., Stasyk, T., de Araujo, M.E., Galluccio, M., Kandasamy, R.K., et al., 2015. SLC38A9 is a component of the lysosomal amino acid sensing machinery that controls mTORC1. Nature 519 (7544), 477–481.

Regazzetti, C., Bost, Fdr, Le Marchand-Brustel, Y., Tanti, J.-F.O., Giorgetti-Peraldi, S., 2010. Insulin induces REDD1 expression through hypoxia-inducible factor 1 activation in adipocytes. J. Biol. Chem. 285, 5157–5164.

Renault, V., Thornell, L.E., Eriksson, P.O., Butler-Browne, G., Mouly, V., 2002. Regenerative potential of human skeletal muscle during aging. Aging Cell 1, 132–139.

Renziehausen, J., Hiebel, C., Nagel, H., Kundu, A., Kins, S., Kogel, D., et al., 2014. The cleavage product of amyloid-beta Protein Precursor sAbetaPPalpha Modulates BAG3-dependent aggresome formation and enhances cellular proteasomal activity. J. Alzheimers Dis. 44 (3), 879–896.

Rieu, I., Balage, M., Sornet, C., Giraudet, C., Pujos, E., Grizard, J., et al., 2006. Leucine supplementation improves muscle protein synthesis in elderly men independently of hyperaminoacidaemia. J. Physiol. 575, 305–315.

Rinnerthaler, M., Duschl, J., Steinbacher, P., Salzmann, M., Bischof, J., Schuller, M., et al., 2013. Age-related changes in the composition of the cornified envelope in human skin. Exp. Dermatol. 22, 329–335.

Rittweger, J., Bareille, M.P., Clement, G., Linnarsson, D., Paloski, W.H., Wuyts, F., et al., 2015. Short-arm centrifugation as a partially effective musculoskeletal countermeasure during 5-day head-down tilt bed rest-results from the BRAG1 study. Eur. J. Appl. Physiol. 115 (6), 1233–1244.

Roccio, M., Bos, J.L., Zwartkruis, F.J., 2006. Regulation of the small GTPase Rheb by amino acids. Oncogene 25, 657–664.

Roebuck, J., 1979. When does old age begin? The evolution of the english definition. J. Soc. Hist. 12, 416–428.

Rooyackers, O.E., Adey, D.B., Ades, P.A., Nair, K.S., 1996. Effect of age on in vivo rates of mitochondrial protein synthesis in human skeletal muscle. Proc. Natl. Acad. Sci. U.S.A. 93, 15364–15369.

Rubinsztein, D.C., Mariño, G., Kroemer, G., 2011. Autophagy and aging. Cell 146, 682–695.

Salahuddin, P., Rabbani, G., Khan, R.H., 2014. The role of advanced glycation end products in various types of neurodegenerative disease: a therapeutic approach. Cell. Mol. Biol. Lett. 19, 407–437.

Salminen, A., Kaarniranta, K., 2011. AMP-activated protein kinase (AMPK) controls the aging process via an integrated signaling network. Ageing Res. Rev. 11, 230–241.

Santilli, V., Bernetti, A., Mangone, M., Paoloni, M., 2015. Clinical definition of sarcopenia. Clin. Cases Miner. Bone Metab. 11, 177–180.

Sculthorpe, N., Solomon, A.M., Sinanan, A.C., Bouloux, P.M., Grace, F., Lewis, M.P., 2011. Androgens affect myogenesis in vitro and increase local IGF-1 expression. Med. Sci. Sports Exerc. 44, 610–615.

Seals, D.R., 2014. Edward F. Adolph distinguished lecture: the remarkable anti-aging effects of aerobic exercise on systemic arteries. J. Appl. Physiol. (1985) 117, 425–439.

Sheffield-Moore, M., 2000. Androgens and the control of skeletal muscle protein synthesis. Ann. Med. 32, 181–186.

Sheffield-Moore, M., Urban, R.J., Wolf, S.E., Jiang, J., Catlin, D.H., Herndon, D.N., et al., 1999. Short-term oxandrolone administration stimulates net muscle protein synthesis in young men. J. Clin. Endocrinol. Metab. 84, 2705–2711.

Sheffield-Moore, M., Wolfe, R.R., Gore, D.C., Wolf, S.E., Ferrer, D.M., Ferrando, A.A., 2000. Combined effects of hyperaminoacidemia and oxandrolone on skeletal muscle protein synthesis. Am. J. Physiol. Endocrinol. Metab. 278, E273–E279.

Sheffield-Moore, M., Yeckel, C.W., Volpi, E., Wolf, S.E., Morio, B., Chinkes, D.L., et al., 2004. Postexercise protein metabolism in older and younger men following moderate-intensity aerobic exercise. Am. J. Physiol. Endocrinol. Metab. 287, E513–E522.

Sheffield-Moore, M., Paddon-Jones, D., Casperson, S.L., Gilkison, C., Volpi, E., Wolf, S.E., et al., 2006. Androgen therapy induces muscle protein anabolism in older women. J. Clin. Endocrinol. Metab. 91, 3844–3849.

Sheffield-Moore, M., Dillon, E.L., Casperson, S.L., Gilkison, C.R., Paddon-Jones, D., Durham, W.J., et al., 2011. A randomized pilot study of monthly cycled testosterone replacement or continuous testosterone replacement versus placebo in older men. J. Clin. Endocrinol. Metab. 96, E1831–E1837.

Sindler, A.L., Delp, M.D., Reyes, R., Wu, G., Muller-Delp, J.M., 2009. Effects of ageing and exercise training on eNOS uncoupling in skeletal muscle resistance arterioles. J. Physiol. 587, 3885–3897.

Sinha, I., Sinha-Hikim, A.P., Wagers, A.J., Sinha-Hikim, I., 2014. Testosterone is essential for skeletal muscle growth in aged mice in a heterochronic parabiosis model. Cell. Tissue Res. 357, 815–821.

Sinha-Hikim, I., Cornford, M., Gaytan, H., Lee, M.L., Bhasin, S., 2006. Effects of testosterone supplementation on skeletal muscle fiber hypertrophy and satellite cells in community-dwelling older men. J. Clin. Endocrinol. Metab. 91, 3024–3033.

Smith, G.I., Yoshino, J., Reeds, D.N., Bradley, D., Burrows, R.E., Heisey, H.D., et al., 2013. Testosterone and progesterone, but not estradiol, stimulate muscle protein synthesis in postmenopausal women. J. Clin. Endocrinol. Metab. 99, 256–265.

Sofer, A., Lei, K., Johannessen, C.M., Ellisen, L.W., 2005. Regulation of mTOR and cell growth in response to energy stress by REDD1. Mol. Cell. Biol. 25, 5834–5845.

Solerte, S.B., Gazzaruso, C., Bonacasa, R., Rondanelli, M., Zamboni, M., Basso, C., et al., 2008. Nutritional supplements with oral amino acid mixtures increases whole-body lean mass and insulin sensitivity in elderly subjects with sarcopenia. Am. J. Cardiol. 101, 69E–77E.

Steinberg, H.O., Brechtel, G., Johnson, A., Fineberg, N., Baron, A.D., 1994. Insulin-mediated skeletal muscle vasodilation is nitric oxide dependent. A novel action of insulin to increase nitric oxide release. J. Clin. Invest. 94, 1172–1179.

Straight, C.R., Ward-Ritacco, C.L., Evans, E.M., 2015. Association between accelerometer-measured physical activity and muscle capacity in middle-aged postmenopausal women. Menopause 22 (11), 1204–1211.

Suetta, C., Frandsen, U., Mackey, A.L., Jensen, L., Hvid, L.G., Bayer, M.L., et al., 2013. Ageing is associated with diminished muscle re-growth and myogenic precursor cell expansion early after immobility-induced atrophy in human skeletal muscle. J. Physiol. 591, 3789–3804.

Suryawan, A., Jeyapalan, A.S., Orellana, R.A., Wilson, F.A., Nguyen, H.V., Davis, T.A., 2008. Leucine stimulates protein synthesis in skeletal muscle of neonatal pigs by enhancing mTORC1 activation. Am. J. Physiol. Endocrinol. Metab. 295, E868–E875.

Symons, T.B., Schutzler, S.E., Cocke, T.L., Chinkes, D.L., Wolfe, R.R., Paddon-Jones, D., 2007. Aging does not impair the anabolic response to a protein-rich meal. Am. J. Clin. Nutr. 86, 451–456.

Symons, T.B., Sheffield-Moore, M., Chinkes, D.L., Ferrando, A.A., Paddon-Jones, D., 2009a. Artificial gravity maintains skeletal muscle protein synthesis during 21 days of simulated microgravity. J. Appl. Physiol. (1985) 107, 34–38.

Symons, T.B., Sheffield-Moore, M., Wolfe, R.R., Paddon-Jones, D., 2009b. A moderate serving of high-quality protein maximally stimulates skeletal muscle protein synthesis in young and elderly subjects. J. Am. Diet. Assoc. 109, 1582–1586.

Taddei, S., Virdis, A., Ghiadoni, L., Salvetti, G., Bernini, G., Magagna, A., et al., 2001. Age-related reduction of NO availability and oxidative stress in humans. Hypertension 38, 274–279.

Takahashi, R., Goto, S., 1990. Alteration of aminoacyl-tRNA synthetase with age: heat-labilization of the enzyme by oxidative damage. Arch. Biochem. Biophys. 277, 228–233.

Thomas, D.R., Ashmen, W., Morley, J.E., Evans, W.J., 2000. Nutritional management in long-term care: development of a clinical guideline. Council for nutritional strategies in long-term care. J. Gerontol. A Biol. Sci. Med. Sci. 55, M725–M734.

Tieland, M., van de Rest, O., Dirks, M.L., van der Zwaluw, N., Mensink, M., van Loon, L.J., et al., 2012. Protein supplementation improves physical performance in frail elderly people: a randomized, double-blind, placebo-controlled trial. J. Am. Med. Dir. Assoc. 13 (8), 720–726.

Timmerman, K.L., Lee, J.L., Fujita, S., Dhanani, S., Dreyer, H.C., Fry, C.S., et al., 2010. Pharmacological vasodilation improves insulin-stimulated muscle protein anabolism but not glucose utilization in older adults. Diabetes 59, 2764–2771.

Timmerman, K.L., Dhanani, S., Glynn, E.L., Fry, C.S., Drummond, M.J., Jennings, K., et al., 2012. A moderate acute increase in physical activity enhances nutritive flow and the muscle protein anabolic response to mixed nutrient intake in older adults. Am. J. Clin. Nutr. 95, 1403–1412.

Toth, C., Martinez, J., Zochodne, D.W., 2007. RAGE, diabetes, and the nervous system. Curr. Mol. Med. 7, 766–776.

Tripathi, D.N., Chowdhury, R., Trudel, L.J., Tee, A.R., Slack, R.S., Walker, C.L., et al., 2013. Reactive nitrogen species regulate autophagy through ATM-AMPK-TSC2-mediated suppression of mTORC1. Proc. Natl. Acad. Sci. U.S.A. 110, E2950–E2957.

Tsai, C.L., Wang, C.H., Pan, C.Y., Chen, F.C., 2015. The effects of long-term resistance exercise on the relationship between neurocognitive performance and GH, IGF-1, and homocysteine levels in the elderly. Front. Behav. Neurosci. 9, 23.

Tyagi, R., Shahani, N., Gorgen, L., Ferretti, M., Pryor, W., Chen, P.Y., et al., 2015. Rheb inhibits protein synthesis by activating the PERK-eIF2alpha signaling cascade. Cell Rep.

Uchiki, T., Weikel, K.A., Jiao, W., Shang, F., Caceres, A., Pawlak, D., et al., 2011. Glycation-altered proteolysis as a pathobiologic mechanism that links dietary glycemic index, aging, and age-related disease (in nondiabetics). Aging Cell 11, 1–13.

Urban, R.J., Dillon, E.L., Choudhary, S., Zhao, Y., Horstman, A.M., Tilton, R.G., et al., 2014a. Translational studies in older men using testosterone to treat sarcopenia. Trans. Am. Clin. Climatol. Assoc. 125, 27–42, discussion 42-24.

Urban, R.J., Dillon, E.L., Choudhary, S., Zhao, Y., Horstman, A.M., Tilton, R.G., et al., 2014b. Translational studies in older men using testosterone to treat sarcopenia. Trans. Am. Clin. Climatol. Assoc. 125, 27–44.

Van Puyvelde, K., Mets, T., Njemini, R., Beyer, I., Bautmans, I., 2014. Effect of advanced glycation end product intake on inflammation and aging: a systematic review. Nutr. Rev. 72, 638–650.

Veldhuis, J.D., 2013. Changes in pituitary function with ageing and implications for patient care. Nat. Rev. Endocrinol. 9, 205–215.

Verdijk, L.B., Snijders, T., Drost, M., Delhaas, T., Kadi, F., van Loon, L.J., 2014. Satellite cells in human skeletal muscle; from birth to old age. Age (Dordr) 36, 545–547.

Verhoeven, S., Vanschoonbeek, K., Verdijk, L.B., Koopman, R., Wodzig, W.K., Dendale, P., et al., 2009. Long-term leucine supplementation does not increase muscle mass or strength in healthy elderly men. Am. J. Clin. Nutr. 89, 1468–1475.

Vermeij, W.P., Backendorf, C., 2013. Reactive oxygen species (ROS) protection via cysteine oxidation in the epidermal cornified cell envelope. Methods Mol. Biol. 1195, 157–169.

Vermeij, W.P., Alia, A., Backendorf, C., 2011. ROS quenching potential of the epidermal cornified cell envelope. J. Invest. Dermatol. 131, 1435–1441.

Vermeij, W.P., Florea, B.I., Isenia, S., Alia, A., Brouwer, J., Backendorf, C., 2012. Proteomic identification of in vivo interactors reveals novel function of skin cornification proteins. J. Proteome Res. 11, 3068–3076.

Vincent, M.A., Clerk, L.H., Lindner, J.R., Price, W.J., Jahn, L.A., Leong-Poi, H., et al., 2006. Mixed meal and light exercise each recruit muscle capillaries in healthy humans. Am. J. Physiol. Endocrinol. Metab. 290, E1191–E1197.

Volek, J.S., Volk, B.M., Gomez, A.L., Kunces, L.J., Kupchak, B.R., Freidenreich, D.J., et al., 2013. Whey protein supplementation during resistance training augments lean body mass. J. Am. Coll. Nutr. 32, 122–135.

Volpi, E., Mittendorfer, B., Wolf, S.E., Wolfe, R.R., 1999. Oral amino acids stimulate muscle protein anabolism in the elderly despite higher first-pass splanchnic extraction. Am. J. Physiol. 277, E513–E520.

Volpi, E., Mittendorfer, B., Rasmussen, B.B., Wolfe, R.R., 2000. The response of muscle protein anabolism to combined hyperaminoacidemia and glucose-induced hyperinsulinemia is impaired in the elderly. J. Clin. Endocrinol. Metab. 85, 4481–4490.

Volpi, E., Sheffield-Moore, M., Rasmussen, B.B., Wolfe, R.R., 2001. Basal muscle amino acid kinetics and protein synthesis in healthy young and older men. J. Am. Med. Assoc. 286, 1206–1212.

Walker, D.K., Dickinson, J.M., Timmerman, K.L., Drummond, M.J., Reidy, P.T., Fry, C.S., et al., 2011. Exercise, amino acids, and aging in the control of human muscle protein synthesis. Med. Sci. Sports Exerc. 43, 2249–2258.

Wall, B.T., Dirks, M.L., Verdijk, L.B., Snijders, T., Hansen, D., Vranckx, P., et al., 2012. Neuromuscular electrical stimulation increases muscle protein synthesis in elderly, type 2 diabetic men. Am. J. Physiol. Endocrinol. Metab. 303 (5), E614–E623.

Walrand, S., Short, K.R., Bigelow, M.L., Sweatt, A.J., Hutson, S.M., Nair, K.S., 2008. Functional impact of high protein intake on healthy elderly people. Am. J. Physiol. Endocrinol. Metab. 295, E921–E928.

Wang, Y., Zhang, L., Zhou, G., Liao, Z., Ahmad, H., Liu, W., et al., 2012. Dietary l-arginine supplementation improves the intestinal development through increasing mucosal Akt and mammalian target of rapamycin signals in intra-uterine growth retarded piglets. Br. J. Nutr.1–11.

Wei, N., Shi, Y., Truong, L.N., Fisch, K.M., Xu, T., Gardiner, E., et al., 2014. Oxidative stress diverts tRNA synthetase to nucleus for protection against DNA damage. Mol. Cell 56, 323–332.

Welle, S., Thornton, C., Jozefowicz, R., Statt, M., 1993. Myofibrillar protein synthesis in young and old men. Am. J. Physiol. 264, E693–E698.

WHO. Definition of an Older or Elderly Person [Online]. Available at <http://www.who.int/healthinfo/survey/ageingdefnolder/en/#>.

Winter, D.L., Paulin, D., Mericskay, M., Li, Z., 2013. Posttranslational modifications of desmin and their implication in biological processes and pathologies. Histochem. Cell Biol. 141, 1–16.

Wolfe, R.R., 2006. The underappreciated role of muscle in health and disease. Am. J. Clin. Nutr. 84, 475–482.

Wood, L.K., Kayupov, E., Gumucio, J.P., Mendias, C.L., Claflin, D.R., Brooks, S.V., 2014. Intrinsic stiffness of extracellular matrix increases with age in skeletal muscles of mice. J. Appl. Physiol. (1985) 117, 363–369.

Wroblewski, A.P., Amati, F., Smiley, M.A., Goodpaster, B., Wright, V., 2011. Chronic exercise preserves lean muscle mass in masters athletes. Phys. Sportsmed. 39, 172–178.

Wu, G., 1998. Intestinal mucosal amino acid catabolism. J. Nutr. 128, 1249–1252.

Wu, G., Meininger, C.J., 2002. Regulation of nitric oxide synthesis by dietary factors. Annu. Rev. Nutr. 22, 61–86.

Wu, G., Jaeger, L.A., Bazer, F.W., Rhoads, J.M., 2004. Arginine deficiency in preterm infants: biochemical mechanisms and nutritional implications. J. Nutr. Biochem. 15, 442–451.

Wu, Y., Bauman, W.A., Blitzer, R.D., Cardozo, C., 2010. Testosterone-induced hypertrophy of L6 myoblasts is dependent upon Erk and mTOR. Biochem. Biophys. Res. Commun. 400, 679–683.

Xi, P., Jiang, Z., Dai, Z., Li, X., Yao, K., Zheng, C., et al., 2011. Regulation of protein turnover by l-glutamine in porcine intestinal epithelial cells. J. Nutr. Biochem. 23, 1012–1017.

Xu, Z.R., Tan, Z.J., Zhang, Q., Gui, Q.F., Yang, Y.M., 2014. Clinical effectiveness of protein and amino acid supplementation on building muscle mass in elderly people: a meta-analysis. PLoS One 9, e109141.

Yang, Z., Klionsky, D.J., 2009. Mammalian autophagy: core molecular machinery and signaling regulation. Curr. Opin. Cell Biol. 22, 124–131.

Yang, Y., Churchward-Venne, T.A., Burd, N.A., Breen, L., Tarnopolsky, M.A., Phillips, S.M., 2012. Myofibrillar protein synthesis following ingestion of soy protein isolate at rest and after resistance exercise in elderly men. Nutr. Metab. (Lond) 9, 57.

Yao, K., Yin, Y.L., Chu, W., Liu, Z., Deng, D., Li, T., et al., 2008. Dietary arginine supplementation increases mTOR signaling activity in skeletal muscle of neonatal pigs. J. Nutr. 138, 867–872.

Yarasheski, K.E., Zachwieja, J.J., Bier, D.M., 1993. Acute effects of resistance exercise on muscle protein synthesis rate in young and elderly men and women. Am. J. Physiol. 265, E210–E214.

Yarasheski, K.E., Castaneda-Sceppa, C., He, J., Kawakubo, M., Bhasin, S., Binder, E.F., et al., 2011. Whole-body and muscle protein metabolism are not affected by acute deviations from habitual protein intake in older men: the Hormonal Regulators of Muscle and Metabolism in Aging (HORMA) Study. Am. J. Clin. Nutr. 94, 172–181.

Yoon, M.S., Chen, J., 2013. Distinct amino acid-sensing mTOR pathways regulate skeletal myogenesis. Mol. Biol. Cell 24, 3754–3763.

Yoshikawa, T., Miyazaki, A., Fujimoto, S., 2009. Decrease in serum levels of advanced glycation end-products by short-term lifestyle modification in non-diabetic middle-aged females. Med. Sci. Monit. 15, PH65–PH73.

Zoncu, R., Bar-Peled, L., Efeyan, A., Wang, S., Sancak, Y., Sabatini, D.M., 2011. mTORC1 senses lysosomal amino acids through an inside-out mechanism that requires the vacuolar H(+)-ATPase. Science 334, 678–683.

CHAPTER

8

Specificity of Amino Acids and Protein Metabolism in Obesity

C. Guillet[1,2], *C. Domingues-Faria*[1,2], *S. Walrand*[1,2] *and Y. Boirie*[1,2,3]

[1]Clermont Université, Université d'Auvergne, Unité de Nutrition Humaine, Clermont-Ferrand, France [2]INRA, UMR 1019, UNH, CRNH Auvergne, Clermont-Ferrand, France [3]CHU Clermont-Ferrand, service de Nutrition Clinique, Clermont-Ferrand, France

8.1. INTRODUCTION: FAT-FREE MASS IN OBESITY

Because aging and obesity are two conditions that represent an important part of health care spending, an increasingly obese elderly population will undoubtedly represent a growing financial problem in health care systems in economically developed countries. Obesity (as defined by a body mass index, BMI $>30\ kg/m^2$) is increasing rapidly and is now recognized as an important global health problem. In Europe, since the 1980s, the prevalence of obesity has tripled (Obepi, 2012; Rennie and Jebb, 2005). In the United States, half of the population is estimated as overweight and around 30% of American people are obese (Flegal et al., 2002; Mokdad et al., 2001). It is remarkable that a strong increase in obesity and overweight among elderly people (defined as person >65 years of age) is reported in both sexes, all ages, all races, and all educational levels, both smokers and nonsmokers (Flegal et al., 2002; Gutierrez-Fisac et al., 2004; Mokdad et al., 2001, 2003; Zamboni et al., 2005). It is estimated that more than 25% of European or US populations will be above 70 years by 2025 (Elia, 2001). In this context, the prevalence of obesity among subjects aged 60–69 years was about 40% in the United States in the period of 1999–2000, 30% in those aged 70–79 years, and 20% in those aged 80 years and older (Flegal et al., 2002; Zamboni et al., 2005). In the United Kingdom, nearly 30% of people aged 55–65 years, 25% in those aged 65–75 years, and 20% in those aged 75 years or older are obese (Seidell, 2002). In France, the ObEpi study has reported that 16% of old people were obese in 2006, that is, significantly more than in the general population (ObEpi, 2012).

Changes in body weight reflect a combination of changes in fat mass (FM) and fat free mass (FFM) which have different impacts on cardiometabolic diseases risk. During the onset of obesity, that is, during weight gain, modifications of body composition are illustrated by the increase of both FM and FFM but the distribution of weight gain between these two compartments varies among individuals: on average, on 10 kg weight gain, 7 kg will be acquired as fat and 3 kg as FFM. So in obese individuals, FFM has been classically reported to be increased or unchanged in comparison with lean subjects (Hulens et al., 2001; Patterson et al., 2002). The higher FFM suggests that exposure to greater weight bearing and greater mechanical loads may have a physical training-like effect (Hulens et al., 2001; Katsanos and Mandarino, 2011; Maffiuletti et al., 2007) and stimulate FFM anabolism. Interestingly, a misinterpretation of changes in body composition has arisen with confusion between FFM and muscle mass compartments: nonmuscle FFM might increase due to an obesity-induced enlargement of several organs like liver, intestine, kidney, and heart tissues (Wang et al., 2012), whereas muscle mass might not be involved. Generally speaking, regarding specific factors linked to individuals (physical activity, diet, genetics, pathology, treatments, etc.), there is a high variability in the evolution of body composition in the context of

The Molecular Nutrition of Amino Acids and Proteins.
DOI: http://dx.doi.org/10.1016/B978-0-12-802167-5.00008-6

weight gain which define individual body trajectories leading to specific phenotypes. For instance, a particular phenotype has been demonstrated in elderly characterized by a frank obesity with a parallel reduction in muscle mass and impairment of strength. Obesity is a strong risk factor for poor health, reduced functional capacity and quality of life in older persons (Alley and Chang, 2007). Analogously, low muscle strength has proven to be a strong predictor of functional disability, institutionalization, and mortality (Rantanen et al., 2000; Visser et al., 2005). This new clinical picture defined as sarcopenic obesity (Baumgartner, 2000) accumulates the risks from changes in each of the two determinants of body composition: the reduction of muscle mass limiting the mobility and involved in the development of metabolic disorders; and the excess of adiposity which generates significant adverse health effects (hypertension, dyslipidemia, insulin resistance, etc.) in addition to mechanical stress effects (Roubenoff, 2004; Stenholm et al., 2008). Sarcopenic obesity is mainly described in the elderly, however this phenotype might concern younger obese, notably in a context of weight loss (Prado et al., 2012). Skeletal muscle plays major roles in locomotive functions and motility and in the regulation of whole body energy homeostasis. Thus, quantitative and qualitative impairments in muscle tissue could affect mobility and quality of life of obese patients.

This chapter will review how protein metabolism could be affected in obesity leading to changes in altered preservation of FFM in obesity.

8.2. INSULIN RESISTANCE AND PROTEIN METABOLISM

In obesity, insulin resistance is characterized by impairments in glucose and lipid metabolisms (Kahn and Flier, 2000). However, insulin is also an important regulator of protein metabolism since this hormone and amino acids are key factors for the regulation of body and muscle protein mass (Biolo and Wolfe, 1993). Since obesity is frequently associated with insulin resistance, the anabolic action of insulin on protein metabolism could be impaired during obesity. Nevertheless, studies that have evaluated this aspect using the euglycemic hyperinsulinemic clamp method in combination with stable isotope tracers have also led to conflicting results. Indeed, the inhibitory action of insulin on whole body protein breakdown is either unchanged (Caballero and Wurtman, 1991; Marchesini et al., 2000; Solini et al., 1997; Welle et al., 1994) or impaired (Chevalier et al., 2005; Jensen and Haymond, 1991; Luzi et al., 1996; Nair et al., 1983). Luzi et al. observed that protein breakdown and leucine oxidation in response to insulin was impaired in obese subjects receiving a low dose of insulin but not in those receiving a high dose of insulin (Luzi et al., 1996). The different results obtained in these studies are probably due to methodological interpretation. For instance, a similar inhibition of protein breakdown during the insulin clamps was obtained with higher plasma insulin levels in obese subjects (Caballero and Wurtman, 1991; Luzi et al., 1996; Welle et al., 1994). Thus, to really estimate the inhibitory action of this hormone on protein breakdown, the magnitude of plasma insulin concentrations during the clamp have to be considered. In these conditions, significant differences between the obese and the nonobese subjects for the inhibition of whole body protein breakdown have been underlined (Guillet et al., 2009). In addition, the effect of insulin alone on whole body protein metabolism has been investigated in previous studies. However, hyperinsulinemia is associated with a decrease in plasma amino acid concentration leading to changes in the regulation of protein metabolism (Castellino et al., 1987). Whole-body protein synthesis is normally stimulated in obese subjects during hyperaminoacidemia with basal insulinemia suggesting that whole-body protein synthesis is still responsive to infusion of amino acid alone during obesity (Luzi et al., 1996). It also suggests that amino acid availability is a key factor for the stimulation of protein synthesis. The specific role of insulin has been considered by clamping plasma amino acid at their postabsorptive concentrations during a hyperinsulinemic clamp in order to discard any effect of plasma amino acids (Chevalier et al., 2005). In these conditions, the blunted whole-body protein anabolic response to the action of insulin results mainly from impaired stimulation of whole-body protein synthesis in obese women. However, the experimental conditions applied in the protocol design did not really match a physiological postprandial state when both insulin and amino acids are elevated in plasma to fully promote protein anabolism. The simulation of a more physiological fed state can be obtained by performing a hyperinsulinemic, hyperaminoacidemic clamp in obese volunteers. Combined infusions of insulin and amino acid failed to induce a normal inhibition of protein breakdown in nondiabetic obese subjects in comparison with nonobese subjects (Chevalier et al., 2015; Guillet et al., 2009). The distribution of fat in the body can affect the response of protein metabolism to nutrients. Indeed, whole-body protein balance following a meal is less positive in obese women with visceral adiposity, and thus insulin-resistant, than in obese women with a fat distribution at lower-body level (Liebau et al., 2014). In middle-aged adults and in the elderly, protein anabolic response to insulin is blunted during obesity (Murphy et al., 2015; Pereira et al., 2008). Thus, inefficient protein anabolism may contribute to the development of sarcopenic obesity with aging.

In obese animals, several studies have observed the inability of insulin to stimulate protein synthesis in muscle. In diabetic obese rats treated with insulin for a short period of time, the increase in muscle protein synthesis rate observed was lower than in lean rats (Chan et al., 1985), suggesting a resistance to the anabolic action of insulin. In obese db/db mice, muscle protein synthesis rate measured in situ in perfused hindquarters was approximately 50% lower than in lean mice (Shargill et al., 1984). On the other side, in obese Zucker rats, in situ insulin infusion in hindlimbs induces stimulation of muscle protein synthesis, despite a significant reduction of muscle mass (Fluckey et al., 2000). This suggests that obese Zucker rats have certainly become resistant to insulin antiproteolytic actions. In mice diet-induced obesity did not affect basal rates of skeletal muscle protein synthesis, but impairs the activation of skeletal muscle protein synthesis in response to nutrient ingestion (Anderson et al., 2008). In an animal model of diet-induced obesity without obesity-associated metabolic alterations (such as insulin resistance, inflammation, or oxidative stress), excessive energy intake has no effect on skeletal muscle and liver proteins synthesis rates in Wistar male rats (Adechian et al., 2009). In obese people, the turnover rate of skeletal muscle proteins has been reported to be reduced in a postabsorptive state, illustrated by a much lower protein synthesis (Guillet et al., 2009) and reduced protein breakdown (Patterson et al., 2002) rates than lean subjects.

In addition, a lack of stimulation of muscle mitochondria protein synthesis in response to insulin and amino acids infusion has been reported in obese subjects (Guillet et al., 2009).

Understanding the metabolic consequences of obesity and insulin resistance on muscle protein turnover may help to recommend appropriate use of dietary protein together with physical activity during energy restriction in promoting protein synthesis to maintain skeletal muscle integrity during obesity. These strategies may optimize mobility and limit obesity-related morbidity.

8.3. LIPOTOXICITY AND MUSCLE PROTEIN METABOLISM

Obesity and aging are both characterized by increased intracellular concentration of fatty acid metabolites in skeletal muscle (Tumova et al., 2015). Most of these lipid compounds have been postulated to activate a serine kinase cascade leading to defects in insulin signaling in muscle (Martins et al., 2012), which results in reduced insulin stimulated glucose transport activity and reduced glycogen synthesis (Tumova et al., 2015). Recent evidence also suggested that muscle lipid infiltration was able to disrupt the regulation of muscle protein synthesis and consequently act as a key modulator of the loss in muscle mass and quality in obese and older subjects.

Lipid deposition is altered in obese and older subjects with an increased liver and skeletal muscle fat infiltration. The increased lipid ectopic deposition in many tissues is associated with the appearance of insulin resistance (Slawik and Vidal-Puig, 2007). Inversely, promoting the storage of fat preferentially in adipose tissue with adiponectin transgenic ob/ob mice or peroxisome proliferator-activated receptor gamma (PPARgamma) agonist treatment improved insulin sensitivity (Kim et al., 2007). Besides increasing the storage capacity of fat in adipocytes, increasing fatty acid oxidation in muscle can also prevent lipotoxicity (Henique et al., 2010). Very old transgenic PGC1a (peroxisome proliferator-activated receptor gamma coactivator 1 alpha) mice showed increased muscle mitochondrial activity and improved metabolic responses as illustrated by decreased circulating lipid levels and increased insulin sensitivity (Tiraby et al., 2007; Wenz et al., 2009). The increased lipid content inside muscle in older persons is independently associated with insulin resistance (Ryan and Nicklas, 1999; Zoico et al., 2010), and a moderate weight loss improves muscle lipid infiltration and insulin resistance in postmenopausal women (Mazzali et al., 2006). Muscle fat accumulation in older people is associated not only with metabolic abnormalities, but also with reduced strength and poorer scores in performance tests and incident mobility disability (Goodpaster et al., 2001; Zoico et al., 2010). Further, the increase in intramuscular fat coincides longitudinally with the progressive muscle weakening in aging (Delmonico et al., 2009). Thus, besides the effect of muscle fat accumulation on insulin sensitivity, muscle lipid accumulation may also promote specific changes in muscle protein metabolism. Anderson and coworkers (Anderson et al., 2008) noted a significant reduction in the ability of muscles from mice fed a high-fat diet to stimulate protein synthesis following a meal, but intramuscular lipid content was not measured. These data suggest that lipid accumulation following a chronic high-fat diet interferes with signaling pathways involved in muscle response to anabolic stimuli, blunting the activation of protein synthesis. We demonstrated (Tardif et al., 2014) that rats exhibited ectopic fat deposition with aging, in particular when a diet-induced obesity was used. In this study, the decreased ability of adipose tissue to store lipids in old rats contributed to the accumulation of intramuscular lipids, which put a lipid burden on mitochondria and created a disconnect between metabolic and regulating pathways in skeletal muscles. Although the precise connection between mitochondrial activity and insulin resistance is still debated, our data also demonstrated that the increased accumulation of lipid metabolism by-products was linked to impaired insulin action in palmitate treated C2C12 myotubes and in old rats fed a high-fat diet. Additionally, this study revealed that intramuscular lipid deposition in old obese rats, together with palmitate-mediated lipid accumulation in C2C12 cells, affected muscle protein synthesis rate via eIF2α phosphorylation. The use of pharmacological inhibitors of ceramide production or the use of ceramide confirmed these effects, showing that the harmful action of reactive lipids on muscle protein synthesis was probably driven by ceramide accumulation inside muscle cells. Therefore, a decrease in the ability of adipose tissue to respond to diet-induced obesity with age could adversely affect protein metabolism in skeletal muscle due to the toxicity of lipid species and thereby accelerate the effects of aging, that is, sarcopenia. Of note in a model of old rats, a high-fat oleate-enriched diet decreased inflammation markers in the plasma and adipose tissue of old rats (Tardif et al., 2011). Oleate treatment also upregulated the expression of enzymes involved in muscle mitochondrial β-oxidation, which likely contributed to the improved muscle insulin sensitivity by purifying cell lipid content. The major effect of feeding old rats a high-oleate diet was to restore muscle protein synthesis in response to amino acid and insulin.

In young obese people, a lower muscle mass or a lower muscle mass gain with increased body weight has been reported (Janssen et al., 2000; Patterson et al., 2002). Noticeably, muscle alterations, such as changes in fiber types (Kemp et al., 2009; Lillioja et al., 1987; Tanner et al., 2002) and in contractile and metabolic functions (Simoneau et al., 1995), have also been described in obesity, suggesting that not only muscle mass but also muscle quality may be impaired in obese patients. In humans, a negative relationship has been highlighted between body fat mass and muscle protein synthesis in young and elderly people (Guillet et al., 2009; Katsanos and Mandarino, 2011), suggesting that an increase in adipose or ectopic fat mass may affect protein kinetics within skeletal muscle. In young obese humans and Zucker rats, studies have reported a decreased (Argiles et al., 1999; Durschlag and Layman, 1983; Fluckey et al., 2004; Guillet et al., 2009; Katsanos and Mandarino, 2011; Nilsson et al., 2010) or unchanged (Fluckey et al., 2000; Nilsson et al., 2010) muscle protein synthesis rate in the postabsorptive state. By contrast, in animal models of diet-induced obesity, studies have reported an increased (Chanseaume et al., 2007) or unchanged (Anderson et al., 2008) protein synthesis rate in the postabsorptive state. In the genetically obese animal or the obese patient, studies have all been performed after long-term obesity. In contrast, diet-induced obesity studies in animals were mostly realized over a relatively short time period and the exposure time to obesity has never been considered with regard to muscle protein metabolism. Therefore, we previously analyzed the sequential changes of protein synthesis in skeletal muscle during the establishment of obesity in a model of prolonged diet-induced obesity in young rats (Masgrau et al., 2012). In this work, muscle and adipose tissue mass, intramyocellular lipids and muscle protein synthesis were differently affected according to the stage of obesity development. Intramyocellular lipid content and muscle protein synthesis were also differently affected depending on muscle typology. The second phase of obesity development was associated with a reduction of adipose tissue expandability and an increase in intramyocellular lipids in glycolytic muscles, which was concomitant with lower protein synthesis stimulation firstly observed with the high-fat, high-sucrose diet. These data suggest that intramyocellular lipids accumulation is deleterious for the incorporation of amino acids in newly synthesized skeletal muscle proteins and could contribute to the loss of skeletal muscle mass and quality during obesity. Moreover, this process, especially if comorbidities of obesity occur, could lead ultimately to sarcopenic obesity, even in young individuals.

8.4. ROLE OF ADIPOSE AND MUSCULAR CYTOKINES IN THE CROSS-TALK BETWEEN MUSCLE AND ADIPOSE TISSUE

A corollary to the diversity of the adipokines is that white adipose tissue communicates extensively with other organs and is involved in multiple metabolic systems. Studies have indicated that adipocytes directly signal to other tissues, such as skeletal muscle (Argiles et al., 2005). To analyze this cross-talk, Dietze et al. (2002) established a system of coculture of human fat and skeletal muscle cells. Insulin-induced phosphorylation of insulin receptor substrate (IRS)-1 and of Akt kinase was completely blocked in myocytes after coculture with adipocytes. Interestingly, addition of TNFα to myocytes reduced IRS1 and Akt phosphorylation comparable to the effect of coculture (Dietze et al., 2002). The authors of this work concluded that the release of fat cell factors was in part responsible for the induction of insulin resistance in skeletal muscle. Interestingly, a similar TNFα-mediated inhibition of the insulin pathway was observed in vivo in adipose tissue and skeletal muscle of the obese and insulin resistant fa/fa rats (Hotamisligil et al., 1993). TNFα has other effects on adipose tissue including the regulation of adipokine expression. In a recent study examining the integrative effect of TNFα on the expression pattern of inflammation-related adipokines in human white adipocytes, the most substantial response occurred with IL6, MCP1, and TNFα itself, expression of each was strongly upregulated by TNFα (Wang et al., 2005).

Pro-inflammatory cytokines (TNFα, IL1β and IL6) have also been shown to induce muscle wasting directly by increasing myofibrillar protein degradation (Fong et al., 1989) and by decreasing protein synthesis (Lang et al., 2002). Enhancement of proteolysis is accomplished by activation of the ubiquitin-dependent proteolytic system (Garcia-Martinez et al., 1993). TNFα activates several serine/threonine kinases and intracellular factors, including the inhibitor of the NF-κB (IκB). This activation contributes to activate the NF-κB which is implicated in the upregulation of myofibrillar proteolysis by the proteasome system and in the suppression of myofibrillar protein synthesis (Guttridge et al., 2000; Li, 2003; Li et al., 1998). Other data (Lang et al., 2002) suggest that TNFα impairs skeletal muscle protein synthesis rate by decreasing translational efficiency resulting from an impairment in translation initiation associated with an alteration in the eukaryotic initiation factor-4E (eIF-4E). Additionally, myotubes treated with TNFα have shown increased reactive oxygen species (ROS) production, evidently released from mitochondria that activated the NF-κB and the FOXO3 forkhead pathways and led to protein loss (Jackman and Kandarian, 2004; Li et al., 1999). Lastly, it has been shown that TNFα activates other intracellular pathways, such as JNK or P38MAP kinase, which are involved in muscle proteolysis (Grivennikov et al., 2006). An indirect effect of TNFα on muscle protein metabolism may also be its capacity to inhibit insulin action since this hormone has been shown to increase muscle protein synthesis and to decrease proteolysis (Biolo and Wolfe, 1993; Guillet et al., 2004). Interestingly, TNFα also activates uncoupling proteins 2 and 3 (UCP2 and UCP3) gene expressions in muscle mitochondria and UCP2 gene expression in adipose tissue (Masaki et al., 1999a,b) which could in return affect mitochondrial function.

It rapidly became clear that many other inflammatory factors exhibit patterns of impact on muscle metabolism in a manner similar to that of TNFα (Wellen and Hotamisligil, 2005). It has been described that muscle cells express the leptin receptor and also that leptin expression in adipose tissue is activated by hyperglycemia or hyperlipidemia (Wang et al., 1998). In addition, leptin seems to have many effects on skeletal muscle (Argiles et al., 2005). Short-term administration of leptin to rats resulted in increased fatty acid oxidation and decreased fatty acid incorporation into triacylglycerols in skeletal muscle clearly demonstrating that leptin alters lipid partitioning in this tissue (Lopez-Soriano et al., 1998). Concerning protein metabolism, administration of leptin results in a decreased rate of myofibrillar protein synthesis in skeletal muscle (Carbo et al., 2000). IL6 and resistin are other well-characterized examples of compounds produced in adipose tissue that may participate in the regulation of muscle metabolism. Moon et al. (2003) have reported that resistin inhibits glucose uptake in L6 muscle cells independently of changes in insulin signaling or GLUT4 translocation. Moreover, IL6 levels are decreased after weight loss in obese subjects, this possibly being related to the improved insulin sensitivity observed in these patients after weight reduction (Bastard et al., 2000). IL6 has been also involved in the regulation of muscle protein turnover and is considered to be a catabolic cytokine (Zoico and Roubenoff, 2002).

It is also interesting to know that muscle cells are able to produced and release cytokines such as IL6 or IL15. These myokines contribute to muscle metabolism, for example, IL15 stimulates contractile protein accumulation, but may also participate in the conversation between skeletal muscle and adipose tissue since the presence of the three IL15 receptor subunits has been demonstrated in fat tissue (Argiles et al., 2005).

8.5. SARCOPENIC OBESITY AND METABOLIC IMPAIRMENTS

Aging is associated with sarcopenia referred to as the universal and involuntary decline in skeletal muscle mass. In young healthy subjects, muscle comprises nearly 60% of the fat-free mass and over half of the total body protein. However, in the elderly, muscle accounts for only 45% of fat-free mass and less than half of body protein, as a result of muscle loss (Proctor et al., 1999). Sarcopenia has major functional and metabolic consequences.

This results in loss of muscle strength and contributes to the eventual inability of the elderly individual to carry out tasks of daily living and to the increased risks of falls and fractures (Rantanen et al., 1999; Short and Nair, 2001). In addition, body proteins are involved in many functional activities such as contractile proteins in muscle, enzymes, hormones, and antibodies. Sarcopenia may therefore contribute to the reduced ability to withstand physical, environmental, or immunological stresses in old age. A body already depleted of proteins because of aging is obviously less capable of withstanding the protein catabolism that comes with chronic inflammation, an acute illness or inadequate intake of proteins. Furthermore, decrease in muscle function can lead to reduced physical activity thereby leading on to osteoporosis, obesity, and impaired glucose tolerance (Dutta et al., 1997; Evans, 1995). The estimated direct healthcare cost attributable to sarcopenia in the United States in 2000 was $18.5 billion, which represented about 1.5% of total healthcare expenditures for that year (Janssen et al., 2004).

The depletion of muscle mass with age does not result in weight loss, suggesting that a corresponding accumulation of body fat occurs; hence, fat mass increases from 18 to 36% in elderly men (Walrand and Boirie, 2005). Aging is also associated with a redistribution of body fat. With aging, there is a greater relative increase in intraabdominal, intrahepatic and intramuscular fat than subcutaneous fat (Beaufrere and Morio, 2000). The size of intraabdominal depots is greater in old than in young adults at any given BMI value. Changes in body fat distribution are associated with an increase in waist circumference in elderly; about 0.7 cm per year between 40 and 70 years with no difference across age-strata, that is, even the oldest continued to have progressive increases in waist circumference (Noppa and Bengtsson, 1980). In addition, abdominal fat accumulation is strongly associated with the risk of metabolic disorders (Farin et al., 2006; Racette et al., 2006). Hence, obesity in the elderly acts synergistically with sarcopenia to maximize disability. Sarcopenic obesity in old age is more strongly associated with disability than either sarcopenia or obesity per se (Baumgartner et al., 2004; Rolland et al., 2009). Older people are particularly susceptible to the adverse effects of excess body weight because it can exacerbate the age-related decline in physical function, that is, the decrease in muscle mass and strength which occur with aging.

Since sarcopenia and visceral obesity may aggravate each other, recent longitudinal studies have revealed that visceral obesity was associated with future loss of skeletal muscle mass in Korean adults (Kim et al., 2014). These new data provide insights into sarcopenic obesity in an aging society.

8.6. BCAA LEVELS AND METABOLISM IN OBESITY

In the late 1960s, impairments of AA metabolism were reported in obesity, suggesting a role of adipose tissue in AA disorders. Obesity and associated insulin resistance have been characterized by elevated blood concentrations of BCAA (Felig et al., 1969). More recently, metabolomic analysis have confirmed these alterations since circulating concentrations of BCAA and some of their metabolite acylcarnitin (AC) derivatives have been shown to be elevated in various human cohorts of obese and/or diabetic individuals of different ethnicity (Huffman et al., 2009; McCormack et al., 2013; Newgard et al., 2009; Tai et al., 2010). Noticeably, concentrations of BCAA in normoglycemic individuals strongly predict type 2 diabetes 12 years later in the offspring of the Framingham cohort (Wang et al., 2011), underscoring the role of AA in the pathogenesis of T2DM. Most obese subjects with insulin resistance do not develop diabetes, so that identifying early markers for the later risk of developing type 2 diabetes have potential clinical implications for targeted interventions.

On the metabolic point of view, BCAAs are poorly metabolized during their hepatic first pass as the liver expresses low levels of the mitochondrial branched-chain aminotransferase (BCAT2 or BCATm). Therefore, BCAA catabolism in the fasted state is primarily done by peripheral tissues, particularly muscle and adipose tissue rather than liver. The capacity of adipose tissue and muscle to catabolize BCAA is about six-fold higher than in the liver taking into account the relative masses of tissues. The decrease in fasted plasma BCAA concentrations after RY-GBP (Laferrere et al., 2011) was shown to be associated with an increase of two key BCAA catabolic enzymes, the BCATm and the branched-chain α-keto acid dehydrogenase E1 (BCKD E1α) in both subcutaneous and visceral fat depots (VAT) (She et al., 2007).

Recently, in animal models the metabolism of BCAA in adipose tissue was shown to exert a significant impact on circulating BCAA levels (Herman et al., 2010). The rate of adipose tissue BCAA oxidation in normal mice was higher than in skeletal muscle and manipulation of the expression of GLUT4 receptor in adipose tissue downregulated the BCAA metabolizing enzymes selectively in adipose tissue not in muscle, in conjunction with increased circulating BCAA levels. These observations suggest that in the case of insulin resistance and lower glucose uptake in muscle, a higher glucose availability for adipose tissue may alter BCAA metabolizing enzymes and contribute to higher BCAA concentrations.

8.7. CONCLUSION

Obesity is associated with many changes in protein metabolism especially at the muscle level. Insulin resistance, lipotoxicity, and inflammation are metabolic abnormalities related to obesity and represent confounding factors in the alteration of protein metabolism in this context. Understanding the metabolic consequences of obesity, insulin resistance, and lipotoxicity on muscle protein turnover is a major issue for recommending the appropriate use of dietary protein together with physical activity during energy restriction in promoting protein synthesis to maintain skeletal muscle integrity during obesity. These strategies may optimize mobility and limit obesity-related morbidity.

References

Adechian, S., Giardina, S., Remond, D., Papet, I., Buonocore, D., Gaudichon, C., et al., 2009. Excessive energy intake does not modify fed-state tissue protein synthesis rates in adult rats. Obesity (Silver Spring) 17, 1348–1355.
Alley, D.E., Chang, V.W., 2007. The changing relationship of obesity and disability, 1988–2004. J. Am. Med. Assoc. 298, 2020–2027.
Anderson, S.R., Gilge, D.A., Steiber, A.L., Previs, S.F., 2008. Diet-induced obesity alters protein synthesis: tissue-specific effects in fasted versus fed mice. Metabolism 57, 347–354.
Argiles, J.M., Busquets, S., Alvarez, B., Lopez-Soriano, F.J., 1999. Mechanism for the increased skeletal muscle protein degradation in the obese Zucker rat. J. Nutr. Biochem. 10, 244–248.
Argiles, J.M., Lopez-Soriano, J., Almendro, V., Busquets, S., Lopez-Soriano, F.J., 2005. Cross-talk between skeletal muscle and adipose tissue: a link with obesity? Med. Res. Rev. 25, 49–65.
Bastard, J.P., Jardel, C., Bruckert, E., Blondy, P., Capeau, J., Laville, M., et al., 2000. Elevated levels of interleukin 6 are reduced in serum and subcutaneous adipose tissue of obese women after weight loss. J. Clin. Endocrinol. Metab. 85, 3338–3342.
Baumgartner, R.N., 2000. Body composition in healthy aging. Ann. N.Y. Acad. Sci. 904, 437–448.
Baumgartner, R.N., Wayne, S.J., Waters, D.L., Janssen, I., Gallagher, D., Morley, J.E., 2004. Sarcopenic obesity predicts instrumental activities of daily living disability in the elderly. Obes. Res. 12, 1995–2004.
Beaufrere, B., Morio, B., 2000. Fat and protein redistribution with aging: metabolic considerations. Eur. J. Clin. Nutr. 54 (Suppl. 3), S48–S53.
Biolo, G., Wolfe, R.R., 1993. Insulin action on protein metabolism. Baillieres Clin. Endocrinol. Metab. 7, 989–1005.
Caballero, B., Wurtman, R.J., 1991. Differential effects of insulin resistance on leucine and glucose kinetics in obesity. Metabolism 40, 51–58.
Carbo, N., Ribas, V., Busquets, S., Alvarez, B., Lopez-Soriano, F.J., Argiles, J.M., 2000. Short-term effects of leptin on skeletal muscle protein metabolism in the rat. J. Nutr. Biochem. 11, 431–435.
Castellino, P., Luzi, L., Simonson, D.C., Haymond, M., DeFronzo, R.A., 1987. Effect of insulin and plasma amino acid concentrations on leucine metabolism in man. Role of substrate availability on estimates of whole body protein synthesis. J. Clin. Invest. 80, 1784–1793.
Chan, C.P., Hansen, R.J., Stern, J.S., 1985. Protein turnover in insulin-treated, alloxan-diabetic lean and obese Zucker rats. J. Nutr. 115, 959–969.
Chanseaume, E., Giraudet, C., Gryson, C., Walrand, S., Rousset, P., Boirie, Y., et al., 2007. Enhanced muscle mixed and mitochondrial protein synthesis rates after a high-fat or high-sucrose diet. Obesity (Silver Spring) 15, 853–859.
Chevalier, S., Marliss, E.B., Morais, J.A., Lamarche, M., Gougeon, R., 2005. Whole-body protein anabolic response is resistant to the action of insulin in obese women. Am. J. Clin. Nutr. 82, 355–365.
Chevalier, S., Burgos, S.A., Morais, J.A., Gougeon, R., Bassil, M., Lamarche, M., et al., 2015. Protein and glucose metabolic responses to hyperinsulinemia, hyperglycemia, and hyperaminoacidemia in obese men. Obesity (Silver Spring) 23, 351–358.
Delmonico, M.J., Harris, T.B., Visser, M., Park, S.W., Conroy, M.B., Velasquez-Mieyer, P., et al., 2009. Longitudinal study of muscle strength, quality, and adipose tissue infiltration. Am. J. Clin. Nutr. 90, 1579–1585.
Dietze, D., Koenen, M., Rohrig, K., Horikoshi, H., Hauner, H., Eckel, J., 2002. Impairment of insulin signaling in human skeletal muscle cells by co-culture with human adipocytes. Diabetes 51, 2369–2376.
Durschlag, R.P., Layman, D.K., 1983. Skeletal muscle growth in lean and obese Zucker rats. Growth 47, 282–291.
Dutta, C., Hadley, E.C., Lexell, J., 1997. Sarcopenia and physical performance in old age: overview. Muscle Nerve Suppl. 5, S5–S9.
Elia, M., 2001. Obesity in the elderly. Obes. Res. 9 (Suppl. 4), 244S–248S.
Evans, W.J., 1995. What is sarcopenia? J. Gerontol. A Biol. Sci. Med. Sci. 50, 5–8, Spec No.
Farin, H.M., Abbasi, F., Reaven, G.M., 2006. Body mass index and waist circumference both contribute to differences in insulin-mediated glucose disposal in nondiabetic adults. Am. J. Clin. Nutr. 83, 47–51.
Felig, P., Marliss, E., Cahill Jr., G.F., 1969. Plasma amino acid levels and insulin secretion in obesity. N. Engl. J. Med. 281, 811–816.
Flegal, K.M., Carroll, M.D., Ogden, C.L., Johnson, C.L., 2002. Prevalence and trends in obesity among US adults, 1999–2000. J. Am. Med. Assoc. 288, 1723–1727.
Fluckey, J.D., Pohnert, S.C., Boyd, S.G., Cortright, R.N., Trappe, T.A., Dohm, G.L., 2000. Insulin stimulation of muscle protein synthesis in obese Zucker rats is not via a rapamycin-sensitive pathway. Am. J. Physiol. Endocrinol. Metab. 279, E182–E187.
Fluckey, J.D., Cortright, R.N., Tapscott, E., Koves, T., Smith, L., Pohnert, S., et al., 2004. Active involvement of PKC for insulin-mediated rates of muscle protein synthesis in Zucker rats. Am. J. Physiol. Endocrinol. Metab. 286, E753–E758.
Fong, Y., Moldawer, L.L., Marano, M., Wei, H., Barber, A., Manogue, K., et al., 1989. Cachectin/TNF or IL-1 alpha induces cachexia with redistribution of body proteins. Am. J. Physiol. 256, R659–R665.
Garcia-Martinez, C., Agell, N., Llovera, M., Lopez-Soriano, F.J., Argiles, J.M., 1993. Tumour necrosis factor-alpha increases the ubiquitinization of rat skeletal muscle proteins. FEBS Lett. 323, 211–214.

Goodpaster, B.H., Carlson, C.L., Visser, M., Kelley, D.E., Scherzinger, A., Harris, T.B., et al., 2001. Attenuation of skeletal muscle and strength in the elderly: the Health ABC Study. J. Appl. Physiol. 1985 90, 2157–2165.

Grivennikov, S.I., Kuprash, D.V., Liu, Z.G., Nedospasov, S.A., 2006. Intracellular signals and events activated by cytokines of the tumor necrosis factor superfamily: from simple paradigms to complex mechanisms. Int. Rev. Cytol. 252, 129–161.

Guillet, C., Zangarelli, A., Gachon, P., Morio, B., Giraudet, C., Rousset, P., et al., 2004. Whole body protein breakdown is less inhibited by insulin, but still responsive to amino acid, in nondiabetic elderly subjects. J. Clin. Endocrinol. Metab. 89, 6017–6024.

Guillet, C., Delcourt, I., Rance, M., Giraudet, C., Walrand, S., Bedu, M., et al., 2009. Changes in basal and insulin and amino acid response of whole body and skeletal muscle proteins in obese men. J. Clin. Endocrinol. Metab. 94, 3044–3050.

Gutierrez-Fisac, J.L., Lopez, E., Banegas, J.R., Graciani, A., Rodriguez-Artalejo, F., 2004. Prevalence of overweight and obesity in elderly people in Spain. Obes. Res. 12, 710–715.

Guttridge, D.C., Mayo, M.W., Madrid, L.V., Wang, C.Y., Baldwin Jr., A.S., 2000. NF-κB-induced loss of MyoD messenger RNA: possible role in muscle decay and cachexia. Science 289, 2363–2366.

Henique, C., Mansouri, A., Fumey, G., Lenoir, V., Girard, J., Bouillaud, F., et al., 2010. Increased mitochondrial fatty acid oxidation is sufficient to protect skeletal muscle cells from palmitate-induced apoptosis. J. Biol. Chem. 285, 36818–36827.

Herman, M.A., She, P., Peroni, O.D., Lynch, C.J., Kahn, B.B., 2010. Adipose tissue branched chain amino acid BCAA metabolism modulates circulating BCAA levels. J. Biol. Chem. 285, 11348–11356.

Hotamisligil, G.S., Shargill, N.S., Spiegelman, B.M., 1993. Adipose expression of tumor necrosis factor-alpha: direct role in obesity-linked insulin resistance. Science 259, 87–91.

Huffman, K.M., Shah, S.H., Stevens, R.D., Bain, J.R., Muehlbauer, M., Slentz, C.A., et al., 2009. Relationships between circulating metabolic intermediates and insulin action in overweight to obese, inactive men and women. Diabetes Care 32, 1678–1683.

Hulens, M., Vansant, G., Lysens, R., Claessens, A.L., Muls, E., 2001. Exercise capacity in lean versus obese women. Scand. J. Med. Sci. Sports 11, 305–309.

Jackman, R.W., Kandarian, S.C., 2004. The molecular basis of skeletal muscle atrophy. Am. J. Physiol. Cell Physiol. 287, C834–C843.

Janssen, I., Heymsfield, S.B., Wang, Z.M., Ross, R., 2000. Skeletal muscle mass and distribution in 468 men and women aged 18–88 yr. J. Appl. Physiol. 1985 89, 81–88.

Janssen, I., Shepard, D.S., Katzmarzyk, P.T., Roubenoff, R., 2004. The healthcare costs of sarcopenia in the United States. J. Am. Geriatr. Soc. 52, 80–85.

Jensen, M.D., Haymond, M.W., 1991. Protein metabolism in obesity: effects of body fat distribution and hyperinsulinemia on leucine turnover. Am. J. Clin. Nutr. 53, 172–176.

Kahn, B.B., Flier, J.S., 2000. Obesity and insulin resistance. J. Clin. Invest. 106, 473–481.

Katsanos, C.S., Mandarino, L.J., 2011. Protein metabolism in human obesity: a shift in focus from whole-body to skeletal muscle. Obesity (Silver Spring) 19, 469–475.

Kemp, J.G., Blazev, R., Stephenson, D.G., Stephenson, G.M., 2009. Morphological and biochemical alterations of skeletal muscles from the genetically obese (ob/ob) mouse. Int. J. Obes. (Lond) 33, 831–841.

Kim, J.Y., van de Wall, E., Laplante, M., Azzara, A., Trujillo, M.E., Hofmann, S.M., et al., 2007. Obesity-associated improvements in metabolic profile through expansion of adipose tissue. J. Clin. Invest. 117, 2621–2637.

Kim, T.N., Park, M.S., Kim, Y.J., Lee, E.J., Kim, M.K., Kim, J.M., et al., 2014. Association of low muscle mass and combined low muscle mass and visceral obesity with low cardiorespiratory fitness. PLoS One 9, e100118.

Laferrere, B., Reilly, D., Arias, S., Swerdlow, N., Gorroochurn, P., Bawa, B., et al., 2011. Differential metabolic impact of gastric bypass surgery versus dietary intervention in obese diabetic subjects despite identical weight loss. Sci. Transl. Med. 3, 80re82.

Lang, C.H., Frost, R.A., Nairn, A.C., MacLean, D.A., Vary, T.C., 2002. TNF-α impairs heart and skeletal muscle protein synthesis by altering translation initiation. Am. J. Physiol. Endocrinol. Metab. 282, E336–E347.

Li, Y.P., 2003. TNF-α is a mitogen in skeletal muscle. Am. J. Physiol. Cell Physiol. 285, C370–C376.

Li, Y.P., Schwartz, R.J., Waddell, I.D., Holloway, B.R., Reid, M.B., 1998. Skeletal muscle myocytes undergo protein loss and reactive oxygen-mediated NF-κB activation in response to tumor necrosis factor alpha. FASEB J. 12, 871–880.

Li, Y.P., Atkins, C.M., Sweatt, J.D., Reid, M.B., 1999. Mitochondria mediate tumor necrosis factor-alpha/NF-κB signaling in skeletal muscle myotubes. Antioxid. Redox Signal. 1, 97–104.

Liebau, F., Jensen, M.D., Nair, K.S., Rooyackers, O., 2014. Upper-body obese women are resistant to postprandial stimulation of protein synthesis. Clin. Nutr. 33, 802–807.

Lillioja, S., Young, A.A., Culter, C.L., Ivy, J.L., Abbott, W.G., Zawadzki, J.K., et al., 1987. Skeletal muscle capillary density and fiber type are possible determinants of in vivo insulin resistance in man. J. Clin. Invest. 80, 415–424.

Lopez-Soriano, J., Carbo, N., Lopez-Soriano, F.J., Argiles, J.M., 1998. Short-term effects of leptin on lipid metabolism in the rat. FEBS Lett. 431, 371–374.

Luzi, L., Castellino, P., DeFronzo, R.A., 1996. Insulin and hyperaminoacidemia regulate by a different mechanism leucine turnover and oxidation in obesity. Am. J. Physiol. 270, E273–E281.

Maffiuletti, N.A., Jubeau, M., Munzinger, U., Bizzini, M., Agosti, F., De Col, A., et al., 2007. Differences in quadriceps muscle strength and fatigue between lean and obese subjects. Eur. J. Appl. Physiol. 101, 51–59.

Marchesini, G., Bianchi, G., Rossi, B., Muggeo, M., Bonora, E., 2000. Effects of hyperglycaemia and hyperinsulinaemia on plasma amino acid levels in obese subjects with normal glucose tolerance. Int. J. Obes. Relat. Metab. Disord. 24, 552–558.

Martins, A.R., Nachbar, R.T., Gorjao, R., Vinolo, M.A., Festuccia, W.T., Lambertucci, R.H., et al., 2012. Mechanisms underlying skeletal muscle insulin resistance induced by fatty acids: importance of the mitochondrial function. Lipids Health Dis. 11, 30.

Masaki, T., Yoshimatsu, H., Chiba, S., Hidaka, S., Tajima, D., Kakuma, T., et al., 1999a. Tumor necrosis factor-alpha regulates in vivo expression of the rat UCP family differentially. Biochim. Biophys. Acta 1436, 585–592.

Masaki, T., Yoshimatsu, H., Chiba, S., Kurokawa, M., Sakata, T., 1999b. Up-regulation of uterine UCP2 and UCP3 in pregnant rats. Biochim. Biophys. Acta 1440, 81–88.

Masgrau, A., Mishellany-Dutour, A., Murakami, H., Beaufrere, A.M., Walrand, S., Giraudet, C., et al., 2012. Time-course changes of muscle protein synthesis associated with obesity-induced lipotoxicity. J. Physiol. 590, 5199–5210.

Mazzali, G., Di Francesco, V., Zoico, E., Fantin, F., Zamboni, G., Benati, C., et al., 2006. Interrelations between fat distribution, muscle lipid content, adipocytokines, and insulin resistance: effect of moderate weight loss in older women. Am. J. Clin. Nutr. 84, 1193–1199.

McCormack, S.E., Shaham, O., McCarthy, M.A., Deik, A.A., Wang, T.J., Gerszten, R.E., et al., 2013. Circulating branched-chain amino acid concentrations are associated with obesity and future insulin resistance in children and adolescents. Pediatr. Obes. 8, 52–61.

Mokdad, A.H., Bowman, B.A., Ford, E.S., Vinicor, F., Marks, J.S., Koplan, J.P., 2001. The continuing epidemics of obesity and diabetes in the United States. J. Am. Med. Assoc. 286, 1195–1200.

Mokdad, A.H., Ford, E.S., Bowman, B.A., Dietz, W.H., Vinicor, F., Bales, V.S., et al., 2003. Prevalence of obesity, diabetes, and obesity-related health risk factors, 2001. J. Am. Med. Assoc. 289, 76–79.

Moon, B., Kwan, J.J., Duddy, N., Sweeney, G., Begum, N., 2003. Resistin inhibits glucose uptake in L6 cells independently of changes in insulin signaling and GLUT4 translocation. Am. J. Physiol. Endocrinol. Metab. 285, E106–E115.

Murphy, J., Chevalier, S., Gougeon, R., Goulet, E.D., Morais, J.A., 2015. Effect of obesity and type 2 diabetes on protein anabolic response to insulin in elderly women. Exp. Gerontol. 69, 20–26.

Nair, K.S., Garrow, J.S., Ford, C., Mahler, R.F., Halliday, D., 1983. Effect of poor diabetic control and obesity on whole body protein metabolism in man. Diabetologia 25, 400–403.

Newgard, C.B., An, J., Bain, J.R., Muehlbauer, M.J., Stevens, R.D., Lien, L.F., et al., 2009. A branched-chain amino acid-related metabolic signature that differentiates obese and lean humans and contributes to insulin resistance. Cell Metab. 9, 311–326.

Nilsson, M.I., Greene, N.P., Dobson, J.P., Wiggs, M.P., Gasier, H.G., Macias, B.R., et al., 2010. Insulin resistance syndrome blunts the mitochondrial anabolic response following resistance exercise. Am. J. Physiol. Endocrinol. Metab. 299, E466–E474.

Noppa, H., Bengtsson, C., 1980. Obesity in relation to socioeconomic status. A population study of women in Goteborg, Sweden. J. Epidemiol. Community Health 34, 139–142.

ObEpi, 2012. Enquête épidémiologique nationale sur le surpoids et l'obésité.

Patterson, B.W., Horowitz, J.F., Wu, G., Watford, M., Coppack, S.W., Klein, S., 2002. Regional muscle and adipose tissue amino acid metabolism in lean and obese women. Am. J. Physiol. Endocrinol. Metab. 282, E931–E936.

Pereira, S., Marliss, E.B., Morais, J.A., Chevalier, S., Gougeon, R., 2008. Insulin resistance of protein metabolism in type 2 diabetes. Diabetes 57, 56–63.

Prado, C.M., Wells, J.C., Smith, S.R., Stephan, B.C., Siervo, M., 2012. Sarcopenic obesity: a critical appraisal of the current evidence. Clin. Nutr. 31, 583–601.

Proctor, D.N., O'Brien, P.C., Atkinson, E.J., Nair, K.S., 1999. Comparison of techniques to estimate total body skeletal muscle mass in people of different age groups. Am. J. Physiol. 277, E489–E495.

Racette, S.B., Evans, E.M., Weiss, E.P., Hagberg, J.M., Holloszy, J.O., 2006. Abdominal adiposity is a stronger predictor of insulin resistance than fitness among 50–95 year olds. Diabetes Care 29, 673–678.

Rantanen, T., Guralnik, J.M., Sakari-Rantala, R., Leveille, S., Simonsick, E.M., Ling, S., et al., 1999. Disability, physical activity, and muscle strength in older women: the Women's Health and Aging Study. Arch. Phys. Med. Rehabil. 80, 130–135.

Rantanen, T., Harris, T., Leveille, S.G., Visser, M., Foley, D., Masaki, K., et al., 2000. Muscle strength and body mass index as long-term predictors of mortality in initially healthy men. J. Gerontol. A Biol. Sci. Med. Sci. 55, M168–M173.

Rennie, K.L., Jebb, S.A., 2005. Prevalence of obesity in Great Britain. Obes. Rev. 6, 11–12.

Rolland, Y., Lauwers-Cances, V., Cristini, C., Abellan van Kan, G., Janssen, I., Morley, J.E., et al., 2009. Difficulties with physical function associated with obesity, sarcopenia, and sarcopenic-obesity in community-dwelling elderly women: the EPIDOS (EPIDemiologie de l'OSteoporose) Study. Am. J. Clin. Nutr. 89, 1895–1900.

Roubenoff, R., 2004. Sarcopenic obesity: the confluence of two epidemics. Obes. Res. 12, 887–888.

Ryan, A.S., Nicklas, B.J., 1999. Age-related changes in fat deposition in mid-thigh muscle in women: relationships with metabolic cardiovascular disease risk factors. Int. J. Obes. Relat. Metab. Disord. 23, 126–132.

Seidell, J.C., 2002. Prevalence and time trends of obesity in Europe. J. Endocrinol. Invest. 25, 816–822.

Shargill, N.S., Ohshima, K., Bray, G.A., Chan, T.M., 1984. Muscle protein turnover in the perfused hindquarters of lean and genetically obese-diabetic (db/db) mice. Diabetes 33, 1160–1164.

She, P., Van Horn, C., Reid, T., Hutson, S.M., Cooney, R.N., Lynch, C.J., 2007. Obesity-related elevations in plasma leucine are associated with alterations in enzymes involved in branched-chain amino acid metabolism. Am. J. Physiol. Endocrinol. Metab. 293, E1552–E1563.

Short, K.R., Nair, K.S., 2001. Muscle protein metabolism and the sarcopenia of aging. Int. J. Sport Nutr. Exerc. Metab. 11 (Suppl.), S119–S127.

Simoneau, J.A., Colberg, S.R., Thaete, F.L., Kelley, D.E., 1995. Skeletal muscle glycolytic and oxidative enzyme capacities are determinants of insulin sensitivity and muscle composition in obese women. FASEB J. 9, 273–278.

Slawik, M., Vidal-Puig, A.J., 2007. Adipose tissue expandability and the metabolic syndrome. Genes Nutr. 2, 41–45.

Solini, A., Bonora, E., Bonadonna, R., Castellino, P., DeFronzo, R.A., 1997. Protein metabolism in human obesity: relationship with glucose and lipid metabolism and with visceral adipose tissue. J. Clin. Endocrinol. Metab. 82, 2552–2558.

Stenholm, S., Harris, T.B., Rantanen, T., Visser, M., Kritchevsky, S.B., Ferrucci, L., 2008. Sarcopenic obesity: definition, cause and consequences. Curr. Opin. Clin. Nutr. Metab. Care 11, 693–700.

Tai, E.S., Tan, M.L., Stevens, R.D., Low, Y.L., Muehlbauer, M.J., Goh, D.L., et al., 2010. Insulin resistance is associated with a metabolic profile of altered protein metabolism in Chinese and Asian-Indian men. Diabetologia 53, 757–767.

Tanner, C.J., Barakat, H.A., Dohm, G.L., Pories, W.J., MacDonald, K.G., Cunningham, P.R., et al., 2002. Muscle fiber type is associated with obesity and weight loss. Am. J. Physiol. Endocrinol. Metab. 282, E1191–E1196.

Tardif, N., Salles, J., Landrier, J.F., Mothe-Satney, I., Guillet, C., Boue-Vaysse, C., et al., 2011. Oleate-enriched diet improves insulin sensitivity and restores muscle protein synthesis in old rats. Clin. Nutr. 30, 799–806.

Tardif, N., Salles, J., Guillet, C., Tordjman, J., Reggio, S., Landrier, J.F., et al., 2014. Muscle ectopic fat deposition contributes to anabolic resistance in obese sarcopenic old rats through eIF2alpha activation. Aging Cell 13, 1001–1011.

Tiraby, C., Tavernier, G., Capel, F., Mairal, A., Crampes, F., Rami, J., et al., 2007. Resistance to high-fat-diet-induced obesity and sexual dimorphism in the metabolic responses of transgenic mice with moderate uncoupling protein 3 overexpression in glycolytic skeletal muscles. Diabetologia 50, 2190–2199.

Tumova, J., Andel, M., Trnka, J., 2015. Excess of free fatty acids as a cause of metabolic dysfunction in skeletal muscle. Physiol. Res. In Press.

Visser, M., Goodpaster, B.H., Kritchevsky, S.B., Newman, A.B., Nevitt, M., Rubin, S.M., et al., 2005. Muscle mass, muscle strength, and muscle fat infiltration as predictors of incident mobility limitations in well-functioning older persons. J. Gerontol. A Biol. Sci. Med. Sci. 60, 324–333.

Walrand, S., Boirie, Y., 2005. Optimizing protein intake in aging. Curr. Opin. Clin. Nutr. Metab. Care 8, 89–94.

Wang, B., Jenkins, J.R., Trayhurn, P., 2005. Expression and secretion of inflammation-related adipokines by human adipocytes differentiated in culture: integrated response to TNF-α. Am. J. Physiol. Endocrinol. Metab. 288, E731–E740.

Wang, M.Y., Koyama, K., Shimabukuro, M., Mangelsdorf, D., Newgard, C.B., Unger, R.H., 1998. Overexpression of leptin receptors in pancreatic islets of Zucker diabetic fatty rats restores GLUT-2, glucokinase, and glucose-stimulated insulin secretion. Proc. Natl. Acad. Sci. U.S.A. 95, 11921–11926.

Wang, T.J., Larson, M.G., Vasan, R.S., Cheng, S., Rhee, E.P., McCabe, E., et al., 2011. Metabolite profiles and the risk of developing diabetes. Nat. Med. 17, 448–453.

Wang, Z., Ying, Z., Bosy-Westphal, A., Zhang, J., Heller, M., Later, W., et al., 2012. Evaluation of specific metabolic rates of major organs and tissues: comparison between nonobese and obese women. Obesity (Silver Spring) 20, 95–100.

Welle, S., Statt, M., Barnard, R., Amatruda, J., 1994. Differential effect of insulin on whole-body proteolysis and glucose metabolism in normal-weight, obese, and reduced-obese women. Metabolism 43, 441–445.

Wellen, K.E., Hotamisligil, G.S., 2005. Inflammation, stress, and diabetes. J. Clin. Invest. 115, 1111–1119.

Wenz, T., Rossi, S.G., Rotundo, R.L., Spiegelman, B.M., Moraes, C.T., 2009. Increased muscle PGC-1alpha expression protects from sarcopenia and metabolic disease during aging. Proc. Natl. Acad. Sci. U.S.A. 106, 20405–20410.

Zamboni, M., Mazzali, G., Zoico, E., Harris, T.B., Meigs, J.B., Di Francesco, V., et al., 2005. Health consequences of obesity in the elderly: a review of four unresolved questions. Int. J. Obes. (Lond) 29, 1011–1029.

Zoico, E., Roubenoff, R., 2002. The role of cytokines in regulating protein metabolism and muscle function. Nutr. Rev. 60, 39–51.

Zoico, E., Rossi, A., Di Francesco, V., Sepe, A., Olioso, D., Pizzini, F., et al., 2010. Adipose tissue infiltration in skeletal muscle of healthy elderly men: relationships with body composition, insulin resistance, and inflammation at the systemic and tissue level. J. Gerontol. A Biol. Sci. Med. Sci. 65, 295–299.

CHAPTER

9

Feeding Modulation of Amino Acid Utilization: Role of Insulin and Amino Acids in Skeletal Muscle

P.J. Atherton, D.J. Wilkinson and K. Smith

MRC-ARUK Centre for Musculoskeletal Ageing Research, School of Medicine, University of Nottingham, Nottingham, United Kingdom

9.1. OVERVIEW OF THE METABOLIC ROLE OF SKELETAL MUSCLE AND AS AN AMINO ACID REPOSITORY

Skeletal muscles perform a number of key functions, but are primarily involved in the maintenance of posture and movement of the limbs, providing both support to the skeleton and a means of intricate acts of locomotion via coordinated muscle contractions. However, skeletal muscle is also extremely important for regulating and maintaining whole-body metabolic homeostasis; for instance, providing a repository for glucose storage in the form of glycogen and intramuscular fat droplets that can be called upon in times of need. Skeletal muscle also acts as the body's largest reservoir of amino acids (AA), concentrating AA intramuscularly against a concentration gradient. Indeed, most AA are found in excess of their levels in plasma, with certain nonessential AA (NEAA), for example, glutamine, alanine and glycine, being ~5–20 fold greater than in the systemic concentrations (Smith and Rennie, 1990). The essential AA (EAA) cannot be de novo synthesized by humans and therefore have to be acquired through the diet. These diet-derived EAA are also concentrated between 1.5- and 2-fold higher within skeletal muscle via a host of active AA transporters (Hundal and Taylor, 2009; Taylor, 2013; see Table 9.1). In contrast, many NEAA undergoing de novo synthesis are concentrated within muscle via specific metabolic pathways generated from other AA, utilizing excess carbon skeletons and nitrogen (Wagenmakers, 1998; Rennie et al., 2006).

Skeletal muscle is the largest organ of the body, comprising ~50% of lean body mass in an average ~75 kg adult male. This pool of protein is continually being turned-over at a rate of approximately 1–1.5%/day in humans. This day-to-day maintenance of skeletal muscle is governed by a dynamic equilibrium between the rates of "building" (ie, MPS) and the rates of "breaking down" (ie, MPB) (Atherton and Smith, 2012). In a healthy weight-bearing adult, the rate at which muscle protein synthesis (MPS) occurs in muscle is balanced by equal rates of muscle protein breakdown (MPB) as part of a diurnal cycle in equilibrium. It follows that during situations of muscle growth, for example, in response to resistance exercise or anabolic steroids, net MPS exceeds net MPB. In contrast, the converse is true in conditions where muscle mass is lost (eg, bed rest, organ failure, cachexia, critical illness). Therefore, knowledge of the control of muscle protein turnover transcends fundamental biological precepts and has great clinical relevance. This is especially true in the context of feeding, which represents a key regulator of muscle maintenance, influencing protein turnover in health and disease.

Before describing the influences of feeding on skeletal muscle (and as will be discussed, particularly the AA constituents of foodstuffs), we must first outline the methodologies that have been applied to determine this. In order to quantify muscle protein turnover, we require sensitive analytical methods and experimental approaches

The Molecular Nutrition of Amino Acids and Proteins.
DOI: http://dx.doi.org/10.1016/B978-0-12-802167-5.00009-8

TABLE 9.1 AA composition of different protein sources in comparison to blood and muscle

	Plasma (μM)	Muscle (μM)	im: plasma ratio	Muscle (μmol/g wet wt)	Muscle (%)	Whey (%)	Casein (%)	Soy (%)	Beef (%)	Egg (%)
EAA										
His	104	420	4.0	34	2.5	2.5	2.8	2.7	3.8	2.1
Ile	67	110	1.6	59	4.3	4.3	4.7	4.9	5.3	5.9
Leu	144	225	1.6	116	8.4	12.1	9.1	8.0	11.0	8.4
Lys	212	1110	5.2	112	8.2	6.5	7.4	6.2	9.5	5.9
Met	29	60	2.1	25	1.8	2.6	2.8	1.3	2.8	3.8
Phe	69	85	1.2	46	3.4	4.8	4.9	5.3	2.8	6
Thr	172	770	4.5	75	5.5	3.6	4.1	3.6	3.5	4.3
Trp	70	nd	–	18	1.3	nd	1.5	0.9	1.5	1.4
Val	271	320	1.2	84	6.1	5.8	6.1	4.9	6.0	7.2
NEAA										
Arg	99	680	6.9	60	4.4	2.8	3.4	7.6	7.3	5.7
Ala	419	2860	6.8	121	8.8	2.7	2.8	4.4	6.0	6.3
Asp	13	1650	126.9	125	9.1	6.5	3.4	11.6	10.0	8.7
Asn	58	420	7.2	Incl amide	Incl amide	Incl amide	3.3	Incl amide	Incl amide	Incl amide
Cys	60	nd	–	nd	nd	nd	0.3	1.3	nd	2.4
Glu	40	3960	99.0	190	13.8	23.9	9.4	19.1	17.8	13.7
Gln	761	19970	26.2	Incl amide	Incl amide	Incl amide	11.8	Incl amide	Incl amide	Incl amide
Gly	303	1660	5.5	134	9.8	1.6	1.7	4.0	4.0	3.6
Pro	231	945	4.1	74	5.4	11.1	10.3	5.3	4.0	3.6
Ser	153	900	5.9	70	5.1	4.0	5.1	5.3	2.8	7.1
Tyr	97	122	1.3	30	2.2	5.0	5.3	3.6	2.0	3.9
ΣEAA (%)	1138 (34)	3100 (9)			41	42	43	38	46	45
ΣBCAA (%)	482 (14)	655 (2)			19	22	20	18	22	22

Data from Dangin et al. (2001), Tang et al. (2009), Tipton et al. (1999), Moore et al. (2009), Smith and Rennie (1990). *nd*, not detected. Glutamine and Asparagine convert to their respective acids upon acid hydrolysis of the protein; therefore, the data is reported as the sum of the two.

to measure the rates of MPS and the rate of MPB. This has been achieved through the application of stable isotopically labeled amino acids (using ^{2}H, ^{13}C, ^{15}N, and ^{18}O) to "trace" the movement of label into the blood and into and out of tissues and proteins—these heavy isotopes are distinguished from their more abundant lighter isotope by mass spectrometric techniques. Traditionally, MPS is determined from sampling, via muscle biopsy, extracting the protein (the product) and measuring the amount of label incorporated over time. From reference to a "surrogate" precursor pool, the true precursor pool for protein synthesis is the amino-acyl tRNA, at which point the AA is committed to peptide bond formation, and the fractional synthesis rate (FSR) can be calculated. However this pool is very small and extremely labile during sample extraction procedures and therefore alternative pools have been chosen, for example, the intracellular pool or in the case of leucine tracers, ketoisocaproate, a metabolite of leucine that is formed intramuscularly and can be sampled easily in venous blood and closely reflects the labeling of intramuscular leucine (Watt et al., 1991; Rennie et al., 1994). The measurement of MPB has been more challenging. Nonetheless methods exist to quantify MPB based on assessing the dilution of a constantly infused EAA tracer across a tissue, organ, or limb is possible through measuring the labeling and concentration of the tracer (usually in a steady state) in the artery supplying the tissue, and measurement of its dilution in the vein draining the organ or limb. Tracer fluxes also require measures of arterial blood flow to the organ or limb (collectively termed the A-V balance approach). From these measures, the rate of appearance (Ra) of endogenous AA, disappearance of endogenous AA (Rd), and net balance (NB) can be used to simultaneously quantify MPB, MPS, and net anabolism/catabolism, respectively. The use of an EAA simplifies the measurement since there is no requirement to account for de novo synthesis of the AA (as the body does not synthesis EAA). It is also helpful if the AA is not further metabolized within the tissue/organ, hence phenylalanine is the AA of choice for skeletal muscle whereas studies using branched chained AA (BCAA), eg, leucine, are complicated by the need to account for intermediary metabolism in muscle, such that the movement of tracer in KIC and CO_2 also needs to be determined (Bennet et al., 1990). Combining these approaches in the study of muscle protein turnover it is possible to investigate the mechanisms by which feeding regulates the magnitude and temporal basis of changes in MPS, MPB and net balance that occur. For more detailed information on these techniques, the reader is referred to Wolfe and Chinkes (2005).

It has been known for over 30 years that feeding a mixed nutrient liquid meal providing around 8 g of protein per hour was capable of doubling the rate of MPS, whilst only increasing whole-body protein turnover by 40% (Rennie et al., 1982). The mechanisms underlying the anabolic effects of this nutrition involve both the stimulation of MPS (Rennie et al., 1982) and suppression of MPB (Wilkes et al., 2009); however, it is generally accepted that increases in MPS are the primary driver behind this anabolic effect (Atherton and Smith, 2012). The absolute stimulation of MPS by nutrition can range from 20% to 300% and is highly dependent on the dosing regimen, the quality of the protein, and most importantly the AA content within the feed (Atherton, 2013). This is because it is AA that represent the "anabolically active" constituents of foodstuffs. In contrast, during periods of fasting both EAA and NEAA are released from muscle, serving to act as substrates to support MPS in other higher turnover tissues within the body, as well as providing carbon to be utilized as fuel, that is, for hepatic gluconeogenesis (Biolo et al., 1995b; Battezzati et al., 1999). These losses of muscle proteins/AA during postabsorptive periods in-between meals can be quantified through the measurement of nitrogen excreted in body fluids (urine, sweat, saliva, and faeces), since the majority of nitrogen is lost as urea—the endpoint as far as AA nitrogen metabolism is concerned (Duggleby and Waterlow, 2005). Daily protein requirements can be estimated based on nitrogen balance. Based on this (note: estimates of protein intake needs vary depending upon the methodologies employed) recent reevaluation of the nitrogen balance method (Humayun et al., 2007) recommended 0.99 g/kg/d of protein, which is slightly lower than the upper limit of 1.2 g/kg/d recommended using the recently introduced indicator amino acid oxidation methods (IAAO) (Elango et al., 2012a). Indeed there is still great debate over the most suitable methods for estimating optimal protein intake (Elango et al., 2012b; Millward and Jackson, 2012) and the current RDA remains 0.8 g/kg/d based on a meta-analysis of nitrogen balance studies (Rand et al., 2003). Exceeding this results in increased nitrogen excretion; conversely failing to take in sufficient protein results in lower excretion rates, and eventually, sacrifice of muscle tissue. Indeed, increased protein intake does not lead to increased protein deposition and muscle growth but increased urea excretion (Quevedo et al., 1994) and AA oxidation (Pacy et al., 1994). This notion is supported by acute protein supplementation studies, where excess protein is metabolized as indicated by increased urea production and AA oxidation (Paddon-Jones et al., 2005; Moore et al., 2009; Witard et al., 2014). The following sections will cover the key steps conferring how anabolic effects to feeding arise in muscle; considering potential influences of protein source, digestion/absorption kinetics, and delivery, all of which can theoretically modulate anabolic responses to feeding.

9.2. IMPACT OF SPLANCHNIC EXTRACTION AND SOURCE OF DIETARY AMINO ACID ON BIOAVAILABILITY AND MUSCLE PROTEIN SYNTHESIS

It is important to consider the potential impact of consumption of different protein source(s) and resulting plasma bioavailability of AA, since this could impact muscle protein metabolism responses to feeding. Dietary proteins are often classified as *high* or *low* "quality" based on their AA composition, with those that are high in EAA being classed as of higher quality (see Table 9.1). In addition, proteins are often deemed to be of "higher quality," based upon their digestibility and rate of appearance in the systemic circulation, which is a function of the rate of digestion, transport across the gut, and extraction by the splanchnic tissues (ie, gut and liver). The protein source of dietary AA can greatly influence the absorption kinetics and plasma availability. Commonly, the kinetics of absorption has been monitored using milk (or animal-based proteins) made up of two distinct fractions: a whey fraction, which is considered soluble, and casein, a micellar fraction, which is much less soluble (Luiking et al., 2015). This means that upon consumption of milk proteins, absorption occurs in two phases: an initial "fast" absorption due to the whey fraction, and a delayed "slow" absorption due to the casein (Boirie et al., 1997a). Thus, it has been shown that whey transits the gut more rapidly than casein (Mahé et al., 1996), greatly impacting the rates at which AA are absorbed from each fraction (Luiking et al., 2015). Whilst whey protein remains largely soluble, casein clots due to precipitation by the acidic media present in the gut (Mahé et al., 1994). Recently, in vitro models of human digestion have shown significant coagulation of casein, with no coagulation of whey (Luiking et al., 2015). This coagulation dictates that casein requires further gastric peptide hydrolysis, slowing emptying rates and plasma AA bioavailability. A similar pattern of digestion, uptake, and appearance in the systemic circulation is seen with beef protein, resulting in a slower, prolonged elevation of AA (Symons et al., 2009). Plant-based proteins on the other hand have been less well investigated, however, it seems that there are significant differences in AA availability when animal- or plant-based proteins are consumed (Van Vliet et al., 2015). If ingestion of milk is compared with soy, despite the more rapid transit of the gut by soy, a greater proportion of soy protein is sequestered by splanchnic tissues and converted to urea, leaving less AA available to peripheral tissues such as muscle (Fouillet et al., 2002, 2009). As such, the source of proteins consumed can greatly impact the rate of appearance and bioavailability of AA in the systemic circulation and for muscle tissue.

Most protein in meals is consumed in the intact protein form, requiring digestion and hydrolysis in the gut before transport of AA into the systemic circulation. However, bioavailability can be improved by provision of protein in its hydrolysate form (Koopman et al., 2009). For example, the rate of appearance of plasma AA is approximately doubled when whey is provided in its hydrolysate form, with plasma AA peaking around ~30 min compared to 60 min post feed in their intact protein form (Morifuji et al., 2010) see Fig. 9.1A); this is coupled to a ~30% greater plasma AA appearance rate and a significantly lower level of splanchnic retention of AA (Koopman et al., 2009). One may therefore expect that if an individual is provided AA in their "free form" then this would increase bioavailability further; yet this is not the case for most AA. If isonitrogenous mixtures of whey hydrolysate and free AA are provided enterally, peak AA concentrations have been shown to be higher for most AA with whey hydrolysate than free AA (Monchi and Rérat, 1993). Furthermore, peak absorption of AA occurs more

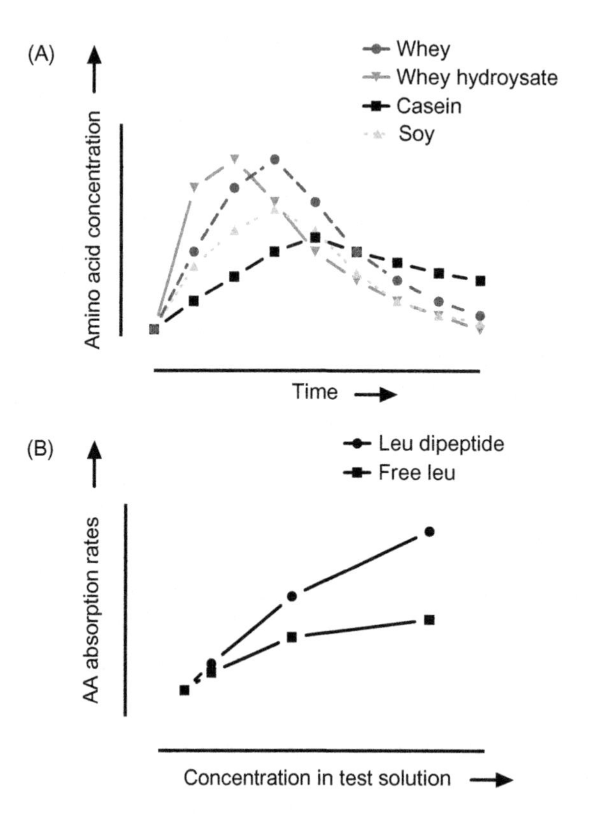

FIGURE 9.1 (A) Graphical representation of plasma amino acid appearance following oral ingestion of ~20 g of protein of differing sources. (B) Gut absorption rates of leucine in its dipeptide and free amino acid forms. *Source: Adapted from Adibi (1971).*

quickly with a hydrolysate than an isonitrogenous AA mixture (Rérat et al., 1988, 1992). This is primarily due to the fact that the large majority of digested protein appears in the gut as di- and tripeptide forms, rather than as free AA. Early research into human digestion and intestinal transport revealed that the human intestine possesses a unique ability to absorb and transport di- and tripeptides from the gut to the systemic circulation using a combination of intact absorption, peptide carrier systems and peptide hydrolysis (Adibi, 1971; Adibi et al., 1975). By perfusing sections of the jejunum of humans with test solutions of the tripeptides, triglycine and trileucine, the dipeptide diglycine and dileucine and the free AA glycine or leucine, the rates of glycine or leucine appearance were shown to be approximately double in peptide solutions than with free AA solutions (Adibi et al., 1975; also see Fig. 9.1B).

The splanchnic bed, inclusive of intestinal mucosa and liver, has a comparatively high rate of turnover (up to ~50%/d in humans; O'keefe et al., 2006). This high rate of turnover requires a large supply of AA to sustain it. In the fasted state, the intestinal mucosa extracts much of its AA needed for the high rates of organ turnover from the systemic circulation (Biolo et al., 1995b). This occurs at the expense of skeletal muscles, that is, AA (principally alanine and glutamine, comprising ~60% of the amino acid leaving muscle despite only making up ~20% of muscle protein) are released via increased fasted state MPB (as previously described), transported to the gut and taken up by the splanchnic tissues (Biolo et al., 1995b). This process is a significant contributor to net negative protein balance and loss of muscle protein during postabsorptive/fasted periods (Atherton and Smith, 2012). Upon intake of a meal however this process is reversed; splanchnic tissues are provided with sufficient supplies of AA via enteral extraction, whilst the remaining AA not sequestered by these tissues appear in the systemic circulation for use by other organs (Capaldo et al., 1999). This replenishes AA lost during postabsorptive cycles, induces positive protein balance, clawing back muscle proteins lost in fasted periods. The plasma availability of these enterally derived AA is dependent on the composition of AA ingested, and the uptake by splanchnic organs. For instance, the majority of NEAA, for example, alanine and glutamate/glutamine, will be sequestered by splanchnic tissues (Matthews et al., 1993; Battezzati et al., 1999) (ie, glutamine as an energy source for enterocytes, with alanine utilized by the liver for gluconeogenesis), with only a small percentage of the EAA being taken up. However, it should be noted that different EAA are taken up in different amounts; for example, in the fed state approximately 25% of dietary leucine (a central skeletal muscle anabolic AA) is retained by splanchnic tissues (Boirie et al., 1997b; Metges et al., 2000), whereas up to 50% of phenylalanine may be retained (Biolo et al., 1992). This likely reflects the fact that the primary site of BCAA metabolism is extrahepatic, being metabolized predominantly in skeletal muscle, with other AA being metabolized by the liver. Further to this splanchnic metabolism effect, plasma postprandial availability of AA is also highly dependent upon macronutrient composition in addition to previously described aspects of protein sources and hydrolysis state (eg, whey, casein, soy, hydrolysate). This aspect will be further discussed in the forthcoming paragraph.

Despite the fact that whey is considered a "fast" protein, which is rapidly digested providing pronounced, early aminoacidemia compared to "slow" casein (Boirie et al., 1997a), whey provides greater amounts of EAA, in particular leucine, than casein per gram of protein (Pennings et al., 2011). This latter aspect may also explain the apparent superior ability of whey protein to stimulate MPS when compared to casein (Tang et al., 2009; Pennings et al., 2011; Burd et al., 2012), with 20 g of whey providing postprandial MPS rates that are approximately double that of 20 g casein (Burd et al., 2012; Pennings et al., 2011). Hence it may be the composition of EAA, and leucine in particular, rather than, for example, total quantity of AA or delivery profile or plasma appearance, that determines the MPS response to feeding. To exemplify this, one recent study showed similar MPS responses to 3 g of 40% leucine-enriched AA in comparison to a 20 g whey bolus (Bukhari et al., 2015), whilst provision of just 3 g leucine to healthy young men resulted in an approximately maximal stimulation of MPS (Wilkinson et al., 2013). Furthermore, recently published data has shown no difference in MPS to equivalent composition 15 g doses of EAA delivered either as a single bolus (providing rapid initial aminoacidemia within ~45 min), or as small regular doses delivered at 45 min intervals (providing low amplitude aminoacidemia spread over ~> 2 h) (Mitchell et al., 2015a,b). These types of AA profiles reflect those commonly observed for slow and fast proteins, such as casein and whey, thus suggesting that when composition and dose are matched, delivery profile may not be a determining factor in anabolism. Together with data on leucine enriched low doses of AA and leucine alone, the acute stimulation of MPS seems to be dependent largely exclusively in accordance to leucine availability. Nonetheless, this remains contentious, since early research suggested improved anabolic response to whey over casein even when leucine compositions were matched, albeit with whole body rather than MPS (Boirie et al., 1997a). Furthermore, in a study comparing 20 g each of whey, casein, and soy (Tang et al., 2009), whey and soy elicited similar rates of MPS that were significantly higher than casein, whereas others have shown identical rates of synthesis when comparing soy and casein (Luiking et al., 2011). In another study comparing 20 g whey, casein, and a casein hydrolysate, the rate of MPS when ingesting whey was significantly higher than either of the casein or casein hydrolysate groups (Pennings et al., 2011), despite the casein hydrolysate being absorbed similarly to the whey. Yang and colleagues compared 20 g and 40 g doses of both whey and soy protein, demonstrating a significant increase only with the whey protein groups with no effect of soy, even at 40 g when the AA AUC profile was greater than that for whey at 20 g (Yang et al., 2012b). This was however at odds with the two previous studies (Luiking et al., 2011), one of which was performed by the same group using a similar approach (Tang et al., 2009). Further confusion over protein sources and composition is provided in studies where both beef (113 g equivalent to 10 g EAA) (Symons et al., 2007, 2009) and egg protein (Moore et al., 2009) produced a robust stimulation of MPS, comparable to those seen with whey. Further studies are required to clarify the benefits or otherwise of different protein sources. Indeed, comparison between different studies is often difficult due to the duration of the measurement (3–6 h post feeding), which impacts determination of MPS, the influence of "muscle full," the tracer employed, the surrogate pool chosen to represent the precursor, and the muscle fraction measured (ie, mixed or myofibrillar), or even lack of comparison to the postabsorptive state or a fasted control group.

9.3. INFLUENCE OF AMINO ACID, MACRONUTRIENT COMPOSITION, AND CALORIC LOAD ON MUSCLE PROTEIN SYNTHESIS

It was elegantly shown more than 25 years ago that AA alone could stimulate MPS (Bennet et al., 1989), with this stimulation almost exclusively driven through the essential AA (EAA). Studies indicated that when given as a large bolus, leucine, phenylalanine, valine, and threonine (all EAA) were all capable of stimulating MPS in man, whilst in contrast, similar large bolus' of arginine, glycine, and serine (all NEAA) did not recapitulate this MPS stimulation, highlighting the importance of the role for EAA in MPS stimulation (Smith et al., 1992, 1998). Given the apparent anabolic effect of individual EAA a significant number of investigations have subsequently attempted to use leucine supplementations to either enhance the stimulatory effect on MPS of protein, AA, or EAA or to maximize the efficiency of MPS when providing suboptimal doses of protein or EAA. In a study providing a suboptimal dose of EAA (6.7 g) where the proportion of leucine was adjusted from 26% (1.7 g) to 41% (2.8 g), there was no additional benefit of increasing the leucine content, at least in the young volunteers; there was however an apparent benefit in an elderly group in whom MPS was only significantly elevated with the higher dose of leucine (Katsanos et al., 2006). The lack of an effect in the young group was confirmed in a similar study in which leucine comprised 18 and 35% of a 10 g EAA mix, the MPS responses were not different between the two groups and there was no additional stimulation of MPS (Glynn et al., 2010). If leucine is indeed the driver for increased MPS (discussed further later) then it appears that doses as low as 1.7 g may be sufficient to maximize the anabolic response of muscle, at least in the young, but to date this has never been fully tested. However, in a recent study comparing the response of MPS to 20 g whey protein or a low dose leucine (1.2 g)-enriched EAA mix (Bukhari et al., 2015) increases in MPS were identical. It is also an intriguing possibility that metabolites of leucine may also be anabolically "active" with previous studies demonstrating the keto-acid of leucine, α-ketoisocaproate (KIC), or beta-hydroxy-beta-methylbutyrate (HMB) downstream metabolites may have roles in regulating MPS. Indeed, infusions of KIC have been shown to stimulate the MPS, however this could simply be due to KIC being reversibly transaminated to leucine, thereby providing this stimulatory effect (Escobar et al., 2010). Nonetheless, there is good evidence of anabolic activities of other more distal leucine metabolites. Indeed ingestion of ~3.5 g of HMB in humans provided a comparable increase in MPS to that of a similar dose of leucine (Wilkinson et al., 2013).

Another important factor affecting kinetics and systemic bioavailability of AA is the composition of the feed and the overall caloric content (Gorissen et al., 2014; Luiking et al., 2015). For example, when 30 g whey protein is consumed with 100 g of carbohydrate (sucrose), gastric emptying is significantly delayed, with the peak appearance of AA from the milk protein (labeled with 15N for measurement) approximately 2 h later than milk protein alone, providing a slower more sustained AA release (Gaudichon et al., 1999); this is not the case with addition of fat. This delay is believed to be due to the higher energy density of the carbohydrates/protein load than protein alone, requiring greater time to transit the gut (Hunt et al., 1985; Gaudichon et al., 1999). Indeed, despite fats adding similar energy density to carbohydrates, these compounds sit at the top of the gastric chyme and are not miscible with the aqueous gastric media where proteins will be present (Meyer et al., 1986) initiating a two phase digestion and ensuring fats do not delay the transit time of AA in the same way as the carbohydrates (which will be miscible in the gastric media alongside protein). There is also evidence to suggest that carbohydrates combined with protein help to retain AA in the system once absorbed (Gaudichon et al., 1999), due to the decreased necessity to utilize carbon skeletons of AA for energy substrates, in addition to the profound insulin response associated with carbohydrates (driving insulin mediated uptake of AA and inhibition of MPB). Indeed, urinary 15N excretion is approximately half that of whey protein alone when consumed alongside carbohydrates (Gaudichon et al., 1999). Despite the fact that there is greater retention of AA and more sustained/gradual release through gastric emptying, this does not seem to benefit muscle protein metabolism (at least in terms of MPS). A number of studies have provided coingestion of protein and carbohydrates without showing any additional benefit for promoting MPS beyond that of protein alone (Staples et al., 2011; Glynn et al., 2013; Gorissen et al., 2014) despite the proposed nitrogen sparing effects. A similar effect can be observed with varying caloric intake, when the same amount of protein is provided with differing amounts of calories (regardless of protein type), the higher calorie protein meal significantly slows digestion based on plasma AA appearance rate (Luiking et al., 2015). An in vitro model of this phenomenon highlighted that with casein protein, higher calories led to an increased coagulation and slower transit and absorption, whilst with whey protein, higher caloric load decreases the overall cumulative release of the AA from the protein source (Luiking et al., 2015). What is directly causing this remains unclear, yet as with addition of carbohydrates, the increased energy need for digestion may contribute to delayed AA release.

9.4. EFFECTS OF DOSE AND DELIVERY PROFILE OF AMINO ACID ON THE FEEDING-INDUCED STIMULATION OF MUSCLE PROTEIN SYNTHESIS

Although still somewhat contentious amongst some, a number of studies appear to indicate that the MPS response to nutrition is dose-dependent and saturable, increasing linearly to a maximum of between 20 and 40 g of high quality protein (meat or whey protein with high leucine/BCAA components; Moore et al., 2009, 2015; Symons et al., 2009; Witard et al., 2014; or around 10 to 20 g of an EAA mix when given orally; Cuthbertson et al., 2005). Similarly when mixed AA infusions are given at increasing doses, MPS is shown to be maximal at 87 mg/kg/h equivalent to 19.2 g of AA over 3 h, during which the concentrations of EAA were effectively double the pre-infusion concentrations (Bohé et al., 2003). Beyond these maximal doses, that is, 10–20 g EAA or 20–40 g protein, there is a tendency toward increased amino acid oxidation (Moore et al., 2009) and urea synthesis (Witard et al., 2014), thereby redistributing carbon skeletons and nitrogen from excess AA, until the AA concentrations return to postabsorptive values (~2–3 h). The time taken to return to baseline is dose-dependent and when in excess of 30 g of protein or AA is administered it can take much longer for the excess AA to be utilized (Dangin et al., 2003; Yang et al., 2012b).

Not only is the response to nutrition dose dependent, it is also temporal in nature. It seems clear that providing excess protein alone does not result in increased muscle mass, at least not without the synergistic stimulation of exercise. Therefore, the stimulation of MPS by feeding must be finite. Due to the relatively low turnover rate of muscle the original AA tracer studies were performed over prolonged periods, anywhere between 4 and 12 h was standard, hence this finite nature was not detected. The first attempt to resolve the time-course of the response of MPS to AA was met with skepticism. However, major improvements in the sensitivity, stability, and accuracy of modern mass spectrometric techniques, and the availability of multiply labeled amino acid stable isotope tracers that is, $^{13}C_6$ or D_5 phenylalanine (see Smith et al., 2011 for review) has dramatically improved the temporal resolution of the measurements that are now possible and confirmed those earlier pioneering studies.

In response to either an intravenous (I.V.) infusion of mixed AA (Bohe et al., 2001; Bohé et al., 2003) or a large 48 g bolus of whey protein (Atherton et al., 2010a), it is now apparent that following an initial lag-period of ~30 min during I.V. infusion (or a bit longer ~45–60 min following oral ingestion—to allow for the digestion, absorption and transport of AA into the systemic circulation), the rate of MPS is increased approximately 2–3-fold reaching a maximum by 90–120 min, subsequently and extremely rapidly the rate of MPS returns to baseline for at least the next 3–4 h and remains refractory to further stimulation, despite the continued availability of increased amino acids (Atherton et al., 2010a; see Fig. 9.2). It appears that the muscle has replenished the AA that were released during fasting and is in fact refilled, hence the coining of the term "muscle full" (Bohe et al., 2001). It should be noted here that without any further stimulus, that is, exercise, that measuring MPS much beyond this period will simply provide a measured rate of MPS that is the average of the period of stimulation diluted by the period of basal turnover and thus the lack of observation of a stimulatory effect of feeding on MPS in some studies may simply be the result of the prolonged measurement period. The duration of this period of refractoriness to further stimulation is not yet known but is likely to be at least 4 h, based on the duration of the original studies (Bohe et al., 2001; Bohé et al., 2003). For example, in a recent study subjects were given a high protein breakfast (~45 g), and approximately 4 h later MPS was measured in response to increasing doses of protein (10, 20, and 40 g) at which time MPS was stimulated at the higher doses (Witard et al., 2014).

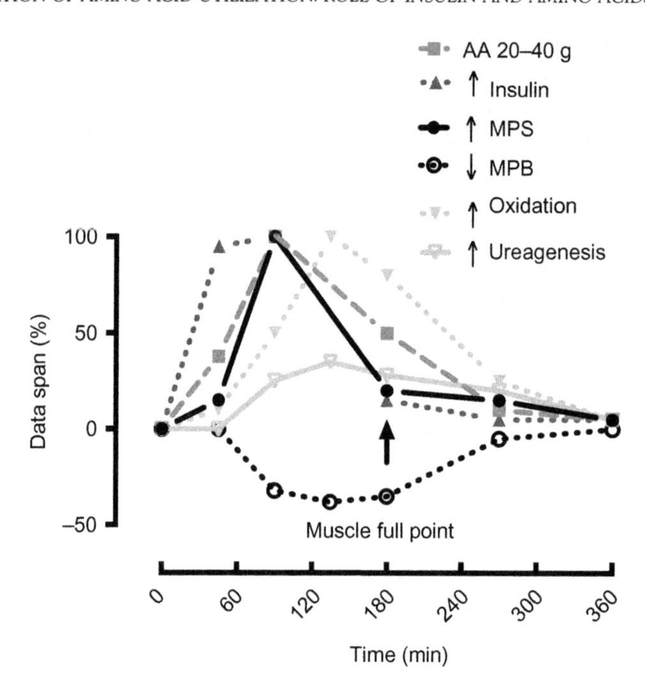

FIGURE 9.2 Graphical representation of the "muscle full" concept. The muscle switches off MPS despite the presence of excess substrate. *Source: Adapted from Atherton and Smith (2012) (Permissions not yet obtained).*

As was discussed previously in the context of protein sources, there has also been extensive focus on the ingestion profile of protein/AA, as a result of the desire to maximize MPS, given the temporal—short-lived nature of the stimulation. Yet the evidence supporting this is not yet convincing. There are a number of studies in which the protein meal (Rennie et al., 1982) or AA solution (provision of the AA solution orally or I.V. had the same doubling effect on MPS) (Rasmussen et al., 2002; Volpi et al., 2003) are divided into regular smaller doses and provided throughout the measurement period and MPS is robustly stimulated over this period, showing the same two-fold elevation in MPS seen with the administration of a whey bolus—or indeed a flooding dose of individual AA (Garlick et al., 1989; Smith et al., 1992, 1998). Similarly when AA are I.V. infused over periods of 3–4 h the rate of MPS is increased significantly (Bennet et al., 1989, 1990; Bohe et al., 2001; Bohé et al., 2003), despite only modest 15–80% increases in BCAA concentrations throughout in some studies (Louis et al., 2003). It appears therefore that neither mode or delivery, oral versus I.V., nor rate, bolus versus small regular doses, impacts on the muscle response but that it is the total dose that drives MPS until the point the "muscle is full." However, there are few studies that have specifically addressed this issue directly, that is, providing the same stimulus (AA amount) as either a bolus or in regular smaller doses thereby providing the same total amount of AA but with a different profile. A recent study addressed this question directly comparing a single 15 g EAA bolus with 4 boluses of 3.75 g every 45 min over 3 h. Although the AA delivery profile was vastly different, the area under the curve for AA over the 3 h period was identical, and increases in MPS were identical peaking 90–180 min before returning to postabsorptive levels over the ensuing 60 min (Mitchell et al., 2015a). Again this supports the idea of an intrinsic "muscle full" response that is unaffected by AA delivery profile or rapid aminoacidemia. Indeed, it is likely that reports of differences between protein sources in terms of impacts on MPS are instead driven by subtle differences in EAA (leucine) composition/bioavailability.

9.5. INFLUENCE OF MICROVASCULAR RESPONSES TO FEEDING IN RELATION TO MUSCLE PROTEIN SYNTHESIS

Following digestion, absorption, and first pass extraction, a large proportion of nutritive AA made available in the plasma will be utilized by skeletal muscle for maintenance of protein homeostasis. This occurs via transport into the muscle and direct stimulation of the translational mechanisms regulating protein metabolism (Hundal and Taylor, 2009; Han et al., 2012). Whilst transmembrane transport of AA is rapid and unlikely to be a rate-limiting factor in protein metabolism (Smith and Rennie, 1990), the provision of AA to the myocytes through recruitment of circulatory mechanisms could be. Indeed, recruitment of nutritive routes of tortuous capillaries contacting myocytes is essential for muscle perfusion and efficient AA uptake. A number of studies have demonstrated increased microvascular recruitment during feeding, with provision of 15 g EAA showing robust increases in microvascular blood flow in young individuals, with early increases in microvascular blood volume followed by later increases in microvascular flow velocity and microvascular blood flow (Mitchell et al., 2013). Similarly, provision of AA combined with dextrose mirrored responses to a mixed-meal feed (Vincent et al., 2006) in causing enhanced limb blood flow, microvascular blood flow, and associated increases in MPS (Phillips et al., 2014), highlighted the important role of the microvasculature in the regulation of skeletal MPS.

The majority of these postprandial increases in microvascular recruitment are driven through insulin. Typically, postprandial nutritive recruitment of the muscle microvasculature is driven through insulin via mechanisms involving nitric-oxide (NO)-dependent vasodilation of precapillary arterioles (Vincent et al., 2004; Rajapakse et al., 2013). For example, clamping insulin at 75 μU/mL led to increased microvascular blood flow (Sjøberg et al., 2011), which combined with insulin's known anabolic effects on muscle (suppressing MPB (see following paragraph) or indirectly stimulating MPS via enhancing delivery of EAA to the capillary-muscle interface; Wilkes et al., 2009) may prove to be a link between muscle microvascular blood flow and muscle protein anabolism. This is supported by work showing that increases in MPS following femoral artery infusions of various insulin titrations were related to AA delivery via enhancement of microvascular blood flow in an insulin availability-dependent manner (Timmerman et al., 2010a); which suggests that altering the delivery of insulin and EAA to muscle could have profound effects on postprandial muscle anabolism. Interestingly however, it has been demonstrated in younger individuals that enhancing limb and microvascular blood flow through intraarterial methacholine infusions did not further enhance muscle anabolic responses to feeding (Phillips et al., 2014). These data suggest that surpassing the subtle vasoactive effects of nutrition via pharmacological enhancement or an altered temporal feeding pattern cannot further increase fed state anabolism; that is, in younger individuals, microvascular perfusion is not a limiting factor for anabolism and nutritive stimulation of microvascular perfusion is sufficient to provide all necessary AA substrate to skeletal muscle for maximal anabolism.

9.6. THE ROLE OF INSULIN IN REGULATING MUSCLE PROTEIN TURNOVER

In response to feeding carbohydrates or protein/AA, insulin secretion is promoted from pancreatic beta cells; due both to direct pancreatic actions or secondary effects facilitated by gut peptides being released in response to gut mucosa L and K cell stimulation upon luminal–systemic nutrient exchange. Moreover, it should be noted that carbohydrate intake is not a prerequisite for nutrition-mediated insulin release, and it is well known that AA can act as potent insulin secretagogues—in particular the NEAA alanine, glutamine and arginine (Newsholme et al., 2007; Newsholme and Krause, 2012) and the BCAA leucine (Yang et al., 2012a; Wilkinson et al., 2013)—independent of glucose, either acting directly on the beta cells to induce insulin exocytosis (Newsholme and Krause, 2012) or indirectly via release of the insulinotropic gut peptides such as GLP-1 and GIP (Abdulla et al., 2014). For example provision of ~50 g whey protein to humans leads to insulin concentrations equivalent to that of a 75 g oral glucose tolerance test (Kulshreshtha et al., 2008; Atherton et al., 2010a).

Insulin is known as an anabolic hormone and in addition to regulating glucose homeostasis (ie, mediating insulin-dependent glucose uptake) insulin has a positive effect on MPS in both cell culture and various pre-clinical models. The role of insulin in adult human muscle is distinct. Following mixed macronutrient ingestion, insulin secretion is stimulated rapidly peaking within 30 min then falling exponentially to postabsorptive values over the next 30–45 min. This response can be recapitulated by oral feeding of protein (Atherton et al., 2010b), mixed amino acids (Dangin et al., 2001) and even EAA (Mitchell et al., 2015b), suggesting insulin has a role to play in the metabolic response to protein or AA alone. Uncovering the role that insulin plays in regulating muscle protein metabolism is difficult, since systemic infusion of insulin to postprandial (30–40 uU/mL) or supra-physiological levels (> 60 uU/mL) without the provision of amino acids leads to significant hypoaminoacidemia and a trend toward a fall in protein synthesis, probably as the result of an inhibition of protein breakdown and a reduction in AA substrate levels for protein synthesis (Bell et al., 2005, 2006). Attempts to offset the hypoaminoacidaemia by infusing the insulin locally into muscle, avoiding any increases in systemic insulin, were successful in that the hypoaminoacidemia was avoided but synthesis remained unchanged. Despite supraphysiological insulin levels within the musculature, the improvement in net muscle protein balance was wholly down to the fall in protein breakdown (Gelfand and Barrett, 1987; Louard et al., 1992). However, in other studies, increased synthesis has been observed only at supraphysiological insulin concentrations (> 100 uU/mL), this AV balance data was supported by small but significant increases in MPS as measured by tracer incorporation (Biolo et al., 1995a; Fujita et al., 2009), though this effect has proven difficult to reproduce in complementary studies by the same group (Timmerman et al., 2010a,b). In an attempt to understand the separate roles that insulin and AA play in the regulation of muscle protein turnover, the dose response of protein turnover was assessed using the precursor product and AV balance approach, under insulin clamp conditions at fasting, postprandial, high physiological, and supraphysiological insulin levels, whilst infusing mixed AA at a high level (18 g/L) (Greenhaff et al., 2008). Muscle protein synthesis was stimulated to the same extent ($\sim$2-fold cf. basal values) under all clamps, whereas MPB was only inhibited at postprandial levels and above, indicating that MPB was unaffected by both high circulating AA concentrations and that insulin was required to induce inhibition of protein breakdown (Greenhaff et al., 2008). With regard to the responsiveness of muscle to the inhibitory role of insulin, in a follow-up study, insulin was clamped both at postabsorptive insulin levels and at 15 uU/mL and both MPS and MPB were measured. MPS was unchanged at $<0.05\%$/h under both insulin clamps, AA concentrations were maintained throughout via an infusion, moreover MPB was reduced by 50% in response to this moderate increase in circulating insulin thereby demonstrating the acute sensitivity of MPB to insulin (Wilkes et al., 2009). Furthermore, despite apparent lack of a consensus on the role of insulin on muscle protein metabolism, a recent meta-analysis concluded that overall insulin appears to play a permissive role in the regulation of MPS in the presence of increased AA substrate, but inhibits MPB independently of the availability of circulating AA (Abdulla et al., 2015). Therefore, at this point in time we can confidently state that EAA, and leucine in particular, are responsible for regulating increases in MPS in response to feeding, while increased insulin concentrations in response to feeding regulate reductions in MPB.

9.7. THE MOLECULAR REGULATION OF SKELETAL MUSCLE PROTEIN SYNTHESIS AND MUSCLE PROTEIN BREAKDOWN BY AMINO ACID AND INSULIN

As has been previously described in this chapter, increased availability of blood AA makes available substrates for cellular energy production, conversion to bioactive metabolites, the stimulation of hormone secretion, and to provide substrate for protein synthesis. In terms of skeletal muscle, one of the most fundamental roles of exogenous AA is to facilitate maintenance of protein balance, via replenishing protein stores broken down to release AA in times of fasting and thereby ensure maintenance of muscle protein homeostasis, or so-called "proteostasis." Crucially, dietary AA do not just act as substrates or building blocks for protein synthesis, but they also act as signals, instructing muscle cells to increase rates of protein synthesis. The mechanisms by which diet-derived AA stimulate intracellular signaling pathways are under construction. AA-mediated increases in muscle protein synthesis (MPS) are initiated after transport of EAAs into the muscle cell (Christie et al., 2002) where leucine in particular (Atherton et al., 2010b) activates mammalian target of rapamycin complex-1 (mTORc1), independently of proximal insulin signaling (phosphatidylinositol 3-kinase (PI3K) pathways; Kimball and Jefferson, 2006). Constructing the signaling pathways governing the detection of intracellular AA in response to feeding included crucial work showing that provision of rapamycin, an immunosuppressant that inhibits mechanistic target of rapamycin (mTOR) ablated anabolic responses to increasing EAA/leucine availability (Gundermann et al., 2014). This led to the supposition that mTOR acts as a central node detecting and responding to cellular AA availability. Indeed, when humans are fed protein, signals in the mTOR pathway rise in synergy with MPS suggesting close links between EAA availability and the stimulation/initiation of MPS (Atherton et al., 2010a). The notion that activation of mTOR might therefore represent a central node communicating AA with the signaling that leads to MPS can be reconciled by the number of substrates associated with its activation. These in particular are involved in translational initiation and peptide elongation processes. Activation of mTOR is thereafter associated with the phosphorylation of multiple translational initiation factors substrates including (4E-binding protein (4EBP1), ribosomal protein S6 kinase (p70S6K1), eukaryotic initiation factors 4 G/A/B (eIF4G/A/B) with ensuing formation of the eIF3F scaffold to promote assembly of a 48S preinitiation complex. In a parallel pathway, activation of the guanine exchange factor, eukaryotic initiation factor 2B (eIF2B) shuttles the initiator tRNA (Met-tRNAi) to the ribosome during formation of the 48S pre-initiation complex, thereby promoting "global" protein synthesis and coordinately enhancing translational efficiency, that is, numbers of mRNAs associated with ribosomes (polysomes) and thus MPS (see Kimball, 2014 for detailed control of mRNA translation). Practically all of these factors have been shown to be stimulated in response to AA, and many are used as a proxy for mTORc1 signaling (eg, S6K1) since kinase activity of mTORc1 has multiple inputs, that is, is governed by a combination of localization, affinity to its binding partners, and phosphorylation events (Efeyan et al., 2012).

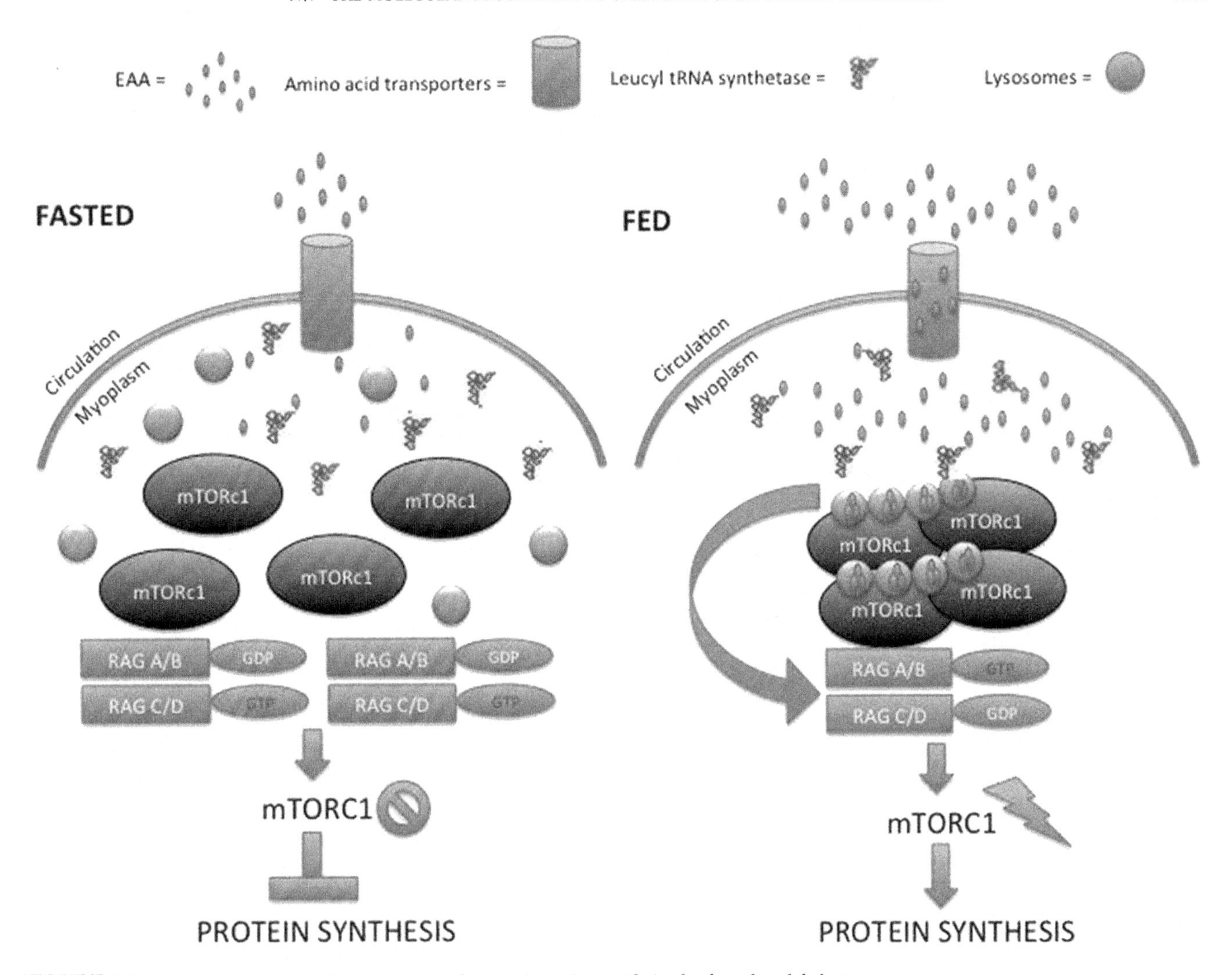

FIGURE 9.3 Intramuscular signaling responses driven via amino acids in the fasted and fed states.

Nonetheless, while mTORc1 is thought an affirmed central regulator of MPS, the proximal mechanisms involved in the activation of mTORc1 by AA/leucine remain incompletely defined. Indeed, it is known MPS responses to increased AA availability are independent of elements of the proximal canonical insulin signaling pathway; at least at the level(s) of PI3K (phosphoinositide (PI) 3-kinase)/protein kinase B since pharmacological inhibition of these constituents does not block AA-stimulated MPS (Bolster et al., 2004). Similarly, other well-characterized signaling pathways upstream of mTOR, such as the tuberous sclerosis complex (TSC), have been investigated. The TSC is a GTPase-activating protein (GAP) for the GTPase RAS-homolog enriched in brain (Rheb) that negatively regulates mTOR by promoting Rheb-GTP hydrolysis, converting Rheb into its inactive GDP-bound state (Zhang et al., 2003). As a result, inhibition of TSC gives rise to GTP-bound Rheb, which is a potent activator of mTORC1 kinase activity. Yet, a number of lines of molecular evidence have shown that this pathway is not crucial for mTOR activity in response to AA (Bolster et al., 2004; Atherton et al., 2010b). The first clues to the regulation of AA-induced MPS arose from another group of GTPases. The presence of AA promotes the formation of the active complex configuration, in which RAGA and RAGB (there are four RAG proteins (A-D)) are GTP-bound and RAGC and RAGD are GDP-bound (note that RAGA/B·GTP–RAGC/D·GDP is the active complex and RAGA/B·GDP–RAGC/D·GTP the inactive). Crucially, under conditions of heightened AA-availability, AA accumulate within lysosomes, activating Vacuolar H^+-ATPase (v-ATPase) where the active RAG complex, via the Ragulator (Sancak et al., 2008; Kim et al., 2008), directly binds to mTORc1 (complex 1) protein raptor, whereby crucially mTORC1 is redistributed to lysosomes. This redistribution is thought to be a critical step in the regulation of cellular AA sensing and ensuing increases in MPS (Fig. 9.3). A second central point of control of MPS in response to AA appears to be at the level of charging of leucyl tRNA synthetase (Han et al., 2012).

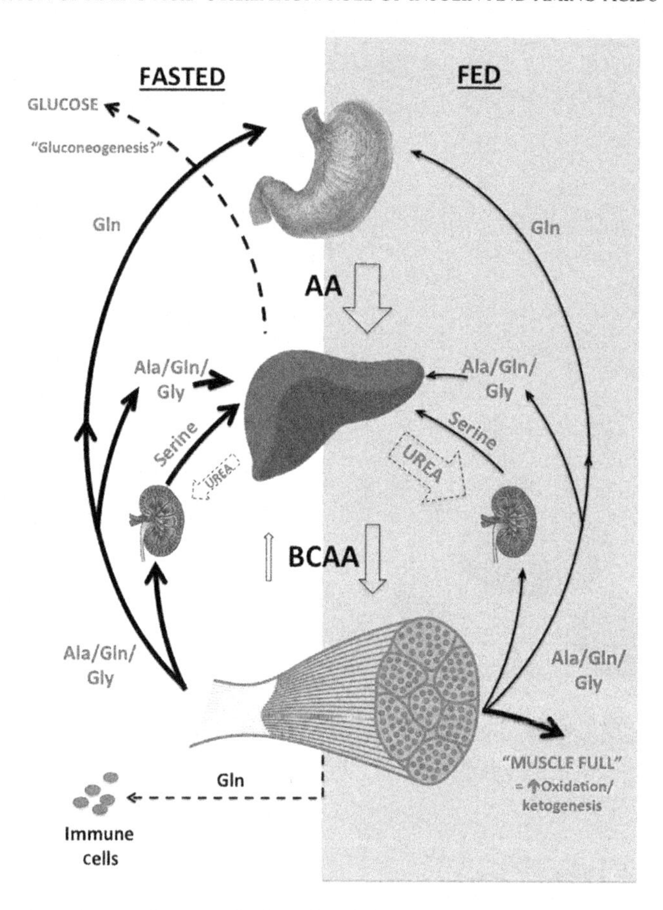

FIGURE 9.4 Summary of interorgan exchange of AA in the fasted and fed states. The bolder the connecting arrow the greater the flow of AA in this direction.

In response to leucine, leucyl tRNA synthetase was found to translocate to the lysosomal membrane, where it bound to and acted as a GAP for RAGD. Whether this is a parallel recognition signal to the intralysosomal mechanisms or is acting somewhere within the same recognition pathway remains to be fully determined.

The regulation of MPB responses to AA is more poorly defined. In humans, MPB is suppressed only in response to increases in insulin availability and does not appear to be in response to increasing AA availability (Chow et al., 2006; Greenhaff et al., 2008) in contrast to many preclinical models (Jefferson et al., 1977; Ashford and Pain, 1986). Indeed, while increased activity of mTOR in response to AA may theoretically suppress MPB, due to direct effects of EAA on MPB ubiquitin proteasome apparatus (Hamel et al., 2003) or concomitant suppression of autophagy through activation of mTORc1 and resulting suppression of uncoordinated like kinases (ULK) (Hosokawa et al., 2009; Kim et al., 2011), this does not lead to a measurable depression of MPB in humans in the absence of insulin (Greenhaff et al., 2008). Therefore, one is left to ponder what are the likely mechanisms are for the suppression of MPB in response to insulin? Previous work has determined that proteasomal activation is inhibited in response to insulin (Hamel et al., 1997), suggesting inhibition of ubiquitin-dependent and or autophagy pathways in response to elevated plasma insulin concentrations are the active mechanisms suppressing MPB in man.

9.8. CONCLUSIONS

In the above sections we have outlined the effects of feeding in relation to key aspects that impact muscle protein metabolism, that is, intake of distinct food composites, protein sources/quantities, systemic bioavailability, microvascular perfusion, and activation of AA (and insulin)-recognition pathways within skeletal muscle cells; the reader should now hopefully grasp each of these key steps in the context of the impact of feeding upon skeletal muscle. The important role of efficient, tightly controlled interorgan AA exchange in both fasted and fed states is summarized in Fig. 9.4. It is clear that amino acids have a variety of important functions in the control of skeletal muscle metabolism, and that providing a diet that includes appropriate doses and composition of EAA is key to this tightly regulated process.

References

Abdulla, H., Phillips, B., Smith, K., Wilkinson, D., Atherton, P.J., Idris, I., 2014. Physiological mechanisms of action of incretin and insulin in regulating skeletal muscle metabolism. Curr. Diabetes Rev. 10, 327–335.

Abdulla, H., Smith, K., Atherton, P.J., Idris, I., 2015. Role of insulin in the regulation of human skeletal muscle protein synthesis and breakdown: a systematic review and meta-analysis. Diabetologia 59 (1), 44–55. Available from: http://dx.doi.org/10.1007/s00125-015-3751-0.

Adibi, S., 1971. Intestinal transport of dipeptides in man: relative importance of hydrolysis and intact protein. J. Clin. Invest. 50, 2266–2275.

Adibi, S., Morse, E., Masilamani, S., Amin, P., 1975. Evidence for two different modes of tripeptide disappearance in human intestine: uptake by peptide carrier systems and hydrolysis by peptide hydrolases. J. Clin. Invest. 56, 1355–1363.

Ashford, A.J., Pain, V.M., 1986. Insulin stimulation of growth in diabetic rats. Synthesis and degradation of ribosomes and total tissue protein in skeletal muscle and heart. J. Biol. Chem. 261, 4066–4070.

Atherton, P.J., 2013. Is there an optimal time for warfighters to supplement with protein? J. Nutr. 143, 1848S–1851S.

Atherton, P.J., Smith, K., 2012. Muscle protein synthesis in response to nutrition and exercise. J. Physiol. 590, 1049–1057.

Atherton, P.J., Etheridge, T., Watt, P.W., Wilkinson, D., Selby, A., Rankin, D., et al., 2010a. Muscle full effect after oral protein: time-dependent concordance and discordance between human muscle protein synthesis and mTORC1 signaling. Am. J. Clin. Nutr. 92, 1080–1088.

Atherton, P.J., Smith, K., Etheridge, T., Rankin, D., Rennie, M.J., 2010b. Distinct anabolic signalling responses to amino acids in C2C12 skeletal muscle cells. Amino Acids 38, 1533–1539.

Battezzati, A., Haisch, M., Brillon, D.J., Matthews, D.E., 1999. Splanchnic utilization of enteral alanine in humans. Metabolism 48, 915–921.

Bell, J.A., Fujita, S., Volpi, E., Cadenas, J.G., Rasmussen, B.B., 2005. Short-term insulin and nutritional energy provision do not stimulate muscle protein synthesis if blood amino acid availability decreases. Am. J. Physiol. Endocrinol. Metab. 289, E999–E1006.

Bell, J.A., Volpi, E., Fujita, S., Cadenas, J.G., Sheffield-Moore, M., Rasmussen, B.B., 2006. Skeletal muscle protein anabolic response to increased energy and insulin is preserved in poorly controlled type 2 diabetes. J. Nutr. 136, 1249–1255.

Bennet, W.M., Connacher, A.A., Scrimgeour, C.M., Smith, K., Rennie, M.J., 1989. Increase in anterior tibialis muscle protein synthesis in healthy man during mixed amino acid infusion: studies of incorporation of [1-13C]leucine. Clin. Sci. (Lond) 76, 447–454.

Bennet, W.M., Connacher, A.A., Scrimgeour, C.M., Rennie, M.J., 1990. The effect of amino acid infusion on leg protein turnover assessed by L-[15N]phenylalanine and L-[1-13C]leucine exchange. Eur J. Clin. Invest. 20, 41–50.

Biolo, G., Tessari, P., Inchiostro, S., Bruttomesso, D., Fongher, C., Sabadin, L., et al., 1992. Leucine and phenylalanine kinetics during mixed meal ingestion: a multiple tracer approach. Am. J. Physiol. 262, E455–E463.

Biolo, G., Declan Fleming, R.Y., Wolfe, R.R., 1995a. Physiologic hyperinsulinemia stimulates protein synthesis and enhances transport of selected amino acids in human skeletal muscle. J. Clin. Invest. 95, 811–819.

Biolo, G., Zhang, X.J., Wolfe, R.R., 1995b. Role of membrane transport in interorgan amino acid flow between muscle and small intestine. Metabolism 44, 719–724.

Bohe, J., Low, J.F., Wolfe, R.R., Rennie, M.J., 2001. Latency and duration of stimulation of human muscle protein synthesis during continuous infusion of amino acids. J. Physiol. 532, 575–579.

Bohé, J., Low, A., Wolfe, R.R., Rennie, M.J., 2003. Human muscle protein synthesis is modulated by extracellular, not intramuscular amino acid availability: a dose-response study. J. Physiol. 552, 315–324.

Boirie, Y., Dangin, M., Gachon, P., Vasson, M.P., Maubois, J.L., Beaufrere, B., 1997a. Slow and fast dietary proteins differently modulate postprandial protein accretion1. Proc. Natl. Acad. Sci. U.S.A. 94, 14930–14935.

Boirie, Y., Gachon, P., Beaufrère, B., 1997b. Splanchnic and whole-body leucine kinetics in young and elderly men. Am. J. Clin. Nutr. 65, 489–495.

Bolster, D.R., Vary, T.C., Kimball, S.R., Jefferson, L.S., 2004. Leucine regulates translation initiation in rat skeletal muscle via enhanced eIF4G phosphorylation. J. Nutr. 134, 1704–1710.

Bukhari, S.S., Phillips, B.E., Wilkinson, D.J., Limb, M.C., Rankin, D., Mitchell, W.K., et al., 2015. Intake of low-dose leucine-rich essential amino acids stimulates muscle anabolism equivalently to bolus whey protein in older women, at rest and after exercise. Am. J. Physiol. Endocrinol. Metab. 308. Available from: http://dx.doi.org/10.1152/ajpendo.00481.2014.

Burd, N.A., Yang, Y., Moore, D.R., Tang, J.E., Tarnopolsky, M.A., Phillips, S.M., 2012. Greater stimulation of myofibrillar protein synthesis with ingestion of whey protein isolate v. micellar casein at rest and after resistance exercise in elderly men. Br. J. Nutr. 108, 958–962.

Capaldo, B., Gastaldelli, A., Antoniello, S., Auletta, M., Pardo, F., Ciociaro, D., et al., 1999. Splanchnic and leg substrate exchange after ingestion of a natural mixed meal in humans. Diabetes 48, 958–966.

Chow, L.S., Albright, R.C., Bigelow, M.L., Toffolo, G., Cobelli, C., Nair, K.S., 2006. Mechanism of insulin's anabolic effect on muscle: measurements of muscle protein synthesis and breakdown using aminoacyl-tRNA and other surrogate measures. AJP Endocrinol. Metab. 291, E729–E736.

Christie, G.R., Hajduch, E., Hundal, H.S., Proud, C.G., Taylor, P.M., 2002. Intracellular sensing of amino acids in *Xenopus laevis* oocytes stimulates p70 S6 kinase in a target of rapamycin-dependent manner. J. Biol. Chem. 277, 9952–9957.

Cuthbertson, D., Smith, K., Babraj, J., Leese, G., Waddell, T., Atherton, P., et al., 2005. Anabolic signaling deficits underlie amino acid resistance of wasting, aging muscle. FASEB J. 19, 422–424.

Dangin, M., Boirie, Y., Garcia-Rodenas, C., Gachon, P., Fauquant, J., Callier, P., et al., 2001. The digestion rate of protein is an independent regulating factor of postprandial protein retention. Am. J. Physiol. Endocrinol. Metab. 280, E340–E348.

Dangin, M., Guillet, C., Garcia-Rodenas, C., Gachon, P., Bouteloup-Demange, C., Reiffers-Magnani, K., et al., 2003. The rate of protein digestion affects protein gain differently during aging in humans. J. Physiol. 549, 635–644.

Duggleby, S.L., Waterlow, J.C., 2005. The end-product method of measuring whole-body protein turnover: a review of published results and a comparison with those obtained by leucine infusion. Br. J. Nutr. 94, 141–153.

Efeyan, A., Zoncu, R., Sabatini, D.M., 2012. Amino acids and mTORC1: from lysosomes to disease. Trends Mol. Med. 18, 524–533.

Elango, R., Ball, R.O., Pencharz, P.B., 2012a. Recent advances in determining protein and amino acid requirements in humans. Br. J. Nutr. 108 (Suppl.), S22–S30.

Elango, R., Humayun, M.A., Ball, R.O., Pencharz, P.B., 2012b. Reply to DJ Millward and AA Jackson. Am. J. Clin. Nutr. 95, 1501–1502.

Escobar, J., Frank, J.W., Suryawan, A., Nguyen, H.V., Van Horn, C.G., Hutson, S.M., et al., 2010. Leucine and -ketoisocaproic acid, but not norleucine, stimulate skeletal muscle protein synthesis in neonatal pigs. J. Nutr. 140, 1418–1424.

Fouillet, H., Mariotti, F., Gaudichon, C., Bos, C., Tomé, D., 2002. Peripheral and splanchnic metabolism of dietary nitrogen are differently affected by the protein source in humans as assessed by compartmental modeling. J. Nutr. 132, 125–133.

Fouillet, H., Juillet, B., Gaudichon, C., Mariotti, F., Tomé, D., Bos, C., 2009. Absorption kinetics are a key factor regulating postprandial protein metabolism in response to qualitative and quantitative variations in protein intake. Am. J. Physiol. Regul. Integr. Comp. Physiol. 297, R1691–R1705.

Fujita, S., Glynn, E.L., Timmerman, K.L., Rasmussen, B.B., Volpi, E., 2009. Supraphysiological hyperinsulinaemia is necessary to stimulate skeletal muscle protein anabolism in older adults: evidence of a true age-related insulin resistance of muscle protein metabolism. Diabetologia 52, 1889–1898.

Garlick, P.J., Wernerman, J., McNurlan, M.A., Essen, P., Lobley, G.E., Milne, E., et al., 1989. Measurement of the rate of protein synthesis in muscle of postabsorptive young men by injection of a "flooding dose" of [1-13C]leucine. Clin. Sci. (Lond) 77, 329–336.

Gaudichon, C., Mahé, S., Benamouzig, R., Luengo, C., Fouillet, H., Daré, S., et al., 1999. Net postprandial utilization of [15N]-labeled milk protein nitrogen is influenced by diet composition in humans. J. Nutr. 129, 890–895.

Gelfand, R.A., Barrett, E.J., 1987. Effect of physiologic hyperinsulinemia on skeletal muscle protein synthesis and breakdown in man. J. Clin. Invest. 80, 1–6.

Glynn, E.L., Fry, C.S., Drummond, M.J., Timmerman, K.L., Dhanani, S., Volpi, E., et al., 2010. Excess leucine intake enhances muscle anabolic signaling but not net protein anabolism in young men and women. J. Nutr. 140, 1970–1976.

Glynn, E.L., Fry, C.S., Timmerman, K.L., Drummond, M.J., Volpi, E., Rasmussen, B.B., 2013. Addition of carbohydrate or alanine to an essential amino acid mixture does not enhance human skeletal muscle protein anabolism. J. Nutr. 143, 307–314.

Gorissen, S.H.M., Burd, N.A., Hamer, H.M., Gijsen, A.P., Groen, B.B., van Loon, L.J.C., 2014. Carbohydrate coingestion delays dietary protein digestion and absorption but does not modulate postprandial muscle protein accretion. J. Clin. Endocrinol. Metab. 99, 2250–2258.

Greenhaff, P.L., Karagounis, L.G., Peirce, N., Simpson, E.J., Hazell, M., Layfield, R., et al., 2008. Disassociation between the effects of amino acids and insulin on signaling, ubiquitin ligases, and protein turnover in human muscle. Am. J. Physiol. Endocrinol. Metab. 295, E595–E604.

Gundermann, D.M., Walker, D.K., Reidy, P.T., Borack, M.S., Dickinson, J.M., Volpi, E., et al., 2014. Activation of mTORC1 signaling and protein synthesis in human muscle following blood flow restriction exercise is inhibited by rapamycin. Am. J. Physiol. Endocrinol. Metab. 306, E1198–E1204.

Hamel, F.G., Bennett, R.G., Harmon, K.S., Duckworth, W.C., 1997. Insulin inhibition of proteasome activity in intact cells. Biochem. Biophys. Res. Commun. 234, 671–674.

Hamel, F.G., Upward, J.L., Siford, G.L., Duckworth, W.C., 2003. Inhibition of proteasome activity by selected amino acids. Metabolism 52, 810–814.

Han, J.M., Jeong, S.J., Park, M.C., Kim, G., Kwon, N.H., Kim, H.K., et al., 2012. Leucyl-tRNA synthetase is an intracellular leucine sensor for the mTORC1-signaling pathway. Cell 149, 410–424.

Hosokawa, N., Hara, T., Kaizuka, T., Kishi, C., Takamura, A., Miura, Y., et al., 2009. Nutrient-dependent mTORC1 association with the ULK1-Atg13-FIP200 complex required for autophagy. Mol. Biol. Cell 20, 1981–1991.

Humayun, M.A., Elango, R., Ball, R.O., Pencharz, P.B., 2007. Reevaluation of the protein requirement in young men with the indicator amino acid oxidation technique. Am. J. Clin. Nutr. 86, 995–1002.

Hundal, H.S., Taylor, P.M., 2009. Amino acid transceptors: gate keepers of nutrient exchange and regulators of nutrient signaling. Am. J. Physiol. Endocrinol. Metab. 296, E603–E613.

Hunt, J.N., Smith, J.L., Jiang, C.L., 1985. Effect of meal volume and energy density on the gastric emptying of carbohydrates. Gastroenterology 89, 1326–1330.

Jefferson, L.S., Li, J.B., Rannels, S.R., 1977. Regulation by insulin of amino acid release and protein turnover in the perfused rat hemicorpus. J. Biol. Chem. 252, 1476–1483.

Katsanos, C.S., Kobayashi, H., Sheffield-Moore, M., Aarsland, A., Wolfe, R.R., 2006. A high proportion of leucine is required for optimal stimulation of the rate of muscle protein synthesis by essential amino acids in the elderly. Am. J. Physiol. Endocrinol. Metab. 291, E381–E387.

Kim, E., Goraksha-Hicks, P., Li, L., Neufeld, T.P., Guan, K.-L., 2008. Regulation of TORC1 by Rag GTPases in nutrient response. Nat. Cell Biol. 10, 935–945.

Kim, J., Kundu, M., Viollet, B., Guan, K.-L., 2011. AMPK and mTOR regulate autophagy through direct phosphorylation of Ulk1. Nat. Cell Biol. 13, 132–141.

Kimball, S.R., 2014. Integration of signals generated by nutrients, hormones, and exercise in skeletal muscle. Am. J. Clin. Nutr. 99, 237S–242S.

Kimball, S.R., Jefferson, L.S., 2006. Signaling pathways and molecular mechanisms through which branched-chain amino acids mediate translational control of protein synthesis. J. Nutr. 136, 227S–231S.

Koopman, R., Walrand, S., Beelen, M., Gijsen, A.P., Kies, A.K., Boirie, Y., et al., 2009. Dietary protein digestion and absorption rates and the subsequent postprandial muscle protein synthetic response do not differ between young and elderly men. J. Nutr. 139, 1707–1713.

Kulshreshtha, B., Ganie, M.A., Praveen, E.P., Gupta, N., Lal Khurana, M., Seith, A., et al., 2008. Insulin response to oral glucose in healthy, lean young women and patients with polycystic ovary syndrome. Gynecol. Endocrinol. 24, 637–643.

Louard, R.J., Fryburg, D.A., Gelfand, R.A., Barrett, E.J., 1992. Insulin sensitivity of protein and glucose metabolism in human forearm skeletal muscle. J. Clin. Invest. 90, 2348–2354.

Louis, M., Poortmans, J.R., Francaux, M., Hultman, E., Berre, J., Boisseau, N., et al., 2003. Creatine supplementation has no effect on human muscle protein turnover at rest in the postabsorptive or fed states. Am. J. Physiol. Endocrinol. Metab. 284, E764–E770.

Luiking, Y.C., Abrahamse, E., Ludwig, T., Boirie, Y., Verlaan, S., 2015. Protein type and caloric density of protein supplements modulate postprandial amino acid profile through changes in gastrointestinal behaviour: a randomized trial. Clin. Nutr. 35 (1), 48–58. Available from: http://dx.doi.org/10.1016/j.clnu.2015.02.013.

Luiking, Y.C., Engelen, M.P.K.J., Soeters, P.B., Boirie, Y., Deutz, N.E.P., 2011. Differential metabolic effects of casein and soy protein meals on skeletal muscle in healthy volunteers. Clin. Nutr. 30, 65–72.

Mahé, S., Roos, N., Benamouzig, R., Sick, H., Baglieri, A., Huneau, J.F., et al., 1994. True exogenous and endogenous nitrogen fractions in the human jejunum after ingestion of small amounts of 15N-labeled casein. J. Nutr. 124, 548–555.

Mahé, S., Roos, N., Benamouzig, R., Davin, L., Luengo, C., Gagnon, L., et al., 1996. Gastrojejunal kinetics and the digestion of [15N]beta-lactoglobulin and casein in humans: the influence of the nature and quantity of the protein. Am. J. Clin. Nutr. 63, 546–552.

Matthews, D.E., Marano, M.A., Campbell, R.G., 1993. Splanchnic bed utilization of glutamine and glutamic acid in humans. Am. J. Physiol. 264, E848–E854.

Metges, C.C., El-Khoury, A.E., Selvaraj, A.B., Tsay, R.H., Atkinson, A., Regan, M.M., et al., 2000. Kinetics of L-[1-(13)C]leucine when ingested with free amino acids, unlabeled or intrinsically labeled casein. Am. J. Physiol. Endocrinol. Metab. 278, E1000–E1009.

Meyer, J.H., Mayer, E.A., Jehn, D., Gu, Y., Fink, A.S., Fried, M., 1986. Gastric processing and emptying of fat. Gastroenterology 90, 1176–1187.

Millward, D.J., Jackson, A.A., 2012. Protein requirements and the indicator amino acid oxidation method. Am. J. Clin. Nutr. 95, 1498–1501.

Mitchell, W.K., Phillips, B.E., Williams, J.P., Rankin, D., Smith, K., Lund, J.N., et al., 2013. Development of a new Sonovue™ contrast-enhanced ultrasound approach reveals temporal and age-related features of muscle microvascular responses to feeding. Physiol. Rep. 1, e00119.

Mitchell, W., Phillips, B.E., Williams, J.P., Rankin, D., Lund, J.N., Smith, K., et al., 2015a. A dose- rather than delivery profile-dependent mechanism regulates the "Muscle-Full" effect in response to oral essential amino acid intake in young men. J. Nutr. 145, 207–214.

Mitchell, W., Phillips, B.E., Williams, J.P., Rankin, D., Lund, J.N., Wilkinson, D.J., et al., 2015b. The impact of delivery profile of essential amino acids upon skeletal muscle protein synthesis in older men: clinical efficacy of pulse vs. bolus supply. Am. J. Physiol. Endocrinol. Metab. 309 (5), E450–E457. Available from: http://dx.doi.org/10.1152/ajpendo.00112.2015.

Monchi, M., Rérat, A.A., 1993. Comparison of net protein utilization of milk protein mild enzymatic hydrolysates and free amino acid mixtures with a close pattern in the rat. J. Parenter. Enteral Nutr. 17, 355–363.

Moore, D.R., Robinson, M.J., Fry, J.L., Tang, J.E., Glover, E.I., Wilkinson, S.B., et al., 2009. Ingested protein dose response of muscle and albumin protein synthesis after resistance exercise in young men. Am. J. Clin. Nutr. 89, 161–168.

Moore, D.R., Churchward-Venne, T.A., Witard, O., Breen, L., Burd, N.A., Tipton, K.D., et al., 2015. Protein ingestion to stimulate myofibrillar protein synthesis requires greater relative protein intakes in healthy older versus younger men. J. Gerontol. Ser. A Biol. Sci. Med. Sci. 70, 57–62.

Morifuji, M., Ishizaka, M., Baba, S., Fukuda, K., Matsumoto, H., Koga, J., et al., 2010. Comparison of different sources and degrees of hydrolysis of dietary protein: effect on plasma amino acids, dipeptides, and insulin responses in human subjects. J. Agric. Food Chem. 58, 8788–8797.

Newsholme, P., Krause, M., 2012. Nutritional regulation of insulin secretion: implications for diabetes. Clin. Biochem. Rev. 33, 35–47.

Newsholme, P., Bender, K., Kiely, A., Brennan, L., 2007. Amino acid metabolism, insulin secretion and diabetes. Biochem. Soc. Trans. 35, 1180–1186.

O'keefe, S.J.D., Lee, R.B., Li, J., Zhou, W., Stoll, B., Dang, Q., 2006. Trypsin and splanchnic protein turnover during feeding and fasting in human subjects. Am. J. Physiol. Gastrointest. Liver Physiol. 290, G213–G221.

Pacy, P.J., Price, G.M., Halliday, D., Quevedo, M.R., Millward, D.J., 1994. Nitrogen homeostasis in man: the diurnal responses of protein synthesis and degradation and amino acid oxidation to diets with increasing protein intakes. Clin. Sci. (Lond) 86, 103–116.

Paddon-Jones, D., Sheffield-Moore, M., Aarsland, A., Wolfe, R.R., Ferrando, A.A., 2005. Exogenous amino acids stimulate human muscle anabolism without interfering with the response to mixed meal ingestion3. Am. J. Physiol. Endocrinol. Metab. 288, E761–E767.

Pennings, B., Boirie, Y., Senden, J.M., Gijsen, A.P., Kuipers, H., van Loon, L.J., 2011. Whey protein stimulates postprandial muscle protein accretion more effectively than do casein and casein hydrolysate in older men. Am. J. Clin. Nutr. 93, 997–1005.

Phillips, B.E., Atherton, P.J., Varadhan, K., Wilkinson, D.J., Limb, M., Selby, A.L., et al., 2014. Pharmacological enhancement of leg and muscle microvascular blood flow does not augment anabolic responses in skeletal muscle of young men under fed conditions. Am. J. Physiol. Endocrinol. Metab. 306, E168–E176.

Quevedo, M.R., Price, G.M., Halliday, D., Pacy, P.J., Millward, D.J., 1994. Nitrogen homoeostasis in man: diurnal changes in nitrogen excretion, leucine oxidation and whole body leucine kinetics during a reduction from a high to a moderate protein intake. Clin. Sci. (Lond) 86, 185–193.

Rajapakse, N.W., Chong, A.L., Zhang, W.-Z., Kaye, D.M., 2013. Insulin-mediated activation of the L-arginine nitric oxide pathway in man, and its impairment in diabetes. PLoS One 8, e61840.

Rand, W.M., Pellett, P.L., Young, V.R., 2003. Meta-analysis of nitrogen balance studies for estimating protein requirements in healthy adults. Am. J. Clin. Nutr. 77, 109–127.

Rasmussen, B.B., Wolfe, R.R., Volpi, E., 2002. Oral and intravenously administered amino acids produce similar effects on muscle protein synthesis in the elderly. J. Nutr. Health Aging 6, 358–362.

Rennie, M.J., Edwards, R.H., Halliday, D., Matthews, D.E., Wolman, S.L., Millward, D.J., 1982. Muscle protein synthesis measured by stable isotope techniques in man: the effects of feeding and fasting. Clin. Sci. (Lond) 63, 519–523.

Rennie, M.J., Smith, K., Watt, P.W., 1994. Measurement of human tissue protein synthesis: an optimal approach. Am. J. Physiol. 266, E298–E307.

Rennie, M.J., Bohé, J., Smith, K., Wackerhage, H., Greenhaff, P., 2006. Branched-chain amino acids as fuels and anabolic signals in human muscle. J. Nutr. 136, 264S–268S.

Rérat, A., Nunes, C.S., Mendy, F., Roger, L., 1988. Amino acid absorption and production of pancreatic hormones in non-anaesthetized pigs after duodenal infusions of a milk enzymic hydrolysate or of free amino acids. Br. J. Nutr. 60, 121–136.

Rérat, A., Simoes-Nuñes, C., Mendy, F., Vaissade, P., Vaugelade, P., 1992. Splanchnic fluxes of amino acids after duodenal infusion of carbohydrate solutions containing free amino acids or oligopeptides in the non-anaesthetized pig. Br. J. Nutr. 68, 111–138.

Sancak, Y., Peterson, T.R., Shaul, Y.D., Lindquist, R.A., Thoreen, C.C., Bar-Peled, L., et al., 2008. The Rag GTPases bind raptor and mediate amino acid signaling to mTORC1. Science 320, 1496–1501.

Sjøberg, K.A., Rattigan, S., Hiscock, N., Richter, E.A., Kiens, B., 2011. A new method to study changes in microvascular blood volume in muscle and adipose tissue: real-time imaging in humans and rat. Am. J. Physiol. Heart Circ. Physiol. 301, H450–H458.

Smith, K., Rennie, M.J., 1990. Protein turnover and amino acid metabolism in human skeletal muscle. Baillieres Clin. Endocrinol. Metab. 4, 461–498.

Smith, G.I., Patterson, B.W., Mittendorfer, B., 2011. Human muscle protein turnover – why is it so variable? J. Appl. Physiol. 110, 480–491.

Smith, K., Barua, J.M., Watt, P.W., Scrimgeour, C.M., Rennie, M.J., 1992. Flooding with L-[1-13C]leucine stimulates human muscle protein incorporation of continuously infused L-[1-13C]valine. Am. J. Physiol. 262, E372–E376.

Smith, K., Reynolds, N., Downie, S., Patel, A., Rennie, M.J., 1998. Effects of flooding amino acids on incorporation of labeled amino acids into human muscle protein. Am. J. Physiol. 275, E73–E78.

Staples, A.W., Burd, N.A., West, D.W.D., Currie, K.D., Atherton, P.J., Moore, D.R., et al., 2011. Carbohydrate does not augment exercise-induced protein accretion versus protein alone. Med. Sci. Sports Exerc. 43, 1154–1161.

Symons, T.B., Schutzler, S.E., Cocke, T.L., Chinkes, D.L., Wolfe, R.R., Paddon-Jones, D., 2007. Aging does not impair the anabolic response to a protein-rich meal. Am. J. Clin. Nutr. 86, 451–456.

Symons, T.B., Sheffield-Moore, M., Wolfe, R.R., Paddon-Jones, D., 2009. A moderate serving of high-quality protein maximally stimulates skeletal muscle protein synthesis in young and elderly subjects. J. Am. Diet Assoc. 109, 1582–1586.

Tang, J.E., Moore, D.R., Kujbida, G.W., Tarnopolsky, M.A., Phillips, S.M., 2009. Ingestion of whey hydrolysate, casein, or soy protein isolate: effects on mixed muscle protein synthesis at rest and following resistance exercise in young men. J. Appl. Physiol. 107, 987–992.

Taylor, P.M., 2013. Role of amino acid transporters in amino acid sensing. Am. J. Clin. Nutr. 99, 223S–230S.

Timmerman, K.L., Lee, J.L., Dreyer, H.C., Dhanani, S., Glynn, E.L., Fry, C.S., et al., 2010a. Insulin stimulates human skeletal muscle protein synthesis via an indirect mechanism involving endothelial-dependent vasodilation and mammalian target of rapamycin complex 1 signaling. J. Clin. Endocrinol. Metab. 95, 3848–3857.

Timmerman, K.L., Lee, J.L., Fujita, S., Dhanani, S., Dreyer, H.C., Fry, C.S., et al., 2010b. Pharmacological vasodilation improves insulin-stimulated muscle protein anabolism but not glucose utilization in older adults. Diabetes 59, 2764–2771.

Tipton, K.D., Ferrando, A.A., Phillips, S.M., Doyle Jr., D., Wolfe, R.R., 1999. Postexercise net protein synthesis in human muscle from orally administered amino acids. Am. J. Physiol. 276, E628–E634.

Vincent, M.A., Clerk, L.H., Lindner, J.R., Klibanov, A.L., Clark, M.G., Rattigan, S., et al., 2004. Microvascular recruitment is an early insulin effect that regulates skeletal muscle glucose uptake in vivo. Diabetes 53, 1418–1423.

Vincent, M.A., Clerk, L.H., Lindner, J.R., Price, W.J., Jahn, L.A., Leong-Poi, H., et al., 2006. Mixed meal and light exercise each recruit muscle capillaries in healthy humans. Am. J. Physiol. Endocrinol. Metab. 290, E1191–E1197.

Van Vliet, S., Burd, N.A., van Loon, L.J., 2015. The skeletal muscle anabolic response to plant- versus animal-based protein consumption. J. Nutr. 145, 1981–1991.

Volpi, E., Kobayashi, H., Sheffield-Moore, M., Mittendorfer, B., Wolfe, R.R., 2003. Essential amino acids are primarily responsible for the amino acid stimulation of muscle protein anabolism in healthy elderly adults. Am. J. Clin. Nutr. 78, 250–258.

Wagenmakers, A.J., 1998. Muscle amino acid metabolism at rest and during exercise: role in human physiology and metabolism. Exerc. Sport Sci. Rev. 26, 287–314.

Watt, P.W., Lindsay, Y., Scrimgeour, C.M., Chien, P.A., Gibson, J.N., Taylor, D.J., et al., 1991. Isolation of aminoacyl-tRNA and its labeling with stable-isotope tracers: Use in studies of human tissue protein synthesis. Proc. Natl. Acad. Sci. U.S.A. 88, 5892–5896.

Wilkes, E.A., Selby, A.L., Atherton, P.J., Patel, R., Rankin, D., Smith, K., et al., 2009. Blunting of insulin inhibition of proteolysis in legs of older subjects may contribute to age-related sarcopenia. Am. J. Clin. Nutr. 90, 1343–1350.

Wilkinson, D.J., Hossain, T., Hill, D.S., Phillips, B.E., Crossland, H., Williams, J., et al., 2013. Effects of Leucine and its metabolite, β-hydroxy-β-methylbutyrate (HMB) on human skeletal muscle protein metabolism. J. Physiol. 591 (11), 2911–2923. Available from: http://dx.doi.org/10.1113/jphysiol.2013.253203.

Witard, O.C., Jackman, S.R., Breen, L., Smith, K., Selby, A., Tipton, K.D., 2014. Myofibrillar muscle protein synthesis rates subsequent to a meal in response to increasing doses of whey protein at rest and after resistance exercise. Am. J. Clin. Nutr. 99, 86–95.

Wolfe, R.R., Chinkes, D.L., 2005. Isotope Tracers in Metabolic Research: Principles and Practice of Kinetic Analysis, second ed. John Wiley & Sons, Inc., Hoboken, NJ.

Yang, J., Dolinger, M., Ritaccio, G., Mazurkiewicz, J., Conti, D., Zhu, X., et al., 2012a. Leucine stimulates insulin secretion via down-regulation of surface expression of adrenergic α2A receptor through the mTOR (mammalian target of rapamycin) pathway: implication in new-onset diabetes in renal transplantation. J. Biol. Chem. 287, 24795–24806.

Yang, Y., Churchward-Venne, T.A., Burd, N.A., Breen, L., Tarnopolsky, M.A., Phillips, S.M., 2012b. Myofibrillar protein synthesis following ingestion of soy protein isolate at rest and after resistance exercise in elderly men. Nutr. Metab. (Lond) 9, 57.

Zhang, Y., Gao, X., Saucedo, L.J., Ru, B., Edgar, B.A., Pan, D., 2003. Rheb is a direct target of the tuberous sclerosis tumour suppressor proteins. Nat. Cell Biol. 5, 578–581.

C H A P T E R

10

Protein Metabolism and Requirement in Intensive Care Units and Septic Patients

P.J.M. Weijs

Department of Nutrition and Dietetics, Internal Medicine, VU University Medical Center; Department of Intensive Care Medicine, VU University Medical Center; Department of Nutrition and Dietetics, School of Sports and Nutrition, Amsterdam University of Applied Sciences, Amsterdam, The Netherlands

10.1. INTRODUCTION

Nutrition in the intensive care unit (ICU) is a highly discussed topic, and the views on feeding protein in particular are rather controversial. For many years, and in fact up to now, protein nutrition has been largely neglected in ICU studies concerning optimal nutrition of the patient. This also means that, in the current text, mention of protein requirements cannot be based on amino acid (AA) level evidence of randomized controlled trials, since there are none published.

However, this chapter will explore the changes in protein metabolism in critically ill patients, and when appropriate, also changes in energy metabolism. Further, we will explore what this means in term of dietary protein feeding, and some suggestions will be also made for protein nutrition in clinical practice, based on descriptive studies amongst some patients in our own institute.

Since findings in sepsis patients may or may not be in line with the general, rather heterogeneous, ICU population, there is specific mention of sepsis patients in this chapter.

The focus of this chapter will be on patients that are admitted to the ICU for more than three days, either medical or surgical.

The message is that protein feeding of critically ill patients in the ICU should be according to the guideline levels of 1.2 g/kg preadmission body weight by day 4. There is no need for aggressive feeding from day 1, and certainly not for caloric provision.

10.2. PROTEIN METABOLISM IN THE CRITICALLY ILL PATIENT

Changes in protein metabolism in critically ill patients appear not to be linear, but more like a phased response. Wischmeyer (2013) describes the following: "There appears to be an acute phase, consisting of the classic ebb and flow phase of shock and sepsis in which the modern ICU patient is undergoing acute resuscitation. ... If the patient survives the acute phase, this is followed by a more chronic phase of critical illness when the patient becomes quite vulnerable to recurrent infection and other complications... If the patient can recover sufficiently, the patient will enter a recovery phase, which often coincides with ICU discharge to a hospital floor or rehabilitation unit." This text will mainly deal with the acute and chronic phase. Since we cannot easily distinguish acute and chronic phase, we will pragmatically consider the first week of ICU admission and deal with patients with an expected ICU stay of more than three days.

The Molecular Nutrition of Amino Acids and Proteins.
DOI: http://dx.doi.org/10.1016/B978-0-12-802167-5.00010-4

It is clear that critical illness affects muscle in a dramatic way, leading to almost immediate and vast muscle wasting during ICU admission (Puthucheary et al., 2013), with an estimate of up to 20% muscle loss after a 10-day ICU stay. In an acute metabolic study, Berg et al. (2013) showed that targeted parenteral nutrition (PN) provides a more positive protein balance, even early in the critically ill patient. In the Berg et al. study two tracers were used of which both indicated an improved protein balance, but only for the phenylalanine tracer, was a clear increased protein synthesis rate demonstrated. No significant effect of targeted parenteral feeding was demonstrated on the protein degradation rate. There was no increased amino acid oxidation up to a feeding level of 1.1 gram protein per kg per day, indicating that requirement levels had probably not been met yet. Putucheary et al. (2013) showed that leg protein degradation was increased in their critically ill cohort, with about half of patients having sepsis. While leg muscle protein synthesis rate was decreased on day 1, it returned to the fed healthy control level at day 7. Klaude et al. (2012) found a large (plus 160%) increase in muscle protein degradation in septic patients, with no change in muscle protein synthesis. The increased muscle protein degradation was further supported by increased (plus 44%) proteasome (proteolytic) activity, the major proteolytic system in skeletal muscle (see chapter: Cellular and Molecular Mechanisms of Protein Synthesis Among Tissues). This study was performed with continuously PN fed critically ill septic patients. While autophagy, the housekeeping system of removing altered proteins, has been supposed to be suppressed by protein feeding, a vastly higher level of protein degradation in fed critically ill patients suggests that housekeeping is still supported. However, autophagy appears to be a bad reason to withhold feeding from critically ill patients who are losing muscle mass fast (McClave and Weijs, 2015; Rooyackers et al., 2015). There may be a timing issue here that is unsolved as yet. Later some evidence for targeted feeding by day 4 will be presented.

10.3. PROTEIN REQUIREMENT OF CRITICALLY ILL PATIENTS: MECHANISTIC STUDIES

While ESPEN and ASPEN guidelines would support an intake of protein of more than 1.2 g/kg per day, there is no randomized study to support these current guidelines (Singer et al., 2009; McClave et al., 2009). One of the most cited studies is the Ishibashi et al. (1998) study with three different levels of protein intake and In Vivo Neutron Activation Analysis for assessment of protein mass at zero and ten days of ICU stay. This study indicates that at a protein intake level of 1.5 g/kg fat-free mass adjusted for hydration, which converts to 1.2 g/kg preadmission body weight, would be most suitable judged by the amount of protein lost in 10 days. Although other studies (eg, Shaw et al., 1987) may point in the same direction, the evidence level is poor and therefore the level of dispute is high. Recently several papers have even suggested to increase the level of protein intake to 2.0–2.5 g/kg for the critically ill (Hoffer and Bistrian, 2012; Dickerson et al., 2012). Almost all studies supporting those high levels of protein intake are based on the nitrogen balance method. However, it is not entirely clear what a nitrogen balance actually means for critically ill patients that seem catabolic by nature. Therefore, we must ask the question of how nitrogen balance relates to patient outcome, but studies up to now are lacking.

Mechanistic studies into the effects of inactivity, such as bed rest, have indicated that muscle wasting in the critically ill could partly be explained by bed rest (Biolo et al., 2008). Bed rest induces insulin resistance (Stuart et al., 1988) and reduces blood flow, leading to the transport of essential amino acids to peripheral leg muscle mass being also reduced. While insulin therapy helps to overcome insulin resistance, including by increasing blood flow, it also improves protein metabolism (Biolo et al., 1995). Also resistance exercise has been shown to increase leg blood flow and muscle protein synthesis (Biolo et al., 1997). The elderly were described as showing increased splanchnic loss of amino acids (Volpi et al., 1999) and muscle protein anabolic resistance (Dardevet et al., 2012), which could contribute to muscle mass loss and may increase the protein requirement (Bauer et al., 2013; Deutz et al., 2014). Critically ill elderly have then lower muscle mass and combined with already existing protein anabolic resistance, they have worse outcomes compared to younger patients (Moisey et al., 2013). Even for the elderly in general we do not have specific protein guidelines; WHO and EFSA still support 0.8 g/kg (Joint WHO/FAO/UNU Expert Consultation, 2007; EFSA Panel on Dietetic Products, Nutrition and Allergies (NDA), 2012), but there is increasing support for a 1.0–1.2 g/kg recommendation for "healthy" elderly and for more than 1.2 g/kg when combined with acute or chronic illness (Bauer et al., 2013; Deutz et al., 2014).

Since there has been a lack of studies into protein needs, we will now explore some of the outcome studies that have been published and what we can learn from them about protein requirements in ICU patients.

10.4. PROTEIN REQUIREMENTS OF CRITICALLY ILL PATIENTS: OUTCOME-BASED STUDIES

10.4.1 Energy

Several small descriptive studies (38–57 patients) showed a relationship between the negative cumulative energy balance of the critically ill and mortality and/or complications (Villet et al., 2005; Dvir et al., 2006; Bartlett et al., 1982; Faisy et al., 2009). However, no such studies appear to be available for protein balance, while it could at least be suggested that a concurrent negative protein balance would have existed in these patients. Larger randomized studies, based on improvement of nutrition guidelines, seemed unsuccessful in improving outcome (Barr et al., 2004; Martin et al., 2004; Doig et al., 2008). However, these studies did improve nutrition only from vastly inadequate to inadequate, therefore it is questionable whether effects on outcome could have been expected (Sauerwein and Strack van Schijndel, 2007). Slightly later larger randomized studies tried to show improved outcome of low energy feeding ("permissive underfeeding," "trophic feeding"), but rather failed to do so (Arabi et al., 2011; Rice et al., 2012; Needham et al., 2013).

10.4.2 Energy and Protein

Alberda et al. (2009) and Strack van Schijndel et al. (2009b) were the first to show a direct relationship between protein intake and patient outcome. They showed a linear improvement with an increase in intake of 30 gram in a large cohort with a mean intake of 0.6 g/kg/d. Strack van Schijndel et al. showed that female patients who reached both the protein target of 1.2 g/kg/d and the measured energy target had lower mortality, whereas those just reaching the measured energy target without reaching the protein target did not. This was later confirmed and extended by Weijs et al. (2012b) and Allingstrup et al. (2012). Weijs et al. showed the benefits of protein intake for all patients and Allingstrup et al. showed that this appeared to be dose-dependent. However, all these studies are descriptive and randomized trials have to be pursued. In the meantime we will also have a look at randomized trials concerning supplemental PN, which were not designed for evaluation of the benefit of increased protein intake per se.

10.4.3 Parenteral Nutrition

In 2011, the TICACOS (Singer et al., 2011) and EPANIC trials (Casaer et al., 2011) were published, and slightly later the Heidegger et al. (2013) and Doig et al. (2013) studies. Although some more trials have been published now, they do not really change the emerging picture. The EPANIC trial is the largest of them all, comparing supplemental PN from day 1 (or 3) and day 8. Casaer et al. showed that starting supplemental PN early was detrimental for critically ill patients based on the outcome 'discharged alive within 8 days' although mortality was not different (Casaer et al., 2011). This study included patients with a short stay of 24 hours, and therefore resulted in very high loss to follow-up by day 3. They provided a high glucose, very low protein feeding on day 1 and 2, combined with a low level of EN (both energy and protein) in general. Furthermore, they estimated energy targets, instead of measuring energy targets (by indirect calorimetry). With targeted feeding strategies, estimated energy targets result in a high level of overfeeding. The TICACOS study was small, with measured energy targets and "aggressive" feeding from day 1 (Singer et al., 2011). Although mortality appeared to be lower in the per protocol analysis, the level of infection and length of stay was higher. Since glucose and propofol were not included into the energy calculation, patients were likely overfed. Heidegger et al. (2013) then showed that if EN failed during days 1–3 in their measured energy target, supplemental PN, however, decreased infection rate in critically ill patients. Finally, Doig et al. showed that in patients with contraindication to EN, the supplemental PN did not change mortality rate, but seemed to decrease the duration of mechanical ventilation, to decrease muscle wasting, and to increase quality of life score (Doig et al., 2013). Taken together, the targeted feeding based on measured energy target appeared to improve outcome. However, "aggressive" feeding may have resulted in overfeeding with a negative effect on outcome, while a higher feeding level in patients with an indication for supplemental PN seemed to have some advantage. Beneficial effects of feeding were not clearly shown in these studies, which is in line with the observation that in all studies protein targets were generally not met.

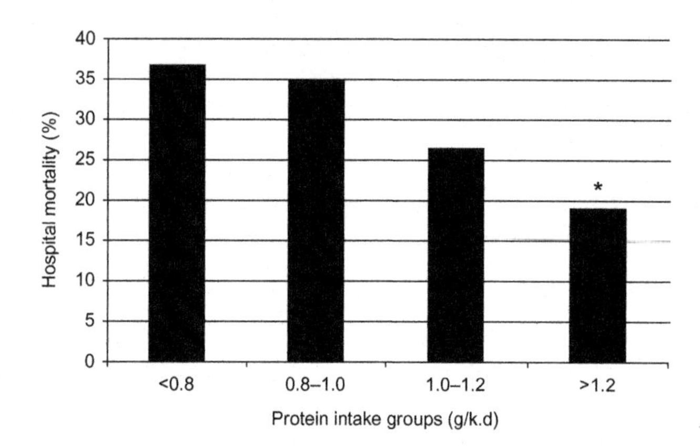

FIGURE 10.1 Hospital mortality (%) for protein intake groups up to more than 1.2 g/kg body weight in nonseptic, nonoverfed (less than 110% measured energy expenditure) critically ill patients on day 4 of ICU admission.

10.4.4 Protein

Since the slow advancement of knowledge about protein requirements in the critically ill was apparent, we reanalyzed our database (Weijs et al., 2014a) that had resulted in earlier publications (Strack van Schijndel et al., 2009b; Weijs et al., 2012b). Although we had observed that reaching protein and energy targets was associated with lower 28 day mortality, hospital mortality appeared to increase. However, hospital mortality was also higher in the energy target met group, while hospital mortality was much lower in the protein target met group. This provided us with a new hypothesis, that protein was indeed beneficial but was obscured by the effect of energy overfeeding. Since, we had been feeding our patients according to a nutrition algorithm using different enteral (and/or parenteral) nutrition formulae, we had more than usual statistical variation in the protein to energy ratio. In multivariate analysis, it became apparent that the protein level of more than 1.2 g/kg was beneficial for outcome, but that higher levels of energy provision were detrimental at the same time. Also, we had observed improved outcome in most admission diagnosis groups, except for sepsis. In sepsis patients meeting energy and protein targets on day 4 of ICU admission, no benefit of protein feeding was observed. Therefore, in the new approach we excluded the sepsis patients, and also those patients with an actual energy intake on day 4 of more than 110% of measured energy expenditure (using indirect calorimetry). Now it became apparent, that in the group of patients without sepsis and without overfeeding on day 4, that increasing the level of protein intake (g/kg/d) was beneficial for outcome (% hospital mortality) in a dose-dependent manner (see Fig. 10.1). The patient group with less than 0.8 g/kg/d was apparently underfed, since mean energy intake was only 49% of measured energy expenditure. However, the higher protein intake groups in Fig. 10.1 (0.8–1.0, 1.0–1.2, and >1.2 g/kg) were all adequately fed in terms of energy (91%, 95%, and 102% of measured energy expenditure). Therefore, it seems unlikely that severity of disease, with associated decreased intake, is an adequate explanation for differences between groups.

10.4.5 Muscle

The above provides us with some evidence that protein provision of more than 1.2 g/kg/d, as stated by current guidelines, appears beneficial for outcome. However, our goal is not only to have the patient to survive the ICU, but rather to become a functional and participating member of society after having to be in an ICU bed for some time. Classic studies by Herridge et al. (2003, 2011) show that the functional status of ICU patients after one year, and also after five years, is extremely low compared to healthy controls. Combined with the information we have on muscle wasting (Puthucheary et al., 2013) and protein intake for the critically ill from an international perspective (0.6 g/kg/d) (Shaw et al., 1987), this suggests that improvement of muscle mass preservation might be of benefit for later functional outcome of patients. We retrospectively investigated our patient records for CT scans of the abdomen at L3 level, which had been taken around the time of admission to the ICU. These CT scans were used to assess muscle mass as described by Mourtzakis et al. (2008). We demonstrated a higher mortality rate for critically ill patients with low muscle mass, and for patients that survived the hospital those with a low muscle mass during admission were more likely admitted to a nursing home after discharge from the hospital (Weijs et al., 2014b). This was also observed for elderly ICU patients by others (Moisey et al., 2013). Our analysis was not yet using information about loss of muscle mass during ICU or hospital stay. Small studies have now also looked at longitudinal changes; however, more work has to be done in this area (Braunschweig et al., 2013; Casaer et al., 2013). Preliminary observations from this database show that the higher level of protein intake of more than 1.2 g/kg is especially beneficial in cases where the muscle mass during admission to the ICU is already low (Looijaard et al., 2015). Further studies will have to demonstrate that feeding in general, and more specifically targeted protein feeding, provides a pathway for the better long-term functional outcome of long-term ICU surviving patients.

10.5. APPLICATION IN CLINICAL PRACTICE

We have been trying to implement nutritional feeding strategies to improve outcome in clinical practice. Critically ill patients have been fed according to both energy as well as protein targets over a period of more than ten years. We first developed a nutritional algorithm that not only accounts for adequate energy supply (based on calculated or measured energy needs) by appropriate pump speed, but also accounts for reaching protein targets (Strack van Schijndel et al., 2007). This algorithm is an important part of our nutrition policy at the ICU (see Fig. 10.2 from Weijs and Wischmeyer, 2013). Since the algorithm is based on calculations and logic steps, we then implemented this algorithm into the ICU patient management computer system (Strack van Schijndel et al., 2009a). The computer system supports both nutrition advice and nutrition monitoring applications. Thus, when the patient is provided with enteral (or parenteral) nutrition, the nutrition advice button provides the nursing staff with the adequate type and amount of EN which is based on patient data on preadmission body weight, energy and protein need, and protein–energy ratio. As we have an obligatory performance indicator of nutrition care in the Netherlands, which asks for protein intake at day 4 of admission to hospital, we monitored how the introduction of the algorithm into the computer system affected this performance indicator (Strack van Schijndel et al., 2009a). We observed that prior to its introduction, about 30% of patients were provided with more than 1.2 g/kg/d protein. Over the period of one year, there was a steady increase to about 60% of patients being provided with more than 1.2 g/kg/d (Strack van Schijndel et al., 2009a). For the implementation to take effect, it therefore took a whole year to reach its full potential on the performance indicator. Considering the hundreds of nursing staff involved, and the complexity of the organization of ICU care, this is still an appreciable speed of change. This again also indicates, since there was no change in patient population within that year, that the feeding levels of patients can be improved by the improvement of the feeding logistics which are certainly not exclusively determined by severity of illness. Organization of care is a key factor to success in the optimal nutrition of critically ill patients.

10.6. PROTEIN–ENERGY RATIO

As we think the protein to energy ratio is important for the outcome of patients, we also have to explain that optimal nutrition of critically ill patients is still limited by the availability of adequate nutritional formulae. When we developed the algorithm, our main goal was a protein target of 1.2 g/kg preadmission weight/d and the calculated or measured energy needs. We defined the need for nutritional formulae of 40, 50, and 60 gram protein per 1000 kcal. However, these nutritional formulae were adequate for the main portion of the ICU patient population, but not for the high and low extremes in kcal per kg body weight. Presently, we are challenged to provide adequate protein intake of even higher levels (1.5 g/kg) and at the same time to provide marginal energy underfeeding (80–90% of measured energy expenditure). This does not even consider yet different approaches to underweight and overweight ICU patients (Weijs et al., 2012a). We have calculated that we would need nutritional formulae up to 100 g protein per 1000 kcal for adequate protein nutrition, which is presently unavailable. Other options, such as feeding extra bolus protein from protein supplements, can be explored to deal with these limitations.

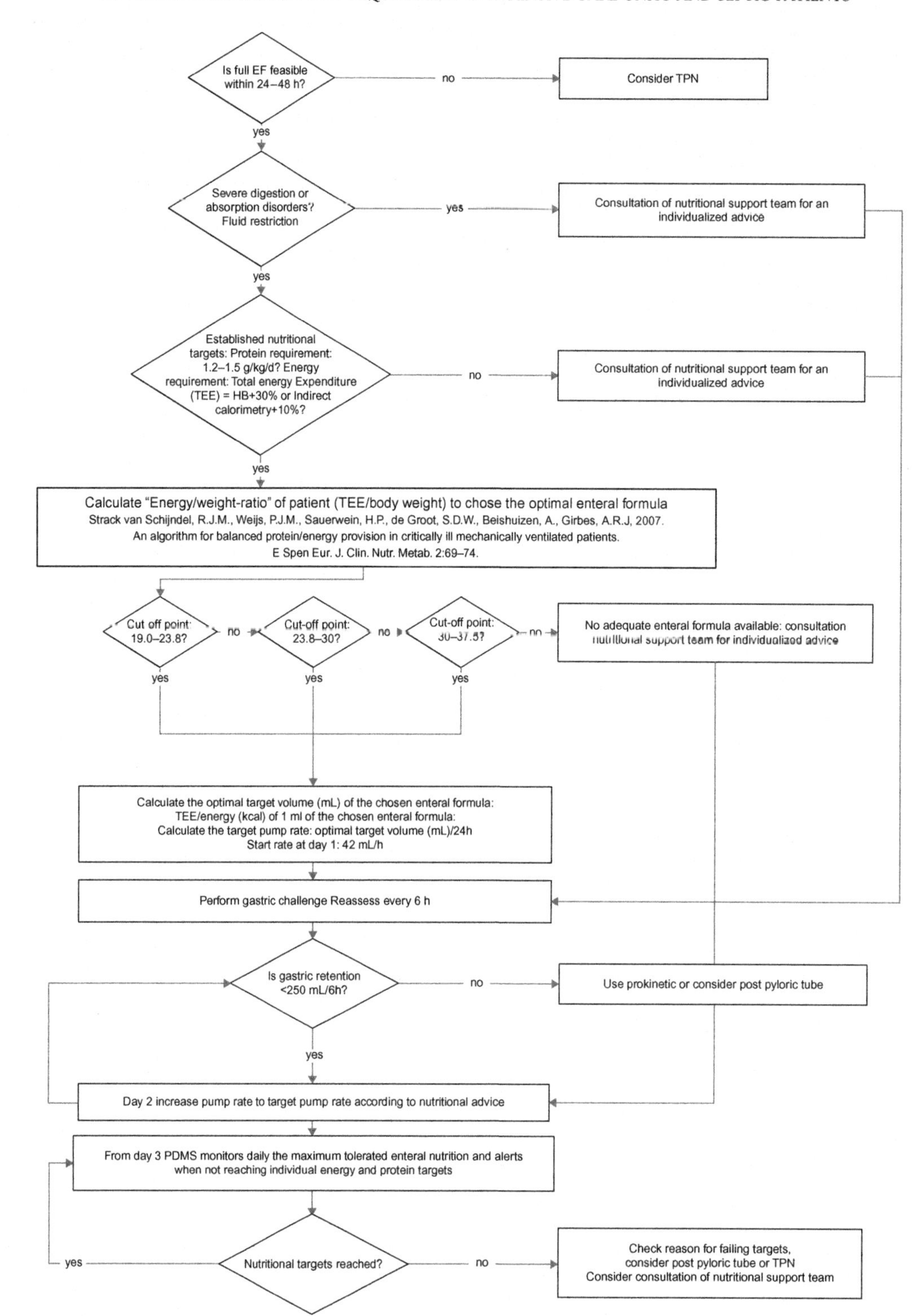

FIGURE 10.2 Feeding algorithm VU University Medical Center. This algorithm has now been adapted toward lower energy targets, in line with recent findings of energy overfeeding (see text). The algorithm still works, but lower energy targets make it even harder to feed patients up to the protein target of 1.2 g/kg.

10.7. CONCLUSION

At present the state of knowledge about protein metabolism in the critically ill is advancing, but it is rather insufficient to adequately address the question of protein requirements in critically ill patients. Randomized studies, preferably including direct measures of muscle mass and/or longer term physical functioning, are needed (Weijs and Wolfe, 2016). Based on available outcome studies, however, the current guidelines for ICU nutrition indicating a protein requirement of more than 1.2 g/kg are still recommended and backed up with some evidence. Concerning the practical application of targeted protein nutrition in critically ill patients, considering the disadvantages of energy overfeeding, clinical practice is in need of higher doses of protein enriched enteral (and parenteral) nutrition formulae.

References

Alberda, C., Gramlich, L., Jones, N., et al., 2009. The relationship between nutritional intake and clinical outcomes in critically ill patients: results of an international multicenter observational study. Intensive Care Med. 35, 1728–1737.

Allingstrup, M.J., Esmailzadeh, N., Wilkens Knudsen, A., et al., 2012. Provision of protein and energy in relation to measured requirements in intensive care patients. Clin. Nutr. 31, 462–468.

Arabi, Y.M., Tamim, H.M., Dhar, G.S., Al-Dawood, A., Al-Sultan, M., Sakkijha, M.H., et al., 2011. Permissive underfeeding and intensive insulin therapy in critically ill patients: a randomized controlled trial. Am. J. Clin. Nutr. 93 (3), 569–577.

Barr, J., Hecht, M., Flavin, K.E., Khorana, A., Gould, M.K., 2004. Outcomes in critically ill patients before and after the implementation of an evidence-based nutritional management protocol. Chest 125 (4), 1446–1457.

Bartlett, R.H., Dechert, R.E., Mault, J.R., et al., 1982. Measurement of metabolism in multiple organ failure. Surgery 92, 771–779.

Bauer, J., Biolo, G., Cederholm, T., Cesari, M., Cruz-Jentoft, A.J., Morley, J.E., et al., 2013. Evidence-based recommendations for optimal dietary protein intake in older people: a position paper from the PROT-AGE Study Group. J. Am. Med. Dir. Assoc. 14 (8), 542–559.

Berg, A., Rooyackers, O., Bellander, B.M., Wernerman, J., 2013. Whole body protein kinetics during hypocaloric and normocaloric feeding in critically ill patients. Crit. Care 17 (4), R158.

Biolo, G., Declan Fleming, R.Y., Wolfe, R.R., 1995. Physiologic hyperinsulinemia stimulates protein synthesis and enhances transport of selected amino acids in human skeletal muscle. J. Clin. Invest. 95 (2), 811–819.

Biolo, G., Tipton, K.D., Klein, S., Wolfe, R.R., 1997. An abundant supply of amino acids enhances the metabolic effect of exercise on muscle protein. Am. J. Physiol. 273 (1 Pt 1), E122–E129.

Biolo, G., Agostini, F., Simunic, B., Sturma, M., Torelli, L., Preiser, J.C., et al., 2008. Positive energy balance is associated with accelerated muscle atrophy and increased erythrocyte glutathione turnover during 5 wk of bed rest. Am. J. Clin. Nutr. 88 (4), 950–958.

Braunschweig, C.A., Sheean, P.M., Peterson, S.J., Perez, S.G., Freels, S., Troy, K.L., et al., 2013. Exploitation of diagnostic computed tomography scans to assess the impact of nutritional support on body composition changes in respiratory failure patients. J. Parenter. Enteral Nutr. 38 (7), 880–885.

Casaer, M.P., Mesotten, D., Hermans, G., Wouters, P.J., Schetz, M., Meyfroidt, G., et al., 2011. Early versus late parenteral nutrition in critically ill adults. N. Engl. J. Med. 365 (6), 506–517.

Casaer, M.P., Langouche, L., Coudyzer, W., Vanbeckevoort, D., De Dobbelaer, B., Güiza, F.G., et al., 2013. Impact of early parenteral nutrition on muscle and adipose tissue compartments during critical illness. Crit. Care Med. 41 (10), 2298–2309.

Dardevet, D., Rémond, D., Peyron, M.A., Papet, I., Savary-Auzeloux, I., Mosoni, L., 2012. Muscle wasting and resistance of muscle anabolism: the "anabolic threshold concept" for adapted nutritional strategies during sarcopenia. ScientificWorldJournal 2012, 269531.

Deutz, N.E., Bauer, J.M., Barazzoni, R., Biolo, G., Boirie, Y., Bosy-Westphal, A., et al., 2014. Protein intake and exercise for optimal muscle function with aging: recommendations from the ESPEN Expert Group. Clin. Nutr. 33 (6), 929–936.

Dickerson, R.N., Pitts, S.L., Maish III, G.O., Schroeppel, T.J., Magnotti, L.J., Croce, M.A., et al., 2012. A reappraisal of nitrogen requirements for patients with critical illness and trauma. J. Trauma Acute Care Surg. 73 (3), 549–557.

Doig, G.S., Simpson, F., Finfer, S., et al., 2008. Effects of evidence-based feeding guidelines on mortality of critically ill patients: a cluster randomized controlled trial. J. Am. Med. Assoc. 300, 2731–2741.

Doig, G.S., Simpson, F., Sweetman, E.A., et al., 2013. Early PN Investigators of the ANZICS Clinical Trials Group. Early parenteral nutrition in critically ill patients with short-term relative contraindications to early enteral nutrition: a randomized controlled trial. J. Am. Med. Assoc. 309, 2130–2138.

Dvir, D., Cohen, J., Singer, P., 2006. Computerized energy balance and complications in critically ill patients: an observational study. Clin. Nutr. 25 (1), 37–44.

EFSA Panel on Dietetic Products, Nutrition and Allergies (NDA), 2012. Scientific opinion on dietary eference values for protein. EFSA J. 10 (2), 2557, 1–66.

Faisy, C., Lerolle, N., Dachraoui, F., Savard, J.F., Abboud, I., Tadie, J.M., et al., 2009. Impact of energy deficit calculated by a predictive method on outcome in medical patients requiring prolonged acute mechanical ventilation. Br. J. Nutr. 101 (7), 1079–1087.

Heidegger, C.P., Berger, M.M., Graf, S., Zingg, W., Darmon, P., Costanza, M.C., et al., 2013. Optimisation of energy provision with supplemental parenteral nutrition in critically ill patients: a randomised controlled clinical trial. Lancet 381 (9864), 385–393.

Herridge, M.S., Cheung, A.M., Tansey, C.M., Matte-Martyn, A., Diaz-Granados, N., Al-Saidi, F., et al., 2003. One-year outcomes in survivors of the acute respiratory distress syndrome. N. Engl. J. Med. 348 (8), 683–693.

Herridge, M.S., Tansey, C.M., Matté, A., Tomlinson, G., Diaz-Granados, N., Cooper, A., et al., 2011. Functional disability 5 years after acute respiratory distress syndrome. N. Engl. J. Med. 364 (14), 1293–1304.

Hoffer, L.J., Bistrian, B.R., 2012. Appropriate protein provision in critical illness: a systematic and narrative review. Am. J. Clin. Nutr. 96 (3), 591–600.

Ishibashi, N., Plank, L.D., Sando, K., Hill, G.L., 1998. Optimal protein requirements during the first 2 weeks after the onset of critical illness. Crit. Care Med. 26, 1529–1535.

Joint WHO/FAO/UNU Expert Consultation, 2007. Protein and amino acid requirements in human nutrition. World Health Organ. Tech. Rep. Ser. 935, 1–265.

Klaude, M., Mori, M., Tjäder, I., Gustafsson, T., Wernerman, J., Rooyackers, O., 2012. Protein metabolism and gene expression in skeletal muscle of critically ill patients with sepsis. Clin. Sci. (Lond) 122 (3), 133–142.

Looijaard, W., Stapel, S., Oudemans-van Straaten, H., Weijs, P.J., 2015. High protein feeding is beneficial for older sarcopenic ICU patients. ESPEN Lisbon, Portugal, 5–8 September 2015.

Martin, C.M., Doig, G.S., Heyland, D.K., et al., 2004. Multicentre, cluster-randomized clinical trial of algorithms for critical-care enteral and parenteral therapy (ACCEPT). Can. Med. Assoc. J. 170, 197–204.

McClave, S.A., Weijs, P.J., 2015. Preservation of autophagy should not direct nutritional therapy. Curr. Opin. Clin. Nutr. Metab. Care 18 (2), 155–161.

McClave, S.A., Martindale, R.G., Vanek, V.W., et al., 2009. Guidelines for the provision and assessment of nutrition support therapy in the adult critically ill patient: Society of Critical Care Medicine (SCCM) and American Society for Parenteral and Enteral Nutrition (A.S.P.E.N.). J. Parenter. Enteral Nutr. 33, 277–316.

Moisey, L.L., Mourtzakis, M., Cotton, B.A., Premji, T., Heyland, D.K., Wade, C.E., et al., 2013. Nutrition and Rehabilitation Investigators Consortium (NUTRIC). Skeletal muscle predicts ventilator-free days, ICU-free days, and mortality in elderly ICU patients. Crit. Care 17 (5), R206.

Mourtzakis, M., Prado, C.M., Lieffers, J.R., Reiman, T., McCargar, L.J., Baracos, V.E., 2008. A practical and precise approach to quantification of body composition in cancer patients using computed tomography images acquired during routine care. Appl. Physiol. Nutr. Metab. 33 (5), 997–1006.

Needham, D.M., Dinglas, V.D., Bienvenu, O.J., Colantuoni, E., Wozniak, A.W., Rice, T.W., et al., 2013. One year outcomes in patients with acute lung injury randomised to initial trophic or full enteral feeding: prospective follow-up of EDEN randomised trial. Br. Med. J. 346, f1532.

Puthucheary, Z.A., Rawal, J., McPhail, M., Connolly, B., Ratnayake, G., Chan, P., et al., 2013. Acute skeletal muscle wasting in critical illness. J. Am. Med. Assoc. 310 (15), 1591–1600, Erratum in: J. Am. Med. Assoc. 12 February 2014; 311 (6), 625.

Rice, T.W., Wheeler, A.P., Thompson, B.T., Steingrub, J., Hite, R.D., Moss, M., et al., 2012. Initial trophic vs full enteral feeding in patients with acute lung injury: the EDEN randomized trial. J. Am. Med. Assoc. 307 (8), 795–803.

Rooyackers, O., Kouchek-Zadeh, R., Tjäder, I., Norberg, Å., Klaude, M., Wernerman, J., 2015. Whole body protein turnover in critically ill patients with multiple organ failure. Clin. Nutr. 34 (1), 95–100.

Sauerwein, H.P., Strack van Schijndel, R.J., 2007. Perspective: how to evaluate studies on peri-operative nutrition? Considerations about the definition of optimal nutrition for patients and its key role in the comparison of the results of studies on nutritional intervention. Clin. Nutr. 26 (1), 154–158.

Shaw, J.H., Wildbore, M., Wolfe, R.R., 1987. Whole body protein kinetics in severely septic patients. The response to glucose infusion and total parenteral nutrition. Ann. Surg. 205 (3), 288–294.

Singer, P., Berger, M.M., Van den Berghe, G., et al., 2009. ESPEN Guidelines on parenteral nutrition: intensive care. Clin. Nutr. 28, 387–400.

Singer, P., Anbar, R., Cohen, J., Shapiro, H., Shalita-Chesner, M., Lev, S., et al., 2011. The tight calorie control study (TICACOS): a prospective, randomized, controlled pilot study of nutritional support in critically ill patients. Intensive Care Med. 37 (4), 601–609.

Strack van Schijndel, R.J.M., Weijs, P.J.M., Sauerwein, H.P., de Groot, S.D.W., Beishuizen, A., Girbes, A.R.J., 2007. An algorithm for balanced protein/energy provision in critically ill mechanically ventilated patients. e-SPEN, European e-Journal Clin. Nutr. Metab. 2, 69–74.

Strack van Schijndel, R.J., de Groot, S.D., Driessen, R.H., Ligthart-Melis, G., Girbes, A.R., Beishuizen, A., et al., 2009a. Computer-aided support improves early and adequate delivery of nutrients in the ICU. Neth. J. Med. 67 (11), 388–393.

Strack van Schijndel, R.J., Weijs, P.J., Koopmans, R.H., et al., 2009b. Optimal nutrition during the period of mechanical ventilation decreases mortality in critically ill, long-term acute female patients: a prospective observational cohort study. Crit. Care 13, R132.

Stuart, C.A., Shangraw, R.E., Prince, M.J., Peters, E.J., Wolfe, R.R., 1988. Bed-rest-induced insulin resistance occurs primarily in muscle. Metabolism 37 (8), 802–806.

Villet, S., Chiolero, R.L., Bollmann, M.D., Revelly, J.P., Cayeux, R.N.M.C., Delarue, J., et al., 2005. Negative impact of hypocaloric feeding and energy balance on clinical outcome in ICU patients. Clin. Nutr. 24 (4), 502–509.

Volpi, E., Mittendorfer, B., Wolf, S.E., Wolfe, R.R., 1999. Oral amino acids stimulate muscle protein anabolism in the elderly despite higher first-pass splanchnic extraction. Am. J. Physiol. 277 (3 Pt 1), E513–E520.

Weijs, P.J., Wischmeyer, P.E., 2013. Optimizing energy and protein balance in the ICU. Curr. Opin. Clin. Nutr. Metab. Care 16 (2), 194–201.

Weijs, P.J., Wolfe, R.R., 2016. Exploration of the protein requirement during weight loss in obese older adults. Clin. Nutr 35 (2), 394–398.

Weijs, P.J., Sauerwein, H.P., Kondrup, J., 2012a. Protein recommendations in the ICU: g protein/kg body weight – which body weight for underweight and obese patients? Clin. Nutr. 31 (5), 774–775.

Weijs, P.J.M., Stapel, S.N., de Groot, S.D.W., Driessen, R.H., de Jong, E., Girbes, A.R.J., et al., 2012b. Optimal protein and energy nutrition decreases mortality in mechanically ventilated, critically ill patients: a prospective observational cohort study. J. Parenter. Enteral Nutr. 36 (1), 60–68.

Weijs, P., Looijaard, W., Beishuizen, A., Girbes, A., Oudemans-van Straaten, H.M., 2014a. Early high protein intake is associated with low mortality and energy overfeeding with high mortality in non-septic mechanically ventilated critically ill patients. Crit. Care 18 (6), 701.

Weijs, P.J., Looijaard, W.G., Dekker, I.M., Stapel, S.N., Girbes, A.R., Oudemans-van Straaten, H.M., et al., 2014b. Low skeletal muscle area is a risk factor for mortality in mechanically ventilated critically ill patients. Crit. Care 18 (1), R12.

Wischmeyer, P.E., 2013. The evolution of nutrition in critical care: how much, how soon? Crit. Care 17 (Suppl. 1), S7.

CHAPTER

11

Muscle Protein Kinetics in Cancer Cachexia

J.M. Argilés[1,2], *S. Busquets*[1,2] *and F.J. López-Soriano*[1,2]

[1]Cancer Research Group, Departament de Bioquímica i Biologia Molecular, Facultat de Biologia, Universitat de Barcelona, Barcelona, Spain [2]Institut de Biomedicina de la Universitat de Barcelona, Barcelona, Spain

11.1. INTRODUCTION: MUSCLE WASTING AS THE MAIN FEATURE OF CANCER CACHEXIA

Cachexia occurs in the majority of terminal cancer patients, and is responsible for the death of 22% of cancer patients. Importantly, the survival of cancer patients suffering from different types of neoplasias is dependent on the amount of weight loss (Dewys et al., 1980). Therefore, cachexia represents an important factor in the treatment of a cancer patient, affecting not only survival, but also the efficacy of anticancer treatment, quality of life, and medical costs (Farkas et al., 2013; Von Haehling and Anker, 2015). Thus, there is a strong pressure to better understand the mechanisms that drive cachexia in order to offer cancer patients more effective care.

Cancer cachexia has been characterized as a syndrome associated with loss of muscle with or without loss of fat mass (Argilés et al., 2010; Evans et al., 2008). Other pathologies associated with cachexia are anorexia, inflammation, insulin resistance, and increased muscle protein degradation. Another defining characteristic is that cachexia cannot be fully reversed by conventional nutritional support and leads to progressive functional impairment. Thus, it can be concluded that cachexia is caused by an energy imbalance which is the result of both decreased food intake, due to marked anorexia, and increased energy expenditure caused by a highly hypermetabolic state. Mitochondrial dysfunction and uncoupling of oxidative phosphorylation together with futile cycling and conversion of white into brown fat contribute to this inefficient energetic state (Argilés et al., 2014b).

Perhaps the most prominent metabolic trend of the cachectic cancer patient is that of muscle wasting. Indeed, protein is lost from skeletal muscle as a consequence of different mechanisms that include an increase in protein degradation, a decrease in protein synthesis, a decrease in amino acid transport, and oxidation of branched chain amino acids, an activation of apoptosis together with changes in muscle regeneration capacity (Argilés et al., 2014b) (Fig. 11.1). If we take into consideration that skeletal muscle represents over 40% of body weight in humans, muscle wasting in cancer accounts for a huge percentage of body weight loss.

11.2. CONTROL OF SKELETAL MASS IN HEALTHY CONDITIONS

Skeletal muscle represents by far the largest tissue in the human body—40% of total body weight including around 640 separate muscles—being involved in many diverse important functions, such as strength, endurance, and physical performance.

An increase in muscle mass occurs during fetal and postnatal development and can be promoted either with mechanical overload, such as strength training, or by anabolic hormonal treatments (beta2 agonists and testosterone) or both. Conversely, muscle mass is decreased as a result of wasting conditions, either healthy, aging, bed-rest or microgravity, or pathological, cancer, diabetes, starvation, sepsis, or loss of neural input (denervation).

The Molecular Nutrition of Amino Acids and Proteins.
DOI: http://dx.doi.org/10.1016/B978-0-12-802167-5.00011-6

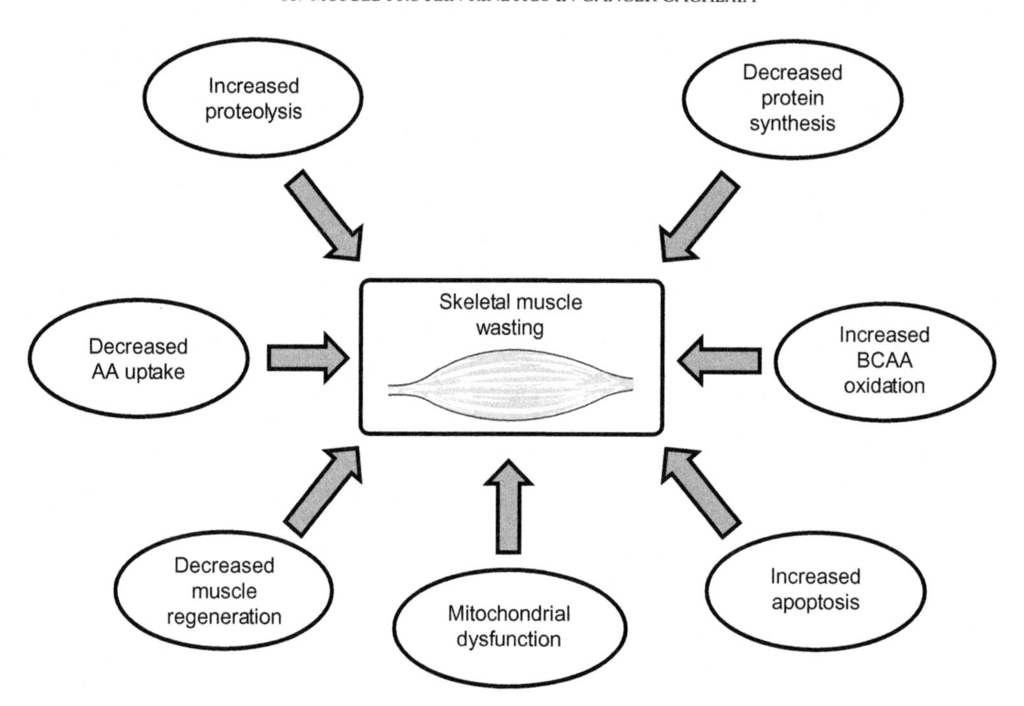

FIGURE 11.1 Different alterations contributing to muscle wasting in cancer. In spite of the fact that protein degradation and protein synthesis are the main factors responsible for muscle wasting, other processes such as mitochondrial dysfunction, apoptosis, regeneration or alterations in uptake and amino acid metabolism also contribute to the syndrome.

The structure of the muscle tissue conditions the control of its mass. Indeed, the regulation of muscle mass depends basically on protein turnover (the balance between protein synthesis and degradation) and because muscle fibers are polynucleated, protein turnover is influenced also by cell or nuclear turnover by either loss (apoptosis) of micronuclei or fusion of satellite cells. Satellite cells are small mononuclear progenitor cells with virtually no cytoplasm found in mature muscle, precursors to skeletal muscle cells, able to give rise to satellite cells or differentiated skeletal muscle cells. Another important factor to bear in mind is that the regulation of fiber mass depends on the type of fiber. Indeed, red (glycolytic, slow, type I) fibers behave very differently from white (lactogenic, fast, type II) fibers in response to denervation (Schiaffino et al., 2013), nutrient deprivation (Li and Goldberg, 1976), or glucocorticoid treatment (Goldberg and Goodman, 1969). In addition, type I and II fibers are also different in relation to their turnover rates, slow ones having faster rates of both protein synthesis and degradation (Li and Goldberg, 1976). Taking all of this into consideration, it becomes clear that the regulation of muscle mass is a fairly complex process that, in addition to the rates of protein synthesis and degradation, including apoptosis, and regeneration that affects the myonuclear domain—all of this results in normal, hypertrophic, or wasted skeletal muscle (Fig. 11.1).

11.3. ANABOLIC SIGNALS

When a shift toward protein synthesis and away from protein degradation takes place, skeletal muscle becomes hypertrophic and, consequently, there is an increase in the size of myofibers. The signals that trigger this anabolic response are varied. Perhaps the most paradigmatic stimulus is that of insulin-like growth factor-1 (IGF-1). Indeed, following the binding of this growth factor to its receptor and ulterior tyrosine phosphorylation, insulin receptor substrate-1 (IRS-1) is recruited. This molecule activates phosphatidylinositol-3 kinase (PI3K) which, in turn, modulates protein kinase B (Akt) to stimulate protein synthesis by activating the mTOR kinase. This enzyme interacts with two multiprotein complexes, mTOR complex-1 (mTORC1), that binds to the regulatory-associated protein of mTOR (RAPTOR) protein, and mTOR complex-2 (mTORC2), that binds to the rapamycin-insensitive companion of mammalian target of rapamycin (RICTOR) protein. mTORC1—which can be inhibited by rapamycin—acts by activating S6 kinase 1 (S6K) and the EIF4E-binding protein (4E-BP), while mTORC2 contributes to Akt activation in a positive feedback loop (see Egerman and Glass, 2014 for review) (Fig. 11.2).

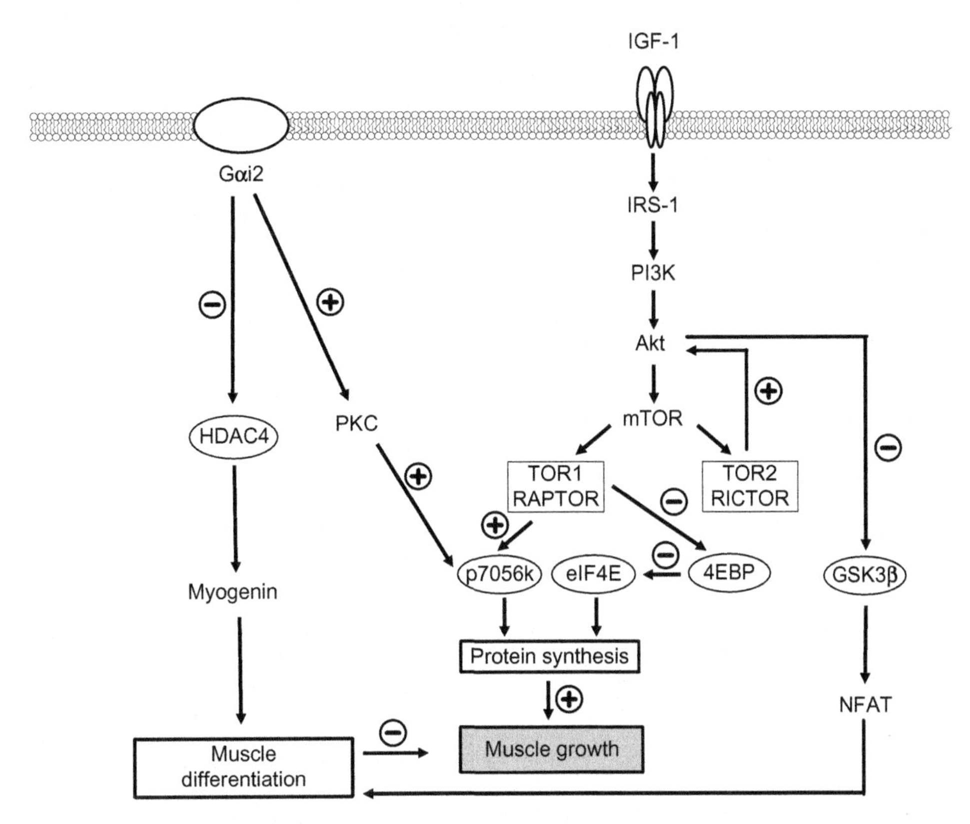

FIGURE 11.2 **Synthetic pathways involved in muscle growth.** Basically, IGF-1 (and also insulin) control muscle protein synthesis by interacting at the level of mTOR through Akt. However, heterotrimeric G proteins can also participate in this process through the activation of PKC. When protein synthesis is enhanced, muscle differentiation is decreased to favor growth—both the inhibition of HDAC4 and GS3kβ contribute to the repressed differentiation.

Among the downstream targets of Akt, glycogen synthase kinase-3β (GSK3β) and the family of Forkhead box O (FOXO) transcription factors are found. These two targets can be inhibited by Akt resulting in activation of protein synthesis through either eIF2B initiation factor or by inhibiting the transcription factor NFAT. This transcription factor activates muscle differentiation (Horsley et al., 2001) via GSK3β or by decreasing the activity of the E3 ubiquitin ligase (Muscle Atrophy Box (MAFbx)—also known as atrogin). This enzyme controls protein synthesis by ubiquitinization of eIF3f through inhibition of FOXO (see (Egerman and Glass, 2014) for review) (Fig. 11.3).

The other major pathway that controls muscle protein synthesis is that of myostatin. Indeed, this molecule—belonging to the TGF-beta superfamily—acts as a negative regulator of muscle growth. Myostatin and activin A (that also belongs to the same family) may interact and bind an heterodimeric receptor that incorporates a type I (activin receptor-like kinase 4 and 5 [ALK4 and ALK5]) and a type II (activin-receptor 2 [ACVR2 and ACVR2B) receptor (Argilés et al., 2012). Downstream signaling includes the phosphorylation and nuclear activation of Smad2 and Smad3 transcription factors and the formation of heterodimers with Smad4. Activation of the Smad proteins leads to inactivation of Akt while activating the expression of E3 ubiquitin ligases by FOXO transcription factors (Fig. 11.4).

In addition to the IGF-1 and myostatin pathways, the Galphai2 (Gαi2) pathway has been shown to induce hypertrophy by the activation of PKC, and, therefore, bypassing Akt. Interestingly, Gαi2—an α subunit of the heterotrimeric G protein complex—is also able to activate protein synthesis in a PKC-independent manner through the inhibition of Histone desacetylase-4 (HDAC4) (Fig. 11.2). This molecule activates atrogenes in a myogenin-dependent fashion. Therefore, Gαi2 signaling participates in skeletal muscle growth, regeneration, satellite cell proliferation, and differentiation (Minetti et al., 2014).

In cancer cachexia, protein synthesis is affected at different levels. IGF-1 is downregulated in experimental cancer cachexia (Costelli et al., 2006). IGF-1 treatment reduces weight loss and improves outcome in a rat model of cancer cachexia (Schmidt et al., 2011). In humans, early perturbations of IGF-1 signaling were observed in gastric cancer patients irrespective of weight loss (Bonetto et al., 2013).

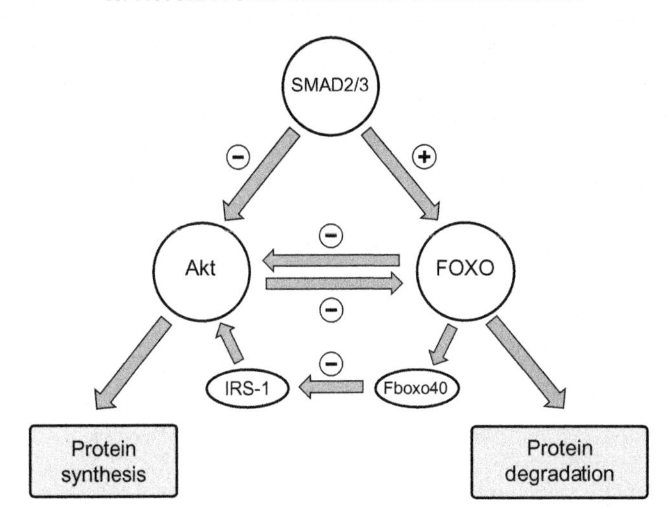

FIGURE 11.3 **Reciprocal regulation of synthetic/degradative transcription factors.** Both FOXO and Akt are reciprocally repressed to avoid a simultaneous increase in synthesis and degradation. On the other hand SMAD2/3 interact by repressing Akt and activating FOXO.

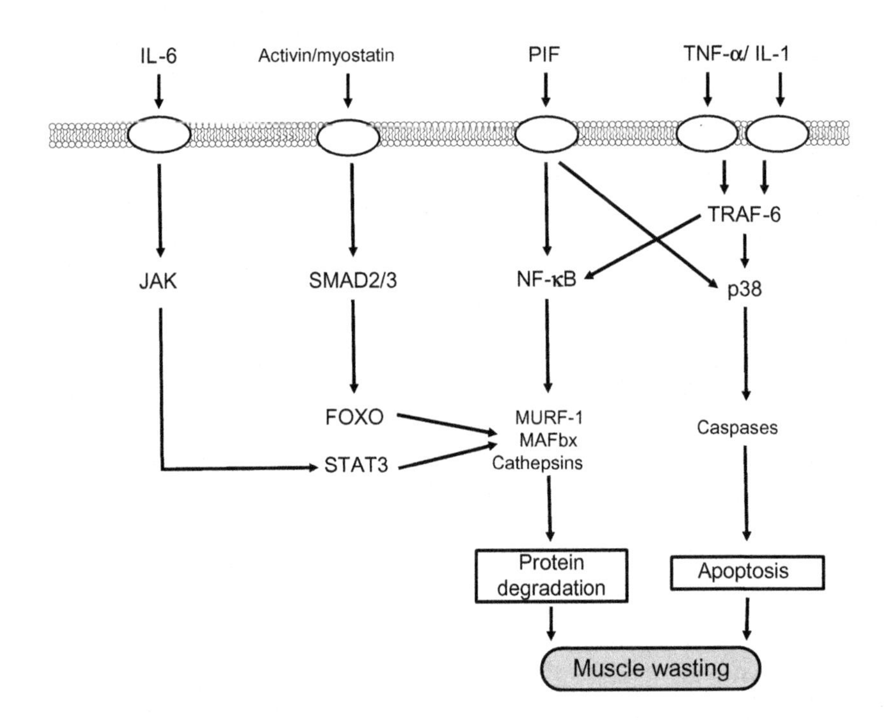

FIGURE 11.4 **Inflammatory pathways contributing to protein degradation.** Different humoral inflammatory stimuli—such as TNF-α, IL-1 and IL-6—and tumoral—PIF—potentiate (via FOXO or STAT3) the activation of proteolytic systems including the ubiquitin-proteasome and the autophagic lysosomal pathways. Myostatin/Activin also participate in the activation of FOXO.

Inflammation certainly contributes to the decreased muscle protein synthesis. Thus, White et al. (2013) reported a dose-dependent suppression of mTOR activity by IL-6 and suppressed mTOR responsiveness to glucose administration in Apc(Min/+) mice. IL-6 suppression of mTOR activity was dependent on AMPK activation and independent of STAT signaling in myotubes.

11.4. INFLAMMATION AND MUSCLE PROTEIN DEGRADATION

Muscle protein degradation is undoubtedly the most important event contributing to muscle wasting associated with cancer cachexia. Different proteolytic systems are involved in the breakdown of skeletal muscle contractile proteins, the ubiquitin-dependent proteolytic system (UPS) one being the most important (Argilés and López-Soriano, 1996). However, recent reports clearly demonstrate an involvement of both the calcium-dependent and the lysosomal proteases.

11.4.1 Ub-Proteasome-Dependent Proteolysis

In animal models of cancer cachexia, there is a clear increase in the expression of muscle-specific E3 ubiquitin ligases—MuRF-1 and MAFbx (Fontes-Oliveira et al., 2013; Hinch et al., 2013). In humans, however, the evidence in favor of UPS is questionable. From this point of view, Stephens et al. (2010) selected from the preclinical literature, candidate genes, including FOXO protein and ubiquitin E3 ligases, to perform a transcriptomic analysis. It was found that these genes were not related to weight loss in a human clinical study involving *rectus abdominis* muscle biopsies obtained from upper gastrointestinal cancer patients. Furthermore, promoter analysis identified that the 83 weight loss-associated genes had fewer FOXO binding sites than expected by chance. In studies where the activity of UPS in quadriceps muscle biopsies was determined, similar levels to healthy controls in patients with lung cancer and weight loss <10% were found (Op den Kamp et al., 2012). In a different study involving low weight loss lung cancer patients, no increases in UPS components resulted, while the lysosomal pathway—cathepsin B—was increased (Jagoe et al., 2002). In favor of the UPS in human cancer cachexia, Bossola et al., using gastric cancer patients, found in skeletal muscle biopsies, both an increase in gene expression (Bossola et al., 2001) and in activity (Bossola et al., 2003) of the UPS—determined by measuring the cleavage of artificial substrates. The authors concluded that an activation of the proteolytic system early in disease before clinical evidence of cachexia, mainly weight loss, was observed (Bossola et al., 2001). In more recent work in gastric cancer patients, the same research group concluded that in skeletal muscle, the gene expression of atrogin-1 and MuRF1 is not affected by the presence of cancer (D'Orlando et al., 2014). The expression of these atrophy-related genes was unaffected by the disease stage and the degree of weight loss. In cancer patient biopsies, the mRNA levels for ubiquitin and the 20S proteasome subunits were two to four times higher in muscle from patients with cancer than in muscle from control patients (Williams et al., 1999). Since the patients did not have a large weight loss, the authors suggest that the enhanced proteolyic gene expression took place before protein breakdown actually took place. Khal et al. measured in cancer patients the skeletal muscle expression of mRNA for proteasome subunits C2 and C5 and the E2 ubiquitin ligase. Their results suggested variations in the expression of these key components of the ubiquitin-proteasome pathway that took place only when the threshold of 10% weight loss was reached (Khal et al., 2005).

Another ubiquitin ligase, Trim32, ubiquitylates thin filament components (actin, tropomyosin, troponins) and Z-band (α-actinin) and promotes their degradation (Cohen et al., 2012). Downregulation of Trim32 during fasting reduced fiber atrophy and the rapid loss of thin filaments (Cohen et al., 2012). Trim32 reduces PI3K-Akt-FoxO signaling in normal and atrophying muscle (Cohen et al., 2014). This mechanism probably contributes to insulin resistance during fasting and catabolic diseases and perhaps also in cancer cachexia, but more studies involving the ligase are necessary to confirm this suggestion.

11.4.2 Lysosomal Proteolysis

For a long time, the involvement of the lysosomal system in muscle proteolysis was not taken much into consideration. However, we now have evidence that the autophagy pathway (necessary to drive substrates to lysosomes and, therefore, cathepsins) certainly contributes to the pathogenesis of muscle wasting not only in myopathies because of intrinsic muscle defects, but also in muscle wasting associated with cancer cachexia (Penna et al., 2013, 2014). Tardif et al. have shown that, whereas other proteolytic systems were unchanged, the autophagic-lysosomal pathway is the main proteolytic system modified in the skeletal muscle of esophageal cancer patients, suggesting an involvement of this proteolytic system during cancer cachexia development in humans (Tardif et al., 2013). In addition, in muscle of patients affected by cancer cachexia, the expression of genes related to autophagy—BNIP3 (*B*cl-2 and *n*ineteen-kilodalton *i*nteracting *p*rotein-3) and GABARAPL1 (GABA(A) receptor-associated protein like 1)—is increased (Stephens et al., 2010).

11.4.3 Calpain-Dependent Proteolysis

In addition to lysosomal-mediated proteolysis, the calpain system constitutes a group of calcium-dependent proteases which is capable of degrading myofibrillar proteins, in addition to cytoskeletal proteins, cell adhesion molecules, and cell receptors (Barta et al., 2005). Calpain activity is increased in skeletal muscle from gastric cancer patients with no or <5% weight loss. This may happen even before changes in molecular markers of muscle wasting and significant weight loss occur (Smith et al., 2011).

11.4.4 Inflammation

Both inflammatory cytokines and glucocorticoids contribute to the activation of proteolysis. However, tumor-derived factors have also been involved (Tisdale, 2004). The cytokines involved in muscle wasting during cancer—TNF-α, TWEAK (TNF-like weak inducer of apoptosis), TRAF6 (tumor necrosis factor receptor (TNFR)-associated factor 6), IL-6, γ-IFN, and LIF (see Argilés et al., 2009 for review)—act via two different intracellular pathways: the NF-κB and the p38 MAP kinase pathways. They are both involved in the upregulation of the expression of E3 ligases MuRF-1 and MAFbx. Inhibition of NF-κB, at least in rodents, results in a decreased muscle loss in tumor-bearing animals (Moore-Carrasco et al., 2009). This pathway acts partly by activating the formation of NO—through induction of the iNOS—therefore increasing nitrosative stress (Barreiro et al., 2005), which can also activate protein degradation in skeletal muscle. These cytokines also activate the Janus kinase/signal transducer and activator of transcription (JAK/STAT) pathway. During cancer, there is an induction of STAT3 phosphorylation in skeletal muscle (Bonetto et al., 2012). However, interestingly, STAT3 also seems to have a role in autophagy resulting in an alteration of the beclin complex and, therefore, blocking autophagy and stopping muscle degeneration (Yamada et al., 2012).

According to White et al., IL-6 suppression of mTOR activity is dependent on AMPK activation and independent of STAT signaling in myotubes (White et al., 2013). AMP kinase is able to phosphorylate FOXO, therefore, stimulating its transcriptional activity. AMPK is also able to activate autophagy genes.

Sun et al. found a positive correlation between TRAF6 and ubiquitin expression in skeletal muscle of gastric cancer patients, suggesting that TRAF6 may up regulates ubiquitin activity in cancer cachexia (Sun et al., 2012). On the same lines, TWEAK, a cytokine belonging to the TNF-α family, is able to induce a cachectic phenotype through the induction of MuRF-1. The same cytokine is also able to induce the autophagy system in skeletal muscle cells and this seems to be dependent on NF-κB activation (Bhatnagar et al., 2012). Furthermore, caspases seem to contribute, at least in part, to the activation of NF-κB in response to TWEAK treatment (Bhatnagar et al., 2012). The cytosolic release of its inhibitory IκB proteins allow the translocation of NF-κB to the nucleus and subsequent transcription of genes involved in proteolysis. Inhibition of NF-κB decreases muscle loss in a rat model of cancer cachexia, in part, by inhibiting the upregulation of MuRF-1 (Moore-Carrasco et al., 2007).

Concerning tumor factors, the so-called proteolysis-inducing factor (PIF) was identified as a circulating tumor-released mediator in mice bearing a cachexia-inducing (MAC16) (Todorov et al., 1996). Moreover, human tumors also express the so-called Proteolysis-inducing factor (PIF) (Cabal-Manzano et al., 2001). PIF—a 24 kDa glycoprotein—causes weight loss by inducing enhanced protein degradation without decreasing the appetite in mice (Fig. 11.4). PIF was also found to be present in the urine of cachectic cancer patients while being absent from normal subjects (Todorov et al., 1996).

It has to be pointed out that some cytokines (IL-4, IL-10) can actually act as anticachectics (Argilés and López-Soriano, 1999). One of them, IL-15, has been shown to have a clear antiproteolytic (Busquets et al., 2005; Carbó et al., 2000, 2001) and antiapoptotic action in skeletal muscle of animals under cancer cachexia (Figueras et al., 2004).

In many cachectic patients, elevated plasma levels of cortisol are found. The mechanisms by which glucocorticoids stimulate muscle protein degradation—particularly type II fibers—are varied. First, glucocorticoids cause defects in insulin-stimulated intracellular signaling that lead to suppression of IRS-1 activity resulting in decreased Akt phosphorylation. Second, low phosphorylated Akt activates caspase-3 thus providing muscle proteins substrates for the ubiquitin-proteasome system. In fact, caspase-3 also stimulates proteasome-dependent proteolysis in muscle. A low phosphorylated Akt also leads to a decrease in the phosphorylation of forkhead transcription factors which enter the nucleus to stimulate the expression of atrogin-1/MAFbx and MuRF1, E3 ubiquitin ligases linked with enhanced proteolysis of skeletal muscle. Unfortunately, glucocorticoids are often given to cancer patients receiving radiotherapy, chemotherapy, or both, and this often results in an exacerbation of their muscle wasting (Schakman et al., 2013).

11.5. CROSS-TALK BETWEEN ANABOLIC AND CATABOLIC MEDIATORS

A cross-talk between anabolic and catabolic mediators and signaling pathways is evident. Thus, IGF1 (and/or insulin), in addition to stimulating protein synthesis, suppresses protein breakdown by regulating both the ubiquitin-proteasome pathway and the autophagic-lysosome pathway by interacting—phosphorylating and therefore promoting their migration from the nucleus to the cytoplasm—with the family of FOXO transcription factors (Latres et al., 2005). It also works the other way round. When FOXO is induced, mTOR is repressed and, therefore, total protein synthesis is reduced (Reed et al., 2012).

FOXO activity can be influenced by different cofactors and by the interaction with other transcription factors. For example, PGC-1α, a molecule involved in mitochondrial biogenesis (Puigserver et al., 2003), is able to reduce proteolysis by inhibiting the transcriptional activity of FOXO and NF-κB without affecting protein synthesis.

Shi et al. demonstrated that the muscle-specific E3 ubiquitin ligase Fbxo40 induces IRS1 ubiquitination and breakdown specifically in skeletal muscle cells and only upon IGF1 stimulation, thus identifying a regulator of IGF1-induced skeletal muscle signaling (Shi et al., 2011). In fact, Fbxo40 seems to be involved in the rapid degradation of IRS1, since IRS1 loss can be rescued by knockdown of Fbxo40—this also induces dramatic hypertrophy of myofibers. Fbxo40 directly ubiquitinates IRS1, and this activity is enhanced by increased tyrosine phosphorylation of IRS1 (Shi et al., 2011).

Both myostatin and activins—proteins belonging to the TGF-β family—bind to the activin receptors leading to a signaling cascade that results not only in activation of atrogin-1 and MuRF1 expression but also in a blockage of the Akt pathway (Argilés et al., 2012). Aversa et al. showed that, in gastric cancer patients, the skeletal muscle expression of myostatin was significantly increased (Aversa et al., 2012). By contrast, in patients with lung cancer, myostatin levels were comparable to controls (Aversa et al., 2012). In a mice model of cancer cachexia, blockage of the myostatin/activin system blocks muscle wasting (Busquets et al., 2012) and prolongs survival (Zhou et al., 2010).

In addition to protein turnover, myogenesis and apoptosis also participate in muscle wasting associated with cancer. Skeletal muscle can be submitted to a repair process involving proliferation and differentiation of the so-called satellite cells—undifferentiated unfused myocells. During cancer cachexia, satellite cells proliferate but their differentiation is blocked resulting in a lack of muscle regeneration (He et al., 2013). On the other hand, evidence for increased apoptosis both in experimental animals (Van Royen et al., 2000) and humans (Busquets et al., 2007) have been associated with skeletal muscle wasting (Fig. 11.1).

11.6. THERAPEUTIC APPROACHES TO INFLUENCE PROTEIN KINETICS

11.6.1 Approaches to Overcome Anabolic Resistance

In cancer patients, insulin resistance of protein anabolism has been speculated to underlie the skeletal muscle anabolic resistance characteristic of cancer cachexia. Therefore, conventional nutritional supplementation is ineffective in stimulating muscle protein synthesis. This anabolic resistance can be partially overcome using specially formulated nutritional supplements that, in a multimodal way, are combined with specific drugs. From this point of view, high-leucine and protein supplements containing ω3-polyunsaturated fatty acids are used as part of a multimodal anabolic approach in long-term trials to confirm their efficacy to sustain anabolism, and attenuate or even reverse muscle wasting (Argilés et al., 2004).

Anamorelin is an oral ghrelin-receptor agonist with appetite-enhancing and anabolic activity. In advanced nonsmall cell lung cancer patients, treatment with the drug resulted in body weight increase and a significant improvement of patient symptoms/concerns regarding anorexia-cachexia over 12 weeks. However, patients receiving anamorelin did not experience improvements in their muscle strength, as measured by hand grip strength (Temel et al., 2014).

11.6.2 Approaches to Target Muscle Protein Turnover

In the last few years, several approaches targeted at either blocking protein degradation or increasing protein synthesis or both have been developed. Among them, the main ones have been beta-2-agonists, ghrelin agonists, SARMs (see Argilés et al., 2004, 2008 for review). Some of these drugs are at present in phase II trials for cachexia (Argilés et al., 2014a). These compounds affect several pathways involved in cachexia by modulating inflammatory activity, anabolic potential, and direct interaction with the muscle at the level of protein degradation (Fig. 11.5).

OHR/AVR118 is a broad-spectrum peptide–nucleic acid immunomodulator drug affecting the action of cytokines. The drug had been evaluated in over 70 AIDS cachectic patients, where it showed beneficial effects in mitigating multiple symptoms of the disease. At the completion of treatment, patients achieved stabilization of body

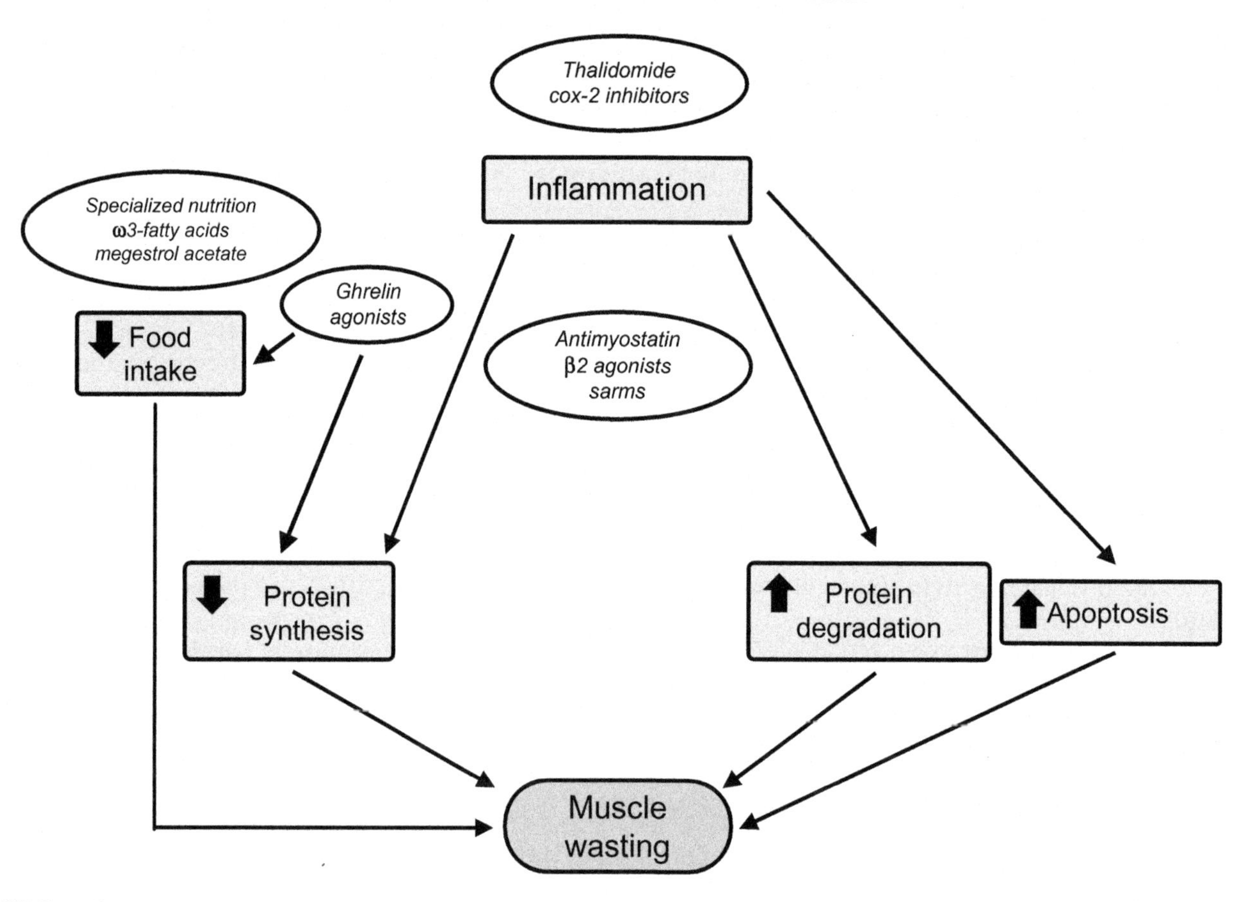

FIGURE 11.5 **Therapeutic approaches based on the interaction with protein turnover in skeletal muscle.** Therapeutic targets are based on either inhibiting the inflammatory response or directly interacting with either protein synthesis or protein degradation.

weight, body fat, and muscle mass, with a significant increase in appetite. Additionally, patients experienced enhanced quality of life as indicated by Patient Generated Subjective Global Assessment (PG-SGA) (Chasen, 2013).

Celecoxib (Celebrex) is a nonsteroidal antiinflammatory drug that reduces the activity of cyclooxygenase 2 (COX-2)—a protein involved in the inflammatory response. Because inflammation appears to be a component in the development of cachexia, cachectic patients with head and neck or gastrointestinal cancer receiving celecoxib gained weight, experienced increased BMI, and demonstrated improved QoL scores. Compliance was good and no adverse events were seen (Lai et al., 2008). Later on, Rogers et al., in a feasibility study, investigated the effect of stimulating the anabolic skeletal muscle pathway with the use of progressive exercise training along with essential amino acids—high in leucine—alongside the combination of EPA and celecoxib in this population (Rogers et al., 2011). So far, no results of this study are yet available.

VT-122 is a novel investigational combination of etodolac and propranolol that also targets pathways associated with cancer-induced systemic inflammation. Bhattacharyya et al. showed the ability to increase lean body mass (muscle) and lower other markers of systemic inflammation in a Phase II trial of patients with advanced lung cancer (Bhattacharyya, 2015).

MT-102 is a beta adrenergic blocker used as therapeutic for the treatment of cancer- and age-related wasting. At present, a phase II multinational, randomized, double blind, placebo controlled clinical study of MT-102 in patients with stage III or IV lung cancer or colorectal cancer that are also suffering severe weight loss and fatigue, is being performed (Stewart Coats et al., 2011). The trial is designed to demonstrate reversal of weight loss following treatment with MT-102 but will also examine improvement in functional ability and quality of life as quantified by a battery of previously validated instruments. The human safety profile of MT-102 has been demonstrated in two phase I/II clinical studies. The preclinical efficacy showed that MT-102 not only significantly improved body weight, muscle mass, and fat mass, but it also significantly improved mobility and survival.

BYM338 (bimagrumab) is a novel, fully human monoclonal antibody developed to treat pathological muscle loss and weakness. BYM338 binds with high affinity to type II activin receptors, preventing natural ligands from binding, including myostatin and activin. BYM338 stimulates muscle growth by blocking signaling from these inhibitory molecules. At present, a clinical trial involving patients with stage IV nonsmall cell lung cancer or stage III/IV adenocarcinoma of the pancreas is being undertaken (Smith and Lin, 2013).

Finally, another clinical trial at present being held involves Ruxolitinib, a Janus kinase 1 (JAK1) and Janus kinase 2 (JAK2) inhibitor (Mesa et al., 2015).

11.7. CONCLUSIONS AND FUTURE DIRECTIONS

Alterations in amino acid protein metabolism are a key feature of the cancer cachexia syndrome. Anabolic resistance—induced by both hormonal changes (that affect insulin sensitivity) and inflammatory mediators—is present in skeletal muscle and this conditions both amino acid uptake and protein synthesis. In addition, skeletal muscle protein turnover is characterized by an exacerbated rate of protein degradation promoted by an activation of different proteolytic systems that include the ubiquitin-proteasome and the autophagic-lysosomal pathways. Changes in the rate of myogenesis/apoptosis also determine skeletal muscle mass during cancer cachexia. Indeed, a decreased skeletal muscle regeneration capacity is observed together with an increased rate of cell death, resulting in muscle wasting. Mitochondrial dysfunction also results in changes in skeletal muscle metabolism and further contributes to the exacerbation of the cancer-wasting syndrome.

Due to the multifactorial aspects of cachexia syndrome, any therapeutic approach based on increasing food intake has to be combined with a pharmacological strategy to counteract skeletal muscle metabolic changes. Moreover, timing is very important and has to be considered seriously when designing the therapeutic approach. A very important aspect to be taken into consideration when treating cancer patients is that any nutritional/metabolic/pharmacological support should be started early in the course of the disease, before severe weight loss occurs. Another important problem associated with the design of the ideal therapeutic approach is that no definite mediators of cachexia have been yet identified. Since the therapy against wasting during cachexia has concentrated on either increasing food intake or normalizing the persistent metabolic alterations that take place in the patient, it is difficult to apply a therapeutic approach based on the neutralization of the potential mediators involved in muscle wasting. Bearing this in mind, it is obvious that a good understanding of the molecular mechanisms involved in the signaling of these mediators may be very positive in the design of the therapeutic strategy. This is especially relevant because different mediators may be sharing the same signaling pathways. At the moment, there are few studies describing the role of cytokines and tumor factors in the signaling associated with muscle wasting.

In conclusion, although both tumoral and humoral (mainly cytokines) factors that trigger cachexia may share common signaling pathways, it is not very likely that a single drug will block the complex processes involved in cachexia. In addition, some of the mediators proposed for the wasting syndrome also play a role in the regulation of body weight in absolutely opposite states such as obesity. In conclusion, the future treatment of the cachectic syndrome will no doubt combine different pharmacological approaches to efficiently reverse the metabolic changes described above and, at the same time, ameliorate the anorexia of the patients. Defining this therapeutic combination of drugs is an exciting project that will stimulate many scientific efforts. Furthermore, future studies should include precachetic patients, since they might be more responsive to treatment. Future studies will, no doubt, benefit from well-defined end points and improved measures of cachexia, providing new insight into the disease. This, in combination with the elucidation of cachexia's underlying mechanisms, will provide new treatment strategies in the near future.

References

Argilés, J.M., López-Soriano, F.J., 1996. The ubiquitin-dependent proteolytic pathway in skeletal muscle: its role in pathological states. Trends Pharmacol. Sci. 17, 223–226.

Argilés, J.M., López-Soriano, F.J., 1999. The role of cytokines in cancer cachexia. Med. Res. Rev. 19, 223–248.

Argilés, J.M., Almendro, V., Busquets, S., López-Soriano, F.J., 2004. The pharmacological treatment of cachexia. Curr. Drug Targets 5, 265–277.

Argilés, J.M., López-Soriano, F.J., Busquets, S., 2008. Novel approaches to the treatment of cachexia. Drug Discov. Today 13, 73–78. Available from: http://dx.doi.org/10.1016/j.drudis.2007.10.008.

Argilés, J.M., Busquets, S., Toledo, M., López-Soriano, F.J., 2009. The role of cytokines in cancer cachexia. Curr. Opin. Support. Palliat. Care 3, 263–268. Available from: http://dx.doi.org/10.1097/SPC.0b013e3283311d09.

Argilés, J.M., Anker, S.D., Evans, W.J., Morley, J.E., Fearon, K.C.H., Strasser, F., et al., 2010. Consensus on cachexia definitions. J. Am. Med. Dir. Assoc. 11, 229–230. Available from: http://dx.doi.org/10.1016/j.jamda.2010.02.004.

Argilés, J.M., Orpí, M., Busquets, S., López-Soriano, F.J., 2012. Myostatin: more than just a regulator of muscle mass. Drug Discov. Today 17, 702–709. Available from: http://dx.doi.org/10.1016/j.drudis.2012.02.001.

Argilés, J.M., López-Soriano, F., Stemmler, B., Busquets, S., 2014a. Recent developments in treatment of Cachexia. In: Folkerts, G., Garssen, J. (Eds.), Pharma-Nutrition SE - 13, AAPS Advances in the Pharmaceutical Sciences Series. Springer International Publishing, pp. 259–273. . Available from: http://dx.doi.org/10.1007/978-3-319-06151-1_13.

Argilés, J.M., Busquets, S., Stemmler, B., López-Soriano, F.J., 2014b. Cancer cachexia: understanding the molecular basis. Nat. Rev. Cancer 14, 754–762. Available from: http://dx.doi.org/10.1038/nrc3829.

Aversa, Z., Bonetto, A., Penna, F., Costelli, P., Di Rienzo, G., Lacitignola, A., et al., 2012. Changes in myostatin signaling in non-weight-losing cancer patients. Ann. Surg. Oncol. 19, 1350–1356. Available from: http://dx.doi.org/10.1245/s10434-011-1720-5.

Barreiro, E., de la Puente, B., Busquets, S., López-Soriano, F.J., Gea, J., Argilés, J.M., 2005. Both oxidative and nitrosative stress are associated with muscle wasting in tumour-bearing rats. FEBS Lett. 579, 1646–1652. Available from: http://dx.doi.org/10.1016/j.febslet.2005.02.017.

Barta, J., Tóth, A., Edes, I., Vaszily, M., Papp, J.G., Varró, A., et al., 2005. Calpain-1-sensitive myofibrillar proteins of the human myocardium. Mol. Cell. Biochem. 278, 1–8. Available from: http://dx.doi.org/10.1007/s11010-005-1370-7.

Bhatnagar, S., Mittal, A., Gupta, S.K., Kumar, A., 2012. TWEAK causes myotube atrophy through coordinated activation of ubiquitin-proteasome system, autophagy, and caspases. J. Cell. Physiol. 227, 1042–1051. Available from: http://dx.doi.org/10.1002/jcp.22821.

Bhattacharyya, G.S., 2015. Vicus Therapeutics Announces Safety and Survival Benefit of VT-122 in Combination with Anti-Cancer Therapies for Advanced Liver and Pancreatic Cancers [WWW Document]. URL <http://www.prnewswire.com/news-releases/vicus-therapeutics-announces-safety-and-survival-benefit-of-vt-122-in-combination-with-anti-cancer-therapies-for-advanced-liver-and-pancreatic-cancers-300021768.html> (accessed 29.05.15.).

Bonetto, A., Aydogdu, T., Jin, X., Zhang, Z., Zhan, R., Puzis, L., et al., 2012. JAK/STAT3 pathway inhibition blocks skeletal muscle wasting downstream of IL-6 and in experimental cancer cachexia. Am. J. Physiol. Endocrinol. Metab. 303, E410–E421. Available from: http://dx.doi.org/10.1152/ajpendo.00039.2012.

Bonetto, A., Penna, F., Aversa, Z., Mercantini, P., Baccino, F.M., Costelli, P., et al., 2013. Early changes of muscle insulin-like growth factor-1 and myostatin gene expression in gastric cancer patients. Muscle Nerve 48, 387–392. Available from: http://dx.doi.org/10.1002/mus.23798.

Bossola, M., Muscaritoli, M., Costelli, P., Bellantone, R., Pacelli, F., Busquets, S., et al., 2001. Increased muscle ubiquitin mRNA levels in gastric cancer patients. Am. J. Physiol. Regul. Integr. Comp. Physiol. 280, R1518–R1523.

Bossola, M., Muscaritoli, M., Costelli, P., Grieco, G., Bonelli, G., Pacelli, F., et al., 2003. Increased muscle proteasome activity correlates with disease severity in gastric cancer patients. Ann. Surg. 237, 384–389. Available from: http://dx.doi.org/10.1097/01.SLA.0000055225.96357.71.

Busquets, S., Figueras, M.T., Meijsing, S., Carbó, N., Quinn, L.S., Almendro, V., et al., 2005. Interleukin-15 decreases proteolysis in skeletal muscle: a direct effect. Int. J. Mol. Med. 16, 471–476.

Busquets, S., Deans, C., Figueras, M., Moore-Carrasco, R., López-Soriano, F.J., Fearon, K.C.H., et al., 2007. Apoptosis is present in skeletal muscle of cachectic gastro-intestinal cancer patients. Clin. Nutr. 26, 614–618. Available from: http://dx.doi.org/10.1016/j.clnu.2007.06.005.

Busquets, S., Toledo, M., Orpí, M., Massa, D., Porta, M., Capdevila, E., et al., 2012. Myostatin blockage using actRIIB antagonism in mice bearing the Lewis lung carcinoma results in the improvement of muscle wasting and physical performance. J. Cachexia Sarcopenia Muscle 3, 37–43. Available from: http://dx.doi.org/10.1007/s13539-011-0049-z.

Cabal-Manzano, R., Bhargava, P., Torres-Duarte, A., Marshall, J., Wainer, I.W., 2001. Proteolysis-inducing factor is expressed in tumours of patients with gastrointestinal cancers and correlates with weight loss. Br. J. Cancer 84, 1599–1601. Available from: http://dx.doi.org/10.1054/bjoc.2001.1830.

Carbó, N., López-Soriano, J., Costelli, P., Busquets, S., Alvarez, B., Baccino, F.M., et al., 2000. Interleukin-15 antagonizes muscle protein waste in tumour-bearing rats. Br. J. Cancer 83, 526–531. Available from: http://dx.doi.org/10.1054/bjoc.2000.1299.

Carbó, N., López-Soriano, J., Costelli, P., Alvarez, B., Busquets, S., Baccino, F.M., et al., 2001. Interleukin-15 mediates reciprocal regulation of adipose and muscle mass: a potential role in body weight control. Biochim. Biophys. Acta 1526, 17–24.

Chasen, M., 2013. Phase II Data on OHR/AVR118 in Advanced Cancer Patients With Cachexia [WWW Document]. Int. Cachexia Conf. Kobe, Japan. URL <http://www.ohrpharmaceutical.com/media-center/press-releases/detail/94/phase-ii-data-on-ohravr118-in-advanced-cancer-patients> (accessed 29.05.15.).

Cohen, S., Zhai, B., Gygi, S.P., Goldberg, A.L., 2012. Ubiquitylation by Trim32 causes coupled loss of desmin, Z-bands, and thin filaments in muscle atrophy. J. Cell Biol. 198, 575–589. Available from: http://dx.doi.org/10.1083/jcb.201110067.

Cohen, S., Lee, D., Zhai, B., Gygi, S.P., Goldberg, A.L., 2014. Trim32 reduces PI3K-Akt-FoxO signaling in muscle atrophy by promoting plakoglobin-PI3K dissociation. J. Cell Biol. 204, 747–758. Available from: http://dx.doi.org/10.1083/jcb.201304167.

Costelli, P., Muscaritoli, M., Bossola, M., Penna, F., Reffo, P., Bonetto, A., et al., 2006. IGF-1 is downregulated in experimental cancer cachexia. Am. J. Physiol. Regul. Integr. Comp. Physiol. 291, R674–R683. Available from: http://dx.doi.org/10.1152/ajpregu.00104.2006.

Dewys, W.D., Begg, C., Lavin, P.T., Band, P.R., Bennett, J.M., Bertino, J.R., et al., 1980. Prognostic effect of weight loss prior to chemotherapy in cancer patients. Eastern Cooperative Oncology Group. Am. J. Med. 69, 491–497.

D'Orlando, C., Marzetti, E., François, S., Lorenzi, M., Conti, V., di Stasio, E., et al., 2014. Gastric cancer does not affect the expression of atrophy-related genes in human skeletal muscle. Muscle Nerve 49, 528–533. Available from: http://dx.doi.org/10.1002/mus.23945.

Egerman, M.A., Glass, D.J., 2014. Signaling pathways controlling skeletal muscle mass. Crit. Rev. Biochem. Mol. Biol. 49, 59–68. Available from: http://dx.doi.org/10.3109/10409238.2013.857291.

Evans, W.J., Morley, J.E., Argiles, J., Bales, C., Baracos, V., Guttridge, D., et al., 2008. Cachexia: a new definition. Clin. Nutr. 27, 793–799, doi: S0261-5614(08)00113-1 [pii] http://dx.doi.org/10.1016/j.clnu.2008.06.013.

Farkas, J., von Haehling, S., Kalantar-Zadeh, K., Morley, J.E., Anker, S.D., Lainscak, M., 2013. Cachexia as a major public health problem: frequent, costly, and deadly. J. Cachexia Sarcopenia Muscle 4, 173–178. Available from: http://dx.doi.org/10.1007/s13539-013-0105-y.

Figueras, M., Busquets, S., Carbó, N., Barreiro, E., Almendro, V., Argilés, J.M., et al., 2004. Interleukin-15 is able to suppress the increased DNA fragmentation associated with muscle wasting in tumour-bearing rats. FEBS Lett. 569, 201–206. Available from: http://dx.doi.org/10.1016/j.febslet.2004.05.066.

Fontes-Oliveira, C.C., Busquets, S., Toledo, M., Penna, F., Aylwin, M.P., Sirisi, S., et al., 2013. Mitochondrial and sarcoplasmic reticulum abnormalities in cancer cachexia: altered energetic efficiency? Biochim. Biophys. Acta 1830, 2770–2778.

Goldberg, A.L., Goodman, H.M., 1969. Relationship between cortisone and muscle work in determining muscle size. J. Physiol. 200, 667–675.

He, W.A., Berardi, E., Cardillo, V.M., Acharyya, S., Aulino, P., Thomas-Ahner, J., et al., 2013. NF-κB-mediated Pax7 dysregulation in the muscle microenvironment promotes cancer cachexia. J. Clin. Invest. 123, 4821–4835. Available from: http://dx.doi.org/10.1172/JCI68523.

Hinch, E.C.A., Sullivan-Gunn, M.J., Vaughan, V.C., McGlynn, M.A., Lewandowski, P.A., 2013. Disruption of pro-oxidant and antioxidant systems with elevated expression of the ubiquitin proteosome system in the cachectic heart muscle of nude mice. J. Cachexia Sarcopenia Muscle 4, 287–293. Available from: http://dx.doi.org/10.1007/s13539-013-0116-8.

Horsley, V., Friday, B.B., Matteson, S., Kegley, K.M., Gephart, J., Pavlath, G.K., 2001. Regulation of the growth of multinucleated muscle cells by an NFATC2-dependent pathway. J. Cell Biol. 153, 329–338.

Jagoe, R.T., Redfern, C.P.F., Roberts, R.G., Gibson, G.J., Goodship, T.H.J., 2002. Skeletal muscle mRNA levels for cathepsin B, but not components of the ubiquitin-proteasome pathway, are increased in patients with lung cancer referred for thoracotomy. Clin. Sci. (Lond) 102, 353–361.

Khal, J., Hine, A.V., Fearon, K.C.H., Dejong, C.H.C., Tisdale, M.J., 2005. Increased expression of proteasome subunits in skeletal muscle of cancer patients with weight loss. Int. J. Biochem. Cell Biol. 37, 2196–2206. Available from: http://dx.doi.org/10.1016/j.biocel.2004.10.017.

Lai, V., George, J., Richey, L., Kim, H.J., Cannon, T., Shores, C., et al., 2008. Results of a pilot study of the effects of celecoxib on cancer cachexia in patients with cancer of the head, neck, and gastrointestinal tract. Head Neck 30, 67–74. Available from: http://dx.doi.org/10.1002/hed.20662.

Latres, E., Amini, A.R., Amini, A.A., Griffiths, J., Martin, F.J., Wei, Y., et al., 2005. Insulin-like growth factor-1 (IGF-1) inversely regulates atrophy-induced genes via the phosphatidylinositol 3-kinase/Akt/mammalian target of rapamycin (PI3K/Akt/mTOR) pathway. J. Biol. Chem. 280, 2737–2744. Available from: http://dx.doi.org/10.1074/jbc.M407517200.

Li, J.B., Goldberg, A.L., 1976. Effects of food deprivation on protein synthesis and degradation in rat skeletal muscles. Am. J. Physiol. 231, 441–448.

Mesa, R.A., Verstovsek, S., Gupta, V., Mascarenhas, J.O., Atallah, E., Burn, T., et al., 2015. Effects of ruxolitinib treatment on metabolic and nutritional parameters in patients with myelofibrosis from COMFORT-I. Clin. Lymphoma Myeloma Leuk. 15, 214–221.e1, http://dx.doi.org/10.1016/j.clml.2014.12.008.

Minetti, G.C., Feige, J.N., Bombard, F., Heier, A., Morvan, F., Nürnberg, B., et al., 2014. Gαi2 signaling is required for skeletal muscle growth, regeneration, and satellite cell proliferation and differentiation. Mol. Cell. Biol. 34, 619–630. Available from: http://dx.doi.org/10.1128/MCB.00957-13.

Moore-Carrasco, R., Busquets, S., Almendro, V., Palanki, M., López-Soriano, F.J., Argilés, J.M., 2007. The AP-1/NF-kappaB double inhibitor SP100030 can revert muscle wasting during experimental cancer cachexia. Int. J. Oncol. 30, 1239–1245.

Moore-Carrasco, R., Busquets, S., Figueras, M., Palanki, M., López-Soriano, F.J., Argilés, J.M., 2009. Both AP-1 and NF-kappaB seem to be involved in tumour growth in an experimental rat hepatoma. Anticancer Res. 29, 1315–1317.

Op den Kamp, C.M., Langen, R.C., Minnaard, R., Kelders, M.C., Snepvangers, F.J., Hesselink, M.K., et al., 2012. Pre-cachexia in patients with stages I-III non-small cell lung cancer: systemic inflammation and functional impairment without activation of skeletal muscle ubiquitin proteasome system. Lung Cancer 76, 112–117. Available from: http://dx.doi.org/10.1016/j.lungcan.2011.09.012.

Penna, F., Costamagna, D., Pin, F., Camperi, A., Fanzani, A., Chiarpotto, E.M., et al., 2013. Autophagic degradation contributes to muscle wasting in cancer cachexia. Am. J. Pathol. 182, 1367–1378. Available from: http://dx.doi.org/10.1016/j.ajpath.2012.12.023.

Penna, F., Baccino, F.M., Costelli, P., 2014. Coming back: autophagy in cachexia. Curr. Opin. Clin. Nutr. Metab. Care 17, 241–246. Available from: http://dx.doi.org/10.1097/MCO.0000000000000048.

Puigserver, P., Rhee, J., Donovan, J., Walkey, C.J., Yoon, J.C., Oriente, F., et al., 2003. Insulin-regulated hepatic gluconeogenesis through FOXO1-PGC-1alpha interaction. Nature 423, 550–555. Available from: http://dx.doi.org/10.1038/nature01667.

Reed, S.A., Sandesara, P.B., Senf, S.M., Judge, A.R., 2012. Inhibition of FoxO transcriptional activity prevents muscle fiber atrophy during cachexia and induces hypertrophy. FASEB J. 26, 987–1000. Available from: http://dx.doi.org/10.1096/fj.11-189977.

Rogers, E.S., MacLeod, R.D., Stewart, J., Bird, S.P., Keogh, J.W.L., 2011. A randomised feasibility study of EPA and Cox-2 inhibitor (Celebrex) versus EPA, Cox-2 inhibitor (Celebrex), resistance training followed by ingestion of essential amino acids high in leucine in NSCLC cachectic patients--ACCeRT study. BMC Cancer 11, 493. Available from: http://dx.doi.org/10.1186/1471-2407-11-493.

Schakman, O., Kalista, S., Barbé, C., Loumaye, A., Thissen, J.P., 2013. Glucocorticoid-induced skeletal muscle atrophy. Int. J. Biochem. Cell Biol. 45, 2163–2172. Available from: http://dx.doi.org/10.1016/j.biocel.2013.05.036.

Schiaffino, S., Dyar, K.A., Ciciliot, S., Blaauw, B., Sandri, M., 2013. Mechanisms regulating skeletal muscle growth and atrophy. FEBS J. 280, 4294–4314. Available from: http://dx.doi.org/10.1111/febs.12253.

Schmidt, K., von Haehling, S., Doehner, W., Palus, S., Anker, S.D., Springer, J., 2011. IGF-1 treatment reduces weight loss and improves outcome in a rat model of cancer cachexia. J. Cachexia Sarcopenia Muscle 2, 105–109. Available from: http://dx.doi.org/10.1007/s13539-011-0029-3.

Shi, J., Luo, L., Eash, J., Ibebunjo, C., Glass, D.J., 2011. The SCF-Fbxo40 complex induces IRS1 ubiquitination in skeletal muscle, limiting IGF1 signaling. Dev. Cell 21, 835–847. Available from: http://dx.doi.org/10.1016/j.devcel.2011.09.011.

Smith, I.J., Aversa, Z., Hasselgren, P.-O., Pacelli, F., Rosa, F., Doglietto, G.B., et al., 2011. Calpain activity is increased in skeletal muscle from gastric cancer patients with no or minimal weight loss. Muscle Nerve 43, 410–414. Available from: http://dx.doi.org/10.1002/mus.21893.

Smith, R.C., Lin, B.K., 2013. Myostatin inhibitors as therapies for muscle wasting associated with cancer and other disorders. Curr. Opin. Support. Palliat. Care 7, 352–360. Available from: http://dx.doi.org/10.1097/SPC.0000000000000013.

Stephens, N.A., Gallagher, I.J., Rooyackers, O., Skipworth, R.J., Tan, B.H., Marstrand, T., et al., 2010. Using transcriptomics to identify and validate novel biomarkers of human skeletal muscle cancer cachexia. Genome Med. 2, 1. Available from: http://dx.doi.org/10.1186/gm122.

Stewart Coats, A.J., Srinivasan, V., Surendran, J., Chiramana, H., Vangipuram, S.R.K.G., Bhatt, N.N., et al., 2011. The ACT-ONE trial, a multicentre, randomised, double-blind, placebo-controlled, dose-finding study of the anabolic/catabolic transforming agent, MT-102 in subjects with cachexia related to stage III and IV non-small cell lung cancer and colorectal cancer. J. Cachexia Sarcopenia Muscle 2, 201–207. Available from: http://dx.doi.org/10.1007/s13539-011-0046-2.

Sun, Y.-S., Ye, Z.-Y., Qian, Z.-Y., Xu, X.-D., Hu, J.-F., 2012. Expression of TRAF6 and ubiquitin mRNA in skeletal muscle of gastric cancer patients. J. Exp. Clin. Cancer Res. 31, 81. Available from: http://dx.doi.org/10.1186/1756-9966-31-81.

Tardif, N., Klaude, M., Lundell, L., Thorell, A., Rooyackers, O., 2013. Autophagic-lysosomal pathway is the main proteolytic system modified in the skeletal muscle of esophageal cancer patients. Am. J. Clin. Nutr. 98, 1485–1492. Available from: http://dx.doi.org/10.3945/ajcn.113.063859.

Temel, J., Currow, D., Fearon, K., Gleich, L., Yan, Y., Friend, J., et al., 2014. Anamorelin for the treatment of cancer anorexia-cachexia in NSCLC: results from the phase 3 studies ROMANA 1 and 2 [WWW Document]. Ann. Oncol.URL <http://oncologypro.esmo.org/Meeting-Resources/ESMO-2014/Supportive-Care/Anamorelin-for-the-treatment-of-cancer-anorexia-cachexia-in-NSCLC-Results-from-the-phase-3-studies-ROMANA-1-and-2> (accessed 29.05.15.).

Tisdale, M.J., 2004. Tumor-host interactions. J. Cell. Biochem. 93, 871–877. Available from: http://dx.doi.org/10.1002/jcb.20246.

Todorov, P., Cariuk, P., McDevitt, T., Coles, B., Fearon, K., Tisdale, M., 1996. Characterization of a cancer cachectic factor. Nature 379, 739–742. Available from: http://dx.doi.org/10.1038/379739a0.

Van Royen, M., Carbó, N., Busquets, S., Alvarez, B., Quinn, L.S., López-Soriano, F.J., et al., 2000. DNA fragmentation occurs in skeletal muscle during tumor growth: A link with cancer cachexia? Biochem. Biophys. Res. Commun. 270, 533–537. Available from: http://dx.doi.org/10.1006/bbrc.2000.2462.

Von Haehling, S., Anker, S.D., 2015. Treatment of cachexia: an overview of recent developments. Int. J. Cardiol. 184, 736–742. Available from: http://dx.doi.org/10.1016/j.ijcard.2014.10.026.

White, J.P., Puppa, M.J., Gao, S., Sato, S., Welle, S.L., Carson, J.A., 2013. Muscle mTORC1 suppression by IL-6 during cancer cachexia: a role for AMPK. Am. J. Physiol. Endocrinol. Metab. 304, E1042–E1052. Available from: http://dx.doi.org/10.1152/ajpendo.00410.2012.

Williams, A., Sun, X., Fischer, J.E., Hasselgren, P.O., 1999. The expression of genes in the ubiquitin-proteasome proteolytic pathway is increased in skeletal muscle from patients with cancer. Surgery 126, 744–749, discussion 749–50.

Yamada, E., Bastie, C.C., Koga, H., Wang, Y., Cuervo, A.M., Pessin, J.E., 2012. Mouse skeletal muscle fiber-type-specific macroautophagy and muscle wasting are regulated by a Fyn/STAT3/Vps34 signaling pathway. Cell Rep. 1, 557–569. Available from: http://dx.doi.org/10.1016/j.celrep.2012.03.014.

Zhou, X., Wang, J.L., Lu, J., Song, Y., Kwak, K.S., Jiao, Q., et al., 2010. Reversal of cancer cachexia and muscle wasting by ActRIIB antagonism leads to prolonged survival. Cell 142, 531–543. Available from: http://dx.doi.org/10.1016/j.cell.2010.07.011.

CHAPTER

12

Amino Acid and Protein Metabolism in Pulmonary Diseases and Nutritional Abnormalities: A Special Focus on Chronic Obstructive Pulmonary Disease

E. Barreiro[1,2] *and J. Gea*[1,2]

[1]Pulmonology Department, Muscle and Lung Cancer Research Group, IMIM-Hospital del Mar, Parc de Salut Mar, Health and Experimental Sciences Department (CEXS), Universitat Pompeu Fabra (UPF), Barcelona Biomedical Research Park (PRBB), Barcelona, Spain [2]Centro de Investigación en Red de Enfermedades Respiratorias (CIBERES), Instituto de Salud Carlos III (ISCIII), Barcelona, Spain

12.1. INTRODUCTION

In the next decade, lung diseases such as Chronic Obstructive Pulmonary Disease (COPD) will be among the main leading causes of death worldwide (Miravitlles et al., 2014; Vestbo et al., 2013). Nutritional abnormalities are common systemic manifestations of patients with chronic respiratory disorders and other diseases such as critical illness, cancer, elective surgery, or prolonged bed rest due to serious disease or trauma. Nutritional abnormalities may further impair other relevant comorbidities that are also present in chronic respiratory patients, such as muscle mass loss and dysfunction, and may also aggravate the number of exacerbations, thus further impairing their quality of life and physical activity and/or exercise performance (Barreiro et al., 2015; Maltais et al., 2014). In addition, nutritional abnormalities and impaired muscle mass may predict morbidity and mortality in patients with COPD, independently of the severity of their lung disease (Marquis et al., 2002; Seymour et al., 2010; Shrikrishna et al., 2012; Swallow et al., 2007). Therefore, in the evaluation of patients with chronic respiratory conditions, a thorough assessment of their nutritional status and muscle mass and performance should always be included. The values of the corresponding parameters will enable health care providers to design specific therapeutic strategies targeted to improve the nutritional status and the overall performance and functional capacity of the chronic respiratory patients. In this regard, nutritional support with and without the administration of anabolic agents that favor protein synthesis will be a priority in the management of patients with chronic respiratory disorders along with the treatment of their respiratory symptoms.

In this chapter, firstly, the epidemiology of nutritional abnormalities in patients with chronic respiratory conditions, with a special emphasis on COPD is explained. Secondly, the definition of nutritional abnormalities in view of the different available diagnostic tools is also described for respiratory patients. Thirdly, the most relevant etiologic factors and biological mechanisms that have been shown to participate in the multifactorial etiology of the nutritional abnormalities of patients with COPD are reviewed. Fourthly, a general overview of protein metabolism, especially focused on protein synthesis and its relationships with exercise, is also given to better understand the

The Molecular Nutrition of Amino Acids and Proteins.
DOI: http://dx.doi.org/10.1016/B978-0-12-802167-5.00012-8

underlying physiology of the protein and amino acid supplementation usually prescribed in patients with COPD. In addition, currently available therapeutic strategies to treat the nutritional status of chronic respiratory patients are also described in the chapter. Finally, a very brief mention of other respiratory-related disorders such as cystic fibrosis, cancer, and critical illness is also provided at the end of the chapter. Nonetheless, we have rather focused the contents of this chapter on COPD as it is nowadays a highly prevalent disabling condition that has drawn the attention of many renowned investigators in the field, thus being the focus of abundant high-class international research over the last two decades. Moreover, as specific chapters intended to review the nutritional abnormalities of patients with critical illness and cancer, have also been allocated in the book, we have decided not to review those conditions in order to avoid potential overlap.

12.2. EPIDEMIOLOGY AND DEFINITION OF NUTRITIONAL ABNORMALITIES IN CHRONIC RESPIRATORY PATIENTS

Nutritional abnormalities are common in patients with COPD, and its prevalence varies according to the employed criteria. Body mass index (BMI) and fat-free mass index (FFMI) are the most commonly used parameters to evaluate nutritional abnormalities in clinical settings. For instance, in the United States and Northern and Eastern European countries, the estimated prevalence ranged from 10 to 30%, and even up to 50% in a series in which nutritional abnormalities were defined using the FFMI (Vermeeren et al., 2006; Wan et al., 2011). Conversely, the prevalence of nutritional abnormalities and poor body composition was substantially much lower in Mediterranean geographical areas (2–3%) (Vermeeren et al., 2006; Wan et al., 2011). Factors related to geographical variations and lifestyle and methodological concerns may account for the significant variability in nutritional abnormalities and body composition reported across countries in COPD. In addition, gender differences in nutritional abnormalities may also exist. For instance, discrepancies between BMI and FFMI have been shown in female COPD, in whom normal body weight may be accompanied by reduced lean muscle mass, while this is not usually the case in male patients (Vermeeren et al., 2006).

In patients with chronic respiratory conditions, nutritional abnormalities and body weight loss are mainly characterized by reduced muscle mass but it can also be the consequence of reduced adipose and bone tissues. Progressive body weight loss leads to cachexia, which is characterized by severe muscle wasting and anorexia in advanced stages, thus further increasing the nutritional abnormalities in the patients. Reduced body weight and muscle mass loss lead to poor exercise tolerance and quality of life and impairs disease prognosis regardless of the severity of the respiratory condition (Marquis et al., 2002; Seymour et al., 2010; Shrikrishna et al., 2012; Swallow et al., 2007).

Importantly, body weight was shown to discriminate for emphysema, cardiovascular morbidity, and musculoskeletal alterations in previous investigations intended to classify patients with COPD into different phenotypes (Burgel et al., 2012; Vanfleteren et al., 2013). The emphysematous phenotype has long been recognized to be associated with severe nutritional abnormalities in patients with COPD. Recently, in these patients, the inclusion of body composition assessment as a marker of nutritional status, such as in the calculation of the BODE index, resulted in the recognition of the paramount importance of the understanding of their systemic manifestations (Divo et al., 2015). Interestingly, in COPD patients, nutritional supplementation has proven to increase exercise capacity through several mechanisms that target body weight loss induced muscle atrophy and poor metabolism (van den Borst et al., 2013). Furthermore, as patients with COPD are also more prone to experience visceral adiposity, nutritional assessment and intervention appears to be a central component of the management of these patients.

12.3. DIAGNOSIS OF NUTRITIONAL ABNORMALITIES IN PATIENTS

Nutritional abnormalities and poor body composition are easily diagnosed in clinical settings. The simplest approach is to estimate the ratio of the patient's body weight to his/her ideal body weight, and low body weight is defined as a ratio lower than 80–85% (Gea et al., 2014). BMI (defined as body weight in kg/(height in meters)^2) below 18.5 kg/m^2 is indicative of low body weight, being considered severe and very severe if values are lower than 16 kg/m^2 and 15 kg/m^2, respectively (Gea et al., 2014). A more precise approach is the measurement of the

Diagnostics tools

Blood analysis (Hemoglobin, Total protein, Albumin, Cholesterol, etc.)
Anthropometry (Weight, IBW,BMI, Triceps Skinfold, Thigh Circumference)
Bioelectrical impedance (FFIM)
Imaging: X-ray, CT, DEXA, MNR, Ultrasounds
Densitometry by immersion in water
Air displacement plethysmography (ADP)
Deuterium dilution
Isotopic methods (in biological fluids)
Spectrophotometry

Abbreviations: IBW, Ideal Body Weight; BMI, Body Mass Index; FFMI, Fat Free Mass Index; CT, computed tomography; DEXA, Dual Energy X-ray Absorptiometry; MNR, Magnetic Nuclear Resonance.

FIGURE 12.1 Schematic representation of the different diagnostics tools employed in clinical settings to define nutritional abnormalities in chronic respiratory patients.

fat-free mass index (FFMI), which is equivalent to fat-free mass in kg/(height in meters)^2. Threshold values commonly accepted to define poor body composition are FFMI lower than 16 kg/m^2 and 15 kg/m^2 in men and women, respectively, although these values may vary across geographical areas (Coin et al., 2008; Fig. 12.1).

Body composition is measured using several tools in clinical settings. The most commonly used for several reasons, for example, its easiness and low cost, is bioelectrical impedance, which is estimated as the resistance of body tissues to an electric current (Gurgun et al., 2013; Schols et al., 1991b). Other more sophisticated and costly techniques that are less frequently used in clinical settings include magnetic resonance, computerized tomography, deuterium dilution, and dual-energy X-ray absorptiometry (DEXA), which is based on the absorption of two different X-ray beams (Schols et al., 1991b; Fig. 12.1).

12.4. ETIOLOGIC FACTORS AND BIOLOGICAL MECHANISMS INVOLVED IN THE NUTRITIONAL ABNORMALITIES OF PATIENTS WITH CHRONIC RESPIRATORY CONDITIONS: COPD AS THE PARADIGM

12.4.1 Cigarette Smoke

Cigarette smoke exerts anorexigenic effects on the central nervous system and favors the release of other molecules that are also involved in body weight loss such as pro-inflammatory cytokines and leptin (Creutzberg et al., 2000; Gea et al., 2014). Additionally, cigarette smoke may also lead to increased muscle oxidative stress, dysfunction, and protein breakdown (Barreiro et al., 2010; Petersen et al., 2007), thus further impairing body composition in the patients (Fig. 12.2).

12.4.2 Physical Inactivity

Poor exercise tolerance, shortness of breath, and depression in patients with chronic respiratory conditions are the three main contributors to reduced physical activity and deconditioning in patients with chronic respiratory diseases. In fact, reduced physical activity negatively impacts on the patients' prognosis and quality of life (Laveneziana and Palange, 2012). In addition, physical activity modulates the release of pro-inflammatory cytokines as well as the expression of structural proteins in the skeletal muscle (Laveneziana and Palange, 2012). Finally in sedentary patients, vigorous exercise may exert deleterious effects on their muscles as levels of oxidative stress were shown to increase (Barreiro et al., 2009), while this was not the case in patients undergoing regular training programs of longer duration (Rodriguez et al., 2012; Fig. 12.2).

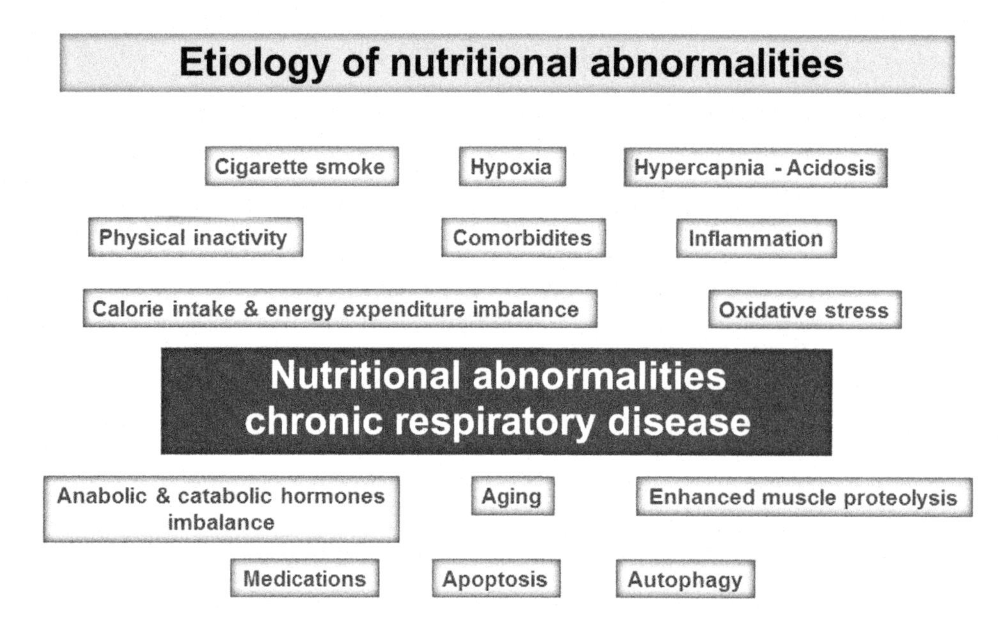

FIGURE 12.2 Schematic representation of the different etiologic factors and biological mechanisms involved in the pathophysiology of the nutritional abnormalities in chronic respiratory patients.

12.4.3 Imbalance Between Calorie Intake and Energy Expenditure

In patients with chronic respiratory conditions, involuntary weight loss takes place when caloric intake is exceeded by energy expenditure. For instance, during acute exacerbations reduced food intake driven by anorexia together with a rise in systemic levels of pro-inflammatory cytokines and leptin leads to a negative energy balance (Creutzberg et al., 2000). Moreover, an increase in physical activity or exercise may further impair the energy imbalance described in patients with chronic respiratory conditions during acute exacerbations (Gea et al., 2014; Fig. 12.2).

In chronic respiratory patients during stable conditions, this scenario is completely different, as dietary intake is maintained or increased even in those with severely impaired body composition (Goris et al., 2003). In this regard, resting energy expenditure (Schols et al., 1991a) and whole-body protein turnover are significantly increased (Engelen et al., 2000a; Kao et al., 2011), probably due to the reported inefficiency of both lower limb and ventilatory muscles to contract and the increased work of breathing shown in patients with respiratory conditions including COPD (Gea et al., 2013; Schols, 2013). In fact, patients who underwent lung volume reduction surgery exhibited an improvement in lung and ventilatory muscle functions together with a gain in body weight (Kim et al., 2012). Hence, the baseline hypermetabolic state of stable patients with chronic respiratory diseases may partly account for their progressive weight loss if that is not counterbalanced by an appropriate energy intake. Thus, dietary supplementation is frequently required in those patients in order to maintain or improve their body weight (Fig. 12.2).

12.4.4 Imbalance Between Anabolic and Catabolic Hormones

Levels of anabolic hormones may be reduced in patients with COPD. For instance, growth hormone (GH), released by the pituitary gland in response to signals from the hypothalamus and the periphery, is a powerful anabolic stimulus. GH favors the synthesis of insulin-like growth factor (IGF)-1, thus promoting calcium resorption and muscle growth and regeneration (Brioche et al., 2014). In patients with COPD, the GH-IGF-1 axis is altered to a certain extent (Creutzberg and Casaburi, 2003). While GH levels were shown to be maintained, or reduced, other investigations showed an increase (Anand et al., 1992; Borghetti et al., 2009). Furthermore, other factors also present in patients with chronic respiratory conditions, for example, aging, several medications, and physical inactivity, also contribute to reducing the levels of GH in the body (Anand et al., 1992; Creutzberg and Casaburi, 2003).

Another hormonal deficiency in patients with COPD is that involving hypogonadism, which decreases testosterone levels, thus leading to muscle mass loss without relevant functional consequences in the affected muscles (Laghi et al., 2005). In any case, hypogonadal deficiency should always be ruled out in patients with chronic respiratory conditions such as COPD (Bhasin et al., 2010; Fig. 12.2).

12.4.5 Comorbidities and Aging

It has been well established that aging is associated with muscle mass loss, also known as sarcopenia. Patients with chronic respiratory diseases are usually elderly subjects in whom sarcopenia may further impair their muscle mass and function. Moreover, elderly patients usually have other comorbid conditions such as chronic heart or kidney failures, diabetes, or tumors that may also negatively influence muscle mass and performance (Sergi et al., 2006). The association of these comorbidities with chronic respiratory conditions will further deteriorate the patients' body composition and muscle mass (Fig. 12.2).

12.4.6 Medications

Several drugs may impair muscle mass and even induce a severe myopathy in patients with respiratory diseases undergoing treatment with corticosteroids (Hasselgren et al., 2010). Other drugs such as beta-blockers and statins, which are common medications taken by patients with chronic respiratory conditions, may also exert deleterious effects on their body composition and muscle mass (Gea et al., 2013; Fig. 12.2).

12.4.7 Blood Gases

Hypoxia and hypercapnia may also contribute to body weight loss and poor muscle mass. Hypoxia induces changes in the concentrations of peptides involved in appetite regulation such as leptin, ghrelin, and AMP-activated kinases, while reducing the levels of anabolic hormones and inducing systemic oxidative stress and inflammation levels (Raguso and Luthy, 2011). Hypoxia may also induce muscle mass loss as a result of decreased mitochondrial biogenesis and accelerated protein breakdown, apoptosis, autophagy, and increased myostatin levels (Raguso and Luthy, 2011). Furthermore, hypercapnia, which is commonly associated with acidosis, may also enhance proteolysis in skeletal muscles of patients with chronic respiratory conditions (England et al., 1991; Jaitovich et al., 2015; Fig. 12.2).

12.4.8 Inflammation and Oxidative Stress

Interestingly, oxidative stress levels are increased in the blood and muscles of patients with COPD (Barreiro et al., 2009, 2010; Fermoselle et al., 2012; Puig-Vilanova et al., 2015; Rodriguez et al., 2012). Oxidative stress may trigger muscle protein loss in these patients, especially in their limb muscles (Fermoselle et al., 2012; Puig-Vilanova et al., 2015; Fig. 12.2). Systemic inflammation also takes place in patients with chronic respiratory conditions (Puig-Vilanova et al., 2015; Rodriguez et al., 2012) as well as in the adipose tissue (van den Borst et al., 2011). Inflammation may trigger tissue degradation via enhanced apoptosis, autophagy, and proteolysis in patients with chronic respiratory diseases in stable conditions, but especially during exacerbations (Karadag et al., 2008; Rodriguez et al., 2012). Moreover, it has been pointed out that systemic inflammatory mediators may be released by the adipose tissue in those patients (Tkacova et al., 2011), potentially leading to increased lipolysis (Franssen et al., 2008). Adipose tissue also synthesizes leptin and adiponectin, which play a crucial role in the regulation of body weight (Engineer and Garcia, 2012; Kirdar et al., 2009). Importantly, despite that chronic respiratory patients with severe nutritional abnormalities and muscle wasting show low levels of leptin and adiponectin, no orexigenic effects were shown in these patients, probably due to the presence of high numbers of receptors in the hypothalamus (Engineer and Garcia, 2012; Kirdar et al., 2009). Besides, adipose tissue loss may also be triggered by hypoxia, systemic inflammation, and medications (Yamamoto et al., 1997; Fig. 12.2).

12.4.9 Enhanced Muscle Proteolysis, Apoptosis, and Autophagy

An imbalance between protein synthesis and degradation exists in muscles of patients with chronic respiratory conditions such as COPD (Fig. 12.3). A reduction in protein synthesis signaling has been demonstrated in the lower limb muscles of these patients (Morrison et al., 1988; Puig-Vilanova et al., 2014), together with an increase in protein catabolism that results in muscle mass loss (Barreiro et al., 2015; Maltais et al., 2014; Puig-Vilanova

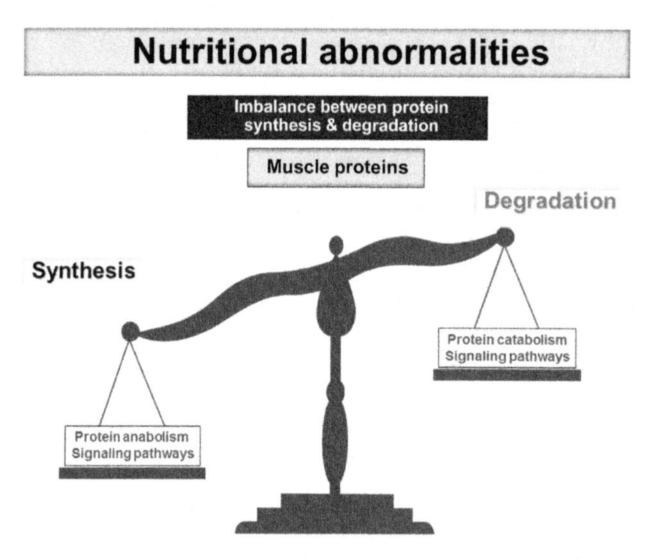

FIGURE 12.3 Schematic representation on the scale of the imbalance between protein synthesis and catabolism that takes place in patients with chronic respiratory diseases and nutritional abnormalities.

et al., 2015). The ubiquitin-proteasome pathway appears to be the major proteolytic system involved in muscle protein degradation in catabolic states, including COPD (Fermoselle et al., 2012; Maltais et al., 2014; Puig-Vilanova et al., 2015) and cancer-induced cachexia models (Puig-Vilanova et al., 2015; Fig. 12.2). Nuclear apoptosis may also take place in the lower limb muscles of patients with advanced respiratory diseases and muscle wasting (Agusti et al., 2002), although its role remains controversial as such findings were not entirely confirmed in another investigation in which apoptotic nuclei were analyzed in respiratory and limb muscles in COPD (Barreiro et al., 2011). Finally, autophagy which degrades organelles and is required for normal muscle mass maintenance (Guo et al., 2013; Puig-Vilanova et al., 2015) has also been shown to increase within the lower limb muscles of patients with COPD especially in those with severe muscle mass loss (Vainshtein et al., 2014). Importantly, systemic inflammation and oxidative stress, through the action of several redox-sensitive signaling mechanisms, such as nuclear factor (NF)-κB and fork head box O (FoxO) families of transcription factors, trigger muscle proteolysis and autophagy in the limb muscles of patients with chronic respiratory conditions (Puig-Vilanova et al., 2015; Fig. 12.2).

Lastly, it should be mentioned that patients with chronic respiratory diseases and nutritional abnormalities also experience decreased bone density, which may further contribute to whole body weight loss and functional capacity impairment (Ionescu and Schoon, 2003). Similarly to the etiology of muscle mass loss, cigarette smoke, reduced physical activity, systemic inflammation, low levels of anabolic hormones and vitamin D, and treatment with corticosteroids are counted among the most relevant factors contributing to bone density abnormalities in those patients (Ionescu and Schoon, 2003; Fig. 12.2).

12.5. PROTEIN METABOLISM, MUSCLES, AND EXERCISE IN HUMANS

12.5.1 Protein Absorption and Synthesis

It has been estimated that 1–2% of skeletal muscle tissue is being renewed every day (van Loon, 2013). Skeletal muscles adapt very rapidly to environmental stimuli such as food intake and exercise. Indeed, physical activity constitutes a very powerful anabolic stimulus in muscles as shown by the rise in protein synthesis following a bout of exercise (van Loon, 2013). Hence, in healthy subjects and patients, daily protein catabolism is counterbalanced through the protein synthesis stimulation induced by protein intake and exercise (van Loon, 2013).

Interestingly, greater rates of muscle protein synthesis and post-exercise-induced protein accretion were observed after the administration of different protein types such as whey (Borsheim et al., 2004), casein (Borsheim et al., 2004), soy (Wilkinson et al., 2007), egg (Moore et al., 2009), whole- and fat-free milk (Elliot et al., 2006; Wilkinson et al., 2007). Importantly, whey protein most effectively stimulated protein synthesis in the muscles after endurance or resistance exercise modalities (van Loon, 2013). As a matter of fact, whey contains a high proportion of branched-chain amino acids (approximately 26%) and is rich in leucine, making this supplement even more interesting to the design of more targeted nutritional strategies in patients (Ha and Zemel, 2003).

Proteins and amino acids are absorbed in the gut. Specifically, food proteins and amino acids are absorbed by the splanchnic tissues, from where they are released to the peripheral organs. When the demands of proteins and amino acids utilized by the splanchnic tissues increase as a result of physiological changes or gastrointestinal disorders their availability to other tissues will consequently decrease. Thus, the first-pass splanchnic extraction of these nutrients is essential to induce modifications in protein anabolism in healthy subjects and patients. In line with this, elderly subjects may experience an increased first-pass splanchnic extraction of the amino acids leucine and phenylalanine, which may jeopardize their availability and flow to the peripheral tissues and organs such as the skeletal muscles (Boirie et al., 1997; Schols, 2013). Nonetheless, Volpi et al. (1999) nicely showed that despite the increased first-pass splanchnic extraction observed in elderly subjects, the delivery to limb muscles of an oral supplement to synthesize proteins was preserved in a similar fashion to that measured in the young controls.

As protein synthesis stimulation depends to a great extent on the availability of amino acids (postprandial stimulation of protein accretion) in the blood of the patients, high levels of these compounds will favor protein anabolism in patients with nutritional abnormalities. More importantly, branched-chain amino acids, for example, leucine, stimulate the anabolic mammalian target of rapamycin (mTOR) signaling pathway that leads to protein synthesis. In fact, in young and elderly subjects, essential amino acids (including leucine) were shown to be mostly responsible for the stimulation of muscle protein synthesis, while nonessential amino acids did not enhance protein anabolism, even if administered at high doses to the participants (Evans, 1995; Ferrando et al., 1995; Volpi et al., 1999). Additionally, leucine has been shown to play a paramount role in the initiation pathway of muscle protein synthesis through the tight regulation of mRNA binding to the ribosomal subunits (Anthony et al., 2001). Furthermore, leucine and other essential amino acids are also substrates for the de novo synthesis of proteins. Therefore, they are key molecules in the control and regulation of protein synthesis, with special attention to skeletal muscles (Ha and Zemel, 2003).

Importantly, in elderly subjects, modifications in protein accretion as a result of alterations in protein digestion and absorption kinetics of certain types of proteins and amino acids have been observed. For instance, whey protein was shown to stimulate postprandial protein accretion in skeletal muscles in a more efficient manner than casein or casein hydrolysate in elderly men (Pennings et al., 2011a,b). Furthermore, the authors also demonstrated that exercise did not impair dietary protein digestion in young or elderly subjects. On the contrary, exercise performed prior to dietary protein intake favored protein synthesis in the muscles of both young and elderly men (Pennings et al., 2011b).

The factors leading to impaired tissue uptake of proteins and amino acids in the elderly can be summarized as follows: a decrease in postprandial hormonal response and microvascular perfusion (Rasmussen et al., 2006; Timmerman et al., 2010), a decline in amino acid uptake and intramuscular signaling from skeletal muscles (Cuthbertson et al., 2005; Drummond et al., 2010; Fry et al., 2011), with the resulting decreased myofibrillar protein synthesis (Cuthbertson et al., 2005). As patients with chronic respiratory diseases, especially COPD are usually old, these factors need to be taken into consideration when designing nutritional interventions in these patients.

12.5.2 Protein Synthesis in Muscles and Exercise

As mentioned above, exercise is a powerful anabolic stimulus in humans. Thus, physical activity performed prior to food intake may counterbalance the anabolic resistance shown in the elderly, favoring the stimulation of protein synthesis from their meals. For a comprehensive review on this specific issue, see van Loon (2013). Additionally, skeletal muscles also become more sensitized to anabolic stimuli in response to exercise. Indeed, basal and postprandial rates of protein synthesis are increased in response to physical activity in healthy subjects (van Loon, 2013). On the contrary, a significant reduction in physical activity and exercise, which are common features in chronic respiratory conditions and other diseases, such as cancer cachexia, critical illness, elective surgery, chronic heart failure, and sarcopenia, clearly results in impaired basal and postprandial rates of muscle protein synthesis and accretion (Burd et al., 2012; Wall and van Loon, 2013). Moreover, other studies have also demonstrated a significant decline in either whole-body (Biolo et al., 2002, 2004) or muscle (Glover et al., 2008) protein synthesis in immobilized patients, suggesting that disuse desensitizes muscles from anabolic stimuli.

12.6. POTENTIAL THERAPEUTIC TARGETS OF NUTRITIONAL ABNORMALITIES IN CHRONIC RESPIRATORY PATIENTS

A multidisciplinary therapeutic approach is absolutely required in patients with chronic respiratory diseases and nutritional abnormalities. On the one hand, several factors need to be targeted beforehand, such as smoking cessation, improvements in physical activity and exercise performance, a reduction in the number of acute exacerbations, and avoidance of the use of systemic corticosteroids as much as possible. On the other hand, anabolic hormones and specific nutritional supplements should also be recommended in those patients (Fig. 12.4).

12.6.1 Energy Balance, Amino Acid, and Protein Supplements

In many patients, dietary counseling and food fortification are sufficient in order to maintain an adequate body weight and muscle mass (Weekes et al., 2009). Nonetheless, in chronic respiratory patients with severe malnutrition, a more aggressive intervention is required such as the prescription of oral supplements containing high levels of amino acids and proteins, which may be reinforced by physical activity or even exercise training (Efthimiou et al., 1988; Fig. 12.4).

In general, nutritional intervention in patients with chronic respiratory diseases should aim to provide a sufficient amount of energy, proteins, and amino acids to promote protein synthesis, which contributes to counterbalancing the increased catabolism and atrophy observed in their muscles. Patients who consume less energy than the required amounts lose muscle mass during a period of bed rest compared to those who receive the appropriate amounts of calorie intake (Biolo et al., 2007). Besides, extracellular rather than intramuscular concentrations of amino acids seem to play a key role in the stimulation of protein synthesis in muscles (Bohe et al., 2003). In fact, protein intake or infusion of a specific solution containing essential and nonessential amino acids increased plasma concentrations of essential amino acids, while no significant changes were induced in the actual intramuscular concentrations (Bohe et al., 2003). Moreover, a higher increase in branched-chain amino acids was achieved in peripheral tissues when milk proteins were administered compared to soy proteins, suggesting that the bioavailability of amino acids is highly dependent on the quality of the nutritional compound (Fouillet et al., 2002). As a matter of fact, the advantages of certain compounds, for example, whey proteins, rely on several other features such as their high content in leucine to initiate protein synthesis, their rapid rate of absorbance in the gut, and their amino acid composition to become a substrate for de novo protein synthesis, which are all key factors

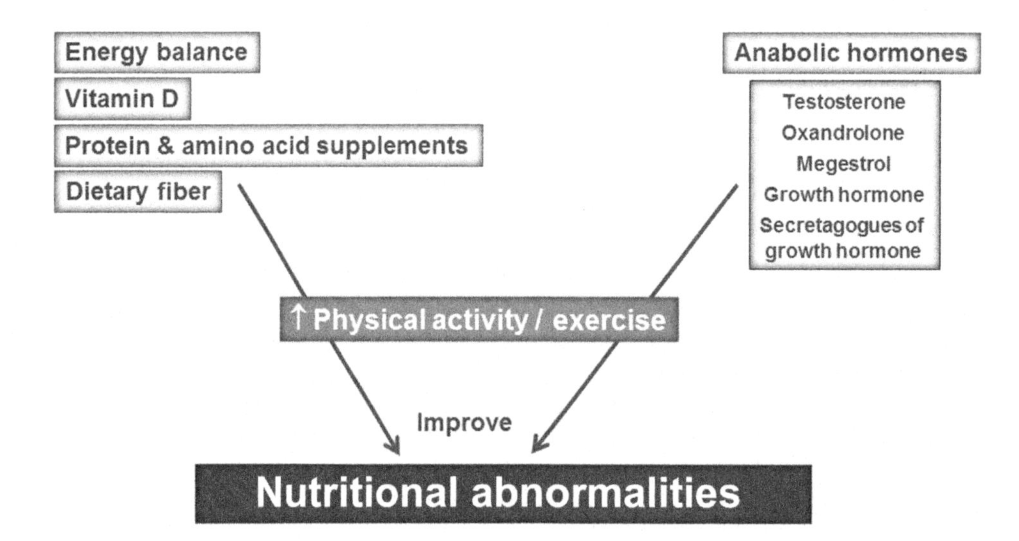

FIGURE 12.4 Schematic representation of the different therapeutic approaches that have been described so far to induce beneficial effects on patients with chronic respiratory patients. Increments in physical activity and/or exercise can be added to the different strategies inasmuch as patients can tolerate them.

that significantly influence the outcome of nutritional support in patients. In this regard, it was clearly demonstrated that dietary supplementation with pressurized whey potentiated the effects of exercise training on exercise tolerance and quality of life in patients with COPD (Laviolette et al., 2010; Fig. 12.4).

Patients with chronic respiratory diseases and muscle wasting have low levels of branched-chain amino acids (Engelen et al., 2000b). Specifically, lower concentrations of leucine, but not of isoleucine or valine, were indeed found in the plasma of the COPD patients, particularly in those with emphysema and severe muscle and adipose tissue loss (Engelen et al., 2000b). As in catabolic states, concentrations of branched-chain amino acids are usually increased, several explanations may account for the reduced levels of these amino acids encountered in COPD patients: (1) greater postabsorptive concentrations of insulin which reduce systemic levels of branched-chain amino acids due to a rise in the uptake of these amino acids in the muscles and adipose tissues (Munro et al., 1975); and (2) systemic inflammation through the action of pro-inflammatory cytokines that may influence adipose tissue content, while they may also favor the liver uptake of amino acids and increase branched-chain amino acid catabolism (Andus et al., 1991; Schols et al., 1996). Hence, supplements containing high levels of branched-chain amino acids will be very beneficial in COPD patients with severe malnutrition and muscle mass loss. In line with this, supplementation with leucine was shown to rather stimulate myofibrillar proteins in vitro, especially myosin heavy chain, than other proteins within the skeletal muscle cells (Haegens et al., 2012). Furthermore, as patients with chronic respiratory diseases, for example, COPD, are usually elderly subjects, they may experience an anabolic resistance to the administration of amino acids and proteins from food intake as also described above (Cuthbertson et al., 2005; Rasmussen et al., 2006; Volpi et al., 2000; Dardevet et al., 2012).

Nonetheless, in elderly patients with COPD and preserved body composition, the increased anabolic response to milk protein sip feeding was the result of a reduced splanchnic extraction of multiple amino acids including all branched-chain amino acids (Engelen et al., 2012). These results suggest that the protein anabolic response of food intake may improve in those patients (Engelen et al., 2012). Furthermore, supplementation with branched-chain amino acids also favored the metabolism of skeletal muscles in elderly subjects and in patients with stable COPD (Engelen et al., 2007; Katsanos et al., 2006; Rieu et al., 2006). Despite the acknowledged beneficial effects, it remains to be answered to what extent a similar response would be found in patients with acute muscle mass loss (eg, during an exacerbation), as well as in those with severe emphysema, since muscle protein synthesis (Morrison et al., 1988) and exercise-induced whole-body protein turnover were significantly reduced in these patients (Engelen et al., 2003). In summary, specific nutritional intervention together with increments in physical activity and/or exercise training are of interest in clinical settings of patients with chronic respiratory diseases (Fig. 12.4).

12.6.2 Other Nutritional Supplements

Vitamin D. The fat-soluble vitamin D favors the gut absorption of calcium, iron, phosphate, zinc, and magnesium. Vitamin D from the diet or dermal synthesis after sunlight exposure requires its activation through hydroxylation (formation of 25(OH)D) in the liver and kidneys. In patients with chronic respiratory conditions, vitamin D deficiency is quite common and was shown to be associated with poor lung and muscle functions, osteoporosis, and reduced immune response (Janssens et al., 2009). As established by international recommendations, vitamin D supplementation exerts beneficial effects on the bones and muscles, while preventing falls in elderly subjects and patients at high risk of vitamin D deficiency such as those with chronic lung diseases (Janssens et al., 2009). Regardless of daily intake, which may vary with age and lifestyle, a dose of 800 International Units vitamin D with 1 g of calcium is recommended for the treatment of vitamin D deficiency in the patients. Furthermore, recent evidence has demonstrated that vitamin D supplementation increased muscle strength in healthy adults with low levels of 25(OH)D (Stockton et al., 2011) and that mitochondrial oxidative phosphorylation also improved in the more susceptible subjects (Sinha et al., 2013; Fig. 12.4).

Dietary fiber. Increasing the dietary fiber content may also represent another possible intervention in patients with chronic respiratory diseases and nutritional abnormalities. In fact, respiratory mortality and greater dietary fiber intake was shown in two different studies (Chuang et al., 2012; Varraso et al., 2010). In this respect, gut immunity and systemic inflammation levels are reduced probably as a result of alterations in the gut microbiome induced by dietary fiber intake (Varraso et al., 2010). A study based on randomized clinical trials have demonstrated that a smaller decline in lung function parameters was observed in those COPD patients who followed a diet containing a greater intake of fruits and vegetables (higher proportions of fiber and antioxidants) for three years (Keranis et al., 2010). In another randomized clinical trial, however, increased intake of vegetables for

12 weeks did not correlate with a decline in oxidative stress or inflammatory markers in patients with COPD (Baldrick et al., 2012). It is likely that the relative short duration of this trial was not sufficient to induce significant changes in the blood parameters analyzed in the study (Baldrick et al., 2012). In summary, it would be possible to conclude that a habitual diet rich in vegetables, fruits, oil fish, wine, and cereals may protect patients from nutritional abnormalities and other systemic manifestations, thus in patients with chronic lung diseases, dietary intervention should be applied together with standard care and medication for their respiratory symptoms (Schols, 2013; Fig. 12.4).

12.6.3 Anabolic Hormones

Testosterone. Testosterone is the best known anabolic hormone, which increases protein synthesis and reduces proteolysis, while it also enhances lipolysis. It can be administered subcutaneously or intramuscularly. Testosterone promotes muscle mass increments together with adipose tissue loss. Its important side effects such as prostate tumor growth and virilization have caused its use to be questioned for the treatment of nutritional abnormalities in patients with chronic lung diseases. Indeed, nowadays the selective androgen receptor modulators are being used, since they have similar effects on tissues to testosterone, but have fewer side effects (Maltais et al., 2014). For instance, oxandrolone was used in patients with COPD and severe nutritional abnormalities for several months, leading to an improvement in body weight and fat-free mass, without improving exercise capacity (Yeh et al., 2002). In another investigation (Schols et al., 1995), treatment with nandrolone and nutritional support for two months induced an improvement in body weight, fat-free mass, and muscle function in patients with COPD. However, the last finding or improvements in the patients' exercise capacity were not demonstrated in a later investigation also conducted on COPD patients by the same group of investigators (Creutzberg et al., 2003; Fig. 12.4).

Megestrol. Megestrol, which is a synthetic compound derived from progesterone, was shown to induce orexigenic effects and improvements in body weight in patients with COPD and low body weight, while increasing fat tissue compartment, with no functional improvement (Weisberg et al., 2002; Fig. 12.4).

GH axis. As abovementioned, GH via IGF-1 activation promotes protein synthesis and decreases protein catabolism, together with an increase in calcium resorption and the immune response in patients (Gea et al., 2014). Apparently, GH exerts beneficial effects on total body and muscle weights and muscle strength of specific muscle groups (Gea et al., 2014). Nonetheless, inconclusive and discrepant results have been shown when administered to patients with COPD. In a study conducted on a small number of COPD patients without a control group of individuals, body weight and muscle mass improved as well as respiratory muscle strength (Pape et al., 1991), whereas in another investigation, patients experienced a significant improvement in body weight and muscle mass with no functional consequences (Burdet et al., 1997). Treatment with growth-hormone-releasing factor in patients with body and muscle mass loss has not yielded conclusive results as the duration of its effects is very short (2 hours) and the individual response shows a great variability (Gea et al., 2014; Fig. 12.4).

Segretagogues of GH. Another group of drugs with potential beneficial effects on muscle mass are the segretagogues of GH that induce its release. Ghrelin is the most widely known segretagogue, which is a natural peptide secreted by several organs including the lungs. The effects of ghrelin are identical to those exerted by GH, such as induction of protein synthesis, prevention of protein oxidation and degradation, and lipolysis (Gea et al., 2014). Ghrelin also stimulates the appetite and may have anti-inflammatory effects (Gea et al., 2014). In patients with COPD and poor body composition, levels of endogenous ghrelin are increased, most likely as a compensatory mechanism (Xu et al., 2012). If administered exogenously, ghrelin induces a rise in total body weight and muscle mass and function in chronic respiratory patients with cachexia as well as in healthy subjects (Nagaya et al., 2005). In another study, the effects on the respiratory patients were moderate as body weight or muscle function did not increase, despite the improvement in symptoms (Miki et al., 2012). Tesamorelin is another GH segretagogue that may improve muscle mass and function in patients with COPD (Gea et al., 2014). The study of the potential beneficial effects of other GH segretagogues (eg, capromorelin, examorelin, sermorelin, ipamorelin, tabimorelin, ibutamoren, and other compounds such as LAB GHRH, GHRP-2) that could be used in patients with chronic respiratory conditions is the focus of ongoing and future research. The ultimate goal of those investigations will be to identify a drug that can be easily and safely administered to the respiratory cachectic patients, while inducing positive effects on their muscle mass and function (Fig. 12.4).

12.7. OTHER CHRONIC RESPIRATORY CONDITIONS

12.7.1 Cystic Fibrosis

Patients with cystic fibrosis, which is an autosomal recessive genetic disorder, experience serious nutritional abnormalities. Cystic fibrosis shortens the patients' life-span and impairs their quality of life as a consequence of the secretion of thick mucus in the respiratory, gastrointestinal, and reproductive tracts and by the excessive losses of sodium and chloride in sweat. These patients usually have frequent respiratory infections that lead to progressive lung fibrosis and destruction as well as severe nutritional abnormalities as a result of impaired absorption of nutrients in the gut (pancreatic insufficiency). Therefore, cystic fibrosis patients do not equally gain weight or growth as subjects without the disease for the same age group. In the last decades, significant improvements in the life expectancy of these patients have been achieved, especially due to recent progress in nutritional strategies and the control of respiratory infections and lose of lung function. In cystic fibrosis care, nutritional support to maintaining an optimal nutritional status is nowadays regarded as a key component. Indeed, significant correlations have been observed between greater body weight and improved lung function and survival through age 18 years in patients with cystic fibrosis (Yen et al., 2013).

As in healthy, full-term toddlers, infants with cystic fibrosis should also receive human milk from breast feeding. After this phase, most of these patients are prescribed a high-calorie diet, containing high-fat and protein intake in order to ensure that the energy demands are met (Powers et al., 2005). However, any high-calorie supplement rather than specific protein energy supplements should be administered to the patients as the latter did not improve their nutritional status (Schindler et al., 2015; Smyth and Rayner, 2014). As diabetes is the most common comorbidity in patients with cystic fibrosis, an adequate control of hyperglycemia and its potential complications in several organs is also required in these patients. Furthermore, fat- and water-soluble vitamins and minerals, for example, zinc, are also administered as nutritional supplements in patients with cystic fibrosis.

12.7.2 Other Respiratory-Related Disorders

Importantly, nutritional abnormalities may also be present in other respiratory diseases, such as asthma and idiopathic pulmonary fibrosis. Nonetheless, little evidence is still available and more research is clearly needed. Moreover, as shown to occur in elderly subjects and COPD patients, physical inactivity, mainly due to shortness of breath, may also be a major trigger of the nutritional abnormalities in patients bearing any of those respiratory conditions, especially in advanced stages.

Finally, nutritional abnormalities are also important complications of other diseases, such as cancer, including lung cancer, and critical illness. Nevertheless, as these two entities are being thoroughly described in specific chapters in this book, no additional information is provided in this chapter in order to avoid any potential conflicting overlap.

12.8. CONCLUSIONS AND FUTURE PERSPECTIVES

Nutritional abnormalities and muscle mass loss are relevant systemic comorbidities in patients with chronic respiratory conditions, especially in COPD. Moreover, such comorbidities may predict mortality in these patients, thus having a great negative impact of the patients' quality of life and survival. Nutritional status assessment should be part of the standard clinical evaluation of patients bearing chronic respiratory conditions. Hence, nutritional abnormalities should be routinely monitored in stable and hospitalized respiratory patients. Several factors and biological mechanisms are involved in the multifactorial etiology of nutritional abnormalities in COPD. Substantial improvements in whole-body and muscle protein anabolism can be achieved by the administration of protein supplements, especially of those with a high content in branched-chain amino acids such as leucine. As physical activity and exercise significantly improve protein synthesis in muscles, they should also be considered as an additional strategy to optimize protein anabolism and accretion from the dietary and supplemental intakes. In summary, specific nutritional intervention together with increments in physical activity and/or exercise training are highly recommended to improving the patients' quality of life and performance in addition to the standard respiratory care of the underlying lung disease.

References

Agusti, A.G., Sauleda, J., Miralles, C., Gomez, C., Togores, B., Sala, E., et al., 2002. Skeletal muscle apoptosis and weight loss in chronic obstructive pulmonary disease. Am. J. Respir. Crit. Care Med. 166, 485–489.

Anand, I.S., Chandrashekhar, Y., Ferrari, R., Sarma, R., Guleria, R., Jindal, S.K., et al., 1992. Pathogenesis of congestive state in chronic obstructive pulmonary disease. Studies of body water and sodium, renal function, hemodynamics, and plasma hormones during edema and after recovery. Circulation 86, 12–21.

Andus, T., Bauer, J., Gerok, W., 1991. Effects of cytokines on the liver. Hepatology 13, 364–375.

Anthony, J.C., Anthony, T.G., Kimball, S.R., Jefferson, L.S., 2001. Signaling pathways involved in translational control of protein synthesis in skeletal muscle by leucine. J. Nutr. 131, 856S–860S.

Baldrick, F.R., Elborn, J.S., Woodside, J.V., Treacy, K., Bradley, J.M., Patterson, C.C., et al., 2012. Effect of fruit and vegetable intake on oxidative stress and inflammation in COPD: a randomised controlled trial. Eur. Respir. J. 39, 1377–1384.

Barreiro, E., Rabinovich, R., Marin-Corral, J., Barbera, J.A., Gea, J., Roca, J., 2009. Chronic endurance exercise induces quadriceps nitrosative stress in patients with severe COPD. Thorax 64, 13–19.

Barreiro, E., Peinado, V.I., Galdiz, J.B., Ferrer, E., Marin-Corral, J., Sanchez, F., et al., 2010. Cigarette smoke-induced oxidative stress: a role in chronic obstructive pulmonary disease skeletal muscle dysfunction. Am. J. Respir. Crit. Care Med. 182, 477–488.

Barreiro, E., Ferrer, D., Sanchez, F., Minguella, J., Marin-Corral, J., Martinez-Llorens, J., et al., 2011. Inflammatory cells and apoptosis in respiratory and limb muscles of patients with COPD. J. Appl. Physiol. (1985) 111, 808–817.

Barreiro, E., Bustamante, V., Cejudo, P., Galdiz, J.B., Gea, J., de, L.P., et al., 2015. Recommendations for the evaluation and treatment of muscle dysfunction in patients with chronic obstructive pulmonary disease. Arch. Bronconeumol. 51 (8), 384–395.

Bhasin, S., Cunningham, G.R., Hayes, F.J., Matsumoto, A.M., Snyder, P.J., Swerdloff, R.S., et al., 2010. Testosterone therapy in men with androgen deficiency syndromes: an endocrine society clinical practice guideline. J. Clin. Endocrinol. Metab. 95, 2536–2559.

Biolo, G., Ciocchi, B., Lebenstedt, M., Heer, M., Guarnieri, G., 2002. Sensitivity of whole body protein synthesis to amino acid administration during short-term bed rest. J. Gravit. Physiol. 9, 197–198.

Biolo, G., Ciocchi, B., Lebenstedt, M., Barazzoni, R., Zanetti, M., Platen, P., et al., 2004. Short-term bed rest impairs amino acid-induced protein anabolism in humans. J. Physiol. 558, 381–388.

Biolo, G., Ciocchi, B., Stulle, M., Bosutti, A., Barazzoni, R., Zanetti, M., et al., 2007. Calorie restriction accelerates the catabolism of lean body mass during 2 wk of bed rest. Am. J. Clin. Nutr. 86, 366–372.

Bohe, J., Low, A., Wolfe, R.R., Rennie, M.J., 2003. Human muscle protein synthesis is modulated by extracellular, not intramuscular amino acid availability: a dose-response study. J. Physiol. 552, 315–324.

Boirie, Y., Gachon, P., Beaufrere, B., 1997. Splanchnic and whole-body leucine kinetics in young and elderly men. Am. J. Clin. Nutr. 65, 489–495.

Borghetti, P., Saleri, R., Mocchegiani, E., Corradi, A., Martelli, P., 2009. Infection, immunity and the neuroendocrine response. Vet. Immunol. Immunopathol. 130, 141–162.

Borsheim, E., Cree, M.G., Tipton, K.D., Elliott, T.A., Aarsland, A., Wolfe, R.R., 2004. Effect of carbohydrate intake on net muscle protein synthesis during recovery from resistance exercise. J. Appl. Physiol. (1985) 96, 674–678.

Brioche, T., Kireev, R.A., Cuesta, S., Gratas-Delamarche, A., Tresguerres, J.A., Gomez-Cabrera, M.C., et al., 2014. Growth hormone replacement therapy prevents sarcopenia by a dual mechanism: improvement of protein balance and of antioxidant defenses. J. Gerontol. A Biol. Sci. Med. Sci. 69, 1186–1198.

Burd, N.A., Wall, B.T., van Loon, L.J., 2012. The curious case of anabolic resistance: old wives' tales or new fables? J. Appl. Physiol. (1985) 112, 1233–1235.

Burdet, L., de, M.B., Schutz, Y., Pichard, C., Fitting, J.W., 1997. Administration of growth hormone to underweight patients with chronic obstructive pulmonary disease. A prospective, randomized, controlled study. Am. J. Respir. Crit. Care Med. 156, 1800–1806.

Burgel, P.R., Paillasseur, J.L., Peene, B., Dusser, D., Roche, N., Coolen, J., et al., 2012. Two distinct chronic obstructive pulmonary disease (COPD) phenotypes are associated with high risk of mortality. PLoS One 7, e51048.

Chuang, S.C., Norat, T., Murphy, N., Olsen, A., Tjonneland, A., Overvad, K., et al., 2012. Fiber intake and total and cause-specific mortality in the European prospective investigation into cancer and nutrition cohort. Am. J. Clin. Nutr. 96, 164–174.

Coin, A., Sergi, G., Minicuci, N., Giannini, S., Barbiero, E., Manzato, E., et al., 2008. Fat-free mass and fat mass reference values by dual-energy X-ray absorptiometry (DEXA) in a 20–80 year-old Italian population. Clin. Nutr. 27, 87–94.

Creutzberg, E.C., Casaburi, R., 2003. Endocrinological disturbances in chronic obstructive pulmonary disease. Eur. Respir. J. Suppl. 46, 76s–80s.

Creutzberg, E.C., Wouters, E.F., Vanderhoven-Augustin, I.M., Dentener, M.A., Schols, A.M., 2000. Disturbances in leptin metabolism are related to energy imbalance during acute exacerbations of chronic obstructive pulmonary disease. Am. J. Respir. Crit. Care Med. 162, 1239–1245.

Creutzberg, E.C., Wouters, E.F., Mostert, R., Pluymers, R.J., Schols, A.M., 2003. A role for anabolic steroids in the rehabilitation of patients with COPD? A double-blind, placebo-controlled, randomized trial. Chest 124, 1733–1742.

Cuthbertson, D., Smith, K., Babraj, J., Leese, G., Waddell, T., Atherton, P., et al., 2005. Anabolic signaling deficits underlie amino acid resistance of wasting, aging muscle. FASEB J. 19, 422–424.

Dardevet, D., Remond, D., Peyron, M.A., Papet, I., Savary-Auzeloux, I., Mosoni, L., 2012. Muscle wasting and resistance of muscle anabolism: the "anabolic threshold concept" for adapted nutritional strategies during sarcopenia. ScientificWorldJournal 2012, 269531.

Divo, M.J., Casanova, C., Marin, J.M., Pinto-Plata, V.M., de-Torres, J.P., Zulueta, J.J., et al., 2015. Chronic obstructive pulmonary disease comorbidities network. Eur. Respir. J. 46 (3), 640–650.

Drummond, M.J., Glynn, E.L., Fry, C.S., Timmerman, K.L., Volpi, E., Rasmussen, B.B., 2010. An increase in essential amino acid availability upregulates amino acid transporter expression in human skeletal muscle. Am. J. Physiol. Endocrinol. Metab. 298, E1011–E1018.

Efthimiou, J., Fleming, J., Gomes, C., Spiro, S.G., 1988. The effect of supplementary oral nutrition in poorly nourished patients with chronic obstructive pulmonary disease. Am. Rev. Respir. Dis. 137, 1075–1082.

Elliot, T.A., Cree, M.G., Sanford, A.P., Wolfe, R.R., Tipton, K.D., 2006. Milk ingestion stimulates net muscle protein synthesis following resistance exercise. Med. Sci. Sports Exerc. 38, 667–674.

Engelen, M.P., Deutz, N.E., Wouters, E.F., Schols, A.M., 2000a. Enhanced levels of whole-body protein turnover in patients with chronic obstructive pulmonary disease. Am. J. Respir. Crit. Care Med. 162, 1488–1492.

Engelen, M.P., Wouters, E.F., Deutz, N.E., Menheere, P.P., Schols, A.M., 2000b. Factors contributing to alterations in skeletal muscle and plasma amino acid profiles in patients with chronic obstructive pulmonary disease. Am. J. Clin. Nutr. 72, 1480–1487.

Engelen, M.P., Deutz, N.E., Mostert, R., Wouters, E.F., Schols, A.M., 2003. Response of whole-body protein and urea turnover to exercise differs between patients with chronic obstructive pulmonary disease with and without emphysema. Am. J. Clin. Nutr. 77, 868–874.

Engelen, M.P., Rutten, E.P., De Castro, C.L., Wouters, E.F., Schols, A.M., Deutz, N.E., 2007. Supplementation of soy protein with branched-chain amino acids alters protein metabolism in healthy elderly and even more in patients with chronic obstructive pulmonary disease. Am. J. Clin. Nutr. 85, 431–439.

Engelen, M.P., De Castro, C.L., Rutten, E.P., Wouters, E.F., Schols, A.M., Deutz, N.E., 2012. Enhanced anabolic response to milk protein sip feeding in elderly subjects with COPD is associated with a reduced splanchnic extraction of multiple amino acids. Clin. Nutr. 31, 616–624.

Engineer, D.R., Garcia, J.M., 2012. Leptin in anorexia and cachexia syndrome. Int. J. Pept. 2012, 287457.

England, B.K., Chastain, J.L., Mitch, W.E., 1991. Abnormalities in protein synthesis and degradation induced by extracellular pH in BC3H1 myocytes. Am. J. Physiol. 260, C277–C282.

Evans, W.J., 1995. What is sarcopenia? J. Gerontol. A Biol. Sci. Med. Sci. 50, 5–8, Spec No.

Fermoselle, C., Rabinovich, R., Ausin, P., Puig-Vilanova, E., Coronell, C., Sanchez, F., et al., 2012. Does oxidative stress modulate limb muscle atrophy in severe COPD patients? Eur. Respir. J. 40, 851–862.

Ferrando, A.A., Stuart, C.A., Brunder, D.G., Hillman, G.R., 1995. Magnetic resonance imaging quantitation of changes in muscle volume during 7 days of strict bed rest. Aviat. Space Environ. Med. 66, 976–981.

Fouillet, H., Mariotti, F., Gaudichon, C., Bos, C., Tome, D., 2002. Peripheral and splanchnic metabolism of dietary nitrogen are differently affected by the protein source in humans as assessed by compartmental modeling. J. Nutr. 132, 125–133.

Franssen, F.M., Sauerwein, H.P., Rutten, E.P., Wouters, E.F., Schols, A.M., 2008. Whole-body resting and exercise-induced lipolysis in sarcopenic [corrected] patients with COPD. Eur. Respir. J. 32, 1466–1471.

Fry, C.S., Drummond, M.J., Glynn, E.L., Dickinson, J.M., Gundermann, D.M., Timmerman, K.L., et al., 2011. Aging impairs contraction-induced human skeletal muscle mTORC1 signaling and protein synthesis. Skelet Muscle 1, 11.

Gea, J., Agusti, A., Roca, J., 2013. Pathophysiology of muscle dysfunction in COPD. J. Appl. Physiol. (1985) 114, 1222–1234.

Gea, J., Martinez-Llorens, J., Barreiro, E., 2014. Nutritional abnormalities in chronic obstructive pulmonary disease. Med. Clin. (Barc) 143, 78–84.

Glover, E.I., Phillips, S.M., Oates, B.R., Tang, J.E., Tarnopolsky, M.A., Selby, A., et al., 2008. Immobilization induces anabolic resistance in human myofibrillar protein synthesis with low and high dose amino acid infusion. J. Physiol. 586, 6049–6061.

Goris, A.H., Vermeeren, M.A., Wouters, E.F., Schols, A.M., Westerterp, K.R., 2003. Energy balance in depleted ambulatory patients with chronic obstructive pulmonary disease: the effect of physical activity and oral nutritional supplementation. Br. J. Nutr. 89, 725–731.

Guo, Y., Gosker, H.R., Schols, A.M., Kapchinsky, S., Bourbeau, J., Sandri, M., et al., 2013. Autophagy in locomotor muscles of patients with chronic obstructive pulmonary disease. Am. J. Respir. Crit. Care Med. 188, 1313–1320.

Gurgun, A., Deniz, S., Argin, M., Karapolat, H., 2013. Effects of nutritional supplementation combined with conventional pulmonary rehabilitation in muscle-wasted chronic obstructive pulmonary disease: a prospective, randomized and controlled study. Respirology 18, 495–500.

Ha, E., Zemel, M.B., 2003. Functional properties of whey, whey components, and essential amino acids: mechanisms underlying health benefits for active people (review). J. Nutr. Biochem. 14, 251–258.

Haegens, A., Schols, A.M., van Essen, A.L., van Loon, L.J., Langen, R.C., 2012. Leucine induces myofibrillar protein accretion in cultured skeletal muscle through mTOR dependent and -independent control of myosin heavy chain mRNA levels. Mol. Nutr. Food Res. 56, 741–752.

Hasselgren, P.O., Alamdari, N., Aversa, Z., Gonnella, P., Smith, I.J., Tizio, S., 2010. Corticosteroids and muscle wasting: role of transcription factors, nuclear cofactors, and hyperacetylation. Curr. Opin. Clin. Nutr. Metab. Care 13, 423–428.

Ionescu, A.A., Schoon, E., 2003. Osteoporosis in chronic obstructive pulmonary disease. Eur. Respir. J. Suppl. 46, 64s–75s.

Jaitovich, A., Angulo, M., Lecuona, E., Dada, L.A., Welch, L.C., Cheng, Y., et al., 2015. High CO_2 levels cause skeletal muscle atrophy via AMP-activated kinase (AMPK), $FoxO_3a$ protein, and muscle-specific Ring finger protein 1 (MuRF1). J. Biol. Chem. 290, 9183–9194.

Janssens, W., Lehouck, A., Carremans, C., Bouillon, R., Mathieu, C., Decramer, M., 2009. Vitamin D beyond bones in chronic obstructive pulmonary disease: time to act. Am. J. Respir. Crit. Care Med. 179, 630–636.

Kao, C.C., Hsu, J.W., Bandi, V., Hanania, N.A., Kheradmand, F., Jahoor, F., 2011. Resting energy expenditure and protein turnover are increased in patients with severe chronic obstructive pulmonary disease. Metabolism 60, 1449–1455.

Karadag, F., Karul, A.B., Cildag, O., Yilmaz, M., Ozcan, H., 2008. Biomarkers of systemic inflammation in stable and exacerbation phases of COPD. Lung 186, 403–409.

Katsanos, C.S., Kobayashi, H., Sheffield-Moore, M., Aarsland, A., Wolfe, R.R., 2006. A high proportion of leucine is required for optimal stimulation of the rate of muscle protein synthesis by essential amino acids in the elderly. Am. J. Physiol. Endocrinol. Metab. 291, E381–E387.

Keranis, E., Makris, D., Rodopoulou, P., Martinou, H., Papamakarios, G., Daniil, Z., et al., 2010. Impact of dietary shift to higher-antioxidant foods in COPD: a randomised trial. Eur. Respir. J. 36, 774–780.

Kim, V., Kretschman, D.M., Sternberg, A.L., DeCamp Jr., M.M., Criner, G.J., 2012. Weight gain after lung reduction surgery is related to improved lung function and ventilatory efficiency. Am. J. Respir. Crit. Care Med. 186, 1109–1116.

Kirdar, S., Serter, M., Ceylan, E., Sener, A.G., Kavak, T., Karadag, F., 2009. Adiponectin as a biomarker of systemic inflammatory response in smoker patients with stable and exacerbation phases of chronic obstructive pulmonary disease. Scand. J. Clin. Lab. Invest. 69, 219–224.

Laghi, F., Langbein, W.E., Antonescu-Turcu, A., Jubran, A., Bammert, C., Tobin, M.J., 2005. Respiratory and skeletal muscles in hypogonadal men with chronic obstructive pulmonary disease. Am. J. Respir. Crit. Care Med. 171, 598–605.

Laveneziana, P., Palange, P., 2012. Physical activity, nutritional status and systemic inflammation in COPD. Eur. Respir. J. 40, 522–529.

Laviolette, L., Lands, L.C., Dauletbaev, N., Saey, D., Milot, J., Provencher, S., et al., 2010. Combined effect of dietary supplementation with pressurized whey and exercise training in chronic obstructive pulmonary disease: a randomized, controlled, double-blind pilot study. J. Med. Food 13, 589–598.

Maltais, F., Decramer, M., Casaburi, R., Barreiro, E., Burelle, Y., Debigare, R., et al., 2014. An official American Thoracic Society/European Respiratory Society statement: update on limb muscle dysfunction in chronic obstructive pulmonary disease. Am. J. Respir. Crit. Care Med. 189, e15–e62.

Marquis, K., Debigare, R., Lacasse, Y., LeBlanc, P., Jobin, J., Carrier, G., et al., 2002. Midthigh muscle cross-sectional area is a better predictor of mortality than body mass index in patients with chronic obstructive pulmonary disease. Am. J. Respir. Crit. Care Med. 166, 809–813.

Miki, K., Maekura, R., Nagaya, N., Nakazato, M., Kimura, H., Murakami, S., et al., 2012. Ghrelin treatment of cachectic patients with chronic obstructive pulmonary disease: a multicenter, randomized, double-blind, placebo-controlled trial. PLoS One 7, e35708.

Miravitlles, M., Soler-Cataluna, J.J., Calle, M., Molina, J., Almagro, P., Antonio, Q.J., et al., 2014. Spanish guideline for COPD (GesEPOC). Update 2014. Arch. Bronconeumol. 50 (Suppl. 1), 1–16.

Moore, D.R., Robinson, M.J., Fry, J.L., Tang, J.E., Glover, E.I., Wilkinson, S.B., et al., 2009. Ingested protein dose response of muscle and albumin protein synthesis after resistance exercise in young men. Am. J. Clin. Nutr. 89, 161–168.

Morrison, W.L., Gibson, J.N., Scrimgeour, C., Rennie, M.J., 1988. Muscle wasting in emphysema. Clin. Sci. (Lond) 75, 415–420.

Munro, H.N., Fernstrom, J.D., Wurtman, R.J., 1975. Insulin, plasma aminoacid imbalance, and hepatic coma. Lancet 1, 722–724.

Nagaya, N., Itoh, T., Murakami, S., Oya, H., Uematsu, M., Miyatake, K., et al., 2005. Treatment of cachexia with ghrelin in patients with COPD. Chest 128, 1187–1193.

Pape, G.S., Friedman, M., Underwood, L.E., Clemmons, D.R., 1991. The effect of growth hormone on weight gain and pulmonary function in patients with chronic obstructive lung disease. Chest 99, 1495–1500.

Pennings, B., Boirie, Y., Senden, J.M., Gijsen, A.P., Kuipers, H., van Loon, L.J., 2011a. Whey protein stimulates postprandial muscle protein accretion more effectively than do casein and casein hydrolysate in older men. Am. J. Clin. Nutr. 93, 997–1005.

Pennings, B., Koopman, R., Beelen, M., Senden, J.M., Saris, W.H., van Loon, L.J., 2011b. Exercising before protein intake allows for greater use of dietary protein-derived amino acids for de novo muscle protein synthesis in both young and elderly men. Am. J. Clin. Nutr. 93, 322–331.

Petersen, A.M., Magkos, F., Atherton, P., Selby, A., Smith, K., Rennie, M.J., et al., 2007. Smoking impairs muscle protein synthesis and increases the expression of myostatin and MAFbx in muscle. Am. J. Physiol. Endocrinol. Metab. 293, E843–E848.

Powers, S.W., Mitchell, M.J., Patton, S.R., Byars, K.C., Jelalian, E., Mulvihill, M.M., et al., 2005. Mealtime behaviors in families of infants and toddlers with cystic fibrosis. J. Cyst. Fibros. 4, 175–182.

Puig-Vilanova, E., Aguilo, R., Rodriguez-Fuster, A., Martinez-Llorens, J., Gea, J., Barreiro, E., 2014. Epigenetic mechanisms in respiratory muscle dysfunction of patients with chronic obstructive pulmonary disease. PLoS One 9, e111514.

Puig-Vilanova, E., Rodriguez, D.A., Lloreta, J., Ausin, P., Pascual-Guardia, S., Broquetas, J., et al., 2015. Oxidative stress, redox signaling pathways, and autophagy in cachectic muscles of male patients with advanced COPD and lung cancer. Free Radic. Biol. Med. 79, 91–108.

Raguso, C.A., Luthy, C., 2011. Nutritional status in chronic obstructive pulmonary disease: role of hypoxia. Nutrition 27, 138–143.

Rasmussen, B.B., Fujita, S., Wolfe, R.R., Mittendorfer, B., Roy, M., Rowe, V.L., et al., 2006. Insulin resistance of muscle protein metabolism in aging. FASEB J. 20, 768–769.

Rieu, I., Balage, M., Sornet, C., Giraudet, C., Pujos, E., Grizard, J., et al., 2006. Leucine supplementation improves muscle protein synthesis in elderly men independently of hyperaminoacidaemia. J. Physiol. 575, 305–315.

Rodriguez, D.A., Kalko, S., Puig-Vilanova, E., Perez-Olabarria, M., Falciani, F., Gea, J., et al., 2012. Muscle and blood redox status after exercise training in severe COPD patients. Free Radic. Biol. Med. 52, 88–94.

Schindler, T., Michel, S., Wilson, A.W., 2015. Nutrition management of cystic fibrosis in the 21st century. Nutr. Clin. Pract. 30 (4), 488–500.

Schols, A.M., 2013. Nutrition as a metabolic modulator in COPD. Chest 144, 1340–1345.

Schols, A.M., Soeters, P.B., Mostert, R., Saris, W.H., Wouters, E.F., 1991a. Energy balance in chronic obstructive pulmonary disease. Am. Rev. Respir. Dis. 143, 1248–1252.

Schols, A.M., Wouters, E.F., Soeters, P.B., Westerterp, K.R., 1991b. Body composition by bioelectrical-impedance analysis compared with deuterium dilution and skinfold anthropometry in patients with chronic obstructive pulmonary disease. Am. J. Clin. Nutr. 53, 421–424.

Schols, A.M., Soeters, P.B., Mostert, R., Pluymers, R.J., Wouters, E.F., 1995. Physiologic effects of nutritional support and anabolic steroids in patients with chronic obstructive pulmonary disease. A placebo-controlled randomized trial. Am. J. Respir. Crit. Care Med. 152, 1268–1274.

Schols, A.M., Buurman, W.A., Staal van den Brekel, A.J., Dentener, M.A., Wouters, E.F., 1996. Evidence for a relation between metabolic derangements and increased levels of inflammatory mediators in a subgroup of patients with chronic obstructive pulmonary disease. Thorax 51, 819–824.

Sergi, G., Coin, A., Marin, S., Vianello, A., Manzan, A., Peruzza, S., et al., 2006. Body composition and resting energy expenditure in elderly male patients with chronic obstructive pulmonary disease. Respir. Med. 100, 1918–1924.

Seymour, J.M., Spruit, M.A., Hopkinson, N.S., Natanek, S.A., Man, W.D., Jackson, A., et al., 2010. The prevalence of quadriceps weakness in COPD and the relationship with disease severity. Eur. Respir. J. 36, 81–88.

Shrikrishna, D., Patel, M., Tanner, R.J., Seymour, J.M., Connolly, B.A., Puthucheary, Z.A., et al., 2012. Quadriceps wasting and physical inactivity in patients with COPD. Eur. Respir. J. 40, 1115–1122.

Sinha, A., Hollingsworth, K.G., Ball, S., Cheetham, T., 2013. Improving the vitamin D status of vitamin D deficient adults is associated with improved mitochondrial oxidative function in skeletal muscle. J. Clin. Endocrinol. Metab. 98, E509–E513.

Smyth, R.L., Rayner, O., 2014. Oral calorie supplements for cystic fibrosis. Cochrane Database Syst. Rev. 11, CD000406.

Stockton, K.A., Mengersen, K., Paratz, J.D., Kandiah, D., Bennell, K.L., 2011. Effect of vitamin D supplementation on muscle strength: a systematic review and meta-analysis. Osteoporos. Int. 22, 859–871.

Swallow, E.B., Reyes, D., Hopkinson, N.S., Man, W.D., Porcher, R., Cetti, E.J., et al., 2007. Quadriceps strength predicts mortality in patients with moderate to severe chronic obstructive pulmonary disease. Thorax 62, 115–120.

Timmerman, K.L., Lee, J.L., Dreyer, H.C., Dhanani, S., Glynn, E.L., Fry, C.S., et al., 2010. Insulin stimulates human skeletal muscle protein synthesis via an indirect mechanism involving endothelial-dependent vasodilation and mammalian target of rapamycin complex 1 signaling. J. Clin. Endocrinol. Metab. 95, 3848–3857.

Tkacova, R., Ukropec, J., Skyba, P., Ukropcova, B., Pobeha, P., Kurdiova, T., et al., 2011. Increased adipose tissue expression of proinflammatory CD40, MKK4 and JNK in patients with very severe chronic obstructive pulmonary disease. Respiration 81, 386–393.

Vainshtein, A., Grumati, P., Sandri, M., Bonaldo, P., 2014. Skeletal muscle, autophagy, and physical activity: the menage a trois of metabolic regulation in health and disease. J. Mol. Med. (Berl) 92, 127–137.

van den Borst, B., Gosker, H.R., Wesseling, G., de, J.W., Hellwig, V.A., Snepvangers, F.J., et al., 2011. Low-grade adipose tissue inflammation in patients with mild-to-moderate chronic obstructive pulmonary disease. Am. J. Clin. Nutr. 94, 1504–1512.

van den Borst, B., Slot, I.G., Hellwig, V.A., Vosse, B.A., Kelders, M.C., Barreiro, E., et al., 2013. Loss of quadriceps muscle oxidative phenotype and decreased endurance in patients with mild-to-moderate COPD. J. Appl. Physiol. (1985) 114, 1319–1328.

van Loon, L.J., 2013. Role of dietary protein in post-exercise muscle reconditioning. Nestle Nutr. Inst. Workshop Ser. 75, 73–83.

Vanfleteren, L.E., Spruit, M.A., Groenen, M., Gaffron, S., van Empel, V.P., Bruijnzeel, P.L., et al., 2013. Clusters of comorbidities based on validated objective measurements and systemic inflammation in patients with chronic obstructive pulmonary disease. Am. J. Respir. Crit. Care Med. 187, 728–735.

Varraso, R., Willett, W.C., Camargo Jr., C.A., 2010. Prospective study of dietary fiber and risk of chronic obstructive pulmonary disease among US women and men. Am. J. Epidemiol. 171, 776–784.

Vermeeren, M.A., Creutzberg, E.C., Schols, A.M., Postma, D.S., Pieters, W.R., Roldaan, A.C., et al., 2006. Prevalence of nutritional depletion in a large out-patient population of patients with COPD. Respir. Med. 100, 1349–1355.

Vestbo, J., Hurd, S.S., Agusti, A.G., Jones, P.W., Vogelmeier, C., Anzueto, A., et al., 2013. Global strategy for the diagnosis, management, and prevention of chronic obstructive pulmonary disease: GOLD executive summary. Am. J. Respir. Crit. Care Med. 187, 347–365.

Volpi, E., Mittendorfer, B., Wolf, S.E., Wolfe, R.R., 1999. Oral amino acids stimulate muscle protein anabolism in the elderly despite higher first-pass splanchnic extraction. Am. J. Physiol. 277, E513–E520.

Volpi, E., Mittendorfer, B., Rasmussen, B.B., Wolfe, R.R., 2000. The response of muscle protein anabolism to combined hyperaminoacidemia and glucose-induced hyperinsulinemia is impaired in the elderly. J. Clin. Endocrinol. Metab. 85, 4481–4490.

Wall, B.T., van Loon, L.J., 2013. Nutritional strategies to attenuate muscle disuse atrophy. Nutr. Rev. 71, 195–208.

Wan, E.S., Cho, M.H., Boutaoui, N., Klanderman, B.J., Sylvia, J.S., Ziniti, J.P., et al., 2011. Genome-wide association analysis of body mass in chronic obstructive pulmonary disease. Am. J. Respir. Cell Mol. Biol. 45, 304–310.

Weekes, C.E., Emery, P.W., Elia, M., 2009. Dietary counselling and food fortification in stable COPD: a randomised trial. Thorax 64, 326–331.

Weisberg, J., Wanger, J., Olson, J., Streit, B., Fogarty, C., Martin, T., et al., 2002. Megestrol acetate stimulates weight gain and ventilation in underweight COPD patients. Chest 121, 1070–1078.

Wilkinson, S.B., Tarnopolsky, M.A., Macdonald, M.J., Macdonald, J.R., Armstrong, D., Phillips, S.M., 2007. Consumption of fluid skim milk promotes greater muscle protein accretion after resistance exercise than does consumption of an isonitrogenous and isoenergetic soy-protein beverage. Am. J. Clin. Nutr. 85, 1031–1040.

Xu, Z.S., Bao, Z.Y., Wang, Z.Y., Yang, G.J., Zhu, D.F., Zhang, L., et al., 2012. The changes of ghrelin, growth hormone, growth hormone releasing hormone and their clinical significances in patients with chronic obstructive pulmonary disease. Zhonghua Nei Ke Za Zhi 51, 536–539.

Yamamoto, C., Yoneda, T., Yoshikawa, M., Fu, A., Takenaka, I., Kobayashi, A., et al., 1997. The relationship between a decrease in fat mass and serum levels of TNF-α in patients with chronic obstructive pulmonary disease. Nihon Kyobu Shikkan Gakkai Zasshi 35, 1191–1195.

Yen, E.H., Quinton, H., Borowitz, D., 2013. Better nutritional status in early childhood is associated with improved clinical outcomes and survival in patients with cystic fibrosis. J. Pediatr. 162, 530–535.

Yeh, S.S., DeGuzman, B., Kramer, T., 2002. Reversal of COPD-associated weight loss using the anabolic agent oxandrolone. Chest 122, 421–428.

CHAPTER

13

Amino Acids, Protein, and the Gastrointestinal Tract

M.J. Bruins[1], *K.V.K. Koelfat*[2] *and P.B. Soeters*[2]

[1]The Hague, The Netherlands [2]Maastricht University Medical Center, Maastricht, The Netherlands

13.1. INTRODUCTION

The gastrointestinal tract has far more functions than just acting as a passive receptacle taking up nutritional components and presenting these to the body. Even this function is very complex and still not completely defined.

The intestine also plays a crucial role in the intermediary metabolism of amino acids and produces, in close conjunction with the liver and peripheral (muscle) tissues, a substrate mix adapted to meet the individual's requirements for maintenance and growth and special requirements under conditions of starvation and metabolic stress (sepsis, trauma, surgery). In addition, the liver, the spleen, and the intestine harbor the largest part of our immune system and it is difficult to disentangle the utilization of our macronutrients by immune cells, hepatocytes, enterocytes, or stroma cells.

Glutamine has been shown to preserve intestinal integrity under conditions of impaired mucosal barrier function, however, not all critical ill patients may benefit from it. In particular, patients with liver and/or renal failure do not tolerate diets with high nitrogen content. Much debate revolves around the role of arginine in severe infection and trauma and whether gut-derived citrulline becomes limiting for renal arginine and, as such, NO production. Similarly glycine and taurine's roles in the conjugation of bile acids and the way in which they participate in enterohepatic cycling has not been completely elucidated.

Finally, it has become clear in recent decades that the proximal intestine produces substantial amounts of ammonia in contradistinction with the previous belief that this was exclusively due to bacterial action in distal parts of the intestine. The role of the quality of protein in the precipitation of liver failure in patients with liver cirrhosis has been elucidated in more detail. In this chapter we will first give a more detailed description of the role of the intestine in the issues mentioned above, discuss the biological value of protein/amino acids in bolus meals, and subsequently outline protein and amino acid metabolism in the whole organism in starvation and stress starvation.

13.2. GASTROINTESTINAL AMINO ACID AND PROTEIN METABOLISM IN HEALTH

Epithelial and immune cells of the intestine have a rapid renewal rate and are largely responsible for the amino acid metabolism observed. Amino acids that are taken up by the enterocytes can be incorporated into protein, or can be converted via transamination or hydroxylation into other amino acids or biosynthetic intermediates. Glucose and glutamine are quantitatively the most important substrates for gut epithelial cells and immune cells (Newsholme and Carrie, 1994; Newsholme, 2001; Stoll et al., 1999). In fasting pigs, glucose, glutamine, and

The Molecular Nutrition of Amino Acids and Proteins.
DOI: http://dx.doi.org/10.1016/B978-0-12-802167-5.00013-X

glutamate were taken up by the portal-drained viscera while lactate and alanine were released, providing substrate for gluconeogenesis in the liver (Bruins et al., 2003). Similarly in postoperative patients during fasting, as measured by arterial and portal venous concentration differences, glucose and glutamine were taken up by the portal-drained viscera and alanine released, while during glucose ingestion, both lactate and alanine were released (Bjorkman et al., 1990).

In all conditions of rapid cell proliferation including the intestine, glucose and glutamine play important roles as building blocks for many cell and matrix components. Glucose in the gut can be transaminated to alanine or converted to pyruvate and lactate. Pyruvate can be introduced into the TCA-cycle via the formation of oxaloacetate and is only partly oxidized in the TCA-cycle after formation of acetyl-coA, when abundant amounts of carbohydrate are ingested (Fig. 13.1).

In starving conditions glucose oxidation is limited but glycolysis is active yielding intermediates that can branch off at several sites. The glycolysis intermediate glucose-6-phosphate can enter the pentose phosphate pathway to produce NADPH to support biosynthesis and to produce ribose-5-phosphate for nucleotides. NADPH is a crucial reducing equivalent, maintaining redox balance, ensuring oxidative bursts when dealing with bacteria or debris and supporting the synthesis of lipids and other products. The pentose phosphate pathway is also a cycle, part of its intermediates also being reintroduced into the glycolytic pathway. Thus, the pentose phosphate pathway of glucose carbon skeletons is essential for safeguarding biosynthetic capacity in proliferating cells. Lactate is produced from glucose in starvation but also in all conditions of rapid cell renewal (as in the gut) and released from the gut to act as a substrate for renewed glucose formation in the liver. At the end of the glycolytic pathway, pyruvate is produced which can be transaminated with glutamic or aspartic acid, producing alanine and α-ketoglutarate or oxaloacetate. Glutamic acid is derived from deamidation of glutamine simultaneously yielding ammonia (Newsholme and Carrie, 1994; Windmueller, 1982).

Both glucose and glutamine have anaplerotic functions during rapid cell proliferation: glucose via pyruvate to oxaloacetate and glutamine via glutamate to α-ketoglutarate. An important function of the TCA cycle is the provision of intermediates that serve as substrates for the biosynthesis of other molecules (eg, bases for

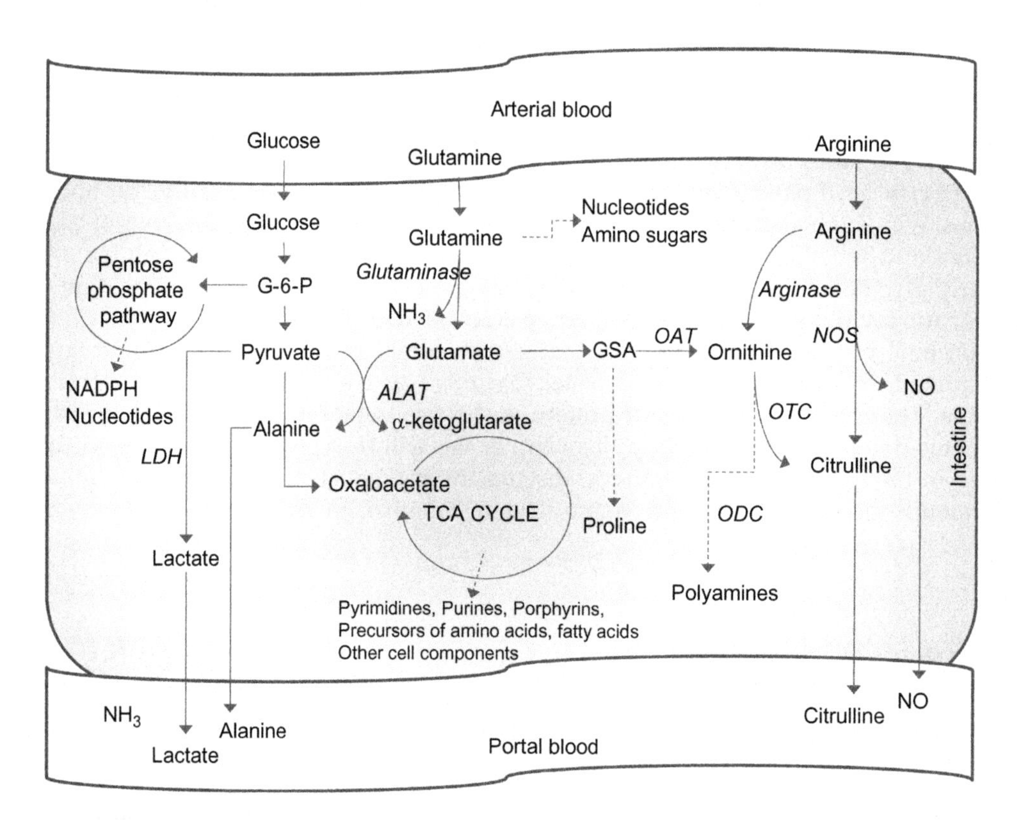

FIGURE 13.1 Main amino acid metabolism and gluconeogenic substrate production in the small intestine. *LDH*, lactate dehydrogenase; *ALAT*, alanine transaminase; *OAT*, ornithine aminotransferase; *ODC*, ornithine decarboxylase; *OTC*, ornithine transcarbamylase; *NOS*, nitric oxide synthase; *TCA*, tricarboxylic acid.

pyrimidines and purines) indispensable in rapidly proliferating enterocytes or immune cells. Moreover, intermediates both from the glycolytic pathway as well as from the TCA-cycle serve as precursors of nonessential amino acids.

Glutamine, via conversion to glutamic acid, can serve as substrate for proline and ornithine production. Ornithine can serve as substrate for polyamines that are essential for the integrity of intestinal epithelial barrier function. High rates of glutamine metabolism in the enterocytes (and immune cells in inflammation) may reflect the substantial requirements of glutamine for synthesis of nucleic acids, proline, and polyamines, due to the rapid turnover of enterocytes and immune cells (Newsholme and Carrie, 1994). In lymphocytes (and very likely in most rapidly proliferating cells), glutamine, via catabolic metabolism involving malate dehydrogenase can generate considerable amounts of NADPH for cell requirements.

Cell renewal and protein turnover rates of the gastrointestinal tract are high, particularly in the mucosa. In piglets, the mucosa constitutes approximately two-thirds of the intestinal mass (Che et al., 2010). Dietary amino acids are the primary source of amino acids for the intestinal mucosa, and enteral feeding is obligatory for maintenance of intestinal mucosal mass and integrity and for recovery after mucosal damage. Absence of enteral nutrition during prolonged fasting leads to reduced protein mass in gut tissues from reduced protein synthesis and increased protein degradation, especially in the small intestine (Burrin et al., 1991a,b; McNurlan and Garlick, 1979; Samuels et al., 1996). In 32-h fasted pigs, decreased protein synthesis in the portal-drained viscera was observed (Bruins et al., 2003). In critically ill patients, fasting was associated with mucosal atrophy, which may be related to their catabolic state (Hernandez et al., 1999). Parenteral nutrition also results in gut atrophy, probably because interaction with luminal nutrients is required to maintain mucosal mass. Transition from enteral to parenteral nutrition in pigs induced a rapid decrease in intestinal blood flow, cell proliferation, and protein synthesis, and led to villous atrophy (Niinikoski et al., 2004).

Notably, the portal-drained viscera (7%), of which 80–90% comprises intestinal tissue, and the liver (3%) together comprise only 10% of whole-body tissue mass as compared to muscle that comprises 36% of body mass (Elowsson and Carlsten, 1997). However, the portal-drained and liver tissues have a relatively high protein synthesis rate *per gram of organ tissue*: 3- and 13-fold higher, respectively, than protein synthesis in muscle (calculated by multiplying protein synthesis in the hindquarter by 2). The turnover of most acute phase proteins, and very likely cellular and matrix proteins, is far more rapid than of myofibrillar protein. The protein synthesis rate in the organs decreased approximately by half during fasting, but the relative contribution of the portal-drained viscera (~23%), liver (~39%), and muscle (~38%), to the whole-body protein synthesis remained relatively constant (Bruins et al., 2000, 2002). A similar high contribution of protein synthesis of splanchnic organs to whole-body protein turnover was observed in fed rats using a primed dose labeled leucine; the liver, stomach, and small intestine together accounted for 43% of total body protein turnover (McNurlan and Garlick, 1980).

13.3. THE FIRST-PASS EFFECT OF A BOLUS MEAL

13.3.1 The Art of the Meal and the Quality of Protein

The quality of a protein has generally been considered to depend on its amino acid composition and its digestibility. However, even when these characteristics are optimal, substantial differences have been found in net protein gain and urea formation, depending on several factors.

When a bolus meal consists exclusively of protein, casein achieves better whole-body net protein gain and less urea formation than whey (Boirie et al., 1997a). Most likely this is caused by clotting of casein (a slow protein) in the stomach, consequently leading to slow digestion and protracted entrance into the duodenum as evidenced by tapered appearance of amino acids in the portal vein, shown in a multicatheterized pig model. In contradistinction, portal amino acid appearance is much faster but of shorter duration after a bolus meal with whey (a fast protein). Simultaneously net nitrogen balance is positive and urea production low with casein and neutral with high urea production with whey (Boirie et al., 1997a). In support of this interpretation is the finding that whey achieves better intestinal protein accumulation and lower urea production when administered as successive small boluses (Dangin et al., 2001). Along similar lines, addition of maltodextrin to a protein meal leads to a slower and tapered release of amino acids in the portal vein and achieves better protein accretion and lower urea formation than a full meal containing casein (Deutz et al., 1995).

A better quality of the amino acid composition of the protein also resulted in a higher retention of amino acid in the portal-drained viscera and a slower rate of amino acid appearance in the portal vein (Deutz et al., 1998). A higher quality of the protein in terms of amino acid profile, has been shown to result in lower liver urea production in animals (Deutz et al., 1998; Ten Have et al., 2012; Tujioka et al., 2011). Proteins lacking an amino acid are poorly retained in the intestine, leading to high urea levels, as shown by ingesting blood meals lacking isoleucine (Deutz et al., 1991; see Section 13.10).

The degree of protein hydrolysis is another factor that affects amino acid retention across the portal-drained viscera; ingestion of amino acids or peptides lead to a more rapid appearance of amino acids in the portal vein, more urea formation, and reduced postprandial peripheral nitrogen availability than when the full protein with similar amino acid composition is administered (Deglaire et al., 2009; Morifuji et al., 2010; Ziegler et al., 1998). Although studies in healthy adults suggest a benefit of intact versus hydrolyzed proteins on the anabolic response in the muscle, more long-term studies are warranted to allow for firm conclusions. Also, more research is required to understand the effects of whole versus hydrolyzed protein on anabolic response under conditions of anabolic resistance. In burn patients, an intact protein diet resulted in less weight loss, more nitrogen retention, and more liver protein than a peptide-based diet (Beaufrère et al., 1996). In patients after abdominal surgery, however, no short-term benefit of intact versus hydrolyzed protein was found (Ziegler et al., 1998).

These findings led to the view that slow absorption of meal protein-derived amino acids and the anabolic stimulus of ingesting carbohydrate promote the utilization of amino acid carbon and nitrogen for anabolic purposes (protein, bases, other nitrogenous products) and limit urea formation in healthy individuals. On the other hand, substantial and rapid digestion and uptake of protein-derived amino acid nitrogen, exceed the capacity of these pathways and therefore lead to increased urea formation.

13.3.2 The Labile Protein Pool Hypothesis

The splanchnic tissues (intestine, stomach, spleen, and pancreas) derive amino acids from both the diet and from the systemic circulation. The first-pass temporary extraction of amino acids across the portal-drained viscera and the liver determines the immediate postprandial availability in the systemic circulation, and thus the supply to other organs. In addition to temporary retention of protein-derived products of a bolus meal by the portal-drained viscera, in the process of continuous turnover of protein, nitrogen-containing products and meal-derived amino acids are taken up in the liver. The amount of amino acids retained from the diet in the intestinal tract may amount to as much as 30–65% depending on the amount of protein in the diet and the composition of the meal. In healthy pigs, the extraction of amino acids by the portal-drained viscera from an enterally infused casein meal was about 40–65% (Deutz et al., 1998). Temporary retention of free amino acids in the gut wall seems unlikely. Alternative hypothetical explanations may be that after initial digestion of the protein, the resulting amino acids are utilized by bacterial proliferation, by synthesis of mucous protein and secretory proteins in the gut wall, or by rapid cell proliferation (mucosa cells, lymphocytes). The dietary amino acids that are retained in the portal-drained viscera without being degraded may function as a temporary protein pool, sometimes called the "labile protein pool." This pool will during and after the meal in turn be degraded ensuring a prolonged delivery of amino acids to the portal vein. Higher amounts lead to a lower proportion retained and vice versa, but maximal absolute amounts will probably be similar when intake is sufficient to reach the maximal synthetic level. This is one potential factor responsible for the so-called first-pass effect. The synthesis of pancreatic secretory proteins or splenic immunocytes cannot account for this phenomenon because these organs do not receive blood and its constituents directly from the lumen of the intestine but exclusively from the arterial circulation. It is also unknown whether amino acids used for the production of secretory proteins in the intestine are derived from the lumen or from the arterial circulation.

Despite these uncertainties, it is clear that spreading the release of meal-derived amino acids into the portal and systemic circulation over time promotes efficient utilization of these amino acids. A complete meal containing intact protein with a balanced amino acid composition and calories ensures efficient utilization whereas administration of exclusively amino acids, small peptides or fast proteins leads to increased degradation and less efficient utilization for whole-body protein synthesis, despite short-term increases in the fractional protein synthesis of protein in muscle biopsies (Ten Have et al., 2007).

In the last decade, much interest has been shown and attributed to this first-pass effect. Initially it was shown that fractional synthesis rates of protein found in vitro in muscle biopsies 3 h after exercise and protein ingestion, were lower in the elderly than in a young age group. This was called anabolic resistance. However, it was also found that in the elderly a larger proportion of the protein in a bolus meal remained in the splanchnic area compared to younger individuals, which most likely reflects an elevated acute phase response in elderly people and is called the first-pass effect (Boirie et al., 1997b; Volpi et al., 1999). This casts doubt on the presence of anabolic resistance, because due to this first-pass effect a lower proportion of the protein meal reaches peripheral (muscle) tissues. This explanation is supported by the finding that increasing the amount of protein ingested in the elderly improves protein synthesis in muscle biopsies (Cuthbertson et al., 2005; Glover et al., 2008; Houston et al., 2008; Katsanos et al., 2006; Timmerman et al., 2012). In another study the role of the first-pass effect was questioned showing that despite increased first-pass extraction, similar amounts of protein reached the peripheral tissues of the elderly compared to the younger age group (Volpi et al., 1999). This view may not be correct because they also show that fat-free mass per length (as an indicator of muscle mass) was lower in the aged group and fat mass higher than in the younger group, while nevertheless receiving an identical amount of amino acids. This means that muscle was exposed to relatively higher quantities of amino acids per unit of weight in the elderly than in the younger age group. This is in line with suggestions in the literature that aged people should consume more protein than younger people to limit muscle losses. Some support for this comes from the Health, Aging, and Body Composition study, showing in a cohort of community-dwelling men and women aged between 70 and 79 years that participants in the highest quintile of protein intake lost approximately 40% less lean mass than did those in the lowest quintile (Houston et al., 2008). Ideally proof for this suggestion should come from randomized controlled intervention trials.

13.4. GASTROINTESTINAL AMINO ACID AND PROTEIN METABOLISM IN STRESS CONDITIONS

Little is known about the effect of stress conditions (infection, trauma, or surgery) on amino acid or protein metabolism in the splanchnic tissues of patients as this requires invasive catheterization techniques. Data on the changes in amino acid and protein metabolism in the gastrointestinal tract of patients in response to stress conditions are therefore scarce. Most data are derived from experimental animal models of infection or trauma although findings are often equivocal due to differences in experimental conditions and the type and phase of injury and the cytokines and hormones produced.

Glucose and glutamine are among the substrates most investigated under stress conditions as they are key substrates for gastrointestinal epithelial cells and immune cells. During experimental hyperdynamic endotoxemia, glucose consumption and alanine and lactate release across the portal-drained viscera were increased (Bruins et al., 2003). This suggests that glucose becomes an important substrate in the gastrointestinal tract, and the pyruvate that is produced may be partly metabolized to lactate as part of the Cori-cycle and partly converted to alanine. Lactate and alanine both serve as substrates for increased gluconeogenesis in the liver under endotoxemic conditions. The uptake of glutamine by the portal-drained viscera was however reduced. Diminished glutamine uptake by the gut and reduced gut glutaminase activity have also been observed in other animal models of endotoxemic conditions (Ardawi et al., 1991; Austgen et al., 1991; Bruins et al., 2003; Souba et al., 1990a). In one of the few studies in which amino acid portal vein-arterial differences were measured, the glutamine extraction in septic patients was diminished by 75% as compared to surgical control patients (Souba et al., 1990b). In sepsis patients, glutamine transport across brushborder membranes from jejunal specimens was decreased in vitro (Salloum et al., 1991). Also in depleted patients with gastrointestinal cancer (van der Hulst et al., 1997) and patients with chronic renal failure (Tizianello et al., 1980), intestinal glutamine extraction, measured by intestinal venous-arterial concentration differences, was decreased. The conditionally indispensable amino acids, arginine, cysteine, tyrosine, and threonine were proposed to become essential in the gastrointestinal tract during intestinal inflammation (Remond et al., 2011). Although intestinal inflammation in an experimental ileitis did not affect cysteine uptake and release by the portal-drained viscera (Remond et al., 2011) the utilization of threonine was increased (Remond et al., 2009). Possibly, the latter may be explained by the high mucin synthesis that was observed, as proteins of mucins secreted in the lumen by the epithelium contain high amounts of threonine.

In an experimental pig model of hyperdynamic endotoxemia, 24-h endotoxin infusion did not significantly affect the protein synthesis rate in the portal-drained viscera (assessed using primed-constant infusion of phenylalanine isotope). In another study with pigs receiving 8-h endotoxin infusion, intestinal protein synthesis (as measured by a flooding dose technique with labeled phenylalanine) increased fractional protein synthesis 14 h after the endotoxin infusion (Orellana et al., 2002). Discrepancies in observations concerning intestinal protein synthesis may result from differences in the experimental model used and the phase of the inflammatory response. In patients with inflammatory bowel disease, where the gastrointestinal tract is the main inflammatory site, fractional protein synthesis rate in colorectal tissue biopsies was increased as compared to normal colon tissue (Heys et al., 1992).

The effect of surgical stress on jejunal protein synthesis was studied in rats (as measured by phenylalanine or leucine isotope flooding dose technique). Immediately after surgical stress, no effect on protein synthesis was observed (Bakic et al., 1988; Preedy et al., 1988). However, in postsurgical patients protein synthesis (as measured using a flooding dose of leucine isotope), the fractional protein synthesis in biopsies of colon and ileal mucosa between 8 to 10 days after surgery was higher as compared to controls (Rittler et al., 2000, 2001). The increased protein synthesis observed in intestinal tissue of postoperative patients was hypothesized to result from temporary ischemia or trauma in the small bowel during surgery, inducing a compensatory response upon restoration of circulation and stimulation of crypt cell proliferation in the small intestine (Rijke et al., 1976).

The response to injury, infection, or other stress conditions is associated with hypermetabolism, and has been described in *"The production of a substrate mix to support host response in stress."* Peripheral protein catabolism and metabolic activity in the hepatosplanchnic region are augmented. Amino acids are released from peripheral tissues and shifted to the liver for promotion of hepatic protein synthesis, gluconeogenesis, and urea synthesis. In fact, a large portion of the hypermetabolism of the septic individual can be attributed to enhancement of splanchnic metabolism. A few studies in patients provided insight into substrate differences across the splanchnic area. Accelerated uptake of lactate, amino acids and free fatty acids across the splanchnic tissue and concomitant splanchnic glucose output were generally observed and considered a hallmark of stress conditions. For instance, these typical alterations in splanchnic metabolism were observed in volunteers receiving a single endotoxin bolus (Fong et al., 1990) but also in burn patients with and without bacteremia (Wilmore et al., 1980) and in cardiac surgery patients (Suojaranta-Ylinen et al., 1998).

A key observation in many stress conditions is the stimulation of net protein breakdown in muscle and the stimulation of protein synthesis in the splanchnic area, mainly liver. In a pig model of endotoxin infusion mimicking systemic inflammation, the liver rather than the portal-drained viscera was accountable for the substantially increased amount of amino acids extracted by the splanchnic area (Bruins et al., 2000, 2003). At 24-h endotoxin infusion, the higher amino acid extraction in the liver was accompanied by both increased glucose production and protein synthesis in the liver (as measured using glucose and valine isotopes) (Bruins et al., 2003). It was calculated that the relative contribution of protein synthesis rate in the portal-drained viscera and muscle to the whole-body protein synthesis decreased during endotoxemia (Bruins et al., 2000, 2002) whereas the contribution of liver protein synthesis to whole-body protein synthesis increased from approximately 39% to 52%. In stressed states, liver protein turnover accounts to a larger extent to total body protein turnover than in non-stressed states, due to activation of the synthesis of acute phase proteins and very likely cellular and matrix proteins accounting for a significant increased proportion of protein turnover compared to portal-drained viscera and peripheral tissues (muscle, skin, bone etc.) (Morais et al., 2000). One to four days after the pigs recovered from endotoxemia and received enteral feeding, both liver amino acid extraction and protein synthesis were still elevated (Bruins et al., 2000). Even though the muscle protein turnover was increased, the release of amino acids from muscle was not elevated as compared to control animals suggesting that amino acids derived from the diet rather than from muscle are used for the liver acute phase response during recovery. During recovery, also a significant accelerated total amino acid release from the diet into the portal vein was observed as compared to control animals (Fig. 13.2).

The observation that dietary amino acids appear faster in the portal vein after inflammatory conditions may imply that endotoxin-induced inflammation interferes with temporary retention of amino acids in the intestine. Although this finding warrants further confirmation, we speculate that the higher portal total amino acid appearance reflects the impaired intestinal function including impaired secretion of enzymes, impaired motility, and dysfunctional utilization of oxygen and nutrients at the cellular level. Consequently, luminal amino acids are poorly retained by the intestine under postseptic feeding conditions which may result in accelerated urea production. This may necessitate slow continuous enteral administration of small amounts of nutrients rather than bolus feeding. In critically ill patients bolus meals are poorly tolerated leading both to secretory and osmotic diarrhea by diminished peristalsis and bacterial overgrowth (Glatzle et al., 2004; Thompson, 1997), insufficient enzyme secretion, secretion of fluids instead of absorption (Field, 2006), internalization of cotransporters (Field, 2006), and possibly other factors.

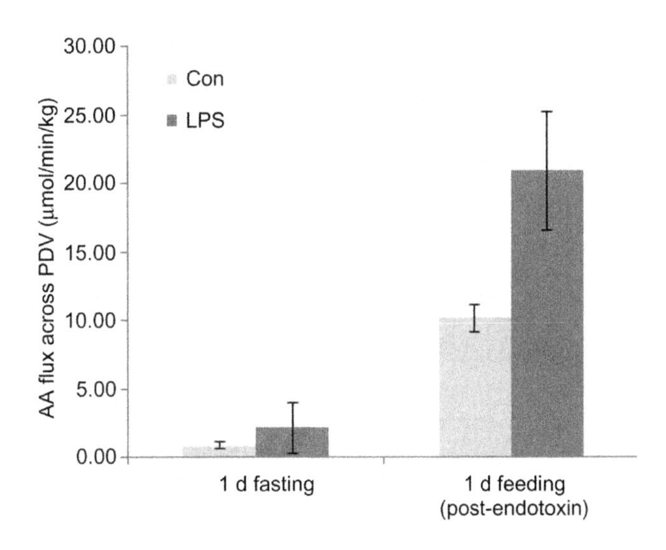

FIGURE 13.2 Net release of total amino acids across the portal-drained viscera during fasting and 24-h endotoxin (LPS) or saline (control) infusion, and subsequent 24-h feeding the following day.

13.5. THE PRODUCTION OF A SUBSTRATE MIX TO SUPPORT HOST RESPONSE IN STRESS

In response to stress (injury, burns, and infection) the changes in protein kinetics (see Section 13.6) lead to a net release of amino acids from peripheral tissues to tissues, involved in host response. During the fed and fasted state, these processes were suggested to be controlled by substrate-mediated hormone responses, whereas during stress conditions, the metabolic responses to hormones and cytokines are no longer substrate controlled (Atkinson and Worthley, 2003). The glucose-lactate-glucose (Cori) cycle and the very similar glucose-alanine-glucose cycle fluxes between peripheral muscle tissues and liver are accelerated but do not produce new glucose. New glucose is derived from amino acids released from peripheral muscle tissues and glycerol released from breakdown of adipose tissue (Bruins et al., 2003; McGuinness and Spitzer, 1984). Although liver gluconeogenesis is increased in stress conditions, most of the glucose produced comes from lactate (derived from glucose) and from alanine (also derived from glucose and a transamination step) and only a modest amount (less than 30%) is newly produced. Lactate is derived from muscle, the intestine. and in fact from almost all tissues, including tissues involved in host response, whereas the liver and kidney consume more lactate than they release (Karlstad et al., 1982; Vary et al., 1995; Wilmore et al., 1980). The kidney also produces glucose from glutamine as part of a complex glucose-glutamine-glucose cycle across several organs. In starvation most of the modest amount of glucose newly formed is produced in the kidney but in stress-starvation it is likely that most of the new production occurs in the liver. All rapidly proliferating tissues (wound, immune system) take up glucose nonoxidatively, utilizing the carbon skeleton for the synthesis cell elements and matrix. This also occurs in the liver and in the intestine, which complicates the interpretation of fluxes of glucose and amino acids across these organs because they play dual roles in intermediary metabolism and cell proliferation during host response. In rapidly proliferating tissues, glucose is therefore used to support proliferation and other activities but part of it is glycolytically degraded to lactate which is exported.

After trauma and infection, the metabolism changes to meet requirements for the host response. A sequence of events can be outlined, although the different elements of the sequence most likely occur simultaneously and are induced by cytokines and damage associated molecular patterns derived from the wounded or infected area.

Fig. 13.3 gives a simplified representation of synthesis of substrate mix in stressed conditions. Data presented are acquired from in vivo data in humans or pigs. It is likely that other nonessential amino acids also produced in excess of their composition in muscle protein.

Animal models of injury and sepsis demonstrate that net peripheral (muscle) amino acid and lactate fluxes are accelerated. This is due to increased protein degradation exceeding protein synthesis (Biolo et al., 2000). Most amino acids are taken up by the liver in a net fashion. Glutamic acid, which is released into the circulation and taken up by the liver in large quantities, is used in part for glucose formation (hand in hand with urea

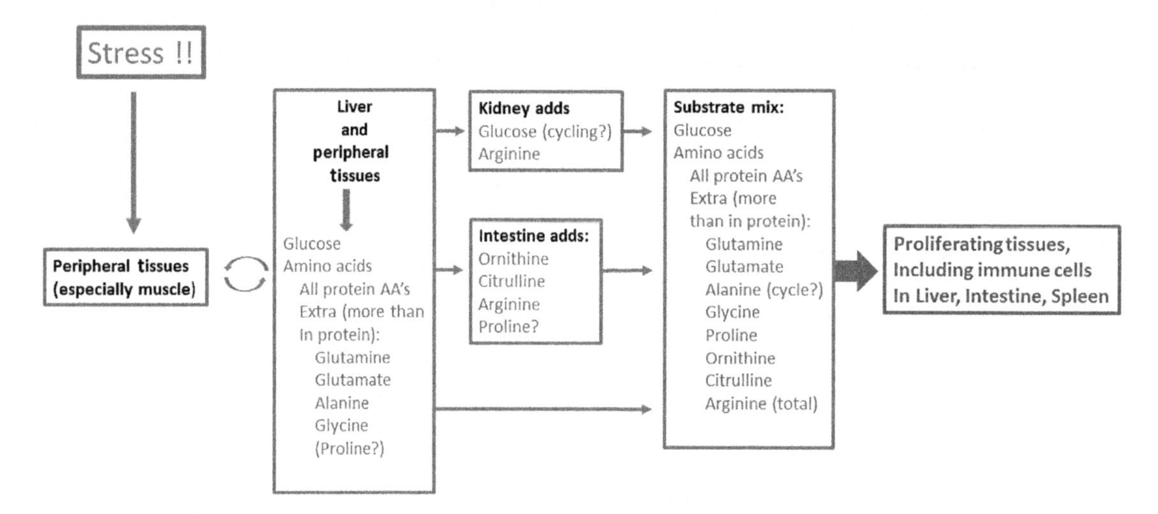

FIGURE 13.3 Simplified representation of synthesis of substrate mix in stressed conditions. Data presented are acquired from in vivo data in humans or pigs. It is likely that other nonessential amino acids also produced in excess of their composition in muscle protein.

formation), which is released in substantially increased amounts into the circulation. Another part is used for liver protein synthesis, most likely consisting of acute phase proteins but also for immune (Kupffer) cell proliferation and activation (Bruins et al., 2003).

Glutamic acid and glucose are taken up in peripheral tissue but quantitatively most importantly in muscle, in which an amino acid mix is produced which is significantly different from the amino acid composition of muscle protein. Especially glutamine, alanine, glycine, and possibly proline are released in quantities exceeding the amounts that could have been derived from muscle protein breakdown and that therefore must have been newly formed. Based on stoichiometric considerations much of the glucose carbon skeleton must have been used for the production of glutamine, alanine, and possibly serine and glycine, whereas glutamic acid may play similar roles but also functions as an amino-group donor, a role that is also played by the branched-chain amino acids derived from muscle protein breakdown. It is not exactly known where and to what degree extra proline is produced. New glutamine production in muscle has received much attention and may be quantitatively more important than other tissues but the brain and the lung may also produce new glutamine. Trans-pulmonary amino acid fluxes have been measured but data are not very reliable due to very high flow and consequently low trans-pulmonary concentration differences, which are therefore difficult to measure reliably (Hulsewe et al., 2003). Net glutamine production by the lung may therefore be modest or negligible but, interestingly significant uptake occurs in the presence of pulmonary infiltrates, most likely supporting the inflammatory process (Hulsewe et al., 2003). Importantly in stressed conditions as part of a complex glucose-glutamine-glucose cycle a substantial amount of glutamine is serving as a precursor of glucose in the kidney but it is not certain to which degree this production is part of a cycle or true new formation.

The efflux of the amino acid mix mentioned in the previous paragraph meets a large part of the requirements of proliferating tissues. Most immune cells and cells in growing tissues require glutamine and glucose as anaplerotic substrates producing in these cells building stones for proliferation and the production of reducing equivalents. Although it is certain that the amino acid mix produced is useful and probably supplies most of the building stones it is not certain whether the cells themselves contribute to an optimal mix. In view of the requirements of glucose and glutamine this seems plausible, because they are the starting point of the pathways described in the previous paragraphs.

Glycine is an important supplier of methyl groups and nitrogen for purines and pyrimidines. Alanine, glycine and proline are present in matrix (collagen) in far higher quantities than in muscle protein although the turnover of collagen is even slower than muscle protein but there also appears to be a rapidly turning over pool (Babraj et al., 2005). Supportive evidence that glutamine and glucose actually serve in vivo to promote the inflammatory and posttraumatic response comes from our observations that after trauma the spleen increases its uptake of glutamine and glucose (Deutz et al., 1992). A similar observation is made in patients with pulmonary infiltrates (Hulsewe et al., 2003) and another observation comes from traumatized pigs in which the liver changes its glutamine uptake from negligible in the nonstressed state to substantial in the stressed state (Bruins et al., 2003). At variance with increased glutamine efflux from muscle (and lung) during injury and infection (Bruins et al., 2003;

Karinch et al., 2001; Remesy et al., 1997) in one publication in very severe stress, decreased de novo glutamine synthesis has been observed (Biolo et al., 2000).

In this paragraph we have outlined the chain of events leading to the production of a substrate mix serving host response. We have described the substrate mix produced in the hindquarter and subsequently what happens in the liver by assessing flow and portal-venous and arterio-venous concentration differences allowing to calculate flux. Although the liver is a crucial organ in host response and receives the result of intestinal metabolism via the portal vein, the role of the portal-drained viscera (intestine, stomach, spleen, and pancreas) is similarly important which has been outlined in previous paragraphs.

13.6. PROTEIN METABOLISM IN STRESS STARVATION

Both skeletal muscle and splanchnic protein metabolism are profoundly affected in metabolically stressed (sepsis, trauma, surgery) patients. One of the hallmark metabolic responses to stress is increased protein turnover, for instance after burns (Prelack et al., 2010), surgery (Tashiro et al., 1985), and trauma (Winthrop et al., 1987). This process involves redistribution of amino acids from peripheral muscle tissues where protein degradation predominates, to central tissues where protein synthesis predominates. Consequently, there is a net release of amino acids from peripheral tissues (which are therefore catabolic) to central tissues like the liver or intestine for acute phase protein synthesis, cell proliferation and other processes, but also to the wound and organs involved in the immune response (which are therefore anabolic).

Although whole-body protein synthesis and degradation are both accelerated during the metabolic stress response in previously well-nourished individuals, the latter predominates. The resulting whole-body nitrogen loss and muscle protein wasting are characteristic for the response to injury, burns, and infections and cannot or only marginally be inhibited by nutrition. During the hypermetabolic state in endotoxemic pigs, these changes in whole-body protein turnover were confirmed leading to a net negative nitrogen balance (Bruins et al., 2003). In the recovery phase, muscle protein turnover is substantially increased, with protein synthesis increased more than degradation, resulting in net muscle protein synthesis (Bruins et al., 2000).

13.7. SUBSTRATE METABOLISM IN STRESS STARVATION TO SPARE PROTEIN

In starvation, the adipose tissue-derived fatty acids are almost exclusively oxidized to furnish energy, which diminishes the necessity to oxidize protein/amino acids. Amino acids are essential to overcome sepsis, trauma, or surgery, producing a multitude of proteins, purines, pyrimidines and other products, crucial for survival. If exclusively protein would be utilized for energy generation and no protein sparing would occur, after one week 40% of protein mass would be lost, which empirically has been shown to lead to mortality. In starvation in a healthy adult with normal body composition, net nitrogen losses decrease after a few days to between 6 and 7 g per day which corresponds to approximately 40 g of dry protein/day (Benedict, 1912). The contribution of gluconeogenesis to glucose production increases from about 50% after 14 h, to almost 100% after 42 h of fasting. The amount of glucose that can be synthesized from this amount of protein is approximately 20 g (nitrogen should be subtracted and not all amino acids have carbon skeletons that can be transformed to glucose). Lipid-derived glycerol also contributes to the formation of new glucose and has been calculated to amount to between 10 and 15 g of new glucose in a normal adult. If these 20 + (10–15) = 30–35 g would be oxidized, this would cover less than 10% of total energy requirement. Therefore, when glycogen stores are depleted after 1–2 days the individual must rely on fatty acid oxidation for approximately 90%. The 30–35 g of glucose produced by gluconeogenesis from amino acids or glycerol would cover only 30–35% of the energy requirement of the brain, but the brain has been shown to be able to utilize ketone bodies (Owen et al., 1967).

In stress starvation the same principles of protein sparing are valid as in starvation. However, nitrogen losses are amounting to roughly double the amount in starvation, that is, 14 g/24 h in adults. This implies that lipid oxidation still covers minimally 85% of energy requirements. Fourteen grams of nitrogen amount to approximately 85 g of (largely muscle) protein, which would deliver approximately 40 g of glucose, which, amounting to 55–60 g when gluconeogenesis from glycerol is taken into account and if oxidized wholly in the brain, would cover approximately half of energy requirement in the brain.

Despite this limited amount of glucose available, the nervous system, but also the renal medulla, bone marrow, and red blood cells have been suggested to be dependent on glucose as their primary energy substrate, which can only be true to a very limited degree in view of lack of glucose and is not necessary, because ketone body formation continues in severely ill starving patients (Beylot et al., 1989). The apparent uptake of glucose by the brain very likely is part of Cori-cycling in which lactate is formed which is exported from the brain, precluding terminal oxidation of glucose.

Another reason for doubt that the small amount of new glucose that is produced is terminally oxidized in a regular manner (via glycolysis, dehydrogenation of pyruvate to acetyl-CoA and introduction of acetyl-CoA into the TCA cycle) for terminal oxidation, is that pyruvate dehydrogenation is inhibited in starvation and in stress. This is, together with inhibition of glycogen synthesis and stimulation of gluconeogenesis, the underlying cause of insulin resistance typical for starvation and stress/starvation, during feeding in conditions of stress and during growth (Soeters and Soeters, 2012).

Less well known is that the little glucose produced may play far more crucial roles than energy generation, including the production of NADPH and ribose in the pentose-phosphate pathway, the introduction of pyruvate into the TCA-cycle via oxaloacetate (anaplerosis) and branching off of intermediates of the cycle (cataplerosis) producing substrates for cell proliferation and redox balance (Fig. 13.4).

These pathways are crucial for supporting cell proliferation and matrix deposition, and to maintain redox balance. In these pathways, glucose acts as a conditionally essential substrate, and glucose is used only for those metabolic processes that solely can be fulfilled by this substrate, thereby limiting terminal oxidation and promoting lipid oxidation to spare protein mass and prolong survival. Also, in the fed state and stress glucose oxidation is preferentially utilized for anabolic purposes, but when abundantly ingested ($\geq$50% of energy requirement) some glucose will also be terminally oxidized.

In all these conditions there is classical insulin resistance to channel glucose derived products into other pathways crucially required to build biomass (cells and matrix), to serve immune function, and to regulate redox balance by the production of NADPH (see above).

13.8. THE ROLE OF INDIVIDUAL AMINO ACIDS IN THE GASTROINTESTINAL TRACT

13.8.1 Citrulline

The intestine is the main source of circulating citrulline, primarily derived from enteral or circulating glutamine (Ligthart-Melis et al., 2008; Tomlinson et al., 2011; Wu et al., 1994). Plasma citrulline has been proposed as a biomarker for small intestinal enterocyte mass, absorptive function, and severity of intestinal failure in short bowel patients (Crenn et al., 2000, 2003). Findings of this pioneering work have later been confirmed for enterocyte mass estimates in small bowel transplantation (Pappas et al., 2001), Crohn disease (Papadia et al., 2007), and enterocyte toxicity resulting from chemotherapy or radiotherapy (Barzal et al., 2014; Lutgens et al., 2004). As a consequence of reduced bowel mass, reduced whole-body turnover of citrulline is also observed in these patients (Vahedi et al., 2001). In rats, 75% enterectomy reduced the renal arginine production from citrulline but whole-body arginine production was not affected (Dejong et al., 1998). However, although citrulline levels and functional small bowel enterocyte mass significantly correlate at a population level, citrulline is less precise at an individual level.

Citrulline uptake concomitantly increased with arginine release in the kidneys, suggesting that the kidney acts as main site of de novo synthesis of arginine (Brosnan, 1987; Buijs et al., 2014; Dhanakoti et al., 1990; van de Poll et al., 2004). However, renal de novo arginine production only contributes approximately 16–20% to the total production of arginine as measured in catheterized pigs, and in healthy humans whole-body de novo arginine production was 17% (Luiking et al., 2009). Despite the relative small contribution of renal arginine production, enteral (Houdijk et al., 1998) and parenteral (Buijs et al., 2014) supplementation of glutamine to patients was able to increase plasma levels of citrulline and arginine. In fasting patients, a positive venous-arterial balance of arginine across the intestine suggests that this organ, besides the kidneys, is capable of producing arginine (Tizianello et al., 1980; van der Hulst et al., 1997). Its magnitude suggests that this arginine cannot be derived from net proteolysis.

13.8.2 Glutamine Supplementation

Some amino acids, particularly glutamine, have been investigated for their effect on the small bowel in starvation or stress. The addition of glutamine to parenteral nutrition was associated with reduced hospital mortality and length-of-stay in patients with diagnoses ranging from pancreatitis, trauma, burns to sepsis (Wischmeyer et al., 2014).

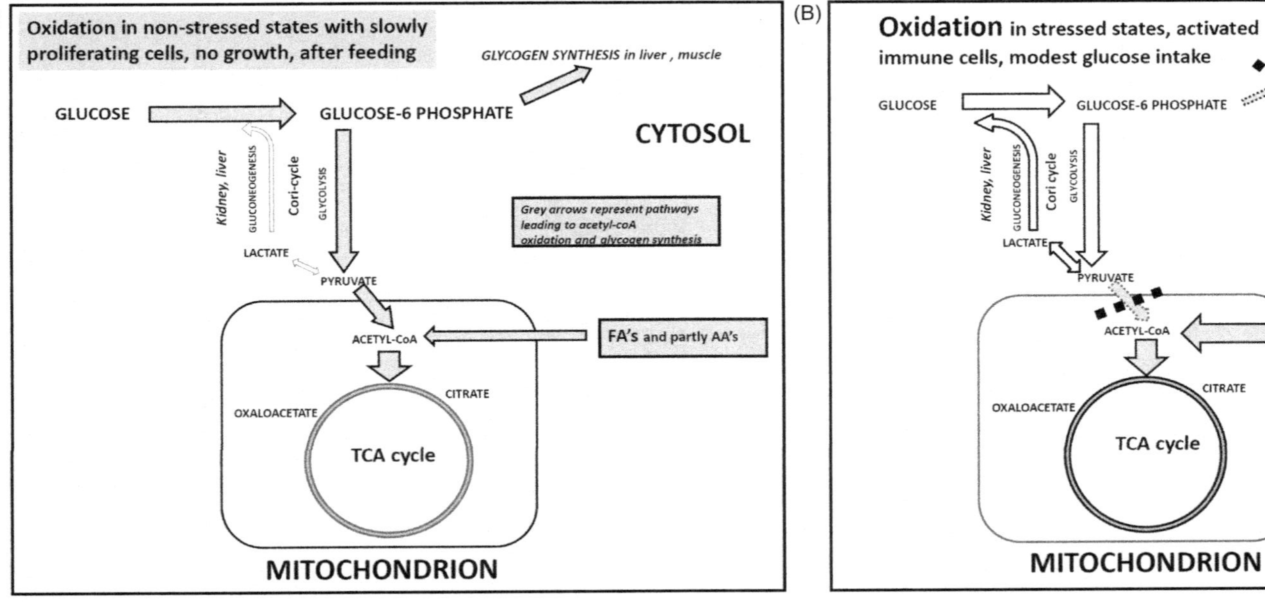

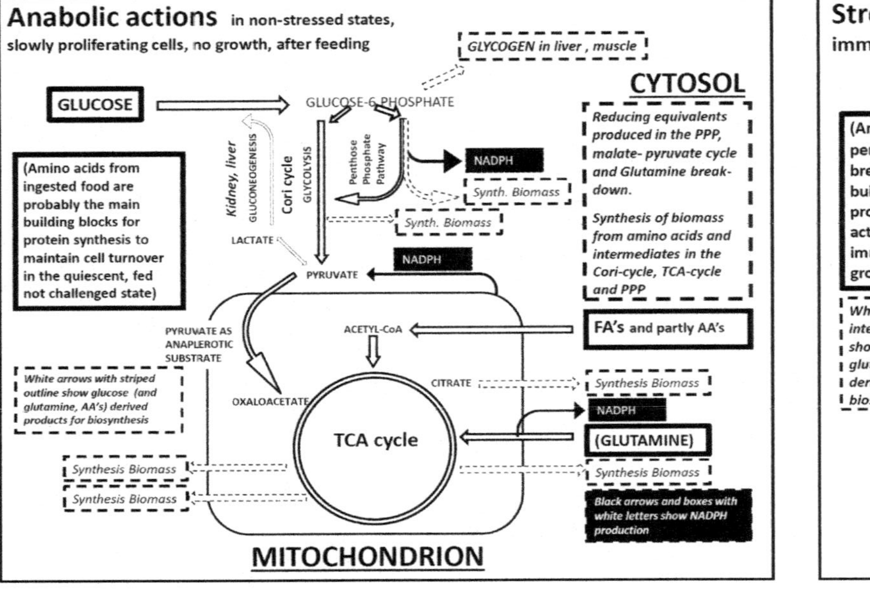

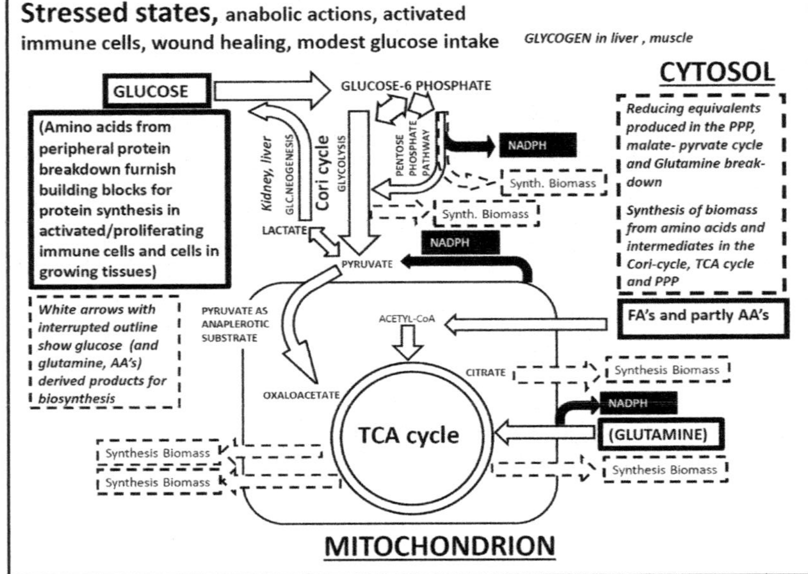

FIGURE 13.4 (A) Schematic representation of glucose utilization in non-stressed states in the whole body after a balanced meal. Upper panel. Oxidation of glucose and fatty acids (Gray arrows, boxes and circle with black outline.). Lower panel. Utilization of glucose to synthesize biomass, immune cells and reducing equivalents. (White arrows, circle and boxes with black border) producing building blocks in Cori-cycle, Pentose Phosphate Pathway and TCA-cycle (White arrows, boxes with dashed outlines). NADPH production indicated with black arrows and boxes, white letters. (B) Schematic representation of glucose utilization in stressed states in the whole body with modest glucose intake. Upper panel. Oxidation of glucose and fatty acids (Grey arrows, boxes and circle with black outline). Very little glucose is oxidized. Energy generation relies mainly on fatty acid oxidation. Instead glucose is utilized for anabolic purposes and redox regulation. Lower panel. Substantial utilization of glucose to synthesize biomass, immune cells. (White arrows, circle and boxes with black border) producing building blocks in Cori-cycle, Pentose Phosphate Pathway and TCA-cycle (White arrows, boxes with dashed outline). Production of reducing equivalents (NADPH) is indicated with black arrows and boxes, white letters.

In trauma patients that were given glutamine-containing enteral nutrition for at least five days following their injuries, glutamine significantly reduced the incidence of pneumonia, bacteremia, and sepsis (Houdijk et al., 1998). Souba et al. proposed that glutamine plays an important role in maintaining a healthy gut and supporting the metabolic response to injury and infection (Souba et al., 1985a, 1990b). Supplemental use of glutamine, either in oral, enteral, or parenteral form, was suggested to maintain mucosal integrity by increasing intestinal villous height, and stimulate gut mucosal cellular proliferation and thereby to prevent hyperpermeability and bacterial translocation, involved in bacteremia and sepsis. Results from numerous animal studies confirm that addition of glutamine to parenteral or enteral nutrition can maintain intestinal function; parenteral and luminal glutamine prevented bacterial translocation in experimental animal models of intestinal resection (Liu et al., 1997) and small-bowel transplantation (Yuzawa et al., 2000) and luminal glutamine maintained the intestinal barrier in animal models of acute pancreatitis (Zou et al., 2010), endotoxemia (Dugan and McBurney, 1995), and sepsis (Gianotti et al., 1995). Evidence for glutamine on bowel integrity in patients has been less well studied; in a study in patients predominantly suffering from exacerbations of inflammatory bowel disease, addition of glutamine to parenteral nutrition improved intestinal integrity as measured by dual sugar test (van der Hulst et al., 1993, 1998). Villus height was negatively influenced by nutritional state (van der Hulst et al., 1993, 1998). In a subsequent study it was found that inflammatory activity was associated with impaired mucosal permeability and plasma glutamine concentrations whereas undernutrition was rather associated with impaired villous height (Hulsewe et al., 2004). These findings suggest that glutamine may particularly affect the gut under conditions of rapid cell proliferation, such as in inflammation, as rapidly dividing immune and mucosa cells utilize increased amounts of glutamine and glucose (Brand et al., 1986). Not in all settings, a beneficial effect of glutamine administration on intestinal function has been observed. For instance, prophylactic glutamine in chemotherapy generally reduced the duration of diarrhea (Sun et al., 2012) but not its severity (Rotovnik Kozjek et al., 2011; Sun et al., 2012). Efficacy of supplemental glutamine on clinical endpoints has been difficult to demonstrate, and glutamine supplementation had no effect on morbidity and mortality in critically ill or major surgery patients (Tao et al., 2014), or on hospital stay and survival in liver-transplanted patients (Langer et al., 2012). The degree of the glutamine deficiency (although not easy to establish) and the presence of inflammation may determine the effect of glutamine supplementation on clinical outcome parameters.

13.9. THE ROLE OF THE INTESTINE IN BILE SALT AND AMINO ACID METABOLISM

The enterohepatic circulation provides an efficient way to eliminate bilirubin and toxic compounds from the body and to conserve bile salts that undergo recycling multiple times (Hofmann, 2009). Bile salts, classically known as emulsifiers, are the major constituents of bile and important for digestion and absorption of dietary lipids and fat soluble vitamins. Bile salts also exert newly identified actions in glucose and lipid metabolism, energy expenditure, thermogenesis, intestinal barrier function, and negative feedback of its own biosynthesis by controlling bile salt-sensing nuclear and membrane-bound receptors (Fang et al., 2015; Schaap et al., 2014). Bile salts are synthesized from cholesterol through multiple enzymatic conversions initiated by the rate-limiting enzyme cholesterol-7-α-hydroxylase (CYP7A1). In humans, the end products of the bile salt biosynthetic pathway are the primary bile salts cholic acid (CA) and chenodeoxycholic acid (CDCA). The final step before entering the enterohepatic cycle is coupling of bile salts with a nonessential amino acid, taurine or glycine by N-acyl amidation (conjugation) (Hofmann and Hagey, 2008).

Conjugation of bile salt is a critical process to increase the solubility of bile salts, requiring Na-linked transport across the enterocyte. Conjugated bile salts cannot passively diffuse across the mucosal border along the biliary and intestinal tract (Hofmann and Hagey, 2008). The majority of recirculating bile salts are in conjugated form. In fact, amino acid conjugation of bile salts is a highly efficient process as infused unconjugated bile salts in isolated rat liver were almost completely conjugated with amino acids after the first-pass through the liver (Gurantz et al., 1991). Moreover, unconjugated bile salts were undetectable in fasting and post prandial jejunal fluid (Tangerman et al., 1986). The conjugation of bile salts with taurine or glycine is catalyzed by one liver-specific enzyme, bile acid-CoA N-acyltransferase (BAAT) (Falany et al., 1994; Pellicoro et al., 2007). Bile salts derived from de novo synthesis and deconjugated bile salts reentering the hepatocyte are required to shuttle to hepatocyte peroxisomes where (re)conjugation with either taurine or glycine occurs before subsequent (re)secretion into bile (Pellicoro et al., 2007; Rembacz et al., 2010). Bile salts control their own conjugation process by regulating biosynthesis of taurine and BAAT expression by activating the nuclear receptor, the farnesoid X receptor (FXR) (Kerr et al., 2014; Pircher et al., 2003). Genetic defects in BAAT have been associated with impaired absorption of fat soluble vitamins and familial hypercholanemia, indicating the essential task of the liver to (re)conjugate the bile salts (Carlton et al., 2003; Setchell et al., 2013).

Bile salts entering the intestinal lumen are reabsorbed primarily in the terminal ileum through active transport facilitated by the apical sodium bile salt transporter (ASBT) (Craddock et al., 1998; Hruz et al., 2006). In a rare human study, venous and arterial blood was sampled from different parts of the intestine, showing significantly increased concentrations of glycine and taurine in the venous effluent in the jejunum (Table 13.1; van der Hulst et al., 1997).

TABLE 13.1 Venous-arterial differences across isolated parts of the human gut

	Tumor		Colon		Ileum		Jejunum	
Ammonia	39.3	(9.9)[a]	30.1	(8.2)[a]	33.2	(7.3)[a]	65.4	(6.5)[a,j]
Glucose	−0.6	(0.3)[a]	−0.3	(0.3)	0.2	(0.2)[c]	0.1	(0.2)[e]
Lactate	0.5	(0.1)[a,g]	0.2	(0.1)[a]	0.1	(0.1)[b]	0.0	(0.1)[i]
Glutamate	0.9	(1.9)	2.0	(1.9)	20.6	(7.5)[a,f]	25.7	(5.6)[a,d]
Asparagine	0.5	(0.7)	−0.9	(0.9)	7.2	(3.5)[b,h]	3.4	(1.5)
Serine	−14.6	(7.8)[a]	−11.2	(8.5)	−4.2	(9.8)[c]	−2.7	(3.6)
Glutamine	−22.8	(5.0)[a]	−32.2	(7.5)[a]	−44.8	(5.6)[a,f]	119.7	(10.9)[a,j]
Glycine	4.63	(6.0)[b]	−2.9	(7.0)	15.2	(8.3)[a,h]	16.0	(3.0)[a,i]
Threonine	−3.4	(1.9)	−5.0	(2.4)[b]	2.3	(3.1)[f]	1.1	(2.7)
Histidine	−1.1	(1.0)	−3.2	(1.7)	−0.7	(1.5)	−2.1	(1.4)
Citrulline	−0.1	(0.4)	−0.2	(0.5)	8.4	(1.7)[a,f]	30.4	(4.0)[a,j]
Alanine	23.8	(4.0)[a]	20.2	(5.0)[a]	44.4	(11.4)[a]	75.3	(10.3)[a,j]
Taurine	7.9	(4.8)[b]	6.8	(2.3)[a]	31.1	(13.0)[a,f]	13.4	(3.7)[a]
Arginine	1.6	(1.2)	0.5	(1.4)	10.5	(3.3)[a,f]	9.3	(2.0)[a,d]
α-Amino butyric acid	0.4	(0.3)	−0.3	(0.5)	1.3	(0.7)[b]	1.4	(0.7)[b]
Tyrosine	−0.8	(1.1)	−1.9	(1.4)	2.2	(1.7)[b,f]	2.0	(1.4)
Valine	−2.5	(2.2)	−7.1	(3.3)[b]	3.6	(2.7)[a,h]	2.1	(3.6)
Methionine	0.1	(0.4)	−0.2	(0.5)	1.3	(0.5)[b,h]	1.3	(0.6)
Isoleucine	−0.9	(1.1)	−2.8	(1.8)	1.8	(1.6)	1.9	(1.4)
Phenylalanine	−0.3	(0.9)	−1.4	(1.2)	1.4	(1.4)	1.2	(1.2)
Tryptophan	−0.4	(0.4)	−0.5	(0.6)	0.7	(0.5)	0.1	(0.6)
Leucine	−1.8	(2.7)	−3.8	(3.3)	4.1	(2.8)	3.8	(2.8)
Lysine	2.1	(5.5)	9.3	(8.4)	2.0	(6.6)	−5.9	(9.7)
Branched-chain amino acids	−5.1	(5.3)	−13.6	(8.0)	9.5	(6.7)[h]	7.7	(7.6)
Sum amino acids	−6.7	(31.4)	−40.2	(45.5)	108.2	(65.4)	57.9	(40.3)

Significant extraction or release (a and b):
[a]*$p < 0.01$.*
[b]*$p < 0.05$.*
Significant differences between groups (b-j):
[c]*$p < 0.05$, ileum vs tumor.*
[d]*$p < 0.01$, jejunum vs colon and tumor.*
[e]*$p < 0.01$, jejunum vs tumor.*
[f]*$p < 0.05$, ileum vs colon and tumor.*
[g]*$p < 0.01$, tumor vs colon, ileum, and jejunum.*
[h]*< 0.05, ileum vs colon.*
[i]*$p < 0.05$, jejunum vs colon.*
[j]*$p < 0.05$, jejunum vs colon, ileum, and tumor.*
Venous minus arterial concentrations of amino acids, ammonia, glucose, and lactate in colon, ileum, jejunum and tumor. Concentrations are given in μmol/L except for lactate and glucose (mmol/L). +, extraction; −, release.
Data as published by van der Hulst et al. 1997

Most likely this reflects total taurine and glycine concentrations including taurine bound in conjugated bile acids. This suggests that already in the jejunum modest but significant reabsorption of conjugated bile acids occurs. A larger part is reabsorbed in the ileum. Bile salts which escape reclamation by the ileocyte (up to five percent) undergo further various microbial processing in the distal part of the small intestine and colon (Hofmann and Hagey, 2008). Bile salts are deconjugated by bacteria catalyzed by bile salt hydrolases (BSH) into deconjugated primary bile salts. Bacterial hydrolysis of conjugated bile salts yields free taurine and glycine, which are released in the portal vein and taken up by the liver completing the enterohepatic cycle of these amino acids, unless metabolized by bacteria. Deconjugated bile salts are largely passively reabsorbed from the colon and recycle back to the liver for reconjugation. Alternatively, bacteria convert primary deconjugated bile salts to secondary bile salts (so-called "damaged bile salts") that also partially recycle back to the liver for subsequent reconjugation (repair) (Hofmann and Hagey, 2014).

In critically ill patients, serum glycine and taurine bile salts conjugates were increased, whereas serum unconjugated bile salts levels remained stable (Vanwijngaerden et al., 2011). Moreover, excretion of conjugated bile salts tended to shift towards systemic blood rather than bile (Vanwijngaerden et al., 2014). These findings indicate that in critical illness conjugation is enhanced to "detoxify" circulating bile salts.

13.10. ROLE OF THE INTESTINE IN AMINO ACID METABOLISM IN LIVER FAILURE

Before the 1970s ammonia generation was considered to take place in the colon by and to have a deleterious influence on the brain leading to neurologic and mental disturbances if present in the systemic circulation in excess. This led to treatment with antibiotics (Neomycin and Bacitracin) to hopefully diminish bacterial action and improve mental state. Some surgeons went so far to perform colectomies to eradicate most of the bacterial flora hopefully diminishing ammonia generation (Singer et al., 1965). Early findings that intestinal ammonia production also occurred in vitro and in germfree animals changed the focus to metabolic generation of ammonia by the mucosa of the intestine (Kline and Nance, 1969; Matsutaka et al., 1973; Nance and Kline, 1971; Windmueller and Spaeth, 1974).

13.10.1 Metabolic (in Contradistinction With Bacterial) Ammonia Generation in Different Parts of the Intestine

Even before ammonia generation in germfree animals had been published, Japanese and German biochemists had found that intestinal tissue produced ammonia in intestinal tissue in vitro especially when incubated with glutamine (Matsutaka et al., 1973; Windmueller and Spaeth, 1974). This led to studies in rats and pigs, confirming in more detail that small bowel very actively metabolized amino acids (Souba et al., 1985b; van Berlo et al., 1989). Glutamine uptake was substantial, leading to ammonia, glutamic acid, alanine, citrulline, and arginine release in the portal circulation (Bruins et al., 2003; Deutz et al., 1992). This was later confirmed by van der Hulst et al. in postabsorptive humans in which during surgery blood was sampled from the venous effluent of different parts of the intestine, showing that especially the jejunum was responsible for these findings (Table 13.1; van der Hulst et al., 1997). In this situation the colon produced substantially less ammonia, only partly corresponding with glutamine uptake, suggesting that a minor part of ammonia production in the colon potentially may be derived from bacterial action (van der Hulst et al., 1997). However, in the nourished state, ammonia production by bacteria may be higher but this could not be measured because at the time of surgery the patients were not fed, supposedly to diminish the risk of aspiration during intubation.

13.10.2 Effect of Portal-Systemic Shunting on Systemic Ammonia Levels in Liver Failure

In healthy individuals all ammonia produced in the gut is completely cleared by the liver either by formation of urea or of glutamine. Ammonia present in the circulation in healthy people is produced in the kidney largely from the breakdown of glutamine. In liver failure, nitrogen homeostasis including hepatic urea synthesis capacity is severely disturbed, leading to an impaired capacity to detoxify gut-derived ammonia contributing to the development of systemic hyperammonemia. In liver disease normal portal flow is often obstructed leading to the formation of collaterals, shunting blood from the portal vein around the liver to the systemic circulation (superior caval vein). Nevertheless, the amount of ammonia produced in the intestine was shown in one study in patients with stable cirrhosis to be of a similar magnitude as the amount cleared by the liver, whereas the ammonia produced by the kidney reaches the systemic circulation and may lead to increased ammonia and glutamine levels in the brain (Olde Damink et al., 2003). On the other hand it appears that when the circulation is overloaded with amino acids in catabolic states or when increased or unbalanced amounts of amino-nitrogen are ingested ammonia production is increased. This leads in patients with severely diminished liver function to encephalopathy (Olde Damink et al., 2007). It is well known that patients with liver insufficiency do not tolerate large amounts of ingested protein well, rising their ammonia levels and risking to develop hepatic encephalopathy.

13.10.3 Effect of the Amount and Quality of Protein on Ammonia Production in Liver Failure

In particular, the amount and quality of protein and the composition of other nutrients in the meal play decisive roles in ammonia generation by the gut. It is well known that patients with liver insufficiency do not tolerate large amounts of ingested protein well, rising their ammonia levels and, in patients with severely compromised liver function, increasing the risk to develop hepatic encephalopathy. There is a hierarchy in the "toxicity" of different proteins, plant proteins being the least, animal protein intermediate, and blood protein being the most toxic. In an effort to elucidate the underlying mechanism aiming to understand why especially variceal bleeding in the intestine often precipitates encephalopathy in patients with liver disease, blood cells were administered in the gut of multicatheterized pigs and the exchange of amino acids, ammonia, and carbohydrate across the gut was assessed. Amino acids derived from blood cells (hemoglobin) appeared much faster in the portal vein (shown by high portal-arterial amino acid concentration differences and extremely high plasma levels of all amino acids except isoleucine) than a normal protein (van Berlo et al., 1989). This was explained by the complete absence of isoleucine in adult hemoglobin, precluding rapid synthesis of protein and its temporary retention in the gut wall and its subsequent slow release. This showed that the normal first-pass effect across the gut (see Section 13.3) was negligible for hemoglobin and that almost all amino acids were immediately absorbed and released in the portal vein. Simultaneously ammonia production was strongly increased due to the abundant presence of amino acids although the precise pathways are not exactly defined. Proof for the deleterious effect of the absence of isoleucine in red blood cells comes from the observation that supplementing isoleucine during a simulated intestinal bleed ameliorates the first-pass effect leading to temporary retention most likely due to rapid protein synthesis in the gut wall and slow release in the circulation, leading to more optimal usage of the amino acid meal for protein synthesis (Olde Damink et al., 2007). However, administering very high quantities of protein of normal composition in a short time period also precipitates encephalopathy in patients with severely diminished liver function.

In conclusion, nutritional measures to improve nutritional state includes the administration of multiple small mixed meals over the day, containing protein with a high biological value, leading to substantial first-pass retention. This spreads appearance of amino-nitrogen in the portal vein and subsequently in part in the systemic circulation, leading to optimal protein synthesis and low and spread urea production and consequently preventing overburdening the liver with amino acids and ammonia. In addition any inflammatory insult should be prevented or acutely treated to diminish inflammation induced catabolism.

References

Ardawi, M.S., Majzoub, M.F., Kateilah, S.M., Newsholme, E.A., 1991. Maximal activity of phosphate-dependent glutaminase and glutamine metabolism in septic rats. J. Lab. Clin. Med. 118, 26–32.

Atkinson, M., Worthley, L.I., 2003. Nutrition in the critically ill patient: part I. Essential physiology and pathophysiology. Crit. Care Resusc. 5, 109–120.

Austgen, T.R., Chen, M.K., Flynn, T.C., Souba, W.W., 1991. The effects of endotoxin on the splanchnic metabolism of glutamine and related substrates. J. Trauma 31, 742–751 (discussion 751–742).

Babraj, J.A., Smith, K., Cuthbertson, D.J.R., Rickhuss, P., Dorling, J.S., Rennie, M.J., 2005. Human bone collagen synthesis is a rapid, nutritionally modulated process. J. Bone Miner. Res. 20, 930–937.

Bakic, V., MacFadyen, B.V., Booth, F.W., 1988. Effect of elective surgical procedures on tissue protein synthesis. J. Surg. Res. 44, 62–66.

Barzal, J.A., Szczylik, C., Rzepecki, P., Jaworska, M., Anuszewska, E., 2014. Plasma citrulline level as a biomarker for cancer therapy-induced small bowel mucosal damage. Acta Biochim. Pol. 61, 615–631.

Beaufrère, B., Collin, C., Chipponi, J., Pezet, D., Ferrier, C., Fauquant, J., 1996. Effects of enteral whole protein and very short chain peptides on protein kinetics in post operative patients. Clin. Nutr. 16, 12.

Benedict, F.G., 1912. An experiment on a fasting man. Science 35, 865.

Beylot, M., Guiraud, M., Grau, G., Bouletreau, P., 1989. Regulation of ketone body flux in septic patients. Am. J. Physiol. 257, E665–E674.

Biolo, G., Fleming, R.Y., Maggi, S.P., Nguyen, T.T., Herndon, D.N., Wolfe, R.R., 2000. Inhibition of muscle glutamine formation in hypercatabolic patients. Clin. Sci. (Lond) 99, 189–194.

Bjorkman, O., Eriksson, L.S., Nyberg, B., Wahren, J., 1990. Gut exchange of glucose and lactate in basal state and after oral glucose ingestion in postoperative patients. Diabetes 39, 747–751.

Boirie, Y., Dangin, M., Gachon, P., Vasson, M.P., Maubois, J.L., Beaufrere, B., 1997a. Slow and fast dietary proteins differently modulate postprandial protein accretion. Proc. Natl. Acad. Sci. U.S.A. 94, 14930–14935.

Boirie, Y., Gachon, P., Beaufrere, B., 1997b. Splanchnic and whole-body leucine kinetics in young and elderly men. Am. J. Clin. Nutr. 65, 489–495.

Brand, K., Leibold, W., Luppa, P., Schoerner, C., Schulz, A., 1986. Metabolic alterations associated with proliferation of mitogen-activated lymphocytes and of lymphoblastoid cell lines: evaluation of glucose and glutamine metabolism. Immunobiology 173, 23–34.

Brosnan, J.T., 1987. The 1986 Borden award lecture. The role of the kidney in amino acid metabolism and nutrition. Can. J. Physiol. Pharmacol. 65, 2355–2362.

Bruins, M.J., Soeters, P.B., Deutz, N.E., 2000. Endotoxemia affects organ protein metabolism differently during prolonged feeding in pigs. J. Nutr. 130, 3003–3013.

Bruins, M.J., Soeters, P.B., Lamers, W.H., Deutz, N.E., 2002. L-arginine supplementation in pigs decreases liver protein turnover and increases hindquarter protein turnover both during and after endotoxemia. Am. J. Clin. Nutr. 75, 1031–1044.

Bruins, M.J., Deutz, N.E., Soeters, P.B., 2003. Aspects of organ protein, amino acid and glucose metabolism in a porcine model of hypermetabolic sepsis. Clin. Sci. (Lond) 104, 127–141.

Buijs, N., Brinkmann, S.J., Oosterink, J.E., Luttikhold, J., Schierbeek, H., Wisselink, W., et al., 2014. Intravenous glutamine supplementation enhances renal de novo arginine synthesis in humans: a stable isotope study. Am. J. Clin. Nutr. 100, 1385–1391.

Burrin, D.G., Davis, T.A., Fiorotto, M.L., Reeds, P.J., 1991a. Stage of development and fasting affect protein synthetic activity in the gastrointestinal tissues of suckling rats. J. Nutr. 121, 1099–1108.

Burrin, D.G., Ferrell, C.L., Eisemann, J.H., Britton, R.A., 1991b. Level of nutrition and splanchnic metabolite flux in young lambs. J. Anim. Sci. 69, 1082–1091.

Carlton, V.E., Harris, B.Z., Puffenberger, E.G., Batta, A.K., Knisely, A.S., Robinson, D.L., et al., 2003. Complex inheritance of familial hypercholanemia with associated mutations in TJP2 and BAAT. Nat. Genet. 34, 91–96.

Che, L., Thymann, T., Bering, S.B., LE Huërou-Luron, I., D'Inca, R., Zhang, K., et al., 2010. IUGR does not predispose to necrotizing enterocolitis or compromise postnatal intestinal adaptation in preterm pigs. Pediatr. Res. 67, 54–59.

Craddock, A.L., Love, M.W., Daniel, R.W., Kirby, L.C., Walters, H.C., Wong, M.H., et al., 1998. Expression and transport properties of the human ileal and renal sodium-dependent bile acid transporter. Am. J. Physiol. 274, G157–G169.

Crenn, P., Coudray-Lucas, C., Thuillier, F., Cynober, L., Messing, B., 2000. Postabsorptive plasma citrulline concentration is a marker of absorptive enterocyte mass and intestinal failure in humans. Gastroenterology 119, 1496–1505.

Crenn, P., Vahedi, K., Lavergne-Slove, A., Cynober, L., Matuchansky, C., Messing, B., 2003. Plasma citrulline: a marker of enterocyte mass in villous atrophy-associated small bowel disease. Gastroenterology 124, 1210–1219.

Cuthbertson, D., Smith, K., Babraj, J., Leese, G., Waddell, T., Atherton, P., et al., 2005. Anabolic signaling deficits underlie amino acid resistance of wasting, aging muscle. FASEB J. 19, 422–424.

Dangin, M., Boirie, Y., Garcia-Rodenas, C., Gachon, P., Fauquant, J., Callier, P., et al., 2001. The digestion rate of protein is an independent regulating factor of postprandial protein retention. Am. J. Physiol. Endocrinol. Metab. 280, E340–E348.

Deglaire, A., Fromentin, C., Fouillet, H., Airinei, G., Gaudichon, C., Boutry, C., et al., 2009. Hydrolyzed dietary casein as compared with the intact protein reduces postprandial peripheral, but not whole-body, uptake of nitrogen in humans. Am. J. Clin. Nutr. 90, 1011–1022.

Dejong, C.H., Welters, C.F., Deutz, N.E., Heineman, E., Soeters, P.B., 1998. Renal arginine metabolism in fasted rats with subacute short bowel syndrome. Clin. Sci. (Lond) 95, 409–418.

Deutz, N.E., Reijven, P.L., Bost, M.C., van Berlo, C.L., Soeters, P.B., 1991. Modification of the effects of blood on amino acid metabolism by intravenous isoleucine. Gastroenterology 101, 1613–1620.

Deutz, N.E., Reijven, P.L., Athanasas, G., Soeters, P.B., 1992. Post-operative changes in hepatic, intestinal, splenic and muscle fluxes of amino acids and ammonia in pigs. Clin. Sci. (Lond) 83, 607–614.

Deutz, N.E., Ten Have, G.A., Soeters, P.B., Moughan, P.J., 1995. Increased intestinal amino-acid retention from the addition of carbohydrates to a meal. Clin. Nutr. 14, 354–364.

Deutz, N.E., Bruins, M.J., Soeters, P.B., 1998. Infusion of soy and casein protein meals affects interorgan amino acid metabolism and urea kinetics differently in pigs. J. Nutr. 128, 2435–2445.

Dhanakoti, S.N., Brosnan, J.T., Herzberg, G.R., Brosnan, M.E., 1990. Renal arginine synthesis: studies in vitro and in vivo. Am. J. Physiol. 259, E437–E442.

Dugan, M.E., McBurney, M.I., 1995. Luminal glutamine perfusion alters endotoxin-related changes in ileal permeability of the piglet. J. Parenter. Enteral Nutr. 19, 83–87.

Elowsson, P., Carlsten, J., 1997. Body composition of the 12-week-old pig studied by dissection. Lab. Anim. Sci. 47, 200–202.

Falany, C.N., Johnson, M.R., Barnes, S., Diasio, R.B., 1994. Glycine and taurine conjugation of bile acids by a single enzyme. Molecular cloning and expression of human liver bile acid CoA:amino acid N-acyltransferase. J. Biol. Chem. 269, 19375–19379.

Fang, S., Suh, J.M., Reilly, S.M., Yu, E., Osborn, O., Lackey, D., et al., 2015. Intestinal FXR agonism promotes adipose tissue browning and reduces obesity and insulin resistance. Nat. Med. 21, 159–165.

Field, M., 2006. T cell activation alters intestinal structure and function. J. Clin. Invest.2580–2582.

Fong, Y.M., Marano, M.A., Moldawer, L.L., Wei, H., Calvano, S.E., Kenney, J.S., et al., 1990. The acute splanchnic and peripheral tissue metabolic response to endotoxin in humans. J. Clin. Invest. 85, 1896–1904.

Gianotti, L., Braga, M., Radaelli, G., Mariani, L., Vignali, A., Di Carlo, V., 1995. Lack of improvement of prognostic performance of weight loss when combined with other parameters [see comments]. Nutrition 11, 12–16.

Glatzle, J., Leutenegger, C.M., Mueller, M.H., Kreis, M.E., Raybould, H.E., Zittel, T.T., 2004. Mesenteric lymph collected during peritonitis or sepsis potently inhibits gastric motility in rats. J. Gastrointest. Surg. 8, 645–652.

Glover, E.I., Phillips, S.M., Oates, B.R., Tang, J.E., Tarnopolsky, M.A., Selby, A., et al., 2008. Immobilization induces anabolic resistance in human myofibrillar protein synthesis with low and high dose amino acid infusion. J. Physiol. 586, 6049–6061.

Gurantz, D., Schteingart, C.D., Hagey, L.R., Steinbach, J.H., Grotmol, T., Hofmann, A.F., 1991. Hypercholeresis induced by unconjugated bile acid infusion correlates with recovery in bile of unconjugated bile acids. Hepatology 13, 540–550.

Hernandez, G., Velasco, N., Wainstein, C., Castillo, L., Bugedo, G., Maiz, A., et al., 1999. Gut mucosal atrophy after a short enteral fasting period in critically ill patients. J. Crit. Care 14, 73–77.

Heys, S.D., Park, K.G., McNurlan, M.A., Keenan, R.A., Miller, J.D., Eremin, O., et al., 1992. Protein synthesis rates in colon and liver: stimulation by gastrointestinal pathologies. Gut 33, 976–981.

Hofmann, A., Hagey, L., 2008. Bile acids: chemistry, pathochemistry, biology, pathobiology, and therapeutics. Cell. Mol. Life Sci. 65, 2461–2483.

Hofmann, A.F., 2009. The enterohepatic circulation of bile acids in mammals: form and functions. Front. Biosci. (Landmark Ed) 14, 2584–2598.

Hofmann, A.F., Hagey, L.R., 2014. Key discoveries in bile acid chemistry and biology and their clinical applications: history of the last eight decades. J. Lipid Res. 55, 1553–1595.

Houdijk, A.P., Rijnsburger, E.R., Jansen, J., Wesdorp, R.I., Weiss, J.K., McCamish, M.A., et al., 1998. Randomised trial of glutamine-enriched enteral nutrition on infectious morbidity in patients with multiple trauma [see comments]. Lancet 352, 772–776.

Houston, D.K., Nicklas, B.J., Ding, J., Harris, T.B., Tylavsky, F.A., Newman, A.B., et al., 2008. Dietary protein intake is associated with lean mass change in older, community-dwelling adults: the Health, Aging, and Body Composition (Health ABC) Study. Am. J. Clin. Nutr. 87, 150–155.

Hruz, P., Zimmermann, C., Gutmann, H., Degen, L., Beuers, U., Terracciano, L., et al., 2006. Adaptive regulation of the ileal apical sodium dependent bile acid transporter (ASBT) in patients with obstructive cholestasis. Gut 55, 395–402.

Hulsewe, K.W., van der Hulst, R.R., Ramsay, G., van Berlo, C.L., Deutz, N.E., Soeters, P.B., 2003. Pulmonary glutamine production: effects of sepsis and pulmonary infiltrates. Intensive Care Med. 29, 1833–1836.

Hulsewe, K.W., van der Hulst, R.W., van Acker, B.A., von Meyenfeldt, M.F., Soeters, P.B., 2004. Inflammation rather than nutritional depletion determines glutamine concentrations and intestinal permeability. Clin. Nutr. 23, 1209–1216.

Karinch, A.M., Pan, M., Lin, C.M., Strange, R., Souba, W.W., 2001. Glutamine metabolism in sepsis and infection. J. Nutr. 131, 2535S–2538S (discussion 2550S–2531S).

Karlstad, M.D., Raymond, R.M., Emerson Jr., T.E., 1982. Mechanism of forelimb skin and skeletal muscle glucose uptake during Escherichia coli endotoxin shock in the dog. Adv. Shock Res. 7, 101–115.

Katsanos, C.S., Kobayashi, H., Sheffield-Moore, M., Aarsland, A., Wolfe, R.R., 2006. A high proportion of leucine is required for optimal stimulation of the rate of muscle protein synthesis by essential amino acids in the elderly. Am. J. Physiol. Endocrinol. Metab. 291, E381–E387.

Kerr, T.A., Matsumoto, Y., Matsumoto, H., Xie, Y., Hirschberger, L.L., Stipanuk, M.H., et al., 2014. Cysteine sulfinic acid decarboxylase regulation: a role for farnesoid X receptor and small heterodimer partner in murine hepatic taurine metabolism. Hepatol. Res. 44, E218–E228.

Kline, D.G., Nance, F.C., 1969. Eck's fistula encephalopathy: long term studies in primates and germfree dogs. Surg. Forum 20, 358–360.

Langer, G., Grossmann, K., Fleischer, S., Berg, A., Grothues, D., Wienke, A., et al., 2012. Nutritional interventions for liver-transplanted patients. Cochrane Database Syst. Rev. 8, CD007605.

Ligthart-Melis, G.C., van de Poll, M.C., Boelens, P.G., Dejong, C.H., Deutz, N.E., van Leeuwen, P.A., 2008. Glutamine is an important precursor for de novo synthesis of arginine in humans. Am. J. Clin. Nutr. 87, 1282–1289.

Liu, Y.W., Bai, M.X., Ma, Y.X., Jiang, Z.M., 1997. Effects of alanyl-glutamine on intestinal adaptation and bacterial translocation in rats after 60% intestinal resection. Clin. Nutr. 16, 75–78.

Luiking, Y.C., Poeze, M., Ramsay, G., Deutz, N.E., 2009. Reduced citrulline production in sepsis is related to diminished de novo arginine and nitric oxide production. Am. J. Clin. Nutr. 89, 142–152.

Lutgens, L.C., Deutz, N., Granzier-Peeters, M., Beets-Tan, R., De Ruysscher, D., Gueulette, J., et al., 2004. Plasma citrulline concentration: a surrogate end point for radiation-induced mucosal atrophy of the small bowel. A feasibility study in 23 patients. Int. J. Radiat. Oncol. Biol. Phys. 60, 275–285.

Matsutaka, H., Aikawa, T., Yamamoto, H., Ishikawa, E., 1973. Gluconeogenesis and amino acid metabolism. 3. Uptake of glutamine and output of alanine and ammonia by non-hepatic splanchnic organs of fasted rats and their metabolic significance. J. Biochem. (Tokyo) 74, 1019–1029.

McGuinness, O.P., Spitzer, J.J., 1984. Hepatic glycerol flux after E. coli endotoxin administration. Am. J. Physiol. 247, R687–R692.

McNurlan, M.A., Garlick, P.J., 1979. Rates of protein synthesis in rat liver and small intestine in protein deprivation and diabetes. Proc. Nutr. Soc. 38, 133A.

McNurlan, M.A., Garlick, P.J., 1980. Contribution of rat liver and gastrointestinal tract to whole-body protein synthesis in the rat. Biochem. J. 186, 381–383.

Morais, J.A., Ross, R., Gougeon, R., Pencharz, P.B., Jones, P.J., Marliss, E.B., 2000. Distribution of protein turnover changes with age in humans as assessed by whole-body magnetic resonance image analysis to quantify tissue volumes. J. Nutr. 130, 784–791.

Morifuji, M., Ishizaka, M., Baba, S., Fukuda, K., Matsumoto, H., Koga, J., et al., 2010. Comparison of different sources and degrees of hydrolysis of dietary protein: effect on plasma amino acids, dipeptides, and insulin responses in human subjects. J. Agric. Food Chem. 58, 8788–8797.

Nance, F.C., Kline, D.G., 1971. Eck's fistula encephalopathy in germfree dogs. Ann. Surg. 174, 856–862.

Newsholme, E.A., Carrie, A.L., 1994. Quantitative aspects of glucose and glutamine metabolism by intestinal cells. Gut 35, S13–S17.

Newsholme, P., 2001. Why is L-glutamine metabolism important to cells of the immune system in health, postinjury, surgery or infection?. J. Nutr. 131, 2515S–2522S (discussion 2523S–2514S).

Niinikoski, H., Stoll, B., Guan, X., Kansagra, K., Lambert, B.D., Stephens, J., et al., 2004. Onset of small intestinal atrophy is associated with reduced intestinal blood flow in TPN-fed neonatal piglets. J. Nutr. 134, 1467–1474.

Olde Damink, S.W., Jalan, R., Deutz, N.E., Redhead, D.N., Dejong, C.H., Hynd, P., et al., 2003. The kidney plays a major role in the hyperammonemia seen after simulated or actual GI bleeding in patients with cirrhosis. Hepatology 37, 1277–1285.

Olde Damink, S.W., Jalan, R., Deutz, N.E., Dejong, C.H., Redhead, D.N., Hynd, P., et al., 2007. Isoleucine infusion during "simulated" upper gastrointestinal bleeding improves liver and muscle protein synthesis in cirrhotic patients. Hepatology 45, 560–568.

Orellana, R.A., O'Connor, P.M., Nguyen, H.V., Bush, J.A., Suryawan, A., Thivierge, M.C., et al., 2002. Endotoxemia reduces skeletal muscle protein synthesis in neonates. Am. J. Physiol. Endocrinol. Metab. 283, E909–E916.

Owen, O.E., Morgan, A.P., Kemp, H.G., Sullivan, J.M., Herrera, M.G., Cahill Jr., G.F., 1967. Brain metabolism during fasting. J. Clin. Invest. 46, 1589–1595.

Papadia, C., Sherwood, R.A., Kalantzis, C., Wallis, K., Volta, U., Fiorini, E., et al., 2007. Plasma citrulline concentration: a reliable marker of small bowel absorptive capacity independent of intestinal inflammation. Am. J. Gastroenterol. 102, 1474–1482.

Pappas, P.A., Saudubray, J.M., Tzakis, A.G., Rabier, D., Carreno, M.R., Gomez-Marin, O., et al., 2001. Serum citrulline and rejection in small bowel transplantation: a preliminary report. Transplantation 72, 1212–1216.

Pellicoro, A., van den Heuvel, F.A., Geuken, M., Moshage, H., Jansen, P.L., Faber, K.N., 2007. Human and rat bile acid-CoA:amino acid N-acyltransferase are liver-specific peroxisomal enzymes: implications for intracellular bile salt transport. Hepatology 45, 340–348.

Pircher, P.C., Kitto, J.L., Petrowski, M.L., Tangirala, R.K., Bischoff, E.D., Schulman, I.G., et al., 2003. Farnesoid X receptor regulates bile acid-amino acid conjugation. J. Biol. Chem. 278, 27703–27711.

Preedy, V.R., Paska, L., Sugden, P.H., Schofield, P.S., Sugden, M.C., 1988. The effects of surgical stress and short-term fasting on protein synthesis in vivo in diverse tissues of the mature rat. Biochem. J. 250, 179–188.

Prelack, K., Yu, Y.M., Dylewski, M., Lydon, M., Sheridan, R.L., Tompkins, R.G., 2010. The contribution of muscle to whole-body protein turnover throughout the course of burn injury in children. J. Burn Care Res. 31, 942–948.

Rembacz, K.P., Woudenberg, J., Hoekstra, M., Jonkers, E.Z., van den Heuvel, F.A., Buist-Homan, M., et al., 2010. Unconjugated bile salts shuttle through hepatocyte peroxisomes for taurine conjugation. Hepatology 52, 2167–2176.

Remesy, C., Moundras, C., Morand, C., Demigne, C., 1997. Glutamine or glutamate release by the liver constitutes a major mechanism for nitrogen salvage. Am. J. Physiol. 272, G257–G264.

Remond, D., Buffiere, C., Godin, J.P., Mirand, P.P., Obled, C., Papet, I., et al., 2009. Intestinal inflammation increases gastrointestinal threonine uptake and mucin synthesis in enterally fed minipigs. J. Nutr. 139, 720–726.

Remond, D., Buffiere, C., Pouyet, C., Papet, I., Dardevet, D., Savary-Auzeloux, I., et al., 2011. Cysteine fluxes across the portal-drained viscera of enterally fed minipigs: effect of an acute intestinal inflammation. Amino Acids 40, 543–552.

Rijke, R.P., Hanson, W.R., Plaisier, H.M., Osborne, J.W., 1976. The effect of ischemic villus cell damage on crypt cell proliferation in the small intestine: evidence for a feedback control mechanism. Gastroenterology 71, 786–792.

Rittler, P., Demmelmair, H., Koletzko, B., Schildberg, F.W., Hartl, W.H., 2000. Determination of protein synthesis in human ileum in situ by continuous [1-(13)C]leucine infusion. Am. J. Physiol. Endocrinol. Metab. 278, E634–E638.

Rittler, P., Demmelmair, H., Koletzko, B., Schildberg, F.W., Hartl, W.H., 2001. Effect of elective abdominal surgery on human colon protein synthesis in situ. Ann. Surg. 233, 39–44.

Rotovnik Kozjek, N., Kompan, L., Soeters, P., Oblak, I., Mlakar Mastnak, D., Mozina, B., et al., 2011. Oral glutamine supplementation during preoperative radiochemotherapy in patients with rectal cancer: a randomised double blinded, placebo controlled pilot study. Clin. Nutr. 30, 567–570.

Salloum, R.M., Copeland, E.M., Souba, W.W., 1991. Brush border transport of glutamine and other substrates during sepsis and endotoxemia. Ann. Surg. 213, 401–409 (discussion 409–410).

Samuels, S.E., Taillandier, D., Aurousseau, E., Cherel, Y., Le Maho, Y., Arnal, M., et al., 1996. Gastrointestinal tract protein synthesis and mRNA levels for proteolytic systems in adult fasted rats. Am. J. Physiol. 271, E232–E238.

Schaap, F.G., Trauner, M., Jansen, P.L., 2014. Bile acid receptors as targets for drug development. Nat. Rev. Gastroenterol. Hepatol. 11, 55–67.

Setchell, K.D., Heubi, J.E., Shah, S., Lavine, J.E., Suskind, D., Al-Edreesi, M., et al., 2013. Genetic defects in bile acid conjugation cause fat-soluble vitamin deficiency. Gastroenterology 144, 945–955 e946; quiz e914–e945.

Singer, H., Harrison, A.W., Aggett, P.W., 1965. The treatment of hepatic encephalopathy by colectomy: case report and review of the literature. Can. Med. Assoc. J. 93, 954–956.

Soeters, M.R., Soeters, P.B., 2012. The evolutionary benefit of insulin resistance. Clin. Nutr. 31, 1002–1007.

Souba, W.W., Herskowitz, K., Klimberg, V.S., Salloum, R.M., Plumley, D.A., Flynn, T.C., et al., 1990a. The effects of sepsis and endotoxemia on gut glutamine metabolism. Ann. Surg. 211, 543–549 (discussion 549–551).

Souba, W.W., Klimberg, V.S., Plumley, D.A., Salloum, R.M., Flynn, T.C., Bland, K.I., et al., 1990b. The role of glutamine in maintaining a healthy gut and supporting the metabolic response to injury and infection. J. Surg. Res. 48, 383–391.

Souba, W.W., Scott, T.E., Wilmore, D.W., 1985a. Intestinal consumption of intravenously administered fuels. J. Parenter. Enteral Nutr. 9, 18–22.

Souba, W.W., Smith, R.J., Wilmore, D.W., 1985b. Glutamine metabolism by the intestinal tract. J. Parenter. Enteral Nutr. 9, 608–617.

Stoll, B., Burrin, D.G., Henry, J., Yu, H., Jahoor, F., Reeds, P.J., 1999. Substrate oxidation by the portal drained viscera of fed piglets. Am. J. Physiol. 277, E168–E175.

Sun, J., Wang, H., Hu, H., 2012. Glutamine for chemotherapy induced diarrhea: a meta-analysis. Asia Pac. J. Clin. Nutr. 21, 380–385.

Suojaranta-Ylinen, R., Poyhonen, M., Takala, J., 1998. Accelerated splanchnic amino acid uptake after cardiac surgery. Clin. Nutr. 17, 51–55.

Tangerman, A., van Schaik, A., van der Hoek, E.W., 1986. Analysis of conjugated and unconjugated bile acids in serum and jejunal fluid of normal subjects. Clin. Chim. Acta 159, 123–132.

Tao, K.M., Li, X.Q., Yang, L.Q., Yu, W.F., Lu, Z.J., Sun, Y.M., et al., 2014. Glutamine supplementation for critically ill adults. Cochrane Database Syst. Rev. 9, CD010050.

Tashiro, T., Yamamori, H., Mashima, Y., Okui, K., 1985. Whole body protein turnover, synthesis, and breakdown in patients receiving total parenteral nutrition before and after recovery from surgical stress. Parenter. Enteral Nutr. 9, 452–455.

Ten Have, G.A., Engelen, M.P., Luiking, Y.C., Deutz, N.E., 2007. Absorption kinetics of amino acids, peptides, and intact proteins. Int. J. Sport Nutr. Exerc. Metab. 17 (Suppl.), S23–S36.

Ten Have, G.A., Engelen, M.P., Soeters, P.B., Deutz, N.E., 2012. Absence of post-prandial gut anabolism after intake of a low quality protein meal. Clin. Nutr. 31, 273–282.

Thompson, J.S., 1997. Can the intestine adapt to a changing environment? [editorial; comment]. Gastroenterology 113, 1402–1405.

Timmerman, K.L., Dhanani, S., Glynn, E.L., Fry, C.S., Drummond, M.J., Jennings, K., et al., 2012. A moderate acute increase in physical activity enhances nutritive flow and the muscle protein anabolic response to mixed nutrient intake in older adults. Am. J. Clin. Nutr. 95, 1403–1412.

Tizianello, A., De Ferrari, G., Garibotto, G., Chiggeri, G.M., Robaudo, C., Motta, G., et al., 1980. Ammonia and amino acid metabolism y the portal-vein-drained viscera in chronic renal insufficiency. Proc. Eur. Dial. Transplant Assoc. 17, 695–699.

Tomlinson, C., Rafii, M., Ball, R.O., Pencharz, P., 2011. Arginine synthesis from enteral glutamine in healthy adults in the fed state. Am. J. Physiol. Endocrinol. Metab. 301, E267–E273.

Tujioka, K., Ohsumi, M., Hayase, K., Yokogoshi, H., 2011. Effect of the quality of dietary amino acids composition on the urea synthesis in rats. J. Nutr. Sci. Vitaminol. (Tokyo) 57, 48–55.

Vahedi, K., Crenn, P., Deutz, N., Messing, B., 2001. Citrulline and arginine turnover in adult short bowel patients: a stable isotope study. In: Presented at ESPEN Research Fellowship Symposium.

van Berlo, C.L., van de Bogaard, A.E., van der Heijden, M.A., van Eijk, H.M., Janssen, M.A., Bost, M.C., et al., 1989. Is increased ammonia liberation after bleeding in the digestive tract the consequence of complete absence of isoleucine in hemoglobin? A study in pigs. Hepatology 10, 315–323.

van de Poll, M.C., Soeters, P.B., Deutz, N.E., Fearon, K.C., Dejong, C.H., 2004. Renal metabolism of amino acids: its role in interorgan amino acid exchange. Am. J. Clin. Nutr. 79, 185–197.

van der Hulst, R.R., van Kreel, B.K., von Meyenfeldt, M.F., Brummer, R.J., Arends, J.W., Deutz, N.E., et al., 1993. Glutamine and the preservation of gut integrity. Lancet 341, 1363–1365.

van der Hulst, R.R., von Meyenfeldt, M.F., Deutz, N.E., Soeters, P.B., 1997. Glutamine extraction by the gut is reduced in depleted [corrected] patients with gastrointestinal cancer. Ann. Surg. 225, 112–121.

van der Hulst, R.R., von Meyenfeldt, M.F., van Kreel, B.K., Thunnissen, F.B., Brummer, R.J., Arends, J.W., et al., 1998. Gut permeability, intestinal morphology, and nutritional depletion. Nutrition 14, 1–6.

Vanwijngaerden, Y.M., Wauters, J., Langouche, L., Vander Perre, S., Liddle, C., Coulter, S., et al., 2011. Critical illness evokes elevated circulating bile acids related to altered hepatic transporter and nuclear receptor expression. Hepatology 54, 1741–1752.

Vanwijngaerden, Y.M., Langouche, L., Derde, S., Liddle, C., Coulter, S., van den Berghe, G., et al., 2014. Impact of parenteral nutrition versus fasting on hepatic bile acid production and transport in a rabbit model of prolonged critical illness. Shock 41, 48–54.

Vary, T.C., Drnevich, D., Jurasinski, C., Brennan Jr., W.A., 1995. Mechanisms regulating skeletal muscle glucose metabolism in sepsis. Shock 3, 403–410.

Volpi, E., Mittendorfer, B., Wolf, S.E., Wolfe, R.R., 1999. Oral amino acids stimulate muscle protein anabolism in the elderly despite higher first-pass splanchnic extraction. Am. J. Physiol. 277, E513–E520.

Wilmore, D.W., Goodwin, C.W., Aulick, L.H., Powanda, M.C., Mason Jr., A.D., Pruitt Jr., B.A., 1980. Effect of injury and infection on visceral metabolism and circulation. Ann. Surg. 192, 491–504.

Windmueller, H.G., 1982. Glutamine utilization by the small intestine. Adv. Enzymol. Relat. Areas Mol. Biol. 53, 201–237.

Windmueller, H.G., Spaeth, A.E., 1974. Uptake and metabolism of plasma glutamine by the small intestine. J. Biol. Chem. 249, 5070–5079.

Winthrop, A.L., Wesson, D.E., Pencharz, P.B., Jacobs, D.G., Heim, T., Filler, R.M., 1987. Injury severity, whole body protein turnover, and energy expenditure in pediatric trauma. J. Pediatr. Surg. 22, 534–537.

Wischmeyer, P.E., Dhaliwal, R., McCall, M., Ziegler, T.R., Heyland, D.K., 2014. Parenteral glutamine supplementation in critical illness: a systematic review. Crit. Care 18, R76.

Wu, G., Knabe, D.A., Flynn, N.E., 1994. Synthesis of citrulline from glutamine in pig enterocytes. Biochem. J. 299 (Pt 1), 115–121.

Yuzawa, H., Azuma, T., Tsutsumi, R., Fujioka, H., Furui, J., Kanematsu, T., 2000. Alanylglutamine-enriched total parenteral nutrition prevents bacterial translocation after small bowel transplantation in pigs. Transplant. Proc. 32, 1662.

Ziegler, F., Nitenberg, G., Coudray-Lucas, C., Lasser, P., Giboudeau, J., Cynober, L., 1998. Pharmacokinetic assessment of an oligopeptide-based enteral formula in abdominal surgery patients. Am. J. Clin. Nutr. 67, 124–128.

Zou, X.P., Chen, M., Wei, W., Cao, J., Chen, L., Tian, M., 2010. Effects of enteral immunonutrition on the maintenance of gut barrier function and immune function in pigs with severe acute pancreatitis. J. Parenter. Enteral Nutr. 34, 554–566.

CHAPTER

14

Regulation of Macroautophagy by Nutrients and Metabolites

S. Lorin[1,2], *S. Pattingre*[3,4,5,6], *A.J. Meijer*[7] *and P. Codogno*[8,9,10]

[1]Faculté de Pharmacie, Université Paris-Saclay, Châtenay-Malabry, France [2]INSERM UMR-S-1193, Châtenay-Malabry, France [3]IRCM, Institut de Recherche en Cancérologie de Montpellier, Montpellier, France [4]INSERM, U1194, Montpellier, France [5]Université de Montpellier, Montpellier, France [6]Institut régional du Cancer de Montpellier, Montpellier, France [7]Department of Medical Biochemistry, Academic Medical Center, University of Amsterdam, Amsterdam, The Netherlands [8]INEM, Institut Necker Enfants-Malades, Paris, France [9]INSERM U1151-CNRS UMR 8253, Paris, France [10]Université Paris Descartes, Paris, France

14.1. INTRODUCTION

Autophagy (from the Greek "auto," oneself, and "phagy," to eat) is an intracellular pathway for degradation of cytoplasmic material within organelles called lysosomes (De Duve and Baudhuin, 1966). This process is the intracellular counterpart of the endocytic pathway that brings extracellular material through the plasma membrane to the lysosome for degradation. The term "autophagy" is used to encompass three distinct mechanisms of intracellular degradation within the lysosome: microautophagy, chaperone-mediated autophagy (CMA), and macroautophagy (Ravikumar et al., 2010). Microautophagy is a nonselective process that enables the trapping and degradation of cytoplasmic molecules in the vicinity of the lysosome by self-invagination of the lysosomal membrane. In contrast, CMA is a selective mechanism in which cytoplasmic proteins bearing the amino acid motif KFERQ are selectively recognized by the chaperone protein HSC70 and targeted to the lysosome. CMA substrates are transported across the lysosomal membrane into the lysosomal lumen by the lysosomal-associated membrane protein 2A (LAMP-2A) (Kaushik and Cuervo, 2008). During macroautophagy (Fig. 14.1), a small membrane (the phagophore) forms and grows in the cytoplasm, sequestering a fraction of cytoplasm containing macromolecules, protein aggregates, lipid droplets, and/or organelles within a double-membrane vesicle, named the autophagosome. The outer membrane of the autophagosome then fuses with the lysosome and the inner membrane contents are released into the lysosomal lumen and the autophagic vesicle components and its contents are degraded by acidic hydrolases (for a literature review of this process, see Meijer and Codogno, 2009). The products of autophagic degradation, such as amino acids, sugars, fatty acids, and nucleotides, are recycled to the cytosol. The third type of autophagy, macroautophagy, can be either nonselective or selective. Macroautophagy can be, specifically directed toward mitochondria (mitophagy), endoplasmic reticulum (ER) components (reticulophagy), peroxisomes (pexophagy), a portion of the nucleus (nucleophagy), ribosomes (ribophagy), protein aggregates (aggrephagy), lipid droplets (lipophagy), or invading pathogens such as viruses and bacteria (xenophagy) (Mizushima and Komatsu, 2011).

During the past decades, macroautophagy (hereafter referred to as autophagy) has been the type of autophagy most extensively studied. These studies began in the 1990s with the identification in *Saccharomyces cerevisiae* of the first members of the autophagy-related (ATG) protein family. The ATG proteins are involved in the regulation, formation, and maturation of autophagosomes (Tsukada and Ohsumi, 1993). To date, 38 *ATG* genes have been identified in yeast. As autophagy is a highly conserved eukaryotic pathway, most of the mammalian orthologs of these 38 genes have been identified (Mizushima et al., 2011; Yang and Klionsky, 2010).

The Molecular Nutrition of Amino Acids and Proteins.
DOI: http://dx.doi.org/10.1016/B978-0-12-802167-5.00014-1

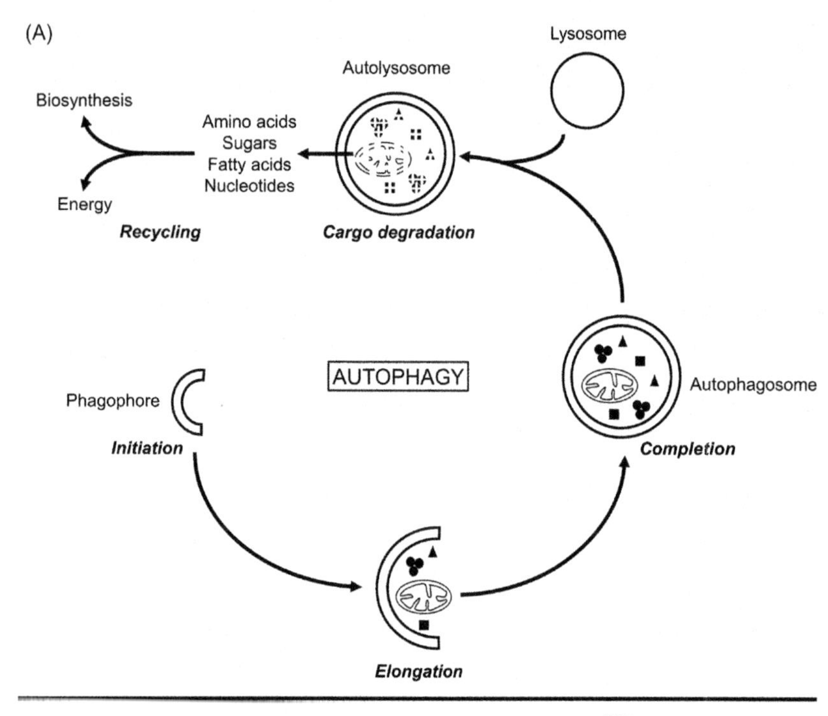

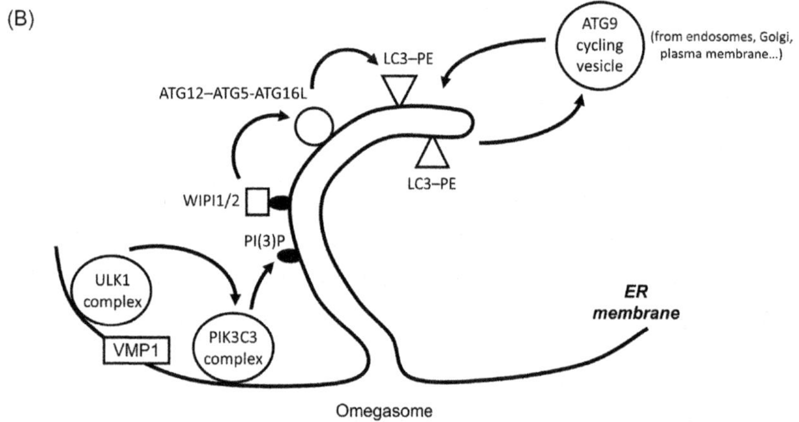

FIGURE 14.1 The autophagy mechanism. (A) Macroautophagy sequesters cytoplasmic material (macromolecules and organelles) in a vesicle called the autophagosome, which then fuses with the lysosome. The hydrolytic enzymes of the lysosome degrade the autophagic cargoes. The degradation products, such as amino acids, fatty acids, sugars, and nucleotides, are exported to the cytosol where they can be re-used for biosynthesis or to generate energy (ATP). (B) The formation of the phagophore takes place at a specific site of the ER membrane called the omegasome. ATG proteins are sequentially involved in the formation of the phagophore that elongates to form an autophagosome.

Autophagy has been of tremendous scientific interest in recent years as its dysregulation is involved in many pathologies including neurodegenerative, heart, liver, and immune diseases, cancer, lysosomal storage disorders, type 2 diabetes, obesity, and aging (Jiang and Mizushima, 2014; Lavallard et al., 2012; Lieberman et al., 2012; Lorin et al., 2013a; Rubinsztein et al., 2012). In most cells, autophagy occurs at a basal rate and serves a housekeeping function by eliminating protein aggregates and damaged organelles and maintaining the quality of the cytoplasm (Mizushima and Komatsu, 2011). Such a mechanism is particularly important in quiescent and differentiated cells, where damaged components are not diluted by cell division. The physiological importance of basal autophagy in maintaining tissue homeostasis has been demonstrated in conditional brain and liver *ATG* knockout mouse models (Mizushima and Komatsu, 2011). These studies have also demonstrated the role of autophagy in preventing the deposition of aggregation-prone proteins in the cytoplasm and in eliminating ubiquitinated proteins that are efficient substrates for the proteasome (Rubinsztein et al., 2011). The antiaging role of autophagy probably depends, at least in part, on this quality control function that limits the deposition of aggregation-prone proteins and the formation of damaging reactive oxygen species (ROS) by mitochondria (Rubinsztein et al., 2011).

Autophagy is also an adaptive response to stress. One of the major autophagic stimuli is nutrient restriction (amino acid or glucose deprivation), which rapidly induces autophagy to provide substrates for metabolism (ie, amino acids, sugars, fatty acids, and nucleotides) in order to maintain protein and ATP synthesis, allowing cells to adapt and survive in nonoptimal conditions (Meijer and Codogno, 2009). In vivo at birth, the sudden interruption of the supply of nutrients via the placenta triggers autophagy in newborn mouse tissues to maintain energy homeostasis and survival (Kuma et al., 2004). Starvation-induced autophagy is also important in adult mammalian muscle and liver as it degrades proteins to produce amino acids for the hepatic synthesis of glucose, which is needed as substrate for energy production in brain and erythrocytes (Meijer and Codogno, 2009).

In this chapter, we will first summarize current knowledge on the molecular mechanisms of the autophagic pathway. Amino acids, glucose, and lipids are important cellular nutrients that regulate autophagy, and mammalian cells have developed diverse mechanisms to sense their abundance. We will then discuss the role of nutrients and intermediate metabolites in the regulation of the autophagic machinery. We will describe the molecular pathways involved both at the transcriptional and posttranslational levels.

14.2. OVERVIEW OF THE AUTOPHAGIC PATHWAY

Induction of autophagy is under the control of several protein complexes composed of *ATG* gene products. Fifteen ATG proteins constitute the core machinery of autophagosome formation. In mammalian cells, these ATG proteins are recruited to form a phagophore, or isolation membrane, that subsequently elongates to form the autophagosome (Fig. 14.1). Recent studies have shown that phagophore elongation takes place at the ER in a structure called the omegasome characterized by the presence of double FYVE-domain-containing protein 1 (DFCP1), a phosphatidylinositol 3-phosphate binding protein (Mizushima and Komatsu, 2011). Membranes of the endosomes and the Golgi apparatus, the contact site between the ER and mitochondria, ER transition elements, and the plasma membrane also contribute to the biogenesis of the autophagosome (Lamb et al., 2013; Sanchez-Wandelmer et al., 2015).

Several functional modules involving ATG proteins have been identified during the different phases of autophagosome formation from the initiation step to the elongation/closure step (Fig. 14.1). The two first modules, the unc-51 like kinase 1 (ULK1) complex and the class 3 phosphatidyl inositol 3-kinase (PIK3C3) complex I, are involved in the initiation of autophagosome formation. The ULK complex is composed of the serine/threonine kinases ULK1 or ULK2 (mammalian orthologs of yeast Atg1), ATG13, focal adhesion kinase family-interacting protein of 200 kDa (FIP200, mammalian ortholog of yeast Atg17) and ATG101. The PIK3C3 complex is composed of three core components: the lipid kinase PIK3C3 (mammalian ortholog of yeast Vps34), phosphatidylinositol 3-kinase regulatory subunit 4 (PIK3R4; mammalian ortholog of yeast Vps15), and Beclin 1 (mammalian ortholog of yeast Atg6). The PIK3 core complex can associate with ATG14 and with (activating molecule in Beclin 1-regulated autophagy protein 1) AMBRA1 (to form PIK3C3 complex I) to drive the first steps of autophagosome formation. Alternatively, the core components of the PIK3 complex can associate with (UV irradiation resistance-associated gene protein) UVRAG to form PIK3C3 complex II, to drive the last steps of the autophagosome maturation.

Several signaling pathways sense the availability of nutrients and growth factors, the cellular energy state, and the presence of ROS to control autophagy by regulating the activity of the ULK complex and/or the activity of the PIK3C3 complex I. The first step of autophagosome formation is the de novo nucleation of an isolation membrane in the cytoplasm, the so-called phagophore. Upon induction of autophagy, the ULK complex localizes to a particular site of the ER specialized in autophagosome formation. The recruitment of the ULK complex is necessary for the subsequent recruitment and activation of the PIK3C3 complex I, and ATG14 may act as an adaptor in the binding of ULK1 to Beclin 1 (Russell et al., 2013). Phosphorylation of Beclin 1 and AMBRA1 by ULK1 enhances the activity of the PIK3C3 complex I, which produces phosphatidylinositol 3-phosphate (PI(3)P). An increase in the local PI(3)P content presumably triggers the formation of a specific structure on the ER, named the omegasome because of its cup-shaded form, creating a cradle where the autophagosomal membrane will elongate and acting as a template for the spherical form of the autophagosome (Mizushima et al., 2011).

At the phagophore, production of PI(3)P also recruits the PI(3)P-binding proteins DFCP1 and WD repeat domain phosphoinositide interacting protein 1/2 (WIPI1/2, mammalian orthologs of yeast Atg18), which constitute the third module of ATG proteins (Simonsen and Tooze, 2009). One of the functions of WIPI2 is to control the transport of the multimembrane spanning ATG9 between the phagophore and a peripheral endosome/Golgi localization. The trafficking of ATG9 to the phagophore is an early event that occurs soon after autophagy induction. Another function of DFCP1 and WIPI1/2 is to allow the expansion, curvation, and closure of the autophagosome by recruitment of two ubiquitin-like conjugation systems; the first is the complex ATG12–ATG5-ATG16L, and the second contains microtubule-associated protein 1A/1B-light chain 3 (LC3, mammalian ortholog of yeast Atg8) conjugated to phosphatidylethanolamine (PE); the LC3–PE conjugate is also known as LC3-II.

Functionally, ATG12–ATG5-ATG16L, which is essential for the elongation of the phagophore, is also required for the conjugation of LC3 to PE. LC3–PE is important for the elongation and/or the complete closure of the

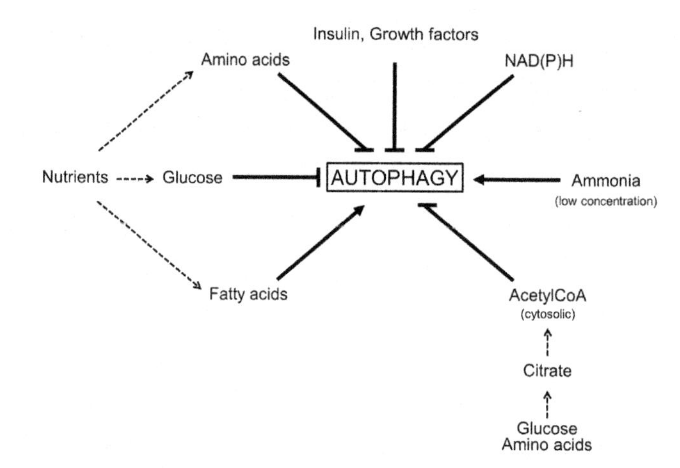

FIGURE 14.2 **Effects of nutrients and metabolites on autophagy.** Most nutrients and their metabolites, except for fatty acids and ammonia (at low concentrations), inhibit autophagy.

autophagosome (Fujita et al., 2008). Unlike ATG12–ATG5-ATG16L, which resides only on the outer membrane of the phagophore, LC3–PE is present on both outer and inner faces of the autophagosomal membrane and remains associated with the complete autophagosome, thus serving as a molecular marker to monitor autophagy (Mizushima et al., 2011). All other ATG proteins associate transiently with the autophagosomal membrane before being recycled to the cytosol. The lipids necessary for autophagosomal membrane extension may be provided by vesicles containing ATG9 and vacuole membrane protein 1 (VMP1); VMP1 is also an ATG protein. These vesicles recycle among the Golgi apparatus, endosomes, and autophagosomes and/or the ER cradle where the autophagosomal membrane is growing (Galluzzi et al., 2014; Rubinsztein et al., 2011). Recently, it has been demonstrated that VMP1 is an ER-resident transmembrane protein that is necessary for the recruitment of other ATG proteins, such as ULK1 and PIK3C3 complex I, through its interaction with Beclin 1 (Koyama-Honda et al., 2013).

The autophagosome engulfs fractions of the cytoplasm in either a nonselective or a selective manner. The recognition of the cargoes to be selectively degraded involves autophagy adaptors such as p62/sequestosome 1 (SQSTM1), neighbor of BRCA1 gene 1 (NBR1), nuclear dot protein 52 (NDP52), and optineurin, that form bridges between the ubiquitinated targets (macromolecules or organelles) and LC3–PE on the growing autophagosomal membrane (Kroemer et al., 2010; Mizushima and Komatsu, 2011).

After formation, most autophagosomes receive input from the endocytic vesicles to form an amphisome. The final fusion with the lysosome requires small Ras-related proteins in brain (Rab) GTPases (such as Rab1, Rab7, Rab8B, and Rab24) and the transmembrane lysosomal protein LAMP-2 (Ao et al., 2014). Acidic hydrolases and cathepsins present in the lysosomal lumen degrade the autophagosomal cargoes. The products of degradation, such as amino acids, sugars, fatty acids, and nucleotides, are then transported back to the cytosol by lysosomal permeases, providing precursors for the synthesis of ATP, proteins, and other essential macromolecules under stress conditions (Meijer and Codogno, 2009). In turn, some products of autophagic degradation, such as amino acids, exert a negative feedback on autophagy initiation.

Induction of autophagy is controlled by hormones, nutrients, the cellular energy state, and by oxidative stress. Withdrawal of nutrients is the most potent known inducer of autophagy. It is then not surprising that molecules that sense nutrients and their metabolites regulate autophagy in response to changes in the cellular microenvironment (Fig. 14.2).

14.3. THE NUTRIENT CODE OF AUTOPHAGY

The availability of amino acids, glucose, and fatty acids controls autophagy through three major protein complexes involved in autophagy initiation: mechanistic target of rapamycin complex 1 (MTORC1), the master regulator of autophagy, and the two initiation complexes, ULK complex and PIK3C3 complex I (Fig. 14.3).

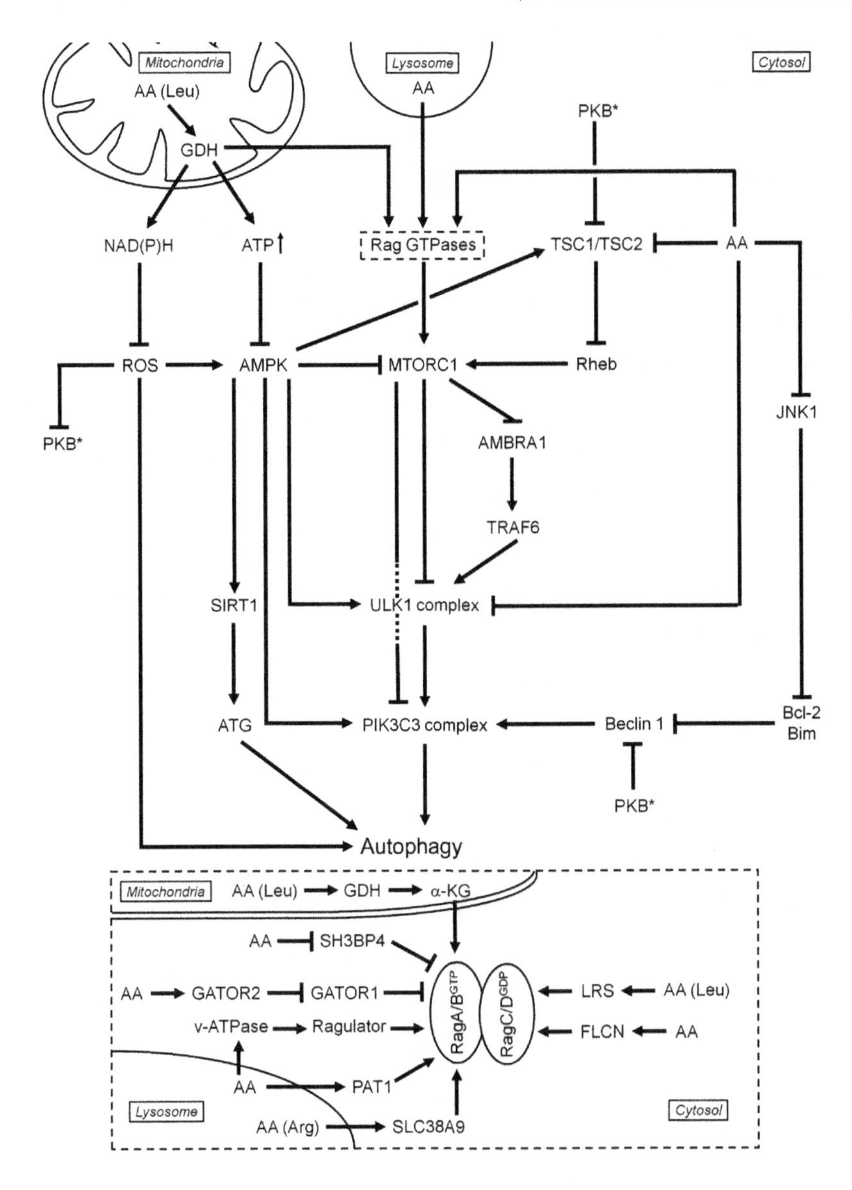

FIGURE 14.3 Main molecular pathways involved in the regulation of autophagy by amino acids. Cytosolic and lysosomal pools of amino acids (AA) inhibit autophagy mainly by regulating MTORC1, the ULK complex, and the PIK3C3 complex I. For clarity, the details of the signaling pathways involved in the regulation of autophagy in response to amino acids are not indicated in the figure. *To simplify the figure, PKB is represented at different places.

14.3.1 Amino Acids

The central role of amino acids in the regulation of autophagy has been known since the 1970s (for detailed reviews, see Lavallard et al., 2012; Meijer et al., 2015). Experiments with the isolated perfused rat liver and with freshly isolated rat hepatocytes demonstrated that deprivation of amino acids resulted in a rapid increase in both the number and volume of autophagosomes and in the rate of proteolysis, with a clear inverse relationship between the concentration of amino acids and the magnitude of autophagic flux (Mortimore and Schworer, 1977; Seglen et al., 1980). Subsequently, it was shown that, among all amino acids, leucine is the most potent inhibitor of autophagy.

14.3.1.1 Regulation of MTORC1 by Amino Acids

MTORC1 is a key complex that is sensitive to growth factors and nutrients. As a central hub of the pathway involving insulin/growth factor-PIK3C1, protein kinase B (PKB), tuberous sclerosis complex (TSC), MTORC1 integrates signals from many stimuli (amino acid availability, energy levels, ROS, insulin, and growth factors). This signaling pathway controls major metabolic processes, such as the metabolism of glucose, lipids, proteins (including autophagy), and ATP production, and is a major controller of cell growth (Meijer et al., 2015).

The central component of MTORC1 complex is the highly conserved serine/threonine kinase MTOR. MTOR is part of two multiprotein complexes: (1) MTORC1, which also includes regulatory-associated protein of MTOR (RAPTOR), mammalian lethal with sec-13 protein 8 (mLST8), DEP domain-containing MTOR-interacting protein

(DEPTOR), and proline-rich AKT substrate 40 (PRAS40); and (2) MTORC2, which also includes rapamycin-insensitive companion of TOR (RICTOR), mLST8, mammalian stress-activated MAP kinase interacting protein 1 (mSin1), protein observed with RICTOR (PROTOR), and DEPTOR (Laplante and Sabatini, 2012). MTORC1 is sensitive to rapamycin, and MTORC2 is rapamycin insensitive.

In the insulin/growth factor-PIK3C1-PKB-TSC-MTORC1 pathway, the proximal regulators of MTORC1 are the heterodimer TSC1/TSC2 and the small G-protein Ras homolog enriched in brain (Rheb). TSC1/TSC2 acts as a GTPase-activating protein (GAP) complex for Rheb. $Rheb^{GTP}$, but not $Rheb^{GDP}$, binds to the lysosomal surface and activates MTORC1, thus inhibiting autophagy. MTORC1 is only activated upon localization at the lysosomal membrane. Several studies have demonstrated that in the presence of amino acids MTORC1 displays a punctate pattern reflecting its lysosomal localization, whereas in the absence of amino acids MTORC1 is diffused in the cytosol (Flinn et al., 2010; Kalender et al., 2010; Korolchuk et al., 2011; Narita et al., 2011; Yoon et al., 2011).

MTORC1 is a key complex coordinating amino acid availability and autophagy. Amino acids are both necessary and sufficient to repress autophagy through MTORC1, and activation of MTORC1 by growth factors requires the presence of amino acids. This allows the coordination of growth-promoting signals with nutrient availability. The first indication that the MTORC1 pathway was involved in this process came from studies showing that inhibition of autophagy by amino acids was accompanied by an increased activity of 70 kDa S6 kinase (S6K), a target of MTORC1 (Blommaart et al., 1995). Insulin was also able to activate S6K and to inhibit autophagy but only in the presence of low concentrations of amino acids. Until recently, the mechanism responsible for the stimulation of MTORC1 activity and repression of autophagy by amino acids was an enigma. As the intracellular level of amino acids is determined by their influx from the extracellular environment, their production by intralysosomal proteolysis, and by their intracellular metabolism, it is not surprising that both cytosolic and lysosomal pools of amino acids are detected by different molecular sensors.

14.3.1.2 Regulation of MTOR by Intracellular Amino Acids

Given that the lysosome is an important source of amino acids and that MTORC1 needs to be localized at its surface to be activated by $Rheb^{GTP}$, this organelle appears to play a central role in the regulation of MTORC1 by amino acids. In fact, the main pathway by which amino acids regulate MTORC1 involves the Ras-related GTP-binding protein (Rag) GTPases, and a panel of proteins that regulate their activities mainly by modulating nucleotide binding (Kim et al., 2008; Sancak et al., 2008). RagA, RagB, RagC, and RagD form heterodimeric complexes comprised of RagA or RagB bound to RagC or RagD. In their active forms, RagA/B is loaded with GTP and RagC/D with GDP, and these complexes tether MTORC1 to the lysosomal membrane, where its activator $Rheb^{GTP}$ resides. It has been proposed that amino acids increase the loading of RagA/B with GTP (Sancak et al., 2008). There is no general agreement on this issue, however (for discussion, see Meijer et al., 2015).

As Rag proteins do not contain lipid modifications commonly used to anchor active small GTPases to target membranes, active Rag is tethered to the lysosome by a multiprotein complex called Ragulator or late endosomal/lysosomal adaptor, MAPK and MTOR activator 1 (LAMTOR). This multiprotein complex is composed of five different proteins: LAMTOR1, the protein that functions as a scaffold for the two heterodimers LAMTOR2/LAMTOR3 and LAMTOR4/LAMTOR5. Ragulator is anchored to the lysosomal surface by myristoylation and palmitoylation of the N-terminus of LAMTOR1 (Bar-Peled and Sabatini, 2014). Ragulator also has a guanine nucleotide exchange (GEF) activity toward RagA/B (GDP dissociation and GTP binding) and thus functions to switch RagA/B to its active conformation (Bar-Peled et al., 2012; Sancak et al., 2010). The binding of Ragulator to the lysosome seems to be regulated by the vacuolar ATPase (v-ATPase), a lysosomal transmembrane protein involved in proton pumping and acidification of the lysosomal interior. It is the v-ATPase that acts as a sensor of the intralysosomal concentration of amino acids. When the amino acid concentration inside the lysosome increases, the v-ATPase changes its conformation and recruits Ragulator to the lysosomal surface to activate MTORC1 (Zoncu et al., 2011). This coupling between intralysosomal amino acid concentration and MTORC1 activity is termed an "inside-out" mechanism. Consistent with this mechanism, nutrient starvation inhibits the interaction between the v-ATPase and Ragulator, leading to MTORC1 inhibition and autophagy induction.

Proton-assisted amino acid transporter 1 (PAT1), a lysosomal proton-assisted amino acid transporter responsible for the efflux of amino acids from the lysosomes, also interacts with the Rag GTPases and regulates MTORC1 localization to the late endosomes (Ögmundsdóttir et al., 2012). That other amino acid transporters participate in regulation of MTORC1 activation is also possible (Meijer et al., 2015). Very recently, a new lysosomal transmembrane protein, member 9 of the solute carrier family 38 (SLC38A9), was identified as an arginine transporter that regulates Rag and MTORC1 (Rebsamen et al., 2015; Wang et al., 2015). Excessive intralysosomal production of amino acids by the degradation of autophagic cargoes may reactivate MTORC1, which then results in inhibition of autophagy.

The fact that MTORC1 activation is determined by the concentration of the intralysosomal pool of amino acids implies that compounds such as the v-ATPase inhibitor bafilomycin or the acidotropic agent chloroquine should not be used to estimate autophagic flux by monitoring the accumulation of the autophagosomal marker LC3-II. This is because inhibition of proteolysis within the lysosomes will directly affect the intralysosomal pool of amino acids, and thus lead to underestimation of MTOR activity and overestimation of autophagic flux (Meijer et al., 2015). In support of this view is the finding that experimental manipulations that result in an increased intralysosomal pH, such as the administration of chloroquine or of inhibitors of the v-ATPase, inhibit MTOR activity (Jewell et al., 2015) and stimulate initiation of autophagy (Jangamreddy et al., 2015; Li et al., 2013b). Of course, high concentrations of such compounds will eventually inhibit autophagic flux because of a total block in intralysosomal degradation when intralysosomal pH reaches high levels. Examples of compounds that completely block autophagic flux are the v-ATPase inhibitor salinomycin (Jangamreddy et al., 2015) and the acidotropic metabolite ammonia (Eng et al., 2010; Polletta et al., 2015).

Ragulator stimulates the loading of GTP on RagA/B, whereas the GTPase-activating protein toward Rags (GATOR) complex inactivates Rag GTPases and inhibits MTORC1 signaling (Bar-Peled et al., 2013). GATOR is composed of two multimolecular subcomplexes, GATOR1 and GATOR2, localized at the lysosomal membrane. GATOR1 interacts directly with Rag proteins and carries the GAP activity, and GATOR2 functions as a negative regulator of GATOR1 (Panchaud et al., 2013; Bar-Peled et al., 2013). In addition to GATOR1, SH3-domain binding protein 4 (SH3BP4) inhibits MTORC1 by interacting with RagB to maintain it in an inactive form (stimulating GTP hydrolysis and preventing GDP dissociation) (Kim et al., 2012). The tumor suppressor protein folliculin (FLCN) is another GAP protein that regulates the Rag complex by stimulating the hydrolysis of GTP to GDP on RagC and RagD (Petit et al., 2013; Tsun et al., 2013). FLCN acts in concert with its binding partner folliculin interacting protein 1 (FNIP1) (Tsun et al., 2013). This regulation is of particular interest as recent in vitro and in vivo studies showed that the nucleotide status of RagC, but not RagA, governs the interaction of Rag with RAPTOR, a subunit of MTORC1.

14.3.1.3 Regulation of MTOR by Nonlysosomal Amino Acids

The regulation of MTORC1 through the Rag GTPase is also controlled by levels of amino acids in the cytosol. When localized at the lysosomal surface, the adaptor protein p62/SQSTM1 interacts with Rag and MTORC1 in the presence of amino acids. p62/SQSTM1 stabilizes the Rag heterodimer and recruits TNF receptor-associated factor 6 (TRAF6) to catalyze MTORC1 ubiquitination; ubiquitination of MTORC1 is necessary for subsequent signaling (Duran et al., 2011; Linares et al., 2013). Amino acids can also activate mitogen-activated protein kinase kinase kinase kinase 3 (MAP4K3), which interacts with the Rag GTPases and stimulates MTORC1 (Bryk et al., 2010; Findlay et al., 2007). During amino acid starvation, SH3BP4 inactivates Rag and inhibits MTORC1 signaling (Kim et al., 2012).

Other mechanisms of amino acid sensing show specificity for leucine, the most potent amino acid inhibitor of autophagy. The tRNA charging enzyme leucyl tRNA synthetase (LRS) functions as a leucine sensor and is important for amino acid-dependent stimulation of MTORC1. In the presence of amino acids, the binding of leucine to LRS induces translocation of LRS to the lysosome where it acts as a GAP for RagD, facilitating the formation of the active Rag complex (Han et al., 2012). Another leucine sensor is the mitochondrial enzyme glutamate dehydrogenase (GDH) (Durán et al., 2012; Lorin et al., 2013b). Leucine binds to GDH and stimulates the deamination of glutamate to α-ketoglutarate (α-KG). α-KG stimulates the loading of GTP on the RagB subunit, which in turn stimulates the translocation of MTORC1 to the lysosome, leading to its activation and to the inhibition of autophagy (Durán et al., 2012).

It is also possible that both α-KG and NADPH, another product of the deamination reaction of glutamate, activate MTORC1 and inhibit autophagy by eliminating ROS (Lavallard et al., 2012; Lorin et al., 2013b; Meijer et al., 2015). The inhibition of autophagy by amino acids other than leucine and glutamine (which generates glutamate for the GDH reaction) does not always correlate with ROS levels (Angcajas et al., 2014). Possible targets of ROS that contribute to stimulation of autophagy are ATG4, Beclin 1, PKB, TSC1/TSC2, and AMP-activated protein kinase (AMPK) (Bolisetty and Jaimes, 2013; Rahman et al., 2014; Scherz-Shouval and Elazar, 2011; Toyoda et al., 2004; Zhang et al., 2013). Moreover, α-KG produced by GDH probably replenishes the tricarboxylic acid (TCA) cycle, generating GTP and cytosolic acetyl-coenzyme A (acetylCoA). These two metabolites of amino acid catabolism regulate autophagy through impacts on small GTPases (such as Rheb and Rag) and protein acetylation (see "AcetylCoA" section), respectively. Involvement of GDH in amino acid sensing has serious consequences for the use of chloroquine in the measurement of autophagic flux. This assay likely leads to overestimation of the autophagic flux, as chloroquine is a potent inhibitor of GDH (Jarzyna et al., 1997).

As glutamine is converted into glutamate by glutaminase in the mitochondria, this enzyme is also indirectly involved in amino acid sensing (Durán et al., 2012). Not surprisingly, mechanisms facilitating cellular leucine uptake, such as glutamine efflux through the plasma membrane amino acid permease member 5 of the solute carrier family 7 (SLC7A5) and member 2 of the solute carrier family 3 (SLC3A2), modulate MTORC1 signaling and autophagy (Nicklin et al., 2009). Of note, effects on MTORC1 seem to be amino acid dependent. In a recent report it was demonstrated that the Rag GTPases are essential for MTORC1 activation by leucine but not by glutamine. Instead, the GTPase ADP-ribosylation factor 1 (ARF1) is necessary for MTORC1 lysosomal localization and activation by glutamine (Jewell et al., 2015). It must be pointed out, however, that the concentration of glutamine used in these experiments (20 mM) was extremely high, and under these conditions cells may undergo extensive osmotic swelling due to massive Na^+-dependent glutamine influx (Baquet et al., 1990; Häussinger et al., 1990).

14.3.1.4 Regulation of MTOR by Other Pathways

Other pathways, not dependent on the Rag GTPases, have been shown to regulate MTORC1 in response to amino acids. Intriguingly, in addition to being involved in autophagy initiation, PIK3C3 is also necessary for amino acid-induced MTORC1 activation (Byfield et al., 2005; Nobukuni et al., 2005). In the presence of amino acids, intracellular Ca^{2+} levels are elevated (contrast Ghislat et al., 2012; cf. Meijer et al., 2015 for discussions of this controversial issue), which would activate PIK3C3 and enhance PI(3)P production on endosomes (Gulati et al., 2008). This may recruit several proteins to the endosomal/lysosomal membranes such as phospholipase D (PLD) (Yoon et al., 2011). The activation of PLD produces phosphatidic acid (PA), which facilitates the association between MTOR and RAPTOR. Thus, this pathway results in MTORC1 activation in presence of amino acids. This process is dependent on PIK3C3 and also on the GTPases RAS-like small GTPase A (RalA) and ARF6 (Xu et al., 2011). Therefore, PIK3C3 is involved in two opposing pathways: one leads to autophagy initiation in the absence of amino acids (see previous section "Overview of the autophagic pathway") and the other leads to activation of MTORC1 and inhibition of autophagy in the presence of amino acids. This puzzling observation could be resolved if PIK3C3 is a shared component of different protein complexes with opposite functions (Duran et al., 2011; Ktistakis et al., 2012). Like PIK3C3 and adaptor protein p62/SQSTM1, Rab1 has a role in autophagy and is also engaged in MTORC1 signaling (Thomas et al., 2014). These apparently opposing functions may be reconciled if one assumes that stimulation of autophagy results in increased proteolytic production of amino acids within the lysosomes, and in this way enhances the activity of MTORC1 on the outside of the lysosomal membrane.

Inositolphosphate multikinase (IMPK) is another component involved in the stimulation of MTORC1 by amino acids. Independent of its catalytic activity, this enzyme appears to stabilize the MTOR–RAPTOR interaction within the MTORC1 complex (Kim et al., 2011b). Ubiquitin protein ligase E3 component N-recognin 2 (UBR2), an E3 ubiquitin ligase, may mediate leucine-induced activation of MTORC1. In vitro, leucine directly binds to the substrate-recognition domain of UBR2 and prevents its degradation, promoting MTORC1 signaling (Dodd and Tee, 2012; Kume et al., 2010a). Finally, the cell surface G protein coupled receptor (GPCR) taste receptor type 1 member 1 (T1R1) and T1R3 can activate MTORC1 and inhibit autophagy in response to extracellular amino acids, but the mechanism involved is still unknown (Wauson et al., 2012).

Amino acids do not appear to activate MTORC1 by regulating upstream components of this pathway, since amino acids do not affect class 1 phosphatidyl 3-phosphate (PIK3C1), PKB, or TSC1/TSC2 activity, nor do they stimulate MTORC1 directly (Kim and Guan, 2011; Laplante and Sabatini, 2012; Van Sluijters et al., 2000). However, a potential mechanism of amino acid sensing was suggested by the observation that amino acids, and leucine in particular, promote the association of Rheb with MTOR (Long et al., 2005). Recently, Demetriades et al. demonstrated that, upon amino acid removal, the GAP TSC1/TSC2 is recruited to the lysosome via its binding to the Rag GTPases. This brings TSC1/TSC2 into the vicinity of Rheb, resulting in stimulation of the GTP hydrolysis of $Rheb^{GTP}$ and leading to MTORC1 inactivation (Demetriades et al., 2014). MTORC1 may be anchored at the lysosomal surface by both Rag and $Rheb^{GTP}$ in the presence of amino acids. In fact, upon starvation in *TSC2*-null cells, MTORC1 remains at the lysosome and autophagy is not induced. In these cells, Rheb, which is constitutively active, seems to be sufficient to retain MTORC1 at the lysosome. The current model proposed by Demetriades et al. suggests that, in the presence of amino acids, Rag in its active conformation binds strongly to MTORC1, recruiting it to the lysosomal surface, and has a low affinity for TSC1/TSC2, which results in localization of this complex in the cytosol. $Rheb^{GTP}$, which is also present on the lysosome, reinforces the positioning of MTORC1 at the lysosome, activating MTORC1. Upon amino acid starvation, Rag shifts to its inactive conformation, which has a high affinity for TSC1/TSC2 and a low affinity for MTORC1. TSC1/TSC2 is recruited to the lysosomal surface and MTORC1 is released both because of the reduction of Rag affinity to MTORC1 and the inhibition of Rheb by TSC1/TSC2.

14.3.1.5 Regulation of ULK1 by Amino Acids

The ULK complex is another key factor in the regulation of autophagy that integrates nutrient availability and energy level. The ULK complex contains one of the kinases ULK1 or ULK2 which are functionally redundant, and FIP200, ATG13, and ATG101. In contrast to yeast, in mammals this complex is constitutively assembled (Wu et al., 2014). ULK1 directly interacts with ATG13 and FIP200 through the C-terminal domain; these interactions stabilize and activate ULK1 kinase activity (Chan et al., 2009; Ganley et al., 2009; Hosokawa et al., 2009; Jung et al., 2009). The ULK1 C-terminal domain is also involved in the binding of ULK1 to membranes (Chan et al., 2009). The ULK complex is regulated by differential phosphorylation, in particular in response to amino acid starvation and energy depletion (Wu et al., 2014). In this section, we will focus on the regulation of the ULK complex by amino acids (Fig. 14.3).

In the presence of amino acids, MTORC1 inhibits the ULK1 kinase activity and autophagy initiation by at least two mechanisms: direct phosphorylation of the ULK complex and indirect destabilization (Dunlop and Tee, 2014). Under nutrient-rich conditions, MTORC1 directly associates with ULK1 through its RAPTOR subunit and phosphorylates both ULK1 (on serine 757) and ATG13, suppressing the kinase activity of the ULK complex and inhibiting autophagy (Hara and Mizushima, 2009; Kim et al., 2011a). These phosphorylations may modulate the localization of the ULK complex and isolate it from its downstream targets (Wong et al., 2013). Interestingly, ULK1 phosphorylation on serine 757 also prevents its association with AMPK, coordinating the regulation of autophagy by amino acids (mainly through MTORC1) and glucose (mainly through AMPK) (see next section "Glucose") (Ganley et al., 2009; Hosokawa et al., 2009; Jung et al., 2009). Under amino acid starvation conditions, MTORC1 is released from the ULK complex, thus relieving the inhibition of ULK1 kinase. ULK1 then phosphorylates itself and its regulatory proteins ATG13 and FIP200, which enhances the ULK1 kinase activity (Kim et al., 2011a). ULK1 also phosphorylates the MTORC1 subunit, RAPTOR, resulting in a reduction of the ability of MTORC1 to bind its substrates. This positive feedback could maintain MTORC1 inhibition during the period of starvation. Lysosomal recycling of amino acids reactivates MTORC1 once an adequate level of cellular nutrients is reached (Russell et al., 2014). In addition to inhibition of autophagy through direct phosphorylation of the ULK complex, MTORC1 also phosphorylates AMBRA1 on serine 52, which indirectly results in the destabilization of the ULK complex (Nazio et al., 2013). The phosphorylation of AMBRA1 prevents the ubiquitination of ULK1 by TNF associated-receptor factor 6 (TRAF6), which reduces the stability and the activity of the ULK1 complex.

AMPK also appears to regulate autophagy in response to amino acid levels (Meijer et al., 2015). Inhibition of AMPK by amino acids has been observed in several studies (Ghislat et al., 2012; Li et al., 2013a; Xiao et al., 2011). This effect could be mediated by GDH, which plays a central role in amino acid catabolism and which is known to be specifically stimulated by leucine (see previous section "Regulation of MTORC1 by amino acids"). By producing α-KG, GDH probably replenishes citric acid cycle intermediates and increases the production of ATP, thus inhibiting AMPK. Moreover, the NADPH produced by glutamate deamination could also inhibit AMPK by reducing intracellular ROS levels (Lorin et al., 2013b; Meijer et al., 2015). As AMPK is involved in autophagy induction through direct and indirect inhibition of MTORC1 and through direct activation of ULK1, AMPK-mediated activities also contribute to induction of autophagy in response to amino acid starvation (see section "Glucose" for more details).

14.3.1.6 Regulation of PIK3C3 by Amino Acids

The core PIK3C3 complex I that promotes autophagy is formed by the kinase PIK3C3, its membrane adaptor VPS15, and Beclin 1, which acts as an adaptor for other regulatory subunits of the protein complex. When associated with ATG14 and AMBRA1, Beclin 1 regulates the early step in the autophagosome formation downstream of the ULK complex. The production of PI(3)P by this complex at the phagophore is necessary for the expansion of the membrane and autophagy initiation. In a separate complex with UVRAG, PIK3C3 controls the formation and maturation of autophagosomes (see previous section "Molecular mechanisms of macroautophagy"). The PIK3C3 complex I is tightly regulated at the level of its formation, localization, and activity.

The activity of the PIK3C3 complex I is controlled by at least the two protein kinases MTORC1 and ULK1, which have opposite effects (Fig. 14.3). Under nutrient-rich conditions, MTORC1 directly phosphorylates ATG14, inhibiting the activity of the PIK3C3 complex I and suppressing autophagosome formation (Yuan et al., 2013). Therefore, in response to amino acids (and other stimuli), MTORC1 inhibits autophagy by phosphorylating and inactivating both autophagy-initiating complexes ULK complex and PIK3C3 complex I.

Upon starvation, the inhibitory effect of MTORC1 on these complexes is relieved. The active ULK complex phosphorylates the two partners of the PIK3C3 complex I, Beclin 1 and AMBRA1, indicating a direct molecular coordination between the PIK3C3 and ULK complexes in the early stage of autophagosome formation. The phosphorylation of Beclin 1 on serine 15 by ULK1 seems to occur specifically at the phagophore and is required for the activation of the PIK3C3 complex I and the induction of autophagy (Russell et al., 2013). ATG14 likely acts as an adaptor to enable Beclin 1 binding to ULK1. The phosphorylation of AMBRA1 by ULK1 regulates the recruitment of the PIK3C3 complex I at the phagophore (Di Bartolomeo et al., 2010). Under nutrient-rich conditions, the PIK3C3 core complex is sequestered on microtubules through an interaction between AMBRA1 and the molecular motor dynein (more precisely, dynein light chain 1 LC8). Upon nutrient starvation, ULK1 phosphorylates AMBRA1 to prevent its interaction with, and to release it from, dynein. AMBRA1 together with Beclin 1 and PIK3C3 are thus translocated to the ER, allowing PIK3C3 to produce PI(3)P at the phagophore. In turn, AMBRA1 exerts a positive feedback loop on ULK1 by recruiting TRAF6 and promoting ULK1 ubiquitination and stabilization (Nazio et al., 2013).

The dynamics of the association between Beclin 1 and PIK3C3 is regulated by amino acids at two levels (Fig. 14.3). Under nutrient-rich conditions, Beclin 1 is associated with the antiapoptotic proteins B cell lymphoma 2 (Bcl-2) and B cell lymphoma extra-large (Bcl-xL) via its Bcl-2 homology domain (BH3) domain. This association inhibits autophagy, and the Beclin 1–Bcl-2 complex must dissociate to allow Beclin 1 binding to PIK3C3 and to initiate autophagy in response to nutrient starvation. The dissociation of the Beclin 1-Bcl-2 complex is promoted by phosphorylation of Bcl-2 (on threonine 69, serine 70, and serine 87) mediated by c-Jun N-terminal kinase 1 (JNK1), by phosphorylation of Beclin 1 mediated by the tumor suppressor death-associated protein kinase (DAPK), or by displacement of Beclin 1 from its complex with Bcl-2 mediated by other BH3-containing proteins such as the pro-apoptotic protein Bcl-2-associated death promoter (Bad) and the pro-autophagic protein Bcl-2/adenovirus E1B 19 kDa interacting protein 3 (BNIP3) (Mariño et al., 2014a; Meijer and Codogno, 2009). During starvation, JNK1 is recruited to microtubules via its interaction with the scaffold protein JNK interacting protein 1 (JIP-1), thus leading to JNK1 activation, Beclin 1-Bcl-2 dissociation, and autophagy induction (Geeraert et al., 2010). The protein Beclin 1 also interacts with microtubules via interactions of Beclin 1, the BH3-only protein Bim, and dynein. In starvation conditions, the phosphorylation of Bim by JNK1 abolishes this interaction and dissociates Beclin 1 from Bim, allowing Beclin 1 to bind PIK3C3 to initiate autophagy (Luo et al., 2012). How activation of the different Beclin 1 complexes is integrated during starvation-induced autophagy remains to be clarified.

14.3.1.7 Regulation of Autophagy by Transcription Factors

In addition to direct regulation of the autophagic machinery, amino acids also regulate autophagy at the transcriptional level by controlling the activity of transcription factors such as C/EBP homologous protein (CHOP), activating transcription factor 4 (ATF4), transcription factor EB (TFEB), transcription factor binding to IGHM enhancer 3 (TFE3), and forkhead box O3 (FoxO3) (Fig. 14.4). Under conditions of amino acid starvation, the drop in the availability of intracellular amino acids leads to an accumulation of uncharged tRNAs. The uncharged tRNAs bind to the serine/threonine kinase general control nondepressible 2 (GCN2). This induces a conformational change, a homodimerization, and an autophosphorylation of GCN2, which thus becomes active. GCN2 subsequently phosphorylates eukaryotic initiation factor 2α (eIF2α) on serine 51, leading to a global reduction of translation and an increase in the translation of a specific set of genes encoding enzymes involved in amino acid biosynthesis, amino acid transport, autophagy regulation, and transcription factors such as CHOP and ATF4. Both CHOP and ATG4 drive the transcription of genes encoding autophagy proteins implicated in the formation (eg, Beclin 1, ATG5, ATG12, and ATG16), elongation (eg, LC3), and function (eg, p62/SQSTM1) of autophagosomes (B'chir et al., 2013). Such a cellular response reduces the demand for amino acids and concomitantly increases their biosynthesis, recycling, and import in order to restore optimal intracellular levels of amino acids and to allow cells to survive under nutritionally scarce conditions.

Amino acids also control autophagy by regulating the transcription factors TFEB and TFE3 in a MTORC1-dependent manner (Settembre et al., 2012). TFEB and TFE3 are master regulators of the synthesis of ATG proteins and of lysosomal biogenesis and upregulate their transcription by binding to the coordinated lysosomal expression and regulation (CLEAR) motif found in the promoter regions of genes that respond to nutrient starvation. Under nutrient-rich conditions, these transcription factors are recruited by active Rag GTPases to the lysosomal membrane where MTORC1 phosphorylates TFEB on serine 142 and TFE3 on serine 311 (Martina et al., 2014; Settembre et al., 2012). Phosphorylated TFEB and TFE3 bind to cytosolic chaperones of 14-3-3 family, leading to their cytoplasmic retention (Martina et al., 2012). During amino acid starvation, TFEB and TFE3 are dephosphorylated and translocate to the nucleus where they upregulate the transcription of their target genes such as *LC3* and

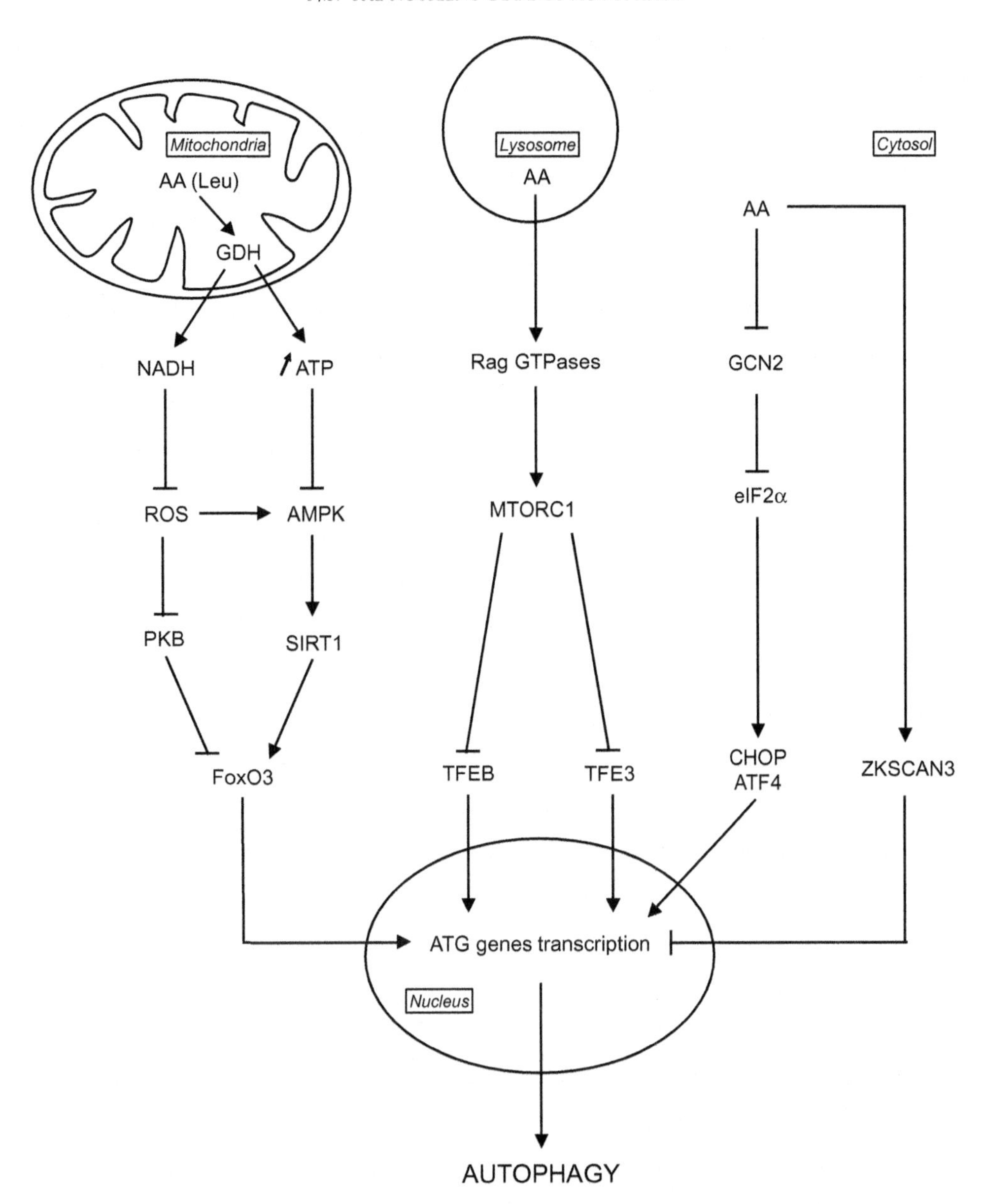

FIGURE 14.4 **Transcriptional regulation of autophagy by amino acids.** Several transcription factors (such as FoxO3, TFEB, TFE3, and CHOP/ATF4) can induce the transcription of autophagy genes in response to amino acid starvation. Each of the transcription factors regulates a different set of *ATG* genes. The transcription factor ZKSCAN3 inhibits the transcription of the same set of *ATG* genes that is activated by TFEB; these transcription factors thus have opposite effects on autophagy.

p62 (Settembre et al., 2011). Very recently, calcineurin, which is activated by a release of calcium from the lysosome through mucolipin 1 (MCOLN1) during starvation and physical exercise, has been identified as a phosphatase that activates TFEB (Medina et al., 2015). Moreover, the transcription factor cAMP-response element binding (CREB) has been identified as a positive regulator of autophagy genes (eg, *ULK1* and *ATG7*) and of the gene encoding transcription factor TFEB during periods of fasting (Seok et al., 2014).

Unlike TFEB, the transcription factor zinc finger with KRAB and SCAN domains 3 (ZKSCAN3) acts as a repressor of autophagy by repressing the transcription of more than 60 TFEB target genes involved in autophagy and lysosomal functions. In response to starvation, ZKSCAN3 translocates from the nucleus to the cytoplasm, removing its inhibitory effect (Settembre et al., 2013). In addition to its role in the regulation of the synthesis of ATG proteins and of the biogenesis of lysosomes, TFEB acts as a key player in the metabolic response to starvation by regulating expression of peroxisome proliferator-activated receptor-γ coactivator 1α (PGC1α, a transcriptional coactivator that stimulates the biogenesis of mitochondria), which, in turn, activates lipid catabolism.

PGC1α might also induce expression TFEB (Tsunemi et al., 2012). This allows a coordinated control of lipid degradation through both autophagy of lipid droplets (ie, lipophagy) and mitochondrial fatty acid oxidation (Settembre et al., 2013).

Amino acids are also involved in the regulation of the transcription factor FoxO3. FoxO3 is a target of the serine/threonine kinase PKB. Under nutrient-rich conditions, FoxO3 is phosphorylated by PKB, allowing its binding to the cytosolic chaperone 14-3-3 and preventing its translocation to the nucleus (Van Der Heide et al., 2004). In response to amino acid starvation, the increase in cellular ROS inactivates PKB (Rahman et al., 2014). FoxO3 then translocates to the nucleus, resulting in the transcription of a set of *ATG* genes (including *ULK2, Beclin 1, ATG4, ATG12,* and *LC3*) that differs from that controlled by TFEB (Mammucari et al., 2007).

14.3.2 Glucose

Glucose is one of the major sources of cellular energy. Glucose is stored as glycogen in liver and muscle cells and is metabolized by glycolysis and oxidative phosphorylation to produce ATP. When ATP is used for energy-requiring intracellular processes, it is hydrolyzed to ADP. Some of the ADP is converted to AMP in a reaction catalyzed by adenylate kinase. Under glucose deprivation, when ATP consumption exceeds its production, the ATP level falls, and levels of ADP and AMP increase. Increased levels of AMP and ADP directly activate the serine/threonine kinase AMPK, a fundamental regulator of cellular metabolism that responds to the cellular energy state (Hardie, 2014). When cells are energy-deprived, AMPK efficiently shuts down the MTORC1 pathway, which stimulates the major ATP-consuming metabolic pathways and simultaneously turns on catabolic pathways, including autophagy, allowing the cells to survive. AMPK acts through several pathways to induce autophagy including inhibition of MTORC1, activation of the ULK1 and PIK3 complexes, and upregulation of sirtuin 1 (SIRT1) (Fig. 14.3).

Upon glucose withdrawal, AMPK inhibits MTORC1 signaling by phosphorylating TSC2, activating its GAP activity to convert Rheb^{GTP} into Rheb^{GDP} to negatively regulate MTORC1. AMPK also phosphorylates RAPTOR on serine 722 and serine 792, stimulating the binding of 14-3-3 and inhibiting MTORC1 (Gwinn et al., 2008). During energy depletion, AMPK binds to Ragulator at the lysosomal membrane through the adaptor protein AXIN. This allows AMPK activation at low concentrations of glucose and prevents MTORC1 recruitment and activation. This process is dependent on the v-ATPase, which, in addition to its role as a sensor of the intralysosomal amino acid pool as discussed earlier (see previous section "Regulation of MTORC1 by amino acids"), also serves as an energy sensor (Zhang et al., 2014).

In addition, AMPK stimulates the activity of the ULK and PIK3C3 complexes to initiate autophagy. At low glucose concentrations, AMPK phosphorylates ULK1 on several residues, leading to autophagy induction (Kim et al., 2011a). ULK1 is thus phosphorylated by both MTORC1 and AMPK, but at different sites, with opposing effects on ULK1 activity and autophagy (Meijer and Codogno, 2011). When MTORC1 is active, AMPK-mediated activation of ULK1 is prevented by MTORC1-mediated phosphorylation of serine 757 of ULK1, which inhibits ULK1 binding to AMPK. The other autophagy complex, PIK3C3, is also activated by AMPK-dependent phosphorylation of Beclin 1 on serines 91 and 94 (Kim et al., 2013). This phosphorylation activates both ATG14- and UVRAG-containing PIK3C3 complexes and is essential for autophagy induction upon glucose withdrawal. Thus, as with ULK1, Beclin 1 is phosphorylated by different kinases with opposing effects: by AMPK leading to PIK3C3 complex I activation and by PKB leading to PIK3C3 complex I inhibition (Wang et al., 2012).

Finally, AMPK can promote autophagy through the activation of the protein deacetylase SIRT1 (see section "AcetylCoA") (Cantó et al., 2009; Ruderman et al., 2010). SIRT1 induces autophagy by deacetylating and activating several ATG proteins and by deacetylating and activating the transcription factor FoxO3, which leads to an increased transcription of *ATG* genes (Cantó et al., 2009; Lee et al., 2008; Ruderman et al., 2010). It is important to note that all mechanisms by which AMPK induces autophagy in response to glucose starvation, as discussed here, could also be involved in the regulation of autophagy in response to amino acid starvation when AMPK is also activated, as discussed earlier.

Apart from being connected to the MTORC1 pathway by AMPK the availability of glucose is also connected to MTORC1 through the glycolytic enzymes hexokinase II (the predominant form of hexokinase present in insulin-sensitive tissues such as muscle, heart, and adipose tissue) and glyceraldehyde 3-phosphate dehydrogenase (GAPDH). When the concentration of glucose falls, substrate-free hexokinase II specifically binds to MTORC1 and inhibits its activity by preventing the binding of RAPTOR; this stimulates autophagy (Roberts et al., 2014). Inhibition of MTORC1 activity by binding of hexokinase II is relieved by glucose 6-phosphate, the product of the hexokinase reaction. Thus, hexokinase II seems to control the switch between two opposing processes, one involved in energy production by glucose degradation and the other involved in energy conservation by autophagy. Glycolytic flux is also signaled to MTORC1 through GAPDH. When the concentration of glucose decreases, glyceraldehyde 3-phosphate-free GAPDH binds to Rheb and prevents its interaction with MTOR (Lee et al., 2009).

14.3.3 Fatty Acids

The capacity of fatty acids to induce autophagy and the mechanisms involved are strongly influenced by the lengths of their carbon chains (Fig. 14.5). Both saturated and *cis*-unsaturated fatty acids stimulate autophagy when the carbon chain lengths are 15–18 and 14–20 carbons, respectively. However, the molecular mechanisms involved in the autophagic response present major functional differences. Saturated fatty acids, such as palmitate, trigger a canonical autophagy by controlling the cytoplasmic association of the transcription factor signal transducer and activator of transcription 3 (STAT3) and the protein kinase RNA-activated (PKR) (Shen et al., 2012). In the absence of palmitate, cytoplasmic STAT3 competitively inhibits PKR activity, probably by binding to its catalytic domain. In the presence of palmitate, the STAT3–PKR complex dissociates, and PKR phosphorylates its substrates eIF2α and JNK1. eIF2α phosphorylation results in inhibition of protein translation, a state necessary for an optimal induction of autophagy. One possibility is that this arrest of protein translation favors the formation of autophagosomes at the ER membrane. Another possibility is that eIF2α phosphorylation activates the transcription factors CHOP and ATF4, thus

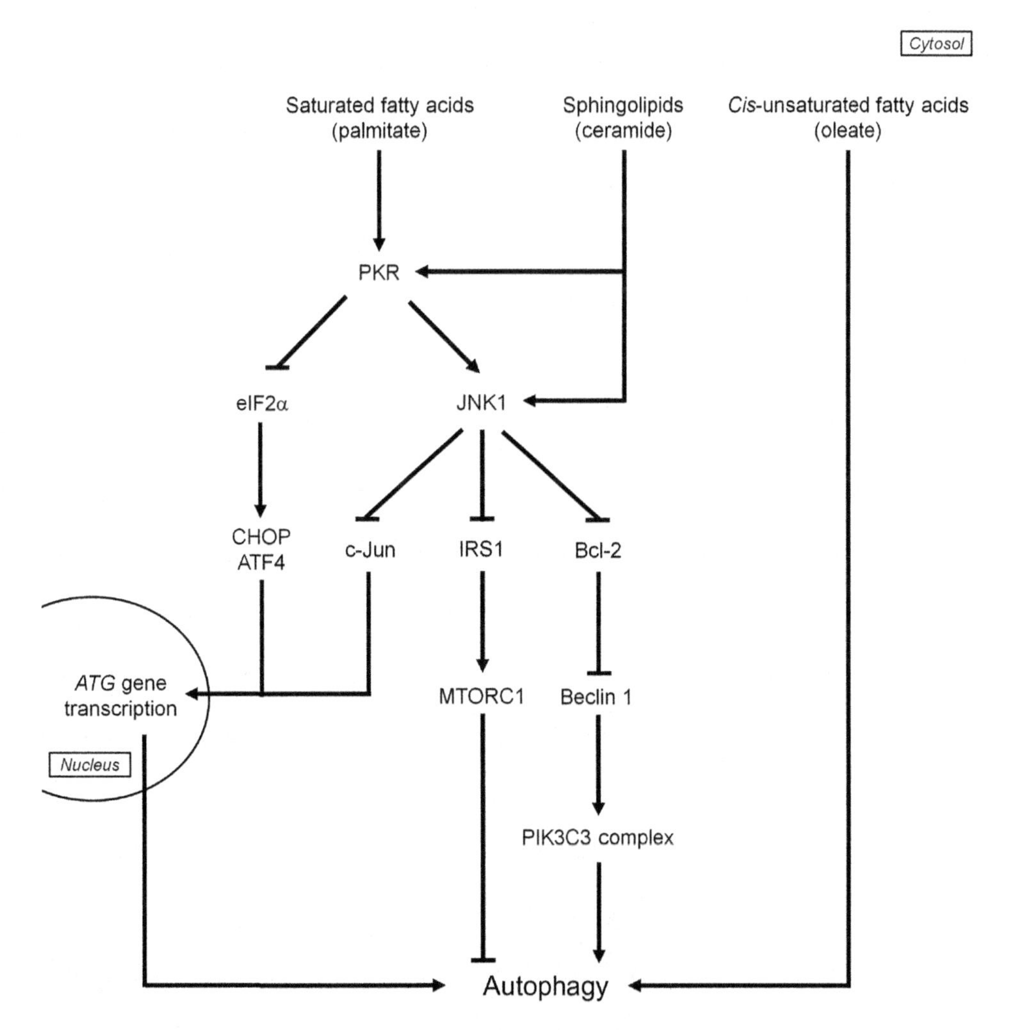

FIGURE 14.5 **Regulation of autophagy by fatty acids.** Sphingolipids and saturated and *cis*-unsaturated fatty acids are able to stimulate autophagy. The molecular mechanisms involved have functional differences: sphingolipids and saturated fatty acids trigger a canonical autophagy, whereas *cis*-unsaturated fatty acids induce a noncanonical autophagy independent of PIK3C3 and Beclin 1.

inducing *ATG* gene transcription and autophagy (see section "Regulation of autophagy by transcription factors"). Moreover, the palmitate-induced dissociation of the STAT3–PKR complex activates JNK1, a kinase that stimulates autophagy by phosphorylating Bcl-2 and Bim, thus freeing Beclin 1 to activate the PIK3C3 complex I. JNK1 can also phosphorylate and inhibit insulin receptor substrate 1 (IRS1), thus leading to MTORC1 inhibition (Yang et al., 2010). The palmitate-induced autophagy is thus dependent on the PIK3C3 complex I.

In contrast to saturated fatty acids, *cis*-unsaturated fatty acids (such as oleate) induce a noncanonical autophagy that is independent of PIK3C3 and Beclin 1 and that does not require PI(3)P for autophagosome formation (Niso-Santano et al., 2015). Oleate-induced autophagosomes colocalize with the Golgi apparatus, which is indispensable for their formation. Interestingly, stearoyl-CoA desaturase (SCD), which converts saturated into unsaturated fatty acids and generates oleate, is required for starvation-induced autophagy (Ogasawara et al., 2014). An SCD inhibitor suppresses the formation of ULK1, WIPI1, ATG16L, p62/SQSTM1, and LC3 puncta, suggesting that SCD is involved in the earliest step of autophagy. Given the opposite effects of saturated and *cis*-unsaturated fatty acids on health, it is tempting to speculate that the beneficial effects of *cis*-unsaturated fatty acids are linked to the induction of a noncanonical autophagy.

Other lipid components such as sphingolipids (eg, ceramide) stimulate autophagy by inhibiting the MTORC1 pathway, activating the dissociation of the Beclin 1–Bcl-2 complex through JNK1 activation, activating transcription of *ATG* genes via c-Jun, and stimulating mitophagy (Bedia et al., 2011; Dall'Armi et al., 2013; Jiang and Ogretmen, 2014). Interestingly, ceramide is known to induce PKR and JNK1-mediated consecutive inhibition of IRS1, thus probably leading to MTORC1 inactivation (Yang et al., 2010). This regulation is reminiscent of the stimulation of autophagy by amino acid depletion probably because ceramide downregulates the membrane expression of amino acid transporters (Guenther et al., 2008).

14.4. METABOLITES AND AUTOPHAGY

14.4.1 NAD^+/NADH

The coenzyme NAD is involved in oxidation–reduction reactions critical for glycolysis, fatty acid oxidation, the TCA cycle, and complex I of the mitochondrial respiratory chain and also is a key regulator of autophagy. At least two different mechanisms are involved. First, the NAD^+-dependent deacetylase SIRT1 activates autophagy by directly deacetylating ATG proteins (Lee et al., 2008). Under starvation conditions, the increased NAD^+/NADH ratio activates SIRT1, which results in stimulation of mitophagy (Jang et al., 2012). Second, the hydrogen of NADH can be transferred to $NADP^+$ to form NADPH via the energy-linked transhydrogenase. In the fed state, when the NAD^+/NADH ratio falls, NADPH inhibits autophagy by scavenging of ROS via the glutathione peroxidase-glutathione reductase system and by preventing the production of ROS at complex 1 of the respiratory chain (Albracht et al., 2011). In addition, NADPH may be formed by the GDH reaction in the mitochondria (Lorin et al., 2013b). This hypothesis is supported by experiments carried out in our laboratories showing that ROS production in starved cells is suppressed by amino acids and that inhibition of GDH expression prevented these effects (Lorin et al., 2013b). Inhibition of GDH expression also stimulates autophagy and inhibits MTORC1-mediated signaling in the presence of amino acids. Cytoplasmic ROS, which can also activate autophagy, may be eliminated by NADPH produced by glucose 6-phosphate dehydrogenase (G6PDH). Inhibition of flux through phosphofructokinase-1 (PFK1), by decreasing the concentration of its essential activator fructose 2,6-diphosphate through genetic means, increases the concentration of glucose 6-phosphate, stimulates NADPH production by G6PDH, stimulates MTORC1 activity, and inhibits autophagy. Opposite changes occur when glycolytic flux is increased (Bensaad et al., 2009; Strohecker et al., 2015).

14.4.2 AcetylCoA

14.4.2.1 Levels of AcetylCoA and Regulation of Autophagy

The metabolite acetylCoA is involved in both catabolism (such as fatty acid oxidation and the TCA cycle) and in biosynthetic pathways (such as the synthesis of fatty acids and cholesterol). Interestingly, acetylCoA is also the sole donor of acetyl residues for acetylation of lysine residues of proteins in the cytoplasm, mitochondria, and nucleus. Acetylation can modify protein activity, stability, or interactome. It has been postulated that the cytoplasmic concentration of acetylCoA reflects the nutritional state of the cell. High acetylCoA levels signify a proliferative or fed state, whereas low acetylCoA levels are indicative of a quiescent or starved state (Cai and Tu, 2011). The acetylation of proteins is governed by the cytoplasmic acetylCoA concentration and allows protein activity to change in accordance with the requirements of cell metabolism. During nutrient starvation, the most potent inducer of autophagy, there is a rapid depletion of acetylCoA from the cytoplasm and a global deacetylation of cytoplasmic proteins in cultured cells, mice, *Drosophila*, and yeast (Eisenberg et al., 2014; Mariño et al., 2014b; Morselli et al., 2011). Additionally, treatments with resveratrol or spermidine, two recently discovered inducers of autophagy, strongly decrease the acetylation of cytoplasmic proteins. Interestingly, analysis of the acetylome also revealed that resveratrol and spermidine decrease the acetylation of 170 intracellular proteins that belong to the human autophagy protein network (Behrends et al., 2010; Morselli et al., 2011). These compounds specifically abrogate the activity of the acetyltransferase E1A binding protein p300 (EP300) (Pietrocola et al., 2015). By using multiple approaches, including genetic deletion of enzymes involved in the generation of acetylCoA in the cytosol, such as ATP citrate lyase (ACLY), it was shown that decreasing the level of acetylCoA induces autophagy in mammals (Mariño et al., 2014b). Altogether, these data strongly suggest that the level of cytosolic acetylCoA is a good indicator of autophagy.

14.4.2.2 Acetylation of ATG Proteins and Regulation of Autophagy

It is now clearly established that posttranslational modifications of ATG proteins modulate autophagosome formation. Among these modifications, recent data strongly support the critical role of acetylation in the regulation of autophagy. The acetylation of critical lysine residues is catalyzed by a specific class of enzymes called acetyltransferases, whereas the removal of the acetyl group is catalyzed by enzymes called deacetylases (histone deacetylases HDAC, sirtuins). In the presence of serum, ATG5, ATG7, ATG12, and LC3 are acetylated by the acetyltransferase EP300, a process that inhibits autophagy (Lee and Finkel, 2009).

As discussed above, the level of acetylCoA is critical for the activity of EP300. The intracellular localization of EP300 itself is controlled by HLA-B-associated transcript 3 (BAT3, also known as BAG6), a protein essential for starvation-induced autophagy. BAT3 is a nucleo-cytoplasmic protein known for its role in apoptosis and in cytoplasmic quality control (Lee and Ye, 2013). During starvation, BAT3 triggers the nuclear translocation of EP300, which causes TP53 acetylation. Depletion of EP300 from the cytosol leads to the deacetylation of ATG7, ATG5, and LC3 in mouse embryonic fibroblasts. Thus, BAT3 guides EP300 and regulates acetylation of both cytosolic and nuclear targets to modulate autophagy (Sebti et al., 2014a,b).

Other acetyltransferases also regulate autophagy. The acetylation of ULK1 by the acetyltransferase Tat interacting protein 60 kDa (TIP60) induces autophagy after growth factor deprivation (Lin et al., 2012). In yeast, essential Sas2-related acetyltransferase 1 (ESA1; TIP60 in mammals) is necessary for acetylation of ATG3, a modification essential for its interaction with, and lipidation of, ATG8 (LC3 in mammals) (Yi et al., 2012). The mechanism of the regulation of autophagy by LC3 acetylation has been carefully explored recently. LC3 intracellular localization is tightly regulated by acetylation. Upon starvation, lysines 49 and 51 of nuclear LC3 are deacetylated by SIRT1, allowing its interaction with the shuttling protein DOR, its cytosolic relocalization, and its pro-autophagic function (this process is inhibited after EP300 overexpression) (Huang et al., 2015). SIRT1 also deacetylates ATG5 and ATG7 during autophagic induction upon starvation, but the mechanism is still poorly understood (Lee et al., 2008). HDAC6 is also able to partially deacetylate LC3-II upon starvation (Liu et al., 2013).

14.4.2.3 Acetylation and Epigenetic Regulation of Autophagy

The first nonhistone protein known to be regulated by acetylation and deacetylation was the protein TP53 (Gu and Roeder, 1997). The role of TP53 in autophagy is still under debate. Acetylation of TP53 on lysine 373 by the acetyltransferase EP300 is essential for starvation-induced autophagy, leading to the expression of pro-autophagic proteins (Sebti et al., 2014b). The acetylation of TP53 is dependent on the nucleo-cytoplasmic shuttling protein BAT3, which transfers a pool of EP300 from the cytosol to the nucleus during autophagy induction. By contrast, another study reported that the nuclear localization of TP53 inhibits autophagy. When lysines 120 and 386 of TP53 are replaced by arginine, TP53 remains located in the nucleus and autophagy is inhibited. When lysine 120 of TP53 is acetylated by the acetyltransferase TIP60, the cytoplasmic localization of TP53 is increased, leading to stimulation of autophagy (Naidu et al., 2012). Another group reported that IFN-γ induces autophagy by suppressing BMF expression through HDAC1-mediated deacetylation of lysine 382 of TP53 (Contreras et al., 2013). It is known that in normal cells, glucose starvation promotes stabilization of wild-type TP53. Interestingly, in cancer cells carrying a TP53 mutation, glucose restriction leads to TP53 mutant deacetylation and degradation by autophagy (Rodriguez et al., 2012).

Acetylation also modulates autophagy through effects on the transcriptome. The deacetylation of the transcription factor FoxO3a by SIRT1 enhances the expression of the pro-autophagic BNIP3 (Kume et al., 2010b). Similarly, FoxO3 also regulates the expression of the GTP-binding protein RAB7 (Kume et al., 2010b). Another sirtuin,

SIRT2, is also involved in the regulation of autophagy. After starvation, SIRT2 dissociates from FoxO1, leading to FoxO1 acetylation. Once acetylated, FoxO1 interacts with ATG7 in the cytosol favoring autophagy (Zhao et al., 2010). FoxO3 promotes the translocation of FoxO1 from the nucleus to the cytoplasm, resulting in an increase in FoxO1-induced autophagy (Zhou et al., 2012). Epigenetic control of the autophagy-relevant transcriptome is also modulated by acetylation. Upon modulation of histone H4 acetylation at lysine 16, cells modified their autophagy-relevant transcriptome (Füllgrabe et al., 2013).

14.4.3 Ammonia

The regulation of autophagy by nutrients is cell autonomous, but it can also occur at a site distant from their production site because of nutrient diffusion through the environment or through an organism. This is the case for ammonia, which, for example, is produced during the degradation of glutamine to glutamate (Eng et al., 2010). After a long-term glucose starvation of yeast, ammonia generated by amino acid catabolism induces autophagy (Cheong et al., 2012). The stimulation of autophagy by ammonia is independent of both MTORC1 and ULK1 (Cheong et al., 2011; Harder et al., 2014). Rather, ammonia seems to trigger autophagy by activating AMPK and favoring the ER-stress response (Harder et al., 2014).

The regulation of autophagy by ammonia is probably important to consider in tumors, which are highly dependent on glutamine for their survival. Due to their high proliferation rate, tumor cells have a high energetic demand, and established cancer cells rely on autophagy for their survival and proliferation (Lorin et al., 2013a). To ensure cancer cell proliferation, stromal cells in the tumor microenvironment are essential partners. Ammonia secreted by cancer cells stimulates autophagy in the adjacent stromal cells. This leads to the stromal generation and secretion of large amounts of glutamine (produced by glutamine synthetase) into the tumor microenvironment. Cancer cells use glutamine and convert it to glutamate, thus releasing ammonia. This positive amplification feedback loop supports tumor cell proliferation.

14.4.4 Nucleotides

A role of autophagy in the regulation of the concentration of deoxyribonucleotides, fundamental for DNA synthesis and repair, has become evident only recently. In human cancer cells, intracellular levels of dNTPs (dCTP, dGTP, dATP, and dTTP) are strongly decreased by starvation- or rapamycin-induced autophagy because of downregulation of ribonucleotide reductase. Conversely, both inhibition of expression of RRM2 (a subunit of ribonucleotide reductase) and a decrease in the level of intracellular dNTP by pharmacological means induces autophagy (Chen et al., 2014). Moreover, high levels of dNTPs desensitize cells to the induction of autophagy by rapamycin, suggesting a direct linkage between nucleotide pools and autophagy.

Among nucleotides, GTP and AMP/ADP appear to have predominant roles in autophagy regulation. As described above, many small GTPases, such as Rheb, Rag, RalA, and Rab, are involved in the regulation or the execution of the autophagic process. The small GTPases require GTP for activation. As discussed earlier (see previous section "Regulation of MTORC1 by amino acids"), a drop in GTP availability has a negative impact on autophagy. AMP and ADP reveal the energetic state of the cell (Cai and Tu, 2011; Hardie, 2011). High levels of AMP and ADP signal to the cell that energy-consuming processes should be turned off and that autophagy should be induced to restore energy stores. Both AMP and ADP are sensed by AMPK, which is allosterically activated by AMP binding to the γ subunit of this trimeric serine/threonine kinase. In addition, AMP and ADP activate AMPK by preventing dephosphorylation of threonine 172 in the activation loop of α subunit of AMPK (Sanders et al., 2007; Suter et al., 2006; Xiao et al., 2007).

14.5. CONCLUSION

As described in this chapter, autophagy is controlled by levels of nutrients and their metabolites. Caloric restriction efficiently induces autophagy in all organisms studied so far. In addition to its role in mobilizing endogenous resources to meet the energetic demands when extracellular nutrients are scarce, autophagy contributes to the maintenance of cellular homeostasis by eliminating damaged organelles. Damaged mitochondria can lead to ROS production and DNA damage. Autophagy also eliminates protein aggregates that are toxic to the cell, thus mediating cytoprotection (Rubinsztein et al., 2011). This cytoprotective effect appears to be involved in the antiaging and health-promoting effects of autophagy. It is important to note that antioxidants have not proven to be successful in combatting aging (Dai et al., 2014; Whelan and Zuckerbraun, 2013). An explanation may be that these compounds strongly interfere with autophagy undermining its protective role (Underwood et al., 2010). Too much (but also too little) activity of MTOR can cause liver injury (Abraham, 2014), perhaps for the same reason.

Interestingly, long-term or intermittent short-term caloric restriction does prolong lifespan in yeast, plants, worms, flies, rodents, and primates (Bergamini et al., 2007; Colman et al., 2009; Fontana et al., 2010; Libert and Guarente, 2013; Mattison et al., 2012; Rubinsztein et al., 2011). In primates, caloric restriction also appears to increase healthspan by reducing the incidence of metabolic diseases, cancer, arteriosclerosis, and neurodegeneration. This beneficial effect of caloric restriction on lifespan appears to be mediated by autophagy, as normal and pathological aging is associated with a diminished autophagic potential in various organs, which correlates with accumulation of ubiquitinated protein aggregates and defective mitochondria. Moreover, manipulations that increase lifespan stimulate autophagy, and suppression of autophagy compromises the extension of longevity induced by caloric restriction (Madeo et al., 2015).

Pharmacologic induction of autophagy has been pursued extensively in the last decade for the treatment of age-associated pathologies (neurodegenerative disease in particular) and for improving healthspan in general. Rapamycin, which has an FDA accreditation, is a potent autophagy inducer (through inhibition of mTORC1) that extends life span in a variety of organisms (Harrison et al., 2009; Li et al., 2014; Neff et al., 2013; Robida-Stubbs et al., 2012; Wilkinson et al., 2012). Unfortunately, rapamycin has severe side effects: it may cause, for example, insulin resistance, glucose intolerance, testicular degeneration, and cataracts, and cannot be used to treat patients long term (Chang et al., 2009; Houde et al., 2010; Wilkinson et al., 2012). Caloric restriction mimetics (CRM), which are molecules that mimic caloric restriction and induce autophagy by deacetylating a large panel of cellular proteins, are promising molecules for antiaging therapy. Interestingly, several of these molecules are currently used in traditional medicines. Examples are anacardic acid (from the nutshell of the cashew), curcumin (an ingredient of curry powder), garcinol (from the fruit of the Kokum tree), epigallocatechin-3-gallate (from green tea), and spermidine (found at particularly high concentrations in durian fruit, fermented soybeans, and wheat germ) (Balasubramanyam et al., 2004; Devipriya and Kumaradhas, 2012, 2013; Eisenberg et al., 2009; Morselli et al., 2010; Niu et al., 2013).

The link between autophagy and nutrition has also had an important impact on the treatment of patients with critical illness. Recent reports suggest that parenteral nutrition should be postponed to day 8 instead of day 3 (as currently recommended) during critical illness (Casaer et al., 2011, 2013; Schetz et al., 2013). In fact, strong evidence links pathological changes at the cell and tissue level with organ failures and mortality in the context of systemic inflammation during critical illness (Yasuhara et al., 2007). In particular, mitochondria are subject to injury during systemic inflammation and impaired mitochondrial function is linked to organ failure and death. In contrast, restoration of mitochondrial normal function is a hallmark of survivors of critical illness (Brealey et al., 2002; Carré et al., 2010). Hyperinsulinemia and hyperglycemia (reflecting insulin resistance) are also common manifestations of critical illness. Correcting this hyperglycemia is associated with enhanced autophagy and a lower 7-day mortality, whereas impaired autophagy correlates with the severity of organ damage. Thus, during critical illness, hyperglycemia suppresses autophagy in vital organs, which promotes organ damage and possibly death (Gunst et al., 2013). Delaying parenteral nutrition during the first days of critical illness could be beneficial as it should induce autophagy and improve patient survival.

Acknowledgments

This work was supported by institutional funding from INSERM (S.L., S.P., P.C.), CNRS (P.C.), University Paris-Saclay (S.L.), and University Paris Descartes (P.C.), and by grants from ANR and INCa (P.C.), the ARC Foundation (S.P.), and La Ligue Régionale contre Le Cancer (Gard and Hérault committee) (S.P.).

References

Abraham, R.T., 2014. Too little mTORC1 activity injures the liver. Cell Metab. 20, 4–6.

Albracht, S.P.J., Meijer, A.J., Rydström, J., 2011. Mammalian NADH: ubiquinone oxidoreductase (Complex I) and nicotinamide nucleotide transhydrogenase (Nnt) together regulate the mitochondrial production of H_2O_2—implications for their role in disease, especially cancer. J. Bioenerg. Biomembr. 43, 541–564.

Angcajas, A.B., Hirai, N., Kaneshiro, K., Karim, M.R., Horii, Y., Kubota, M., et al., 2014. Diversity of amino acid signaling pathways on autophagy regulation: a novel pathway for arginine. Biochem. Biophys. Res. Commun. 446, 8–14.

Ao, X., Zou, L., Wu, Y., 2014. Regulation of autophagy by the Rab GTPase network. Cell Death Differ. 21, 348–358.

Balasubramanyam, K., Altaf, M., Varier, R.A., Swaminathan, V., Ravindran, A., Sadhale, P.P., et al., 2004. Polyisoprenylated benzophenone, garcinol, a natural histone acetyltransferase inhibitor, represses chromatin transcription and alters global gene expression. J. Biol. Chem. 279, 33716–33726.

Baquet, A., Hue, L., Meijer, A.J., van Woerkom, G.M., Plomp, P.J., 1990. Swelling of rat hepatocytes stimulates glycogen synthesis. J. Biol. Chem. 265, 955–959.

Bar-Peled, L., Sabatini, D.M., 2014. Regulation of mTORC1 by amino acids. Trends Cell Biol. 24, 400–406.

Bar-Peled, L., Schweitzer, L.D., Zoncu, R., Sabatini, D.M., 2012. Ragulator is a GEF for the rag GTPases that signal amino acid levels to mTORC1. Cell 150, 1196–1208.

Bar-Peled, L., Chantranupong, L., Cherniack, A.D., Chen, W.W., Ottina, K.A., Grabiner, B.C., et al., 2013. A Tumor suppressor complex with GAP activity for the Rag GTPases that signal amino acid sufficiency to mTORC1. Science 340, 1100–1106.

B'chir, W., Maurin, A.-C., Carraro, V., Averous, J., Jousse, C., Muranishi, Y., et al., 2013. The eIF2α/ATF4 pathway is essential for stress-induced autophagy gene expression. Nucleic Acids Res. 41, 7683–7699.

Bedia, C., Levade, T., Codogno, P., 2011. Regulation of autophagy by sphingolipids. Anticancer Agents Med. Chem. 11, 844–853.

Behrends, C., Sowa, M.E., Gygi, S.P., Harper, J.W., 2010. Network organization of the human autophagy system. Nature 466, 68–76.

Bensaad, K., Cheung, E.C., Vousden, K.H., 2009. Modulation of intracellular ROS levels by TIGAR controls autophagy. EMBO J. 28, 3015–3026.

Bergamini, E., Cavallini, G., Donati, A., Gori, Z., 2007. The role of autophagy in aging: its essential part in the anti-aging mechanism of caloric restriction. Ann. N. Y. Acad. Sci. 1114, 69–78.

Blommaart, E.F., Luiken, J.J., Blommaart, P.J., van Woerkom, G.M., Meijer, A.J., 1995. Phosphorylation of ribosomal protein S6 is inhibitory for autophagy in isolated rat hepatocytes. J. Biol. Chem. 270, 2320–2326.

Bolisetty, S., Jaimes, E.A., 2013. Mitochondria and reactive oxygen species: physiology and pathophysiology. Int. J. Mol. Sci. 14, 6306–6344.

Brealey, D., Brand, M., Hargreaves, I., Heales, S., Land, J., Smolenski, R., et al., 2002. Association between mitochondrial dysfunction and severity and outcome of septic shock. Lancet 360, 219–223.

Bryk, B., Hahn, K., Cohen, S.M., Teleman, A.A., 2010. MAP4K3 regulates body size and metabolism in Drosophila. Dev. Biol. 344, 150–157.

Byfield, M.P., Murray, J.T., Backer, J.M., 2005. hVps34 is a nutrient-regulated lipid kinase required for activation of p70 S6 kinase. J. Biol. Chem. 280, 33076–33082.

Cai, L., Tu, B.P., 2011. On acetyl-CoA as a gauge of cellular metabolic state. Cold Spring Harb. Symp. Quant. Biol. 76, 195–202.

Cantó, C., Gerhart-Hines, Z., Feige, J.N., Lagouge, M., Noriega, L., Milne, J.C., et al., 2009. AMPK regulates energy expenditure by modulating NAD+ metabolism and SIRT1 activity. Nature 458, 1056–1060.

Carré, J.E., Orban, J.-C., Re, L., Felsmann, K., Iffert, W., Bauer, M., et al., 2010. Survival in critical illness is associated with early activation of mitochondrial biogenesis. Am. J. Respir. Crit. Care Med. 182, 745–751.

Casaer, M.P., Mesotten, D., Hermans, G., Wouters, P.J., Schetz, M., Meyfroidt, G., et al., 2011. Early versus late parenteral nutrition in critically ill adults. N. Engl. J. Med. 365, 506–517.

Casaer, M.P., Wilmer, A., Hermans, G., Wouters, P.J., Mesotten, D., Van den Berghe, G., 2013. Role of disease and macronutrient dose in the randomized controlled EPaNIC trial: a post hoc analysis. Am. J. Respir. Crit. Care Med. 187, 247–255.

Chan, E.Y.W., Longatti, A., McKnight, N.C., Tooze, S.A., 2009. Kinase-inactivated ULK proteins inhibit autophagy via their conserved C-terminal domains using an Atg13-independent mechanism. Mol. Cell. Biol. 29, 157–171.

Chang, G.-R., Wu, Y.-Y., Chiu, Y.-S., Chen, W.-Y., Liao, J.-W., Hsu, H.-M., et al., 2009. Long-term administration of rapamycin reduces adiposity, but impairs glucose tolerance in high-fat diet-fed KK/HlJ mice. Basic Clin. Pharmacol. Toxicol. 105, 188–198.

Chen, W., Zhang, L., Zhang, K., Zhou, B., Kuo, M.-L., Hu, S., et al., 2014. Reciprocal regulation of autophagy and dNTP pools in human cancer cells. Autophagy 10, 1272–1284.

Cheong, H., Lindsten, T., Wu, J., Lu, C., Thompson, C.B., 2011. Ammonia-induced autophagy is independent of ULK1/ULK2 kinases. Proc. Natl. Acad. Sci. U.S.A 108, 11121–11126.

Cheong, H., Lindsten, T., Thompson, C.B., 2012. Autophagy and ammonia. Autophagy 8, 122–123.

Colman, R.J., Anderson, R.M., Johnson, S.C., Kastman, E.K., Kosmatka, K.J., Beasley, T.M., et al., 2009. Caloric restriction delays disease onset and mortality in rhesus monkeys. Science 325, 201–204.

Contreras, A.U., Mebratu, Y., Delgado, M., Montano, G., Hu, C.-A.A., Ryter, S.W., et al., 2013. Deacetylation of p53 induces autophagy by suppressing Bmf expression. J. Cell Biol. 201, 427–437.

Dai, D.-F., Chiao, Y.A., Marcinek, D.J., Szeto, H.H., Rabinovitch, P.S., 2014. Mitochondrial oxidative stress in aging and healthspan. Longev. Healthspan 3, 6.

Dall'Armi, C., Devereaux, K.A., Di Paolo, G., 2013. The role of lipids in the control of autophagy. Curr. Biol. 23, R33–R45.

Demetriades, C., Doumpas, N., Teleman, A.A., 2014. Regulation of TORC1 in response to amino acid starvation via lysosomal recruitment of TSC2. Cell 156, 786–799.

De Duve, C., Baudhuin, P., 1966. Peroxisomes (microbodies and related particles). Physiol. Rev. 46, 323–357.

Devipriya, B., Kumaradhas, P., 2012. Probing the effect of intermolecular interaction and understanding the electrostatic moments of anacardic acid in the active site of p300 enzyme via DFT and charge density analysis. J. Mol. Graph. Model. 34, 57–66.

Devipriya, B., Kumaradhas, P., 2013. Molecular flexibility and the electrostatic moments of curcumin and its derivatives in the active site of p300: a theoretical charge density study. Chem. Biol. Interact. 204, 153–165.

Di Bartolomeo, S., Corazzari, M., Nazio, F., Oliverio, S., Lisi, G., Antonioli, M., et al., 2010. The dynamic interaction of AMBRA1 with the dynein motor complex regulates mammalian autophagy. J. Cell Biol. 191, 155–168.

Dodd, K.M., Tee, A.R., 2012. Leucine and mTORC1: a complex relationship. Am. J. Physiol. Endocrinol. Metab. 302, E1329–E1342.

Dunlop, E.A., Tee, A.R., 2014. mTOR and autophagy: a dynamic relationship governed by nutrients and energy. Semin. Cell Dev. Biol. 36, 121–129.

Duran, A., Amanchy, R., Linares, J.F., Joshi, J., Abu-Baker, S., Porollo, A., et al., 2011. p62 is a key regulator of nutrient sensing in the mTORC1 pathway. Mol. Cell 44, 134–146.

Durán, R.V., Oppliger, W., Robitaille, A.M., Heiserich, L., Skendaj, R., Gottlieb, E., et al., 2012. Glutaminolysis activates Rag-mTORC1 signaling. Mol. Cell 47, 349–358.
Eisenberg, T., Knauer, H., Schauer, A., Büttner, S., Ruckenstuhl, C., Carmona-Gutierrez, D., et al., 2009. Induction of autophagy by spermidine promotes longevity. Nat. Cell Biol. 11, 1305–1314.
Eisenberg, T., Schroeder, S., Andryushkova, A., Pendl, T., Küttner, V., Bhukel, A., et al., 2014. Nucleocytosolic depletion of the energy metabolite acetyl-coenzyme a stimulates autophagy and prolongs lifespan. Cell Metab. 19, 431–444.
Eng, C.H., Yu, K., Lucas, J., White, E., Abraham, R.T., 2010. Ammonia derived from glutaminolysis is a diffusible regulator of autophagy. Sci. Signal. 3, ra31.
Findlay, G.M., Yan, L., Procter, J., Mieulet, V., Lamb, R.F., 2007. A MAP4 kinase related to Ste20 is a nutrient-sensitive regulator of mTOR signalling. Biochem. J. 403, 13–20.
Flinn, R.J., Yan, Y., Goswami, S., Parker, P.J., Backer, J.M., 2010. The late endosome is essential for mTORC1 signaling. Mol. Biol. Cell 21, 833–841.
Fontana, L., Partridge, L., Longo, V.D., 2010. Extending healthy life span—from yeast to humans. Science 328, 321–326.
Fujita, N., Hayashi-Nishino, M., Fukumoto, H., Omori, H., Yamamoto, A., Noda, T., et al., 2008. An Atg4B mutant hampers the lipidation of LC3 paralogues and causes defects in autophagosome closure. Mol. Biol. Cell 19, 4651–4659.
Füllgrabe, J., Lynch-Day, M.A., Heldring, N., Li, W., Struijk, R.B., Ma, Q., et al., 2013. The histone H4 lysine 16 acetyltransferase hMOF regulates the outcome of autophagy. Nature 500, 468–471.
Galluzzi, L., Pietrocola, F., Levine, B., Kroemer, G., 2014. Metabolic control of autophagy. Cell 159, 1263–1276.
Ganley, I.G., Lam, D.H., Wang, J., Ding, X., Chen, S., Jiang, X., 2009. ULK1.ATG13.FIP200 complex mediates mTOR signaling and is essential for autophagy. J. Biol. Chem. 284, 12297–12305.
Geeraert, C., Ratier, A., Pfisterer, S.G., Perdiz, D., Cantaloube, I., Rouault, A., et al., 2010. Starvation-induced hyperacetylation of tubulin is required for the stimulation of autophagy by nutrient deprivation. J. Biol. Chem. 285, 24184–24194.
Ghislat, G., Patron, M., Rizzuto, R., Knecht, E., 2012. Withdrawal of essential amino acids increases autophagy by a pathway involving Ca2 + /calmodulin-dependent kinase kinase-β (CaMKK-β). J. Biol. Chem. 287, 38625–38636.
Gu, W., Roeder, R.G., 1997. Activation of p53 sequence-specific DNA binding by acetylation of the p53 C-terminal domain. Cell 90, 595–606.
Guenther, G.G., Peralta, E.R., Rosales, K.R., Wong, S.Y., Siskind, L.J., Edinger, A.L., 2008. Ceramide starves cells to death by downregulating nutrient transporter proteins. Proc. Natl. Acad. Sci. U.S.A. 105, 17402–17407.
Gulati, P., Gaspers, L.D., Dann, S.G., Joaquin, M., Nobukuni, T., Natt, F., et al., 2008. Amino acids activate mTOR complex 1 via Ca2 + /CaM signaling to hVps34. Cell Metab. 7, 456–465.
Gunst, J., Derese, I., Aertgeerts, A., Ververs, E.-J., Wauters, A., Van den Berghe, G., et al., 2013. Insufficient autophagy contributes to mitochondrial dysfunction, organ failure, and adverse outcome in an animal model of critical illness. Crit. Care Med. 41, 182–194.
Gwinn, D.M., Shackelford, D.B., Egan, D.F., Mihaylova, M.M., Mery, A., Vasquez, D.S., et al., 2008. AMPK phosphorylation of raptor mediates a metabolic checkpoint. Mol. Cell 30, 214–226.
Han, J.M., Jeong, S.J., Park, M.C., Kim, G., Kwon, N.H., Kim, H.K., et al., 2012. Leucyl-tRNA synthetase is an intracellular leucine sensor for the mTORC1-signaling pathway. Cell 149, 410–424.
Hara, T., Mizushima, N., 2009. Role of ULK-FIP200 complex in mammalian autophagy: FIP200, a counterpart of yeast Atg17?. Autophagy 5, 85–87.
Harder, L.M., Bunkenborg, J., Andersen, J.S., 2014. Inducing autophagy: a comparative phosphoproteomic study of the cellular response to ammonia and rapamycin. Autophagy 10, 339–355.
Hardie, D.G., 2011. AMP-activated protein kinase: an energy sensor that regulates all aspects of cell function. Genes Dev. 25, 1895–1908.
Hardie, D.G., 2014. AMPK—sensing energy while talking to other signaling pathways. Cell Metab. 20, 939–952.
Harrison, D.E., Strong, R., Sharp, Z.D., Nelson, J.F., Astle, C.M., Flurkey, K., et al., 2009. Rapamycin fed late in life extends lifespan in genetically heterogeneous mice. Nature 460, 392–395.
Häussinger, D., Lang, F., Bauers, K., Gerok, W., 1990. Interactions between glutamine metabolism and cell-volume regulation in perfused rat liver. Eur. J. Biochem. 188, 689–695.
Hosokawa, N., Hara, T., Kaizuka, T., Kishi, C., Takamura, A., Miura, Y., et al., 2009. Nutrient-dependent mTORC1 association with the ULK1-Atg13-FIP200 complex required for autophagy. Mol. Biol. Cell 20, 1981–1991.
Houde, V.P., Brûlé, S., Festuccia, W.T., Blanchard, P.-G., Bellmann, K., Deshaies, Y., et al., 2010. Chronic rapamycin treatment causes glucose intolerance and hyperlipidemia by upregulating hepatic gluconeogenesis and impairing lipid deposition in adipose tissue. Diabetes 59, 1338–1348.
Huang, R., Xu, Y., Wan, W., Shou, X., Qian, J., You, Z., et al., 2015. Deacetylation of nuclear LC3 drives autophagy initiation under starvation. Mol. Cell 57, 456–466.
Jang, S., Kang, H.T., Hwang, E.S., 2012. Nicotinamide-induced mitophagy: event mediated by high NAD + /NADH ratio and SIRT1 protein activation. J. Biol. Chem. 287, 19304–19314.
Jangamreddy, J.R., Panigrahi, S., Łos, M.J., 2015. Monitoring of autophagy is complicated-salinomycin as an example. Biochim. Biophys. Acta 1854, 604–610.
Jarzyna, R., Lenarcik, E., Bryła, J., 1997. Chloroquine is a potent inhibitor of glutamate dehydrogenase in liver and kidney-cortex of rabbit. Pharmacol. Res. 35, 79–84.
Jewell, J.L., Kim, Y.C., Russell, R.C., Yu, F.-X., Park, H.W., Plouffe, S.W., et al., 2015. Metabolism. Differential regulation of mTORC1 by leucine and glutamine. Science 347, 194–198.
Jiang, P., Mizushima, N., 2014. Autophagy and human diseases. Cell Res. 24, 69–79.
Jiang, W., Ogretmen, B., 2014. Autophagy paradox and ceramide. Biochim. Biophys. Acta 1841, 783–792.
Jung, C.H., Jun, C.B., Ro, S.-H., Kim, Y.-M., Otto, N.M., Cao, J., et al., 2009. ULK-Atg13-FIP200 complexes mediate mTOR signaling to the autophagy machinery. Mol. Biol. Cell 20, 1992–2003.
Kalender, A., Selvaraj, A., Kim, S.Y., Gulati, P., Brûlé, S., Viollet, B., et al., 2010. Metformin, independent of AMPK, inhibits mTORC1 in a rag GTPase-dependent manner. Cell Metab. 11, 390–401.

Kaushik, S., Cuervo, A.M., 2008. Chaperone-mediated autophagy. Methods Mol. Biol. 445, 227–244.

Kim, J., Guan, K.-L., 2011. Amino acid signaling in TOR activation. Annu. Rev. Biochem. 80, 1001–1032.

Kim, E., Goraksha-Hicks, P., Li, L., Neufeld, T.P., Guan, K.-L., 2008. Regulation of TORC1 by Rag GTPases in nutrient response. Nat. Cell Biol. 10, 935–945.

Kim, J., Kundu, M., Viollet, B., Guan, K.-L., 2011a. AMPK and mTOR regulate autophagy through direct phosphorylation of Ulk1. Nat. Cell Biol. 13, 132–141.

Kim, S., Kim, S.F., Maag, D., Maxwell, M.J., Resnick, A.C., Juluri, K.R., et al., 2011b. Amino acid signaling to mTOR mediated by inositol polyphosphate multikinase. Cell Metab. 13, 215–221.

Kim, Y.-M., Stone, M., Hwang, T.H., Kim, Y.-G., Dunlevy, J.R., Griffin, T.J., et al., 2012. SH3BP4 is a negative regulator of amino acid-Rag GTPase-mTORC1 signaling. Mol. Cell 46, 833–846.

Kim, J., Kim, Y.C., Fang, C., Russell, R.C., Kim, J.H., Fan, W., et al., 2013. Differential regulation of distinct Vps34 complexes by AMPK in nutrient stress and autophagy. Cell 152, 290–303.

Korolchuk, V.I., Saiki, S., Lichtenberg, M., Siddiqi, F.H., Roberts, E.A., Imarisio, S., et al., 2011. Lysosomal positioning coordinates cellular nutrient responses. Nat. Cell Biol. 13, 453–460.

Koyama-Honda, I., Itakura, E., Fujiwara, T.K., Mizushima, N., 2013. Temporal analysis of recruitment of mammalian ATG proteins to the autophagosome formation site. Autophagy 9, 1491–1499.

Kroemer, G., Mariño, G., Levine, B., 2010. Autophagy and the integrated stress response. Mol. Cell 40, 280–293.

Ktistakis, N.T., Manifava, M., Schoenfelder, P., Rotondo, S., 2012. How phosphoinositide 3-phosphate controls growth downstream of amino acids and autophagy downstream of amino acid withdrawal. Biochem. Soc. Trans. 40, 37–43.

Kuma, A., Hatano, M., Matsui, M., Yamamoto, A., Nakaya, H., Yoshimori, T., et al., 2004. The role of autophagy during the early neonatal starvation period. Nature 432, 1032–1036.

Kume, K., Iizumi, Y., Shimada, M., Ito, Y., Kishi, T., Yamaguchi, Y., et al., 2010a. Role of N-end rule ubiquitin ligases UBR1 and UBR2 in regulating the leucine-mTOR signaling pathway. Genes Cells 15, 339–349.

Kume, S., Uzu, T., Horiike, K., Chin-Kanasaki, M., Isshiki, K., Araki, S.-I., et al., 2010b. Calorie restriction enhances cell adaptation to hypoxia through Sirt1-dependent mitochondrial autophagy in mouse aged kidney. J. Clin. Invest. 120, 1043–1055.

Lamb, C.A., Yoshimori, T., Tooze, S.A., 2013. The autophagosome: origins unknown, biogenesis complex. Nat. Rev. Mol. Cell Biol. 14, 759–774.

Laplante, M., Sabatini, D.M., 2012. mTOR signaling in growth control and disease. Cell 149, 274–293.

Lavallard, V.J., Meijer, A.J., Codogno, P., Gual, P., 2012. Autophagy, signaling and obesity. Pharmacol. Res. 66, 513–525.

Lee, I.H., Finkel, T., 2009. Regulation of autophagy by the p300 acetyltransferase. J. Biol. Chem. 284, 6322–6328.

Lee, J.-G., Ye, Y., 2013. Bag6/Bat3/Scythe: a novel chaperone activity with diverse regulatory functions in protein biogenesis and degradation. Bioessays 35, 377–385.

Lee, I.H., Cao, L., Mostoslavsky, R., Lombard, D.B., Liu, J., Bruns, N.E., et al., 2008. A role for the NAD-dependent deacetylase Sirt1 in the regulation of autophagy. Proc. Natl. Acad. Sci. U.S.A. 105, 3374–3379.

Lee, M.N., Ha, S.H., Kim, J., Koh, A., Lee, C.S., Kim, J.H., et al., 2009. Glycolytic flux signals to mTOR through glyceraldehyde-3-phosphate dehydrogenase-mediated regulation of Rheb. Mol. Cell. Biol. 29, 3991–4001.

Li, L., Chen, Y., Gibson, S.B., 2013a. Starvation-induced autophagy is regulated by mitochondrial reactive oxygen species leading to AMPK activation. Cell. Signal. 25, 50–65.

Li, M., Khambu, B., Zhang, H., Kang, J.-H., Chen, X., Chen, D., et al., 2013b. Suppression of lysosome function induces autophagy via a feedback down-regulation of MTOR complex 1 (MTORC1) activity. J. Biol. Chem. 288, 35769–35780.

Li, J., Kim, S.G., Blenis, J., 2014. Rapamycin: one drug, many effects. Cell Metab. 19, 373–379.

Libert, S., Guarente, L., 2013. Metabolic and neuropsychiatric effects of calorie restriction and sirtuins. Annu. Rev. Physiol. 75, 669–684.

Lieberman, A.P., Puertollano, R., Raben, N., Slaugenhaupt, S., Walkley, S.U., Ballabio, A., 2012. Autophagy in lysosomal storage disorders. Autophagy 8, 719–730.

Lin, S.-Y., Li, T.Y., Liu, Q., Zhang, C., Li, X., Chen, Y., et al., 2012. GSK3-TIP60-ULK1 signaling pathway links growth factor deprivation to autophagy. Science 336, 477–481.

Linares, J.F., Duran, A., Yajima, T., Pasparakis, M., Moscat, J., Diaz-Meco, M.T., 2013. K63 polyubiquitination and activation of mTOR by the p62-TRAF6 complex in nutrient-activated cells. Mol. Cell 51, 283–296.

Liu, K.-P., Zhou, D., Ouyang, D.-Y., Xu, L.-H., Wang, Y., Wang, L.-X., et al., 2013. LC3B-II deacetylation by histone deacetylase 6 is involved in serum-starvation-induced autophagic degradation. Biochem. Biophys. Res. Commun. 441, 970–975.

Long, X., Ortiz-Vega, S., Lin, Y., Avruch, J., 2005. Rheb binding to mammalian target of rapamycin (mTOR) is regulated by amino acid sufficiency. J. Biol. Chem. 280, 23433–23436.

Lorin, S., Hamaï, A., Mehrpour, M., Codogno, P., 2013a. Autophagy regulation and its role in cancer. Semin. Cancer Biol. 23, 361–379.

Lorin, S., Tol, M.J., Bauvy, C., Strijland, A., Poüs, C., Verhoeven, A.J., et al., 2013b. Glutamate dehydrogenase contributes to leucine sensing in the regulation of autophagy. Autophagy 9, 850–860.

Luo, S., Garcia-Arencibia, M., Zhao, R., Puri, C., Toh, P.P.C., Sadiq, O., et al., 2012. Bim inhibits autophagy by recruiting Beclin 1 to microtubules. Mol. Cell 47, 359–370.

Madeo, F., Zimmermann, A., Maiuri, M.C., Kroemer, G., 2015. Essential role for autophagy in life span extension. J. Clin. Invest. 125, 85–93.

Mammucari, C., Milan, G., Romanello, V., Masiero, E., Rudolf, R., Del Piccolo, P., et al., 2007. FoxO3 controls autophagy in skeletal muscle *in vivo*. Cell Metab. 6, 458–471.

Mariño, G., Niso-Santano, M., Baehrecke, E.H., Kroemer, G., 2014a. Self-consumption: the interplay of autophagy and apoptosis. Nat. Rev. Mol. Cell Biol. 15, 81–94.

Mariño, G., Pietrocola, F., Eisenberg, T., Kong, Y., Malik, S.A., Andryushkova, A., et al., 2014b. Regulation of autophagy by cytosolic acetyl-coenzyme A. Mol. Cell 53, 710–725.

Martina, J.A., Chen, Y., Gucek, M., Puertollano, R., 2012. MTORC1 functions as a transcriptional regulator of autophagy by preventing nuclear transport of TFEB. Autophagy 8, 903–914.

Martina, J.A., Diab, H.I., Lishu, L., Jeong-A, L., Patange, S., Raben, N., et al., 2014. The nutrient-responsive transcription factor TFE3 promotes autophagy, lysosomal biogenesis, and clearance of cellular debris. Sci. Signal. 7, ra9.

Mattison, J.A., Roth, G.S., Beasley, T.M., Tilmont, E.M., Handy, A.M., Herbert, R.L., et al., 2012. Impact of caloric restriction on health and survival in rhesus monkeys from the NIA study. Nature 489, 318–321.

Medina, D.L., Di Paola, S., Peluso, I., Armani, A., De Stefani, D., Venditti, R., et al., 2015. Lysosomal calcium signalling regulates autophagy through calcineurin and TFEB. Nat. Cell Biol. 17, 288–299.

Meijer, A.J., Codogno, P., 2009. Autophagy: regulation and role in disease. Crit. Rev. Clin. Lab. Sci. 46, 210–240.

Meijer, A.J., Codogno, P., 2011. Autophagy: regulation by energy sensing. Curr. Biol. 21, R227–R229.

Meijer, A.J., Lorin, S., Blommaart, E.F., Codogno, P., 2015. Regulation of autophagy by amino acids and MTOR-dependent signal transduction. Amino Acids 47, 2037–2063.

Mizushima, N., Komatsu, M., 2011. Autophagy: renovation of cells and tissues. Cell 147, 728–741.

Mizushima, N., Yoshimori, T., Ohsumi, Y., 2011. The role of Atg proteins in autophagosome formation. Annu. Rev. Cell Dev. Biol. 27, 107–132.

Morselli, E., Maiuri, M.C., Markaki, M., Megalou, E., Pasparaki, A., Palikaras, K., et al., 2010. Caloric restriction and resveratrol promote longevity through the Sirtuin-1-dependent induction of autophagy. Cell Death Dis. 1, e10.

Morselli, E., Mariño, G., Bennetzen, M.V., Eisenberg, T., Megalou, E., Schroeder, S., et al., 2011. Spermidine and resveratrol induce autophagy by distinct pathways converging on the acetylproteome. J. Cell Biol. 192, 615–629.

Mortimore, G.E., Schworer, C.M., 1977. Induction of autophagy by amino-acid deprivation in perfused rat liver. Nature 270, 174–176.

Naidu, S.R., Lakhter, A.J., Androphy, E.J., 2012. PIASy-mediated Tip60 sumoylation regulates p53-induced autophagy. Cell Cycle 11, 2717–2728.

Narita, M., Young, A.R.J., Arakawa, S., Samarajiwa, S.A., Nakashima, T., Yoshida, S., et al., 2011. Spatial coupling of mTOR and autophagy augments secretory phenotypes. Science 332, 966–970.

Nazio, F., Strappazzon, F., Antonioli, M., Bielli, P., Cianfanelli, V., Bordi, M., et al., 2013. mTOR inhibits autophagy by controlling ULK1 ubiquitylation, self-association and function through AMBRA1 and TRAF6. Nat. Cell Biol. 15, 406–416.

Neff, F., Flores-Dominguez, D., Ryan, D.P., Horsch, M., Schröder, S., Adler, T., et al., 2013. Rapamycin extends murine lifespan but has limited effects on aging. J. Clin. Invest. 123, 3272–3291.

Nicklin, P., Bergman, P., Zhang, B., Triantafellow, E., Wang, H., Nyfeler, B., et al., 2009. Bidirectional transport of amino acids regulates mTOR and autophagy. Cell 136, 521–534.

Niso-Santano, M., Malik, S.A., Pietrocola, F., Bravo-San Pedro, J.M., Mariño, G., Cianfanelli, V., et al., 2015. Unsaturated fatty acids induce non-canonical autophagy. EMBO J.

Niu, Y., Na, L., Feng, R., Gong, L., Zhao, Y., Li, Q., et al., 2013. The phytochemical, EGCG, extends lifespan by reducing liver and kidney function damage and improving age-associated inflammation and oxidative stress in healthy rats. Aging Cell 12, 1041–1049.

Nobukuni, T., Joaquin, M., Roccio, M., Dann, S.G., Kim, S.Y., Gulati, P., et al., 2005. Amino acids mediate mTOR/raptor signaling through activation of class 3 phosphatidylinositol 3OH-kinase. Proc. Natl. Acad. Sci. U.S.A. 102, 14238–14243.

Ogasawara, Y., Itakura, E., Kono, N., Mizushima, N., Arai, H., Nara, A., et al., 2014. Stearoyl-CoA desaturase 1 activity is required for autophagosome formation. J. Biol. Chem. 289, 23938–23950.

Ögmundsdóttir, M.H., Heublein, S., Kazi, S., Reynolds, B., Visvalingam, S.M., Shaw, M.K., et al., 2012. Proton-assisted amino acid transporter PAT1 complexes with Rag GTPases and activates TORC1 on late endosomal and lysosomal membranes. PLoS ONE 7, e36616.

Panchaud, N., Péli-Gulli, M.-P., De Virgilio, C., 2013. Amino acid deprivation inhibits TORC1 through a GTPase-activating protein complex for the Rag family GTPase Gtr1. Sci. Signal. 6, ra42.

Petit, C.S., Roczniak-Ferguson, A., Ferguson, S.M., 2013. Recruitment of folliculin to lysosomes supports the amino acid-dependent activation of Rag GTPases. J. Cell Biol. 202, 1107–1122.

Pietrocola, F., Lachkar, S., Enot, D.P., Niso-Santano, M., Bravo-San Pedro, J.M., Sica, V., et al., 2015. Spermidine induces autophagy by inhibiting the acetyltransferase EP300. Cell Death Differ. 22, 509–516.

Polletta, L., Vernucci, E., Carnevale, I., Arcangeli, T., Rotili, D., Palmerio, S., et al., 2015. SIRT5 regulation of ammonia-induced autophagy and mitophagy. Autophagy 11, 253–270.

Rahman, M., Mofarrahi, M., Kristof, A.S., Nkengfac, B., Harel, S., Hussain, S.N.A., 2014. Reactive oxygen species regulation of autophagy in skeletal muscles. Antioxid. Redox Signal. 20, 443–459.

Ravikumar, B., Sarkar, S., Davies, J.E., Futter, M., Garcia-Arencibia, M., Green-Thompson, Z.W., et al., 2010. Regulation of mammalian autophagy in physiology and pathophysiology. Physiol. Rev. 90, 1383–1435.

Rebsamen, M., Pochini, L., Stasyk, T., de Araújo, M.E.G., Galluccio, M., Kandasamy, R.K., et al., 2015. SLC38A9 is a component of the lysosomal amino acid sensing machinery that controls mTORC1. Nature 519, 477–481.

Roberts, D.J., Tan-Sah, V.P., Ding, E.Y., Smith, J.M., Miyamoto, S., 2014. Hexokinase-II positively regulates glucose starvation-induced autophagy through TORC1 inhibition. Mol. Cell 53, 521–533.

Robida-Stubbs, S., Glover-Cutter, K., Lamming, D.W., Mizunuma, M., Narasimhan, S.D., Neumann-Haefelin, E., et al., 2012. TOR signaling and rapamycin influence longevity by regulating SKN-1/Nrf and DAF-16/FoxO. Cell Metab. 15, 713–724.

Rodriguez, O.C., Choudhury, S., Kolukula, V., Vietsch, E.E., Catania, J., Preet, A., et al., 2012. Dietary downregulation of mutant p53 levels via glucose restriction: mechanisms and implications for tumor therapy. Cell Cycle 11, 4436–4446.

Rubinsztein, D.C., Mariño, G., Kroemer, G., 2011. Autophagy and aging. Cell 146, 682–695.

Rubinsztein, D.C., Codogno, P., Levine, B., 2012. Autophagy modulation as a potential therapeutic target for diverse diseases. Nat. Rev. Drug Discov. 11, 709–730.

Ruderman, N.B., Xu, X.J., Nelson, L., Cacicedo, J.M., Saha, A.K., Lan, F., et al., 2010. AMPK and SIRT1: a long-standing partnership? Am. J. Physiol. Endocrinol. Metab. 298, E751–E760.

Russell, R.C., Tian, Y., Yuan, H., Park, H.W., Chang, Y.-Y., Kim, J., et al., 2013. ULK1 induces autophagy by phosphorylating Beclin-1 and activating VPS34 lipid kinase. Nat. Cell Biol. 15, 741–750.

Russell, R.C., Yuan, H.-X., Guan, K.-L., 2014. Autophagy regulation by nutrient signaling. Cell Res. 24, 42–57.

Sancak, Y., Peterson, T.R., Shaul, Y.D., Lindquist, R.A., Thoreen, C.C., Bar-Peled, L., et al., 2008. The Rag GTPases bind raptor and mediate amino acid signaling to mTORC1. Science 320, 1496–1501.

Sancak, Y., Bar-Peled, L., Zoncu, R., Markhard, A.L., Nada, S., Sabatini, D.M., 2010. Ragulator-Rag complex targets mTORC1 to the lysosomal surface and is necessary for its activation by amino acids. Cell 141, 290–303.

Sanchez-Wandelmer, J., Ktistakis, N.T., Reggiori, F., 2015. ERES: sites for autophagosome biogenesis and maturation? J. Cell. Sci. 128, 185–192.

Sanders, M.J., Grondin, P.O., Hegarty, B.D., Snowden, M.A., Carling, D., 2007. Investigating the mechanism for AMP activation of the AMP-activated protein kinase cascade. Biochem. J. 403, 139–148.

Scherz-Shouval, R., Elazar, Z., 2011. Regulation of autophagy by ROS: physiology and pathology. Trends Biochem. Sci. 36, 30–38.

Schetz, M., Casaer, M.P., Van den Berghe, G., 2013. Does artificial nutrition improve outcome of critical illness? Crit. Care 17, 302.

Sebti, S., Prébois, C., Pérez-Gracia, E., Bauvy, C., Desmots, F., Pirot, N., et al., 2014a. BAG6/BAT3 modulates autophagy by affecting EP300/p300 intracellular localization. Autophagy 10, 1341–1342.

Sebti, S., Prébois, C., Pérez-Gracia, E., Bauvy, C., Desmots, F., Pirot, N., et al., 2014b. BAT3 modulates p300-dependent acetylation of p53 and autophagy-related protein 7 (ATG7) during autophagy. Proc. Natl. Acad. Sci. U.S.A. 111, 4115–4120.

Seglen, P.O., Gordon, P.B., Poli, A., 1980. Amino acid inhibition of the autophagic/lysosomal pathway of protein degradation in isolated rat hepatocytes. Biochim. Biophys. Acta 630, 103–118.

Seok, S., Fu, T., Choi, S.-E., Li, Y., Zhu, R., Kumar, S., et al., 2014. Transcriptional regulation of autophagy by an FXR-CREB axis. Nature 516, 108–111.

Settembre, C., Di Malta, C., Polito, V.A., Garcia Arencibia, M., Vetrini, F., Erdin, S., et al., 2011. TFEB links autophagy to lysosomal biogenesis. Science 332, 1429–1433.

Settembre, C., Zoncu, R., Medina, D.L., Vetrini, F., Erdin, S., Erdin, S., et al., 2012. A lysosome-to-nucleus signalling mechanism senses and regulates the lysosome via mTOR and TFEB. EMBO J. 31, 1095–1108.

Settembre, C., De Cegli, R., Mansueto, G., Saha, P.K., Vetrini, F., Visvikis, O., et al., 2013. TFEB controls cellular lipid metabolism through a starvation-induced autoregulatory loop. Nat. Cell Biol. 15, 647–658.

Shen, S., Niso-Santano, M., Adjemian, S., Takehara, T., Malik, S.A., Minoux, H., et al., 2012. Cytoplasmic STAT3 represses autophagy by inhibiting PKR activity. Mol. Cell 48, 667–680.

Simonsen, A., Tooze, S.A., 2009. Coordination of membrane events during autophagy by multiple class III PI3-kinase complexes. J. Cell Biol. 186, 773–782.

Strohecker, A.M., Joshi, S., Possemato, R., Abraham, R.T., Sabatini, D.M., White, E., 2015. Identification of 6-phosphofructo-2-kinase/fructose-2,6-bisphosphatase as a novel autophagy regulator by high content shRNA screening. Oncogene 34, 5662–5676.

Suter, M., Riek, U., Tuerk, R., Schlattner, U., Wallimann, T., Neumann, D., 2006. Dissecting the role of 5′-AMP for allosteric stimulation, activation, and deactivation of AMP-activated protein kinase. J. Biol. Chem. 281, 32207–32216.

Thomas, J.D., Zhang, Y.-J., Wei, Y.-H., Cho, J.-H., Morris, L.E., Wang, H.-Y., et al., 2014. Rab1A is an mTORC1 activator and a colorectal oncogene. Cancer Cell 26, 754–769.

Toyoda, T., Hayashi, T., Miyamoto, L., Yonemitsu, S., Nakano, M., Tanaka, S., et al., 2004. Possible involvement of the alpha1 isoform of 5′AMP-activated protein kinase in oxidative stress-stimulated glucose transport in skeletal muscle. Am. J. Physiol. Endocrinol. Metab. 287, E166–E173.

Tsukada, M., Ohsumi, Y., 1993. Isolation and characterization of autophagy-defective mutants of *Saccharomyces cerevisiae*. FEBS Lett. 333, 169–174.

Tsun, Z.-Y., Bar-Peled, L., Chantranupong, L., Zoncu, R., Wang, T., Kim, C., et al., 2013. The folliculin tumor suppressor is a GAP for the RagC/D GTPases that signal amino acid levels to mTORC1. Mol. Cell 52, 495–505.

Tsunemi, T., Ashe, T.D., Morrison, B.E., Soriano, K.R., Au, J., Roque, R.A.V., et al., 2012. PGC-1α rescues Huntington's disease proteotoxicity by preventing oxidative stress and promoting TFEB function. Sci. Transl. Med. 4, 142ra97.

Underwood, B.R., Imarisio, S., Fleming, A., Rose, C., Krishna, G., Heard, P., et al., 2010. Antioxidants can inhibit basal autophagy and enhance neurodegeneration in models of polyglutamine disease. Hum. Mol. Genet. 19, 3413–3429.

Van Der Heide, L.P., Hoekman, M.F.M., Smidt, M.P., 2004. The ins and outs of FoxO shuttling: mechanisms of FoxO translocation and transcriptional regulation. Biochem. J. 380, 297–309.

Van Sluijters, D.A., Dubbelhuis, P.F., Blommaart, E.F., Meijer, A.J., 2000. Amino-acid-dependent signal transduction. Biochem. J. 351 (Pt 3), 545–550.

Wang, R.C., Wei, Y., An, Z., Zou, Z., Xiao, G., Bhagat, G., et al., 2012. Akt-mediated regulation of autophagy and tumorigenesis through Beclin 1 phosphorylation. Science 338, 956–959.

Wang, S., Tsun, Z.-Y., Wolfson, R.L., Shen, K., Wyant, G.A., Plovanich, M.E., et al., 2015. Metabolism. Lysosomal amino acid transporter SLC38A9 signals arginine sufficiency to mTORC1. Science 347, 188–194.

Wauson, E.M., Zaganjor, E., Lee, A.-Y., Guerra, M.L., Ghosh, A.B., Bookout, A.L., et al., 2012. The G protein-coupled taste receptor T1R1/T1R3 regulates mTORC1 and autophagy. Mol. Cell 47, 851–862.

Whelan, S.P., Zuckerbraun, B.S., 2013. Mitochondrial signaling: forwards, backwards, and in between. Oxid. Med. Cell Longev. 2013, 351613.

Wilkinson, J.E., Burmeister, L., Brooks, S.V., Chan, C.-C., Friedline, S., Harrison, D.E., et al., 2012. Rapamycin slows aging in mice. Aging Cell 11, 675–682.

Wong, P.-M., Puente, C., Ganley, I.G., Jiang, X., 2013. The ULK1 complex: sensing nutrient signals for autophagy activation. Autophagy 9, 124–137.

Wu, S.-B., Wu, Y.-T., Wu, T.-P., Wei, Y.-H., 2014. Role of AMPK-mediated adaptive responses in human cells with mitochondrial dysfunction to oxidative stress. Biochim. Biophys. Acta 1840, 1331–1344.

Xiao, B., Heath, R., Saiu, P., Leiper, F.C., Leone, P., Jing, C., et al., 2007. Structural basis for AMP binding to mammalian AMP-activated protein kinase. Nature 449, 496–500.

Xiao, F., Huang, Z., Li, H., Yu, J., Wang, C., Chen, S., et al., 2011. Leucine deprivation increases hepatic insulin sensitivity via GCN2/mTOR/S6K1 and AMPK pathways. Diabetes 60, 746–756.

Xu, L., Salloum, D., Medlin, P.S., Saqcena, M., Yellen, P., Perrella, B., et al., 2011. Phospholipase D mediates nutrient input to mammalian target of rapamycin complex 1 (mTORC1). J. Biol. Chem. 286, 25477–25486.

Yang, Z., Klionsky, D.J., 2010. Eaten alive: a history of macroautophagy. Nat. Cell Biol. 12, 814–822.

Yang, X., Nath, A., Opperman, M.J., Chan, C., 2010. The double-stranded RNA-dependent protein kinase differentially regulates insulin receptor substrates 1 and 2 in HepG2 cells. Mol. Biol. Cell 21, 3449–3458.

Yasuhara, S., Asai, A., Sahani, N.D., Martyn, J.A.J., 2007. Mitochondria, endoplasmic reticulum, and alternative pathways of cell death in critical illness. Crit. Care Med. 35, S488–S495.

Yi, C., Ma, M., Ran, L., Zheng, J., Tong, J., Zhu, J., et al., 2012. Function and molecular mechanism of acetylation in autophagy regulation. Science 336, 474–477.

Yoon, M.-S., Du, G., Backer, J.M., Frohman, M.A., Chen, J., 2011. Class III PI-3-kinase activates phospholipase D in an amino acid-sensing mTORC1 pathway. J. Cell Biol. 195, 435–447.

Yuan, H.-X., Russell, R.C., Guan, K.-L., 2013. Regulation of PIK3C3/VPS34 complexes by MTOR in nutrient stress-induced autophagy. Autophagy 9, 1983–1995.

Zhang, J., Kim, J., Alexander, A., Cai, S., Tripathi, D.N., Dere, R., et al., 2013. A tuberous sclerosis complex signalling node at the peroxisome regulates mTORC1 and autophagy in response to ROS. Nat. Cell Biol. 15, 1186–1196.

Zhang, C.-S., Jiang, B., Li, M., Zhu, M., Peng, Y., Zhang, Y.-L., et al., 2014. The lysosomal v-ATPase-Ragulator complex is a common activator for AMPK and mTORC1, acting as a switch between catabolism and anabolism. Cell Metab. 20, 526–540.

Zhao, Y., Yang, J., Liao, W., Liu, X., Zhang, H., Wang, S., et al., 2010. Cytosolic FoxO1 is essential for the induction of autophagy and tumour suppressor activity. Nat. Cell Biol. 12, 665–675.

Zhou, J., Liao, W., Yang, J., Ma, K., Li, X., Wang, Y., et al., 2012. FOXO3 induces FOXO1-dependent autophagy by activating the AKT1 signaling pathway. Autophagy 8, 1712–1723.

Zoncu, R., Bar-Peled, L., Efeyan, A., Wang, S., Sancak, Y., Sabatini, D.M., 2011. mTORC1 senses lysosomal amino acids through an inside-out mechanism that requires the vacuolar H(+)-ATPase. Science 334, 678–683.

SECTION III

CELLULAR AND MOLECULAR ACTIONS OF AMINO ACIDS IN NON PROTEIN METABOLISM

CHAPTER

15

Dietary Protein and Colonic Microbiota: Molecular Aspects

G. Boudry[1], *I. Le Huërou-Luron*[1] *and C. Michel*[2]

[1]INRA UR1341 ADNC, St-Gilles, France [2]INRA UMR1280 PhAN, Nantes, France

15.1. INTRODUCTION

The human gastrointestinal tract hosts more than 100 trillion bacteria and archae, with maximal bacterial densities occurring in the colon, reaching 10^{11} bacteria per gram of content (Doré and Corthier, 2010; Lepage et al., 2013). Approximately 1000 different bacterial species have been detected as potential component of the intestinal microbiota. Each individual microbiota is composed of only some of them (150 to 400), with less than 100 being shared by most individuals (Bäckhed et al., 2005; Lepage et al., 2013; Tap et al., 2009). The vast majority of these bacteria belongs to Firmicutes and Bacteroidetes phyla while Actinobacteria, Proteobacteria, and Verrucomicrobia phyla represent a minority. Approximately 30 genera are represented with *Bacteroides* sp. and genera related to clostridial clusters IV and XIVa being the most abundant (Arumugam et al., 2011).

Clustering by diet (herbivore, omnivore, and carnivore) of bacterial 16S rRNA gene sequences from the fecal microbiota of 60 mammalian species suggests that the type of nutrients the microbiota encounters is a major determinant of gut microbiotal community (Ley et al., 2008). Newly developed metagenomic analyses reveal that the intestinal microbiota harbors 10s of millions of genes (Li et al., 2014), corresponding to an overall potential of 19,000 metabolic functions (Qin et al., 2010). This is consistent with the considerable metabolic activity of the intestinal microbiota which is estimated as comparable to, or even significantly higher than that of the liver (Blottière et al., 2013; Possemiers et al., 2011). Half of these functions are still unknown but among the identified ones, "Amino Acid (AA) metabolism and transport" is one of the most represented (Arumugam et al., 2011; Kurokawa et al., 2007). This suggests that the intestinal microbiota is particularly adapted to protein metabolism. However, and somewhat surprisingly, microbiotal proteolysis and AA fermentation has not been extensively considered by the scientific community (Macfarlane and Macfarlane, 2012). This chapter aims to give an overview of dietary protein metabolism by the gut microbiota, focusing first on the protein, peptides, and AA available for the gut microbiota, second on bacterial protein fermentation mechanisms, and finally on the effect of protein-derived bacterial metabolites on the host.

15.1.1 Protein Available for the Gut Microbiota

Dietary protein digestion is a complex process resulting from the concerted action of several digestive enzymes and depending on factors such as the type and amount of dietary proteins, endogenous secretions, motility or luminal pH. Although digestion of dietary proteins along the upper gastrointestinal tract is considered as an efficient process, substantial amounts of peptides and AA reach the distal part of the small intestine (ileum) and enter the large intestine (Davila et al., 2013). Other important sources of protein reaching the distal intestine are endogenous protein, made up of a mixture of gastric and pancreatic secretory enzymes, exfoliated intestinal epithelial cells and mucus proteins (Moughan and Rutherfurd, 2012). The amount of nitrogen reaching the large

The Molecular Nutrition of Amino Acids and Proteins.
DOI: http://dx.doi.org/10.1016/B978-0-12-802167-5.00015-3

intestine in adults is estimated to be between 2 and 5 g/day, with the endogenous fraction contributing to about 1 g/day (Chacko and Cummings, 1988; Gaudichon et al., 2002). Yet, the assessment of the relative contribution of dietary versus endogenous sources to the ileal protein content is not easy since they are both dependent of the diet composition (Chacko and Cummings, 1988; Gaudichon et al., 2002; Hodgkinson et al., 2000).

15.1.1.1 *Digestibility of Dietary Proteins*

In an adult human, the nitrogen fraction that enters the large intestine every day is composed of 10–15% urea, NH_4^+, nitrate and AA, 48–51% proteins, and 34–42% peptides (Chacko and Cummings, 1988). Among food proteins, milk constitutes the highest digestible source of protein with ileal true digestibility of milk proteins estimated to be around 95% in humans (Gaudichon et al., 2002). Meat proteins are also highly digestible; the amount of meat protein entering the large intestine is very low, with 90 and 95% of ileal digestibility in human subjects with ileostomies (Silvester and Cummings, 1995) and in adult minipigs (Bax et al., 2013), respectively. It is noteworthy that the amount of meat protein entering the large intestine increases with the level of meat intake although values are always low (less than 5% of ingested protein) (Bax et al., 2013). Protein digestibility differs among legumes. For example, soy-protein digestibility is lower than milk protein (92 vs 95%) in humans (Gaudichon et al., 2002). Regarding peas, 31% of ingested proteins escape proximal digestion in healthy volunteers; two-thirds of them being further absorbed between the proximal jejunum and the distal ileum while one-third enter the large intestine (Gausserès et al., 1996).

The composition and the technological process of the food matrix impact protein digestion and the subsequent nutrient bioavailability for the host and the microbiota. In patients with high jejunostomy, β-lactoglobulin was found completely intact in the jejunum when ingested alone whereas only 64% remained intact when ingested as skimmed milk (Mahé et al., 1991). Caseins were almost entirely degraded in the proximal jejunum (Mahé et al., 1991). Comparing heat treatments and gelation that modify both the micro- and the macrostructure of milk matrices, Barbe et al. confirmed the strong impact of technological processes on the different steps of milk protein digestion in pigs, that is, gastric emptying rate, sensitivity to hydrolysis, and subsequent kinetics of AA absorption (Barbé et al., 2014). Transformation of milk to yogurt delays gastric emptying, and subsequently the kinetics of digestion and nitrogen absorption, without affecting the overall digestibility of milk protein. The differences between milk and yogurt nitrogen transit rate probably explain the accumulation of dietary protein in cecum and colon observed 8 h after the yogurt meal (Gaudichon et al., 1994). Only moderate effects of cooking temperature were recently reported for meat protein digestibility in minipigs (Bax et al., 2013). Cooking temperature modulated the speed of protein digestion, without changing intestinal digestion and the amount of meat protein residues reaching the colon. Finally, the physical form of vegetables, for example, whole or pureed soy-beans, was also shown to influence protein digestibility in adult humans (Chacko and Cummings, 1988).

15.1.1.2 *Variable Amount of Endogenous Proteins*

The importance of endogenous protein flow in the distal small intestine was first demonstrated in adult patients with ileostomy fed a protein-free diet with a mean daily dry matter output of 19 g accounting for 0.7 g nitrogen losses (Fuller et al., 1994). The average total nitrogen flow in the distal small intestine ranged from 2 to 5 g/d, with endogenous and dietary nitrogen losses ranging from 0.7 to 4 and 0.3 to 1 g/d, respectively, in adult humans (Gaudichon et al., 2002; Gausserès et al., 1996; Mahé et al., 1992; Mariotti et al., 1999; Moughan and Rutherfurd, 2012). Endogenous nitrogen losses seem lower in infants (0.4 g/d in infant pigs, Reis de Souza et al., 2013). Mucus was the most abundant endogenous component within the ileal digesta of human adults fed a casein-based diet (Miner-Williams et al., 2012). Endogenous proteins are made also of digestive enzymes, mainly from pancreas, ensuring food digestion up to the ileal part and themselves resistant to digestion to achieve this role. Endogenous protein losses are affected by the type of dietary protein and their AA composition (Daenzer et al., 2001; Gaudichon et al., 2002). For example, casein- or enzyme-hydrolyzed-casein-based diets increase by 41% the ileal mucin flows compared to a free AA-based diet (Miner-Williams et al., 2014).

Collectively, these data indicate that although dietary protein digestibility is quite high, undigested dietary protein, fractions of these proteins, and endogenous proteins are present in the distal intestine and enter the colon. The nature, quantity and kinetic of arrival of this nitrogen fraction are dependent upon the quantity, quality, and food processing of the initial dietary protein. Yet this N fraction is fully available for the colonic microbiota.

15.1.2 Protein Fermentation by Intestinal Microbiota

15.1.2.1 Site of Protein and AA Metabolism by the Gut Microbiota

Analysis of gut contents obtained from people who died suddenly reveals that concentration of typical end-products of AA fermentation (eg, branched short-chain fatty acids (BCFA), or p-cresol and phenol) is higher in distal than proximal colon (Cummings et al., 1987; Macfarlane et al., 1992; Smith and Macfarlane, 1996a), although this was not corroborated for amines (Smith and Macfarlane, 1996b). Protein fermentation is therefore considered to be more efficient in distal than proximal digestive segments (Macfarlane and Macfarlane, 2012; Windey et al., 2012b), probably due to a more optimal pH for bacterial proteases and peptidases (Gibson et al., 1989) and/or differences in the bacterial composition of the microbiota along the large intestine (Marteau et al., 2001).

Although concentrations of protein-derived bacterial end-products, such as NH_4^+ and BCFA, are low in ileal contents obtained from sudden death victims (Macfarlane et al., 1986), the capacity of small intestinal bacteria for protein metabolism has been demonstrated ex vivo using pig luminal contents (Dai et al., 2010; Yang et al., 2014). The contribution of ileal microbiota to protein metabolism is also supported by comparisons of apparent AA absorption measured in vivo with the AA metabolic capacities of isolated enterocytes measured in vitro (Davila et al., 2013). Protein metabolism by the small intestinal microbiota deserves therefore further investigation. Moreover, the current perception of the contribution of gastrointestinal microbiota to protein degradation and the possible microbiota–host interactions in the proximal part of the intestine should be reevaluated.

15.1.2.2 Metabolic Pathways Involved in Protein and AA Fermentation

15.1.2.2.1 General Overview of Protein and AA Fermentation

Degradation of proteins in the colon starts with hydrolysis of the proteins to smaller peptides and AA by bacterial proteases and peptidases. In the large intestine, this depolymerization is catalyzed by a mixture of residual pancreatic endopeptidases and bacterial proteases and peptidases (Macfarlane and Macfarlane, 1995). Proteases of bacterial origin may be extracellular or cell-bound. Those detected in human fecal samples include trypsin, chymotrypsin, elastase, serine, cysteine, and metalloproteinases (Gibson et al., 1989). The released AA can then be incorporated into bacterial cells; this process varies with AA structure, involves different strategies based on sodium- or phosphate-dependent transporters or facilitated diffusion, and is affected by extracellular pH (Dai et al., 2011). Amino acids can then either be deaminated by intestinal bacteria which form a variety of products, including H_2, NH_4, carboxylic acids such as short-chain fatty acids (SCFA) and BCFA, indoles and phenols, or decarboxylated to produce amines and CO_2 or metabolized to release H_2S (Fig. 15.1). Proteins, peptides, AA, and some of the products from AA metabolism may also be used as nitrosatable substrates. Many different electron donors and acceptors participate in these reactions, including keto acids, molecular H_2, unsaturated fatty acids, and other AA (Macfarlane and Macfarlane, 2012).

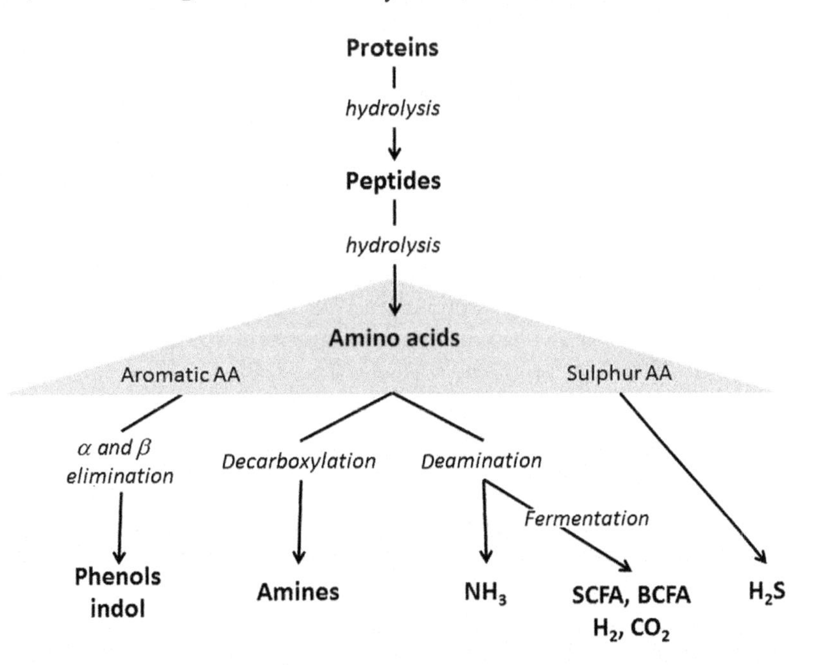

FIGURE 15.1 Overview of metabolism of protein by gut microbiota and resulting end-products.

15.1.2.2.2 End-Products

Short-chain fatty acids (mainly acetic, propionic, and butyric acids) are known as the major end-products of carbohydrate fermentation, but they are also the major metabolites produced from many AA by reductive deamination (Macfarlane and Macfarlane, 2012; Smith and Macfarlane, 1996b). On the other hand, BCFA exclusively originate from the deamination of branched AA and their excretion is thus often considered as a marker of colonic protein fermentation (Windey et al., 2012b). Isobutyrate, 2-methylbutyrate, and isovalerate are produced from the fermentation of valine, isoleucine, and leucine, respectively. These metabolites are usually one carbon shorter than their AA precursors. They are formed in large amounts through the Stickland reactions involving the coupled oxidation and reduction of AA to organic acids (Rist et al., 2013). Isocaproate, a minor BCFA in the large intestine, is the product of leucine oxidation (Macfarlane and Macfarlane, 2012).

Ammonia originates either from microbiotal AA deamination or from urea hydrolysis due to the action of bacterial urease (Rist et al., 2013). Yet, the actual involvement of the latter pathway in the human colon is debated (Macfarlane and Macfarlane, 1995). In human feces, NH_4^+ concentrations range from 12 to 30 mM depending on initial protein uptake (Hughes et al., 2000).

Amines detected in gut contents include monoamines, such as tyramine, dimethylamine, pyrrolidine, and piperidine, as well as polyamines, such as cadaverine, agmatine, histamine, putrescine, and spermine (Macfarlane and Macfarlane, 1995; Rist et al., 2013). These amines not only arise from microbiotal metabolism of AA but also from endogenous secretion, unabsorbed dietary polyamines, or release by desquamated cells (Blachier et al., 2007). Amines produced by intestinal bacteria mainly result from decarboxylation of AA but other biochemical pathways are also involved (demethylation, N-dealkylation). Amines concentrations in the human colon exhibit large interindividual variations but can reach 10 mmol/kg (Smith and Macfarlane, 1996b).

Phenolic compounds specifically result from bacterial degradation of the aromatic AA phenylalanine, tyrosine, and tryptophan and are thus possible markers of colonic protein fermentation (Windey et al., 2012b). Degradation products from tyrosine include 4-hydroxyphenylpyruvate, 4-hydroxyphenyllactate, 4-hydroxyphenylpropionate, and 4-hydroxyphenylacetate as well as phenol, p-cresol, and 4-ethylphenol. Phenylalanine bacterial metabolism leads to similar derivatives, that is, phenylpyruvate, phenyllactate, phenylacetate, and phenylpropionate, and benzoic acid. Tryptophan degradation generates indole, indole acetate, indole propionate, indole lactate, and 3-methylindole (skatole) (Macfarlane and Macfarlane, 1995; Windey et al., 2012b). Intracolonic concentrations of these end-products are usually low since the complete metabolism of aromatic AA by bacteria is thermodynamically unfavorable in the absence of an inorganic electron acceptor, which can nonetheless be partly alleviated by some other bacterial activities such as nitrate reduction (Hughes et al., 2000; Macfarlane and Macfarlane, 1995).

Most of the hydrogen sulfide in the gut is produced via metabolism of sulfur AA and taurine (Magee et al., 2000). Bacteria responsible for this production harbor a desulfhydrase, which converts cysteine to pyruvate, NH_4^+ and hydrogen sulfide (Davila et al., 2013; Smith and Macfarlane, 1997) or reduces the sulfonic acid moiety of taurine to hydrogen sulfide after deconjugation of bile acid (Louis et al., 2014). Mercaptans can also be produced, as for example from methionine which leads to the production of α-ketobutyrate, NH_4^+, and methanethiol (Davila et al., 2013; Smith and Macfarlane, 1997). Sulfide concentrations fluctuate with protein intake and can vary from 1.0 to 2.4 mM in luminal colonic contents (Macfarlane et al., 1992) and from 0.2 to 3.4 mmol/kg in feces (Magee et al., 2000).

N-nitroso compounds result from the reaction of nitrosating agents with nitrosatable substrates. This reaction can be catalyzed by some bacteria, and may occur in the colon where nitrosating agents from dissimilatory nitrate metabolism and nitrogenous residues from endogenous and dietary sources as well as nitrosatable substrates (ie, dietary proteins and peptides, AA, secondary amines, indoles, and phenols) are present (Hughes et al., 2000). However, colonic N-nitrosation has been poorly studied and its actual significance is not established, although nitrosamines have been detected in feces of healthy human volunteers in amounts related to dietary nitrate and red meat consumption (Hughes et al., 2000; Macfarlane and Macfarlane, 1995).

15.1.2.3 Bacteria Involved in Protein and AA Fermentation

15.1.2.3.1 Bacterial Densities

Enumerations by culture-based methods reveal that fecal bacteria able to grow on peptides or AA as the sole N and energy source account for 8 $\log_{10}$ per g of wet weight feces (Richardson et al., 2013; Smith and Macfarlane, 1996b). However, similar experiments carried out with more complex media (eg, enriched in yeast extract) showed that bacteria capable of hydrolyzing protein (Macfarlane et al., 1986) or to produce SCFA from peptides (Smith and Macfarlane, 1998) reach 10 to 11 $\log_{10}$ per g of wet weight feces and are therefore largely predominant in the fecal microbiota.

To our knowledge, specific enumerations of bacteria producing BCFA and NH_4^+ have not been carried out. Since these abilities are widespread among intestinal bacteria (Holdeman et al., 1977; Macfarlane and Macfarlane,

2012, 1995), it is likely that such numbers equate with total anaerobe counts. Similarly, precise enumeration of bacteria producing H_2S from AA and N-nitroso compounds is not documented. However, according to the taxonomic distribution of these functions (Davila et al., 2013; Hughes et al., 2000), they are likely to account for approximately 8 $\log_{10}$ per g of wet weight feces. Total amines producing bacteria account for 11.4 $\log_{10}$ per g of wet weight feces, similar to total anaerobe counts (Smith and Macfarlane, 1996b). However, a huge variability was observed for specific amine: while such densities were observed for bacteria producing methylamine, dimethylamine, or propylamine, only 3 $\log_{10}$ per g wet weight feces bacteria were capable of producing butylamine (Smith and Macfarlane, 1996b). Phenol and indole producing bacteria are estimated at densities between 9.7 to 11.4 $\log_{10}$ per g of wet weight feces (Smith and Macfarlane, 1996a) while only 8.4 and 8.0 $\log_{10}$ per g of wet weight feces bacteria grew on phenylalanine and tyrosine as the sole N and energy source, respectively, and as few as 4.8 $\log_{10}$ per g of wet weight feces bacteria were able to utilize tryptophan as a sole energy source (Smith and Macfarlane, 1998).

15.1.2.3.2 Bacterial Genera

Proteolysis has a wide taxonomic distribution since it is found in many different genera including both saccharolytic bacteria and obligate AA fermenters, such as *Bacteroides* sp., *Clostridium* sp., *Propionibacterium* sp., *Fusobacterium* sp., *Prevotella* sp., *Lactobacillus* sp., and *Enterococcus* sp. (Dai et al., 2011; Macfarlane and Macfarlane, 1995).

Predominance of bacteria belonging to *Clostridium* sp., *Enterococcus* sp., and *Bacteroides* sp among peptidolytic bacteria has been recently confirmed by identification based on 16s rRNA gene sequencing (Richardson et al., 2013) but this study also identified some enterobacteria as involved in peptides fermentation.

Clostridia and peptostreptococci were the most prevalent isolates from media containing AA as the sole energy source, although bacteria belonging to the genera *Fusobacterium* sp., *Bacteroides* sp., *Actinomyces* sp., *Propionibacterium* sp., and a variety of Gram-positive cocci including micrococci, peptococci, and ruminococci were also identified (Smith and Macfarlane, 1998). Conversely, AA fermenting species identified by Richardson et al. were mainly related to enterobacteria and also included *Eggerthela lenta*, belonging to *Coriobacteriales* (Richardson et al., 2013).

More specifically, bacteria able to produce NH_4^+ from urea belong to *Bacteroides* sp., *Bifidobacterium* sp., *Clostridium* sp., *Proteus* sp., and *Klebsiella* sp. (Vince et al., 1973), while the main NH_4^+ producers from AA are related to bacteroides, clostridia, and enterobacteria (Vince and Burridge, 1980). Production of BCFA is widely distributed among intestinal bacteria: many members of *Bacteroides* sp., and some species from *Peptococcus* sp., *Peptostreptococcus* sp., *Clostridium* sp., *Fusobacterium* sp., *Eubacterium* sp., and *Propionibacterium* sp. (Holdeman et al., 1977). The main amines producing bacteria belong to enterobacteria and clostridia (Allison and Macfarlane, 1989) but this ability is also present in bacteria related to *Bacteroides* sp., *Bifidobacterium* sp., *Lactobacillus* sp., *Peptostreptococcus* sp., and *Streptococcus* sp. (Macfarlane and Macfarlane, 1995; Smith and Macfarlane, 1996b).

Intestinal bacteria involved in the production of phenolic compounds include clostridia, enterobacteria, bifidobacteria, and lactobacilli, whereas bacteria from *Bacteroides* sp., *Clostridium* sp., *Peptostreptococcus* sp., *Fusobacterium* sp., *Porphyromonas* sp., and *Propionibacterium* sp. are indole producers (Hughes et al., 2000; Macfarlane and Macfarlane, 1995). Metabolism of sulfur-containing AA has not been related to any particular taxonomic group. Yet, numerous bacterial groups which are commonly found in the large intestine are known to ferment sulfur-containing AA (ie, enterococci, enterobacteria, clostridia, peptostreptococci, fusobacteria, and eubacteria) (Davila et al., 2013). From studies carried out using biological fluids other than intestinal contents, bacteria able to perform nitrosation have been identified mainly among facultative anaerobes such as *Escherichia* sp, *Pseudomonas* sp., *Proteus* sp., and *Klebsiella* sp. (Hughes et al., 2000). Whether some of those bacteria, which are components of the intestinal microbiota (eg., enterobacteria) and which encode nitroreductases and nitrate reductases, contribute to nitrosation reactions in the colon has still to be established (Louis et al., 2014).

Both the high densities and the large taxonomic diversity of the bacteria responsible for the different steps of protein metabolism suggest that intestinal microbiota is particularly adapted to putrefaction. Yet, protein metabolism by bacteria at different levels of the gut and from different protein sources is still largely overlooked, despite the fact that it could be a major media for host–microbiota interactions, taking into account both the variety of end-products generated and their biological/physiological properties.

15.1.3 Physiological and Pathophysiological Effects of Protein-Derived Bacterial Metabolites

15.1.3.1 Effect Upon the Microbiota

The microbiota is a complex, competitive, but also cooperative ecosystem whereby by-products or end-products metabolites released by bacteria can benefit other species which will use them for their own metabolism (Freilich et al., 2011). Ammonia and H_2S can be used by bacteria for AA and protein synthesis (Carbonero et al., 2012; Takahashi et al., 1980) provided that an energy source (ie, carbohydrates) is available (Rist et al., 2013; Windey et al., 2012b). Whether these metabolites, when released into the luminal content by some bacterial species are used and metabolized by other species for their own metabolism is however not known. Beside this conceivable mutualism, bacteria have also developed intercellular signaling to adapt and survive in the face of nutritional limitation, competition with other bacteria, and the host defense system. Amongst these cell-to-cell signal molecules, indole has recently emerged. It controls diverse aspects of bacterial physiology, such as spore formation, plasmid stability, drug resistance, biofilm formation, and virulence (Lee and Lee, 2010). Recent evidences also suggest that polyamines play a crucial role in bacterial pathogenesis: in some species, polyamines favor bacterial survival within the host; in some other pathogens, they are required to promote the expression of virulence determinants following infection of the host and in some others, they are crucial for triggering the genes involved in biofilm formation (Di Martino et al., 2013).

15.1.3.2 *Effect Upon the Gut*

Beside impact of protein-derived bacterial metabolites on the gut microbiota itself, a vast array of literature evaluated the impact of protein-derived bacterial metabolites on various functions of the colon, and especially colonocytes. Note that there is already a large body of literature on the effect of SCFA on these functions that will not be discussed here as they are not specific to protein metabolism.

15.1.3.2.1 Transport and Metabolism of Protein-Derived Bacterial Metabolites Into Colonocytes

Several studies identified protein-derived bacterial metabolites or their derivatives in the plasma in several species, suggesting intestinal absorption and host transformation (Nyangale et al., 2012). Polyamines can be transported through several pathways in several mammalian cell types (Abdulhussein and Wallace, 2014). In colon cancer-derived HCT116 cells, both endocytic and solute carrier-dependent (probably *SLC3A2*) mechanisms have been described for polyamine uptake (Uemura et al., 2010). They are degraded by diamine oxidase and spermidine/spermine-N acetyltransferase, with a tight intracellular polyamine concentration regulation. Amines produced by colonic bacteria are detoxified by monoamine and diamine oxidases present in the gut mucosa and liver. However some amines appear to escape such detoxification since dimethylamine has been detected in human urine samples (Hughes et al., 2000).

Large amounts of NH_4^+ absorption through the human large intestine mucosa has been originally described (Summerskill and Wolpert, 1970). Yet, permeability for NH_4^+ at the apical border of rabbit crypt colonocytes was described as very low (Singh et al., 1995). A later study identified two members of the NH_4^+ transporter family, RhBG and RhCG, expression in the mouse intestinal tract (Handlogten et al., 2005), suggesting transporter-mediated absorption of NH_4^+. Part of NH_4^+ is metabolized within the colonocytes: NH_4^+ contributes to colonocyte glutamine synthesis through the activity of the glutamine synthetase in pig (Eklou-Lawson et al., 2009) and participates to colonocyte citrulline production in rat (Mouillé et al., 1999). In the colonic environment, H_2S is found in gaseous, dissolved, and anionic forms depending on the pH. In the human descending colon with a pH of 6, sulfide likely occurs as H_2S, either dissolved or as a gas (Magee et al., 2000). H_2S is lipophilic and penetrates biological membranes (Reiffenstein et al., 1992), yet its specific transport into colonocytes has not been studied. Sulfide in low concentrations can be oxidized in HT-29 colonocyte mitochondria, providing fuel to the colonocytes (Goubern et al., 2007). Phenolic compounds are largely absorbed from the human colonic lumen and are partly metabolized within the mucosa by glucuronide and sulfate conjugation (Windey et al., 2012a) and excreted via urine, principally as p-cresols ($>90\%$ of urinary phenolic compounds) but also as phenol and 4-ethylphenol (Hughes et al., 2000; Rist et al., 2013).

15.1.3.2.2 Genotoxicity of Protein-Derived Bacterial Metabolites

Several protein-derived bacterial metabolites have shown genotoxicity even at doses representative of their luminal concentrations. Slight genomic damages were observed with low concentration of H_2S in HT-29 colonic epithelial cells (Attene-Ramos et al., 2006). Phenol has also been categorized as a carcinogen based on its ability to enhance N-nitrosation to form nitrosamines which are carcinogens (Nyangale et al., 2012). It also directly impairs HT-29 viability in vitro (Pedersen et al., 2002). Finally, in a recent study of geno- and cytotoxicity of fecal water from high dietary protein-fed volunteers, indole was associated with the cytotoxicity of the fecal water (Windey et al., 2012a).

15.1.3.2.3 Impact of Protein-Derived Bacterial Metabolites on Colonocyte Metabolism

Several protein-derived bacterial metabolites are considered as colonocyte metabolic trouble-makers. For example, NH_4^+ interferes with the oxidative metabolism of colonocytes and dose-dependently inhibits mitochondrial oxygen consumption in rats (Andriamihaja et al., 2010). High concentrations of NH_4^+ decrease metabolism of butyrate and acetate to CO_2 and ketones by rat colonocytes, yet increase oxidation of glucose and glutamine in pig colonocytes (Cremin et al., 2003; Darcy-Vrillon et al., 1996). Likewise, sulfide can impair colonocyte metabolism depending on its concentration: low concentrations can be used as energetic substrate by HT-29 colonic cells (Goubern et al., 2007), high concentrations inhibit HT-29 glutamine, butyrate, and acetate oxidation in a dose-dependent manner (Leschelle et al., 2005).

15.1.3.2.4 Impact of Protein-Derived Bacterial Metabolites on Epithelial Cell Proliferation, Differentiation, and Apoptosis

Maintenance of intestinal epithelial homeostasis depends on a complex interplay between processes involving proliferation, differentiation, migration, and apoptosis. Protein-derived bacterial metabolites can interfere with these mechanisms. The most studied ones are polyamines since those compounds, either endogenous or luminally derived, are absolutely required for epithelial cell division. Cellular polyamine content increases rapidly when mammalian cells are stimulated to grow and divide, whereas decreased cellular polyamine content represses intestinal cell proliferation both in vivo and in vitro (Timmons et al., 2012). The other side of the coin is that polyamines seem involved in the growth of chemically induced preneoplastic colonic lesions in animal models, consistent with the fact that colonocytes isolated from human cancerous areas have higher polyamine content than healthy colonocytes (Timmons et al., 2012). Beside these well studied metabolites, reports indicated that other protein-derived bacterial metabolites may interfere with intestinal epithelial cell turnover. Hydrogen sulfide induces mucosal hyperproliferation with an expansion of the proliferative zone to the upper crypt in human mucosal biopsies (Christl et al., 1996). This was confirmed in vitro in intestinal epithelial IEC-18 cells where H_2S increased cell proliferation (Deplancke et al., 2003). Conversely, NH_4^+ decreases epithelial cell proliferation rate with a slowdown of all cell cycle phases in HT-29 cells (Mouillé et al., 2003) but had not effect on cell apoptosis in isolated pig colonic crypts (Leschelle et al., 2002).

15.1.3.2.5 Impact of Protein-Derived Bacterial Metabolites on Electrolyte and Water Absorption or Secretion

In the colon, hydrolysis and absorption of nutrients is generally completed and the main role of colonocytes is to re-absorb electrolytes and water that have been secreted in the small intestine during the digestion process. Several studies reported that protein-derived bacterial metabolites can interfere with electrolytes and water transport across colonocytes. Ammonia inhibits cAMP-induced Cl^- secretion in human T84 intestinal epithelial cells (Prasad et al., 1995). It also interferes with the mechanisms involved in intracellular (Ca^{2+}) increase, a preliminary step in Ca^{2+} mediated Cl^- secretion (Mayol et al., 1997). Beside this antisecretory effect, NH_4^+ has also been shown to decrease basal Na^+ absorption in the rat proximal colon (Cermak et al., 2000). Similarly to NH_4^+, isobutyrate prevents cAMP-triggered Cl^- secretion (Dagher et al., 1996) and reverses the cGMP-induced Cl^- secretion in both proximal and distal colon of the rat (Charney et al., 1999). It also stimulates Na^+ absorption in the rat proximal colon (Zaharia et al., 2001). Likewise, luminal or basolateral spermine dose-dependently increases net fluid absorption and inhibits cAMP-stimulated fluid secretion in rat isolated perfused colonic crypts (Cheng et al., 2004). Conversely to these pro-absorptive or antisecretory effects, H_2S induces colonic ion secretion by affecting basolateral K^+ channels, the basolateral Na^+-K^+-ATPase as well as Ca^{2+} transporters in rats (Hennig and Diener, 2009; Pouokam and Diener, 2011).

15.1.3.2.6 Impact of Protein-Derived Metabolites on Colonocyte Barrier Function

Intestinal epithelial cells form a physical barrier limiting the entry of bacteria and unwanted luminal products. This is crucial in the colon which displays the highest concentration of luminal bacteria. This barrier function is supported by the presence of tight junctions, specialized protein complexes anchored in the lateral membrane of adjacent cells, forming size- and charge-selective pores between cells. This paracellular permeability is a dynamic phenomenon; permeability can change within minutes or longer time scales depending on the stimulus (Bischoff et al., 2014). Several reports indicate an impact of protein-derived bacterial metabolites upon intestinal permeability or expression of tight junction proteins. Polyamines are necessary for the maintenance of intestinal barrier function as suggested by polyamine depletion experiments in IEC-6 cells (Guo et al., 2003, 2005). Spermidine protects T84 and HT-29 intestinal epithelial cells from inflammatory challenge by reducing the increase in epithelial permeability induced by IFN-γ (Penrose et al., 2013). Yet, all polyamines do not seem to affect intestinal barrier function equally. Spermine effect on intestinal permeability to different-sized molecules seems to depend on the intestinal region and on the polyamine concentration; high spermine concentration enhanced, while low concentration decreased intestinal permeability in rats (Osman et al., 1998). Like spermidine, BCFA dose-dependently protect Caco-2 cells from pro-inflammatory cytokine-induced intestinal barrier function defaults (Boudry et al., 2013). Ammonia increases intestinal permeability in Caco-2 cells (Hughes et al., 2008). No direct report of the effect of H_2S on intestinal permeability is available. However, as oxidation of butyrate supports the epithelial barrier function of colonocytes, the fact that H_2S impairs butyrate metabolism suggests that it may affect colonic mucosal integrity when in large concentrations. Indole significantly increases the expression of several genes involved in the maintenance of epithelial cell structure and function in vitro in HCT-8 cells, including genes responsible for tight junction organization, actin cytoskeleton and adherens junction. This translates phenotypically in strengthened intestinal epithelial cell barrier properties and increased resistance to pathogen colonization in mouse (Bansal et al., 2010; Shimada et al., 2013; Venkatesh et al., 2014). Conversely, phenol increases intestinal epithelial cell permeability in various intestinal cell in vitro models (Hughes et al., 2008; McCall et al., 2009).

15.1.3.2.7 Impact of Protein-Derived Bacterial Metabolites on Nutrient Sensing and Gastrointestinal Hormone Release

Within the intestinal epithelium, some cells have more specialized functions. Entero-endocrine cells sense nutrients present in the luminal content. In response to these luminal stimuli, they secrete various gastrointestinal hormones that are released locally in the lamina propria or in the bloodstream, signaling to several organs (eg, vagus afferent neurons, brain, and pancreas) to coordinate feeding behavior, digestion process, and insulin release (Breen et al., 2013). Indole is able to modulate the secretion of GLP-1 from immortalized and primary mouse colonic L cells (Chimerel et al., 2014). Conversely, H_2S dose-dependently decrease bile acid-induced GLP-1 and PYY secretion from STC-1 cells, a cellular model of entero-endocrine cells, likely by inhibition of the intracellular signaling cascade (Bala et al., 2014). The effect of other protein-derived bacterial metabolites like BCFA, NH_4^+, and other indolic and phenolic compounds upon entero-endocrine cell activation is not documented.

15.1.3.2.8 Impact of Protein-Derived Bacterial Metabolites Upon Goblet Cells and Mucin Secretion

Goblet cells are mucin-secreting cells that secrete various types of mucins and factors like trefoil factor (TFF) in the lumen, forming a dense mucus layer covering the apical surface of the epithelium (Pelaseyed et al., 2014). The impact of protein-derived bacterial metabolites upon goblet cells and mucin secretion is poorly studied. Yet, high protein diet consumption induces goblet cell hyperplasia and greater intestinal mucus content in ileum, and modifies goblet cell distribution in rat colonic epithelium (Lan et al., 2015). However, since AA, like threonine, are precursors for mucin synthesis, it is difficult to decipher if the effect of the high protein diet is mediated directly by larger amounts of AA or by bacterial metabolites. However, some reports indicate that protein-derived bacterial metabolites modulate goblet cell differentiation or mucin and mucus properties and secretion. Indeed, H_2S reduces heteromer complexes TFF3- IgG Fc binding protein in vitro to release TFF3 monomer or dimers in human colonic biopsies (Albert et al., 2010). Likewise, indole upregulates the expression of several MUC genes as well as of TFF2 in HCT-8 cells (Bansal et al., 2010).

15.1.3.2.9 Impact of Protein-Derived Bacterial Metabolites on Enteric Nerves

Several molecules with neuroactive functions such as gamma-aminobutyric acid (GABA from glutamate), serotonin (5-HT from tryptophan), catecholamines (dopamine and norepinephrine, from phenylalanine and tyrosine, respectively), histamine (from histidine), indole-3-propionic acid, and H_2S have been reported to be produced by intestinal bacteria. The exact role of these neuroactive compounds is not known. Norepinephrine has been shown to enhance growth and pathogenicity of several bacteria (Lyte et al., 2011). Secreted neurotransmitters from bacteria in the intestinal lumen may impact colonocytes directly as demonstrated for dopamine in mice (Asano et al., 2012). They may also induce intestinal epithelial cells to release molecules that in turn modulate neural signaling within the enteric nervous system or signal directly to the enteric neurons and modulate intestinal function (motility, electrolyte secretion, epithelial layer permeability) or even signal to the brain to modulate host behavior (Wall et al., 2014).

15.1.3.2.10 Impact of Protein-Derived Bacterial Metabolites on Intestinal Immune Cells

Recent evidence suggests that some protein-derived bacterial metabolites signal to intestinal immune cells and contribute to gut mucosal homeostasis. For example, under high tryptophan availability some *Lactobacilli* sp. produce indole- 3-aldehyde that contributes to aryl hydrocarbon receptor (AhR)-dependent IL-22 transcription, balancing mucosal response and providing colonization resistance to the fungus *Candida albicans* and mucosal protection from inflammation in mice (Zelante et al., 2013). Physiological concentrations of isovalerate stimulates the secretion of TNF-α and suppresses the secretion of IL-10 by macrophages in vitro, while NH_4^+ suppresses the release of both cytokines (Van Nuenen et al., 2005). Finally, H_2S enhances mouse T-cell activation and dose-dependently enhances TCR-stimulated proliferation (Miller et al., 2012).

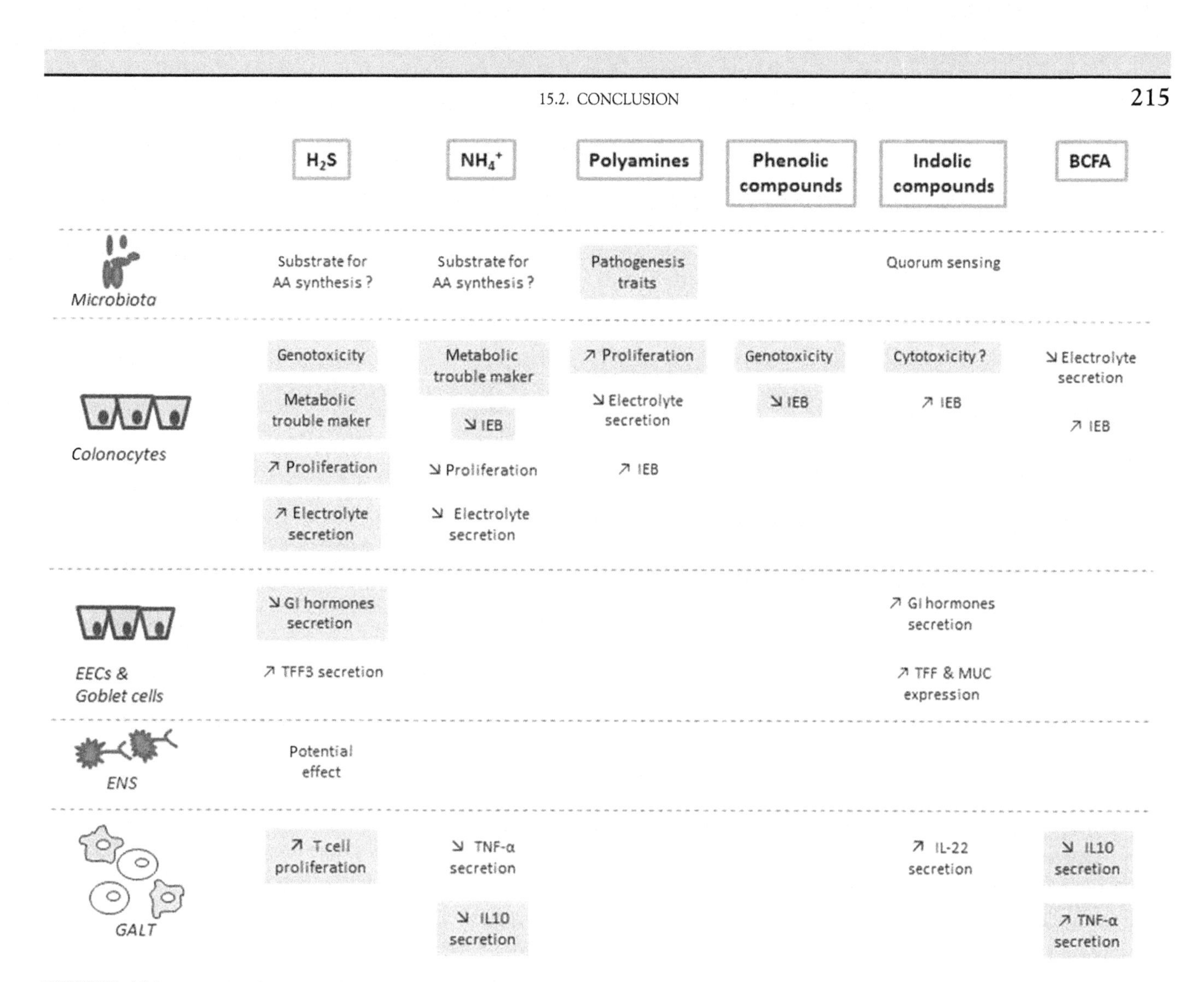

FIGURE 15.2 **Impact of protein-derived bacterial metabolites upon the microbiota and the host colonic mucosa.** The impact of protein-derived metabolites (upper line) on each component of the colonic mucosa (left column) is described when documented. Shade: detrimental effect upon the host. *EEC*, entero-endocrine cell; *ENS*, enteric nervous system; *GALT*, gut-associated lymphoid tissue; *GI*, gastrointestinal; *IEB*, intestinal epithelial barrier; *TFF*, trefoil factor.

15.1.3.3 *Effect Beyond the Gut*

As stated earlier, several protein-derived bacterial metabolites can be found in the host plasma. Ammonia concentration is well regulated in the body with the majority being converted to urea in the liver of healthy individuals. However, in individuals with impaired liver function, hyperammonemia can develop, leading to encephalopathy and eventually death (Nyangale et al., 2012). Phenol, indole, and p-cresol have an impact on kidney failure and endothelial function. Prebiotic and probiotic therapy are even suggested to reduce bacterial putrefaction and reduce p-cresol plasma level in individuals suffering kidney failure (Ramezani and Raj, 2014). Besides these pathological states, it is very conceivable that protein-derived bacterial metabolites impact diverse host functions. Molecules with neuroactive functions may reach the brain and modulate behavior and metabolism. Other metabolites may be involved in host metabolism and physiology.

15.2. CONCLUSION

The diversity of protein and peptides entering the colon is widely dependent upon protein source and quantity but also food process. Yet, the large variety of bacteria able to metabolize dietary protein and AA and their complementary and overlapping metabolic pathways ensure efficient use by the gut microbiota of these dietary "left-over" or endogenous compound. The resulting protein-derived bacterial metabolites exert diverse effects upon the gut microbiota itself but also on various cell types within the colonic mucosa (Fig. 15.2). Those metabolites may well also exert some effects beyond the gut. Protein-derived metabolites have long been described as harmful compounds with genotoxic and procarcinogenic functions. Although several reports do confirm this statement, recent evidence also suggests that metabolites arising from protein fermentation participates in the host–commensal microbiota interactions. New experimental work is therefore needed to understand the interplay of dietary protein and gut microbiota in health and disease.

References

Abdulhussein, A.A., Wallace, H.M., 2014. Polyamines and membrane transporters. Amino Acids 46, 655–660. Available from: http://dx.doi.org/10.1007/s00726-013-1553-6.

Albert, T.K., Laubinger, W., Müller, S., Hanisch, F.-G., Kalinski, T., Meyer, F., et al., 2010. Human intestinal TFF3 forms disulfide-linked heteromers with the mucus-associated FCGBP protein and is released by hydrogen sulfide. J. Proteome Res. 9, 3108–3117. Available from: http://dx.doi.org/10.1021/pr100020c.

Allison, C., Macfarlane, G.T., 1989. Influence of pH, nutrient availability, and growth rate on amine production by *Bacteroides fragilis* and *Clostridium perfringens*. Appl. Environ. Microbiol. 55, 2894–2898.

Andriamihaja, M., Davila, A.-M., Eklou-Lawson, M., Petit, N., Delpal, S., Allek, F., et al., 2010. Colon luminal content and epithelial cell morphology are markedly modified in rats fed with a high-protein diet. Am. J. Physiol. Gastrointest. Liver Physiol. 299, G1030–G1037. Available from: http://dx.doi.org/10.1152/ajpgi.00149.2010.

Arumugam, M., Raes, J., Pelletier, E., Le Paslier, D., Yamada, T., Mende, D.R., et al., 2011. Enterotypes of the human gut microbiome. Nature 473, 174–180. Available from: http://dx.doi.org/10.1038/nature09944.

Asano, Y., Hiramoto, T., Nishino, R., Aiba, Y., Kimura, T., Yoshihara, K., et al., 2012. Critical role of gut microbiota in the production of biologically active, free catecholamines in the gut lumen of mice. AJP Gastrointest. Liver Physiol. 303, G1288–G1295. Available from: http://dx.doi.org/10.1152/ajpgi.00341.2012.

Attene-Ramos, M.S., Wagner, E.D., Plewa, M.J., Gaskins, H.R., 2006. Evidence that hydrogen sulfide is a genotoxic agent. Mol. Cancer Res. 4, 9–14. Available from: http://dx.doi.org/10.1158/1541-7786.MCR-05-0126.

Bäckhed, F., Ley, R.E., Sonnenburg, J.L., Peterson, D.A., Gordon, J.I., 2005. Host-bacterial mutualism in the human intestine. Science 307, 1915–1920. Available from: http://dx.doi.org/10.1126/science.1104816.

Bala, V., Rajagopal, S., Kumar, D.P., Nalli, A.D., Mahavadi, S., Sanyal, A.J., et al., 2014. Release of GLP-1 and PYY in response to the activation of G protein-coupled bile acid receptor TGR5 is mediated by Epac/PLC-ε pathway and modulated by endogenous H2S. Front. Physiol. 5, 420. Available from: http://dx.doi.org/10.3389/fphys.2014.00420.

Bansal, T., Alaniz, R.C., Wood, T.K., Jayaraman, A., 2010. The bacterial signal indole increases epithelial-cell tight-junction resistance and attenuates indicators of inflammation. Proc. Natl. Acad. Sci. U.S.A. 107, 228–233. Available from: http://dx.doi.org/10.1073/pnas.0906112107.

Barbé, F., Ménard, O., Le Gouar, Y., Buffière, C., Famelart, M.-H., Laroche, B., et al., 2014. Acid and rennet gels exhibit strong differences in the kinetics of milk protein digestion and amino acid bioavailability. Food Chem. 143, 1–8. Available from: http://dx.doi.org/10.1016/j.foodchem.2013.07.100.

Bax, M.-L., Buffière, C., Hafnaoui, N., Gaudichon, C., Savary-Auzeloux, I., Dardevet, D., et al., 2013. Effects of meat cooking, and of ingested amount, on protein digestion speed and entry of residual proteins into the colon: a study in minipigs. PloS One 8, e61252. Available from: http://dx.doi.org/10.1371/journal.pone.0061252.

Bischoff, S.C., Barbara, G., Buurman, W., Ockhuizen, T., Schulzke, J.-D., Serino, M., et al., 2014. Intestinal permeability – a new target for disease prevention and therapy. BMC Gastroenterol. 14, 189. Available from: http://dx.doi.org/10.1186/s12876-014-0189-7.

Blachier, F., Mariotti, F., Huneau, J.F., Tomé, D., 2007. Effects of amino acid-derived luminal metabolites on the colonic epithelium and physiopathological consequences. Amino Acids 33, 547–562. Available from: http://dx.doi.org/10.1007/s00726-006-0477-9.

Blottière, H.M., de Vos, W.M., Ehrlich, S.D., Doré, J., 2013. Human intestinal metagenomics: state of the art and future. Curr. Opin. Microbiol. 16, 232–239. Available from: http://dx.doi.org/10.1016/j.mib.2013.06.006.

Boudry, G., Jamin, A., Chatelais, L., Gras-Le Guen, C., Michel, C., Le Huërou-Luron, I., 2013. Dietary protein excess during neonatal life alters colonic microbiota and mucosal response to inflammatory mediators later in life in female pigs. J. Nutr. 143, 1225–1232. Available from: http://dx.doi.org/10.3945/jn.113.175828.

Breen, D.M., Rasmussen, B.A., Côté, C.D., Jackson, V.M., Lam, T.K.T., 2013. Nutrient-sensing mechanisms in the gut as therapeutic targets for diabetes. Diabetes 62, 3005–3013. Available from: http://dx.doi.org/10.2337/db13-0523.

Carbonero, F., Benefiel, A.C., Alizadeh-Ghamsari, A.H., Gaskins, H.R., 2012. Microbial pathways in colonic sulfur metabolism and links with health and disease. Front. Physiol. 3, 448. Available from: http://dx.doi.org/10.3389/fphys.2012.00448.

Cermak, R., Lawnitzak, C., Scharrer, E., 2000. Influence of ammonia on sodium absorption in rat proximal colon. Pflüg. Arch. 440, 619–626.

Chacko, A., Cummings, J.H., 1988. Nitrogen losses from the human small bowel: obligatory losses and the effect of physical form of food. Gut 29, 809–815.

Charney, A.N., Giannella, R.A., Egnor, R.W., 1999. Effect of short-chain fatty acids on cyclic 3′,5′-guanosine monophosphate-mediated colonic secretion. Comp. Biochem. Physiol. A. Mol. Integr. Physiol. 124, 169–178.

Cheng, S.X., Geibel, J.P., Hebert, S.C., 2004. Extracellular polyamines regulate fluid secretion in rat colonic crypts via the extracellular calcium-sensing receptor. Gastroenterology 126, 148–158.

Chimerel, C., Emery, E., Summers, D.K., Keyser, U., Gribble, F.M., Reimann, F., 2014. Bacterial metabolite indole modulates incretin secretion from intestinal enteroendocrine L cells. Cell Rep. 9, 1202–1208. Available from: http://dx.doi.org/10.1016/j.celrep.2014.10.032.

Christl, S.U., Eisner, H.D., Dusel, G., Kasper, H., Scheppach, W., 1996. Antagonistic effects of sulfide and butyrate on proliferation of colonic mucosa: a potential role for these agents in the pathogenesis of ulcerative colitis. Dig. Dis. Sci. 41, 2477–2481.

Cremin, J.D., Fitch, M.D., Fleming, S.E., 2003. Glucose alleviates ammonia-induced inhibition of short-chain fatty acid metabolism in rat colonic epithelial cells. Am. J. Physiol. Gastrointest. Liver Physiol. 285, G105–114. Available from: http://dx.doi.org/10.1152/ajpgi.00437.2002.

Cummings, J.H., Pomare, E.W., Branch, W.J., Naylor, C.P., Macfarlane, G.T., 1987. Short chain fatty acids in human large intestine, portal, hepatic and venous blood. Gut 28, 1221–1227.

Daenzer, M., Petzke, K.J., Bequette, B.J., Metges, C.C., 2001. Whole-body nitrogen and splanchnic amino acid metabolism differ in rats fed mixed diets containing casein or its corresponding amino acid mixture. J. Nutr. 131, 1965–1972.

Dagher, P.C., Egnor, R.W., Taglietta-Kohlbrecher, A., Charney, A.N., 1996. Short-chain fatty acids inhibit cAMP-mediated chloride secretion in rat colon. Am. J. Physiol. 271, C1853–1860.

Dai, Z.-L., Zhang, J., Wu, G., Zhu, W.-Y., 2010. Utilization of amino acids by bacteria from the pig small intestine. Amino Acids 39, 1201–1215. Available from: http://dx.doi.org/10.1007/s00726-010-0556-9.

Dai, Z.-L., Wu, G., Zhu, W.-Y., 2011. Amino acid metabolism in intestinal bacteria: links between gut ecology and host health. Front. Biosci. Landmark Ed. 16, 1768–1786.

Darcy-Vrillon, B., Cherbuy, C., Morel, M.T., Durand, M., Duée, P.H., 1996. Short chain fatty acid and glucose metabolism in isolated pig colonocytes: modulation by NH_4^+. Mol. Cell. Biochem. 156, 145–151.

Davila, A.-M., Blachier, F., Gotteland, M., Andriamihaja, M., Benetti, P.-H., Sanz, Y., et al., 2013. Re-print of "Intestinal luminal nitrogen metabolism: role of the gut microbiota and consequences for the host.". Pharmacol. Res. 69, 114–126. Available from: http://dx.doi.org/10.1016/j.phrs.2013.01.003.

Deplancke, B., Finster, K., Graham, W.V., Collier, C.T., Thurmond, J.E., Gaskins, H.R., 2003. Gastrointestinal and microbial responses to sulfate-supplemented drinking water in mice. Exp. Biol. Med. (Maywood NJ) 228, 424–433.

Di Martino, M.L., Campilongo, R., Casalino, M., Micheli, G., Colonna, B., Prosseda, G., 2013. Polyamines: emerging players in bacteria–host interactions. Int. J. Med. Microbiol. 303, 484–491. Available from: http://dx.doi.org/10.1016/j.ijmm.2013.06.008.

Doré, J., Corthier, G., 2010. The human intestinal microbiota. Gastroentérologie Clin. Biol. 34 (Suppl. 1), S7–15, http://dx.doi.org/10.1016/S0399-8320(10)70015-4.

Eklou-Lawson, M., Bernard, F., Neveux, N., Chaumontet, C., Bos, C., Davila-Gay, A.-M., et al., 2009. Colonic luminal ammonia and portal blood L-glutamine and L-arginine concentrations: a possible link between colon mucosa and liver ureagenesis. Amino Acids 37, 751–760. Available from: http://dx.doi.org/10.1007/s00726-008-0218-3.

Freilich, S., Zarecki, R., Eilam, O., Segal, E.S., Henry, C.S., Kupiec, M., et al., 2011. Competitive and cooperative metabolic interactions in bacterial communities. Nat. Commun. 2, 589. Available from: http://dx.doi.org/10.1038/ncomms1597.

Fuller, M.F., Milne, A., Harris, C.I., Reid, T.M., Keenan, R., 1994. Amino acid losses in ileostomy fluid on a protein-free diet. Am. J. Clin. Nutr. 59, 70–73.

Gaudichon, C., Roos, N., Mahé, S., Sick, H., Bouley, C., Tomé, D., 1994. Gastric emptying regulates the kinetics of nitrogen absorption from 15N-labeled milk and 15N-labeled yogurt in miniature pigs. J. Nutr. 124, 1970–1977.

Gaudichon, C., Bos, C., Morens, C., Petzke, K.J., Mariotti, F., Everwand, J., et al., 2002. Ileal losses of nitrogen and amino acids in humans and their importance to the assessment of amino acid requirements. Gastroenterology 123, 50–59.

Gausserès, N., Mahè, S., Benamouzig, R., Luengo, C., Drouet, H., Rautureau, J., et al., 1996. The gastro-ileal digestion of 15N-labelled pea nitrogen in adult humans. Br. J. Nutr. 76, 75–85.

Gibson, S.A., McFarlan, C., Hay, S., Macfarlane, G.T., 1989. Significance of microflora in proteolysis in the colon. Appl. Environ. Microbiol. 55, 679–683.

Goubern, M., Andriamihaja, M., Nübel, T., Blachier, F., Bouillaud, F., 2007. Sulfide, the first inorganic substrate for human cells. FASEB J. 21, 1699–1706. Available from: http://dx.doi.org/10.1096/fj.06-7407com.

Guo, X., Rao, J.N., Liu, L., Zou, T.-T., Turner, D.J., Bass, B.L., et al., 2003. Regulation of adherens junctions and epithelial paracellular permeability: a novel function for polyamines. AJP Cell Physiol. 285, C1174–C1187. Available from: http://dx.doi.org/10.1152/ajpcell.00015.2003.

Guo, X., Rao, J.N., Liu, L., Zou, T., Keledjian, K.M., Boneva, D., et al., 2005. Polyamines are necessary for synthesis and stability of occludin protein in intestinal epithelial cells. Am. J. Physiol. Gastrointest. Liver Physiol. 288, G1159–1169. Available from: http://dx.doi.org/10.1152/ajpgi.00407.2004.

Handlogten, M.E., Hong, S.-P., Zhang, L., Vander, A.W., Steinbaum, M.L., Campbell-Thompson, M., et al., 2005. Expression of the ammonia transporter proteins Rh B glycoprotein and Rh C glycoprotein in the intestinal tract. Am. J. Physiol. Gastrointest. Liver Physiol. 288, G1036–1047. Available from: http://dx.doi.org/10.1152/ajpgi.00418.2004.

Hennig, B., Diener, M., 2009. Actions of hydrogen sulphide on ion transport across rat distal colon. Br. J. Pharmacol. 158, 1263–1275. Available from: http://dx.doi.org/10.1111/j.1476-5381.2009.00385.x.

Hodgkinson, S.M., Moughan, P.J., Reynolds, G.W., James, K.A., 2000. The effect of dietary peptide concentration on endogenous ileal amino acid loss in the growing pig. Br. J. Nutr. 83, 421–430.

Holdeman, L.V., Cato, E.P., Moore, W., 1977. Anaerobe Laboratory Manual, fourth ed. Polytechnic Institute and State University Press, Backsburg.

Hughes, R., Kurth, M.J., McGilligan, V., McGlynn, H., Rowland, I., 2008. Effect of colonic bacterial metabolites on Caco-2 cell paracellular permeability in vitro. Nutr. Cancer 60, 259–266. Available from: http://dx.doi.org/10.1080/01635580701649644.

Hughes, R., Magee, E.A., Bingham, S., 2000. Protein degradation in the large intestine: relevance to colorectal cancer. Curr. Issues Intest. Microbiol. 1, 51–58.

Kurokawa, K., Itoh, T., Kuwahara, T., Oshima, K., Toh, H., Toyoda, A., et al., 2007. Comparative metagenomics revealed commonly enriched gene sets in human gut microbiomes. DNA Res. 14, 169–181. Available from: http://dx.doi.org/10.1093/dnares/dsm018.

Lan, A., Andriamihaja, M., Blouin, J.-M., Liu, X., Descatoire, V., Desclée de Maredsous, C., et al., 2015. High-protein diet differently modifies intestinal goblet cell characteristics and mucosal cytokine expression in ileum and colon. J. Nutr. Biochem. 26, 91–98. Available from: http://dx.doi.org/10.1016/j.jnutbio.2014.09.007.

Lee, J.-H., Lee, J., 2010. Indole as an intercellular signal in microbial communities. FEMS Microbiol. Rev. 34, 426–444. Available from: http://dx.doi.org/10.1111/j.1574-6976.2009.00204.x.

Lepage, P., Leclerc, M.C., Joossens, M., Mondot, S., Blottière, H.M., Raes, J., et al., 2013. A metagenomic insight into our gut's microbiome. Gut 62, 146–158. Available from: http://dx.doi.org/10.1136/gutjnl-2011-301805.

Leschelle, X., Robert, V., Delpal, S., Mouillé, B., Mayeur, C., Martel, P., et al., 2002. Isolation of pig colonic crypts for cytotoxic assay of luminal compounds: effects of hydrogen sulfide, ammonia, and deoxycholic acid. Cell Biol. Toxicol. 18, 193–203.

Leschelle, X., Goubern, M., Andriamihaja, M., Blottière, H.M., Couplan, E., Gonzalez-Barroso, M.-D.-M., et al., 2005. Adaptative metabolic response of human colonic epithelial cells to the adverse effects of the luminal compound sulfide. Biochim. Biophys. Acta 1725, 201–212. Available from: http://dx.doi.org/10.1016/j.bbagen.2005.06.002.

Ley, R.E., Hamady, M., Lozupone, C., Turnbaugh, P., Ramey, R.R., Bircher, J.S., et al., 2008. Evolution of mammals and their gut microbes. Science 320, 1647–1651. Available from: http://dx.doi.org/10.1126/science.1155725.

Li, J., Jia, H., Cai, X., Zhong, H., Feng, Q., Sunagawa, S., et al., 2014. An integrated catalog of reference genes in the human gut microbiome. Nat. Biotechnol. 32, 834–841. Available from: http://dx.doi.org/10.1038/nbt.2942.

Louis, P., Hold, G.L., Flint, H.J., 2014. The gut microbiota, bacterial metabolites and colorectal cancer. Nat. Rev. Microbiol. 12, 661–672. Available from: http://dx.doi.org/10.1038/nrmicro3344.

Lyte, M., Vulchanova, L., Brown, D.R., 2011. Stress at the intestinal surface: catecholamines and mucosa-bacteria interactions. Cell Tissue Res. 343, 23–32. Available from: http://dx.doi.org/10.1007/s00441-010-1050-0.

Macfarlane, S., Macfarlane, G.T., 1995. Proteolysis and amino acid fermentation. In: Gibson, G.R., Macfarlane, G.T. (Eds.), Human Colonic Bacteria, Role in Nutrition, Physiology, and Pathology. CRC Press, Boca Raton.

Macfarlane, G.T., Macfarlane, S., 2012. Bacteria, colonic fermentation, and gastrointestinal health. J. AOAC Int. 95, 50–60.

Macfarlane, G.T., Cummings, J.H., Allison, C., 1986. Protein degradation by human intestinal bacteria. J. Gen. Microbiol. 132, 1647–1656.

Macfarlane, G.T., Gibson, G.R., Cummings, J.H., 1992. Comparison of fermentation reactions in different regions of the human colon. J. Appl. Bacteriol. 72, 57–64.

Magee, E.A., Richardson, C.J., Hughes, R., Cummings, J.H., 2000. Contribution of dietary protein to sulfide production in the large intestine: an in vitro and a controlled feeding study in humans. Am. J. Clin. Nutr. 72, 1488–1494.

Mahé, S., Messing, B., Thuillier, F., Tomé, D., 1991. Digestion of bovine milk proteins in patients with a high jejunostomy. Am. J. Clin. Nutr. 54, 534–538.

Mahé, S., Huneau, J.F., Marteau, P., Thuillier, F., Tomé, D., 1992. Gastroileal nitrogen and electrolyte movements after bovine milk ingestion in humans. Am. J. Clin. Nutr. 56, 410–416.

Mariotti, F., Mahé, S., Benamouzig, R., Luengo, C., Daré, S., Gaudichon, C., et al., 1999. Nutritional value of [15N]-soy protein isolate assessed from ileal digestibility and postprandial protein utilization in humans. J. Nutr. 129, 1992–1997.

Marteau, P., Pochart, P., Doré, J., Béra-Maillet, C., Bernalier, A., Corthier, G., 2001. Comparative study of bacterial groups within the human cecal and fecal microbiota. Appl. Environ. Microbiol. 67, 4939–4942.

Mayol, J.M., Hrnjez, B.J., Akbarali, H.I., Song, J.C., Smith, J.A., Matthews, J.B., 1997. Ammonia effect on calcium-activated chloride secretion in T84 intestinal epithelial monolayers. Am. J. Physiol. 273, C634–642.

McCall, I.C., Betanzos, A., Weber, D.A., Nava, P., Miller, G.W., Parkos, C.A., 2009. Effects of phenol on barrier function of a human intestinal epithelial cell line correlate with altered tight junction protein localization. Toxicol. Appl. Pharmacol. 241, 61–70. Available from: http://dx.doi.org/10.1016/j.taap.2009.08.002.

Miller, T.W., Wang, E.A., Gould, S., Stein, E.V., Kaur, S., Lim, L., et al., 2012. Hydrogen sulfide is an endogenous potentiator of T cell activation. J. Biol. Chem. 287, 4211–4221. Available from: http://dx.doi.org/10.1074/jbc.M111.307819.

Miner-Williams, W., Deglaire, A., Benamouzig, R., Fuller, M.F., Tomé, D., Moughan, P.J., 2012. Endogenous proteins in terminal ileal digesta of adult subjects fed a casein-based diet. Am. J. Clin. Nutr. 96, 508–515. Available from: http://dx.doi.org/10.3945/ajcn.111.033472.

Miner-Williams, W., Deglaire, A., Benamouzig, R., Fuller, M.F., Tomé, D., Moughan, P.J., 2014. Endogenous proteins in the ileal digesta of adult humans given casein-, enzyme-hydrolyzed casein- or crystalline amino-acid-based diets in an acute feeding study. Eur. J. Clin. Nutr. 68, 363–369. Available from: http://dx.doi.org/10.1038/ejcn.2013.270.

Moughan, P.J., Rutherfurd, S.M., 2012. Gut luminal endogenous protein: implications for the determination of ileal amino acid digestibility in humans. Br. J. Nutr. 108 (Suppl. 2), S258–263. Available from: http://dx.doi.org/10.1017/S0007114512002474.

Mouillé, B., Morel, E., Robert, V., Guihot-Joubrel, G., Blachier, F., 1999. Metabolic capacity for L-citrulline synthesis from ammonia in rat isolated colonocytes. Biochim. Biophys. Acta 1427, 401–407.

Mouillé, B., Delpal, S., Mayeur, C., Blachier, F., 2003. Inhibition of human colon carcinoma cell growth by ammonia: a non-cytotoxic process associated with polyamine synthesis reduction. Biochim. Biophys. Acta 1624, 88–97.

Nyangale, E.P., Mottram, D.S., Gibson, G.R., 2012. Gut microbial activity, implications for health and disease: the potential role of metabolite analysis. J. Proteome Res. 11, 5573–5585. Available from: http://dx.doi.org/10.1021/pr300637d.

Osman, N.E., Weström, B., Wang, Q., Persson, L., Karlsson, B., 1998. Spermine affects intestinal in vitro permeability to different-sized molecules in rats. Comp. Biochem. Physiol. C Pharmacol. Toxicol. Endocrinol. 120, 211–216.

Pedersen, G., Brynskov, J., Saermark, T., 2002. Phenol toxicity and conjugation in human colonic epithelial cells. Scand. J. Gastroenterol. 37, 74–79.

Pelaseyed, T., Bergström, J.H., Gustafsson, J.K., Ermund, A., Birchenough, G.M.H., Schütte, A., et al., 2014. The mucus and mucins of the goblet cells and enterocytes provide the first defense line of the gastrointestinal tract and interact with the immune system. Immunol. Rev. 260, 8–20. Available from: http://dx.doi.org/10.1111/imr.12182.

Penrose, H.M., Marchelletta, R.R., Krishnan, M., McCole, D.F., 2013. Spermidine stimulates T cell protein-tyrosine phosphatase-mediated protection of intestinal epithelial barrier function. J. Biol. Chem. 288, 32651–32662. Available from: http://dx.doi.org/10.1074/jbc.M113.475962.

Possemiers, S., Bolca, S., Verstraete, W., Heyerick, A., 2011. The intestinal microbiome: a separate organ inside the body with the metabolic potential to influence the bioactivity of botanicals. Fitoterapia 82, 53–66. Available from: http://dx.doi.org/10.1016/j.fitote.2010.07.012.

Pouokam, E., Diener, M., 2011. Mechanisms of actions of hydrogen sulphide on rat distal colonic epithelium. Br. J. Pharmacol. 162, 392–404. Available from: http://dx.doi.org/10.1111/j.1476-5381.2010.01026.x.

Prasad, M., Smith, J.A., Resnick, A., Awtrey, C.S., Hrnjez, B.J., Matthews, J.B., 1995. Ammonia inhibits cAMP-regulated intestinal Cl- transport. Asymmetric effects of apical and basolateral exposure and implications for epithelial barrier function. J. Clin. Invest. 96, 2142–2151. Available from: http://dx.doi.org/10.1172/JCI118268.

Qin, J., Li, R., Raes, J., Arumugam, M., Burgdorf, K.S., Manichanh, C., et al., 2010. A human gut microbial gene catalogue established by metagenomic sequencing. Nature 464, 59–65. Available from: http://dx.doi.org/10.1038/nature08821.

Ramezani, A., Raj, D.S., 2014. The gut microbiome, kidney disease, and targeted interventions. J. Am. Soc. Nephrol. 25, 657–670. Available from: http://dx.doi.org/10.1681/ASN.2013080905.

Reiffenstein, R.J., Hulbert, W.C., Roth, S.H., 1992. Toxicology of hydrogen sulfide. Annu. Rev. Pharmacol. Toxicol. 32, 109–134. Available from: http://dx.doi.org/10.1146/annurev.pa.32.040192.000545.

Reis de Souza, T.C., Barreyro, A.A., Mariscal-Landín, G., 2013. Estimation of endogenous protein and amino acid ileal losses in weaned piglets by regression analysis using diets with graded levels of casein. J. Anim. Sci. Biotechnol. 4, 36. Available from: http://dx.doi.org/10.1186/2049-1891-4-36.

Richardson, A.J., McKain, N., Wallace, R.J., 2013. Ammonia production by human faecal bacteria, and the enumeration, isolation and characterization of bacteria capable of growth on peptides and amino acids. BMC Microbiol. 13, 6. Available from: http://dx.doi.org/10.1186/1471-2180-13-6.

Rist, V.T.S., Weiss, E., Eklund, M., Mosenthin, R., 2013. Impact of dietary protein on microbiota composition and activity in the gastrointestinal tract of piglets in relation to gut health: a review. Animal 7, 1067–1078. Available from: http://dx.doi.org/10.1017/S1751731113000062.

Shimada, Y., Kinoshita, M., Harada, K., Mizutani, M., Masahata, K., Kayama, H., et al., 2013. Commensal bacteria-dependent indole production enhances epithelial barrier function in the colon. PloS One 8, e80604. Available from: http://dx.doi.org/10.1371/journal.pone.0080604.

Silvester, K.R., Cummings, J.H., 1995. Does digestibility of meat protein help explain large bowel cancer risk? Nutr. Cancer 24, 279–288. Available from: http://dx.doi.org/10.1080/01635589509514417.

Singh, S.K., Binder, H.J., Geibel, J.P., Boron, W.F., 1995. An apical permeability barrier to NH_3/NH_4^+ in isolated, perfused colonic crypts. Proc. Natl. Acad. Sci. U.S.A. 92, 11573–11577.

Smith, E.A., Macfarlane, G.T., 1996a. Enumeration of human colonic bacteria producing phenolic and indolic compounds: effects of pH, carbohydrate availability and retention time on dissimilatory aromatic amino acid metabolism. J. Appl. Bacteriol. 81, 288–302.

Smith, E.A., Macfarlane, G.T., 1996b. Studies on amine production in the human colon: enumeration of amine forming bacteria and physiological effects of carbohydrate and pH. Anaerobe285–297.

Smith, E.A., Macfarlane, G.T., 1997. Dissimilatory amino acid metabolism in human colonic bacteria. Anaerobe 3, 327–337. Available from: http://dx.doi.org/10.1006/anae.1997.0121.

Smith, E.A., Macfarlane, G.T., 1998. Enumeration of amino acid fermenting bacteria in the human large intestine: effects of pH and starch on peptide metabolism and dissimilation of amino acids. FEMS Microbiol. Ecol.355–368.

Summerskill, W.H., Wolpert, E., 1970. Ammonia metabolism in the gut. Am. J. Clin. Nutr. 23, 633–639.

Takahashi, M., Benno, Y., Mitsuoka, T., 1980. Utilization of ammonia nitrogen by intestinal bacteria isolated from pigs. Appl. Environ. Microbiol. 39, 30–35.

Tap, J., Mondot, S., Levenez, F., Pelletier, E., Caron, C., Furet, J.-P., et al., 2009. Towards the human intestinal microbiota phylogenetic core. Environ. Microbiol. 11, 2574–2584. Available from: http://dx.doi.org/10.1111/j.1462-2920.2009.01982.x.

Timmons, J., Chang, E.T., Wang, J.-Y., Rao, J.N., 2012. Polyamines and gut mucosal homeostasis. J. Gastrointest. Dig. Syst. 2, piii:2.

Uemura, T., Stringer, D.E., Blohm-Mangone, K.A., Gerner, E.W., 2010. Polyamine transport is mediated by both endocytic and solute carrier transport mechanisms in the gastrointestinal tract. Am. J. Physiol. Gastrointest. Liver Physiol. 299, G517–G522. Available from: http://dx.doi.org/10.1152/ajpgi.00169.2010.

Van Nuenen, M.H.M.C., de Ligt, R.A.F., Doornbos, R.P., van der Woude, J.C.J., Kuipers, E.J., Venema, K., 2005. The influence of microbial metabolites on human intestinal epithelial cells and macrophages in vitro. FEMS Immunol. Med. Microbiol. 45, 183–189. Available from: http://dx.doi.org/10.1016/j.femsim.2005.03.010.

Venkatesh, M., Mukherjee, S., Wang, H., Li, H., Sun, K., Benechet, A.P., et al., 2014. Symbiotic bacterial metabolites regulate gastrointestinal barrier function via the xenobiotic sensor PXR and toll-like receptor 4. Immunity 41, 296–310. Available from: http://dx.doi.org/10.1016/j.immuni.2014.06.014.

Vince, A.J., Burridge, S.M., 1980. Ammonia production by intestinal bacteria: the effects of lactose, lactulose and glucose. J. Med. Microbiol. 13, 177–191.

Vince, A.J., Dawson, A.M., Park, N., O'Grady, F., 1973. Ammonia production by intestinal bacteria. Gut171–177.

Wall, R., Cryan, J.F., Ross, R.P., Fitzgerald, G.F., Dinan, T.G., Stanton, C., 2014. Bacterial neuroactive compounds produced by psychobiotics. Adv. Exp. Med. Biol. 817, 221–239. Available from: http://dx.doi.org/10.1007/978-1-4939-0897-4_10.

Windey, K., De Preter, V., Louat, T., Schuit, F., Herman, J., Vansant, G., et al., 2012a. Modulation of protein fermentation does not affect fecal water toxicity: a randomized cross-over study in healthy subjects. PLoS One 7, e52387. Available from: http://dx.doi.org/10.1371/journal.pone.0052387.

Windey, K., De Preter, V., Verbeke, K., 2012b. Relevance of protein fermentation to gut health. Mol. Nutr. Food Res. 56, 184–196. Available from: http://dx.doi.org/10.1002/mnfr.201100542.

Yang, Y.-X., Dai, Z.-L., Zhu, W.-Y., 2014. Important impacts of intestinal bacteria on utilization of dietary amino acids in pigs. Amino Acids 46, 2489–2501. Available from: http://dx.doi.org/10.1007/s00726-014-1807-y.

Zaharia, V., Varzescu, M., Djavadi, I., Newman, E., Egnor, R.W., Alexander-Chacko, J., et al., 2001. Effects of short chain fatty acids on colonic Na^+ absorption and enzyme activity. Comp. Biochem. Physiol. A. Mol. Integr. Physiol. 128, 335–347.

Zelante, T., Iannitti, R.G., Cunha, C., De Luca, A., Giovannini, G., Pieraccini, G., et al., 2013. Tryptophan catabolites from microbiota engage aryl hydrocarbon receptor and balance mucosal reactivity via interleukin-22. Immunity 39, 372–385. Available from: http://dx.doi.org/10.1016/j.immuni.2013.08.003.

CHAPTER

16

Control of Food Intake by Dietary Amino Acids and Proteins: Molecular and Cellular Aspects

G. Fromentin, N. Darcel, C. Chaumontet, P. Even, D. Tomé and C. Gaudichon

UMR Physiologie de la Nutrition et du Comportement Alimentaire, AgroParisTech, INRA, Université Paris-Saclay, Paris, France

16.1. INTRODUCTION

It is well known that the macronutrient composition influences energy intake and metabolism and can induce long-term changes in body weight and body composition. Protein is a mandatory constituent of the diet, as it is a source of nitrogen and essential amino acids (Bensaid et al., 2002; Tome, 2004). Insufficient intake of dietary protein is incompatible with growth and life. As a consequence, a control of the regulation of protein intake is critical for the organism.

Decades of research in various animal species (Morrison et al., 2012) have shown that both the quantity and the quality of dietary protein are regulated, and that this regulation may influence feeding behavior by promoting or inhibiting protein intake. High protein diets have been shown to decrease food intake, at least in the short-term, both in animal models and in humans (Jean et al., 2001). Animals also learn to avoid very low protein diets (Du et al., 2000; Peters and Harper, 1985) whereas some studies suggest that animals with moderately low protein diets tend to increase their food intake, presumably in order to fulfill their protein needs (White et al., 2000). It is well-established that eating a diet with an imbalance in the protein or indispensable amino acid content induces a decrease in food intake through conditioned food aversion, and animals will avoid this diet if any other possibility is available (Anthony and Gietzen, 2013; Fromentin et al., 2000; Gietzen et al., 2007). These nutritional manipulation studies show that diets with a protein content that differs from required levels may influence food intake and feeding behavior, and that protein intake is regulated separately, but not necessarily independently, of other dietary macronutrients (carbohydrate and fat), as well as total energy intake. It is unclear whether organisms meet their protein needs by striving a unique "protein target," or if regulatory processes ensure that protein deficiency and/or toxicity do not occur by maintaining protein intake within a "safe" range. In addition, the signaling pathways that convey information regarding protein intake to the central nervous system (CNS) and the regulatory pathways controlling subsequent food intake are not well understood.

The main purpose of this review is to present current views and knowledge concerning the effect of eating high protein meals or a high protein diet on food intake and the neuronal mechanisms associated with peripheral and central signaling processes by which high dietary protein intake influences overall energy intake. Recent findings regarding essential amino acid deficiencies and low protein diets have been summarized in other up-to-date reviews (Anthony and Gietzen, 2013; Morrison and Laeger, 2015). Numerous partially redundant hypotheses have been proposed to explain how dietary protein and amino acids decrease food intake (Fromentin et al., 2012; Westerterp-Plantenga et al., 2009). These theories mostly differ on the putative location where the reduction in eating signals is initiated. After nutrients are ingested, preabsorptive signals originating from the intestine are transmitted to the brain's satiety centers via the action of gut peptides on peripheral nerves or via the

The Molecular Nutrition of Amino Acids and Proteins.
DOI: http://dx.doi.org/10.1016/B978-0-12-802167-5.00016-5

bloodstream. Postabsorptive signals, which occur after nutrients and/or gut peptides cross the gut wall and enter circulation, are initiated in the hepatic-portal zone. Finally, signals indicating the status of energy stores (leptin and insulin levels) may also be potent mediators of the effect of protein consumption on eating behavior (Schwartz et al., 2000). In addition to the location where the eating signals originate, the brain areas and neuronal populations responsible for integrating these sensory signals are not yet fully understood. These signals are integrated in the CNS, where structures in the reward system, the hypothalamus and brainstem regulate energy homeostasis as well as the onset of appetite and satiety (Berthoud, 2011). Meal size is controlled via negative feedback transmitted by gastrointestinal signals and the bloodstream to the dorsal vagal complex in the brainstem and the hypothalamic nuclei. Increasing evidence from recent studies suggests that the reward system may be modulated, with the main consequence being a decrease in the desire for food, which is known to be a major driver of eating behavior.

16.2. THE EFFECT OF PROTEIN INTAKE AND OVERALL ENERGY INTAKE ON BODY WEIGHT AND BODY COMPOSITION

Dietary protein intake has frequently been associated with food intake regulation because of its effects on energy balance. However, these effects depend on a wide array of factors, such as the protein content of the meal and/or the diet, other diet constituents, and the species studied.

16.2.1 Protein Snacks/Meals and Food Intake

In the short-term, dietary protein appears to be a strong appetite inhibitor and reduces food intake in subsequent meals more than would be expected based on energy content alone, both in rats (Bensaid et al., 2002) and in humans (Porrini et al., 1997). Moreover, in humans (Marsset-Baglieri et al., 2014) and in rats (Burton-Freeman et al., 1997), consumption of a high protein snack delayed the request for the following meal (ie, increased the intermeal interval). It is currently thought that, in rats (Bensaid et al., 2002) and humans (Porrini et al., 1997), protein has the greatest effect on suppressing appetite of the three macronutrients. In human studies, ingesting a high protein load decreased the total meal or daily food intake differently than carbohydrate (Bertenshaw et al., 2008) or lipid loads (Porrini et al., 1997) or both (Latner and Schwartz, 1999). Moreover, high protein meals suppress energy intake in lean and obese subjects (Brennan et al., 2012). However, the commonly acknowledged hierarchy of proteins > carbohydrates > lipids with respect to satiety is not always observed in rats (Geliebter et al., 1984) or in humans (Potier et al., 2010; Raben et al., 2003). Many of these discrepancies originated from the nature of the load, the method of nutrient administration (ie, oral, intragastric, or intravenous), the duration of eating, and the physiological status of the subjects (Reid and Hetherington, 1997). It is important to pay special attention to the characteristics of the protein load, as the amount of protein used, protein type, food form, volume, energy density, and palatability can influence the degree of satiation or satiety that it induces. In addition, subjects should be selected carefully based on their physiological status and function (such as body mass index) in conjunction with their cognitive restraint and disinhibition scores (Fromentin et al., 2012). For example, protein type can modulate satiety in rats and in humans, as shown in studies of fish versus beef protein (Borzoei et al., 2006; Faipoux et al., 2006; Jung et al., 2014; Uhe et al., 1992), whey or soy protein versus albumin (Anderson et al., 2004), and pork versus soy protein (Mikkelsen et al., 2000). However, these effects based on the nature of the proteins have not been detected in all experiments. Indeed, various proteins (egg albumin, casein, soy, whey, and pea protein) have been used in other human studies, and it is still not clear whether the nature of the protein significantly affects feeding responses (Lang et al., 1998, 1999).

Even though several human studies of low calorie diets have shown that one or two high protein meal replacements per day may decrease energy intake and body weight (Ashley et al., 2001; Noakes et al., 2004), the long-term consequences of protein meal consumption in restricted or nonrestricted subjects have not yet been determined (Li et al., 2010). It has been shown that eating 0.5 g of yeast protein hydrolysate twice a day reduces body weight and the accumulation of abdominal fat in obese adults by reducing energy intake, and has no adverse effects on lean body mass, regardless of sex (Jung et al., 2014). Consuming snacks containing whey protein and polydextrose induced a sustained reduction in daily energy intake in lean men; however, the authors did not assess the separate effects of protein versus polydextrose consumption on food intake (Astbury et al., 2014). We carried out an experiment in rats fed a normal protein (NP) diet and showed that long-term (3 weeks) administration of a high protein load one hour before the lunch in a "humanized" three meal feeding sequence decreased the size of the following meal, but did not reduce total daily energy intake and did not modify body composition (Davidenko et al., 2013). Moreover, substituting an ad libitum high protein meal for an NP meal in a "humanized" three meal feeding sequence reduced energy intake at lunch but did not reduce total daily energy intake. The test meal was the second meal of the day (lunch). At the end of the study, there was no difference in mean body weight and body composition between the two groups (Davidenko et al., 2013).

16.2.2 High Protein Diet and Food Intake

The consumption of a high protein diet induces a strong and immediate decrease in food intake followed by a progressive, but not complete, return to the level of energy intake of the control diet in animals (Jean et al., 2001). Harper described this phenomenon in rats that consumed diets with protein contents ranging from 5% to 75% (Harper and Peters, 1989). The reduction of food intake occurred when the protein content of the diet was greater than 40% (protein/energy [P/E] ratio), and became even more pronounced as the protein content increased (Davidenko et al., 2013).

Given the decrease in food intake and the active avoidance of food intake that occurs, it has been proposed that eating a high protein diet induces a conditioned food aversion (Tews et al., 1992). Our own experiments (Bensaid et al., 2003; L'Heureux-Bouron et al., 2004a) showed that rats adapted to a high protein diet did not acquire conditioned food aversion, but exhibited satiety and a normal behavioral satiety sequence. Similar to our results, other studies (Kinzig et al., 2007; Ropelle et al., 2007) do not confirm the conditioned food aversion hypothesis. Regardless of the protein type used in the high protein diet, these diets result in a greater decrease in the average energy intake compared to NP diets. This effect is modulated by the protein type: total milk protein, whey protein, or beta-lactoglobulin in (Pichon et al., 2008); soy or gluten or total milk protein in (L'Heureux-Bouron et al., 2004b); and also by the ratio of carbohydrate and fat (Marsset-Baglieri et al., 2004). The interaction between these two factors is still not well characterized.

It has been suggested that protein sensing in animals might be linked to the L-glutamate content of foods (free + protein-bound) (Kondoh et al., 2009). This could provide a reasonable index for protein ingestion because L-glutamate is the most abundant amino acid in almost all dietary proteins. Nevertheless, eating an NP diet (14% total milk protein, P/E ratio) enriched in glutamate (2%) for 15 days did not induce a decrease in food intake or weight gain (Boutry et al., 2011). Glutamic acid is well-tolerated by the rat in amounts as high as 5%, and probably up to 10%, in diets with low to moderate protein content (Harper et al., 1970).

The poor palatability of high protein diets has been documented (McArthur et al., 1993), but the relative importance of palatability with respect to protein intake remains unclear. It is possible that the appetite suppressing effect of dietary protein is partly due to poor palatability (L'Heureux-Bouron et al., 2004b).

The usefulness of high protein diets in relation to weight loss and caloric restriction is questionable. The effect of diet composition on weight loss during energy restriction has been widely studied, and there has been no clear demonstration that the macronutrient composition of the diet has an additive effect when combined with calorie restriction. The absence of any effect of diet composition in humans (apart from calorie restriction) has been reported by several authors (Aldrich et al., 2011; Das et al., 2007; Due et al., 2004; Gilbert et al., 2011; Sacks et al., 2009), although not all (Layman et al., 2009; Treyzon et al., 2008). The loss of lean body mass that accompanies caloric restriction is a major obstacle to the success of this strategy as a long-term diet. Importantly, increasing dietary protein during energy restriction may prevent the loss of lean body mass (Chaston et al., 2007). This effect is likely modulated by the protein type, but also by the ratio between carbohydrate and fat. In a 10-week randomized trial of two energy-restricted diets that were low or high in fat, Stocks et al. (2013) reported that articipants on a low-fat diet who modified their diet to increase the amount of energy obtained from protein exhibited the greatest weight loss (although not in lean body mass). In rats, eating a high protein diet ad libitum mainly induced a reduction in the fat mass, but also a higher ratio of lean body mass to fat mass.

16.2.3 Low Protein Diet and Food Intake

Rats and mice develop hyperphagia in response to moderately low protein diets (Morrison and Brannigan, 2015), but reduce their food intake if the protein content is extremely low (Du et al., 2000; Malta et al., 2014). It has been shown that food intake in young adult rats, when assessed over a range of low dietary protein content, showed a quasi bell-shaped response curve with peak intake occurring at or just below the estimated protein requirement. Studies of low protein intake in humans are scarce. Individuals that consumed a low protein diet (5% P/E ratio) did not increase their diet intake (Martens et al., 2014). In another study, subjects with a moderately restricted protein intake, when given access to protein, ate more protein but did not alter their overall energy intake (Griffioen-Roose et al., 2012). However, there was an important difference between these two experiments: in Griffioen-Roose's study, the individuals were allowed to choose their foods, but in Martens' study they were not. Indeed, specific hunger for protein has already been demonstrated, as people can acquire food flavor preferences when specific flavors are associated with the reversal of a mild protein deficit (Gibson et al., 1995). According to the protein leverage hypothesis (Simpson and Raubenheimer, 2005), energy intake may be adjusted to maintain a relatively constant protein intake, and protein intake is regulated more precisely than energy intake. However, whether this self-selection precisely regulates protein intake is still a matter of debate (Leidy et al., 2015; Martens et al., 2014).

16.3. DETECTION OF PROTEIN AND AMINO ACIDS DURING DIGESTION AND CONTROL OF FOOD INTAKE BY FEEDBACK SIGNALING

16.3.1 Oral Sensing

Taste is a major element of the food experience that plays a key role in food intake. Like many food constituents, protein may be sensed initially in the oral cavity. Amino acid detection occurs as early as the oral cavity. Indeed, several amino acids taste sweet, bitter, or umami to humans and are attractive to rodents and other animals (Beauchamp, 2009). Taste receptors (namely T1R1 + T1R3 heterodimers) are present on the tongue and can detect most of the 20 amino acids, including glutamate. Some amino acids, such as glutamate, can also be detected specifically by metabotropic glutamate receptors (mGluR1 and mGluR4). The heterodimeric T1R1 + T1R3 receptor also mediates the sensing of umami flavors (Nelson et al., 2002), although a knockout study suggests that additional umami receptors are likely to exist (Chaudhari and Kinnamon, 2009). Umami, one of the five basic tastes, is mainly attributed to free glutamate, and it has been suggested that it represents "protein taste." Umami is elicited by many small molecules, including amino acids (glutamate and aspartate) and nucleotides. In addition to amino acids, certain peptides can also elicit and enhance umami, suggesting that protein breakdown products may contribute to this taste (Haid et al., 2013). L-glutamate in food can be either protein-bound or in free form. When glutamate is protein-bound, it is tasteless and does not impart an umami flavor to food. L-glutamate is present as a free amino acid in various food products, not all of which are rich in proteins (for instance, corn and green peas), and ripening or maturation increases the glutamate level of products (such as tomatoes or cheese). In contrast, not all products that are rich in protein are also rich in glutamate: for instance, human milk is quite high in free glutamate, in contrast with cow's milk. Moreover, whereas the nutritional state of children affected casein hydrolysate intake (malnourished infants preferred soup plus casein hydrolysate, while well-nourished infants preferred unsupplemented soup), there was no effect on glutamate preference, as both malnourished and well-nourished infants consumed more glutamate-containing soup than soup containing no glutamate. These results therefore do not support the hypothesis that glutamate signals the presence of protein (Beauchamp, 2009). In the absence of a unique "protein taste" system, the potential role of oral protein sensing shifts toward learned associations between orosensory food characteristics and postoral signals.

16.3.2 Gastric and Gut Signals

Satiation feedback signals originating from the stomach are the result of volumetric signals produced by mechanoreceptors (Powley and Phillips, 2004). The initial increase in gastric volume after ingesting dietary protein is probably due to increased gastric secretions and increased water intake (L'Heureux-Bouron et al., 2004b). As explained below, dietary proteins and amino acids are detected within the duodenum, and this detection delays gastric emptying (Ma et al., 2009), hence prolonging gastric distension satiation signals. However, it is unlikely that these mechanisms play a central role in protein-induced satiety. Although intraduodenal whey protein, infused in amounts reflecting average gastric emptying, suppressed energy intake in young lean adults, it exhibited only weak effects on gastrointestinal motility and hormone release, suggesting that the effects of whey on energy intake may be mediated by mechanisms other than nutrient emptying sensing (Ryan et al., 2012). Moreover, we showed that a liquid protein snack containing 30 g of carbohydrates and 30 g of protein induces short-term satiety efficiently in overweight subjects, but that this satiety effect is independent of digestive and plasma amino acid kinetics (Marsset-Baglieri et al., 2014).

The upper intestine is a strong contributor to the effect of proteins and amino acids on food intake. There is some evidence that protein digestion products such as amino acids and oligopeptides are detected within the lumen of the duodenum (Rasoamanana et al., 2012). While the mechanism for this detection was initially thought to be linked to protein and amino acid absorption and processing by enterocytes (Raybould et al., 2006), other groups have shown that nutrient-specific receptors that are similar to either lingual taste receptors (Lindemann, 2001) or functional oligopeptide transporters (Darcel et al., 2005b) exist on the apical surface of enterocytes and enteroendocrine cells. It has also been shown that intestinal transport mechanisms involving the peptide transporter PEPT1 participate in controlling food intake on a high protein diet (Nassl et al., 2011). PEPT1$^{-/-}$ animals showed a much more pronounced reduction in food intake than wild type (WT) animals in response to a high protein (45% of the total energy) diet only. Ultimately, amino acid and oligopeptide detection by the intestinal wall is dependent upon the release of cholecystokinin (CCK) by enteroendocrine cells (Conigrave et al., 2000). Duodenal CCK then increases the firing rate of afferent vagus nerves, whose terminals extend close to the brush border, which convey information to the nucleus of the solitary tract (NTS) in the brainstem. The effect of this detection on food intake seems to be restricted to short-term food intake control (Raybould, 1991).

G protein-coupled receptor family C group 6 member A (GPRC6A) is activated by proteinogenic amino acids and may sense amino acids in the gastrointestinal tract and the brain. To investigate whether GPRC6A is necessary to mediate the effects of low- and high protein diets on body weight and food intake, GPRC6a knockout (GPRC6a-KO) and WT mice were fed a control diet or a low protein diet or a high protein diet (Kinsey-Jones et al., 2015). Their results showed that GPRC6A was not necessary for mediating the effects of a low- or high protein diet on body weight and likely does not play a role in protein-induced satiety.

Food intake was reduced most potently by oral L-arginine, L-lysine, and L-glutamic acid gavage compared to the other 17 amino acids in rats (Jordi et al., 2013). These three amino acids induced neuronal activity in the area postrema and the NTS. Surgical lesion of the area postrema abolished the anorectic response to L-arginine and L-glutamic acid whereas vagal afferent lesion prevented the response to lysine.

Within the lower intestine, the ileal brake is a feedback mechanism that inhibits proximal gastrointestinal motility and secretion. Studies in animals and humans have shown that activation of the ileal brake by local nutrient perfusion increases feelings of satiety and reduces ad libitum food intake (Maljaars et al., 2008). Ileal perfusion of amino acids or proteins activates the ileal brake in both humans and animals (Meyer et al., 1998; van Avesaat et al., 2015). Implementation of the ileal brake is mediated notably by peptide YY (PYY), which is released by L cells located on the ileal mucosa. According to Moran and Dailey (Moran and Dailey, 2011), PYY plasma levels can remain elevated for up to 6 h following meal termination. The long duration of this release pattern suggests the PYY may have roles that extend beyond the meal that originally stimulated its release. According to Batterham et al. (2006), consuming a high protein diet induced the greatest release of PYY and the most pronounced satiety in normal-weight and obese human subjects. Also, dietary protein is also a very strong stimulus compared to other macronutrients for glucagon like peptide-1 (GLP-1) release by the small intestine (Bowen et al., 2006). Since GLP-1 receptors are expressed in afferent vagal neurons, it is likely that GLP-1 acts on vagal afferent terminals in close vicinity to the enteroendocrine L cells (Moran and Dailey, 2011). As de Lartigue et al. (2010) describe, there may be complex and cooperative interplay between ghrelin and CCK, but also between CCK and GLP-1, suggesting critical synergistic effects (Rehfeld, 2011).

The participation of vagal afferent fibers in the hepatic portal vein in protein sensing and signaling to the brain is supported by electrophysiological recordings, which show that hepatic portal vein perfusion of amino acids activates vagal afferent fibers (Darcel et al., 2005b). However, hepatic vagal afferent fibers do not appear to be essential for peripheral detection of a high protein diet (Tome et al., 2009). Others have proposed that the hepatic portal vein is critical for detecting intestinal gluconeogenesis induced by the ingestion of high protein diets, and therefore responsible for the decrease in high protein diet intake (Penhoat et al., 2011), but this view seems rather unlikely considering the high level of redundancy present in gut-brain axis satiety signaling (Berthoud et al., 2011). These findings have also been challenged by a lack of evidence for significant intestinal gluconeogenesis, based on a study using stable isotope-labeled glutamine in two strains of fasted rats. When a 15N-labeled yeast extract was administrated to wild-type and PEPT1-deficient animals by gavage, no detectable differences were found between the two groups in the level of gluconeogenesis induced by the relevant amino acids (Nassl et al., 2011).

Signals generated in response to gut peptides and circulating amino acids act on the CNS in two ways: (1) directly at the hypothalamus (arcuate nucleus of the hypothalamus (ARC)) and brainstem (area postrema

(AP)) via the bloodstream (notably through hormones and circulating nutrients acting as mediators); and (2) indirectly through gastrointestinal tract innervation, namely the vagus nerve and splanchnic nerves, though subdiaphragmatic vagotomy fails to suppress the decrease in food intake induced by the administration of a high protein diet in rats (L'Heureux-Bouron et al., 2003).

16.3.3 Post Absorptive Signals

Since the 2000s, many human studies have been conducted to examine differences in postprandial hormone profiles that could be the cause of satiety induced by protein consumption. Studies have focused on CCK, GLP-1, amylin, ghrelin, leptin, insulin, and glucagon, but in general no clear correlation was detected between these hormones and satiety (Bowen et al., 2006; Marsset-Baglieri et al., 2014, 2015; Ryan et al., 2012, 2013; Veldhorst et al., 2008). As for individual mediators, studies have shown that ghrelin is primarily blood-borne and alters hypothalamic function, whereas the effects of CCK and GLP-1 are primarily vagally mediated (Moran and Dailey, 2011). Plasma CCK levels are unlikely to be a relevant signal (Brenner et al., 1993). Ghrelin, an orexigenic peptide, decreased after consumption of a high protein breakfast by lean subjects (Blom et al., 2006). This result was confirmed by Bowen in lean and overweight subjects (Bowen et al., 2006), but not by Westerterp as reviewed in (Veldhorst et al., 2008). It is also possible that a high protein diet, but not a single high protein meal, influences ghrelin responses (Smeets et al., 2008).

It is currently accepted that, in humans, protein stimulates diet-induced thermogenesis to a greater extent than other macronutrients (Westerterp-Plantenga, 2008). This is due in part to the energetic cost required to incorporate each amino acid into a protein and to catabolize the excess amino acids. However, it has never been shown that increased postprandial thermogenesis is the direct cause of satiety induced by protein ingestion, and central mechanisms linking increased thermogenesis to postprandial satiety have yet to be identified.

Amino acid-induced gluconeogenesis may prevent a decrease in glycemia, which could contribute to satiety. Whether gluconeogenesis in humans remains relatively stable despite varying metabolic conditions is still a matter of debate. It has been shown that gluconeogenesis increased and appetite decreased in response to a high protein diet compared with an NP diet; however, these two factors were unrelated to each other (Veldhorst et al., 2012).

The high concentration of amino acids in the plasma after protein ingestion could cause peripheral signals that would be detected in specific regions of the hypothalamus (Choi et al., 2001) in agreement with the aminostatic theory (Mellinkoff et al., 1956).

The general hypothesis that certain amino acid precursors of neurotransmitters are responsible for the decrease in food intake induced by proteins is still a matter of debate. Tryptophan, the precursor of serotonin, is known to inhibit appetite. It has been suggested that the high tryptophan content of α-lactalbumin accounts for its satiating effect (Nieuwenhuizen et al., 2009). Indeed, ingesting α-lactalbumin increases the level of tryptophan in the plasma and the ratio of tryptophan to branched chain amino acids more than other proteins, such as gelatin (which does not contain tryptophan). However, the addition of free tryptophan to gelatin, a protein devoid of tryptophan, at the same level as α-lactalbumin does not affect hunger levels (Veldhorst et al., 2009). Similarly, adding tyrosine, a precursor of dopamine, to the diet at a level of 5% had no influence on the level of intake by rats (Bassil et al., 2007). Because the ingestion of foods high in protein (and thus L-glutamate) does not lead to appreciable changes in plasma L-glutamate concentrations (Kondoh et al., 2009), the body (ie, the brain) is unlikely to monitor protein intake via meal- or diet-related variations in plasma L-glutamate. The data concerning the addition of histidine, a precursor of histamine, to the diet are also contradictory. Chronic ingestion of 5% histidine to the diet had no influence on the level of intake in rats (Bassil et al., 2007). In contrast, adding 1, 2.5, or 5% of histidine in the diet for 8 days decreased food intake in rats (Kasaoka et al., 2004).

16.4. PROTEIN-INDUCED REDUCTION IN EATING AND CENTRAL NEURONAL PATHWAYS

Ingesting a well-equilibrated level of dietary protein induces a reduction in food intake during the subsequent meal which activates neuronal populations in the NTS and the hypothalamus. In addition, consumption of a high protein diet inhibits the activation of opioid and GABAergic neurons in the nucleus accumbens (NAcc), and thus inhibits food intake by reducing the hedonic response to food, presumably because of its low palatability.

The NTS is the main entry point of the vagus nerve into the CNS and thus receives afferent fiber projections from most of the organs of the gastrointestinal tract (Berthoud, 2002). In addition, the NTS receives some afferent cranial nerves that convey extensive information from the orosensory area on food texture, taste, smell, appearance, and palatability. The involvement of vagal afferent pathways in protein sensing and signaling to the brain is supported by studies showing that intraduodenal protein activates vagal afferent fibers, and that consumption of high protein induces c-Fos expression in neurons within the NTS (Darcel et al., 2005a; Phifer and Berthoud, 1998). A reduction in food intake after administration of a high protein load (versus an NP load) resulted from activation of the noradrenergic neurons associated with CCK-induced anorexia (Faipoux et al., 2008). Recently, the NTS was shown to integrate forebrain descending melanocortinergic input and leptin signals with gut-derived satiety signals to determine food intake (Zhao et al., 2012). It has been shown that L-leucine detection by the caudomedial NTS regulates food intake and body weight gain through the activation of the amino acid-sensing p70 S6 kinase 1 (S6K1) signaling pathway, with NTS catecholaminergic and proopiomelanocortin (POMC) neurons acting as L-leucine-sensing cells (Blouet and Schwartz, 2012). These effects are diet-dependent and can be blunted by consumption of high fat foods (Cavanaugh et al., 2015). Finally, the AP could be involved in leucine chemoreception by the brain (Zampieri et al., 2013).

The hypothalamus is the focus of much peripheral information sensing and helps to control body energy homeostasis and food intake. This vast area contains many nuclei that are in constant interaction (Berthoud, 2002). The hypothalamus contains a number of discrete neuronal populations or nuclei, including the ARC, the paraventricular nucleus (PVN), the ventromedial nucleus, the dorsomedial nucleus (DMH), and the lateral hypothalamic area (LH). These nuclei contain and are connected by energy homeostasis-regulating circuits. In the ARC, POMC neuron activation induces a reduction in food intake. The activation of POMC neurons is also inseparable from the behavior of another population in the ARC, neuropeptide Y (NPY)/Agouti-related peptide (AgRP) neurons. The activation of these neurons has a potent effect on increasing food intake and inhibiting POMC neuron activation (Cowley et al., 2001). We previously showed that after the ingestion of a high protein meal, the number of cells that were double-labeled with c-Fos and α-Melanocyte Stimulating Hormone (a marker for POMC neuron activation) increased concomitantly with a reduction in the activation of non-POMC neurons (Faipoux et al., 2008). This result was less pronounced after ingestion of a high protein diet for 21 days than when subjects had only consumed one high protein meal. Moreover, because arcuate neurons mainly exhibit a POMC or NPY phenotype, it could be hypothesized that NPY neurons are less strongly activated after high protein meals. Similarly, rats or obese mice fed a high protein diet ad libitum exhibit a greater decrease in hypothalamic NPY mRNA levels and increase in hypothalamic POMC mRNA levels compared to animals on an NP diet (Ropelle et al., 2007).

AMP-activated protein kinase (AMPK) and the mammalian target of rapamycin (mTOR) are involved in the reduction in eating induced by high protein diets. Increasing dietary leucine (Ropelle et al., 2007) or intracerebroventricular (ICV) administration of amino acids (or leucine only) (Cota et al., 2006; Morrison et al., 2007) reduces food intake and body weight. AMPK is the downstream component of a kinase cascade that acts as a sensor for cellular energy charge, and is activated by an increase in the AMP/ATP ratio. Once activated, AMPK phosphorylates acetyl-CoA carboxylase (ACC) and switches on energy-producing pathways at the expense of energy-depleting processes. L-leucine sensing by the mediobasal hypothalamus regulates energy balance and presented evidence suggesting that activation of the S6K1 signaling pathway may be involved in L-leucine detection (Blouet et al., 2008). Moreover both a high protein diet and ICV leucine administration resulted in a decrease in AMPK and ACC phosphorylation in the rat hypothalamus, as well as a reduced AMP/ATP ratio (Ropelle et al., 2007). At the same time, there has been growing interest in mTOR, an intracellular signaling molecule that is sensitive to both amino acids and growth factors, and has also been described as a metabolic sensor. Both a high protein diet and the ICV administration of free amino acids, or leucine only, led to mTOR activation in the hypothalamus (Cota et al., 2006; Morrison et al., 2007). Moreover, high protein diets modulate AMPK and mTOR expression in the same specific neuronal subsets: the ARC and PVN of the hypothalamus. AMPK and mTOR may have overlapping and reciprocal functions. Finally, mTOR activation and the suppression of AMPK phosphorylation activity appear to modulate the expression of hypothalamic neuropeptides: they reduce levels of orexigenic NPY and AgRP neuropeptides and increase the expression of POMC, which exerts an anorexigenic effect (Morrison et al., 2007; Ropelle et al., 2007).

This effect of high protein diets appears to be leucine-specific, as Laeger et al. (2014) have shown that the anorectic effect produced by ICV leucine injection does not occur in response to injection with other amino acids, suggesting that leucine is unique in acting as a signaling molecule in the brain. Moreover a high protein diet or an NP diet enriched in leucine to the same level found in a high protein diet produced the same results without inducing a conditioned taste aversion (Ropelle et al., 2007). Nevertheless, this comparison is somewhat limited, because NP diets containing an amino acid in excess and high protein diets have different peripheral catabolism (Harper et al., 1970). Moreover the brain can detect changes in leucine availability based on oral intake (Zampieri et al., 2013). Despite the fact that the brain can sense leucine, chronic oral leucine supplementation (in drinking water) does not induce any changes in food intake, nor does it cause an anorectic pattern of gene expression in the hypothalamus.

The NAcc forms part of the structure of the ventral striatum and is pivotal for modulating the hedonic control of eating behavior. It is divided into two parts: the core (nucleus accumbens core (AccCo)) and the shell (nucleus accumbens shell (AccSh)) (Berthoud, 2002). It was identified early on as the primary interface between motivation and action in the brain, because it receives many afferents (mostly glutamatergic) from regions involved in cognitive processes and learning and directs efferent fibers (mostly GABAergic) to areas of motor control. The AccCo is involved in learning and implementing adaptive mechanical actions, while the AccSh acts as a relay between cortical regions and other regions of the brain to help regulate behavioral aspects, and particularly food behavior (Kelley, 2004). As previously discussed, dietary proteins have a low palatability. Palatability has a major impact on the reward system and food liking, but cannot solely account for the anorexigenic effect of proteins (Bensaid et al., 2003). Proteins inhibit the activation of opioid and GABAergic neurons in the NAcc, and could thus inhibit food intake by reducing the hedonic response to food (Davidenko et al., 2013). It is well known that neurons in the AccSh project directly into the LH (Kirouac and Ganguly, 1995), and that the stimulation of orexigenic LH neurons can induce robust feeding (Stratford and Kelley, 1999). Chemical manipulation of the AccSh elicits robust feeding, as well as c-Fos expression in the hypothalamus, particularly in the ARC. c-Fos activation was significantly lower in POMC/cocaine and amphetamine-regulated transcript (CART) neurons and higher in NPY neurons (Zheng et al., 2003). Recently, we showed that postprandial activation of NAcc is decreased by a high protein diet administered for 2 days or after a 15-day period (Chaumontet et al., 2011; Davidenko et al., 2013), and could in turn inhibit NPY neurons and activate POMC/CART neurons in the ARC.

16.5. CONCLUSION

Although it is generally accepted that dietary protein intake is a key factor influencing eating behavior and is closely regulated, it is difficult to evaluate the importance of each individual signaling pathway that has been proposed as a mechanisms for this regulation. It seems clear that protein sensing takes place in several locations, including the oral cavity, gastrointestinal tract, and brain. Sensing occurs during protein ingestion and digestion, and after absorption. Several results support the importance of CCK-mediated vagus nerve signaling, as well as blood and brain amino acid concentrations. However, homeostatic control of protein intake does not appear to be the only regulatory pathway. Food reward circuits influence food intake, both by inhibiting it (in the case of a high protein diet) and by inciting a protein-specific appetite (in cases where protein intake is low). However, the role of reward in protein intake is still unclear. In response to the consumption of a high protein diet, the protein component of food intake control is downregulated, and food intake control is mainly based on energy homeostasis. It is difficult to assess whether protein intake is regulated to achieve a certain optimal level, or to avoid dangerously low and/or high intakes. Studies seem to show a general avoidance (and even learned aversion) toward very low protein diets, which would argue in favor of a protein deficiency sensor. This type of avoidance does not develop in response to very high protein diets, even though reinforced satiety contributes to earlier meal cessation. However, as protein is naturally quite a rare nutrient, there could be an evolutionary reason for regulatory mechanisms that would be stricter in regards to avoiding deficiency and permissive with respect to high intake.

Acknowledgments

This review was supported by funds from our laboratory. The authors declare that they have no conflicts of interest.

References

Aldrich, N.D., Reicks, M.M., Sibley, S.D., Redmon, J.B., Thomas, W., Raatz, S.K., 2011. Varying protein source and quantity do not significantly improve weight loss, fat loss, or satiety in reduced energy diets among midlife adults. Nutr. Res. 31, 104–112.

Anderson, G.H., Tecimer, S.N., Shah, D., Zafar, T.A., 2004. Protein source, quantity, and time of consumption determine the effect of proteins on short-term food intake in young men. J. Nutr. 134, 3011–3015.

Anthony, T.G., Gietzen, D.W., 2013. Detection of amino acid deprivation in the central nervous system. Curr. Opin. Clin. Nutr. Metab. Care 16, 96–101.

Ashley, J.M., St Jeor, S.T., Perumean-Chaney, S., Schrage, J., Bovee, V., 2001. Meal replacements in weight intervention. Obes. Res. 9 (Suppl. 4), 312S–320S.

Astbury, N.M., Taylor, M.A., French, S.J., Macdonald, I.A., 2014. Snacks containing whey protein and polydextrose induce a sustained reduction in daily energy intake over 2 wk under free-living conditions. Am. J. Clin. Nutr. 99, 1131–1140.

Bassil, M.S., Hwalla, N., Obeid, O.A., 2007. Meal pattern of male rats maintained on histidine-, leucine-, or tyrosine-supplemented diet. Obesity (Silver Spring) 15, 616–623.

Batterham, R.L., Heffron, H., Kapoor, S., Chivers, J.E., Chandarana, K., Herzog, H., et al., 2006. Critical role for peptide YY in protein-mediated satiation and body-weight regulation. Cell Metab. 4, 223–233.

Beauchamp, G.K., 2009. Sensory and receptor responses to umami: an overview of pioneering work. Am. J. Clin. Nutr. 90, 723S–727S.

Bensaid, A., Tome, D., Gietzen, D., Even, P., Morens, C., Gausseres, N., et al., 2002. Protein is more potent than carbohydrate for reducing appetite in rats. Physiol. Behav. 75, 577–582.

Bensaid, A., Tome, D., L'Heureux-Bourdon, D., Even, P., Gietzen, D., Morens, C., et al., 2003. A high-protein diet enhances satiety without conditioned taste aversion in the rat. Physiol. Behav. 78, 311–320.

Bertenshaw, E.J., Lluch, A., Yeomans, M.R., 2008. Satiating effects of protein but not carbohydrate consumed in a between-meal beverage context. Physiol. Behav. 93, 427–436.

Berthoud, H.R., 2002. Multiple neural systems controlling food intake and body weight. Neurosci. Biobehav. Rev. 26, 393–428.

Berthoud, H.R., 2011. Metabolic and hedonic drives in the neural control of appetite: who is the boss? Curr. Opin. Neurobiol.

Berthoud, H.R., Shin, A.C., Zheng, H., 2011. Obesity surgery and gut-brain communication. Physiol. Behav. 105 (1), 106–119.

Blom, W.A., Lluch, A., Stafleu, A., Vinoy, S., Holst, J.J., Schaafsma, G., et al., 2006. Effect of a high-protein breakfast on the postprandial ghrelin response. Am. J. Clin. Nutr. 83, 211–220.

Blouet, C., Schwartz, G.J., 2012. Brainstem nutrient sensing in the nucleus of the solitary tract inhibits feeding. Cell Metab. 16, 579–587.

Blouet, C., Ono, H., Schwartz, G.J., 2008. Mediobasal hypothalamic p70 S6 kinase 1 modulates the control of energy homeostasis. Cell Metab. 8, 459–467.

Borzoei, S., Neovius, M., Barkeling, B., Teixeira-Pinto, A., Rossner, S., 2006. A comparison of effects of fish and beef protein on satiety in normal weight men. Eur. J. Clin. Nutr. 60, 897–902.

Boutry, C., Bos, C., Matsumoto, H., Even, P., Azzout-Marniche, D., Tome, D., et al., 2011. Effects of monosodium glutamate supplementation on glutamine metabolism in adult rats. Front. Biosci. (Elite Ed) 3, 279–290.

Bowen, J., Noakes, M., Clifton, P.M., 2006. Appetite regulatory hormone responses to various dietary proteins differ by body mass index status despite similar reductions in ad libitum energy intake. J. Clin. Endocrinol. Metab. 91, 2913–2919.

Brennan, I.M., Luscombe-Marsh, N.D., Seimon, R.V., Otto, B., Horowitz, M., Wishart, J.M., et al., 2012. Effects of fat, protein, and carbohydrate and protein load on appetite, plasma cholecystokinin, peptide YY, and ghrelin, and energy intake in lean and obese men. Am. J. Physiol. Gastrointest. Liver Physiol. 303, G129–G140.

Brenner, L., Yox, D.P., Ritter, R.C., 1993. Suppression of sham feeding by intraintestinal nutrients is not correlated with plasma cholecystokinin elevation. Am. J. Physiol. 264, R972–R976.

Burton-Freeman, B., Gietzen, D.W., Schneeman, B.O., 1997. Meal pattern analysis to investigate the satiating potential of fat, carbohydrate, and protein in rats. Am. J. Physiol. 273, R1916–R1922.

Cavanaugh, A.R., Schwartz, G.J., Blouet, C., 2015. High-fat feeding impairs nutrient sensing and gut brain integration in the caudomedial nucleus of the solitary tract in mice. PLoS One 10, e0118888.

Chaston, T.B., Dixon, J.B., O'Brien, P.E., 2007. Changes in fat-free mass during significant weight loss: a systematic review. Int. J. Obes. (Lond) 31, 743–750.

Chaudhari, N., Kinnamon, S.C., 2009. Symposium overview: sweet taste: receptors, transduction, and hormonal modulation. Ann. N. Y. Acad. Sci. 1170, 95–97.

Chaumontet, C., Darcel, N., Tome, D., Fromentin, G., 2011. Postprandial activation of accumbens nucleus, brain area involved in hedonism, is decreased by high protein diet. FASEB J. 25, 328.1.

Choi, Y.H., Fletcher, P.J., Anderson, G.H., 2001. Extracellular amino acid profiles in the paraventricular nucleus of the rat hypothalamus are influenced by diet composition. Brain Res. 892, 320–328.

Conigrave, A.D., Quinn, S.J., Brown, E.M., 2000. L-amino acid sensing by the extracellular Ca2 + -sensing receptor. Proc. Natl. Acad. Sci. U.S. A. 97, 4814–4819.

Cota, D., Proulx, K., Smith, K.A., Kozma, S.C., Thomas, G., Woods, S.C., et al., 2006. Hypothalamic mTOR signaling regulates food intake. Science 312, 927–930.

Cowley, M.A., Smart, J.L., Rubinstein, M., Cerdan, M.G., Diano, S., Horvath, T.L., et al., 2001. Leptin activates anorexigenic POMC neurons through a neural network in the arcuate nucleus. Nature 411, 480–484.

Darcel, N., Fromentin, G., Raybould, H.E., Gougis, S., Gietzen, D.W., Tome, D., 2005a. Fos-positive neurons are increased in the nucleus of the solitary tract and decreased in the ventromedial hypothalamus and amygdala by a high-protein diet in rats. J. Nutr. 135, 1486–1490.

Darcel, N.P., Liou, A.P., Tome, D., Raybould, H.E., 2005b. Activation of vagal afferents in the rat duodenum by protein digests requires PepT1. J. Nutr. 135, 1491–1495.

Das, S.K., Gilhooly, C.H., Golden, J.K., Pittas, A.G., Fuss, P.J., Cheatham, R.A., et al., 2007. Long-term effects of 2 energy-restricted diets differing in glycemic load on dietary adherence, body composition, and metabolism in CALERIE: a 1-y randomized controlled trial. Am. J. Clin. Nutr. 85, 1023–1030.

Davidenko, O., Darcel, N., Fromentin, G., Tome, D., 2013. Control of protein and energy intake—brain mechanisms. Eur. J. Clin. Nutr. 67, 455–461.

de Lartigue, G., Lur, G., Dimaline, R., Varro, A., Raybould, H., Dockray, G.J., 2010. EGR1 Is a target for cooperative interactions between cholecystokinin and leptin, and inhibition by ghrelin, in vagal afferent neurons. Endocrinology 151, 3589–3599.

Du, F., Higginbotham, D.A., White, B.D., 2000. Food intake, energy balance and serum leptin concentrations in rats fed low-protein diets. J. Nutr. 130, 514–521.

Due, A., Toubro, S., Skov, A.R., Astrup, A., 2004. Effect of normal-fat diets, either medium or high in protein, on body weight in overweight subjects: a randomised 1-year trial. Int. J. Obes. Relat. Metab. Disord. 28, 1283–1290.

Faipoux, R., Tome, D., Bensaid, A., Morens, C., Oriol, E., Bonnano, L.M., et al., 2006. Yeast proteins enhance satiety in rats. J. Nutr. 136, 2350–2356.

Faipoux, R., Tome, D., Gougis, S., Darcel, N., Fromentin, G., 2008. Proteins activate satiety-related neuronal pathways in the brainstem and hypothalamus of rats. J. Nutr. 138, 1172–1178.

Fromentin, G., Feurte, S., Nicolaidis, S., Norgren, R., 2000. Parabrachial lesions disrupt responses of rats to amino acid devoid diets, to protein-free diets, but not to high-protein diets. Physiol. Behav. 70, 381–389.

Fromentin, G., Darcel, N., Chaumontet, C., Marsset-Baglieri, A., Nadkarni, N., Tome, D., 2012. Peripheral and central mechanisms involved in the control of food intake by dietary amino acids and proteins. Nutr. Res. Rev. 25, 29–39.

Geliebter, A., Liang, J.T., Van Itallie, T.B., 1984. Effects of repeated isocaloric macronutrient loads on daily food intake of rats. Am. J. Physiol. 247, R387–R392.

Gibson, E.L., Wainwright, C.J., Booth, D.A., 1995. Disguised protein in lunch after low-protein breakfast conditions food-flavor preferences dependent on recent lack of protein intake. Physiol. Behav. 58, 363–371.

Gietzen, D.W., Hao, S., Anthony, T.G., 2007. Mechanisms of food intake repression in indispensable amino acid deficiency. Annu. Rev. Nutr. 27, 63–78.

Gilbert, J.A., Bendsen, N.T., Tremblay, A., Astrup, A., 2011. Effect of proteins from different sources on body composition. Nutr. Metab. Cardiovasc. Dis. 21 (Suppl. 2), B16–B31.

Griffioen-Roose, S., Mars, M., Siebelink, E., Finlayson, G., Tome, D., de Graaf, C., 2012. Protein status elicits compensatory changes in food intake and food preferences. Am. J. Clin. Nutr. 95, 32–38.

Haid, D., Widmayer, P., Voigt, A., Chaudhari, N., Boehm, U., Breer, H., 2013. Gustatory sensory cells express a receptor responsive to protein breakdown products (GPR92). Histochem. Cell Biol. 140, 137–145.

Harper, A.E., Peters, J.C., 1989. Protein intake, brain amino acid and serotonin concentrations and protein self-selection. J. Nutr. 119, 677–689.

Harper, A.E., Benevenga, N.J., Wohlhueter, R.M., 1970. Effects of ingestion of disproportionate amounts of amino acids. Physiol. Rev. 50, 428–558.

Jean, C., Rome, S., Mathe, V., Huneau, J.F., Aattouri, N., Fromentin, G., et al., 2001. Metabolic evidence for adaptation to a high protein diet in rats. J. Nutr. 131, 91–98.

Jordi, J., Herzog, B., Camargo, S.M., Boyle, C.N., Lutz, T.A., Verrey, F., 2013. Specific amino acids inhibit food intake via the area postrema or vagal afferents. J. Physiol. 591, 5611–5621.

Jung, E.Y., Cho, M.K., Hong, Y.H., Kim, J.H., Park, Y., Chang, U.J., et al., 2014. Yeast hydrolysate can reduce body weight and abdominal fat accumulation in obese adults. Nutrition 30, 25–32.

Kasaoka, S., Tsuboyama-Kasaoka, N., Kawahara, Y., Inoue, S., Tsuji, M., Ezaki, O., et al., 2004. Histidine supplementation suppresses food intake and fat accumulation in rats. Nutrition 20, 991–996.

Kelley, A.E., 2004. Ventral striatal control of appetitive motivation: role in ingestive behavior and reward-related learning. Neurosci. Biobehav. Rev. 27, 765–776.

Kinsey-Jones, J.S., Alamshah, A., McGavigan, A.K., Spreckley, E., Banks, K., Cereceda Monteoliva, N., et al., 2015. GPRC6a is not required for the effects of a high-protein diet on body weight in mice. Obesity (Silver Spring) 23, 1194–1200.

Kinzig, K.P., Hargrave, S.L., Hyun, J., Moran, T.H., 2007. Energy balance and hypothalamic effects of a high-protein/low-carbohydrate diet. Physiol. Behav. 92, 454–460.

Kirouac, G.J., Ganguly, P.K., 1995. Topographical organization in the nucleus accumbens of afferents from the basolateral amygdala and efferents to the lateral hypothalamus. Neuroscience 67, 625–630.

Kondoh, T., Mallick, H.N., Torii, K., 2009. Activation of the gut-brain axis by dietary glutamate and physiologic significance in energy homeostasis. Am. J. Clin. Nutr. 90, 832S–837S.

L'Heureux-Bouron, D., Tome, D., Rampin, O., Even, P.C., Larue-Achagiotis, C., Fromentin, G., 2003. Total subdiaphragmatic vagotomy does not suppress high protein diet-induced food intake depression in rats. J. Nutr. 133, 2639–2642.

L'Heureux-Bouron, D., Tome, D., Bensaid, A., Morens, C., Gaudichon, C., Fromentin, G., 2004a. A very high 70%-protein diet does not induce conditioned taste aversion in rats. J. Nutr. 134, 1512–1515.

L'Heureux-Bouron, D., Tome, D., Bensaid, A., Morens, C., Lacroix, M., Huneau, J.F., et al., 2004b. Preabsorptive factors are not the main determinants of intake depression induced by a high-protein diet in the rat. Physiol. Behav. 81, 499–504.

Laeger, T., Reed, S.D., Henagan, T.M., Fernandez, D.H., Taghavi, M., Addington, A., Münzberg, H., Martin, R.J., Hutson, S.M., Morrison, C.D., 2014. Leucine acts in the brain to suppress food intake but does not function as a physiological signal of low dietary protein. Am. J. Physiol. Regul. Integr. Comp. Physiol 307 (3), R310–312.

Lang, V., Bellisle, F., Oppert, J.M., Craplet, C., Bornet, F.R., Slama, G., et al., 1998. Satiating effect of proteins in healthy subjects: a comparison of egg albumin, casein, gelatin, soy protein, pea protein, and wheat gluten. Am. J. Clin. Nutr. 67, 1197–1204.

Lang, V., Bellisle, F., Alamowitch, C., Craplet, C., Bornet, F.R., Slama, G., et al., 1999. Varying the protein source in mixed meal modifies glucose, insulin and glucagon kinetics in healthy men, has weak effects on subjective satiety and fails to affect food intake. Eur. J. Clin. Nutr. 53, 959–965.

Latner, J.D., Schwartz, M., 1999. The effects of a high-carbohydrate, high-protein or balanced lunch upon later food intake and hunger ratings. Appetite 33, 119–128.

Layman, D.K., Evans, E.M., Erickson, D., Seyler, J., Weber, J., Bagshaw, D., et al., 2009. A moderate-protein diet produces sustained weight loss and long-term changes in body composition and blood lipids in obese adults. J. Nutr. 139, 514–521.

Leidy, H.J., Clifton, P.M., Astrup, A., Wycherley, T.P., Westerterp-Plantenga, M.S., Luscombe-Marsh, N.D., et al., 2015. The role of protein in weight loss and maintenance. Am. J. Clin. Nutr. 101 (Suppl.), 1320S–1329SS.

Li, Z., Treyzon, L., Chen, S., Yan, E., Thames, G., Carpenter, C.L., 2010. Protein-enriched meal replacements do not adversely affect liver, kidney or bone density: an outpatient randomized controlled trial. Nutr. J. 9, 72.

Lindemann, B., 2001. Receptors and transduction in taste. Nature 413, 219–225.
Ma, J., Stevens, J.E., Cukier, K., Maddox, A.F., Wishart, J.M., Jones, K.L., et al., 2009. Effects of a protein preload on gastric emptying, glycemia, and gut hormones after a carbohydrate meal in diet-controlled type 2 diabetes. Diabetes Care 32, 1600–1602.
Maljaars, P.W., Peters, H.P., Mela, D.J., Masclee, A.A., 2008. Ileal brake: a sensible food target for appetite control. A review. Physiol. Behav. 95, 271–281.
Malta, A., de Oliveira, J.C., Ribeiro, T.A., Tofolo, L.P., Barella, L.F., Prates, K.V., et al., 2014. Low-protein diet in adult male rats has long-term effects on metabolism. J. Endocrinol. 221, 285–295.
Marsset-Baglieri, A., Fromentin, G., Tome, D., Bensaid, A., Makkarios, L., Even, P.C., 2004. Increasing the protein content in a carbohydrate-free diet enhances fat loss during 35% but not 75% energy restriction in rats. J. Nutr. 134, 2646–2652.
Marsset-Baglieri, A., Fromentin, G., Airinei, G., Pedersen, C., Leonil, J., Piedcoq, J., et al., 2014. Milk protein fractions moderately extend the duration of satiety compared with carbohydrates independently of their digestive kinetics in overweight subjects. Br. J. Nutr. 112, 557–564.
Marsset-Baglieri, A., Fromentin, G., Nau, F., Airinei, G., Piedcoq, J., Remond, D., et al., 2015. The satiating effects of eggs or cottage cheese are similar in healthy subjects despite differences in postprandial kinetics. Appetite 90, 136–143.
Martens, E.A., Tan, S.Y., Dunlop, M.V., Mattes, R.D., Westerterp-Plantenga, M.S., 2014. Protein leverage effects of beef protein on energy intake in humans. Am. J. Clin. Nutr. 99, 1397–1406.
McArthur, L.H., Kelly, W.F., Gietzen, D.W., Rogers, Q.R., 1993. The role of palatability in the food intake response of rats fed high-protein diets. Appetite 20, 181–196.
Mellinkoff, S.M., Frankland, M., Boyle, D., Greipel, M., 1956. Relationship between serum amino acid concentration and fluctuations in appetite. J. Appl. Physiol. 8, 535–538.
Meyer, J.H., Hlinka, M., Tabrizi, Y., DiMaso, N., Raybould, H.E., 1998. Chemical specificities and intestinal distributions of nutrient-driven satiety. Am. J. Physiol. 275, R1293–R1307.
Mikkelsen, P.B., Toubro, S., Astrup, A., 2000. Effect of fat-reduced diets on 24-h energy expenditure: comparisons between animal protein, vegetable protein, and carbohydrate. Am. J. Clin. Nutr. 72, 1135–1141.
Moran, T.H., Dailey, M.J., 2011. Intestinal feedback signaling and satiety. Physiol. Behav. 105, 77–81.
Morrison, C.D., Brannigan, R.E., 2015. Metabolic syndrome and infertility in men. Best Pract. Res. Clin. Obstet. Gynaecol. 29, 507–515.
Morrison, C.D., Laeger, T., 2015. Protein-dependent regulation of feeding and metabolism. Trends Endocrinol. Metab. 26, 256–262.
Morrison, C.D., Xi, X., White, C.L., Ye, J., Martin, R.J., 2007. Amino acids inhibit Agrp gene expression via an mTOR-dependent mechanism. Am. J. Physiol. Endocrinol. Metab. 293, E165–E171.
Morrison, C.D., Reed, S.D., Henagan, T.M., 2012. Homeostatic regulation of protein intake: in search of a mechanism. Am. J. Physiol. Regul. Integr. Comp. Physiol. 302, R917–R928.
Nassl, A.M., Rubio-Aliaga, I., Sailer, M., Daniel, H., 2011. The intestinal peptide transporter PEPT1 is involved in food intake regulation in mice fed a high-protein diet. PLoS One 6, e26407.
Nelson, G., Chandrashekar, J., Hoon, M.A., Feng, L., Zhao, G., Ryba, N.J., et al., 2002. An amino-acid taste receptor. Nature 416, 199–202.
Nieuwenhuizen, A.G., Hochstenbach-Waelen, A., Veldhorst, M.A., Westerterp, K.R., Engelen, M.P., Brummer, R.J., et al., 2009. Acute effects of breakfasts containing alpha-lactalbumin, or gelatin with or without added tryptophan, on hunger, 'satiety' hormones and amino acid profiles. Br. J. Nutr. 101, 1859–1866.
Noakes, M., Foster, P.R., Keogh, J.B., Clifton, P.M., 2004. Meal replacements are as effective as structured weight-loss diets for treating obesity in adults with features of metabolic syndrome. J. Nutr. 134, 1894–1899.
Penhoat, A., Mutel, E., Correig, M.A., Pillot, B., Stefanutti, A., Rajas, F., et al., 2011. Protein-induced satiety is abolished in the absence of intestinal gluconeogenesis. Physiol. Behav. 105 (1), 89–93.
Peters, J.C., Harper, A.E., 1985. Adaptation of rats to diets containing different levels of protein: effects on food intake, plasma and brain amino acid concentrations and brain neurotransmitter metabolism. J. Nutr. 115, 382–398.
Phifer, C.B., Berthoud, H.R., 1998. Duodenal nutrient infusions differentially affect sham feeding and Fos expression in rat brain stem. Am. J. Physiol. 274, R1725–R1733.
Pichon, L., Potier, M., Tome, D., Mikogami, T., Laplaize, B., Martin-Rouas, C., et al., 2008. High-protein diets containing different milk protein fractions differently influence energy intake and adiposity in the rat. Br. J. Nutr. 99, 739–748.
Porrini, M., Santangelo, A., Crovetti, R., Riso, P., Testolin, G., Blundell, J.E., 1997. Weight, protein, fat, and timing of preloads affect food intake. Physiol. Behav. 62, 563–570.
Potier, M., Fromentin, G., Lesdema, A., Benamouzig, R., Tome, D., Marsset-Baglieri, A., 2010. The satiety effect of disguised liquid preloads administered acutely and differing only in their nutrient content tended to be weaker for lipids but did not differ between proteins and carbohydrates in human subjects. Br. J. Nutr. 104, 1406–1414.
Powley, T.L., Phillips, R.J., 2004. Gastric satiation is volumetric, intestinal satiation is nutritive. Physiol. Behav. 82, 69–74.
Raben, A., Agerholm-Larsen, L., Flint, A., Holst, J.J., Astrup, A., 2003. Meals with similar energy densities but rich in protein, fat, carbohydrate, or alcohol have different effects on energy expenditure and substrate metabolism but not on appetite and energy intake. Am. J. Clin. Nutr. 77, 91–100.
Rasoamanana, R., Darcel, N., Fromentin, G., Tome, D., 2012. Nutrient sensing and signalling by the gut. Proc. Nutr. Soc.1–10.
Raybould, H.E., 1991. Capsaicin-sensitive vagal afferents and CCK in inhibition of gastric motor function induced by intestinal nutrients. Peptides 12, 1279–1283.
Raybould, H.E., Glatzle, J., Freeman, S.L., Whited, K., Darcel, N., Liou, A., et al., 2006. Detection of macronutrients in the intestinal wall. Auton. Neurosci. 125, 28–33.
Rehfeld, J.F., 2011. Incretin physiology beyond glucagon-like peptide 1 and glucose-dependent insulinotropic polypeptide: cholecystokinin and gastrin peptides. Acta Physiol. (Oxf) 201, 405–411.
Reid, M., Hetherington, M., 1997. Relative effects of carbohydrates and protein on satiety -- a review of methodology. Neurosci. Biobehav. Rev. 21, 295–308.

Ropelle, E.R., Pauli, J.R., Fernandes, M.F., Rocco, S.A., Marin, R.M., Morari, J., et al., 2007. A centre role for neuronal AMPK and mTOR in high protein diet-induced weight loss. Diabetes 57 (3), 594–605.

Ryan, A.T., Feinle-Bisset, C., Kallas, A., Wishart, J.M., Clifton, P.M., Horowitz, M., et al., 2012. Intraduodenal protein modulates antropyloroduodenal motility, hormone release, glycemia, appetite, and energy intake in lean men. Am. J. Clin. Nutr. 96, 474–482.

Ryan, A.T., Luscombe-Marsh, N.D., Saies, A.A., Little, T.J., Standfield, S., Horowitz, M., et al., 2013. Effects of intraduodenal lipid and protein on gut motility and hormone release, glycemia, appetite, and energy intake in lean men. Am. J. Clin. Nutr. 98, 300–311.

Sacks, F.M., Bray, G.A., Carey, V.J., Smith, S.R., Ryan, D.H., Anton, S.D., et al., 2009. Comparison of weight-loss diets with different compositions of fat, protein, and carbohydrates. N. Engl. J. Med. 360, 859–873.

Schwartz, M.W., Woods, S.C., Porte Jr., D., Seeley, R.J., Baskin, D.G., 2000. Central nervous system control of food intake. Nature 404, 661–671.

Simpson, S.J., Raubenheimer, D., 2005. Obesity: the protein leverage hypothesis. Obes. Rev. 6, 133–142.

Smeets, A.J., Soenen, S., Luscombe-Marsh, N.D., Ueland, O., Westerterp-Plantenga, M.S., 2008. Energy expenditure, satiety, and plasma ghrelin, glucagon-like peptide 1, and peptide tyrosine-tyrosine concentrations following a single high-protein lunch. J. Nutr. 138, 698–702.

Stocks, T., Taylor, M.A., Angquist, L., Macdonald, I.A., Arner, P., Holst, C., et al., 2013. Change in proportional protein intake in a 10-week energy-restricted low- or high-fat diet, in relation to changes in body size and metabolic factors. Obes. Facts 6, 217–227.

Stratford, T.R., Kelley, A.E., 1999. Evidence of a functional relationship between the nucleus accumbens shell and lateral hypothalamus subserving the control of feeding behavior. J. Neurosci. 19, 11040–11048.

Tews, J.K., Repa, J.J., Harper, A.E., 1992. Protein selection by rats adapted to high or moderately low levels of dietary protein. Physiol. Behav. 51, 699–712.

Tome, D., 2004. Protein, amino acids and the control of food intake. Br. J. Nutr. 92 (Suppl. 1), S27–S30.

Tome, D., Schwarz, J., Darcel, N., Fromentin, G., 2009. Protein, amino acids, vagus nerve signaling, and the brain. Am. J. Clin. Nutr. 90, 838S–843S.

Treyzon, L., Chen, S., Hong, K., Yan, E., Carpenter, C.L., Thames, G., et al., 2008. A controlled trial of protein enrichment of meal replacements for weight reduction with retention of lean body mass. Nutr. J. 7, 23.

Uhe, A.M., Collier, G.R., O'Dea, K., 1992. A comparison of the effects of beef, chicken and fish protein on satiety and amino acid profiles in lean male subjects. J. Nutr. 122, 467–472.

van Avesaat, M., Troost, F.J., Ripken, D., Hendriks, H.F., Masclee, A.A., 2015. Ileal brake activation: macronutrient-specific effects on eating behavior?. Int. J. Obes. (Lond) 39, 235–243.

Veldhorst, M., Smeets, A., Soenen, S., Hochstenbach-Waelen, A., Hursel, R., Diepvens, K., et al., 2008. Protein-induced satiety: effects and mechanisms of different proteins. Physiol. Behav. 94, 300–307.

Veldhorst, M.A., Nieuwenhuizen, A.G., Hochstenbach-Waelen, A., Westerterp, K.R., Engelen, M.P., Brummer, R.J., et al., 2009. A breakfast with alpha-lactalbumin, gelatin, or gelatin + TRP lowers energy intake at lunch compared with a breakfast with casein, soy, whey, or whey-GMP. Clin. Nutr. 28, 147–155.

Veldhorst, M.A., Westerterp, K.R., Westerterp-Plantenga, M.S., 2012. Gluconeogenesis and protein-induced satiety. Br. J. Nutr. 107 (4), 595–600.

Westerterp-Plantenga, M.S., 2008. Protein intake and energy balance. Regul. Pept. 149, 67–69.

Westerterp-Plantenga, M.S., Nieuwenhuizen, A., Tome, D., Soenen, S., Westerterp, K.R., 2009. Dietary protein, weight loss, and weight maintenance. Annu. Rev. Nutr. 29, 21–41.

White, B.D., Porter, M.H., Martin, R.J., 2000. Protein selection, food intake, and body composition in response to the amount of dietary protein. Physiol. Behav. 69, 383–389.

Zampieri, T.T., Pedroso, J.A., Furigo, I.C., Tirapegui, J., Donato Jr., J., 2013. Oral leucine supplementation is sensed by the brain but neither reduces food intake nor induces an anorectic pattern of gene expression in the hypothalamus. PLoS One 8, e84094.

Zhao, S., Kanoski, S.E., Yan, J., Grill, H.J., Hayes, M.R., 2012. Hindbrain leptin and glucagon-like-peptide-1 receptor signaling interact to suppress food intake in an additive manner. Int. J. Obes. (Lond) 36, 1522–1528.

Zheng, H., Corkern, M., Stoyanova, I., Patterson, L.M., Tian, R., Berthoud, H.R., 2003. Peptides that regulate food intake: appetite-inducing accumbens manipulation activates hypothalamic orexin neurons and inhibits POMC neurons. Am. J. Physiol. Regul. Integr. Comp. Physiol. 284, R1436–R1444.

CHAPTER

17

Dietary Protein and Hepatic Glucose Production

C. Gaudichon, D. Azzout-Marniche and D. Tomé

UMR Physiologie de la Nutrition et du Comportement Alimentaire, AgroParisTech, INRA, Université Paris Saclay, Paris, France

17.1. INTRODUCTION

Proteins are a source of gluconeogenic substrates that can be used to produce glucose under some nutritional conditions, such as fasting or low carbohydrate (CHO) intake. Except leucine and lysine, all amino acids (AA) are potentially gluconeogenic, especially those directly converted to pyruvate, namely alanine, glycine, serine, cysteine, and threonine. Other AAs enter at different levels of the Krebs cycle and can generate glucose from oxaloacetate (Fig. 17.1).

Consequentially, proteins can be used to sustain the glucose demand of organs, especially the brain, in several nutritional and physiopathological conditions. Proteins are also claimed to maintain glycemia and, in this way, to promote satiety (Westerterp-Plantenga et al., 2009). However, the role of dietary proteins in hepatic glucose production is still controversial and few teams have attempted to clarify this question.

Hepatic glucose is produced from two pathways: gluconeogenesis and glycogenolysis. Even if the endogenous glucose production can easily be assessed using an euglycemic clamp or tracer methods, the differentiation between both pathways remains complex. Therefore, the role of proteins in hepatic glucose production is not fully understood due to fragmentary information from the literature.

17.2. AMINO ACIDS AS GLUCOSE PRECURSORS AND EFFECT OF PROTEIN INTAKE

If the gluconeogenic role of some AAs has been well described, there is to date no general knowledge concerning the efficiency of the different AAs to be converted to glucose.

Alanine has been reported as the main gluconeogenic AA taken up by the liver (Fafournoux et al., 1983; MacDonald et al., 1976; Yamamoto et al., 1974). Alanine catabolism in liver cells plays a role in the utilization of this AA in respect to the adaptation of alanine aminotransferase (ALAT) (Azzout et al., 1987). In addition to alanine, glycine, threonine, and serine are important precursors of gluconeogenesis. Their hepatic use has also been shown to be enhanced in rats fed a high protein (HP) diet, even in the fasted state (Remesy et al., 1983), in accordance with the increase of serine dehydratase activity (Coleman, 1980).

Kaloyianni and Freedland (1990) intended to quantify the relative contribution of different AAs to gluconeogenesis. Using ^{14}C AAs, they showed in isolated rat hepatocytes that alanine and glutamine each contributed about 10% of the glucose production. This was approximately fourfold the production from serine, glycine, and threonine, and 10-fold that of other AAs. However, they reported that lactate was the major precursor for gluconeogenesis (over 60%), regardless of the nutritional conditions.

The Molecular Nutrition of Amino Acids and Proteins.
DOI: http://dx.doi.org/10.1016/B978-0-12-802167-5.00017-7

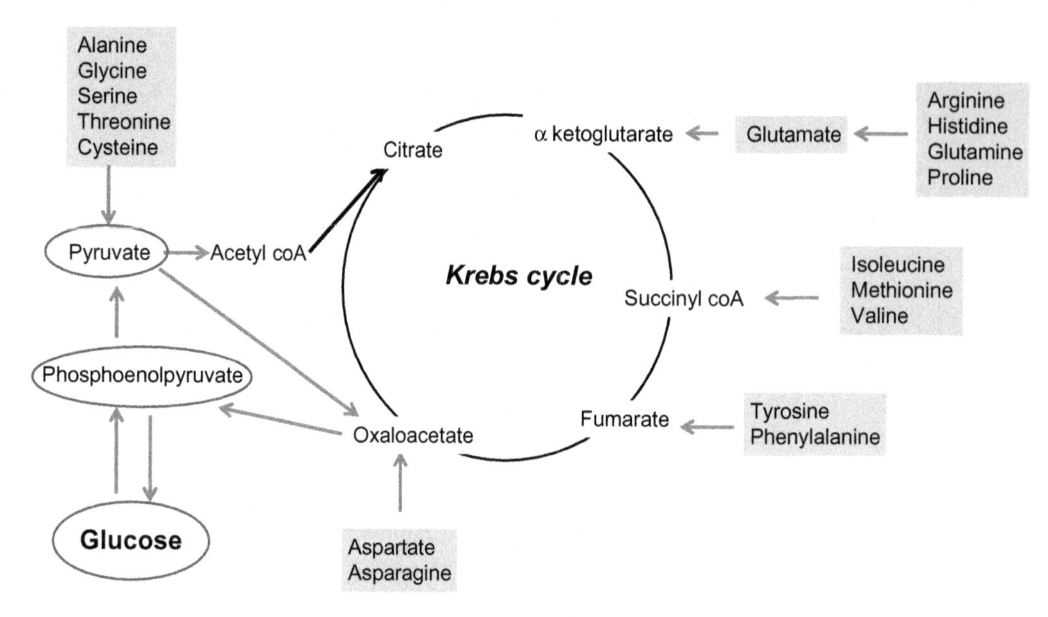

FIGURE 17.1 Biochemical pathways of glucose production from amino acids. *Source: Adapted from Häggström, Mikael. "Medical gallery of Mikael Häggström 2014". Wikiversity J. Med. 1 (2). doi:10.15347/wjm/2014.008. ISSN 20018762.*

Aspartate is another gluconeogenic AA. It is converted to oxaloacetate by the cytosolic aspartate aminotransferase (ASAT). In fasted rats adapted to an HP diet (80% protein), the mRNA and the activity of the liver ASAT were increased twofold compared to rats fed a diet containing 20% protein (Horio et al., 1988). It has been shown in Fao hepatoma cells that cytosolic ASAT was mainly regulated at the transcriptional level by glucagon positively and negatively, both by glucocorticoids and insulin (Aggerbeck et al., 1993). The activity of the mitochondrial ASAT does not seem to be affected by the nutritional environment (Horio et al., 1988).

Studies from our laboratory reported that the liver uptake of proline was enhanced in rats fed an HP diet (unpublished data). Indeed, we observed differences between portal and venous proline concentrations of 21 and 34% in the fasted and fed state, respectively. This apparent uptake was similar to that of alanine: 45 and 19% (Fig. 17.2). Moreover, proline oxidase (PO) gene expression was twofold higher in rats given an HP diet compared to an normal protein (NP) diet. These variations, being similar to proline and alanine uptake by the liver and to ALAT gene expression, suggest that proline could be an efficient hepatic gluconeogenic substrate.

Some studies have suggested that gluconeogenic substrates may contribute to glycogen accumulation more by exerting regulatory effects on hepatic glycogen metabolism than by being used as substrates (Bode et al., 1992; Bode and Nordlie, 1993; Katz et al., 1976; Riesenfeld et al., 1981). Surprisingly, glucose alone has not been reported as a good precursor for hepatic glycogen synthesis in perfused livers or in incubated hepatocytes (Boyd et al., 1981; Youn et al., 1986). The addition of modest amounts of gluconeogenic substrates into the culture medium (fructose, lactate, or AAs) resulted in markedly enhanced glycogenesis (Bode et al., 1992; Bode and Nordlie, 1993; Youn et al., 1986). Therefore, glucose and gluconeogenic precursors act synergistically to promote glycogen synthesis.

17.3. INSULIN AND GLUCAGON MEDIATED EFFECTS OF AMINO ACIDS AND PROTEINS ON GLUCOSE PRODUCTION

High postprandial AA levels were shown to stimulate insulin and glucagon secretion (Calbet and MacLean, 2002; Krebs et al., 2003). Many studies have also reported the insulinotropic effect of intravenous infusion of AAs in healthy individuals (Azzout-Marniche et al., 2014). Combined infusion of leucine, arginine and glucose were particularly insulinotropic (El Khoury and Hwalla, 2010; Manders et al., 2012). In respect to their AA composition and digestion speed, dietary proteins such as whey, casein, soy, and their respective hydrolysates triggered different insulin responses (Claessens et al., 2008; Power et al., 2009). Intake of whey and soy protein with or without supplementation with AAs (isoleucine, leucine, lysine, threonine, and valine) induced an increase in

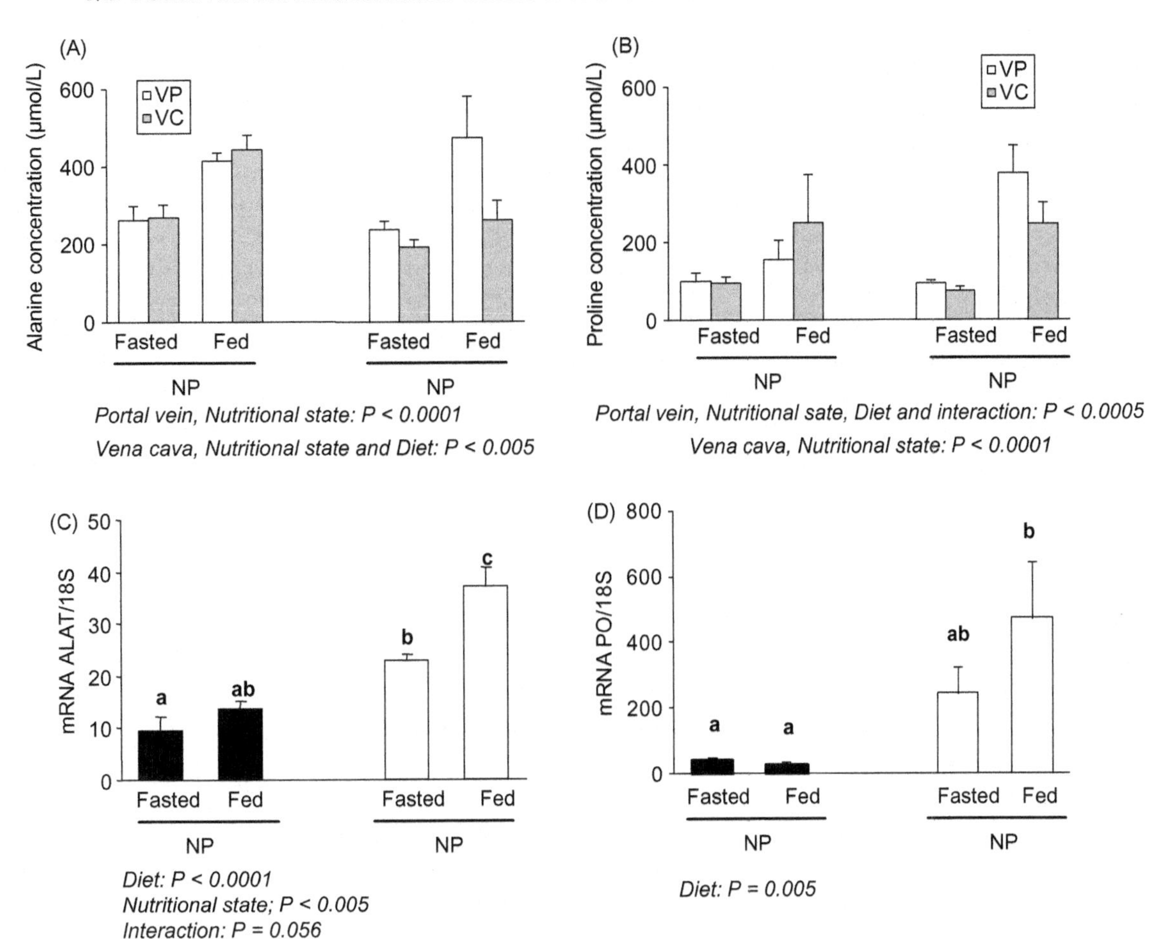

FIGURE 17.2 Portal and venous concentration of Alanine (A) and Proline (B), and hepatic expression of alanine amino transferase (C) and proline oxidase (D) in rats fed NP or HP diet, in the fasted and the fed state. For ALAT and PO gene expression, total RNA was extracted from liver and analyzed by real time PCR. Ribosomal 18 S RNA amplification was used as internal control. The results are expressed as means ± SE of four different animals. The effects of the nutritional state (fed or fasted), the diet and interaction were tested as fixed factors (ANOVA), and associated P values are indicated under each panel. Significant differences between groups are indicated by different letters (Post hoc Tukey test).

postprandial insulinemia and plasma glucagon-like peptide-1 responses associated with reduced glycemia in healthy individuals (Gunnerud et al., 2012). The addition of free leucine to protein mixtures has also been shown to enhance insulin secretion in Type 2 Diabetes patients (Manders et al., 2009). The effect of branched-chain amino acids (BCAA), and especially leucine, on insulin sensitivity is a matter of debate because their circulating concentrations have been shown to increase in obese and insulin resistant subjects (Menni et al., 2013; Newgard et al., 2009). However, it is not known whether this is a cause or a consequence of metabolic dysfunctions. Additionally, several studies reported either no effect or an improvement of insulin sensitivity associated to a higher leucine intake (Qin et al., 2011; Xiao et al., 2011; Zanchi et al., 2012; Zeanandin et al., 2012). Additionally, BCAA were shown to act synergistically with glucose to enhanced muscle glycogen production in rats at exercise, compared with glucose alone (Morifuji et al., 2009), but their effect on hepatic glucose production is unknown.

The insulinotropic action of AAs may lower the endogenous production of glucose by downregulating the key enzymes of gluconeogenesis, that is, phosphoenol pyruvate carboxykinase (PEPCK), glucose 6 phosphatase (G6PC1), and upregulating glycogen synthase (GS). Accordingly, we observed that the expression of GS progressively increased throughout the adaptation to a high protein diet (Stepien et al., 2011). Additionally, we observed in isolated rat hepatocytes that high concentrations of AAs, equivalent to the concentrations measured in vivo in the portal vein, reinforced the GS dephosphorylation induced by insulin (unpublished data).

Adversely, dietary proteins also potentiate glucagon secretion (Calbet and MacLean, 2002), which in turn can activate glycogenolysis through glycogen phosphorylase. It has been shown in rats that the addition of proteins to an oral glucose load increased the glycogen phosphorylase activity (Gannon and Nuttall, 1987).

Overall, it is unclear how the presence of both insulin and glucagon modulate the glycolytic and gluconeogenic fluxes after a protein intake. We have reported that glycolysis enzyme (glucokinase and L-pyruvate kinase) mRNA decreased in rats fed an HP diet, whereas PEPCK increased but G6PC1 mRNA did not (Stepien et al., 2011). These effects may be partly due to the ability of high physiological concentrations of portal blood AAs to counteract glucagon-induced liver G6PC1 but not PEPCK gene expression (Azzout-Marniche et al., 2007). This suggests that AAs, independent of their effect on insulin and glucagon secretion, modulate their metabolic orientation by themselves. They can act by signaling in several pathways that are connected to gluconeogenesis. For instance, high AA levels were reported to downregulate 5′AMP-activated protein kinase (AMPK) and general control nonderepressible 2 (GCN2) and to upregulate mammalian target of rapamycin (mTOR) in hepatocytes (Chotechuang et al., 2009). Therefore, there are several lines of evidence suggesting that the AMPK signaling pathway can lead to the inhibition of PEPCK gene expression (He et al., 2009; Horike et al., 2008; Lochhead et al., 2000; Zhou et al., 2001). GCN2 also probably downregulates PEPCK. In GCN2 knockout mice, PEPCK expression in the fed state was abnormally elevated and failed to be further induced during fasting (Xu et al., 2013).

17.4. PROTEIN MEAL AND HEPATIC GLUCOSE PRODUCTION

There are few studies on the effect of protein or AAs on endogenous glucose production (EGP). Using either an euglycemic clamp (Boden and Tappy, 1990) or stable isotopes to trace glucose (Tappy et al., 1992), Tappy and colleagues reported that the intravenous infusion of AAs in humans increased the endogenous glucose production. In the second study, the amount of infused AAs was huge (about 1 g/kg in 3 h), and it resulted in a drastic increase of gluconeogenesis (+235%). After the intravenous infusion of about 0.5 g/kg of AAs in 5 h, Krebs et al. (2003) also reported a 100% increase of gluconeogenesis but not of EGP.

Nuttall and Gannon conducted several studies addressing the impact of protein meal on glucose homeostasis. In one, they measured the ingestion effects of 210 g of cottage cheese (50 g of proteins, 5 g of carbohydrates and 1.5 g of lipids) on postprandial gluconeogenesis attributed to ingested proteins (Khan et al., 1992). The results indicated that about 10 g of glucose were produced from proteins during the 8 h after the meal. In Type 2 diabetes patients, they studied the glucose appearance rate after the ingestion of beef (50 g protein) or water (Gannon et al., 2001). Over 8 h, the EGP was 71 g and 68 g after the ingestion of beef or water, respectively, leading to the assessment that 3 g of glucose were produced from beef protein.

More recently, we evaluated the effect of egg ingestion (23 g protein and 27 g fat) on EGP using a multitracer approach (Fromentin et al., 2013). Healthy volunteers were continuously infused with deuterated glucose, and they ingested eggs in which the proteins were uniformly and intrinsically labeled with ^{15}N and ^{13}C. We found that the EGP was stable during 5 h after the meal ingestion and then decreased. Thanks to the intrinsic labeling of egg proteins, we were able to assess the contribution of proteins to glucose production. In the 50 g of glucose produced during the 8 h after the meal, only 4 g originated from proteins, a value that is consistent with that found in diabetes patients (Gannon et al., 2001). Although this contribution appeared to be low, it is compatible with the gluconeogenic potential of the proteins. If we consider that all AAs except leucine and lysine can provide glucose, the total amount of gluconeogenic acids was 19 g for 23 g of egg proteins. The intrinsic labeling of proteins with ^{15}N allowed determination that 18% of dietary AAs were deaminated postprandially (Fromentin et al., 2013). Thus, 3.4 g of AAs were used to synthetize glucose. Knowing the AA composition of dietary proteins and their postprandial deamination rate, it is possible to calculate a gluneocogenic potential of the proteins (Table 17.1). This reveals that the maximum amount would be from 15% for milk protein to 22% for wheat proteins. Taking into account the probable relative contribution of AAs to glucose production, based on the study of Kaloyianni and Freedland (1990), only 5 to 10% could be used for gluconeogenesis of milk and wheat proteins, respectively. Wheat proteins are the more gluconeogenic because of their high deamination rate (Bos et al., 2005) and their high glutamine content. However, it must be kept in mind that wheat proteins are rarely consumed in a purified form but in starchy products, and consequently in low gluconeogenesis conditions.

TABLE 17.1 Gluconeogenic potential of different protein sources

	Milk	Eggs	Meat	Soy	Wheat
AA COMPOSITION (G/100 G)					
Lysine	7.7	7	5.9	6.4	3.1
Leucine	9.5	8.6	7.3	8	7.1
Threonine	4.5	4.7	4.1	4	3.1
Serine	5.2	7.6	5	5.2	4.9
Glycine	2.3	8.6	11.4	4.3	4.4
Alanine	3.8	5.4	7.1	4.4	3.9
Methionine	2	2.8	1.8	1.4	1.6
Cystine	1.2	2	1.4	1.4	2.3
Tryptophan	1.5	1.6	1	1.4	1.4
Valine	5.8	6.1	5.2	4.9	4.6
Isoleucine	5.1	5.2	4	4.7	3.6
Arginine	3.7	6.6	7.4	7.4	5.1
Phenylalanine	4.5	4.7	4.1	5.2	4.8
Tyrosine	4.1	4.6	3.1	3.8	2.9
Histidine	2.5	2.4	1.8	2.6	2.4
Aspartic acid + asparagine	8.7	7.7	8.5	11.6	5.5
Glutamic acid + glutamine	19.6	10.8	13.5	18.3	29.6
Proline	8.3	3.8	7.6	5.1	9.7
Total gluconeogenic AA (g/100 g)[a]	83	85	87	86	90
Weighted gluconeogenic AA (g/100 g)[b]	31	26	30	31	40
Postprandial deamination (%)[c]	18	18	20	23	25
Total AA disposal for gluconeogenesis (g/100 g)[d]	15	15	17	20	22
Gluconeogenic potential (g/100 g)[e]	5	5	6	7	10

[a]*Sum of all AAs except lysine and leucine.*
[b]*Weighted sum of gluneocogenic AAs considering that alanine and glutamine are the more gluconeogenic; serine, threonine, and glycine are fourfold less gluconeogenic; the others AA are 10-fold less gluconeogenic. These coefficients were adapted from the study of Kaloyianni and Freedland (1990).*
[c]*Postprandial deamination rate in humans 8 h after the ingestion of ^{15}N labeled proteins (Bos et al., 2005; Fromentin et al., 2013; Lacroix et al., 2006; Mariotti et al., 1999; Oberli et al., 2015).*
[d]*Total neoglucogenic AA × deamination/100.*
[e]*Weighted neoglucogenic AA × deamination/100.*

17.5. HIGH PROTEIN DIET AND HEPATIC GLUCOSE PRODUCTION

HP diets are generally poor in carbohydrates and are assumed to promote gluconeogenesis. However, the evidence from literature remains unclear because metabolic investigations are generally performed in the fasting state. For instance, Bisschop et al. (2000) compared the effect of low- and rich-CHO diets on glucose fluxes in healthy subjects, using deuterated water to assess the relative contribution of gluconeogenesis and glycogenesis. After fasting overnight, the EGP decreased with the low CHO diet when compared with the high CHO diet. This was due to an approximate 55% decrease in glycogenolysis, while gluconeogenesis increased by only 14% with the low CHO diet. Veldhorst et al. (2009) compared the glucose fluxes in the fasting state of subjects after 2 days of a balanced or a CHO-free diet. The EGP decreased by 20% after the CHO-free diet, while gluconeogenesis represented the main flux (95%) of EGP. Linn et al. (2000) also reported an increase fasting gluconeogenesis rate in subjects fed an HP diet for 6 months, together with a moderate increase of EGP. These studies show that the increase in proteins at the expense of CHO in the diet moderately increases gluconeogenesis but drastically decreases glycogen storage due to the lack of CHO substrate.

Nevertheless, the existence of an indirect pathway from de novo synthesized glucose to liver glycogen has already been shown. Using specific ^{13}C labeling of glucose on C_1 or C_6 positions, it was reported that the indirect pathway contributed 80% to glycogen synthesis in rats fed a high-protein, low-CHO diet, while this contribution was 48% with a low-protein, high-CHO diet (Rossetti et al., 1989). Liver glycogen content was similar between groups, but it must be noted that measurements were made in the fasting state. We also suspected that in rats fed an HP diet, the indirect pathway was stimulated since as described above, PEPCK expression increased while G6PC1 did not, suggesting a channeling of de novo synthesized glucose to glycogen synthesis (Peret et al., 1975; Azzout-Marniche et al., 2007). However, we did not further confirm this hypothesis. After the administration of ^{13}C labeled dietary AAs, the ^{13}C enrichment in liver glycogen was low and did not differ between rats fed an HP or a normal protein diet (Fromentin et al., 2011). Additionally, the postprandial glycogen content in the liver was lower in HP than in NP fed rats (Fromentin et al., 2011; Stepien et al., 2011).

Finally, the stimulating effect of HP diets on gluconeogenesis is mainly the result of the increase in protein turnover and, consequently, protein breakdown during the nocturne periods (Forslund et al., 1998; Pacy et al., 1994). Consistently, Chevalier et al. (2006) have reported a significant correlation between gluconeogenic flux and whole body protein breakdown in lean and obese subjects. We thus cannot exclude activation of the indirect pathway using endogenous AAs or other gluconeogenic substrates as precursors.

17.6. CONCLUSION

Overall, many studies have shown that dietary proteins interact with hepatic glucose production due to their role as glucose precursors, their secretagogue effect, as well as their signaling role in metabolic pathways. Tracer studies have indicated that the dietary AA contribution to glucose production, even in optimal gluconeogenic conditions, is moderate. Additionally, the contribution of endogenous AAs to gluconeogenesis may be considerable under HP diet conditions, but this was never quantified. The use of AAs in glycogen synthesis, namely the indirect pathway, is also questioned, with conflicting results between cellular/molecular studies and flux data. Proteins can interact with many pathways involved in glucose homeostasis, either indirectly via the modulation of pancreatic and gastrointestinal secretions, or directly by activating or downregulating some molecular effectors in energy and protein sensing pathways. Overall, there are only parceled data on the effect of dietary proteins on the liver production of glucose. Further integrative studies are needed to clarify how the protein intake better interacts, in a direct or indirect manner, with gluconeogenesis and glycogenolysis.

References

Aggerbeck, M., Garlatti, M., Feilleux-Duche, S., Veyssier, C., Daheshia, M., Hanoune, J., et al., 1993. Regulation of the cytosolic aspartate aminotransferase housekeeping gene promoter by glucocorticoids, cAMP, and insulin. Biochemistry 32 (35), 9065–9072.

Azzout-Marniche, D., Gaudichon, C., Blouet, C., Bos, C., Mathe, V., Huneau, J.F., et al., 2007. Liver glyconeogenesis: a pathway to cope with postprandial amino acid excess in high-protein fed rats?. Am. J. Physiol. Regul. Integr. Comp. Physiol. 292 (4), R1400–R1407.

Azzout-Marniche, D., Gaudichon, C., Tome, D., 2014. Dietary protein and blood glucose control. Curr. Opin. Clin. Nutr. Metab. Care 17 (4), 349–354.

Azzout, B., Bois-Joyeux, B., Chanez, M., Peret, J., 1987. Development of gluconeogenesis from various precursors in isolated rat hepatocytes during starvation or after feeding a high protein, carbohydrate-free diet. J. Nutr. 117 (1), 164–169.

Bisschop, P.H., Pereira Arias, A.M., Ackermans, M.T., Endert, E., Pijl, H., Kuipers, F., et al., 2000. The effects of carbohydrate variation in isocaloric diets on glycogenolysis and gluconeogenesis in healthy men. J. Clin. Endocrinol. Metab. 85 (5), 1963–1967.

Bode, A.M., Nordlie, R.C., 1993. Reciprocal effects of proline and glutamine on glycogenesis from glucose and ureagenesis in isolated, perfused rat livers. J. Biol. Chem. 268 (22), 16298–16301.

Bode, A.M., Foster, J.D., Nordlie, R.C., 1992. Glyconeogenesis from L-proline involves metabolite inhibition of the glucose-6-phosphatase system. J. Biol. Chem. 267 (5), 2860–2863.

Boden, G., Tappy, L., 1990. Effects of amino acids on glucose disposal. Diabetes 39 (9), 1079–1084.

Bos, C., Juillet, B., Fouillet, H., Turlan, L., Dare, S., Luengo, C., et al., 2005. Postprandial metabolic utilization of wheat protein in humans. Am. J. Clin. Nutr. 81 (1), 87–94.

Boyd, M.E., Albright, E.B., Foster, D.W., McGarry, J.D., 1981. In vitro reversal of the fasting state of liver metabolism in the rat. Reevaluation of the roles of insulin and glucose. J. Clin. Invest. 68 (1), 142–152.

Calbet, J.A., MacLean, D.A., 2002. Plasma glucagon and insulin responses depend on the rate of appearance of amino acids after ingestion of different protein solutions in humans. J. Nutr. 132 (8), 2174–2182.

Chevalier, S., Burgess, S.C., Malloy, C.R., Gougeon, R., Marliss, E.B., Morais, J.A., 2006. The greater contribution of gluconeogenesis to glucose production in obesity is related to increased whole-body protein catabolism. Diabetes 55 (3), 675–681.

Chotechuang, N., Azzout-Marniche, D., Bos, C., Chaumontet, C., Gausseres, N., Steiler, T., et al., 2009. mTOR, AMPK, and GCN2 coordinate the adaptation of hepatic energy metabolic pathways in response to protein intake in the rat. Am. J. Physiol. Endocrinol. Metab. 297 (6), E1313–E1323.

Claessens, M., Saris, W.H., van Baak, M.A., 2008. Glucagon and insulin responses after ingestion of different amounts of intact and hydrolysed proteins. Br. J. Nutr. 100 (1), 61–69.

Coleman, D.L., 1980. Genetic control of serine dehydratase and phosphoenolpyruvate carboxykinase in mice. Biochem. Genet. 18 (9–10), 969–979.

El Khoury, D., Hwalla, N., 2010. Metabolic and appetite hormone responses of hyperinsulinemic normoglycemic males to meals with varied macronutrient compositions. Ann. Nutr. Metab. 57 (1), 59–67.

Fafournoux, P., Remesy, C., Demigne, C., 1983. Control of alanine metabolism in rat liver by transport processes or cellular metabolism. Biochem. J. 210 (3), 645–652.

Forslund, A.H., Hambraeus, L., Olsson, R.M., El-Khoury, A.E., Yu, Y.M., Young, V.R., 1998. The 24-h whole body leucine and urea kinetics at normal and high protein intakes with exercise in healthy adults. Am. J. Physiol. 275 (2 Pt 1), E310–E320.

Fromentin, C., Azzout-Marniche, D., Tome, D., Even, P., Luengo, C., Piedcoq, J., et al., 2011. The postprandial use of dietary amino acids as an energy substrate is delayed after the deamination process in rats adapted for 2 weeks to a high protein diet. Amino Acids 40 (5), 1461–1472.

Fromentin, C., Tome, D., Nau, F., Flet, L., Luengo, C., Azzout-Marniche, D., et al., 2013. Dietary proteins contribute little to glucose production, even under optimal gluconeogenic conditions in healthy humans. Diabetes 62 (5), 1435–1442.

Gannon, M.C., Nuttall, F.Q., 1987. Oral protein hydrolysate causes liver glycogen depletion in fasted rats pretreated with glucose. Diabetes 36 (1), 52–58.

Gannon, M.C., Nuttall, J.A., Damberg, G., Gupta, V., Nuttall, F.Q., 2001. Effect of protein ingestion on the glucose appearance rate in people with type 2 diabetes. J. Clin. Endocrinol. Metab. 86 (3), 1040–1047.

Gunnerud, U.J., Heinzle, C., Holst, J.J., Ostman, E.M., Bjorck, I.M., 2012. Effects of pre-meal drinks with protein and amino acids on glycemic and metabolic responses at a subsequent composite meal. PLoS One 7 (9), e44731.

He, L., Sabet, A., Djedjos, S., Miller, R., Sun, X., Hussain, M.A., et al., 2009. Metformin and insulin suppress hepatic gluconeogenesis through phosphorylation of CREB binding protein. Cell 137 (4), 635–646.

Horike, N., Sakoda, H., Kushiyama, A., Ono, H., Fujishiro, M., Kamata, H., et al., 2008. AMP-activated protein kinase activation increases phosphorylation of glycogen synthase kinase 3beta and thereby reduces cAMP-responsive element transcriptional activity and phosphoenolpyruvate carboxykinase C gene expression in the liver. J. Biol. Chem. 283 (49), 33902–33910.

Horio, Y., Nishida, Y., Sakakibara, R., Inagaki, S., Kamisaki, Y., Wada, H., 1988. Induction of cytosolic aspartate aminotransferase by a high-protein diet. Biochem. Int. 16 (3), 579–586.

Kaloyianni, M., Freedland, R.A., 1990. Contribution of several amino acids and lactate to gluconeogenesis in hepatocytes isolated from rats fed various diets. J. Nutr. 120 (1), 116–122.

Katz, J., Golden, S., Wals, P.A., 1976. Stimulation of hepatic glycogen synthesis by amino acids. Proc. Natl. Acad. Sci. U.S.A. 73 (10), 3433–3437.

Khan, M.A., Gannon, M.C., Nuttall, F.Q., 1992. Glucose appearance rate following protein ingestion in normal subjects. J. Am. Coll. Nutr. 11 (6), 701–706.

Krebs, M., Brehm, A., Krssak, M., Anderwald, C., Bernroider, E., Nowotny, P., et al., 2003. Direct and indirect effects of amino acids on hepatic glucose metabolism in humans. Diabetologia 46 (7), 917–925.

Lacroix, M., Bos, C., Leonil, J., Airinei, G., Luengo, C., Dare, S., et al., 2006. Compared with casein or total milk protein, digestion of milk soluble proteins is too rapid to sustain the anabolic postprandial amino acid requirement. Am. J. Clin. Nutr. 84 (5), 1070–1079.

Linn, T., Santosa, B., Gronemeyer, D., Aygen, S., Scholz, N., Busch, M., et al., 2000. Effect of long-term dietary protein intake on glucose metabolism in humans. Diabetologia 43 (10), 1257–1265.

Lochhead, P.A., Salt, I.P., Walker, K.S., Hardie, D.G., Sutherland, C., 2000. 5-aminoimidazole-4-carboxamide riboside mimics the effects of insulin on the expression of the 2 key gluconeogenic genes PEPCK and glucose-6-phosphatase. Diabetes 49 (6), 896–903.

MacDonald, M., Neufeldt, N., Park, B.N., Berger, M., Ruderman, N., 1976. Alanine metabolism and gluconeogenesis in the rat. Am. J. Physiol. 231 (2), 619–626.

Manders, R.J., Praet, S.F., Vikstrom, M.H., Saris, W.H., van Loon, L.J., 2009. Protein hydrolysate co-ingestion does not modulate 24 h glycemic control in long-standing type 2 diabetes patients. Eur. J. Clin. Nutr. 63 (1), 121–126.

Manders, R.J., Little, J.P., Forbes, S.C., Candow, D.G., 2012. Insulinotropic and muscle protein synthetic effects of branched-chain amino acids: potential therapy for type 2 diabetes and sarcopenia. Nutrients 4 (11), 1664–1678.

Mariotti, F., Mahe, S., Benamouzig, R., Luengo, C., Dare, S., Gaudichon, C., et al., 1999. Nutritional value of [15N]-soy protein isolate assessed from ileal digestibility and postprandial protein utilization in humans. J. Nutr. 129 (11), 1992–1997.

Menni, C., Fauman, E., Erte, I., Perry, J.R., Kastenmuller, G., Shin, S.Y., et al., 2013. Biomarkers for type 2 diabetes and impaired fasting glucose using a nontargeted metabolomics approach. Diabetes 62 (12), 4270–4276.

Morifuji, M., Koga, J., Kawanaka, K., Higuchi, M., 2009. Branched-chain amino acid-containing dipeptides, identified from whey protein hydrolysates, stimulate glucose uptake rate in L6 myotubes and isolated skeletal muscles. J. Nutr. Sci. Vitaminol. (Tokyo) 55 (1), 81–86.

Newgard, C.B., An, J., Bain, J.R., Muehlbauer, M.J., Stevens, R.D., Lien, L.F., et al., 2009. A branched-chain amino acid-related metabolic signature that differentiates obese and lean humans and contributes to insulin resistance. Cell Metab. 9 (4), 311–326.

Oberli, M., Marsset-Baglieri, A., Airinei, G., Santé-Lhoutellier, V., Khodorova, N., Rémond, D., et al., 2015. High true ileal digestibility but not postprandial utilization of nitrogen from bovine meat protein in humans is moderately decreased by high-temperature, long-duration cooking. J. Nutr. 145(10), 2221–2228.

Pacy, P.J., Price, G.M., Halliday, D., Quevedo, M.R., Millward, D.J., 1994. Nitrogen homeostasis in man: the diurnal responses of protein synthesis and degradation and amino acid oxidation to diets with increasing protein intakes. Clin. Sci. (Lond) 86 (1), 103–116.

Peret, J., Chanez, M., Cota, J., Macaire, I., 1975. Effects of quantity and quality of dietary protein and variation in certain enzyme activities on glucose metabolism in the rat. J. Nutr. 105 (12), 1525–1534.

Power, O., Hallihan, A., Jakeman, P., 2009. Human insulinotropic response to oral ingestion of native and hydrolysed whey protein. Amino Acids 37 (2), 333–339.

Qin, L.Q., Xun, P., Bujnowski, D., Daviglus, M.L., Van Horn, L., Stamler, J., et al., 2011. Higher branched-chain amino acid intake is associated with a lower prevalence of being overweight or obese in middle-aged East Asian and Western adults. J. Nutr. 141 (2), 249–254.

Remesy, C., Fafournoux, P., Demigne, C., 1983. Control of hepatic utilization of serine, glycine and threonine in fed and starved rats. J. Nutr. 113 (1), 28–39.

Riesenfeld, G., Wals, P.A., Golden, S., Katz, J., 1981. Glucose, amino acids, and lipogenesis in hepatocytes of Japanese quail. J. Biol. Chem. 256 (19), 9973–9980.

Rossetti, L., Rothman, D.L., DeFronzo, R.A., Shulman, G.I., 1989. Effect of dietary protein on in vivo insulin action and liver glycogen repletion. Am. J. Physiol. 257 (2 Pt 1), E212–E219.

Stepien, M., Gaudichon, C., Fromentin, G., Even, P., Tome, D., Azzout-Marniche, D., 2011. Increasing protein at the expense of carbohydrate in the diet down-regulates glucose utilization as glucose sparing effect in rats. PLoS One 6 (2), e14664.

Tappy, L., Acheson, K., Normand, S., Schneeberger, D., Thelin, A., Pachiaudi, C., et al., 1992. Effects of infused amino acids on glucose production and utilization in healthy human subjects. Am. J. Physiol. 262 (6 Pt 1), E826–E833.

Veldhorst, M.A., Westerterp-Plantenga, M.S., Westerterp, K.R., 2009. Gluconeogenesis and energy expenditure after a high-protein, carbohydrate-free diet. Am. J. Clin. Nutr. 90 (3), 519–526.

Westerterp-Plantenga, M.S., Nieuwenhuizen, A., Tome, D., Soenen, S., Westerterp, K.R., 2009. Dietary protein, weight loss, and weight maintenance. Annu. Rev. Nutr. 29, 21–41.

Xiao, F., Huang, Z., Li, H., Yu, J., Wang, C., Chen, S., et al., 2011. Leucine deprivation increases hepatic insulin sensitivity via GCN2/mTOR/S6K1 and AMPK pathways. Diabetes 60 (3), 746–756.

Xu, X., Hu, J., McGrath, B.C., Cavener, D.R., 2013. GCN2 regulates the CCAAT enhancer binding protein beta and hepatic gluconeogenesis. Am. J. Physiol. Endocrinol. Metab. 305 (8), E1007–E1017.

Yamamoto, H., Aikawa, T., Matsutaka, H., Okuda, T., Ishikawa, E., 1974. Interorganal relationships of amino acid metabolism in fed rats. Am. J. Physiol. 226 (6), 1428–1433.

Youn, J.H., Youn, M.S., Bergman, R.N., 1986. Synergism of glucose and fructose in net glycogen synthesis in perfused rat livers. J. Biol. Chem. 261 (34), 15960–15969.

Zanchi, N.E., Guimaraes-Ferreira, L., Siqueira-Filho, M.A., Gabriel Camporez, J.P., Nicastro, H., Seixas Chaves, D.F., et al., 2012. The possible role of leucine in modulating glucose homeostasis under distinct catabolic conditions. Med. Hypotheses 79 (6), 883–888.

Zeanandin, G., Balage, M., Schneider, S.M., Dupont, J., Hebuterne, X., Mothe-Satney, I., et al., 2012. Differential effect of long-term leucine supplementation on skeletal muscle and adipose tissue in old rats: an insulin signaling pathway approach. Age (Dordr) 34 (2), 371–387.

Zhou, G., Myers, R., Li, Y., Chen, Y., Shen, X., Fenyk-Melody, J., et al., 2001. Role of AMP-activated protein kinase in mechanism of metformin action. J. Clin. Invest. 108 (8), 1167–1174.

CHAPTER

18

Impact of Dietary Proteins on Energy Balance, Insulin Sensitivity and Glucose Homeostasis: From Proteins to Peptides to Amino Acids

G. Chevrier[1,2], *P. Mitchell*[1,2], *M.-S. Beaudoin*[1,2] *and A. Marette*[1,2]

[1]Department of Medicine, Faculty of Medicine, Cardiology Axis of the Québec Heart and Lung Institute, Québec, QC, Canada [2]Institute of Nutrition and Functional Foods, Laval University, Québec, QC, Canada

18.1. INTRODUCTION

As "westernization" of the diet becomes a global phenomenon, the incidences of obesity, insulin resistance, and type 2 diabetes (T2D) are the new epidemics in countries that once faced infectious disease and starvation. Much has been made about our current unhealthy lifestyle and how it promotes the development of chronic societal diseases. The incidence of obesity and associated cardiometabolic diseases has exponentially increased in the past few decades. Recent data from the World Health Organisation (WHO) shows that global obesity has more than doubled since 1980 and now affects 600 million people (WHO, 2015). Obesity and T2D affect, respectively, 13% and 9% of the adult population, and the disease is projected to be one of the leading causes of death in people by 2030 (WHO, 2015). While public health strategies and intervention are put into place to reduce the risk factors and the associated economic burden of obesity-associated diseases, personal strategies such as increasing level of physical activity and improving nutritional habits should be encouraged. Indeed, data gathered from several epidemiological and clinical studies indicate that in addition to genetic factors, an unhealthy diet represents the most important determinant of the rapid progression of metabolic diseases.

While it is clear that poor dietary habits are contributing to the obesity epidemic, we still face major gaps in our understanding of the role of individual foods and nutrients in the development of insulin resistance and T2D. Research in the last decade has provided mounting evidence that dietary proteins are key modulators of energy balance and metabolic health, which is in large part related to their amino acid composition. Dietary proteins are needed for maintenance of multiple body functions and are particularly important to support increased needs during specific physiological conditions (eg, lactation, pregnancy and growth) (WHO, 2002).

Expanding the role of foods from merely providing nutritional requirements to being a key contributing factor to health led to the conceptual development of nutraceuticals and functional foods. For instance, our understanding of the role of proteins in the diet is no longer centered on the provision of adequate nitrogen sources and amino acids (AA) as building blocks. There is accumulated evidence that bioactive peptides derived from dietary proteins have key physiological functions (Erdmann et al., 2008) and a growing number of peptides with therapeutic benefits have been identified (Chalamaiah et al., 2012; Harnedy and FitzGerald, 2012; Meisel, 2004), including many from marine origin (Kristinsson and Rasco, 2000).

While the health impact of consuming proteins and bioactive peptides is increasingly being recognized, we are also making great progress in our understanding of the mechanisms by which AA modulates insulin action

The Molecular Nutrition of Amino Acids and Proteins.
DOI: http://dx.doi.org/10.1016/B978-0-12-802167-5.00018-9

and glucose metabolism and how this can play a role in the development of obesity, insulin resistance, and T2D. We will also discuss the recent evidence that circulating levels of AA are reliable predictors of T2D risk through the modulation of key cellular signaling pathways involved in glucose and lipid metabolism. Finally, the newly established role of the gut microbiota in modulating the immunometabolic effects of proteins and AA and its potential impact on the development of the MetS will be reviewed.

18.1.1 Effects of Dietary Proteins on Energy Balance and Body Weight

Protein supplementation was found to be a popular aid to weight control after the failure of low-fat diets popularized in the 1970s and 1980s. In fact, many years of research had showed that excessive fat, and more particularly saturated fats, was the main cause of cardiovascular diseases (CVD) (Gifford, 2002). However, it was reported that despite reducing their consumption of calories from fat from 40% to 34% between 1977 and 1995, the US population still was found to have increased their body mass index (BMI) and body weight (Ogden et al., 2004), total caloric intake by 21% (Gifford, 2002), and calories from sugar by 17% (USDA). During these years, new diet trends have emerged and among them, many diets rich in protein (>20–35% of energy) such as Atkins, South Beach, Zone and the Paleolithic diets, remain popular today (Eaton and Eaton, 2000; Konner and Eaton, 2010; Pesta and Samuel, 2014). In comparison, the WHO recommends consuming 10–15% of daily energy intake as protein (WHO, 2003) with a recommended dietary allowance of 0.83 g/kg body weight (WHO, 2007).

Potential mechanisms to explain the advantages of high-protein (HP) diets include the stimulation of energy expenditure, the sparing of fat-free mass at the expense of fat mass, the reduction of appetite, and improvement of the metabolic profile (reviewed in Westerterp-Plantenga et al., 2009). Protein is also known as the most effective satiating macronutrient as compared to fat and carbohydrate (Soenen and Westerterp-Plantenga, 2008). Some of these effects may be explained by the release of gut hormones, as will be discussed in the following section.

18.1.1.1 Protein-Induced Incretin Release and Satiety

The secretion of several gut hormones such as cholecystokinin (CCK), Peptide YY (PYY), glucagon-like peptide 1 (GLP-1), and possibly ghrelin, is modulated by the levels of proteins and AA in the gastrointestinal tract. Many factors may influence the release of anorexigenic and the inhibition of orexigenic hormones, as their circulating levels do not necessarily correlate with satiety (Diepvens et al., 2008; Smeets et al., 2008). Incretin response to protein and subsequent satiety may also be influenced by a complex interplay between other peptidic hormones such as leptin, insulin, or gastric inhibitory peptide (Veldhorst et al., 2008), while overweight and obesity are known to alter incretin levels or response to protein intake (Newgard et al., 2009; Small and Bloom, 2004). Changes in incretin levels and satiety will also depend on the protein source consumed (eg, whey, casein, soy, fish), the matrix of food (eg, liquid, solid), the rate of gastric emptying, and the type of peptides and AA present in food (Diepvens et al., 2008; Hall et al., 2003).

AA and proteins contribute to the release of CCK, a gut hormone synthetized from mucosal enteroendocrine cells in the duodenum (Davidenko et al., 2013). CCK was first shown to inhibit food intake in rats (Gibbs et al., 1973) and is known since then to induce satiety and participate in many physiological functions, such as activation of gallbladder contraction, pancreatic secretion and intestinal motility, and delaying of gastric emptying (Moran, 2000). Release of CCK activates vagal feedback to the nucleus of the solitary track in the brainstem, thus providing information on intestinal content and inducing protein-satiating effects and reducing food intake (Faipoux et al., 2008). The importance of functional CCK signaling pathway was demonstrated using Otsuka Long Evans Tokushima (OLEFT) rats, which have no CKK_A receptors, and thus display hyperphagia and obesity (Moran, 2000). In clinical settings, lean and obese humans receiving either an HP or a normal protein (NP) meal (1.35 vs 0.8 g/kg body weight) had elevated perception of fullness and CCK levels when compared to consumption of a lower-protein meal (Brennan et al., 2012). Similar results were obtained in lean and overweight male subjects after casein, whey, soy, and gluten protein preloads (Bowen et al., 2006a,b) and the same authors later found that CCK-mediated satiety by protein was dose-dependent (Bowen et al., 2007).

Another gut hormone, PYY is an important contributor of the so-called ileal break, an inhibitory feedback-loop mechanism that maximizes absorption and digestion of food in the gastrointestinal track (Davidenko et al., 2013). It is released by L cells in the ileum and colon and is recognized by hypothalamic Y2 receptors to reduce feeding

(Badman and Flier, 2005). PYY response to protein may be dose-dependent in human subjects (Smeets et al., 2008). An elegant study showed that PYY3-36, the most anorectic form of PYY, was elevated in lean and obese subjects upon an HP meal (Batterham et al., 2006). Then the same authors demonstrated that mice lacking the *PYY* gene displayed hyperphagia and obesity, which could be normalized by exogenous PYY treatment, thus showing that PYY is involved in the satiety action of dietary proteins (Batterham et al., 2006).

Another key gut peptide is GLP-1, which is cleaved by proglucagon and produced by intestinal L cells. It is inactivated by human neutral endopeptidase 24.11, by renal clearance, but most importantly by DPP-4 (Mansour et al., 2013). Aside from fulfilling satiety function just like other anorexic incretins, it stimulates glucose-dependant insulin release and inhibits hepatic glucose production and β-cell apoptosis (Mansour et al., 2013). It was suggested that gut AA activates p38 mitogen activated protein kinase and ERK1/2 pathways and could thus promote GLP-1 secretion (Reimer, 2006). While release of GLP-1 is triggered by several types of nutrients (Mansour et al., 2013), some evidence shows that a 4-day HP diet increases its levels in healthy women (Lejeune et al., 2006) and after a liquid whey preload (Hall et al., 2003). However, another group showed that an HP meal induced a higher level of satiety without changes in GLP-1 and PYY levels as compared to an NP meal (Smeets et al., 2008). The presence of carbohydrate and the type and proportion of AA in proteins may also be important factors that influence GLP-1 release, although the findings are discordant between studies (Mansour et al., 2013; Smeets et al., 2008).

Ghrelin is a key orexigenic hormone known to increase appetite and food intake in humans, which is primarily secreted by the stomach and in small concentrations by the duodenum and the small intestine (reviewed in Cummings, 2006). Ghrelin can also be released just before food intake as a cephalic response to the anticipatory process of food ingestion (Cummings, 2006). Ghrelin increases motility of the gastrointestinal tractus and decreases insulin secretion after a meal (Cummings, 2006). It plays an important role in the regulation of body weight, as it increases during weight loss and decreases in weight gain (Cummings, 2006). Its response to nutrients is also dependent on gender and the BMI status (Greenman et al., 2004). Indeed obese subjects have lower fasting ghrelin levels (Greenman et al., 2004) and a lower suppression of ghrelin upon food intake, perhaps indicating an orexigenic drive failing to respond to food intake (English et al., 2002). Simple and complex carbohydrates suppress ghrelin secretion in a dose-dependent manner (Karhunen et al., 2008), no matter the mode of administration (orally or intravenously). Carbohydrates also suppress ghrelin better than lipids when calorie-equivalent (Monteleone et al., 2003), perhaps explained by the high insulin- and glucose-suppressing effects on ghrelin (Nakagawa et al., 2002; Saad et al., 2002). However, the impact of protein on ghrelin release is less clear. While an HP meal induced a high suppression of ghrelin levels in lean and obese subjects over time (Brennan et al., 2012), no changes in ghrelin were observed after an isocaloric high-carbohydrate, high-fat, or high-protein load (Greenman et al., 2004) or meal (Batterham et al., 2006). The absence of changes in ghrelin levels are in line with other studies looking at the acute response of two types of meals different in their protein content (Smeets et al., 2008) or after a chronic 4-day HP- or NP-diet equivalent in their fat content (Lejeune et al., 2006). These results may indicate that carbohydrates and proteins could equally suppress ghrelin response although not necessarily with the same kinetics (Cummings, 2006).

In summary, the superior satiating effects of proteins, as compared to carbohydrates and fats (Westerterp-Plantenga et al., 2009) is explained by a complex interplay between protein-induced incretin release as well as involvement of other hormones. While enhanced satiation is an important contributor to the popularity of HP diets, protein is also a very thermogenic nutrient that may help in weight loss (Westerterp-Plantenga et al., 2009).

18.1.1.2 Protein-Induced Thermogenesis

Much evidence shows that protein intake increases thermogenesis through diet-induced energy expenditure (DEE) and ultimately raises sleeping metabolic rate (SMR) and resting metabolic rate (RMR) (reviewed in Westerterp-Plantenga et al., 2009). DEE represents about 10% of total energy intake in healthy subjects and varies according to nutrient availability, meal size, BMI status, and ATP requirements for metabolism and storage (Robinson et al., 1990). In this regard, protein-metabolizing processes such as protein synthesis, protein turnover, ureogenesis, and gluconeogenesis have high energetic requirements for ATP synthesis (Westerterp-Plantenga et al., 2009), that is, 20–30% of energy consumed as protein is dissipated as DEE, as compared to 5–10% of that for carbohydrate and 0–3% of that for fat metabolism, respectively (Tappy, 1996). The unique thermogenic effect of protein comes from the difference between its ingested energy value (IE) of 23.6 kJ/g and its net metabolizable energy value (NME), the food energy available for body functions that require ATP (UN, 2002), of 13.3 kJ/g. In comparison, IE and NME is 15.7 kJ/g and 15.7 kJ/g for carbohydrates, and 39.3 kJ/g and 36.6 kJ/g for fat, respectively (Livesey, 2001).

The SMR, which is part of daily energy expenditure, was shown to be increased by an HP diet as compared to NP diet (Lejeune et al., 2006), and was also increased by 2% following consumption of animal protein derived from pork as compared to vegetable protein from soy (Mikkelsen et al., 2000). The authors suggested that the balanced mixture of AA and the higher biological value of animal protein may be contributing to thermogenesis (Mikkelsen et al., 2000). This may be due to the cost of ATP synthesis for AA degradation, some generating more energy than others (van Milgen, 2002). For instance, catabolism of cysteine yields 153 kJ/ATP and glutamate, 99.2 kJ/ATP (van Milgen, 2002). This thermogenic effect of protein is also partly explained by the inability of the body to store excess protein and its obligation to catabolize protein into peptides and AA, as opposed to carbohydrate and lipids, which are stored in the liver and adipose tissue. This was demonstrated in the $BCATm^{-/-}$ mouse model, which displays high levels of branched-chain amino acids (BCAA) due to its inability to catabolize them. These mice had an increased protein turnover resulting in higher energy expenditure, VO_2 consumption, and lower body weight despite higher energy intake as compared to their normal counterparts (She et al., 2007a). Protein turnover may also account for higher SMR and thermogenesis in different conditions, such as after an HP diet that followed a fast (Robinson et al., 1990) and after exercise, since protein synthesis may be promoted for up to 48 hours (Koopman et al., 2007; Welle and Nair, 1990).

18.1.1.3 Long-Term Health Effects of HP Diets

Diets rich in proteins were once thought to negatively alter renal function and promote bone loss. However, recent advances from epidemiological studies show that long-term high-protein diets increase bone mineral density and do not contribute to development of fracture and osteoporosis (reviewed in Cao and Nielsen, 2010). Even if this type of diet favors renal acid load and increases urinary calcium excretion, they are offset by the intestinal calcium uptake, the increase in plasma of insulin-like growth factor 1, and the decrease in serum parathyroid hormone. The consumption of alkali buffers such as fruits and vegetables when on an HP diet may be beneficial to bone health (Cao and Nielsen, 2010).

While there is no evidence that HP diets initiate renal disease in healthy subjects, those with obesity-associated conditions such as the metabolic syndrome and T2D should express some caution, as these conditions are associated with preexisting renal malfunction and chronic kidney disease. Indeed, while some AA are involved in processes such as ureagenesis and gluconeogenesis and may have a blood pressure lowering effect, other AA such as cysteine, homocysteine, methionine, and taurine may alter the acid–base homeostasis and contribute to a raise of blood pressure by reducing the mass of nephrons, particularly in obesity (Friedman, 2004; Veldhorst et al., 2008).

Many epidemiological and intervention studies have also assessed the association between protein consumption, protein source, and obesity-associated factors and diseases. It was shown that high intake of red meat, processed meat, and chicken, but not fish, dairy, or plant protein, was associated with weight gain (Halkjaer et al., 2011; Vergnaud et al., 2010). The future risk of coronary heart disease (CHD) was lower with high intakes of poultry, fish and nuts, whereas the opposite was true with high consumption of red meat. In this study it was calculated that fish was the best replacement option for meat to reduce the risk of CHD (Bernstein et al., 2010). Furthermore, an animal versus vegetable low-carbohydrate diet is associated with higher versus lower all-cause mortality, respectively (Fung et al., 2010). Many epidemiological studies of various sizes have shown that high consumption of red meat (and especially processed meat) (Aune et al., 2009; Pan et al., 2011), total protein intake (Tinker et al., 2011; Wang et al., 2010), and total protein intake from animal origin, but not from plant (Sluijs et al., 2010; van Nielen et al., 2014), were associated with an increased risk of developing T2D. These studies need to be interpreted with caution, however, as sometimes high meat consumption is associated with an unhealthier lifestyle and other confounding behaviors thus making it hard to tell with certainty that this specific eating habit is causing disease or premature mortality. While some studies have controlled for some confounding effects, such as for consumption of fruits and vegetables, smoking, or exercise habits, other parameters associated with meat consumption, such as the male gender, smoking, and BMI status, may also represent key confounders for T2D risk (Joost, 2013; Sluijs et al., 2010). A limitation of epidemiological studies is that it can never take into account all possible confounding factors especially when the studies are conducted with heterogeneous large community-based populations.

Interestingly, a study (Pan et al., 2013) analyzed subjects for their voluntary increase in red meat consumption instead of their habitual intake per se. They had a 1.48-fold increase risk of T2D in the 4-year observation period as compared to a reference group that had not made such changes, suggesting a dose-dependent effect and stronger evidence that high intakes of red meat should be limited (Pan et al., 2013).

On the opposite, other sources of protein may also contribute to decreasing the risk of T2D. For instance, research shows that a high consumption of plant protein, such as nuts, soy and legumes (Halton et al., 2008; Kendall et al., 2010; Villegas et al., 2008), and dairy products (Fumeron et al., 2011; Malik et al., 2011) lowers the long-term risk of T2D. The association with fish and/or seafood intake is less conclusive, some showing beneficial effects (Feskens et al., 1991; Patel et al., 2009) after long-term consumption while others did not (Kaushik et al., 2009; Patel et al., 2012; van Woudenbergh et al., 2009; Wallin et al., 2012; Wu et al., 2012; Xun and He, 2012; Zhou et al., 2012).

HP diets may have positive outcomes associated with body weight regulation and appetite as seen previously. However the AA surplus may cause detrimental effects to kidneys as well as increase the risk of obesity-associated disorders in some populations. Choosing plant-, dairy- and likely fish-derived proteins may be a safer alternative over animal-derived protein to prevent mortality, T2D, and CVD over the long-term. Body weight management may also be improved by increasing the consumption of these sources of proteins instead of meat-derived proteins.

18.1.2 Impact of Dietary Protein Sources and Derived Peptides on the Metabolic Syndrome

Several foods and compounds from marine, plant, or dairy origin have shown therapeutic properties and could help to prevent or alleviate obesity-related diseases, or can be developed as functional foods to be consumed as part of a normal diet. Marine-derived compounds, such as oligopeptides and peptides, have been extensively studied, especially those from fish, with their AA sequence identified, or already commercialized as natural health products (Harnedy and FitzGerald, 2012). Several other peptides cleaved from milk compounds are bioactive and were shown to have beneficial health properties, that is, antihypertensive, antithrombotic, immunomodulatory, metal-biocarrier, antimicrobial, cytomodulatory among others (reviewed in Mills et al., 2011). Legume-derived protein, especially those from soy, have also attracted some attention for their lipid-lowering effects and other metabolic properties. The following sections explore the added value of marine, dairy, and plant-derived proteins and bioactive peptides and their therapeutic targets for combating obesity and associated cardiometabolic complications.

18.1.2.1 Impact of Marine-Derived Proteins and Peptides

For a long time the beneficial effects of fish consumption were attributed to their content in ω-3 polyunsaturated fatty acids (PUFA). The lipid-lowering effects of ω-3 PUFA supplementation are usually accepted by the scientific community (Jacobson et al., 2012; Lopez-Huertas, 2012) as well as their positive impact on the incidence of CVD (Baum et al., 2012). However, the effect of ω-3 PUFA on insulin sensitivity and glucose homeostasis remains unclear, especially in diabetic subjects (Delarue et al., 2004; Flachs et al., 2009; Kopecky et al., 2009). While some meta-analyses have failed to establish a link between incidence of T2D and consumption of fish and/or ω-3 rich fish oil in various populations (Patel et al., 2012; Wallin et al., 2012; Wu et al., 2012; Xun and He, 2012; Zhou et al., 2012), others have shown that consumption of fish in an elderly population (Feskens et al., 1991) and in a large population-based prospective cohort (Patel et al., 2012) could protect against the development of T2D. We also reported in a systematic review with meta-analyses that although no significant effect of fish/seafood or marine ω-3 PUFA intake was observed on risk of T2D, that a significant effect was obtained when measuring the impact of oily fish intake on T2D risk, suggesting that ω-3 PUFA in foods may exert more protective effects than when extracted and given as supplements (Zhang et al., 2013). Moreover, dose-response analysis suggested that every 80 g/day intake of oily fish may reduce the risk of T2D by up to 20%.

Whereas the protective effect of ω-3 and/or fish consumption on the future incidence of T2D remains unclear, chronic dietary treatments with fish proteins in rodents and human subjects were shown to exert many beneficial effects on the lipid profile, insulin sensitivity, and inflammation. Dietary intervention with human subjects showed that the substitution of most proteins by fish fillets have resulted in decreased very-low density lipoproteins triglycerides (TG) in women (Gascon et al., 1996) and increased HDL2 in healthy and hypercholesterolemic men (Beauchesne-Rondeau et al., 2003; Lacaille et al., 2000). In a large multicentric study, an 8-week daily white fish consumption was reported to decrease waist circumference, serum low-density lipoproteins (LDL) levels, and diastolic blood pressure in subjects with MetS when compared with no-fish consumption (Vazquez et al., 2014). A fish protein supplement of 3–6 g/day was also reported to improve glucose homeostasis, body composition, and reduce LDL cholesterol in overweight adults (Vikoren et al., 2013). The impact of fish proteins and peptides on blood pressure may be linked to inhibition of angiotensin I-converting enzyme (ACE) activity.

Indeed, several studies have shown that fish protein hydrolysates and peptides derived from shrimp, sardine, bonito, salmon, and other marine sources can inhibit ACE activity in vitro (Astawan et al., 1995; Ewart et al., 2009; Fujita and Yoshikawa, 1999; Nii et al., 2008) and, in some cases, these in vitro data were confirmed in vivo as determined by blood pressure lowering effects in spontaneously hypertensive rats (SHR) or Wistar rats (Ait-Yahia et al., 2003; Ait-Yahia et al., 2005; Astawan et al., 1995; Ewart et al., 2009; Fujita and Yoshikawa, 1999; Nii et al., 2008; Otani et al., 2009).

In overweight/obese insulin-resistant men and women, a significant improvement in insulin sensitivity and reduced low-grade inflammation, as revealed by C-reactive protein (CRP) levels, was observed after a 4-week consumption of cod protein from cod fillets (Ouellet et al., 2007, 2008). Our group also showed that the combined effect of fish gelatin hydrolysate (FGH) and ω-3 PUFAs in vitro decreases tumor necrosis factor-α (TNF-α) expression as compared to ω-3 PUFAs or FGH alone in human macrophages (Rudkowska et al., 2010). We further reported that FGH potentiated the effects of ω-3 PUFAs on plasma TG in insulin-resistant women and on CRP levels in insulin-resistant men, suggesting sex-dependant lipid and anti-inflammatory responses to fish proteins (Picard-Deland et al., 2012). We further showed that various fish hydrolysates isolated from mackerel, bonito, herring, and salmon had anti-inflammatory properties, as revealed by decreased levels of the pro-inflammatory cytokines interleukin 6 and TNF-α in visceral adipose tissue of high-fat fed rats (Pilon et al., 2011). However, the salmon protein hydrolysate also reduced body weight gain and adiposity and improved insulin sensitivity, and the former effect was possibly due to its calcitonin content (Pilon et al., 2011). Also, dietary cod, as compared to soy or casein, reduced fasting and postprandial glucose and insulin responses in rats and peripheral insulin sensitivity during a clamp (Lavigne et al., 2000). The mechanisms of action behind these positive physiological effects were later shown to be mediated by improved skeletal muscle insulin resistance by normalizing insulin activation of the phosphatidylinositol 3-kinase/Akt-protein kinase B (PI3K/Akt) pathway and by improving the cell surface recruitment of GLUT4 transporters in skeletal muscle (Lavigne et al., 2001; Tremblay et al., 2003). The unique AA content of dietary cod protein may explain the augmented skeletal muscle glucose uptake in vivo as it was also shown to enhance insulin-stimulated glucose uptake in myocytes (Lavigne et al., 2001).

Dietary cod also improved triglyceride clearance in these rats (Lavigne et al., 2000), similarly to the improved lipid profile in rabbits fed with the same dietary protein sources (Bergeron et al., 1992a,b; Bergeron and Jacques, 1989). A study that examined the combination of either source of protein (casein or cod) with lipid (beef tallow or menhaden oil) showed that the combination cod protein-menhaden oil resulted in better lipid homeostasis, that is, lower plasma TG, triglyceride secretion rates, hepatic TG, plasma cholesterol, and hepatic cholesterol, as compared to the casein-beef tallow diet (Demonty et al., 2003). The potential mechanisms of action behind these lipid-lowering effects of fish hydrolysates are numerous. Some suggest modulation of acyl-CoA:cholesterol acyltransferase activity (Wergedahl et al., 2004), bile acid metabolism (Hosomi et al., 2011; Liaset et al., 2009, 2011; Matsumoto et al., 2007), or changes in the expression of gene involved in hepatic lipid synthesis (Bjorndal et al., 2013).

Growing evidence that fish proteins and hydrolysates can improve the MetS in animal models and humans lead us to search for bioactive peptides that may contribute to explaining these beneficial effects on insulin sensitivity, inflammation, and other aspects of the MetS. We recently showed that the substitution of casein hydrolysate (CH) for a low-molecular weight (<1 kDa) bioactive salmon peptide fraction (SPF) improved glucose tolerance and hepatic insulin signaling through Akt activation and the mechanistic target of rapamycin (mTOR)/S6K1/IRS1 pathway in a mouse model of the metabolic syndrome. Moreover, combination of SPF to dietary ω-3 PUFA-rich was shown to exert additive effects to reduce visceral adipose tissue inflammation, plasma nonesterified fatty acid levels, and hepatic insulin signaling (Chevrier et al., 2015). Moreover, the glucoregulatory and anti-inflammatory effects observed in the mouse model in vivo were confirmed in vitro, using relevant cell lines, implicating some cell-autonomous activity of this low-molecular weight peptide fraction for preventing the MetS. The overall impact of fish proteins, hydrolysates and peptides in preclinical models and human subjects are summarized in Fig. 18.1.

18.1.2.2 *Vegetable-Derived Proteins and Peptides: The Case of Legumes, Pulses and Soy*

Widely consumed in many regions of the world where they are a diet staple, legumes consist of seeds such as peanuts, soybeans, lupins, fresh peas, and beans, as well as pulses, which are defined as "dry seeds of leguminous plants which are distinguished from leguminous oil seeds by their low fat content" according to the Food and Agricultural Organization (FAO) (FAO, 2007). All are a good source of complex carbohydrates, dietary soluble and insoluble fibers, vitamins, minerals, and other nutritional compounds in the phytochemical family

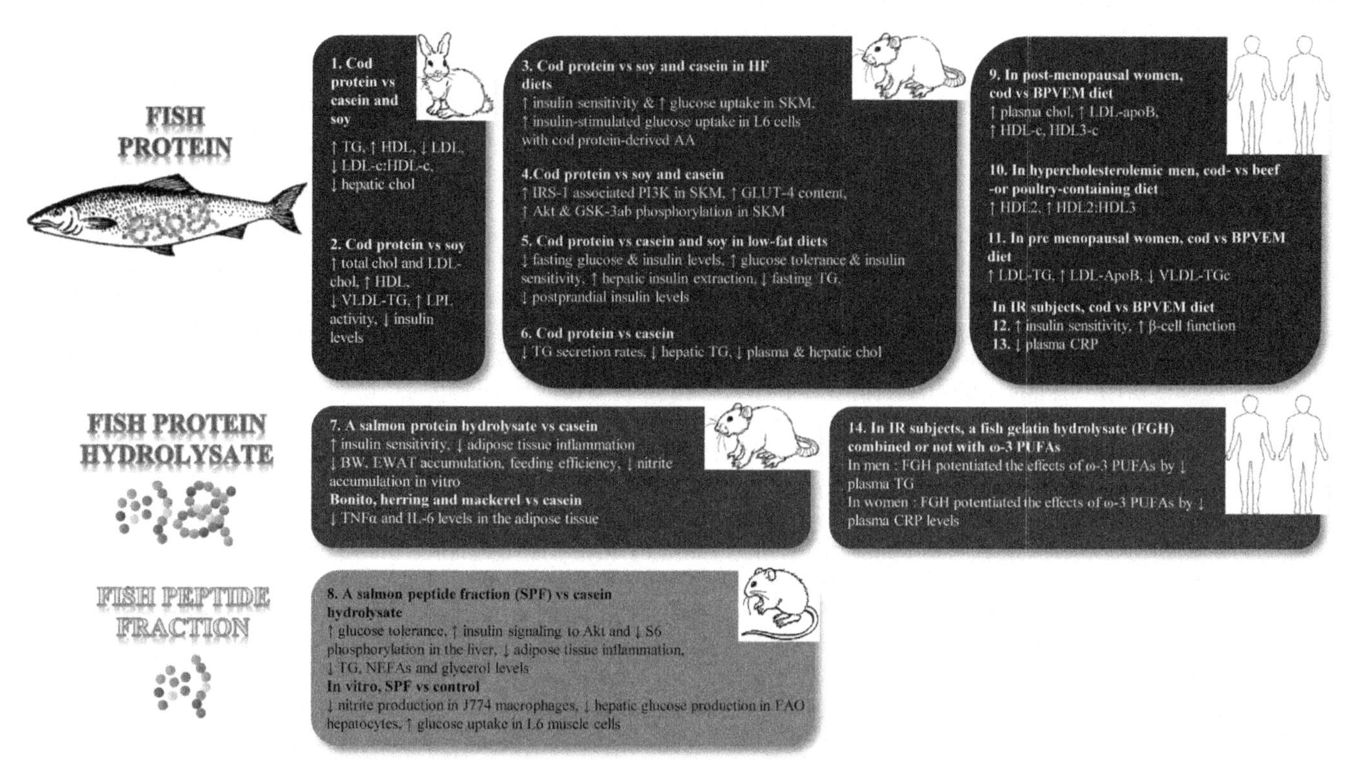

FIGURE 18.1 Summary of in vivo and clinical studies carried out with fish protein, fish protein hydrolysate, and fish peptide fraction in our extended research team. AA, Amino acids; Akt, protein kinase B; ApoB, apolipoprotein B; BPVEM diet, diet containing beef, pork, veal, eggs, and milk products; BW, body weight; chol, cholesterol; CRP, C-Reactive protein; EWAT, epidydimal adipose tissue; GLUT-4, glucose transporter 4; HDL, high-density lipoproteins; IL-6, interleukin 6; IR, Insulin resistant; IRS-1, insulin-receptor substrate 1; LDL, low-density lipoproteins; LPL, lipoprotein lipase; NEFAs, nonesterified fatty acids; PI3K, phosphoinositide 3-kinase; PUFA, polyunsaturated fatty acids; S6, S6 protein; SKM, skeletal muscle; TG, triglycerides; TNFα, tumor necrosis factor alpha; VLDL, very-low density lipoproteins. *Source: From relevant references for Fig. 18.1: (1) Bergeron et al., Atherosclerosis 1989 Aug;78(2–3):113–21; (2) Bergeron et al., J. Nutr. 1992 Aug;122(8):1731–71992; (3) Lavigne et al., Am. J. Physiol. Endocrinol. Metab. 2001 Jul;281(1):E62–714; (4) Tremblay et al., Diabetes 2003 Jan;52(1):29–37; (5) Lavigne et al., Am. J. Physiol. Endocrinol. Metab. 2000 Mar;278(3):E491–500; (6) Demonty et al., J. Nutr. 2003 May;133(5):1398–402; (7) Pilon et al., Metabolism 2011 Aug;60 (8):1122–30; (8) Chevrier et al., J. Nutr. 2015 Jul;145(7):1415–22; (9) Jacques et al., Am. J. Clin. Nutr. 1992 Apr;55(4):896–901; (10) Beauchesne-Rondeau et al., Am. J. Clin. Nutr. 2003 Mar;77(3):587–93; (11) Gascon et al., Am. J. Clin. Nutr. 1996 Mar;63(3):315–21; (12) Ouellet et al., Diabetes Care. 2007 Nov;30(11):2816–21; (13) Ouellet et al., J. Nutr. 2008 Dec;138(12):2386–91; (14) Picard-Deland et al., J. Nutr. Sci. 2012 Oct 23;1:e15.*

(Bouchenak and Lamri-Senhadji, 2013), and have drawn a lot of attention in the past few years because of the consumer's demand for healthy foods.

As a distinct part of the legume family, pulses have long been studied in human subjects for their beneficial properties against CVD, hypertension, cancer, obesity, and diseases within the digestive tract effects likely mediated by their high fiber-, resistant starch-, and nonnutritional phytochemical content as well as low-glycemic index (reviewed in Duranti, 2006). Pulses also contain low amounts of fat and are of interest for their excellent protein content, which varies between 22% and 38% of energy or 20% to 40% of dry weight depending on the species, making them one of the richest source of protein among foods along with meat (Lopez-Barrios et al., 2014; Messina, 1999). In a diet where pulses are properly combined with grains or nuts, they are a complete source of protein since their AA are complimentary (Duranti, 2006). Storage proteins such as globulins and glutelins represent the two major sources of protein in pulses with 70% and 10–20% of total protein, respectively (Roy et al., 2010). Other types of so-called antinutritional proteins and compounds, present in the seeds as a result of evolution for protection against a number of predators and environmental stresses, are found in uncooked beans, such as lectins, or hydrolysase or protease inhibitors. They may impact on nutrient absorption by inactivating digestive enzymes and they were even reported to induce growth retardation in animals or other negative and to cause unpleasant digestive symptoms when injected in humans (reviewed in Duranti, 2006; Roy et al., 2010). Fortunately, these antinutritional proteins and compounds have a positive impact on disease prevention when properly cooked, and may act as possible anticancer treatments via various mechanisms of action (Duranti, 2006). Contrary to soybean-derived peptides which have drawn a lot of attention in the past few years, research on peptides from pulses and legumes is scarce and has focused on characterization and in vitro therapeutic potential of hydrolysates and peptides from pea, mung bean and chickpea, which have been investigated for their inhibitory effects on ACE activity, calmodulin-dependent enzyme, and copper-chelating activity (Aluko, 2008; Humiski and Aluko, 2007; Vermeirssen et al., 2005), as well as antioxidant capacity (Girgih et al., 2015).

Soy protein was previously believed to have a positive impact on risk factors for cardiovascular diseases. However in 2006, the American Heart Association Nutrition Committee compiled 22 randomized trials and concluded that soy proteins with isoflavones only slightly reduced LDL cholesterol but showed no impact on HDL cholesterol, TG, lipoprotein(a), or blood pressure (Clifton, 2011; Sacks et al., 2006). Even with these new findings, the FDA approved food-labeling health claims in the United States and several other countries have followed since then, leading to huge increases in consumption and production of soy-derived foods (Xiao, 2008). It is the synergistic effect of soy protein with other compounds, such as isoflavones, fiber, saponins, minerals, and phytic acid, that may explain its mild effects on cholesterolemia (reviewed in Blachier et al., 2010). Their mechanisms of action include an increased cholesterol clearance by hepatic LDL receptors, changes in hepatic cholesterol metabolism, and a decrease in intestinal cholesterol absorption and biliary salts (Blachier et al., 2010).

Furthermore, while some evidence shows that soybean products could lower the risk of developing T2D (Villegas et al., 2008), fermented soybean products generally have a more favorable impact on obesity-linked features such as glucose and lipid homeostasis and body fat accumulation in rodents (Kim et al., 2013; Kwak et al., 2012; Oh et al., 2014) and human subjects (Cha et al., 2012, 2014), effects likely mediated by their enhanced bioactive peptide and isoflovanoid content generated during fermentation (Kwon et al., 2010). These fermented products are largely consumed in Asian countries and are traditionally produced by letting cooked soybeans form into a block ferment outdoors for 20–60 days with natural microorganisms. It appears that functional and nutritional properties have been attributed to the small molecules produced by the enzymatic hydrolysis during fermentation, that is, peptides, AA, fatty acids, and sugars, as well as from two major isoflavones, genistein and daidzein, that result from long-term fermentation.

Another way to yield bioactive peptides, derived or not from soy, is by using the electrodialysis technology (EDUF), which gives an electrical charge to the peptides to make them either anionic or cationic and separate them according to their molecular weight (reviewed in Roblet et al., 2014). With previous work showing that the net charge, the size, and the recovery of peptides were important factors for the bioactivity of peptides, our group aimed to isolate soy-derived bioactive peptides with the EDUF method. We found that the recovered 300–500 Da soybean fractions increased insulin-stimulated glucose uptake in muscle cells and the phosphorylation of adenosine monophosphate activated protein kinase (AMPK), an important protein involved in proper insulin signaling (Roblet et al., 2014).

With their high content of fibers, minerals, unsaturated fatty acids, and high-quality protein, fermented soy products and legumes are nourishing and attractive foods to put in our plates.

18.1.2.3 *Dairy Proteins and Peptides*

Milk protein is comprised of 80% (wt/wt) casein along with ~20% whey proteins and minerals. The components of casein include alpha-s1, alpha-s2, beta and kappa-casein while whey has many globular proteins, enzymes, and growth factors (McGregor and Poppitt, 2013). The AA profile of casein and whey differ substantially as well: casein is rich in several nonessential AA while whey contains a high proportion of BCAA, aromatic amino acids (histidine and phenylalanine), and methionine (McGregor and Poppitt, 2013). Milk also has other valuable components such as oligosaccharides, immunoglobulins, specific lipids, vitamins, and minerals.

Milk proteins are of excellent nutritional value as they have a high metabolic utilization by the organism (Bos et al., 2000). We could take the example of a fermented dairy product, yogurt, which has gained considerable attention in the past 30 years. Already full of nutritional benefits, yogurt has gained even more attention since the introduction in 1998 of the Greek or Greek-like yogurt types, on the shelves of supermarkets in the United States (Palmer, 2011), partly due to its high-protein content as well as its purported superior satiety effects compared to other foods and beverages (Chapelot and Payen, 2010; Douglas et al., 2013; Tsuchiya et al., 2006). Bacterial predigestion of milk proteins by proteolytic enzymes and peptidases and a finer coagulation of casein during fermentation of yogurt both increase its content of peptides and AA during shelf time, making its proteins easily digestible even when compared to milk (Adolfsson et al., 2004). While the intestinal availability of nitrogen of milk and yogurt is similar, yogurt lowers the gastric emptying rate, a phenomenon possibly due to its viscosity and consistency (Gaudichon et al., 1994, 1995).

The milk-derived peptides can be released either during gastrointestinal digestion from intact proteins or during microbial fermentation with starter bacteria used in yogurt and cheese processes. Once liberated and absorbed, these bioactive peptides may exert a physiological effect on various systems of the body including the cardiovascular, gastointestinal, musculo-skeletal, endocrine, and immune and nervous systems (Agyei and Danquah, 2012; Phelan and Kerins, 2011). In particular, the sequence of several ACE-inhibitory bioactive peptides derived from milk has been identified and some have been commercialized after their effects were demonstrated in SH rats and, in some cases, in hypertensive humans (Hsieh et al., 2015). Other bioactive peptides have been isolated from fermented milks and commercial dairy products, through the action of lactic acid bacteria (LAB) or their proteinases (Phelan and Kerins, 2011). Strain selection is one of the main factors that influence their release in fermented foods. Among LAB, *Lactobacillus helveticus* has been shown to exhibit strong proteolytic activity in milk-based media. A study showed that casein-enriched milk fermented by *Lb. helveticus* contained higher ACE-inhibitory activity and significantly decreased mean arterial blood pressure in SHR rats. These effects were related to peptides released from casein during milk fermentation, which were possibly further hydrolyzed by digestive enzymes in gastrointestinal tract (Leclerc et al., 2002). The presence of ACE-inhibitory peptides was also reported in several ripened cheeses (Butikofer et al., 2008; Saito et al., 2000) and some studies suggested that ACE-inhibiting activity and the concentration of the bioactive peptides increased with cheese ripening (Haque and Chand, 2008).

While some bacterial species acting as probiotics in yogurt have already been shown to diversify the gut microbiota (Alvaro et al., 2007; Garcia-Albiach et al., 2008) and have antiobesity potential (reviewed in Arora et al., 2013), dairy-derived peptides, hydrolysates and AA mixtures may also benefit intestinal health by increasing goblet cell counts as well as mucin gene expression and production (reviewed in Hsieh et al., 2015), that is, epithelial glucoproteins capable of forming a gel-like substance that protects the epithelium against assaults from different toxins (Faure et al., 2006). These results were also observed in a study after administration of threonine, proline, serine, and cysteine, the limiting AA for mucin production, into dextran sodium sulfate-treated rats, a model of intestinal inflammation (Faure et al., 2006). Moreover, it was speculated that the synergy of dairy-containing bioactive peptides, BCAA, and calcium may explain the superior weight-reducing effects of high-dairy diets as compared to high-calcium consumption, as it was demonstrated in several randomized clinical trials and in vivo studies (reviewed in Zemel, 2005).

While randomized controlled trials have shown discordant findings on the impact of yogurt on weight loss and weight maintenance (reviewed in Jacques and Wang, 2014), yogurt consumption is generally associated with better diet quality score/dietary patterns and improved metabolic profile (Cormier et al., 2016; Wang et al., 2013). These findings are concordant with a prospective study showing that consumption of yogurt, whole grain, fruits, nuts, and vegetables were associated with a lesser weight gain over a 4-year period in nonobese subjects. The authors suggest that these individuals may have weight-influencing behaviors; alternatively, they may indirectly lower intake of other energy-dense or processed foods (Mozaffarian et al., 2011). Another prospective study showed that high intake of yogurt was associated with lower systolic blood pressure and Homeostatis model assessment of insulin resistance (HOMA-IR), and lower fasting glucose, insulin, and TG levels, even when adjusted for diet quality score (Wang et al., 2013). In addition, the beneficial effect of yogurt consumption on weight management and metabolic syndrome features may be linked to beneficial components such as bioactive peptides released by fermentation, minerals as well as probiotics that may help prevent obesity-related dysmetabolism through modulation of the gut microbiota and gut health.

18.1.3 The Role of Bioactive Peptides in the Metabolic Effects of Dietary Proteins

Several laboratories have spent considerable time and effort to identify new bioactive peptides which may have potent physiological actions beyond the proteins that they are derived from, and that can also provide a means to concentrate the benefits of a given food not otherwise much consumed or appreciated (for instance, fish) into a functional ingredient form. Peptides can also be added to specific foods by alterations of their manufacturing processes to provide a nutritional added value for the consumer. Food-derived peptides may also be low-cost alternatives to drugs, without the usual side effects (Rutherfurd-Markwick, 2012).

However, several precautions need to be taken when evaluating the potency of so-called bioactive peptides, especially with the use of in vitro cell lines. Their efficacy at the cell level in an animal model or a human subject may be very different than in a petri dish, because of the influence of many factors during digestion (Rutherfurd-Markwick, 2012). For instance, the enzymatic degradation in the gastrointestinal tract may further hydrolyze peptides into smaller peptides or AA. The gut bacteria can also breakdown some peptides, use AA for their own needs, or synthetize AA from molecules already generated in the gut (see section on gut microbiota). Moreover, interindividual differences may modulate the bioactivity of the ingested peptides (Sato et al., 2008), thus not yielding the same expected physiological response. Conversely, inactive peptides in the petri dish could alternatively become bioactive once in the circulation (Sato et al., 2008). Thus, to make sure that bioactivity of a given peptide is the same in vitro and in vivo or in a clinical setting, characterization of the peptide sequence and identification of the same peptides in the circulation after ingestion is required using sophisticated analytical techniques (Sato et al., 2008). A valid alternative is to determine the AA plasma profile of animals and reproduce their relative concentrations on cultured cell models as a way to assess their in vivo effects (Lavigne et al., 2001).

Notwithstanding the above, some bioactive peptides already identified in vitro may be absorbed intact by the intestine, partly due to their resistance to proteolysis by digestive enzymes (Rutherfurd-Markwick, 2012). Furthermore, several early studies have shown that absorption of AA and nitrogen depends on the chain length of the peptide mixture (reviewed in Grimble et al., 1987). Di- and tripeptides have indeed a greater intact absorption as compared to higher molecular weight peptides, possibly because their hydrolysis into di- and tripeptides at the intestinal brush border is mandatory for absorption (Grimble et al., 1987). Moreover, the transport capacity of di- and tripeptides is greater than that of AA carrier system, making AA absorption rate faster with ingestion of low-molecular-weight protein hydrolysates than from AA mixtures. Absorption of di- and tripeptides is mediated by the PEPT1 and PEPT2 transporters in the intestinal and renal cells (Daniel and Kottra, 2004). Oral administration of 50 g of fish hydrolysate in the form of small peptides induced a higher and faster increment of plasma AA than in the form of a mixture of AA (Silk et al., 1979). Similarly, hospitalized patients receiving either a diet containing 60% small peptides or an isocaloric diet of nondegraded proteins had higher absorption of AA with the first diet (Ziegler et al., 1990). We have also observed that the higher the content of CH in a diet (ie, its content of small peptides) as compared to whole casein, the better was the metabolic profile of these animals. As illustrated in Figs. 18.2 and 18.3, mice fed a high-fat plus sucrose diet containing 20% CH had lower body weight gain, liver weight, food efficiency ratio, and significant improvements in insulin sensitivity as determined during an hyperinsulinemic-euglycemic clamp, as compared to those fed 20% nonhydrolyzed casein.

18.1.4 A New Role for the Gut Microbiota in AA Metabolism and the Modulation of Immunometabolism

We have seen that many factors can modulate or even compromise the bioactivity of peptides and the availability of AA after passing through the digestive system. The gut microbiota is certainly a major player in this regard, and its implication in the catabolism and metabolism of AA is likely contributing to the pathogenesis of obesity-associated diseases and in particular T2D.

The gut microbiota is now well recognized for its implication in the development of insulin resistance and obesity-linked inflammation (reviewed in Cani et al., 2012). Perhaps the most compelling evidence for its key role in metabolic diseases came from the finding that transplantation of cecum bacterial content from obese mice into lean, germ-free (GF) mice was found to induce obesity in the recipient mice (Backhed et al., 2004; Turnbaugh et al., 2006, 2008). Conversely, GF mice lost weight after receiving a transplantation of the gut microbiota from their counterparts who had a Roux-en-Y gastric bypass surgery (Liou et al., 2013). It is also well known that gut dysbiosis induced by feeding a HF diet raises levels of *lipopolysaccharides* and induces metabolic endotoxemia and systemic inflammation (Cani et al., 2008). On the other hand, dietary fibers (Neyrinck et al., 2012; Serino et al., 2012), prebiotics (Cani et al., 2009; Everard et al., 2013), and polyphenol-rich extracts from various sources (Anhe et al., 2015; Neyrinck et al., 2013; Roopchand et al., 2015) were reported to positively modulate the gut microbiota and were able to prevent obesity and associated features of the metabolic syndrome.

The gut microbiota is also known to play an important role in the metabolism of peptides and AA (Davila et al., 2013). Indeed the three big families (*phyla*) of intestinal bacteria, *Actinobacteria*, *Bacteroidetes*, and *Firmicutes*, all have a role to play in the degradation of AA (Turnbaugh et al., 2009). Once hydrolyzed by proteases, then peptidases from pancreatic secretions, proteins become polypeptides, then peptides and AA (Do et al., 2014). The AA may be transferred across the brush membrane border by several transporters, or used by the enterocytes and the gut bacteria to fulfill their own metabolic requirements (Davila et al., 2013; Do et al., 2014). While limited evidence suggests that gut bacteria from the small intestine may play a role in AA metabolism, the large intestine has a higher bacterial density and a better capacity to utilize remaining luminal AA (Davila et al., 2013). Indeed, several bacterial species in the distal gut express genes required for AA synthesis (Gill et al., 2006). Some

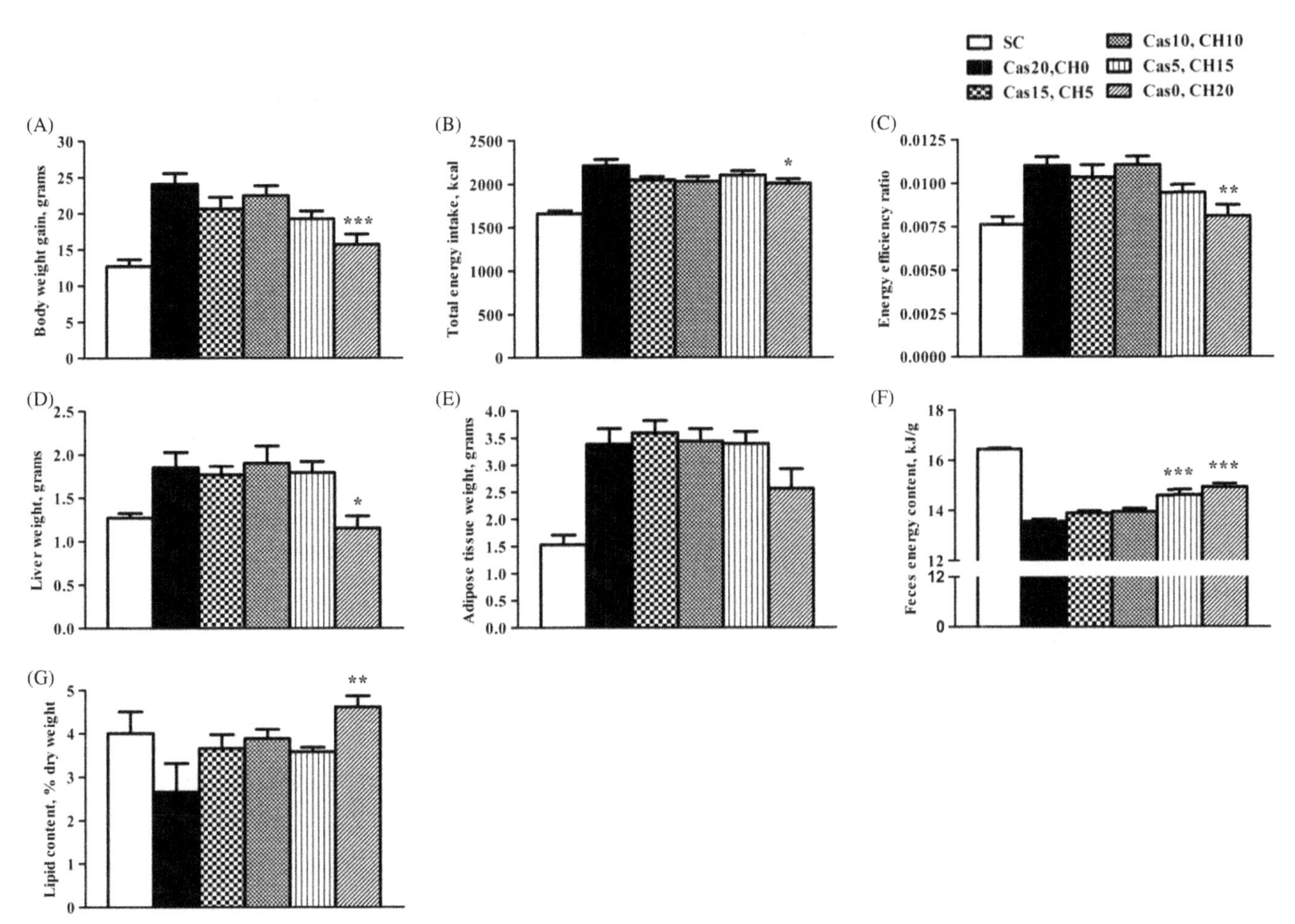

FIGURE 18.2 Effects of casein hydrolyzation on body weight gain, energy intake, energy efficiency, organ weight, and feces energy and lipid content after 24 weeks of dietary treatment in male $LDLr^{-/-}/ApoB^{100/100}$ mice. $LDLr^{-/-}/ApoB^{100/100}$ mice were fed either a standard chow (SC) diet or one of five high-fat, high-sucrose diets (HFS) for 6 months. All HFS diet contained the same amount of protein (200 g/kg), although they varied in the hydrolyzation of their casein. Indeed the HFS diets differ in their casein (Cas: 20, 15, 10, 5 or 0% protein wt/wt) and casein hydrolysate (CH: 0, 5, 10, 15, 20% protein wt/wt) content. The dietary groups are: SC; HFS control (Cas20, CH0); Cas15, CH5; Cas10, CH10; Cas5, CH15 and Cas0, CH20. Several physiological parameters such as body weight gain (A), total energy intake (B), energy efficiency ratio (body weight gain/total energy intake) (C), and liver weight (D) were decreased in the group fed the highest casein hydrolysate content (Cas0, CH20), while adipose tissue accumulation (E) showed the same tendency but did not reach statistical significance. These results may be explained in part by the higher energy and lipid content in the feces of these mice (F-G). $n = 10-13$ for all except for feces energy content (F) and feces lipid content (G), $n = 3-5$. Statistical analysis was performed with GraphPad Prism, and data were analyzed using a one-way ANONA with a Dunnet post hoc test. The SC group was not included in the statistical analysis since it represented a reference group to show the presence of diet-induced obesity. Data are means ± SEM. Results were considered significant when $P < 0.05$. $^{*}P < 0.05$, $^{**}P < 0.01$, $^{***}P < 0.001$ versus CAS20, CH0.

evidence also shows that distal microbiota synthetizes essential AA with the help of glutamate and glutamine, as well as with molecules generated from the intermediary metabolism such as oxaloacetate, pyruvate, α-oxoglutarate, and urea (Davila et al., 2013). Depending on their type, AA also undergo processes in the large intestine, such as transamination, deamination, decarboxylation, and desulfhydration, yielding various compounds, such as short-chain fatty acids (SCFAs) (propionate, butyrate and acetate), branched-chain fatty acids (BCFAs) (valerate, isobutyrate and isovalerate), ketoacids, saturated fatty acids, organic acids (lactate, formate and succinate), amines, polyamines, indolic and phenolic compounds, ethanol, gazes (H_2 and CO_2), all of which can have an impact on gut homeostasis and metabolic health (Davila et al., 2013).

Di- and tripeptides are taken up by the transporter PEPT1 in the small intestine and colon, which is also responsible for most nitrogen absorption in the body as described earlier (Do et al., 2014). Interestingly, an increase in the expression of PEPT1 and other AA transporters after a HF diet was linked to the modulation of the gut microbiota, an increase in intestinal permeability, transit time, and plasma AA levels (Do et al., 2014).

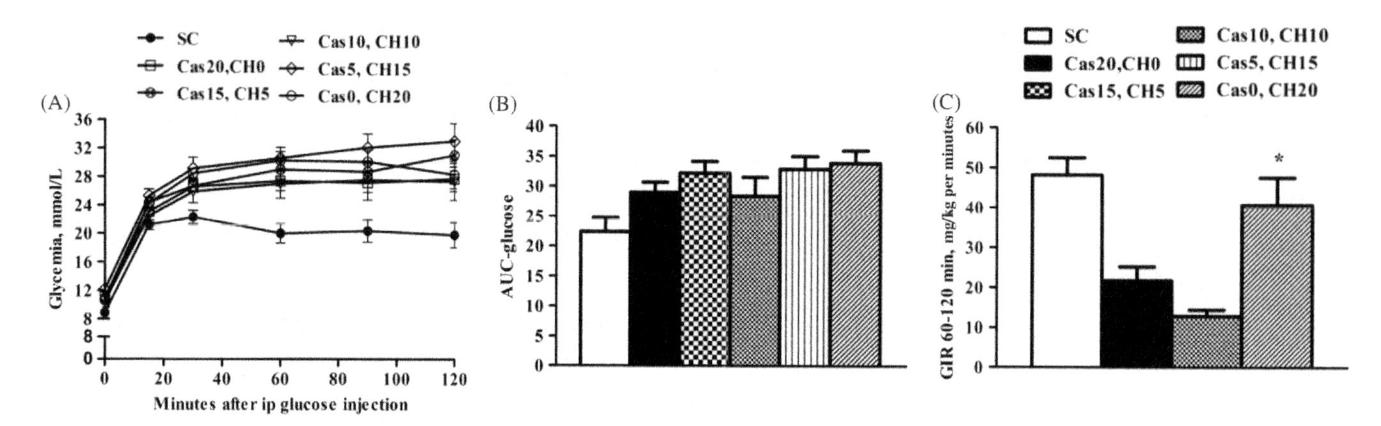

FIGURE 18.3 Effects of casein hydrolyzation on glucose homeostasis and insulin sensitivity after 24 wks of dietary treatment in male LDLr$^{-/-}$/ApoB$^{100/100}$ mice. Despite lower body weight gain, glucose tolerance tests (GTT) and total area under the GTT curves (A-B) were not different between the groups ($n = 10-13$). The hyperinsulinemic-euglycemic clamp technique was performed on a subset of animals ($n = 4-5$), and the results show that the glucose infusion rate (GIR), an index of insulin sensitivity, was increased in the Cas0, CH20 diet group, for which the animals consumed their entire protein consumption under the form of hydrolyzed casein (CH). Statistical analysis was performed with GraphPad Prism, and data were analyzed using a one-way ANONA with a Dunnet post hoc test. The SC group was not included in the statistical analysis since it represented a reference group to show the presence of diet-induced obesity. Data are means ± SEM. Results were considered significant when $P < 0.05$. *$P < 0.05$ versus CAS20, CH0.

Few studies have explored the effects of an HP diet or of different sources of proteins on the gut microbiota. It has been reported that an HP diet (53% protein) induces higher colonic levels of SCFAs, BCFAs, organic acids, and ethanol in rats compared to their counterparts fed an NP diet (14% protein) (Liu et al., 2014). The HP diet also reduced the richness of gut microbiota as well as the gene copy numbers of several cecal and colonic groups and species of bacteria (Liu et al., 2014). In another study, rats fed diets made of hydrolyzed casein were shown to have a lower incidence of type 1 diabetes which was associated with a shift in the gut microbiota, as revealed by increased abundance of *Lactobacilli* and decreased presence of *Bacteroides* (Visser et al., 2012).

Altogether, these findings support the concept that peptide and AA metabolism by the gut microbiota have an impact on health outcomes which is partly dependent on AA-generating molecules. There is clearly a need for further research in this field in order to better understand the influence of dietary proteins, peptides, and AA on the abundance and diversity of the gut microbiota, but also to advance our knowledge on how the gut microbiota can influence the metabolic faith of dietary proteins and peptides on the bioavailability and bioactivity of several AA in health and diseases.

18.1.5 Effects of AA on Metabolic Control and Cellular Signaling Pathways

18.1.5.1 The Effects of AA on Insulin and Glucagon Secretion

Once in circulation, AA may exert several functions: they can modulate specific actions at the level of the pancreas or participate in the modulation of metabolic processes such as ketogenesis, gluconeogenesis, or glycogenolysis. They can also bind to specific receptors at the surface of cells and induce profound changes in various signaling pathways.

An early study from the 1950s showed for the first time that leucine could be an insulin secretagogue as its administration induced hypoglycemia in patients with familial idiopathic hypoglycemia (Cochrane et al., 1956). While glucose is obviously the most powerful insulin secretagogue, studies have demonstrated that proteins and AA can also exert insulinotropic actions. Early studies have first demonstrated the insulin-producing effect of protein meals (Floyd et al., 1966) or the enhanced insulinotropic effects of combined protein and glucose in healthy (Krezowski et al., 1986; Pallotta and Kennedy, 1968; Rabinowitz et al., 1966; van Loon et al., 2000) and type 2 diabetic patients (Nuttall et al., 1984), over that of either protein or glucose alone. Since then AA have been recognized as activators of insulin secretion in the β-cell via several mechanisms of action, for instance by activating glutamate dehydrogenase or the mTOR signaling pathway, or by modulating the opening of ion channels (reviewed in Azzout-Marniche et al., 2014).

While low plasma glucose is the main contributor to glucagon release by α-cells, some AA (eg, alanine, glutamine, arginine, and glycine) can also stimulate glucagon release (Gannon et al., 2002; Marroqui et al., 2014;

Pipeleers et al., 1985). AA are especially effective at provoking glucagon response when used in combination (Cheng-Xue et al., 2013; Pipeleers et al., 1985; Quoix et al., 2009). When released, glucagon binds to its receptor in the liver and induces gluconeogenesis or glycogenolysis (Marroqui et al., 2014). It was demonstrated in an elegant study that, in healthy males, infusion of AA stimulated endogenous glucose production while increasing insulin stimulated glucose uptake, thereby maintaining stable glucose levels (Krebs et al., 2003). Furthermore, glucagon and somatostatin infusion induced significant elevation in plasma glucose, which was associated with increased glycogenolysis and gluconeogenesis (Krebs et al., 2003).

Insulin resistance, but more specifically T2D, are associated with elevations in glucagon and insulin secretion, which concomitantly dysregulate protein metabolism (Marroqui et al., 2014). Early stage T2D patients have increased pancreatic β-cell and α-cell masses, thus present hyperglucagonemia, hyperinsulinemia, and an impaired capacity to tightly regulate glucose levels. A defective hepatic insulin signaling underlies insulin resistance in liver, leading to impaired inhibition of hepatic glucose production, but a lack of suppression of plasma glucagon levels by hyperglycemia is also pathognomonic to T2D.

18.1.5.2 Altered BCAA Levels and Metabolism in Obesity and T2D

The BCAA valine, isoleucine, and leucine are essential AA and compose ~20–25% of dietary proteins (Harper et al., 1984), and are involved in the regulation of protein synthesis, insulin secretion, satiety, and glucose homeostasis. However, an early study revealed more than four decades ago that plasma levels of BCAA were higher in obese individuals as compared to age- and sex-matched lean controls and that they correlated with insulin levels (Felig et al., 1969). More recent studies using sophisticated metabolomics analyses have shown similar results not only in adult obese individuals but also in obese children, as well as in other insulin-resistant conditions, such as in MetS or in polycystic ovarian syndrome (Felig et al., 1974; McCormack et al., 2013; Newgard et al., 2009; Perng et al., 2014; Serralde-Zuniga et al., 2014; Tai et al., 2010; Wiklund et al., 2014; Zhao et al., 2012). Furthermore, prospective studies have established a link between BCAA levels and the future risk of developing T2D (Cheng et al., 2012; Lee et al., 2014; McCormack et al., 2013; Wang et al., 2011).

Several mechanisms of action may contribute to the increase in BCAA levels in obesity-related metabolic disorders. Possible hypotheses include the rate of appearance (food intake and proteolysis) and disappearance (protein synthesis and catabolism) of BCAA and other metabolically related AA. It is very pertinent to question whether overnutrition or a high consumption of BCAA and protein could contribute to the circulating plasma levels of BCAA and impair glucose metabolism as seen in obese and T2D subjects. However, it was reported that protein and BCAA intakes were not associated with BCAA levels in insulin-resistant subjects (Shah et al., 2012; Tai et al., 2010; Wang et al., 2011). Long-term epidemiological studies further revealed that high intake levels of BCAA were not related to a higher risk of developing diabetes later in life (Qin et al., 2011) and may even be protective (Nagata et al., 2013). When looking at the general impact of protein-rich diets (>20% of energy intake) on insulin resistance or T2D outcomes, the results from interventional and longitudinal studies remain inconclusive (reviewed in Rietman et al., 2014). On the one hand, protein-rich diets may help obese people lose weight, while on the other hand they may increase insulin secretion over time (Rietman et al., 2014). Moreover, many key variables vary between studies, that is, the type of population studied, the presence or absence of weight loss and the length of the study, making it difficult to assess with certainty whether protein-rich diets can increase the risk for insulin resistance and T2D (Rietman et al., 2014). An interesting hypothesis first proposed by some authors (Newgard et al., 2009) is the "BCAA overload," which suggests that overnutrition, comprising of high levels of fat and protein in obese humans, could overwhelm the BCAA catabolic system and lead to higher than normal BCAA levels in this condition.

On the other hand, plasma BCAA levels have been studied in the context of lower food intake, such as in starvation or protein malnutrition (Shimomura et al., 2001) and after gastric bypass surgeries (GBS), or also upon diet intervention and lifestyle modification programs (Kamaura et al., 2010; Laferrere et al., 2011; Lips et al., 2014; Magkos et al., 2013; She et al., 2007b). While weight loss-induced reductions in BCAA levels by diet intervention have been reported (Kamaura et al., 2010; Shah et al., 2012), others have not observed this effect despite improvement in glucose homeostasis (Lips et al., 2014). However, it seems that GBS, especially Roux-en-Y surgery, have more profound effects on BCAA levels and/or their derivatives, which even appear to be independent from weight loss (Laferrere et al., 2011; Lips et al., 2014).

Defective proteolysis and protein synthesis mechanisms have also been studied in the context of elevated BCAA levels. While it is well known that insulin is a key modulator of glucose and lipid metabolism (Sesti, 2006), it is also a key regulator of protein metabolism, through promoting protein synthesis and inhibiting proteolysis in healthy subjects (Liu et al., 2006). Whether alterations in both proteolysis and protein synthesis contribute to the usually higher BCAA and AA levels in T2D patients, however, remain to be firmly established. Whole-body protein synthesis is effective in hyperaminoacidemic conditions, in the presence or absence of hyperinsulinemia, in both nondiabetic and T2D patients (reviewed in Tessari et al., 2011). However, others have highlighted a higher protein turnover, a reduced protein synthesis and the restoration of protein metabolism at insulin concentrations required to normalize glycemia in obese and/or T2D patients (Gougeon et al., 1994, 1997, 1998, 2000, 2008; Pereira et al., 2008). As for proteolysis, it was found to be inhibited only when using higher insulin concentrations in T2D patients (reviewed in Tessari et al., 2011). Thus impairments in proteolysis and protein synthesis rates certainly contribute but may not fully explain the raised BCAA levels in obese, insulin-resistant, or T2D subjects.

Growing evidence indicates that BCAA catabolism may also account for altered BCAA levels in obesity and T2D subjects. This system is regulated by a specific set of enzymes and proteins: BCATm or BCATc (branched-chain amino acid aminotransferase mitochondrial or cytosolic), BCKDHC (branched-chain α-ketoacid dehydrogenase complex), BCKDK (branched-chain α-ketoacid dehydrogenase kinase), and protein phosphatase 1K (PPM1K or PP2CM). Unlike other AA, the BCAA bypass the liver (Brosnan and Brosnan, 2006) but are then predominantly catabolized by skeletal muscles and adipose tissues, via BCATm and BCKDHC (Brosnan and Brosnan, 2006). Their transamination leads to synthesis of BCKAs (branched-chain keto-acids) by BCATm or BCATc, generating α-ketoisovalerate (KIV) from valine, α-ketoisocaproate (KIC) from leucine, and α-keto-β-methylvalerate (KMV) from isoleucine. The second step of BCAA catabolism yields the irreversible decarboxylation of BCKAs by the BCKDHC, expressed in liver, adipose tissue, and skeletal muscle (Brosnan and Brosnan, 2006), and comprising of several components (Harris et al., 2005). BCKDHC is the rate-limited enzyme of the pathway and is inactivated by BCKDK by phosphorylation of BCKDHC on E1-α Ser293, thereby stopping BCKA catabolism and conserving BCAA for protein synthesis. Conversely, the BCKDH phosphatase (PP2cm) activates BCKDHC by dephosphorylation when BCAA are in excess (Shimomura et al., 2001). Oxidation of BCKAs in the second step leads to formation of either acetyl-CoA by leucine or propionyl-CoA or succinyl-CoA by isoleucine and valine, and these metabolites then participate in the citric acid cycle (Valerio et al., 2011). These acyl-CoAs generate acylcarnitines in the mitochondria that can be measured in the urine or in the plasma when mitochondrial oxidative capacity (or activity) is compromised, as reported in obese and T2D subjects (Huffman et al., 2009; Mihalik et al., 2010; Newgard et al., 2009).

Changes in BCAA catabolism have been studied in various nutritional and disease conditions and in different tissues (Shimomura et al., 2001). For instance, in maple syrup urine disease, defective genes encoding BCKDHC lead to reduced capacity for BCAA oxidation, raising levels of plasma BCAA, brain damage and seizures (Zimmerman et al., 2013). Interestingly, reduction in the gene or protein expression, and/or phosphorylation of several molecular components of BCKDHC were observed in the adipose tissue of genetic rodent models of obesity and in animals fed a high-fat diet (Herman et al., 2010; Lackey et al., 2013; She et al., 2007b, 2013). Conversely, bariatric surgery was found to normalize BCAA levels and adipose tissue protein expression in human subjects (She et al., 2007b). In this regard, obesity-related dysfunctions of adipose tissue appear to be involved in altered BCAA catabolism. This was demonstrated by transplanting adipose tissue from healthy, wild-type mice into BCAT2$^{-/-}$ mice characterized with defective BCAA catabolism capacity and thus with lower BCKAs and higher BCAA levels (Herman et al., 2010). The adipose tissue transplant lowered BCAA levels by 31 and 46% in the fast and the fed state, respectively, without changes in glycemia, insulin, or leptin levels (Herman et al., 2010). These results are in line with another study (Zimmerman et al., 2013), in which a subcutaneous transplantation of adipose tissue resulted in 51–82% reductions of BCAA levels in BCAT2$^{-/-}$ and PP2Cm$^{-/-}$ mice, the latter characterized by elevated BCKAs and BCAA (Zimmerman et al., 2013). These results demonstrate that adipose tissue is significant to BCAA oxidation and thus a major determinant of plasma BCAA levels.

It should also be mentioned that regional variations exist regarding BCAA metabolism by the adipose tissue. Indeed it has been demonstrated that the BCAA catabolism is altered in omental but not subcutaneous adipose tissues of obese adult female subjects (Boulet et al., 2015) and with MetS (Lackey et al., 2013), suggesting that visceral obesity promotes alterations in BCAA catabolism. In this regard, C57BL/6 mice treated with the PPARγ ligand rosiglitazone for 14 days exhibited elevated BCKD E1α protein abundance in retroperitoneal (visceral) fat whereas this effect was reduced in obese db/db mice. PPARγ agonists were also shown to increase adipose BCAA gene expression in adipocytes (Lackey et al., 2013) as well as in vivo in rats and human subjects treated with PPARγ ligands (Hsiao et al., 2011; Sears et al., 2009). Fig. 18.4 summarizes how different physiological, nutritional, genetic, and pharmaceutical conditions may play a role in modulating BCAA catabolism through the rate of appearance and disappearance of plasma BCAA.

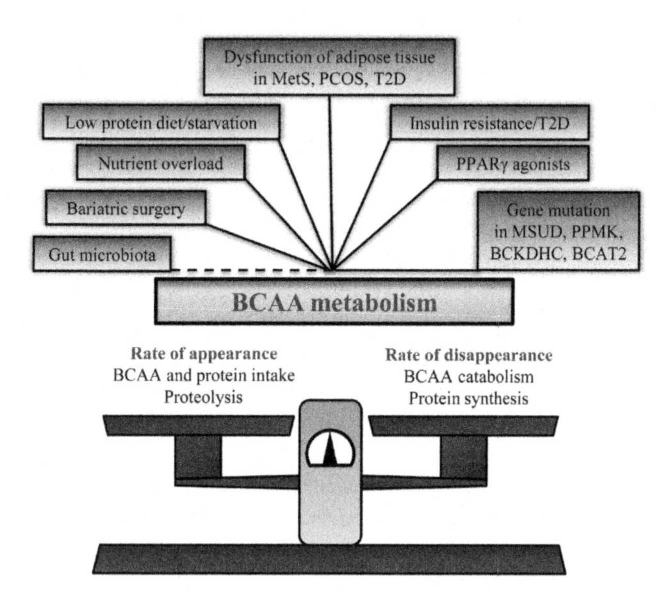

FIGURE 18.4 Different physiological, nutritional, genetic and pharmaceutical conditions that may play a role in modulating BCAA metabolism through the rate of appearance and disappearance of plasma BCAA. BCAA, Branched-chain amino acids; BCAT2, Branched-chain amino acid transferase 2; BCKDHC, branched-chain α-ketoacid dehydrogenase complex; MetS, Metabolic syndrome; MSUD, Maple syrup urine disease; PCOS, Polycystic ovarian syndrome; PPARγ, Peroxisome proliferator-activated receptor gamma; PPMK, Protein phosphatase K; T2D, Type 2 diabetes. The dotted line indicates that gut microbiota may possibly be involved in BCAA metabolism, but this question remains to be explored.

18.1.5.3 Role of AA in the Activation of Nutrient Sensing Pathways and Obesity-Linked Insulin Resistance and T2D

AA are transported into cells by the class-specific transporters located at the surface of the cell. While some of them act solely as transporters, others are referred to as transceptors since their binding with an AA activates an intracellular signaling cascade (reviewed in Taylor, 2014). The transceptors help "sense" the pool of AA outside and inside the cell, determining AA abundance leading to activation of two important sensing systems: the mechanistic target of rapamycin complex 1 (mTORC1), and the general control nonderepressible (GNC) pathways. While the GNC pathway is activated when intracellular AA are scarce, activation of the mTOR pathway occurs with abundance of certain AA, especially BCAA.

mTOR is a serine/threonine protein kinase that is involved in the operation of two signaling complexes, mTORC1 and mTORC2. The two mTOR complexes differ by their components, cellular regulation, their inhibition sensitivity to rapamycin, and thus control different yet overlapping functions. mTORC1 is composed of several components, including Raptor (the regulatory-associated protein of mTOR), mLST8 (mammalian lethal with Sec13 protein 8), PRAS40 (proline-rich AKT substrate 40 kDa), as well as Deptor (DEP-domain-containing mTOR-interacting protein). Most common mTORC1 downstream effectors include S6K (p70 S6 kinase), 4E-BP1 (eukaryotic initiation factor 4E-binding protein), and ULK1 that regulate proteins, nucleotides, and lipids synthesis, lysosome biogenesis and autophagy, as well as cellular growth and proliferation. mTORC2 includes mTOR, Rictor (rapamycin-insensitive companion of mTOR), mSIN1 (mammalian stress-activated protein kinase interacting protein), Protor-1 (protein observed with Rictor-1), mLST8, and Deptor, that regulate Akt and SGK1 (serum and glucocorticoid-induced protein kinase 1) (reviewed in Jewell et al., 2013; Laplante and Sabatini, 2012).

While the role of mTORC1 has been well-characterized, much less is known about mTORC2. The most known effect of mTORC2 is to regulate Akt Ser473 phosphorylation, giving it an important role in mediating growth factors action on cellular proliferation, growth, and metabolism. mTORC1 plays a major role in nutrient-sensing and protein synthesis. It is why this chapter focuses more on mTORC1 function since mTORC2 has been reviewed elsewhere (Oh and Jacinto, 2011).

mTORC1 is a master regulator of growth since it senses multiple stimuli, such as growth factors, stress, hypoxia, mechanical strain, and nutrients, notably AA, glucose, lipids, and also energy status. mTORC1 coordinates the nutrient and endocrine signals in the postprandial state and promotes protein turnover and cellular growth. Both AAs and growth factors activating mTORC1 involve small GTPases, respectively, Rag

GTPases and Rheb GTPase (Dibble and Cantley, 2015; Jewell et al., 2013). Rag (Ras-related GTPase) proteins are heterodimers of Rag A/B combined with Rag C/D. The RAG GTPase complex is inactive, with RagA or RagB binding with GDP (RagA/B-GDP) and RagC or RagD binding with GTP (RagC/D-GTP) and become activated with the inverted nucleotide form (RagA/B-GTP; RagC/D-GDP). AA promote the formation of the active configuration of the Rag GTPase complex that leads to the recruitment of inactive cytosolic mTORC1 and causes the localization of mTORC1 to the surface of lysosomes. Once recruited to the lysosome, mTORC is fully activated by binding Rheb-GTP, which is activated by growth factor stimuli.

Nutrient overload and high levels of insulin and AA lead to a sustained activation of mTORC1 (Huang and Manning, 2009), increasing the activity of the two key downstream effectors of mTORC1 involved in mRNA translation, that is, ribosomal S6K (S6 kinase) and eukaryotic translation initiation factor 4E-binding protein (4E-BP1). In a series of studies involving cellular and animal models of obesity- and nutrient satiation-linked insulin resistance, as well as AA-infused human subjects, we have previously demonstrated that overactivation of S6K1 is involved in a negative feedback loop to inhibit insulin signaling through inhibitory phosphorylation of IRS-1 on Ser1101 and other Ser sites, leading to impaired activation of PI3K/Akt and insulin resistance in skeletal muscle and liver (Khamzina et al., 2005; Tremblay et al., 2005, 2007; Tremblay and Marette, 2001; Veilleux et al., 2010). An acute AA infusion during a clamp in healthy men induced insulin resistance through reduction in glucose disposal and glycogen synthesis rates (Krebs et al., 2002), increased skeletal muscle activation of S6K1 and IRS1 (Krebs et al., 2007; Tremblay et al., 2005, 2007), and blunted PI 3-kinase activity (Tremblay et al., 2005).

Paradoxally, others have shown beneficial effects of certain AA and particularly leucine supplementation on metabolic disorders in HF-fed or in rodent models of obesity and insulin resistance. Many have reported significant improvements in blood glucose (Nairizi et al., 2009; Zhang et al., 2007), cholesterol levels (Torres-Leal et al., 2011; Zhang et al., 2007), adiposity (Li et al., 2012; Zhang et al., 2007), HbA1c (Guo et al., 2010), macrophage infiltration (Guo et al., 2010), adiponectin (Torres-Leal et al., 2011), and insulin sensitivity (Li et al., 2012) with leucine supplementation. Isoleucine was also reported in a few studies to lower blood glucose in normal and diabetic rats, and this could be linked to an insulin-independent effect on muscle glucose uptake as seen in vitro in myocytes (Doi et al., 2003). Importantly, while supplementation with BCAA of chow-fed animals increases C3 and C5 acylcarnitines levels, this was not accompanied by insulin resistance suggesting that BCAA reduce insulin action on glucose metabolism particularly in the context of an oversupply of plasma lipids when fed a HF diet or driven by a genetic mutation (Newgard et al., 2009). When it comes to the regulation of energy balance, leucine administration was shown to reduce food intake and promote body weight loss in rats by acting on the mediobasal hypothalamus (Blouet et al., 2009) and on the caudomedial nucleus of the solitary tract of the brainstem through the activation of the S6K1 signaling pathway (Blouet and Schwartz, 2012; Cota et al., 2006). Just as in rodent studies, a BCAA or AA mixture may help improve glucose homeostasis in clinical settings and reduce HbA1c levels in subjects with T2D, T1D, and/or with chronic viral liver disease (Kawaguchi et al., 2008; Solerte et al., 2008a,b). Infusion of a balanced cocktail of AA can also trigger hepatic glucose production but at the same time enhance insulin-stimulated glucose uptake, thereby maintaining glycemia (Krebs et al., 2003). These studies highlight some discrepancies between the effect of various individual AA or combinations of AA on insulin sensitivity and glucose metabolism and, clearly, more research is needed to fully understand the underlying mechanisms.

18.2. CONCLUSION

This chapter focused on the physiological and molecular impact of dietary proteins, peptides, and AA. Overall the review of the literature provides extensive and well-documented evidence that the amount of proteins in the diet as well as the type of dietary proteins modulate several features of the metabolic syndrome. Several studies also suggest that the metabolic effects of dietary proteins are mediated by small peptides and AA, involving a complex interaction with the gut microbiota and cross-talk between central and peripheral metabolic tissues. Further understanding of the underlying mechanisms of action by which dietary protein composition, bioactive peptides, and AA levels control endocrine functions, energy metabolism, and inflammation will help the design of novel diets and/or functional foods to prevent or alleviate obesity-linked diseases.

References

Adolfsson, O., Meydani, S.N., Russell, R.M., 2004. Yogurt and gut function. Am. J. Clin. Nutr. 80, 245–256.

Agyei, D., Danquah, M.K., 2012. Rethinking food-derived bioactive peptides for antimicrobial and immunomodulatory activities. Trends Food Sci. Technol. 23, 62–69.

Ait-Yahia, D., Madani, S., Savelli, J.L., Prost, J., Bouchenak, M., Belleville, J., 2003. Dietary fish protein lowers blood pressure and alters tissue polyunsaturated fatty acid composition in spontaneously hypertensive rats. Nutrition 19, 342–346.

Ait-Yahia, D., Madani, S., Prost, J., Bouchenak, M., Belleville, J., 2005. Fish protein improves blood pressure but alters HDL2 and HDL3 composition and tissue lipoprotein lipase activities in spontaneously hypertensive rats. Eur. J. Nutr. 44, 10–17.

Aluko, R.E., 2008. Determination of nutritional and bioactive properties of peptides in enzymatic pea, chickpea, and mung bean protein hydrolysates. J. AOAC Int. 91, 947–956.

Alvaro, E., Andrieux, C., Rochet, V., Rigottier-Gois, L., Lepercq, P., Sutren, M., et al., 2007. Composition and metabolism of the intestinal microbiota in consumers and non-consumers of yogurt. Br. J. Nutr. 97, 126–133.

Anhe, F.F., Roy, D., Pilon, G., Dudonne, S., Matamoros, S., Varin, T.V., et al., 2015. A polyphenol-rich cranberry extract protects from diet-induced obesity, insulin resistance and intestinal inflammation in association with increased Akkermansia spp. population in the gut microbiota of mice. Gut 64, 872–883.

Arora, T., Singh, S., Sharma, R.K., 2013. Probiotics: interaction with gut microbiome and antiobesity potential. Nutrition 29, 591–596.

Astawan, M., Wahyuni, M., Yasuhara, T., Yamada, K., Tadokoro, T., Maekawa, A., 1995. Effects of angiotensin I-converting enzyme inhibitory substances derived from Indonesian dried-salted fish on blood pressure of rats. Biosci. Biotechnol. Biochem. 59, 425–429.

Aune, D., Ursin, G., Veierod, M.B., 2009. Meat consumption and the risk of type 2 diabetes: a systematic review and meta-analysis of cohort studies. Diabetologia 52, 2277–2287.

Azzout-Marniche, D., Gaudichon, C., Tome, D., 2014. Dietary protein and blood glucose control. Curr. Opin. Clin. Nutr. Metab. Care 17, 349–354.

Backhed, F., Ding, H., Wang, T., Hooper, L.V., Koh, G.Y., Nagy, A., et al., 2004. The gut microbiota as an environmental factor that regulates fat storage. Proc. Natl. Acad. Sci. U.S.A. 101, 15718–15723.

Badman, M.K., Flier, J.S., 2005. The gut and energy balance: visceral allies in the obesity wars. Science 307, 1909–1914.

Batterham, R.L., Heffron, H., Kapoor, S., Chivers, J.E., Chandarana, K., Herzog, H., et al., 2006. Critical role for peptide YY in protein-mediated satiation and body-weight regulation. Cell Metab. 4, 223–233.

Baum, S.J., Kris-Etherton, P.M., Willett, W.C., Lichtenstein, A.H., Rudel, L.L., Maki, K.C., et al., 2012. Fatty acids in cardiovascular health and disease: a comprehensive update. J. Clin. Lipidol. 6, 216–234.

Beauchesne-Rondeau, E., Gascon, A., Bergeron, J., Jacques, H., 2003. Plasma lipids and lipoproteins in hypercholesterolemic men fed a lipid-lowering diet containing lean beef, lean fish, or poultry. Am. J. Clin. Nutr. 77, 587–593.

Bergeron, N., Jacques, H., 1989. Influence of fish protein as compared to casein and soy protein on serum and liver lipids, and serum lipoprotein cholesterol levels in the rabbit. Atherosclerosis 78, 113–121.

Bergeron, N., Deshaies, Y., Jacques, H., 1992a. Dietary fish protein modulates high density lipoprotein cholesterol and lipoprotein lipase activity in rabbits. J. Nutr. 122, 1731–1737.

Bergeron, N., Deshaies, Y., Jacques, H., 1992b. Factorial experiment to determine influence of fish protein and fish oil on serum and liver lipids in rabbits. Nutrition 8, 354–358.

Bernstein, A.M., Sun, Q., Hu, F.B., Stampfer, M.J., Manson, J.E., Willett, W.C., 2010. Major dietary protein sources and risk of coronary heart disease in women. Circulation 122, 876–883.

Bjorndal, B., Berge, C., Ramsvik, M.S., Svardal, A., Bohov, P., Skorve, J., et al., 2013. A fish protein hydrolysate alters fatty acid composition in liver and adipose tissue and increases plasma carnitine levels in a mouse model of chronic inflammation. Lipids Health Dis. 12, 143.

Blachier, F., Lancha Jr., A.H., Boutry, C., Tome, D., 2010. Alimentary proteins, amino acids and cholesterolemia. Amino Acids 38, 15–22.

Blouet, C., Jo, Y.H., Li, X., Schwartz, G.J., 2009. Mediobasal hypothalamic leucine sensing regulates food intake through activation of a hypothalamus-brainstem circuit. J. Neurosci. 29, 8302–8311.

Blouet, C., Schwartz, G.J., 2012. Brainstem nutrient sensing in the nucleus of the solitary tract inhibits feeding. Cell Metab. 16, 579–587.

Bos, C., Gaudichon, C., Tome, D., 2000. Nutritional and physiological criteria in the assessment of milk protein quality for humans. J. Am. Coll. Nutr. 19, 191S–205S.

Bouchenak, M., Lamri-Senhadji, M., 2013. Nutritional quality of legumes, and their role in cardiometabolic risk prevention: a review. J. Med. Food 16, 185–198.

Boulet, M.M., Chevrier, G., Grenier-Larouche, T., Pelletier, M., Nadeau, M., Scarpa, J., et al., 2015. Alterations of plasma metabolite profiles related to adipose tissue distribution and cardiometabolic risk. Am. J. Physiol. Endocrinol. Metab. 309, E736–746.

Bowen, J., Noakes, M., Clifton, P.M., 2006a. Appetite regulatory hormone responses to various dietary proteins differ by body mass index status despite similar reductions in ad libitum energy intake. J. Clin. Endocrinol. Metab. 91, 2913–2919.

Bowen, J., Noakes, M., Trenerry, C., Clifton, P.M., 2006b. Energy intake, ghrelin, and cholecystokinin after different carbohydrate and protein preloads in overweight men. J. Clin. Endocrinol. Metab. 91, 1477–1483.

Bowen, J., Noakes, M., Clifton, P.M., 2007. Appetite hormones and energy intake in obese men after consumption of fructose, glucose and whey protein beverages. Int. J. Obes. (Lond) 31, 1696–1703.

Brennan, I.M., Luscombe-Marsh, N.D., Seimon, R.V., Otto, B., Horowitz, M., Wishart, J.M., et al., 2012. Effects of fat, protein, and carbohydrate and protein load on appetite, plasma cholecystokinin, peptide YY, and ghrelin, and energy intake in lean and obese men. Am. J. Physiol. Gastrointest. Liver Physiol. 303, G129–140.

Brosnan, J.T., Brosnan, M.E., 2006. Branched-chain amino acids: enzyme and substrate regulation. J. Nutr. 136, 207S–211S.

Butikofer, U., Meyer, J., Sieber, R., Walther, B., Wechsler, D., 2008. Occurrence of the angiotensin-converting enzyme inhibiting tripeptides Val-Pro-Pro and Ile-Pro-Pro in different cheese varieties of Swiss origin. J. Dairy Sci. 91, 29–38.

Cani, P.D., Bibiloni, R., Knauf, C., Waget, A., Neyrinck, A.M., Delzenne, N.M., et al., 2008. Changes in gut microbiota control metabolic endotoxemia-induced inflammation in high-fat diet-induced obesity and diabetes in mice. Diabetes 57, 1470–1481.

Cani, P.D., Possemiers, S., Van de Wiele, T., Guiot, Y., Everard, A., Rottier, O., et al., 2009. Changes in gut microbiota control inflammation in obese mice through a mechanism involving GLP-2-driven improvement of gut permeability. Gut 58, 1091–1103.

Cani, P.D., Osto, M., Geurts, L., Everard, A., 2012. Involvement of gut microbiota in the development of low-grade inflammation and type 2 diabetes associated with obesity. Gut Microbes 3, 279–288.

Cao, J.J., Nielsen, F.H., 2010. Acid diet (high-meat protein) effects on calcium metabolism and bone health. Curr. Opin. Clin. Nutr. Metab. Care 13, 698–702.

Cha, Y.S., Yang, J.A., Back, H.I., Kim, S.R., Kim, M.G., Jung, S.J., et al., 2012. Visceral fat and body weight are reduced in overweight adults by the supplementation of Doenjang, a fermented soybean paste. Nutr. Res. Pract. 6, 520–526.

Cha, Y.S., Park, Y., Lee, M., Chae, S.W., Park, K., Kim, Y., et al., 2014. Doenjang, a Korean fermented soy food, exerts antiobesity and antioxidative activities in overweight subjects with the PPAR-gamma2 C1431T polymorphism: 12-week, double-blind randomized clinical trial. J. Med. Food 17, 119–127.

Chalamaiah, M., Dinesh Kumar, B., Hemalatha, R., Jyothirmayi, T., 2012. Fish protein hydrolysates: proximate composition, amino acid composition, antioxidant activities and applications: a review. Food Chem. 135, 3020–3038.

Chapelot, D., Payen, F., 2010. Comparison of the effects of a liquid yogurt and chocolate bars on satiety: a multidimensional approach. Br. J. Nutr. 103, 760–767.

Cheng, S., Rhee, E.P., Larson, M.G., Lewis, G.D., McCabe, E.L., Shen, D., et al., 2012. Metabolite profiling identifies pathways associated with metabolic risk in humans. Circulation 125, 2222–2231.

Cheng-Xue, R., Gomez-Ruiz, A., Antoine, N., Noel, L.A., Chae, H.Y., Ravier, M.A., et al., 2013. Tolbutamide controls glucagon release from mouse islets differently than glucose: involvement of K(ATP) channels from both alpha-cells and delta-cells. Diabetes 62, 1612–1622.

Chevrier, G., Mitchell, P.L., Rioux, L.E., Hasan, F., Jin, T., Roblet, C.R., et al., 2015. Low-molecular-weight peptides from salmon protein prevent obesity-linked glucose intolerance, inflammation and dyslipidemia in LDLR − / − /ApoB100/100 Mice. J. Nutr. 145 (7), 1415–1422.

Clifton, P.M., 2011. Protein and coronary heart disease: the role of different protein sources. Curr. Atheroscler. Rep. 13, 493–498.

Cochrane, W.A., Payne, W.W., Simpkiss, M.J., Woolf, L.I., 1956. Familial hypoglycemia precipitated by amino acids. J. Clin. Invest. 35, 411–422.

Cormier, H., Thifault, E., Garneau, V., Tremblay, A., Drapeau, V., Perusse, L., Vohl, M.C., 2016. Association between yogurt consumption, dietary patterns, and cardio-metabolic risk factors. Eur J Nutr 55, 577–587.

Cota, D., Proulx, K., Smith, K.A., Kozma, S.C., Thomas, G., Woods, S.C., et al., 2006. Hypothalamic mTOR signaling regulates food intake. Science 312, 927–930.

Cummings, D.E., 2006. Ghrelin and the short- and long-term regulation of appetite and body weight. Physiol. Behav. 89, 71–84.

Daniel, H., Kottra, G., 2004. The proton oligopeptide cotransporter family SLC15 in physiology and pharmacology. Pflugers Arch. 447, 610–618.

Davidenko, O., Darcel, N., Fromentin, G., Tome, D., 2013. Control of protein and energy intake—brain mechanisms. Eur. J. Clin. Nutr. 67, 455–461.

Davila, A.M., Blachier, F., Gotteland, M., Andriamihaja, M., Benetti, P.H., Sanz, Y., et al., 2013. Re-print of "Intestinal luminal nitrogen metabolism: role of the gut microbiota and consequences for the host". Pharmacol. Res. 69, 114–126.

Delarue, J., LeFoll, C., Corporeau, C., Lucas, D., 2004. N-3 long chain polyunsaturated fatty acids: a nutritional tool to prevent insulin resistance associated to type 2 diabetes and obesity? Reprod. Nutr. Dev. 44, 289–299.

Demonty, I., Deshaies, Y., Lamarche, B., Jacques, H., 2003. Cod protein lowers the hepatic triglyceride secretion rate in rats. J. Nutr. 133, 1398–1402.

Dibble, C.C., Cantley, L.C., 2015. Regulation of mTORC1 by PI3K signaling. Trends Cell Biol. 25, 545–555.

Diepvens, K., Haberer, D., Westerterp-Plantenga, M., 2008. Different proteins and biopeptides differently affect satiety and anorexigenic/orexigenic hormones in healthy humans. Int. J. Obes. (Lond) 32, 510–518.

Do, T.T., Hindlet, P., Waligora-Dupriet, A.J., Kapel, N., Neveux, N., Mignon, V., et al., 2014. Disturbed intestinal nitrogen homeostasis in a mouse model of high-fat diet-induced obesity and glucose intolerance. Am. J. Physiol. Endocrinol. Metab. 306, E668–680.

Doi, M., Yamaoka, I., Fukunaga, T., Nakayama, M., 2003. Isoleucine, a potent plasma glucose-lowering amino acid, stimulates glucose uptake in C2C12 myotubes. Biochem. Biophys. Res. Commun. 312, 1111–1117.

Douglas, S.M., Ortinau, L.C., Hoertel, H.A., Leidy, H.J., 2013. Low, moderate, or high protein yogurt snacks on appetite control and subsequent eating in healthy women. Appetite 60, 117–122.

Duranti, M., 2006. Grain legume proteins and nutraceutical properties. Fitoterapia 77, 67–82.

Eaton, S.B., Eaton III, S.B., 2000. Paleolithic vs. modern diets—selected pathophysiological implications. Eur. J. Nutr. 39, 67–70.

English, P.J., Ghatei, M.A., Malik, I.A., Bloom, S.R., Wilding, J.P., 2002. Food fails to suppress ghrelin levels in obese humans. J. Clin. Endocrinol. Metab. 87, 2984.

Erdmann, K., Cheung, B.W., Schroder, H., 2008. The possible roles of food-derived bioactive peptides in reducing the risk of cardiovascular disease. J. Nutr. Biochem. 19, 643–654.

Everard, A., Belzer, C., Geurts, L., Ouwerkerk, J.P., Druart, C., Bindels, L.B., et al., 2013. Cross-talk between Akkermansia muciniphila and intestinal epithelium controls diet-induced obesity. Proc. Natl. Acad. Sci. U.S.A. 110, 9066–9071.

Ewart, H.S., Dennis, D., Potvin, M., Tiller, C., Fang, L.-h, Zhang, R., et al., 2009. Development of a salmon protein hydrolysate that lowers blood pressure. Eur. Food Res. Technol. 229, 561–569.

Faipoux, R., Tome, D., Gougis, S., Darcel, N., Fromentin, G., 2008. Proteins activate satiety-related neuronal pathways in the brainstem and hypothalamus of rats. J. Nutr. 138, 1172–1178.

FAO, 2007. Cereals, Pulses, Legumes and Vegetable Proteins. Food and Agriculture Organization of the United Nations, Rome, pp. 1–116.

Faure, M., Mettraux, C., Moennoz, D., Godin, J.P., Vuichoud, J., Rochat, F., et al., 2006. Specific amino acids increase mucin synthesis and microbiota in dextran sulfate sodium-treated rats. J. Nutr. 136, 1558–1564.

Felig, P., Marliss, E., Cahill Jr., G.F., 1969. Plasma amino acid levels and insulin secretion in obesity. N. Engl. J. Med. 281, 811–816.

Felig, P., Wahren, J., Hendler, R., Brundin, T., 1974. Splanchnic glucose and amino acid metabolism in obesity. J. Clin. Invest. 53, 582–590.

Feskens, E.J., Bowles, C.H., Kromhout, D., 1991. Inverse association between fish intake and risk of glucose intolerance in normoglycemic elderly men and women. Diabetes Care 14, 935–941.

Flachs, P., Rossmeisl, M., Bryhn, M., Kopecky, J., 2009. Cellular and molecular effects of n-3 polyunsaturated fatty acids on adipose tissue biology and metabolism. Clin. Sci. (Lond) 116, 1–16.

Floyd Jr., J.C., Fajans, S.S., Conn, J.W., Knopf, R.F., Rull, J., 1966. Insulin secretion in response to protein ingestion. J. Clin. Invest. 45, 1479–1486.

Friedman, A.N., 2004. High-protein diets: potential effects on the kidney in renal health and disease. Am. J. Kidney Dis. 44, 950–962.

Fujita, H., Yoshikawa, M., 1999. LKPNM: a prodrug-type ACE-inhibitory peptide derived from fish protein. Immunopharmacology 44, 123–127.

Fumeron, F., Lamri, A., Abi Khalil, C., Jaziri, R., Porchay-Balderelli, I., Lantieri, O., et al., 2011. Dairy consumption and the incidence of hyperglycemia and the metabolic syndrome: results from a french prospective study, Data from the Epidemiological Study on the Insulin Resistance Syndrome (DESIR). Diabetes Care 34, 813–817.

Fung, T.T., van Dam, R.M., Hankinson, S.E., Stampfer, M., Willett, W.C., Hu, F.B., 2010. Low-carbohydrate diets and all-cause and cause-specific mortality: two cohort studies. Ann. Intern. Med. 153, 289–298.

Gannon, M.C., Nuttall, J.A., Nuttall, F.Q., 2002. The metabolic response to ingested glycine. Am. J. Clin. Nutr. 76, 1302–1307.

Garcia-Albiach, R., Pozuelo de Felipe, M.J., Angulo, S., Morosini, M.I., Bravo, D., Baquero, F., et al., 2008. Molecular analysis of yogurt containing Lactobacillus delbrueckii subsp. bulgaricus and Streptococcus thermophilus in human intestinal microbiota. Am. J. Clin. Nutr. 87, 91–96.

Gascon, A., Jacques, H., Moorjani, S., Deshaies, Y., Brun, L.D., Julien, P., 1996. Plasma lipoprotein profile and lipolytic activities in response to the substitution of lean white fish for other animal protein sources in premenopausal women. Am. J. Clin. Nutr. 63, 315–321.

Gaudichon, C., Roos, N., Mahe, S., Sick, H., Bouley, C., Tome, D., 1994. Gastric emptying regulates the kinetics of nitrogen absorption from 15N-labeled milk and 15N-labeled yogurt in miniature pigs. J. Nutr. 124, 1970–1977.

Gaudichon, C., Mahe, S., Roos, N., Benamouzig, R., Luengo, C., Huneau, J.F., et al., 1995. Exogenous and endogenous nitrogen flow rates and level of protein hydrolysis in the human jejunum after [15N]milk and [15N]yoghurt ingestion. Br. J. Nutr. 74, 251–260.

Gibbs, J., Young, R.C., Smith, G.P., 1973. Cholecystokinin decreases food intake in rats. J. Comp. Physiol. Psychol. 84, 488–495.

Gifford, K.D., 2002. Dietary fats, eating guides, and public policy: history, critique, and recommendations. Am. J. Med. 113 (Suppl. 9B), 89S–106S.

Gill, S.R., Pop, M., Deboy, R.T., Eckburg, P.B., Turnbaugh, P.J., Samuel, B.S., et al., 2006. Metagenomic analysis of the human distal gut microbiome. Science 312, 1355–1359.

Girgih, A.T., Chao, D., Lin, L., He, R., Jung, S., Aluko, R.E., 2015. Enzymatic protein hydrolysates from high pressure-pretreated isolated pea proteins have better antioxidant properties than similar hydrolysates produced from heat pretreatment. Food Chem. 188, 510–516.

Gougeon, R., Pencharz, P.B., Marliss, E.B., 1994. Effect of NIDDM on the kinetics of whole-body protein metabolism. Diabetes 43, 318–328.

Gougeon, R., Pencharz, P.B., Sigal, R.J., 1997. Effect of glycemic control on the kinetics of whole-body protein metabolism in obese subjects with non-insulin-dependent diabetes mellitus during iso- and hypoenergetic feeding. Am. J. Clin. Nutr. 65, 861–870.

Gougeon, R., Marliss, E.B., Jones, P.J., Pencharz, P.B., Morais, J.A., 1998. Effect of exogenous insulin on protein metabolism with differing nonprotein energy intakes in Type 2 diabetes mellitus. Int. J. Obes. Relat. Metab. Disord. 22, 250–261.

Gougeon, R., Styhler, K., Morais, J.A., Jones, P.J., Marliss, E.B., 2000. Effects of oral hypoglycemic agents and diet on protein metabolism in type 2 diabetes. Diabetes Care 23, 1–8.

Gougeon, R., Morais, J.A., Chevalier, S., Pereira, S., Lamarche, M., Marliss, E.B., 2008. Determinants of whole-body protein metabolism in subjects with and without type 2 diabetes. Diabetes Care 31, 128–133.

Greenman, Y., Golani, N., Gilad, S., Yaron, M., Limor, R., Stern, N., 2004. Ghrelin secretion is modulated in a nutrient- and gender-specific manner. Clin. Endocrinol. (Oxf) 60, 382–388.

Grimble, G.K., Rees, R.G., Keohane, P.P., Cartwright, T., Desreumaux, M., Silk, D.B., 1987. Effect of peptide chain length on absorption of egg protein hydrolysates in the normal human jejunum. Gastroenterology 92, 136–142.

Guo, K., Yu, Y.H., Hou, J., Zhang, Y., 2010. Chronic leucine supplementation improves glycemic control in etiologically distinct mouse models of obesity and diabetes mellitus. Nutr. Metab. (Lond) 7, 57.

Halkjaer, J., Olsen, A., Overvad, K., Jakobsen, M.U., Boeing, H., Buijsse, B., et al., 2011. Intake of total, animal and plant protein and subsequent changes in weight or waist circumference in European men and women: the Diogenes project. Int. J. Obes. (Lond) 35, 1104–1113.

Hall, W.L., Millward, D.J., Long, S.J., Morgan, L.M., 2003. Casein and whey exert different effects on plasma amino acid profiles, gastrointestinal hormone secretion and appetite. Br. J. Nutr. 89, 239–248.

Halton, T.L., Liu, S., Manson, J.E., Hu, F.B., 2008. Low-carbohydrate-diet score and risk of type 2 diabetes in women. Am. J. Clin. Nutr. 87, 339–346.

Haque, E., Chand, R., 2008. Antihypertensive and antimicrobial bioactive peptides from milk proteins. Eur. Food Res. Technol. 227, 7–15.

Harnedy, P.A., FitzGerald, R.J., 2012. Bioactive peptides from marine processing waste and shellfish: a review. J. Funct. Foods 4, 6–24.

Harper, A.E., Miller, R.H., Block, K.P., 1984. Branched-chain amino acid metabolism. Annu. Rev. Nutr. 4, 409–454.

Harris, R.A., Joshi, M., Jeoung, N.H., Obayashi, M., 2005. Overview of the molecular and biochemical basis of branched-chain amino acid catabolism. J. Nutr. 135, 1527S–1530S.

Herman, M.A., She, P., Peroni, O.D., Lynch, C.J., Kahn, B.B., 2010. Adipose tissue branched chain amino acid (BCAA) metabolism modulates circulating BCAA levels. J. Biol. Chem. 285, 11348–11356.

Hosomi, R., Fukunaga, K., Arai, H., Kanda, S., Nishiyama, T., Yoshida, M., 2011. Fish protein decreases serum cholesterol in rats by inhibition of cholesterol and bile acid absorption. J. Food Sci. 76, H116–121.

Hsiao, G., Chapman, J., Ofrecio, J.M., Wilkes, J., Resnik, J.L., Thapar, D., et al., 2011. Multi-tissue, selective PPARgamma modulation of insulin sensitivity and metabolic pathways in obese rats. Am. J. Physiol. Endocrinol. Metab. 300, E164–174.

Hsieh, C.C., Hernandez-Ledesma, B., Fernandez-Tome, S., Weinborn, V., Barile, D., de Moura Bell, J.M., 2015. Milk proteins, peptides, and oligosaccharides: effects against the 21st century disorders. Biomed. Res. Int. 2015, 146840.

Huang, J., Manning, B.D., 2009. A complex interplay between Akt, TSC2 and the two mTOR complexes. Biochem. Soc. Trans. 37, 217–222.

Huffman, K.M., Shah, S.H., Stevens, R.D., Bain, J.R., Muehlbauer, M., Slentz, C.A., et al., 2009. Relationships between circulating metabolic intermediates and insulin action in overweight to obese, inactive men and women. Diabetes Care 32, 1678–1683.

Humiski, L.M., Aluko, R.E., 2007. Physicochemical and bitterness properties of enzymatic pea protein hydrolysates. J. Food Sci. 72, S605–611.

Jacobson, T.A., Glickstein, S.B., Rowe, J.D., Soni, P.N., 2012. Effects of eicosapentaenoic acid and docosahexaenoic acid on low-density lipoprotein cholesterol and other lipids: a review. J. Clin. Lipidol. 6, 5–18.

Jacques, P.F., Wang, H., 2014. Yogurt and weight management. Am. J. Clin. Nutr. 99, 1229S–1234S.

Jewell, J.L., Russell, R.C., Guan, K.L., 2013. Amino acid signalling upstream of mTOR. Nat. Rev. Mol. Cell Biol. 14, 133–139.

Joost, H.G., 2013. Nutrition: red meat and T2DM—the difficult path to a proof of causality. Nat. Rev. Endocrinol. 9, 509–511.

Kamaura, M., Nishijima, K., Takahashi, M., Ando, T., Mizushima, S., Tochikubo, O., 2010. Lifestyle modification in metabolic syndrome and associated changes in plasma amino acid profiles. Circ. J. 74, 2434–2440.

Karhunen, L.J., Juvonen, K.R., Huotari, A., Purhonen, A.K., Herzig, K.H., 2008. Effect of protein, fat, carbohydrate and fibre on gastrointestinal peptide release in humans. Regul. Pept. 149, 70–78.

Kaushik, M., Mozaffarian, D., Spiegelman, D., Manson, J.E., Willett, W.C., Hu, F.B., 2009. Long-chain omega-3 fatty acids, fish intake, and the risk of type 2 diabetes mellitus. Am. J. Clin. Nutr. 90, 613–620.

Kawaguchi, T., Nagao, Y., Matsuoka, H., Ide, T., Sata, M., 2008. Branched-chain amino acid-enriched supplementation improves insulin resistance in patients with chronic liver disease. Int. J. Mol. Med. 22, 105–112.

Kendall, C.W., Josse, A.R., Esfahani, A., Jenkins, D.J., 2010. Nuts, metabolic syndrome and diabetes. Br. J. Nutr. 104, 465–473.

Khamzina, L., Veilleux, A., Bergeron, S., Marette, A., 2005. Increased activation of the mammalian target of rapamycin pathway in liver and skeletal muscle of obese rats: possible involvement in obesity-linked insulin resistance. Endocrinology 146, 1473–1481.

Kim, J., Choi, J.N., Choi, J.H., Cha, Y.S., Muthaiya, M.J., Lee, C.H., 2013. Effect of fermented soybean product (Cheonggukjang) intake on metabolic parameters in mice fed a high-fat diet. Mol. Nutr. Food Res. 57, 1886–1891.

Konner, M., Eaton, S.B., 2010. Paleolithic nutrition: twenty-five years later. Nutr. Clin. Pract. 25, 594–602.

Koopman, R., Saris, W.H., Wagenmakers, A.J., van Loon, L.J., 2007. Nutritional interventions to promote post-exercise muscle protein synthesis. Sports Med. 37, 895–906.

Kopecky, J., Rossmeisl, M., Flachs, P., Kuda, O., Brauner, P., Jilkova, Z., et al., 2009. n-3 PUFA: bioavailability and modulation of adipose tissue function. Proc. Nutr. Soc. 68, 361–369.

Krebs, M., Krssak, M., Bernroider, E., Anderwald, C., Brehm, A., Meyerspeer, M., et al., 2002. Mechanism of amino acid-induced skeletal muscle insulin resistance in humans. Diabetes 51, 599–605.

Krebs, M., Brehm, A., Krssak, M., Anderwald, C., Bernroider, E., Nowotny, P., et al., 2003. Direct and indirect effects of amino acids on hepatic glucose metabolism in humans. Diabetologia 46, 917–925.

Krebs, M., Brunmair, B., Brehm, A., Artwohl, M., Szendroedi, J., Nowotny, P., et al., 2007. The Mammalian target of rapamycin pathway regulates nutrient-sensitive glucose uptake in man. Diabetes 56, 1600–1607.

Krezowski, P.A., Nuttall, F.Q., Gannon, M.C., Bartosh, N.H., 1986. The effect of protein ingestion on the metabolic response to oral glucose in normal individuals. Am. J. Clin. Nutr. 44, 847–856.

Kristinsson, H.G., Rasco, B.A., 2000. Fish protein hydrolysates: production, biochemical, and functional properties. Crit. Rev. Food Sci. Nutr. 40, 43–81.

Kwak, C.S., Park, S.C., Song, K.Y., 2012. Doenjang, a fermented soybean paste, decreased visceral fat accumulation and adipocyte size in rats fed with high fat diet more effectively than nonfermented soybeans. J. Med. Food 15, 1–9.

Kwon, D.Y., Daily III, J.W., Kim, H.J., Park, S., 2010. Antidiabetic effects of fermented soybean products on type 2 diabetes. Nutr. Res. 30, 1–13.

Lacaille, B., Julien, P., Deshaies, Y., Lavigne, C., Brun, L.D., Jacques, H., 2000. Responses of plasma lipoproteins and sex hormones to the consumption of lean fish incorporated in a prudent-type diet in normolipidemic men. J. Am. Coll. Nutr. 19, 745–753.

Lackey, D.E., Lynch, C.J., Olson, K.C., Mostaedi, R., Ali, M., Smith, W.H., et al., 2013. Regulation of adipose branched-chain amino acid catabolism enzyme expression and cross-adipose amino acid flux in human obesity. Am. J. Physiol. Endocrinol. Metab. 304, E1175–1187.

Laferrere, B., Reilly, D., Arias, S., Swerdlow, N., Gorroochurn, P., Bawa, B., et al., 2011. Differential metabolic impact of gastric bypass surgery versus dietary intervention in obese diabetic subjects despite identical weight loss. Sci. Transl. Med. 3, 80re82.

Laplante, M., Sabatini, D.M., 2012. mTOR signaling in growth control and disease. Cell 149, 274–293.

Lavigne, C., Marette, A., Jacques, H., 2000. Cod and soy proteins compared with casein improve glucose tolerance and insulin sensitivity in rats. Am. J. Physiol. Endocrinol. Metab. 278, E491–500.

Lavigne, C., Tremblay, F., Asselin, G., Jacques, H., Marette, A., 2001. Prevention of skeletal muscle insulin resistance by dietary cod protein in high fat-fed rats. Am. J. Physiol. Endocrinol. Metab. 281, E62–71.

Leclerc, P.-L., Gauthier, S.F., Bachelard, H., Santure, M., Roy, D., 2002. Antihypertensive activity of casein-enriched milk fermented by *Lactobacillus helveticus*. Int. Dairy J. 12, 995–1004.

Lee, A., Jang, H.B., Ra, M., Choi, Y., Lee, H.J., Park, J.Y., et al., 2014. Prediction of future risk of insulin resistance and metabolic syndrome based on Korean boy's metabolite profiling. Obes. Res. Clin. Pract.

Lejeune, M.P., Westerterp, K.R., Adam, T.C., Luscombe-Marsh, N.D., Westerterp-Plantenga, M.S., 2006. Ghrelin and glucagon-like peptide 1 concentrations, 24-h satiety, and energy and substrate metabolism during a high-protein diet and measured in a respiration chamber. Am. J. Clin. Nutr. 83, 89–94.

Li, H., Xu, M., Lee, J., He, C., Xie, Z., 2012. Leucine supplementation increases SIRT1 expression and prevents mitochondrial dysfunction and metabolic disorders in high-fat diet-induced obese mice. Am. J. Physiol. Endocrinol. Metab. 303, E1234–1244.

Liaset, B., Madsen, L., Hao, Q., Criales, G., Mellgren, G., Marschall, H.U., et al., 2009. Fish protein hydrolysate elevates plasma bile acids and reduces visceral adipose tissue mass in rats. Biochim. Biophys. Acta 1791, 254–262.

Liaset, B., Hao, Q., Jorgensen, H., Hallenborg, P., Du, Z.Y., Ma, T., et al., 2011. Nutritional regulation of bile acid metabolism is associated with improved pathological characteristics of the metabolic syndrome. J. Biol. Chem. 286, 28382–28395.

Liou, A.P., Paziuk, M., Luevano Jr., J.M., Machineni, S., Turnbaugh, P.J., Kaplan, L.M., 2013. Conserved shifts in the gut microbiota due to gastric bypass reduce host weight and adiposity. Sci. Transl. Med. 5, 178ra141.

Lips, M.A., Van Klinken, J.B., van Harmelen, V., Dharuri, H.K., t Hoen, P.A., Laros, J.F., et al., 2014. Roux-en-Y gastric bypass surgery, but not calorie restriction, reduces plasma branched-chain amino acids in obese women independent of weight loss or the presence of type 2 Diabetes. Diabetes Care 37, 3150–3156.

Liu, X., Blouin, J.M., Santacruz, A., Lan, A., Andriamihaja, M., Wilkanowicz, S., et al., 2014. High-protein diet modifies colonic microbiota and luminal environment but not colonocyte metabolism in the rat model: the increased luminal bulk connection. Am. J. Physiol. Gastrointest. Liver Physiol. 307, G459–470.

Liu, Z., Long, W., Fryburg, D.A., Barrett, E.J., 2006. The regulation of body and skeletal muscle protein metabolism by hormones and amino acids. J. Nutr. 136, 212S–217S.

Livesey, G., 2001. A perspective on food energy standards for nutrition labelling. Br. J. Nutr. 85, 271–287.

Lopez-Barrios, L., Gutierrez-Uribe, J.A., Serna-Saldivar, S.O., 2014. Bioactive peptides and hydrolysates from pulses and their potential use as functional ingredients. J. Food Sci. 79, R273–283.

Lopez-Huertas, E., 2012. The effect of EPA and DHA on metabolic syndrome patients: a systematic review of randomised controlled trials. Br. J. Nutr. 107 (Suppl. 2), S185–194.

Magkos, F., Bradley, D., Schweitzer, G.G., Finck, B.N., Eagon, J.C., Ilkayeva, O., et al., 2013. Effect of Roux-en-Y Gastric Bypass and laparoscopic adjustable gastric banding on branched-chain amino acid metabolism. Diabetes 62, 2757–2761.

Malik, V.S., Sun, Q., van Dam, R.M., Rimm, E.B., Willett, W.C., Rosner, B., et al., 2011. Adolescent dairy product consumption and risk of type 2 diabetes in middle-aged women. Am. J. Clin. Nutr. 94, 854–861.

Mansour, A., Hosseini, S., Larijani, B., Pajouhi, M., Mohajeri-Tehrani, M.R., 2013. Nutrients related to GLP1 secretory responses. Nutrition 29, 813–820.

Marroqui, L., Alonso-Magdalena, P., Merino, B., Fuentes, E., Nadal, A., Quesada, I., 2014. Nutrient regulation of glucagon secretion: involvement in metabolism and diabetes. Nutr. Res. Rev. 27, 48–62.

Matsumoto, J., Enami, K., Doi, M., Kishida, T., Ebihara, K., 2007. Hypocholesterolemic effect of katsuobushi, smoke-dried bonito, prevents ovarian hormone deficiency-induced hypercholesterolemia. J. Nutr. Sci. Vitaminol. (Tokyo) 53, 225–231.

McCormack, S.E., Shaham, O., McCarthy, M.A., Deik, A.A., Wang, T.J., Gerszten, R.E., et al., 2013. Circulating branched-chain amino acid concentrations are associated with obesity and future insulin resistance in children and adolescents. Pediatr. Obes. 8, 52–61.

McGregor, R.A., Poppitt, S.D., 2013. Milk protein for improved metabolic health: a review of the evidence. Nutr. Metab. (Lond) 10, 46.

Meisel, H., 2004. Multifunctional peptides encrypted in milk proteins. Biofactors 21, 55–61.

Messina, M.J., 1999. Legumes and soybeans: overview of their nutritional profiles and health effects. Am. J. Clin. Nutr. 70, 439S–450S.

Mihalik, S.J., Goodpaster, B.H., Kelley, D.E., Chace, D.H., Vockley, J., Toledo, F.G., et al., 2010. Increased levels of plasma acylcarnitines in obesity and type 2 diabetes and identification of a marker of glucolipotoxicity. Obesity (Silver Spring) 18, 1695–1700.

Mikkelsen, P.B., Toubro, S., Astrup, A., 2000. Effect of fat-reduced diets on 24-h energy expenditure: comparisons between animal protein, vegetable protein, and carbohydrate. Am. J. Clin. Nutr. 72, 1135–1141.

Mills, S., Ross, R.P., Hill, C., Fitzgerald, G.F., Stanton, C., 2011. Milk intelligence: mining milk for bioactive substances associated with human health. Int. Dairy J. 21, 377–401.

Monteleone, P., Bencivenga, R., Longobardi, N., Serritella, C., Maj, M., 2003. Differential responses of circulating ghrelin to high-fat or high-carbohydrate meal in healthy women. J. Clin. Endocrinol. Metab. 88, 5510–5514.

Moran, T.H., 2000. Cholecystokinin and satiety: current perspectives. Nutrition 16, 858–865.

Mozaffarian, D., Hao, T., Rimm, E.B., Willett, W.C., Hu, F.B., 2011. Changes in diet and lifestyle and long-term weight gain in women and men. N. Engl. J. Med. 364, 2392–2404.

Nagata, C., Nakamura, K., Wada, K., Tsuji, M., Tamai, Y., Kawachi, T., 2013. Branched-chain amino acid intake and the risk of diabetes in a Japanese community: the Takayama study. Am. J. Epidemiol. 178, 1226–1232.

Nairizi, A., She, P., Vary, T.C., Lynch, C.J., 2009. Leucine supplementation of drinking water does not alter susceptibility to diet-induced obesity in mice. J. Nutr. 139, 715–719.

Nakagawa, E., Nagaya, N., Okumura, H., Enomoto, M., Oya, H., Ono, F., et al., 2002. Hyperglycaemia suppresses the secretion of ghrelin, a novel growth-hormone-releasing peptide: responses to the intravenous and oral administration of glucose. Clin. Sci. (Lond) 103, 325–328.

Newgard, C.B., An, J., Bain, J.R., Muehlbauer, M.J., Stevens, R.D., Lien, L.F., et al., 2009. A branched-chain amino acid-related metabolic signature that differentiates obese and lean humans and contributes to insulin resistance. Cell Metab. 9, 311–326.

Neyrinck, A.M., Van Hee, V.F., Piront, N., De Backer, F., Toussaint, O., Cani, P.D., et al., 2012. Wheat-derived arabinoxylan oligosaccharides with prebiotic effect increase satietogenic gut peptides and reduce metabolic endotoxemia in diet-induced obese mice. Nutr. Diabetes 2, e28.

Neyrinck, A.M., Van Hee, V.F., Bindels, L.B., De Backer, F., Cani, P.D., Delzenne, N.M., 2013. Polyphenol-rich extract of pomegranate peel alleviates tissue inflammation and hypercholesterolaemia in high-fat diet-induced obese mice: potential implication of the gut microbiota. Br. J. Nutr. 109, 802–809.

Nii, Y., Fukuta, K., Yoshimoto, R., Sakai, K., Ogawa, T., 2008. Determination of antihypertensive peptides from an izumi shrimp hydrolysate. Biosci. Biotechnol. Biochem. 72, 861–864.

Nuttall, F.Q., Mooradian, A.D., Gannon, M.C., Billington, C., Krezowski, P., 1984. Effect of protein ingestion on the glucose and insulin response to a standardized oral glucose load. Diabetes Care 7, 465–470.

Ogden, C.L., Fryar, C.D., Carroll, M.D., Flegal, K.M., 2004. Mean body weight, height, and body mass index, United States 1960–2002. Adv. Data1–17.

Oh, H.G., Kang, Y.R., Lee, H.Y., Kim, J.H., Shin, E.H., Lee, B.G., et al., 2014. Ameliorative effects of Monascus pilosus-fermented black soybean (Glycine max L. Merrill) on high-fat diet-induced obesity. J. Med. Food 17, 972–978.

Oh, W.J., Jacinto, E., 2011. mTOR complex 2 signaling and functions. Cell Cycle 10, 2305–2316.

Otani, L., Ninomiya, T., Murakami, M., Osajima, K., Kato, H., Murakami, T., 2009. Sardine peptide with angiotensin I-converting enzyme inhibitory activity improves glucose tolerance in stroke-prone spontaneously hypertensive rats. Biosci. Biotechnol. Biochem. 73, 2203–2209.

Ouellet, V., Marois, J., Weisnagel, S.J., Jacques, H., 2007. Dietary cod protein improves insulin sensitivity in insulin-resistant men and women: a randomized controlled trial. Diabetes Care 30, 2816–2821.

Ouellet, V., Weisnagel, S.J., Marois, J., Bergeron, J., Julien, P., Gougeon, R., et al., 2008. Dietary cod protein reduces plasma C-reactive protein in insulin-resistant men and women. J. Nutr. 138, 2386–2391.

Pallotta, J.A., Kennedy, P.J., 1968. Response of plasma insulin and growth hormone to carbohydrate and protein feeding. Metabolism 17, 901–908.

Palmer, D., 2011. The Rise of Greek. UBS Investment Research, UBS Securities, LLC.

Pan, A., Sun, Q., Bernstein, A.M., Schulze, M.B., Manson, J.E., Willett, W.C., et al., 2011. Red meat consumption and risk of type 2 diabetes: 3 cohorts of US adults and an updated meta-analysis. Am. J. Clin. Nutr. 94, 1088–1096.

Pan, A., Sun, Q., Bernstein, A.M., Manson, J.E., Willett, W.C., Hu, F.B., 2013. Changes in red meat consumption and subsequent risk of type 2 diabetes mellitus: three cohorts of US men and women. JAMA Intern. Med. 173, 1328–1335.

Patel, P.S., Sharp, S.J., Luben, R.N., Khaw, K.T., Bingham, S.A., Wareham, N.J., et al., 2009. Association between type of dietary fish and seafood intake and the risk of incident type 2 diabetes: the European prospective investigation of cancer (EPIC)-Norfolk cohort study. Diabetes Care 32, 1857–1863.

Patel, P.S., Forouhi, N.G., Kuijsten, A., Schulze, M.B., van Woudenbergh, G.J., Ardanaz, E., et al., 2012. The prospective association between total and type of fish intake and type 2 diabetes in 8 European countries: EPIC-InterAct Study. Am. J. Clin. Nutr. 95, 1445–1453.

Pereira, S., Marliss, E.B., Morais, J.A., Chevalier, S., Gougeon, R., 2008. Insulin resistance of protein metabolism in type 2 diabetes. Diabetes 57, 56–63.

Perng, W., Gillman, M.W., Fleisch, A.F., Michalek, R.D., Watkins, S.M., Isganaitis, E., et al., 2014. Metabolomic profiles and childhood obesity. Obesity (Silver Spring) 22, 2570–2578.

Pesta, D.H., Samuel, V.T., 2014. A high-protein diet for reducing body fat: mechanisms and possible caveats. Nutr. Metab. (Lond) 11, 53.

Phelan, M., Kerins, D., 2011. The potential role of milk-derived peptides in cardiovascular disease. Food Funct. 2, 153–167.

Picard-Deland, E., Lavigne, C., Marois, J., Bisson, J., Weisnagel, S.J., Marette, A., et al., 2012. Dietary supplementation with fish gelatine modifies nutrient intake and leads to sex-dependent responses in TAG and C-reactive protein levels of insulin-resistant subjects. J. Nutr. Sci. 1, e15.

Pilon, G., Ruzzin, J., Rioux, L.E., Lavigne, C., White, P.J., Froyland, L., et al., 2011. Differential effects of various fish proteins in altering body weight, adiposity, inflammatory status, and insulin sensitivity in high-fat-fed rats. Metabolism 60, 1122–1130.

Pipeleers, D.G., Schuit, F.C., Van Schravendijk, C.F., Van de Winkel, M., 1985. Interplay of nutrients and hormones in the regulation of glucagon release. Endocrinology 117, 817–823.

Qin, L.Q., Xun, P., Bujnowski, D., Daviglus, M.L., Van Horn, L., Stamler, J., et al., 2011. Higher branched-chain amino acid intake is associated with a lower prevalence of being overweight or obese in middle-aged East Asian and Western adults. J. Nutr. 141, 249–254.

Quoix, N., Cheng-Xue, R., Mattart, L., Zeinoun, Z., Guiot, Y., Beauvois, M.C., et al., 2009. Glucose and pharmacological modulators of ATP-sensitive K + channels control [Ca2 +]c by different mechanisms in isolated mouse alpha-cells. Diabetes 58, 412–421.

Rabinowitz, D., Merimee, T.J., Maffezzoli, R., Burgess, J.A., 1966. Patterns of hormonal release after glucose, protein, and glucose plus protein. Lancet 2, 454–456.

Reimer, R.A., 2006. Meat hydrolysate and essential amino acid-induced glucagon-like peptide-1 secretion, in the human NCI-H716 enteroendocrine cell line, is regulated by extracellular signal-regulated kinase1/2 and p38 mitogen-activated protein kinases. J. Endocrinol. 191, 159–170.

Rietman, A., Schwarz, J., Tome, D., Kok, F.J., Mensink, M., 2014. High dietary protein intake, reducing or eliciting insulin resistance? Eur. J. Clin. Nutr. 68, 973–979.

Robinson, S.M., Jaccard, C., Persaud, C., Jackson, A.A., Jequier, E., Schutz, Y., 1990. Protein turnover and thermogenesis in response to high-protein and high-carbohydrate feeding in men. Am. J. Clin. Nutr. 52, 72–80.

Roblet, C., Doyen, A., Amiot, J., Pilon, G., Marette, A., Bazinet, L., 2014. Enhancement of glucose uptake in muscular cell by soybean charged peptides isolated by electrodialysis with ultrafiltration membranes (EDUF): activation of the AMPK pathway. Food Chem. 147, 124–130.

Roopchand, D.E., Carmody, R.N., Kuhn, P., Moskal, K., Rojas-Silva, P., Turnbaugh, P.J., et al., 2015. Dietary Polyphenols Promote Growth of the Gut Bacterium Akkermansia muciniphila and Attenuate High-Fat Diet-Induced Metabolic Syndrome. Diabetes 64, 2847–2858.

Roy, F., Boye, J.I., Simpson, B.K., 2010. Bioactive proteins and peptides in pulse crops: pea, chickpea and lentil. Food Res. Int. 43, 432–442.

Rudkowska, I., Marcotte, B., Pilon, G., Lavigne, C., Marette, A., Vohl, M.C., 2010. Fish nutrients decrease expression levels of tumor necrosis factor-{alpha} in cultured human macrophages. Physiol. Genomics 40, 189–194.

Rutherfurd-Markwick, K.J., 2012. Food proteins as a source of bioactive peptides with diverse functions. Br. J. Nutr. 108 (Suppl. 2), S149–157.

Saad, M.F., Bernaba, B., Hwu, C.M., Jinagouda, S., Fahmi, S., Kogosov, E., et al., 2002. Insulin regulates plasma ghrelin concentration. J. Clin. Endocrinol. Metab. 87, 3997–4000.

Sacks, F.M., Lichtenstein, A., Van Horn, L., Harris, W., Kris-Etherton, P., Winston, M., et al., 2006. Soy protein, isoflavones, and cardiovascular health: an American Heart Association Science Advisory for professionals from the Nutrition Committee. Circulation 113, 1034–1044.

Saito, T., Nakamura, T., Kitazawa, H., Kawai, Y., Itoh, T., 2000. Isolation and structural analysis of antihypertensive peptides that exist naturally in Gouda cheese. J. Dairy Sci. 83, 1434–1440.

Sato, K., Iwai, K., Aito-Inoue, M., 2008. Identification of food-derived bioactive peptides in blood and other biological samples. J. AOAC Int. 91, 995–1001.

Sears, D.D., Hsiao, G., Hsiao, A., Yu, J.G., Courtney, C.H., Ofrecio, J.M., et al., 2009. Mechanisms of human insulin resistance and thiazolidinedione-mediated insulin sensitization. Proc. Natl. Acad. Sci. U. S. A. 106, 18745–18750.

Serino, M., Luche, E., Gres, S., Baylac, A., Berge, M., Cenac, C., et al., 2012. Metabolic adaptation to a high-fat diet is associated with a change in the gut microbiota. Gut 61, 543–553.

Serralde-Zuniga, A.E., Guevara-Cruz, M., Tovar, A.R., Herrera-Hernandez, M.F., Noriega, L.G., Granados, O., et al., 2014. Omental adipose tissue gene expression, gene variants, branched-chain amino acids, and their relationship with metabolic syndrome and insulin resistance in humans. Genes Nutr. 9, 431.

Sesti, G., 2006. Pathophysiology of insulin resistance. Best Pract. Res. Clin. Endocrinol. Metab. 20, 665–679.

Shah, S.H., Crosslin, D.R., Haynes, C.S., Nelson, S., Turer, C.B., Stevens, R.D., et al., 2012. Branched-chain amino acid levels are associated with improvement in insulin resistance with weight loss. Diabetologia 55, 321–330.

She, P., Reid, T.M., Bronson, S.K., Vary, T.C., Hajnal, A., Lynch, C.J., et al., 2007a. Disruption of BCATm in mice leads to increased energy expenditure associated with the activation of a futile protein turnover cycle. Cell Metab. 6, 181–194.

She, P., Van Horn, C., Reid, T., Hutson, S.M., Cooney, R.N., Lynch, C.J., 2007b. Obesity-related elevations in plasma leucine are associated with alterations in enzymes involved in branched-chain amino acid metabolism. Am. J. Physiol. Endocrinol. Metab. 293, E1552–1563.

She, P., Olson, K.C., Kadota, Y., Inukai, A., Shimomura, Y., Hoppel, C.L., et al., 2013. Leucine and protein metabolism in obese Zucker rats. PLoS One 8, e59443.

Shimomura, Y., Obayashi, M., Murakami, T., Harris, R.A., 2001. Regulation of branched-chain amino acid catabolism: nutritional and hormonal regulation of activity and expression of the branched-chain alpha-keto acid dehydrogenase kinase. Curr. Opin. Clin. Nutr. Metab. Care 4, 419–423.

Silk, D.B., Chung, Y.C., Berger, K.L., Conley, K., Beigler, M., Sleisenger, M.H., et al., 1979. Comparison of oral feeding of peptide and amino acid meals to normal human subjects. Gut 20, 291–299.

Sluijs, I., Beulens, J.W., van der, A.D., Spijkerman, A.M., Grobbee, D.E., van der Schouw, Y.T., 2010. Dietary intake of total, animal, and vegetable protein and risk of type 2 diabetes in the European Prospective Investigation into Cancer and Nutrition (EPIC)-NL study. Diabetes Care 33, 43–48.

Small, C.J., Bloom, S.R., 2004. Gut hormones as peripheral anti obesity targets. Curr. Drug Targets CNS Neurol. Disord. 3, 379–388.

Smeets, A.J., Soenen, S., Luscombe-Marsh, N.D., Ueland, O., Westerterp-Plantenga, M.S., 2008. Energy expenditure, satiety, and plasma ghrelin, glucagon-like peptide 1, and peptide tyrosine-tyrosine concentrations following a single high-protein lunch. J. Nutr. 138, 698–702.

Soenen, S., Westerterp-Plantenga, M.S., 2008. Proteins and satiety: implications for weight management. Curr. Opin. Clin. Nutr. Metab. Care 11, 747–751.

Solerte, S.B., Fioravanti, M., Locatelli, E., Bonacasa, R., Zamboni, M., Basso, C., et al., 2008a. Improvement of blood glucose control and insulin sensitivity during a long-term (60 weeks) randomized study with amino acid dietary supplements in elderly subjects with type 2 diabetes mellitus. Am. J. Cardiol. 101, 82E–88E.

Solerte, S.B., Gazzaruso, C., Bonacasa, R., Rondanelli, M., Zamboni, M., Basso, C., et al., 2008b. Nutritional supplements with oral amino acid mixtures increases whole-body lean mass and insulin sensitivity in elderly subjects with sarcopenia. Am. J. Cardiol. 101, 69E–77E.

Tai, E.S., Tan, M.L., Stevens, R.D., Low, Y.L., Muehlbauer, M.J., Goh, D.L., et al., 2010. Insulin resistance is associated with a metabolic profile of altered protein metabolism in Chinese and Asian-Indian men. Diabetologia 53, 757–767.

Tappy, L., 1996. Thermic effect of food and sympathetic nervous system activity in humans. Reprod. Nutr. Dev. 36, 391–397.

Taylor, P.M., 2014. Role of amino acid transporters in amino acid sensing. Am. J. Clin. Nutr. 99, 223S–230S.

Tessari, P., Cecchet, D., Cosma, A., Puricelli, L., Millioni, R., Vedovato, M., et al., 2011. Insulin resistance of amino acid and protein metabolism in type 2 diabetes. Clin. Nutr. 30, 267–272.

Tinker, L.F., Sarto, G.E., Howard, B.V., Huang, Y., Neuhouser, M.L., Mossavar-Rahmani, Y., et al., 2011. Biomarker-calibrated dietary energy and protein intake associations with diabetes risk among postmenopausal women from the Women's Health Initiative. Am. J. Clin. Nutr. 94, 1600–1606.

Torres-Leal, F.L., Fonseca-Alaniz, M.H., Teodoro, G.F., de Capitani, M.D., Vianna, D., Pantaleao, L.C., et al., 2011. Leucine supplementation improves adiponectin and total cholesterol concentrations despite the lack of changes in adiposity or glucose homeostasis in rats previously exposed to a high-fat diet. Nutr. Metab. (Lond) 8, 62.

Tremblay, F., Marette, A., 2001. Amino acid and insulin signaling via the mTOR/p70 S6 kinase pathway. A negative feedback mechanism leading to insulin resistance in skeletal muscle cells. J. Biol. Chem. 276, 38052–38060.

Tremblay, F., Lavigne, C., Jacques, H., Marette, A., 2003. Dietary cod protein restores insulin-induced activation of phosphatidylinositol 3-kinase/Akt and GLUT4 translocation to the T-tubules in skeletal muscle of high-fat-fed obese rats. Diabetes 52, 29–37.

Tremblay, F., Krebs, M., Dombrowski, L., Brehm, A., Bernroider, E., Roth, E., et al., 2005. Overactivation of S6 kinase 1 as a cause of human insulin resistance during increased amino acid availability. Diabetes 54, 2674–2684.

Tremblay, F., Brule, S., Hee Um, S., Li, Y., Masuda, K., Roden, M., et al., 2007. Identification of IRS-1 Ser-1101 as a target of S6K1 in nutrient- and obesity-induced insulin resistance. Proc. Natl. Acad. Sci. U.S.A. 104, 14056–14061.

Tsuchiya, A., Almiron-Roig, E., Lluch, A., Guyonnet, D., Drewnowski, A., 2006. Higher satiety ratings following yogurt consumption relative to fruit drink or dairy fruit drink. J. Am. Diet Assoc. 106, 550–557.

Turnbaugh, P.J., Ley, R.E., Mahowald, M.A., Magrini, V., Mardis, E.R., Gordon, J.I., 2006. An obesity-associated gut microbiome with increased capacity for energy harvest. Nature 444, 1027–1031.

Turnbaugh, P.J., Backhed, F., Fulton, L., Gordon, J.I., 2008. Diet-induced obesity is linked to marked but reversible alterations in the mouse distal gut microbiome. Cell Host Microbe 3, 213–223.

Turnbaugh, P.J., Hamady, M., Yatsunenko, T., Cantarel, B.L., Duncan, A., Ley, R.E., et al., 2009. A core gut microbiome in obese and lean twins. Nature 457, 480–484.

UN, F.a.a.o.o.t., 2002. Food Energy—Methods of Analysis and Conversion Factors. Food and agriculture organization of the UN, Rome.

USDA, Caloric sweeteners: per capita availability adjusted for loss USDA.

Valerio, A., D'Antona, G., Nisoli, E., 2011. Branched-chain amino acids, mitochondrial biogenesis, and healthspan: an evolutionary perspective. Aging (Albany NY) 3, 464–478.

van Loon, L.J., Saris, W.H., Verhagen, H., Wagenmakers, A.J., 2000. Plasma insulin responses after ingestion of different amino acid or protein mixtures with carbohydrate. Am. J. Clin. Nutr. 72, 96–105.

van Milgen, J., 2002. Modeling biochemical aspects of energy metabolism in mammals. J. Nutr. 132, 3195–3202.

van Nielen, M., Feskens, E.J., Mensink, M., Sluijs, I., Molina, E., Amiano, P., et al., 2014. Dietary protein intake and incidence of type 2 diabetes in Europe: the EPIC-InterAct Case-Cohort Study. Diabetes Care 37, 1854–1862.

van Woudenbergh, G.J., van Ballegooijen, A.J., Kuijsten, A., Sijbrands, E.J., van Rooij, F.J., Geleijnse, J.M., et al., 2009. Eating fish and risk of type 2 diabetes: a population-based, prospective follow-up study. Diabetes Care 32, 2021–2026.

Vazquez, C., Botella-Carretero, J.I., Corella, D., Fiol, M., Lage, M., Lurbe, E., et al., 2014. White fish reduces cardiovascular risk factors in patients with metabolic syndrome: the WISH-CARE study, a multicenter randomized clinical trial. Nutr. Metab. Cardiovasc. Dis. 24, 328–335.

Veilleux, A., Houde, V.P., Bellmann, K., Marette, A., 2010. Chronic inhibition of the mTORC1/S6K1 pathway increases insulin-induced PI3K activity but inhibits Akt2 and glucose transport stimulation in 3T3-L1 adipocytes. Mol. Endocrinol. 24, 766–778.

Veldhorst, M., Smeets, A., Soenen, S., Hochstenbach-Waelen, A., Hursel, R., Diepvens, K., et al., 2008. Protein-induced satiety: effects and mechanisms of different proteins. Physiol. Behav. 94, 300–307.

Vergnaud, A.C., Norat, T., Romaguera, D., Mouw, T., May, A.M., Travier, N., et al., 2010. Meat consumption and prospective weight change in participants of the EPIC-PANACEA study. Am. J. Clin. Nutr. 92, 398–407.

Vermeirssen, V., Augustijns, P., Morel, N., Van Camp, J., Opsomer, A., Verstraete, W., 2005. In vitro intestinal transport and antihypertensive activity of ACE inhibitory pea and whey digests. Int J. Food Sci. Nutr. 56, 415–430.

Vikoren, L.A., Nygard, O.K., Lied, E., Rostrup, E., Gudbrandsen, O.A., 2013. A randomised study on the effects of fish protein supplement on glucose tolerance, lipids and body composition in overweight adults. Br. J. Nutr. 109, 648–657.

Villegas, R., Gao, Y.T., Yang, G., Li, H.L., Elasy, T.A., Zheng, W., et al., 2008. Legume and soy food intake and the incidence of type 2 diabetes in the Shanghai Women's Health Study. Am. J. Clin. Nutr. 87, 162–167.

Visser, J.T., Bos, N.A., Harthoorn, L.F., Stellaard, F., Beijer-Liefers, S., Rozing, J., et al., 2012. Potential mechanisms explaining why hydrolyzed casein-based diets outclass single amino acid-based diets in the prevention of autoimmune diabetes in diabetes-prone BB rats. Diabetes Metab. Res. Rev. 28, 505–513.

Wallin, A., Di Giuseppe, D., Orsini, N., Patel, P.S., Forouhi, N.G., Wolk, A., 2012. Fish consumption, dietary long-chain n-3 fatty acids, and risk of type 2 diabetes: systematic review and meta-analysis of prospective studies. Diabetes Care 35, 918–929.

Wang, E.T., de Koning, L., Kanaya, A.M., 2010. Higher protein intake is associated with diabetes risk in South Asian Indians: the Metabolic Syndrome and Atherosclerosis in South Asians Living in America (MASALA) study. J. Am. Coll. Nutr. 29, 130–135.

Wang, H., Livingston, K.A., Fox, C.S., Meigs, J.B., Jacques, P.F., 2013. Yogurt consumption is associated with better diet quality and metabolic profile in American men and women. Nutr. Res. 33, 18–26.

Wang, T.J., Larson, M.G., Vasan, R.S., Cheng, S., Rhee, E.P., McCabe, E., et al., 2011. Metabolite profiles and the risk of developing diabetes. Nat. Med. 17, 448–453.

Welle, S., Nair, K.S., 1990. Relationship of resting metabolic rate to body composition and protein turnover. Am. J. Physiol. 258, E990–998.

Wergedahl, H., Liaset, B., Gudbrandsen, O.A., Lied, E., Espe, M., Muna, Z., et al., 2004. Fish protein hydrolysate reduces plasma total cholesterol, increases the proportion of HDL cholesterol, and lowers acyl-CoA:cholesterol acyltransferase activity in liver of Zucker rats. J. Nutr. 134, 1320–1327.

Westerterp-Plantenga, M.S., Nieuwenhuizen, A., Tome, D., Soenen, S., Westerterp, K.R., 2009. Dietary protein, weight loss, and weight maintenance. Annu. Rev. Nutr. 29, 21–41.

WHO, 2002. Protein and Amino Acid Requirements in Human Nutrition. World Health Organization, Geneva, Switzerland.

WHO, 2003. Diet, Nutrition and the Prevention of Chronic Diseases, WHO Technical Report Series. World Health Organization, Geneva.

WHO, 2007. Protein and Amino Acid Requirements in Human Nutrition, WHO Technical Report Series. World Health Organization, Geneva.

WHO, 2015. Obesity and Overweight, Fact sheet No 311. World Health Organization, Geneva.

Wiklund, P.K., Pekkala, S., Autio, R., Munukka, E., Xu, L., Saltevo, J., et al., 2014. Serum metabolic profiles in overweight and obese women with and without metabolic syndrome. Diabetol. Metab. Syndr. 6, 40.

Wu, J.H., Micha, R., Imamura, F., Pan, A., Biggs, M.L., Ajaz, O., et al., 2012. Omega-3 fatty acids and incident type 2 diabetes: a systematic review and meta-analysis. Br. J. Nutr. 107 (Suppl. 2), S214–227.

Xiao, C.W., 2008. Health effects of soy protein and isoflavones in humans. J. Nutr. 138, 1244S–1249S.

Xun, P., He, K., 2012. Fish Consumption and Incidence of Diabetes: meta-analysis of data from 438,000 individuals in 12 independent prospective cohorts with an average 11-year follow-up. Diabetes Care 35, 930–938.

Zemel, M.B., 2005. The role of dairy foods in weight management. J. Am. Coll. Nutr. 24, 537S–546S.

Zhang, M., Picard-Deland, E., Marette, A., 2013. Fish and marine omega-3 polyunsatured Fatty Acid consumption and incidence of type 2 diabetes: a systematic review and meta-analysis. Int. J. Endocrinol. 2013, 501015.

Zhang, Y., Guo, K., LeBlanc, R.E., Loh, D., Schwartz, G.J., Yu, Y.H., 2007. Increasing dietary leucine intake reduces diet-induced obesity and improves glucose and cholesterol metabolism in mice via multimechanisms. Diabetes 56, 1647–1654.

Zhao, Y., Fu, L., Li, R., Wang, L.N., Yang, Y., Liu, N.N., et al., 2012. Metabolic profiles characterizing different phenotypes of polycystic ovary syndrome: plasma metabolomics analysis. BMC Med. 10, 153.

Zhou, Y., Tian, C., Jia, C., 2012. Association of fish and n-3 fatty acid intake with the risk of type 2 diabetes: a meta-analysis of prospective studies. Br. J. Nutr. 108, 408–417.

Ziegler, F., Ollivier, J.M., Cynober, L., Masini, J.P., Coudray-Lucas, C., Levy, E., et al., 1990. Efficiency of enteral nitrogen support in surgical patients: small peptides v non-degraded proteins. Gut 31, 1277–1283.

Zimmerman, H.A., Olson, K.C., Chen, G., Lynch, C.J., 2013. Adipose transplant for inborn errors of branched chain amino acid metabolism in mice. Mol. Genet. Metab. 109, 345–353.

CHAPTER

19

Sulfur Amino Acids Metabolism From Protein Synthesis to Glutathione

G. Courtney-Martin[1] *and P.B. Pencharz*[2]

[1]Faculty of Kinesiology & Physical Education, Department of Clinical Dietetics, University of Toronto, The Hospital for Sick Children, Toronto, ON, Canada [2]Department of Paediatrics and Nutritional Sciences (Emeritus), Senior Scientist Research Institute, University of Toronto, The Hospital for Sick Children, Toronto, ON, Canada

19.1. INTRODUCTION

The SAAs are methionine and cysteine. They are so named because of the presence of a sulfur atom in their molecule (Fig. 19.1). Methionine is indispensable (Snyderman et al., 1964; Fomon et al., 1986; Holt and Snyderman, 1961; Rose, 1937), which means it must be obtained preformed in the diet because it cannot be synthesized by the body. Cysteine is dispensable because it is synthesized in the body from methionine and serine (De Vigneaud et al., 1944), although it is also obtained preformed in the diet. In the synthesis of cysteine, only the sulfur atom is donated by methionine. The carbon skeleton is donated by serine (De Vigneaud et al., 1944).

The increased scientific and clinical attention on the SAAs in recent years is due to interest in their metabolites; homocysteine (Hcy), glutathione (GSH) (Fig. 19.2), and more recently hydrogen sulfide (H_2S). Hcy, a central product in the metabolic pathway of methionine metabolism and hyperhomocysteinemia (Hhcy) (plasma Hcy $>15\,\mu M$), was a well-recognized risk factor for cardiovascular disease (Refsum et al., 1998), ischemic and hemorrhagic stroke in newborn infants and children (Hogeveen et al., 2002; van Beynum et al., 1999), cognitive deterioration (Troen, 2005; Ingenbleek and Kimura, 2013), and Alzheimer's disease (Stipanuk, 2004; Selhub, 1999). A recent Cochrane Review evaluating the impact of Hcy-lowering interventions on cardiovascular events showed that B-complex vitamins (B6, B9 or B12, given alone or in combination) failed to prevent cardiovascular events (Marti-Carvajal et al., 2015). The authors concluded that prescription of these interventions are not justified but stated the need to further investigate other Hcy pathways.

GSH is the most prevalent intracellular thiol (Meister and Anderson, 1983) and most important endogenous antioxidant and scavenger (Wernerman and Hammarqvist, 1999). It is synthesized de novo within all cells (Reid and Jahoor, 2000) and cysteine is the rate-limiting amino acid for its synthesis (Lyons et al., 2000; Jackson et al., 2004; Badaloo et al., 2002). GSH concentration is reduced in several disease states including HIV (Jahoor et al., 1999), liver cirrhosis (Bianchi et al., 1997, 2000), diabetes (Ghosh et al., 2004), and Alzheimer's disease (Liu et al., 2005). GSH concentration is also reduced in surgical trauma patients (Luo et al., 1998), septic patients (Lyons et al., 2001), premature infants (Vina et al., 1995), and in children with severe protein energy malnutrition (Badaloo et al., 2002; Reid et al., 2000). Clear understanding of sulfur amino acid (SAA) metabolism, knowledge of their requirement and the factors that impact, are advantageous to human health.

Emerging experimental evidence suggests that H_2S plays an important role in normal physiological and pathophysiological. Cysteine is the major source of H_2S in mammals and its formation from cysteine is catalyzed by the enzymes cystathionine beta-synthase (CBS), cystathionine gamma-lyase (CSE), and 3-mercaptopyruvate sulfurtransferase (3-MST) (Predmore et al., 2012). Until recently H_2S was viewed as a toxic gas and an

The Molecular Nutrition of Amino Acids and Proteins.
DOI: http://dx.doi.org/10.1016/B978-0-12-802167-5.00019-0

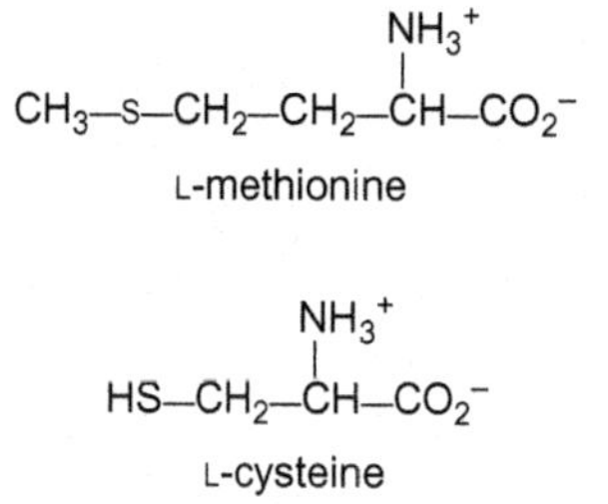

$$CH_3{-}S{-}CH_2{-}CH_2{-}CH(NH_3^+){-}CO_2^-$$

L-methionine

$$HS{-}CH_2{-}CH(NH_3^+){-}CO_2^-$$

L-cysteine

FIGURE 19.1 Structure of the sulfur amino acids methionine and cysteine.

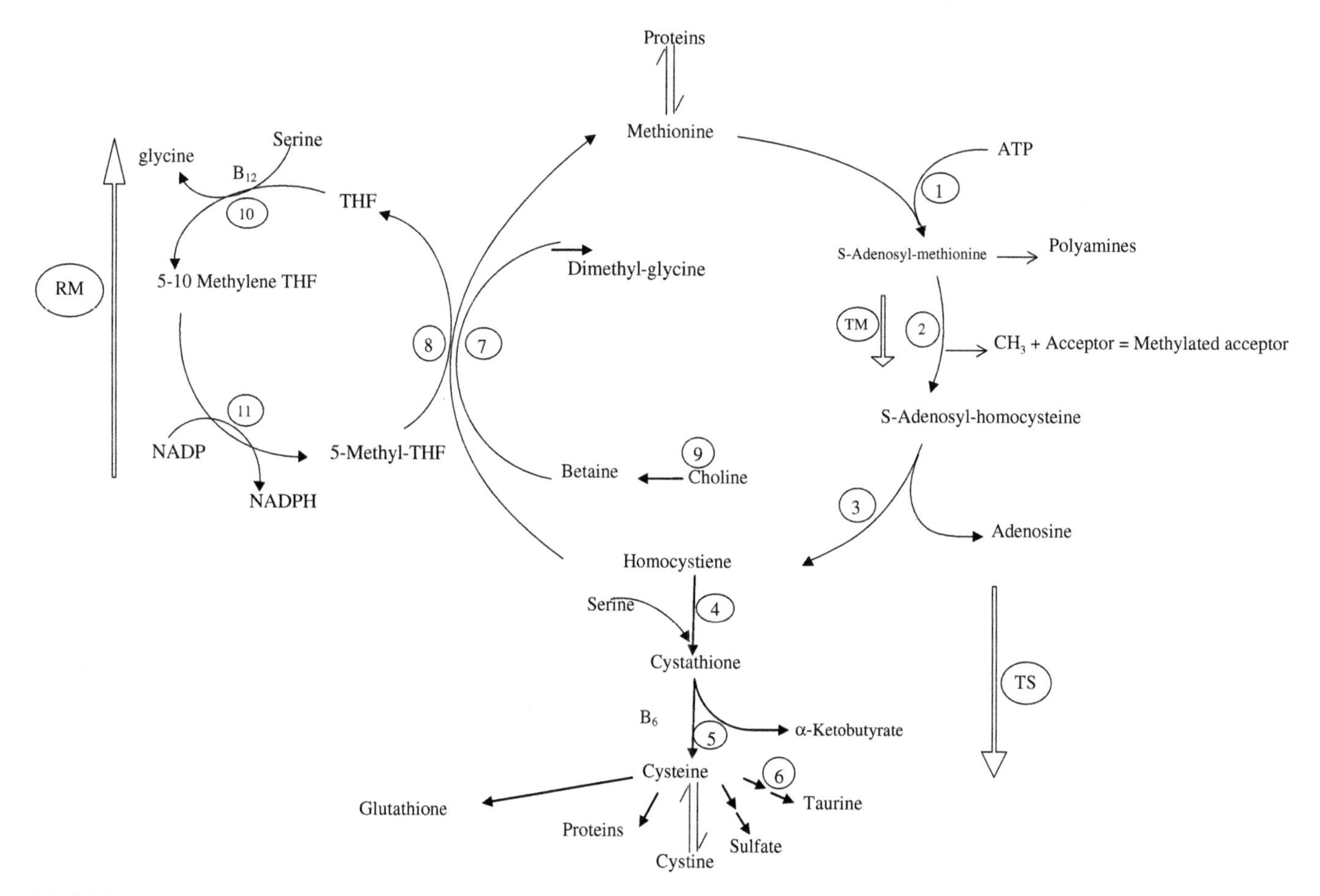

FIGURE 19.2 Pathways to methionine metabolism. Illustration of the pathways of methionine metabolism in mammalian tissue via transmethylation (TM), transsulfuration (TS), and remethylation (RM). The numbers represent the following enzyme or reaction sequence: (1) L-methionine-s-adenosyl-transferase (EC 2.5.1.6); (2) Transmethylation reaction; (3) Adenosylhomocysteinase (EC 3.3.1.1); (4) Cystathionine-β-synthase (EC 4.2.1.22); (5) Cystathionase (EC 4.4.1.1); (6) Multiple reactions leading from cysteine to sulfate or taurine, cysteine used for protein or GSH or protein synthesis or the reversible conversion between cysteine and cystine; (7) Betaine-homocysteine methyl transferase (EC 2.1.1.5); (8) Methyltetrahydrofolate homocysteine methyltransferase (EC 2.1.1.13); (9) Choline dehydrogenase (EC 1.1.3.17) and betaine aldehyde dehydrogenase (EC 1.2.1.8); (10) Serine hydroxymethylase (EC 2.1.2.1); (11) Methylene tetrahydrofolate reductase (EC 1.5.1.20).

environmental hazard (Szabo et al., 2014; Li and Lancaster, 2013). However, recent evidence points to its beneficial effect on vascular tone (blood pressure regulation) and inflammation (Polhemus and Lefer, 2014). It is now considered a potent antiinflammatory molecule with vasodilator actions. In addition, H_2S is being studied as a potential therapeutic agent in the treatment of ischemic reperfusion injury in the heart, brain, lungs, and liver (Polhemus and Lefer, 2014). Understanding the precise pathophysiological signaling mechanism, metabolism, and bioavailability of H_2S are areas of active research.

19.2. FUNCTIONS OF THE SAAS

19.2.1 Methionine

Methionine is a dietary indispensable amino acid (AA) which means it cannot be synthesized in the body and must be obtained from food. The functions of methionine are as follows:

1. It is required for normal growth and development of humans (Snyderman et al., 1964; Fomon et al., 1986; Holt and Snyderman, 1961; Holt, 1968), other mammals (Finkelstein et al., 1988), and avian species (Baker, 2006).
2. It is a substrate for protein synthesis.
3. It serves as the major methyl group donor in vivo via its metabolite adenosylmethionine (Griffith, 1987; Stipanuk, 1986); serving as a source of the methyl group for synthesis of polyamines, methylation of phospholipids, and DNA and RNA intermediates.
4. Its metabolism via the remethylation (RM) pathway in which Hcy accepts a methyl group from 5-methyl tetrahydrofolate in a reaction catalyzed by 5-methyltetrahydrofolate homocysteine methyl transferase (methionine synthase), or from betaine or choline in a reaction catalyzed by betaine homocysteine methyltransferase, is the only reaction which allows for the recycling of this form of folate and for the catabolism of betaine and choline in vivo (Fig. 19.2).
5. It is a precursor for cysteine: donates its sulfur atom to cysteine through the process of transsulfuration (TS) (De Vigneaud et al., 1944) (Fig. 19.2).

19.2.2 Cysteine

1. It is a substrate for protein synthesis.
2. It is required for the synthesis of taurine and sulfate (Griffith, 1987)
3. It is the rate-limiting substrate for the synthesis of GSH (Lyons et al., 2000; Jackson et al., 2004; Badaloo et al., 2002)
4. It lowers the dietary methionine requirement by increasing the RM of Hcy to methionine and decreasing the TS of Hcy to cysteine (Baker, 2006; Ball et al., 2006; Di Buono et al., 2001a).

19.3. PHYSIOLOGICAL ASPECTS OF SAA METABOLISM

The metabolism of the SAA's methionine and cysteine is intractably linked to that of their metabolites Hcy and GSH (Fig. 19.2). While both methionine and cysteine are required for protein synthesis, Hcy, although frequently referred to as an AA, is not incorporated into proteins. Hcy is a central product in the metabolism of methionine while the tri-peptide GSH, the most abundant intracellular antioxidant and scavenger, is a related metabolite of cysteine. Because cysteine is also synthesized from methionine, GSH can be regarded as a related metabolite of methionine, although indirectly.

19.3.1 Methionine

The indispensable AA methionine is metabolized via three major pathways (Fig. 19.2); transmethylation (TM), TS, and RM (Stipanuk, 2004). In the TM pathway, methionine is degraded to form Hcy. The TM pathway begins with the activation of methionine by ATP to form S-adenosylmethionine (SAM) in a reaction catalyzed by methionine adenosyltransferase (MAT). SAM serves as the methyl donor for all known biological methylation reactions with the exception of those involved in the RM of Hcy. SAM donates its methyl group and in the process forms S-adenosylhomocysteine (SAH), which is hydrolyzed to yield adenosine and Hcy. Hcy can then be remethylated to form methionine or enter the TS pathway to form cysteine (Stipanuk, 2004).

The RM pathway allows for the regeneration of methionine from Hcy using methyl groups from one of two pathways (Stipanuk, 2004; Finkelstein et al., 1988). In the folate coenzyme system, methionine synthase enzyme catalyzes the transfer of a methyl group from methylated folic acid (N^5-Methyl-tetrahydro-folate, MTHF) to Hcy assisted by vitamin B12. This is the only reaction which allows for the recycling of this form of folate in the body (Finkelstein et al., 1988). SAM is the inhibitor whereas SAH is the activator of this reaction (Stipanuk, 2004). The methyl group is donated by the dispensable amino acid serine. The final step of MTHF synthesis is the irreversible reduction of $N^{5,10}$-methylene-THF using NADH as the electron donor. In another pathway (only in the liver, kidney, and lens of humans), betaine is the source of the methyl group transferred to Hcy in a reaction catalyzed by betaine-Hcy methyltransferase (Stipanuk, 2004). Betaine is derived either from preformed dietary betaine, dietary choline or choline synthesized from SAM-dependent methylation of phosphatidylethanolamine (Stipanuk, 2004). This enzyme is feedback inhibited by its product dimethyl-glycine. In folate-replete tissue, nearly all of the one-carbon units used for the RM of Hcy occur via the folate-dependent RM. In folate-deficient states or when betaine or choline levels are high, betaine becomes an important methyl donor (Stipanuk, 2004).

The TS pathway allows for the irreversible conversion of Hcy to cystathionine and subsequently to cysteine (Finkelstein et al., 1988). In the TS pathway, methionine and serine are the precursors for cysteine and the derivatives of that AA (Finkelstein et al., 1988). The first step of the TS pathway is the irreversible condensation of Hcy and serine to form cystathionine in a reaction catalyzed by the enzyme cystathionine β-synthase (Stipanuk, 2004). Cysteine is not a precursor for methionine because of the irreversibility of the cystathionase β-synthase reaction (Rose, 1937) (Fig. 19.2). In this reaction, only the sulfur atom from methionine is transferred to cysteine since the carbon skeleton is donated by serine (De Vigneaud et al., 1944). Cystathionase β-synthase is a highly regulated enzyme. It is located at a branched point between the recycling and irreversible catabolism of methionine. SAM is activator of this enzyme which results in increased removal of Hcy by TS, rather than by RM to methionine (Stipanuk, 2004). In the second reaction, cystathionine is hydrolyzed by the enzyme cystathionase (cystathionine γ-lyase) to form cysteine, α-ketobutyrate and ammonia (Stipanuk, 2004). The α-ketobutyrate undergoes oxidative decarboxylation to propionyl-CoA which enters the tricarboxylic aid cycle at the level of succinyl-CoA. The TS pathway is responsible for the oxidative catabolism of the carbon chain of methionine and for the formation of cysteine via the transfer of the sulfur atom of methionine to serine (Stipanuk, 2004). Therefore, TS allows the methionine sulfur to be used for GSH synthesis. It also allows the syntheses of taurine, sulfate, and coenzyme A (Daugherty et al., 2002).

All cells are capable of TM and RM. The TS pathway has a limited tissue distribution. TS occurs only in liver, intestine, kidney, pancreas, and adrenals (Zlotkin and Anderson, 1982b; Brosnan and Brosnan, 2006). Therefore, tissues not capable of TS need an exogenous source of cysteine. The liver is the main organ of TS and maximum hepatic activity is reached at about 3 months in the human infant (Zlotkin and Anderson, 1982b). However, pancreas, kidney, small intestine, brain, and adipose tissues are also capable of significant TS (Stipanuk, 2004). The traffic of methionine through the TM, RM, and TS pathways are regulated by SAM in response to available methionine (Selhub, 1999; Selhub and Miller, 1992). When methionine concentration is high, SAM concentration is increased, which in turn increases Cystathionase β-synthase. This increases TS with the irreversible oxidation of the excess methionine. At the same time, high SAM reduces MTHF, thereby decreasing RM and preventing excess synthesis of methionine. Low methionine concentration as would occur with a diet low in methionine, decreases intracellular concentrations of SAM. This reduces Cystathionase β-synthase activity and stimulates MTHF activity resulting in decreasing TS and increasing RM. This conserves methionine for its essential TM functions. An increase in TM reaction is interpreted by the cell as a deficiency in SAM. This increases the RM rates further, in order to supply methionine for SAM, with a low constant TS rate resulting in the preservation of methionine and maintenance of methionine balance (Selhub, 1999; Finkelstein and Martin, 1984).

19.3.2 Cysteine

Cysteine is a dispensable AA required for protein synthesis and for the synthesis of nonprotein compounds including taurine, sulfate, coenzyme A, and GSH. The key regulatory enzymes of cysteine metabolism are cysteine dioxygenase (CDO), γ-glutamyl-cysteine synthetase (GCS), cysteine-sulfinate decarboxylase (CSDC), and aspartate aminotransferase (Bella et al., 1999a; Stipanuk et al., 2006) (Fig. 19.3). Because cysteine is synthesized from methionine as well as obtained preformed in the diet, an important question is what source of cysteine is preferred by the cell for synthesis of its metabolites; sulfate, taurine and GSH. Current data from in vitro studies show that the source of cysteine is not an important factor for the synthesis of is metabolites (Stipanuk et al., 1992). Both cysteine supplied preformed in the diet and cysteine formed from methionine are equally partitioned toward the synthesis of sulfate, taurine, and GSH (Stipanuk et al., 1992). This demonstrates that methionine is not a superior substrate to preformed cysteine. On the other hand, the concentration of cysteine is an important factor in determining the type of metabolite synthesized (Kwon and Stipanuk, 2001). Sulfate and taurine synthesis are favored when cysteine concentration is high, whereas GSH synthesis is favored when cysteine concentration is low (Stipanuk et al., 1992). Within a given tissue, synthesis of taurine or sulfate is dependent on the concentration of cysteinesulfinate. In general the K_m of CSDC for cysteinesulfinate is lower than the K_m of aspartate

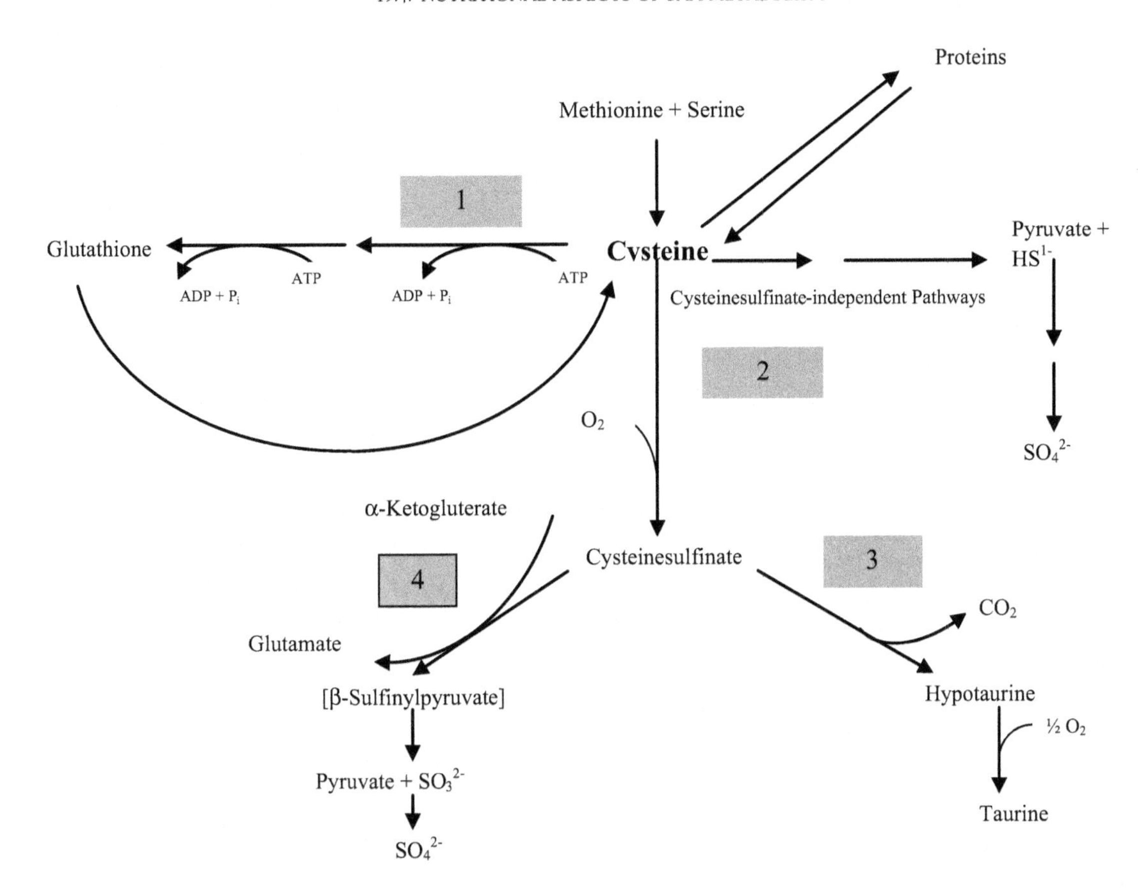

FIGURE 19.3 Pathways of cysteine metabolism. Pathways of cysteine metabolism: Reaction 1, γ-Glutamylcysteine Synthetase (GCS); Reaction 2, Cysteine Dioxygenase (CDO); Reaction 3, Cysteinesulfinate Decarboxylase (CSDC); Reaction 4, Aspartate Aminotransferase.
The enzymes of the Cysteinesulfinate-independent Pathway are Cystathionine γ-lyase and Cystathionine β-synthase responsible for HS production from cysteine.

aminotransferase for cysteinesulfinate. Therefore, the synthesis of sulfate is favored when cysteinesulfinate concentration is high and the synthesis of taurine is favored under more physiological conditions (Bella et al., 1999a; Stipanuk and Ueki, 2011) (Fig. 19.3).

CDO plays a dominant role in cysteine catabolism. It catalyzes the oxidation of the sulfhydryl group of cysteine to form cysteinesulfinate which is the precursor for synthesis of taurine as well as substrate for transamination to yield pyruvate and inorganic sulfate (Fig. 19.3). γ-Glutamylcysteine synthetase (GCS) catalyzes the rate-limiting step in GSH synthesis and therefore competes with CDO for cysteine as a substrate (Stipanuk et al., 2002).

The liver is the main organ responsible for the regulation of cysteine homeostasis (Stipanuk et al., 2002). The key enzymes involved in cysteine metabolism act to keep cysteine concentrations within a tightly regulated normal range in order to support the need for protein synthesis, for the production of other essential molecules including GSH, and to keep cysteine below the level of cytotoxicity (Stipanuk et al., 2006).

In vivo, whole body cysteine metabolism responds to changes in protein, methionine, and cysteine intake, and the enzymes of central importance CDO, GCS, and CSDC respond based on cysteine availability. Greater proportions are partitioned toward oxidation and sulfate production in the presence of high cysteine concentrations. Conversely, a relatively larger proportion is channeled toward GSH synthesis when cysteine concentrations are lower (Stipanuk et al., 2002; Cresenzi et al., 2003; Lee et al., 2004; Bella et al., 1999b). GSH acts as a storage form of cysteine and can be broken down to supply cysteine to the free AA pool when cysteine intake is low (Cho et al., 1981).

19.4. NUTRITIONAL ASPECTS OF SAA METABOLISM

The metabolic significance of the SAAs began with the classic work of William C. Rose and his colleagues in the first half of the 20th century. Using positive nitrogen balance as the criterion of adequacy, Rose et al. (1950)

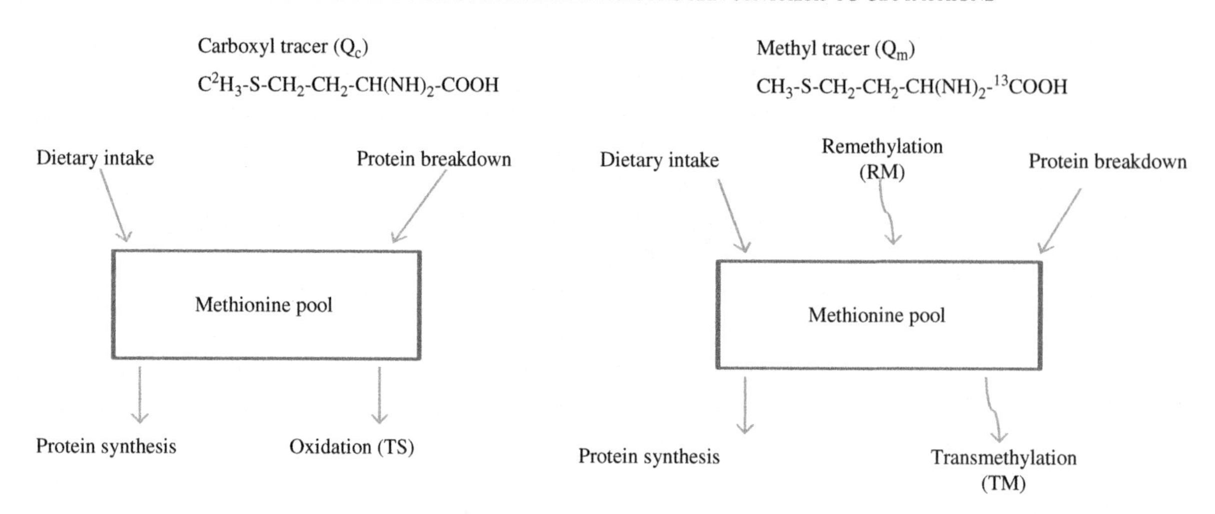

FIGURE 19.4 Schematic outline of carboxyl and methyl tracer models of methionine kinetics. Dilution of the [^{13}C] carboxyl moiety occurs via methionine entry from dietary intake (I) and from protein breakdown (B) whereas loss of moiety occurs via protein synthesis (S) and TS (oxidation). Dilution of the methyl—$^2H_3$ moiety occurs via entry of methionine into the pool from dietary intake (I), protein breakdown (B) and RM of Hcy whereas loss of label from the pool occurs via protein synthesis (S) and TM reactions. Under steady state conditions, differences between fluxes of methyl (Q_m) and carboxyl (Q_c) labeled methionine equals rates of RM.

were the first to demonstrate qualitatively that methionine was an indispensable AA in humans. In a follow-up study, the quantitative minimum and safe methionine requirements was estimated at 1.2 and 2.2 g/kg per day, respectively (Rose et al., 1955). Less than 1 year later, Rose and Wixom (1955) published their report which has led to much controversy in the ensuing years. That report summarized the results of three experiments in which the methionine requirement was first determined in the absence of cysteine followed by the requirement determination in the presence of a set dietary excess of cysteine. Those results showed that cysteine was capable of replacing 80–89% of the methionine requirement of adult men and provided evidence for a regulatory mechanism not only in rats (Womack and Rose, 1941), but also in humans. Those studies however, despite being ground breaking in nature, did not delineate the mechanisms, quantify substrate, or isolate precursors in vivo.

Mudd et al. (1980), Mudd and Poole (1975) were the first to establish a method (the methyl balance approach method) with which to identify and quantify different aspects of methionine metabolism in humans. The method provides a noninvasive, indirect approach for estimating the rates of methylneogenesis and Hcy recycling. The determination is based on the difference between the total daily utilization of methyl groups (SAM) and the dietary intake of preformed methyl groups mainly from methionine, lecithin, choline, and betaine. Differences between the inputs and outputs are attributed to de novo methyl group formation. Outputs include urinary excretion of creatine, creatinine, *n*-methyl nicotinic acid, carnitine, and methylated AAs; polyamines; and the terminal oxidation of methyl groups, including sarcosine (N-methylglycine) (Mudd et al., 1980). Although useful, the methyl balance approach has many limitations. The total body pools of the methylated compounds creatine and phosphatidylcholine are large and turn over relatively slowly (Crim et al., 1975). This meant the method was insensitive as acute changes could go undetected. There were also problems with the use of the method to quantify fluxes through the TM and TS pathways (Storch et al., 1988). Therefore the method was abandoned.

Subsequently, Storch et al. (1988) successfully developed a stable isotope tracer method for quantifying the various aspects of methionine metabolism in humans (Fig. 19.4). In this model whole body rates of TM, RM, and TS can be measured in the steady state by means of a primed constant infusion of tracers [methyl – $^2H_3$] (C $^2\mathbf{H}_3$-S-CH_2-CH_2-CH$(NH)_2$- COOH) and [1-^{13}C]methionine (C $^2H_3$-S-CH_2-CH_2-CH$(NH)_2$- ^{13}COOH). At steady state, metabolic pool sizes are constant, thus the sum of inputs is equal to the sum of outputs. The following two equations depict the methionine turnover rates (flux (Q)) measured using the methyl tracer Q_m and carboxyl tracer (Q_c) of methionine. **I** is dietary intake; **B** is appearance of methionine from protein breakdown; RM is appearance of methionine from Hcy RM; **S** is disappearance of methionine via protein synthesis; TM is utilization of methionine for TM; and TS is oxidation of methionine. At steady state

$$Q = \text{sum of inputs} = \text{sum of outputs}$$

hence

$$Q_m = I + B + RM = S + TM \tag{19.1}$$

and

$$Q_c = I + B = S + TS \tag{19.2}$$

There is not an input from RM in the carboxyl flux equation because the carboxyl label is conserved during each cycle of TM and RM. RM and TM are calculated as follows

$$RM = Q_m - Q_c \tag{19.3}$$

and

$$TM = RM + TS \tag{19.4}$$

Synthesis (S) and breakdown (B) are determined as follows:

$$S = Q_c - TS \tag{19.5}$$

$$B = Q_c - I \tag{19.6}$$

TS is calculated from the rate of $^{13}CO_2$ appearance ($V^{13}CO_2$) using equations of Matthews et al. (1980).

$TS = V^{13}CO_2 \times (1/[^{13}C]$ methionine pool enrichment$) - (1/[^{13}C]$ methionine tracer enrichment$)$.

Using this model, various aspects of methionine metabolism have been examined in both the fed and fasted states (Storch et al., 1988). In the fed state, there is a significant increase in plasma methionine concentration compared to fasting. Feeding increases whole body methionine flux, methionine TM, RM, and TS. Although TS is increased, this is accompanied by an increased efficiency of methionine recycling via RM relative to TS. Feeding also results in a decrease in methionine release from protein breakdown. Thus in the fed state, methionine metabolism is regulated toward anabolism with an increased flux partly accounted for by increased dietary intake, enhanced flow of methionine into TM and RM relative to the fasted state.

In the fasted state, the rate of methionine utilization for protein synthesis is increased relative to its use for TM. Thus in the fasted state, methionine is conserved through the process of protein synthesis and protein synthesis takes precedence when SAA availability is low.

The same model has been used to explore regulatory aspects of methionine metabolism (Storch et al., 1990). In particular, the mechanism underlying the cysteine sparing effect on the methionine requirement was explored using different diets; 25 mg/kg per day methionine without cysteine (adequate diet), a TSAA free diet, or zero methionine plus 20 mg/kg per day cysteine. These data show that plasma methionine concentration are significantly lower when an SAA-free diet is ingested compared with a diet adequate in methionine (Storch et al., 1990). However, plasma cysteine concentration does not differ across diets. All aspects of methionine metabolism are decreased on a SAA free diet and methionine incorporation into protein synthesis is increased relative to TM which confirms previous results that methionine is conserved via protein synthesis when SAA intakes are low or absent (Storch et al., 1988). The addition of cysteine to the SAA-free diet, results in a significant decline in TS and a trend toward increased RM relative to TS. Therefore, the mechanism underlying the sparing effect of cysteine on the methionine requirement is through a decrease in TS (Storch et al., 1990).

These studies (Storch et al., 1988, 1990) were central to the understanding of the various aspects of SAA metabolism in vivo. In the presence of an SAA-free diet, methionine is highly directed toward protein synthesis relative to TM, and Hcy is preferentially partitioned toward RM relative to TS. These all serve to conserve methionine by decreasing oxidation (Ball et al., 2006).

The elegant animal work of Finkelstein et al. (1986, 1988) provides clarification of the mechanisms involved in the SAA metabolism. Using in vivo (Finkelstein et al., 1988), as well as in vitro (Finkelstein et al., 1986) experimental models Finkelstein et al. (1988) demonstrated the following:

1. In growing animals, the net flow of methionine is in the direction of protein synthesis, which removes methionine from the cycle.
2. The utilization of SAM in the formation of polyamines is the second outlet.
3. The irreversible Cystathionase β-synthase reaction is the final outlet since Hcy used in this way is committed TS—these three outlets represent the three essential functions of methionine.

4. The reactions of the cycle itself fulfill three additional requirements: (1) TM reaction; (2) the recycling of methyltetrahydrofolate; and (3) the catabolism of choline (betaine) via RM.
5. Cysteine can spare methionine in only one of these functions; the synthesis of cysteine and its derivatives by means of TS.
6. The residual methionine requirement after cysteine supplementation represents the need for protein synthesis, the obligatory synthesis of cystathionine (if relevant) and methionine used in the process of RM secondary to inefficient conservation (since these two enzymes are utilized in Hcy conservation).

The methionine-sparing effect of cysteine is based on the redistribution of Hcy between competing reactions, notably an increase in RM relative to TS. While the absolute rates of RM remain unchanged, there is a marked decrease in TS as the rates of flow of metabolites through Cystathionase β-synthase reaction decrease. The determinant of this metabolic pattern is represented by a reduction in the liver enzymes together with a decrease in SAM which is an effector of Cystathionase β-synthase (Finkelstein et al., 1988).

19.5. SAA REQUIREMENT

19.5.1 Definitions of Dietary Requirements With Respect to the SAA

Methionine is clearly accepted as a dietary indispensable amino acid and cysteine as a dietary dispensable AA, which can be entirely replaced with dietary methionine. This has been confirmed in all animal species examined to date.

19.5.2 Total SAA Requirement

Because methionine can be converted to cysteine as required and can meet 100% of the metabolic requirement for cysteine (Rose et al., 1955; Di Buono et al., 2001b), the Total SAA (TSAA) requirement is defined as:

> the methionine intake, in the absence of cysteine, that satisfies all of the physiological requirement for both methionine and cysteine (eg, growth, nitrogen balance, TS, methyl donation, glutathione synthesis, taurine synthesis etc) (Ball et al., 2006).

19.5.3 Minimum Obligatory Requirement for Methionine

The sparing effect of cysteine on the methionine requirement means that cysteine is capable of replacing some proportion of the TSAA requirement. The indispensable nature of methionine means that there is a *Minimum Obligatory Requirement* for methionine, representing the quantity of the methionine requirement that cannot be replaced by cysteine. This can be defined as

> the intake of methionine that cannot be replaced by cysteine and that will not be reduced by addition of any methyl donor, cofactor or any other metabolite (Ball et al., 2006).

19.5.4 Cysteine Sparing of Methionine

The definition of cysteine sparing of methionine can be derived from the definition of TSAA as follows:

> Cysteine sparing is:
> The proportion of the TSAA (as described above) that can be met with dietary cysteine.

Determination of the above three sets of requirements requires the conduction of a minimum of three experiments: (1) the TSAA requirement (measured by feeding graded intakes of methionine and zero dietary cysteine); (2) The minimum obligatory methionine requirement, (measured by feeding an excess of dietary cysteine and graded intakes of methionine); and (3) the cysteine sparing effect (measured by feeding the minimum obligatory methionine requirement and graded intakes of cysteine; Ball et al., 2006).

19.5.5 SAA Requirement Using Nitrogen Balance

Using nitrogen balance technique, Rose et al. (1950) were the first to determine the TSAA requirement in young adult males. The minimal and safe requirements were estimated at 1.1 and 2.2 g/day respectively, representing a requirement of 13.25 mg/kg per day to keep all subjects in positive nitrogen balance. The minimum methionine requirement was also estimated in young men (Rose and Wixom, 1955), and found to be 0.1–0.2 g/day. This demonstrated a sparing effect of cysteine on the methionine requirement of 89% and 80%.

The TSAA requirement of the human infant was estimated using the nitrogen balance technique with nitrogen balance and growth as the criterion of adequacy (Snyderman et al., 1964; Fomon et al., 1986; Albanese et al., 1948). Estimates of 80–88 and 44–49 mg/kg per day were derived for TSAA and minimum methionine requirement respectively. This represents a 55–60% sparing effect of cysteine on the methionine requirement of infants.

The nitrogen balance method has many practical limitations and flaws (Fuller and Garlick, 1994; Young and Bier, 1987). Nitrogen balance is a relatively small value, obtained by subtracting a relatively large value of nitrogen losses from a similarly large value of nitrogen intake. This result in overestimation in the prediction of nitrogen balances and a tendency toward a falsely positive nitrogen balance because of overestimation of intake, and underestimation of losses. This leads to underestimation of the requirement. Due to the slow turnover of the body urea pool, nitrogen balance also require a prolonged adaptation time (minimum of 7 days) (Fuller and Garlick, 1994; and Young and Bier, 1987), to the test diets for the dietary change to be reflected in the urinary nitrogen excretion. This makes the method unfit for use in vulnerable populations like children and pregnant women because it is unethical to maintain them on deficient intakes for such prolonged periods. The result is that nitrogen balance studies do not allow for the evaluation of sufficient levels of intake of the AA in order to determine the requirement. This is problematic because since the requirement is defined for the individual, each individual needs to be studied at several levels of intake (at least three) both above and below the predicted requirement in order to be able to estimate the individual requirement (Rand et al., 1976; Zello et al., 1990). Nitrogen balance calculations and estimation of nitrogen can be significantly affected by miscellaneous and dermal losses, therefore, they must be included in the calculation. However, they are very difficult to measure and vary with environmental conditions (eg, ambient temperature) (Calloway et al., 1971; Rand and Young, 1999).

Thus the nitrogen balance method is cumbersome, and lacks precision. In addition it is burdensome to subjects and totally unsuitable for use in vulnerable populations. These disadvantages pointed to a need for more sensitive, less cumbersome, and minimally invasive methods to be developed.

19.5.6 SAA Requirement Using Stable Isotope Tracer Kinetics

19.5.6.1 Indicator Amino Acid Oxidation Technique

The most important contribution to our current knowledge of the SAA requirement using stable isotope tracer kinetics have been by the combined Toronto/Alberta group headed by P. Pencharz and R. Ball and the MIT group headed by the late V.R. Young and his collaborators in India, namely Kurpad et al. These two groups have employed the techniques of indicator amino acid oxidation (IAAO) and IAAO-balance technique to determine the TSAA and minimum methionine requirement. In contrast to nitrogen balance, only minimal adaptation time to each test level of AA intake (6–8 h) is needed (Elango et al., 2009), which is likely due to the time needed for equilibration in the acyl-tRNA pools. This means that several levels of AA intakes can be studied without putting the subjects at risk because they are only exposed to a deficient or excess level of intake for hours rather than days.

The IAAO technique is based on the principle that the partitioning of any indispensable AA between oxidation and protein synthesis is sensitive to the level of the most limiting AA in the diet. When an indispensable AA is limiting in the diet for protein synthesis, all other (AAs) are in excess and therefore must be oxidized (Zello et al., 1995). It follows that as the dietary level of the limiting AA is increased in graded amounts, the uptake of all other (AAs) for protein synthesis increases, leading to a decrease in their oxidation, including the oxidation of the indicator AA. This decrease in oxidation occurs until the requirement is met, after which further increase in the limiting AA (test AA) will have no effect on the uptake of other indispensable (AAs) (the indicator amino acid) for protein synthesis or oxidation (Fig. 19.5) (Ball and Bayley, 1984; Kim et al., 1983a,b).

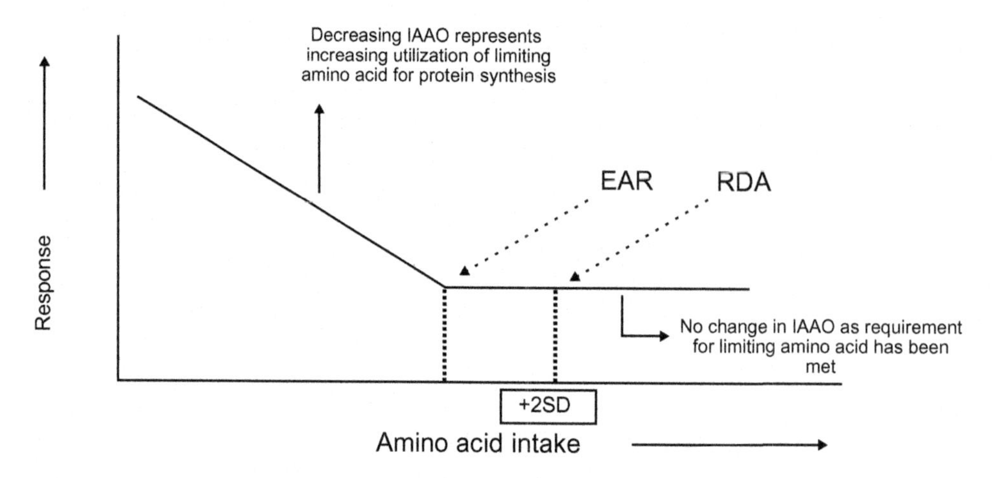

FIGURE 19.5 Schematic representation of the pattern of oxidation of amino acids in studies of amino acid requirement using the indicator amino acid oxidation (IAAO) technique. Oxidation of the indicator amino acid decreases as intake of the test amino acid increases until the requirement (breakpoint) is reached. After the breakpoint further increase in the test amino acid has no effect on the oxidation of the indicator amino acid and oxidation remains constant. *IAAO*; Indicator amino acid oxidation, *EAR*; Estimated average requirement, *RDA*; Recommended dietary allowance.

IAAO provides a functional approach to measuring AA requirements. It is safe, noninvasive, and can be used in vulnerable groups like children and pregnant women. It applies the stochastic modeling technique (Waterlow et al., 1978) for the measurement of amino acid kinetics. This model makes the following assumptions:

1. the size of the metabolic pool with respect to both labeled and unlabeled indicator AA is constant during the course of the study (steady state)
2. there is no significant reentry of isotope into the metabolic pool—at steady state ^{13}C is treated in the same way as ^{12}C
3. AAs derived from the diet and from the catabolism of protein are handled in the same way (Picou and Taylor-Roberts, 1969).

The IAAO method was used to determine the TSAA requirement, and the sparing effect of cysteine in young adult males (Di Buono et al., 2001a,b). The mean TSSA requirement estimated by the IAAO method is 12.6 mg/kg per day. This mean requirement estimate is the breakpoint of the estimate in the oxidation curve of the indicator AA (Fig. 19.5) (Di Buono et al., 2001b). The population safe requirement calculated from the variance of the individual requirement at two standard deviations above the mean was calculated at 21 mg/kg per day (Di Buono et al., 2001b). The sparing effect of cysteine on the methionine requirement was determined by providing methionine at the safe intake of 21 mg/kg per day and varying the intakes of cysteine. This results show that cysteine is capable of sparing 64% of the methionine requirement of adult humans (Di Buono et al., 2001a).

Various aspects of SAA metabolism were studies by varying the ratio of methionine and cysteine to represent the ratios present in common foods (Di Buono et al., 2003). Using the model of Storch et al. (1988), healthy men were fed three different diets in random order; diet A: 24 mg/kg methionine without cysteine, diet B: 13 mg methionine plus 11 mg/kg cysteine, and diet C: 5 mg/kg methionine plus 19 mg/kg cysteine. Methionine kinetics was measured in the fed state using an orally administered L-[1-^{13}C, methyl-$^{2}H_3$]methionine.

Based on the results it is apparent that the ratio of cysteine to methionine regulates whole body SAA metabolism in adult humans. When TSAA intake is adequate and held constant at 24 mg/kg, replacement of methionine with cysteine results in increased RM at the expense of TS, whereas at high methionine intakes, the methionine pool is regulated by high rates of TS.

The development of the piglet model as a surrogate for the human neonate (Wykes et al., 1993) to study AA requirement and metabolism led to increased knowledge on SAA requirements. With the piglet model, Shoveller et al. (2003b) determined the TSAA as methionine only (methionine in the absence of cysteine) of the enterally and parenterally fed neonatal piglet. Using IAAO technique the mean methionine requirement of the enterally and parenterally fed neonatal piglet was estimated at 0.42 and 0.26 g/kg respectively. The methionine requirement of the total parenterally (TPN) fed piglet was 30% lower than the enteral requirement. The splanchnic tissue (gut and liver) is very important in AA metabolism (Stoll et al., 1998; Bertolo et al., 1999, 2000) with approximately one-third of dietary essential AA being consumed by the splanchnic tissue on first-pass metabolism (Stoll et al., 1998). When the gastrointestinal track is bypassed with intravenous feeding, whole body nitrogen metabolism is decreased in intravenously fed animals compared with those fed enterally despite similar weight gain (Bertolo et al., 1999). In addition, intestinal atrophy, characterized by decreased villous height and crypt depth occurs in intravenously fed animals. These alterations in AA metabolism and decreased nitrogen retention have been used as possible explanation for the decrease AA requirement observed in TPN feeding.

Using these requirement estimates derived from the piglet model the TSAA requirement of the TPN fed human neonate was predicted. Since piglets grow at five times the rate of the human infant, the TSAA requirement estimates of the TPN fed human neonate was predicted to be 52 mg/kg per day. When the IAAO technique was applied to the determination of the TSAA of the TPN fed human neonate the mean and safe estimates were 49.0 and 58.0 mg/kg, respectively (Courtney-Martin et al., 2008a). This study provides evidence that the piglet model is valid for the study of AA requirement of the human neonate.

The minimum methionine requirement (methionine in the presence of excess cysteine) of the enteral and TPN fed neonatal piglet was also estimated using the IAAO method. A reduction of the requirement to 0.25 and 0.18 g/kg per day for enterally and parenterally fed piglets respectively (Shoveller et al., 2003a) demonstrated a 40% sparing effect of cysteine on the methionine requirement when cysteine was added to the diet. The sparing effect was not affected by route of feeding (Shoveller et al., 2003a). This shows that dietary cysteine is equally effective in producing a sparing effect on the TSAA requirement whether fed enterally or parenterally.

In healthy school-age children, the TSAA requirement and minimal methionine requirement estimated using the IAAO method was 12.9 and 5.8 mg/kg, respectively. This represents a 55% cysteine sparing effect on the methionine requirement. These requirements are similar to that of healthy adults (Di Buono et al., 2001a,b) which shows that the maintenance requirements of children and adults are similar and represents the predominant part of SAA requirement in school-aged children.

19.5.6.2 *Twenty-Four Hour IAAO and Balance Technique*

The 24 h IAAO and balance technique is based on the same fundamental principles of the IAAO method. In addition, the indicator AA balance is calculated as an absolute balance or as a percentage of doses of the indicator AA oxidized in response to the intake of the test AA. This method combines the feature of 24 h direct balance with indicator oxidation approach. It measures indicator oxidation and balance in the fed and fasted states over a 24 h period. The point at which the indicator AA balance is closest to zero in response to the intake of the test AA is deemed the mean AA requirement. The 24 h IAAO and balance method is favored over the IAAO method because measurements are conducted over the full 24 h in the fed and fasted states whereas the IAAO protocol is done only in the fed state. This method is therefore the chosen method of determining AA requirement when data are available. This method however, has only been used in healthy adults because of the prolonged adaptation time of 5–7 days to the test level of AA intake.

The TSAA requirement was estimated using the 24-h IAAO and balance technique in healthy (Kurpad et al., 2003), and chronically undernourished (Kurpad et al., 2004b) Indian men. The requirement estimates were 15 and 16 mg/kg, respectively. These estimates are in agreement with the estimates of 12.6 mg/kg per day derived using the IAAO method (Di Buono et al., 2001b). These requirement estimates suggest that chronic undernutrition in the absence of infection does not increase the requirement for SAAs.

The sparing effect of cysteine on the methionine requirement was assessed at two different cysteine intakes (5 and 12 mg/kg) with seven different intakes of methionine ranging from 3 to 21 mg/kg per day (Kurpad et al., 2004a). With a cysteine intake of 5 mg/kg per day, methionine breakpoint was 20 mg/kg per day. At a cysteine intake 12 mg/kg per day, methionine breakpoint was 10 mg/kg per day. The authors concluded based on the overall results that cysteine may spare methionine requirement in healthy men but that the amount of sparing is difficult to quantify.

Based on the definitions under the heading "*Definitions of dietary requirements with respect to the SA*," it is apparent that the intake of 5 mg/kg per day of cysteine was less than that required to arrest the flow of methionine through the TS pathway and therefore the sparing effect was too small to measure relative to the total potential effect of cysteine to replace methionine in providing the total SAA requirement (Ball et al., 2006). The decrease in the methionine breakpoint to 10 mg/kg per day when cysteine was provided at 12 mg/kg per day suggest a sparing effect of cysteine on the methionine requirement. However, the estimate of total SAA requirement (corrected for molecular weight) was 25 mg/kg per day when 12 mg/kg per day of cysteine was fed. Since the

TABLE 19.1 Summary of studies on estimates of methionine requirement and cysteine sparing

References	Sulfur AA intake			Number of diets used	Estimated requirements mean			Comments
	Mg/kg		Method of estimation		TSAA	Minimum Meth	Cys Sparing (%)	
	Met	Cys						
Rose et al. (1955)	10–33[a]	0	N Balance	6	12.5 mg/kg	ND	ND	Mean calorie intake of 56 kcal/kg. 123–150 mg/kg of N
Rose and Wixom (1955)	1.2–5[a]	9.5–13.5	N Balance	4–9	12.7 mg/kg	1.9 mg/kg	87–89	
Di Buono et al. (2001b)	0,6.5,13.0,19.5, 26.0,32.0	0	IAAO	6	12.6	ND	ND	
Di Buono et al. (2001a)	0,2.5,5.0,7.5, 10.0,13.0	21	IAAO	6	ND	4.5	64	
Kurpad et al. (2003)	3,6,9,13,18,21,24	0	IAAO and Balance Technique	7	15	ND	ND	
Kurpad et al. (2004a)	3,6,9,13,18,21,24	5 or 12	IAAO and Balance Technique	7	ND	20 and 10	ND	

[a]*DL-methionine used in study.*
ND, not detected; *IAAO*, indicator amino acid oxidation; N balance, nitrogen balance.

addition of cysteine to the diet should not result in a higher estimate of total SAA requirement compared with methionine alone, a possible explanation for this unusual result is that glycine and serine were used to balance the nitrogen and AA in this experiment. Glycine and serine are involved in the RM of methionine and serine donates the carbon skeleton for the synthesis of cysteine, therefore difference in intake of these AAs could affect the rates of TS and or RM.

Table 19.1 presents a summary of some of the studies which have evaluated the sparing effect of cysteine on the methionine requirement in humans.

19.5.7 SAA Metabolism: Effect of Route of Feeding

Enteral nutrition results in higher plasma cysteine and Hcy concentrations than TPN (Shoveller et al., 2004; Stegink and Den Besten, 1972; Miller et al., 1995a) in piglets, human neonates, and adult humans. These data show that the gut is a significant site of methionine TS and TM producing significant amounts of Hcy and cysteine for net release into the circulation in neonates as well as in adults.

Data on splanchnic metabolism of SAAs are generated mostly from studies conducted in the pig and piglet models because they have similar physiology to man, especially gut physiology (Bauchart-Thevret et al., 2009). The importance of the gut in SAA metabolism was demonstrated in piglets when the SAA requirement of enterally fed piglets was 30% higher than piglets fed parenterally (Shoveller et al., 2003b). In addition 20% of the dietary methionine intake and 25% of whole body methionine is metabolized by the gut via TM and TS (Riedijk et al., 2007a).

With regards to cysteine, studies in pigs indicate the gastrointestinal tract is an important site for its utilization (Stoll et al., 1998; Bos et al., 2003). Approximately 40% of cysteine intake is metabolized in the splanchnic tissue during first-pass metabolism (Bauchart-Thevret et al., 2011). Of this, the gastrointestinal tract utilizes about 25% of the dietary cysteine intake, which represents 53% of the splanchnic first-pass uptake (Bauchart-Thevret et al., 2011), with approximately 25% consumed in nonoxidative pathways.

In the neonatal piglet, plasma Hcy is highest when methionine is provided enterally in the absence of cysteine, compared to when methionine is provided enterally with cysteine, or parenterally (Shoveller et al., 2004). Increases in Hcy also occurs in the TPN-fed human neonate when methionine was fed in the absence of cysteine

(Courtney-Martin et al., 2008a). These data show that both routes of feeding and dietary supply of methionine and cysteine affect plasma Hcy concentration. Since high plasma Hcy is a predictive marker for hemorrhagic and ischemic stroke in infants and children (Hogeveen et al., 2002; van Beynum et al., 1999), the provision of the SAA as a balance between methionine and cysteine with the minimum amount of methionine for its obligatory requirements would be beneficial in TPN and enteral feeding of infants and children.

19.5.8 Is Cysteine a Conditionally Essential AA in Human Neonates?

Sturman and Gaull were among the first to report on the absence of cystathionase activity in the livers of premature and newborn infants (Sturman et al., 1970), and to make the suggestion that cysteine is a conditionally essential AA in the newborn until sometime after birth. Cystathionase is the second enzyme in the TS pathway (Fig. 19.2). When that enzyme is absent or underdeveloped, cystathionine concentrations are elevated. This idea was in question when it was observed that premature infants on cysteine-free TPN had adequate growth and nitrogen retention which was not improved by adding cysteine to cysteine-free TPN formulation (Zlotkin et al., 1981).

In later work, cystathionase activity was measured in samples of human liver tissue during postmortem examination of infants who died prior to 1 year of age (Zlotkin and Anderson, 1982b). Based on those results, cystathionase activity in the liver is dependent of both gestational age and postnatal age. In addition, kidney and adrenals have considerable activity which is not affected by postnatal age. In the full-term infant there is a gradual increase in liver cystathionase activity during the first 3 months of life whereas in the premature infant there is a more marked increase during the first 2 weeks of life. This provided further evidence that cysteine is not conditionally essential in the human neonate.

Nevertheless, the existence of several reports of low plasma cysteine concentration in neonates on TPN served as existing evidence in the minds of clinicians and researchers alike that cysteine may be conditionally essential in the human neonate (Winters et al., 1977; Kanaya et al., 1984; Pohlandt, 1974; Wu et al., 1986; Stegink and Baker, 1971).

Using growth and nitrogen balance, the essentiality of cysteine was tested in term and preterm infants on cysteine-free and cysteine-supplemented TPN (Zlotkin et al., 1981). There was no difference in the nitrogen retention between the unsupplemented and cysteine-supplemented group. Both groups showed similar positive nitrogen retention of 282 mg/kg per day which was 56% of nitrogen infused. These retentions paralleled the expected nitrogen in utero retention. There was no difference in the weight change between groups. As expected, plasma cysteine concentrations were higher in the cysteine supplemented group.

Still others have shown that cysteine supplementation to cysteine-free TPN (Malloy et al., 1984) did not improve nitrogen retention or weight gain in the cysteine supplemented group when compared with the cysteine unsupplemented group. These studies provided further evidence that cysteine is not indispensable in the human neonate.

In 1995, Miller et al. published a stable isotope tracer technique to assess human neonatal AA synthesis using D-[U-^{13}C]glucose (Miller et al., 1995b). With this technique the conversion of glucose carbon into seven nonessential (AAs) was assessed by measuring their isotopic enrichments in plasma using gas chromatography/mass spectrometry (GC/MS). Using this technique they were unable to detect significant ^{13}C enrichment in plasma cysteine (Miller et al., 1995a,b). This led them to suggest that cysteine is an essential AA in parenterally fed premature neonates.

However, using the same method but a more sensitive end-point, apo B-100, ^{13}C labeled cysteine was detected in hepatically derived apo B-100 (Shew et al., 2005). The tracer/tracee ratios of the M + 1 isotopomer of cysteine derived from apo B-100 were significantly greater after the [$^{13}C_6$]glucose than at baseline. There was a direct correlation between the increase in cysteine synthesis and birth weight. They concluded that a functional pathway exists for cysteine synthesis in premature neonates and that the minimum synthetic capacity of this pathway is directly related to neonatal maturity.

Previous studies using growth and nitrogen balance in neonates fed cysteine-free, methionine-adequate TPN provided evidence that the neonate is able to synthesize adequate cysteine from methionine, at least for protein synthesis (Zlotkin et al., 1981; Malloy et al., 1984; Zlotkin and Anderson, 1982a). The results of a study using IAAO method in enterally fed preterm neonate (Riedijk et al., 2007b) also revealed that cysteine is not a conditionally essential AA in the preterm neonate. Neonates were fed methionine-adequate formula, with graded intakes of cysteine. There was no change in oxidation of the indicator AA in response to changes in the intake of cysteine.

A more recent report using stable isotope tracers provided evidence that the neonate is able to synthesize adequate cysteine from methionine not only for protein synthesis but also for GSH synthesis (Courtney-Martin et al., 2010). GSH synthesis was measured in erythrocytes of neonates fed TPN in which the TSAA was provided as methionine only, or as methionine plus supplemental cysteine. Cysteine supplementation had no effect on erythrocyte GSH synthesis (Courtney-Martin et al., 2010). Notwithstanding, the increase in plasma Hcy when SAAs is provided as methionine only (Courtney-Martin et al., 2008a; Shoveller et al., 2004), especially in enteral feeding, is evidence that provision of the SAAs as a balance of methionine and cysteine is advantageous over the provision as methionine only.

19.6. GLUTATHIONE

19.6.1 Introduction to GSH Metabolism

The tripeptide GSH (gamma-glutamyl-cysteinyl-glycine:GSH) is synthesized de novo within all cells from glycine, cysteine, and glutamate (Reid and Jahoor, 2000). Although synthesized within all cells, the liver is the major producer and exporter of GSH. It is synthesized primarily if not exclusively in the cytoplasm (Smith et al., 1996). Therefore, most of the cellular GSH (85–90%) is present in the cytosol. Unlike the synthesis of larger peptides, no RNA template is involved in GSH synthesis (Beutler, 1989).

There are two steps in the synthesis of GSH: first the enzyme γ-glutamyl cysteine synthetase catalyzes the formation of a peptide bond between the γ-carboxyl group of glutamic acid and the amino group of cysteine (Fig. 19.6). This is the rate-limiting step in GSH synthesis (Meister and Anderson, 1983), and because the peptide bond formed between glutamic acid and cysteine is at the γ rather than the α carboxyl group of glutamic acid, the tripeptide is resistant to peptidase (Beutler, 1989). In the next step, glycine is joined to γ-glutamyl cysteine

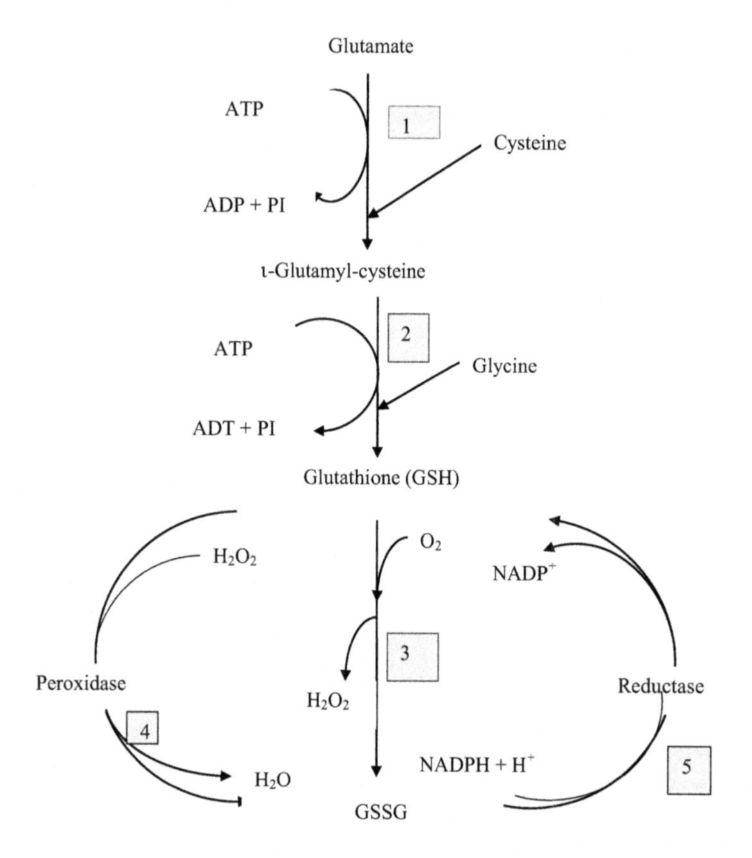

FIGURE 19.6 Glutathione metabolism. Reaction 1. γ-Glutamylcysteine synthetase; Reaction 2. Glutathione synthetase; Reaction 3. Oxidation of GSH by O_2; conversion of GSH to GSSG is also modified by free radicals; Reaction 4. GSH peroxidase; Reaction 5. GSSG reductase.

(Fig. 19.6) to form GSH. This reaction is catalyzed by glutathione synthetase (Beutler, 1989). GSH has an inhibitory influence on the first enzyme γ-glutamyl cysteine synthetase, which acts as feedback control for the regulation of GSH synthesis. In rare cases of hereditary deficiency of glutathione synthetase, the second reaction in GSH synthesis is halted and large amounts of γ-glutamyl cysteine accumulates, which is then catabolized to 5-oxyproline and excreted in the urine (Beutler, 1989).

Most of the functions of GSH require its reduced form, (Fig. 19.6) in which state it has a free sulfhydryl group and is designated GSH. However, the cysteine residue is easily oxidized nonenzymatically to glutathione disulfide (GSSG). Because most of the functions of GSH require its reduced form, an active enzyme mechanism exists in the form of glutathione reductase for the reduction of GSSG to GSH (Fig. 19.6). This enzyme uses NADPH or NADH as the hydrogen donor. Hence the activity of GSH is very dependent on the intake of riboflavin (Beutler, 1989). In addition, the concentration of GSH, as well as the enzymes involved in its metabolism, are markedly influenced by diet (Beutler, 1989).

19.6.2 Functions of GSH

The intracellular concentration of GSH in mammalian cells is in the millimolar range (0.5–10 mM) with 85–90% being present in the cytosol. The extracellular concentration (with the exception of bile acids which contain up to 10 mM) is typically in the micromolar range, for example, 2–20 μM (Meister and Anderson, 1983). GSH is therefore regarded as the most prevalent intracellular thiol (Meister and Anderson, 1983) and the most important endogenous antioxidant and scavenger (Wernerman and Hammarqvist, 1999). The [GSH]:[GSSG] ratio is often used as an indicator of the cellular redox state and is >10 under normal physiological conditions.

1. GSH is consumed in the detoxification of electrophylic metabolites and xenobiotics, converting the first step in the conversion of formaldehyde (a toxic product of methanol oxidation) to formic acid (Beutler, 1989). It is an effective free radical scavenger, protecting cells from the toxic effects of reactive oxygen compounds (Reid and Jahoor, 2000).
2. Through the enzyme glutathione peroxidase (Fig. 19.6), GSH removes peroxides that could oxidize sulfhydrils and participates in several reactions that serve to prevent oxidation of SH groups or to reduce them once they have become oxidized (Beutler, 1989). This function is important to promote and protect the normal functioning of proteins.
3. GSH is needed for the synthesis of leukotrienes, making GSH an important mediator of inflammation (Beutler, 1989).
4. GSH plays an important role in AA transport and is a source of cysteine reserve during food deprivation, and a major source of cysteine for lymphocytes (Cho et al., 1981; Fukagawa et al., 1996; Malmezat et al., 2000b).

Irreversible cell damage occurs when the cell is no longer able to maintain its content of GSH (Reid and Jahoor, 2000). Indeed poor prognosis is associated with decreasing GSH concentration in certain disease states. Consequently an understanding of GSH metabolism and kinetics with particular emphasis/knowledge of substrate needs for its synthesis is of importance in health as well as in disease states.

19.6.3 Physiological Aspects of GSH

19.6.3.1 Concentration Measurement

In the biological compartments, changes in GSH concentration are affected if there is a difference between the rates of synthesis and the rates of disposal of GSH (Reid and Jahoor, 2000). A single concentration measure, therefore, while it gives a static measure of previous kinetics and of amounts of GSH available or lacking in that compartment and possible surrounding tissues, tells nothing about the rates of synthesis and loss of GSH. Consequently, kinetic measures using radio and stable isotope tracers provide an opportunity to make meaningful interpretation of concentration measurements.

19.6.3.2 Kinetic Measurement

The first report in which GSH kinetics was measured (Dimant et al., 1955) used the rate of incorporation of orally administered ^{15}N glycine to estimate GSH synthesis in erythrocytes. This provided the first evidence that erythrocytes synthesize GSH de novo. Recent reports suggest that erythrocyte contribute up to 10% of whole body GSH synthesis in humans (Wu et al., 2004). Subsequent kinetic studies used intravenous (IV) injections of supraphysiologic doses of GSH with measurement of loss from the plasma compartment; still others measured incorporation of radio labeled precursors of GSH into in vitro systems. These methods were flawed because they suggested that plasma GSH reflected interorgan, particularly hepatic GSH efflux. Venous plasma GSH concentration is higher than arterial glutathione concentration suggesting a limited role for plasma GSH in interorgan GSH homeostasis (Reid and Jahoor, 2000). In addition, plasma GSH in vivo is much less than intracellular concentration (μmol vs mmol) (Reid and Jahoor, 2000), and plasma GSH is highly unstable, readily undergoing autooxidation to GSSG or protein GSH disulfides (Reid and Jahoor, 2000).

In 1995, with the development of a stable isotope precursor product model for measuring GSH synthesis in vivo, (Jahoor et al., 1995), there began an opportunity for the more effective characterization of various aspects of GSH metabolism and improvement in this body of knowledge.

19.6.3.3 The Precursor Product Model

The precursor product model for measuring GSH kinetics was developed by Jahoor et al. (1995). The minimum requirement for the calculation of the rate of synthesis of a protein or peptide with this model is the measurement of the isotopic enrichment at two time points during the quasilinear portion of the exponential increase in peptide-bound AA labeling (Reid and Jahoor, 2000). In addition, an estimate of the enrichment of AA tracer in the precursor pool (the free pool of the tissue being studied) is necessary since the AA tracer should be at isotopic steady state before a measurement of its incorporation into the protein/peptide is made.

19.6.3.4 Infusion Protocol

A primed continuous infusion (CI) of either $^{13}C_2$ or $^{2}H_2$ glycine is administered intravenously or intragastrically for 6 h in neonates and 7–8 h in adults (Reid and Jahoor, 2000). $^{13}C_2$ glycine is used for glycine flux measurements because of the loss of one or two of the deuterium when $^{2}H_2$ glycine is used as the tracer (Reid and Jahoor, 2000). Blood samples are collected at baseline and hourly during the infusion, with sampling restricted to the last 3 h of infusion in neonates and small children (Reid and Jahoor, 2000). The rate of synthesis of erythrocyte GSH is obtained from the rate of incorporation of $^{13}C_2$ or $^{2}H_2$ glycine into the GSH. Erythrocyte-free glycine isotopic enrichment is used to represent the enrichment of the glycine precursor pool from which erythrocytes make GSH (Reid and Jahoor, 2000).

19.6.3.5 Calculations

- The fractional synthesis rate (FSR) of erythrocyte GSH is calculated as follows:
  - FSR_{GSH} (%/h) = $PE_{t_2} - PE_{t_1}/IEP_1 \times (t_2 - t_1) \times 100$
  - Where $PE_{t_2} - PE_{t_1}$ = the increase in the enrichment of GSH bound glycine over the period $t_2 - t_1$ of the infusion.
  - IE_{pl} = isotopic enrichment at plateau of erythrocyte free glycine
- Absolute synthesis rate (ASR) of GSH is calculated as follows:
  - $ASR = GSH_{mass} \times FSR_{GSH}$
  - Where GSH_{mass} = the product of the cell volume (or cell number or cell protein) and the concentration of GSH in the cell.

19.6.4 GSH Metabolism and Synthesis Rates

19.6.4.1 In Healthy States

Cysteine is the rate-limiting factor for GSH synthesis. When healthy adults were fed a diet containing protein of 1.0 g/kg per day or the same protein intake but devoid of methionine and cysteine (Lyons et al., 2000), their erythrocyte FSR and ASR of GSH fell significantly when the diet was devoid of methionine and cysteine. However, GSH concentration did not differ between groups. Therefore, GSH concentration alone cannot be used as an isolated marker of GSH metabolism, since it does not reflect changes in GSH synthesis within the cell.

A decrease in GSH synthesis has also been observed in response to a 30% decrease in protein intake in healthy adults (Jackson et al., 2004). Erythrocyte GSH synthesis was measured in healthy adult males while consuming their habitual protein intake (1.13 g/kg per day), and while consuming a diet providing the safe WHO-recommended amount of dietary protein of 0.75 g/kg per day (FAO/WHO/UNU, 1985). Both FSR and ASR decreased significantly from baseline despite maintenance of nitrogen balance.

Animals are capable of retaining sulfur beyond that required for protein synthesis when fed a diet low in protein, and infused with methionine. Pigs fed an adequate protein (AP) diet and low protein (LP) diet have similar weight gain (Hou et al., 2003). After receiving a methionine infusion, sulfate excretion was significantly lower in the LP pigs than the pigs who received AP resulting in a more positive sulfur balance after methionine infusion but lower nitrogen to sulfur ratio compared to the AP group. In the AP piglets the entire methionine load was catabolized and excreted in the urine whereas only 69% of the infused methionine was excreted in the LP piglets. This sulfur retention was not due to increased methionine uptake for protein synthesis because there was no increase in nitrogen balance in the face of the increased sulfur balance (Hou et al., 2003).

On the other hand, when the TSAA requirement is provided as methionine only in the presence of an adequate protein intake, additional graded intakes of supplemental cysteine do not increase erythrocyte GSH synthesis in healthy adults (Courtney-Martin et al., 2008b). Rather, the increased cysteine intake results in a linear increase in sulfate excretion (Courtney-Martin et al., 2008b). The protein intake in that study was provided as crystalline (AAs) and the methionine was provided at 14 mg/kg per day.

In healthy states therefore, consumption of protein at intakes of 1.0–13 g/kg per day provided in the form of native protein in subjects habitual diet, is adequate for the maintenance of nitrogen balance as well as for synthesis of the most abundant intracellular antioxidant, GSH. The same level of protein intake from a crystalline AA mixture designed to provide the TSAA at its requirement of 14 mg/kg per day, is also adequate for maintenance of GSH synthesis. A protein intake of 0.75 mg/kg per day, although sufficient to maintain nitrogen balance, is inadequate for maintenance of GSH synthesis.

19.6.4.2 In Stress/Disease/Aging

GSH concentration is reduced in several disease states, including HIV infection (Jahoor et al., 1999), liver cirrhosis (Bianchi et al., 1997, 2000), diabetes (Ghosh et al., 2004), Sickle cell disease (Reid et al., 2006), and Alzheimer's disease (Liu et al., 2005). GSH concentration is also found to be reduced in surgical trauma (Luo et al., 1998), septic patients (Lyons et al., 2001), premature infants (Vina et al., 1995), as well as in children with severe childhood undernutrition (SCU) (Badaloo et al., 2002; Reid et al., 2000). The mechanism surrounding this decreased concentration was believed to be increased utilization. However, protein-deficient animals subject to the stress of inflammation are unable to maintain GSH homeostasis and ASR, while piglets fed adequate protein maintained GSH homeostasis even when subjected to the stress of inflammation (Jahoor et al., 1995). In addition, cysteine and methionine supplementation was shown to modulate the effect of TNF-α on protein and GSH synthesis in animals fed a low protein diet (Hunter and Grimble, 1994). Survival of guinea pig pups subjected to oxidative stress was improved by feeding nutritional substrate for GSH synthesis (Chessex et al., 1999).

In HIV-infected individuals, GSH deficiency is due in part, to reduced synthesis, secondary to cysteine deficiency (Jahoor et al., 1999). GSH fractional and ASR were measured in symptom-free HIV-infected subjects before and after a 1-week supplementation with N-acetylcysteine (NAC) (Jahoor et al., 1999). After NAC supplementation, both fractional and ASR of GSH experienced a significant increase in the HIV-infected individuals which was similar to that observed in healthy control subjects.

Earlier publications had shown that children with SCU have lower total free plasma concentrations of (AAs) including glutamine and cysteine (Edozien et al., 1960) and a 60% lower plasma methionine concentration than healthy children (Roediger, 1995). This knowledge led to a series of investigations in children with edematous and nonedematous SCU in which the precursor product model was used to measure erythrocyte GSH synthesis at three time points during hospitalization: shortly after admission when the children were infected and malnourished (phase 1); 8 days postadmission when they were no longer infected (phase 2); and when they had recovered (phase 3). Children with edematous (SCU) have significantly lower erythrocyte GSH concentrations and lower ASR of erythrocyte GSH than those with nonedematous SCU (Reid et al., 2000) during phases 1 and 2 of study. During those two phases, the group with edematous SCU had lower erythrocyte GSH concentrations and lower ASR than at recovery (phase 3). Plasma and erythrocyte free cysteine concentration were lower in the children with edematous SCU during phases 1 and 2 than at recovery (phase 3). On the other hand, erythrocyte GSH concentration, rates of GSH synthesis, and plasma and erythrocyte free cysteine concentrations of the nonedematous group were similar at all three points and were greater at phase 1 and 2 than in the edematous group. Those results led to the conclusion that GSH deficiency in children with edematous SCU is due to decreased synthesis as a result of cysteine deficiency both as preformed cysteine and from methionine (Reid et al., 2000; Jahoor, 2012).

To test this hypothesis, two groups of children with edematous SCU were supplemented with either NAC or alanine for 7 days (Badaloo et al., 2002). Supplementation with NAC resulted in higher erythrocyte cysteine and GSH concentration as well as ASR of GSH. Importantly, in the cysteine supplemented group edema resolved approximately 5 days sooner than in the control group. However, both groups had similar rates of weight gain, yet the clinically important loss of edema occurred quicker in the cysteine supplemented group. This observation is supportive of prior reports in healthy subjects that SAA is partitioned more toward protein synthesis when intakes are low (Jackson et al., 2004; Storch et al., 1988). Similar results are obtained in protein or SAA-deficient animals subjected to the stress of inflammation (Jahoor et al., 1995; Hunter and Grimble, 1994) that SAA is partitioned more into proteins than into GSH when SAA intake is low.

Recent evidence shows that GSH concentration and synthesis are lower in elderly subjects than younger adults and that cysteine and glycine supplementation lowers both oxidative stress and oxidative damage in the elderly (Sekhar et al., 2011). The suggestion is that cysteine deficiency is a characteristic of aging, and can be alleviated by supplementation. The mechanism underlying the deficiency is unknown but decreased intake as a result of inadequate dietary protein intake is a possible explanation.

The results from all the above studies suggest an increased need for SAA particularly cysteine during illness, infection, disease, and even aging. This increased need for cysteine can be partly explained by the increased need for GSH synthesis; a lack of which can lead to slower recovery time and prolonged illness (Badaloo et al., 2002).

Animal studies demonstrate an increased requirement for cysteine in infection (Malmezat et al., 2000a,b). Cysteine flux, as well as the activity of the enzymes of GSH synthesis (γ-glutamyl cysteine synthetase, and glutathione reductase) are increased in infected rats compared to controls. GSH synthesis accounts for at least 40% of the enhanced cysteine flux during inflammation. In addition methionine TS and methionine flux are increased during sepsis but to a less extent than cysteine flux (Malmezat et al., 2000b). The increased cysteine flux observed in inflammation is higher than predicted from estimates of protein turnover suggesting that the increase cysteine flux is due to increased GSH synthesis. This is believed to be the case because previous work in humans have shown that up 50% of cysteine flux in the fasted state is due to GSH breakdown (Fukagawa et al., 1996).

In addition, human (Sekhar et al., 2011; Mercier et al., 2006) and animal (Vidal et al., 2014) studies support an increased requirement for cysteine with aging. Elderly humans have lower cysteine and GSH concentrations as well as lower fractional and ASR of GSH (Sekhar et al., 2011) when compared to younger controls. After cysteine and glycine supplementation, all markers of GSH homeostasis improved with no observed differences between elderly and young subjects.

19.7. CONCLUSIONS

The SAAs are important nutrient substrates for protein synthesis. In addition, they have important roles outside of protein synthesis; methionine via its metabolite SAM is the most important methyl donor in vivo and cysteine is the rate-limiting substrate for synthesis of the most abundant intracellular antioxidant GSH. Deficiency of SAA therefore, results in impaired growth and protein synthesis but also compromises methylation reactions and host antioxidant status. Deficient SAA and low protein intake results in decrease in GSH synthesis in healthy adults. In addition, GSH deficiency is a characteristic feature of multiple disease states including HIV, diabetes, sepsis, and severe childhood undernutrition. The mechanism underlying GSH deficiency is cysteine deficiency. Existing evidence points to hierarchy regarding cysteine utilization by the body; cysteine is preferentially used for protein synthesis when intakes are low but intakes in excess of that required for protein synthesis are partitioned into GSH synthesis, then taurine, and finally oxidized to sulfate when intakes are high.

References

Albanese, A.A., Holt Jr., L.E., et al., 1948. The sulfur amino acid requirement of the infant. Fed. Proc. 7 (1 Pt 1), 141.

Badaloo, A., Reid, M., Forrester, T., Heird, W.C., Jahoor, F., 2002. Cysteine supplementation improves the erythrocyte glutathione synthesis rate in children with severe edematous malnutrition. Am. J. Clin. Nutr. 76 (3), 646–652.

Baker, D.H., 2006. Comparative species utilization and toxicity of sulfur amino acids. J. Nutr. 136 (6 Suppl), 1670S–1675S.

Ball, R.O., Bayley, H.S., 1984. Tryptophan requirement of the 2.5-kg piglet determined by the oxidation of an indicator amino acid. J. Nutr. 114 (10), 1741–1746.

Ball, R.O., Courtney-Martin, G., Pencharz, P.B., 2006. The in vivo sparing of methionine by cysteine in sulfur amino acid requirements in animal models and adult humans. J. Nutr. 136 (6 Suppl), 1682S–1693S.

Bauchart-Thevret, C., Stoll, B., Burrin, D.G., 2009. Intestinal metabolism of sulfur amino acids. Nutr. Res. Rev. 22 (2), 175–187.

Bauchart-Thevret, C., Cottrell, J., Stoll, B., Burrin, D.G., 2011. First-pass splanchnic metabolism of dietary cysteine in weanling pigs. J. Anim. Sci. 89 (12), 4093–4099.

Bella, D.L., Hahn, C., Stipanuk, M.H., 1999a. Effects of nonsulfur and sulfur amino acids on the regulation of hepatic enzymes of cysteine metabolism. Am. J. Physiol. 277 (1 Pt 1), E144–153.

Bella, D.L., Hirschberger, L.L., Hosokawa, Y., Stipanuk, M.H., 1999b. Mechanisms involved in the regulation of key enzymes of cysteine metabolism in rat liver in vivo. Am. J. Physiol. 276 (2 Pt 1), E326–335.

Bertolo, R.F., Chen, C.Z., Pencharz, P.B., Ball, R.O., 1999. Intestinal atrophy has a greater impact on nitrogen metabolism than liver by-pass in piglets fed identical diets via gastric, central venous or portal venous routes. J. Nutr. 129 (5), 1045–1052.

Bertolo, R.F., Pencharz, P.B., Ball, R.O., 2000. Organ and plasma amino acid concentrations are profoundly different in piglets fed identical diets via gastric, central venous or portal venous routes. J. Nutr. 130 (5), 1261–1266.

Beutler, E., 1989. Nutritional and metabolic aspects of glutathione. Annu. Rev. Nutr. 9, 287–302.

Bianchi, G., Bugianesi, E., Ronchi, M., Fabbri, A., Zoli, M., Marchesini, G., 1997. Glutathione kinetics in normal man and in patients with liver cirrhosis. J. Hepatol. 26 (3), 606–613.

Bianchi, G., Brizi, M., Rossi, B., Ronchi, M., Grossi, G., Marchesini, G., 2000. Synthesis of glutathione in response to methionine load in control subjects and in patients with cirrhosis. Metabolism 49 (11), 1434–1439.

Bos, C., Stoll, B., Fouillet, H., et al., 2003. Intestinal lysine metabolism is driven by the enteral availability of dietary lysine in piglets fed a bolus meal. Am. J. Physiol. Endocrinol. Metab. 285 (6), E1246–1257.

Brosnan, J.T., Brosnan, M.E., 2006. The sulfur-containing amino acids: an overview. J. Nutr. 136 (6 Suppl), 1636S–1640S.

Calloway, D.H., Odell, A.C., Margen, S., 1971. Sweat and miscellaneous nitrogen losses in human balance studies. J. Nutr. 101 (6), 775–786.

Chessex, P., Lavoie, J.C., Laborie, S., Vallee, J., 1999. Survival of guinea pig pups in hyperoxia is improved by enhanced nutritional substrate availability for glutathione production. Pediatr. Res. 46 (3), 305–310.

Cho, E.S., Sahyoun, N., Stegink, L.D., 1981. Tissue glutathione as a cyst(e)ine reservoir during fasting and refeeding of rats. J. Nutr. 111 (5), 914–922.

Courtney-Martin, G., Chapman, K.P., Moore, A.M., Kim, J.H., Ball, R.O., Pencharz, P.B., 2008a. Total sulfur amino acid requirement and metabolism in parenterally fed postsurgical human neonates. Am. J. Clin. Nutr. 88 (1), 115–124.

Courtney-Martin, G., Rafii, M., Wykes, L.J., Ball, R.O., Pencharz, P.B., 2008b. Methionine-adequate cysteine-free diet does not limit erythrocyte glutathione synthesis in young healthy adult men. J. Nutr. 138 (11), 2172–2178.

Courtney-Martin, G., Moore, A.M., Ball, R.O., Pencharz, P.B., 2010. The addition of cysteine to the total sulphur amino acid requirement as methionine does not increase erythrocytes glutathione synthesis in the parenterally fed human neonate. Pediatr. Res. 67 (3), 320–324.

Cresenzi, C.L., Lee, J.I., Stipanuk, M.H., 2003. Cysteine is the metabolic signal responsible for dietary regulation of hepatic cysteine dioxygenase and glutamate cysteine ligase in intact rats. J. Nutr. 133 (9), 2697–2702.

Crim, M.C., Calloway, D.H., Margen, S., 1975. Creatine metabolism in men: urinary creatine and creatinine excretions with creatine feeding. J. Nutr. 105 (4), 428–438.

Daugherty, M., Polanuyer, B., Farrell, M., et al., 2002. Complete reconstitution of the human coenzyme A biosynthetic pathway via comparative genomics. J. Biol. Chem. 277 (24), 21431–21439.

De Vigneaud, V., Kilmer, G.W., Rachele, J.R., Cohn, M., 1944. On the mechanism of the conversion in vivo of methionine to cystine. J. Biol. Chem. 155, 645–651.

Di Buono, M., Wykes, L.J., Ball, R.O., Pencharz, P.B., 2001a. Dietary cysteine reduces the methionine requirement in men. Am. J. Clin. Nutr. 74 (6), 761–766.

Di Buono, M., Wykes, L.J., Ball, R.O., Pencharz, P.B., 2001b. Total sulfur amino acid requirement in young men as determined by indicator amino acid oxidation with L-[1-13C]phenylalanine. Am. J. Clin. Nutr. 74 (6), 756–760.

Di Buono, M., Wykes, L.J., Cole, D.E., Ball, R.O., Pencharz, P.B., 2003. Regulation of sulfur amino acid metabolism in men in response to changes in sulfur amino acid intakes. J. Nutr. 133 (3), 733–739.

Dimant, E., Landsberg, E., London, I.M., 1955. The metabolic behavior of reduced glutathione in human and avian erythrocytes. J. Biol. Chem. 213 (2), 769–776.

Edozien, J.C., Phillips, E.J., Collis, W.R., 1960. The free amino acids of plasma and urine in kwashiorkor. Lancet 1 (7125), 615–618.

Elango, R., Humayun, M.A., Ball, R.O., Pencharz, P.B., 2009. Indicator amino acid oxidation is not affected by period of adaptation to a wide range of lysine intake in healthy young men. J. Nutr. 139 (6), 1082–1087.

FAO/WHO/UNU. Energy and protein requirements. Report of a joint FAO/WHO/UNU expert consultation. World Health Organization Technical Report Service. 1985;724:1–206.

Finkelstein, J.D., Martin, J.J., 1984. Methionine metabolism in mammals. Distribution of homocysteine between competing pathways. J. Biol. Chem. 259 (15), 9508–9513.

Finkelstein, J.D., Martin, J.J., Harris, B.J., 1986. Effect of dietary cystine on methionine metabolism in rat liver. J. Nutr. 116 (6), 985–990.

Finkelstein, J.D., Martin, J.J., Harris, B.J., 1988. Methionine metabolism in mammals. The methionine-sparing effect of cystine. J. Biol. Chem. 263 (24), 11750–11754.

Fomon, S.J., Ziegler, E.E., Nelson, S.E., Edwards, B.B., 1986. Requirement for sulfur-containing amino acids in infancy. J. Nutr. 116 (8), 1405–1422.

Fukagawa, N.K., Ajami, A.M., Young, V.R., 1996. Plasma methionine and cysteine kinetics in response to an intravenous glutathione infusion in adult humans. Am. J. Physiol. 270 (2 Pt 1), E209–214.

Fuller, M.F., Garlick, P.J., 1994. Human amino acid requirements: can the controversy be resolved? Annu. Rev. Nutr. 14, 217–241.

Ghosh, S., Ting, S., Lau, H., et al., 2004. Increased efflux of glutathione conjugate in acutely diabetic cardiomyocytes. Can. J. Physiol. Pharmacol. 82 (10), 879–887.

Griffith, O.W., 1987. Mammalian sulfur amino acid metabolism: an overview. Methods Enzymol. 143, 366–376.

Hogeveen, M., Blom, H.J., Van Amerongen, M., Boogmans, B., Van Beynum, I.M., Van De Bor, M., 2002. Hyperhomocysteinemia as risk factor for ischemic and hemorrhagic stroke in newborn infants. J. Pediatr. 141 (3), 429–431.

Holt Jr., L.E., 1968. Some problems in dietary amino acid requirements. Am. J. Clin. Nutr. 21 (5), 367–375.

Holt Jr., L.E., Snyderman, S.E., 1961. The amino acid requirements of infants. JAMA 175, 100–103.
Hou, C., Wykes, L.J., Hoffer, L.J., 2003. Urinary sulfur excretion and the nitrogen/sulfur balance ratio reveal nonprotein sulfur amino acid retention in piglets. J. Nutr. 133 (3), 766–772.
Hunter, E.A., Grimble, R.F., 1994. Cysteine and methionine supplementation modulate the effect of tumor necrosis factor alpha on protein synthesis, glutathione and zinc concentration of liver and lung in rats fed a low protein diet. J. Nutr. 124 (12), 2319–2328.
Ingenbleek, Y., Kimura, H., 2013. Nutritional essentiality of sulfur in health and disease. Nutr. Rev. 71 (7), 413–432.
Jackson, A.A., Gibson, N.R., Lu, Y., Jahoor, F., 2004. Synthesis of erythrocyte glutathione in healthy adults consuming the safe amount of dietary protein. Am. J. Clin. Nutr. 80 (1), 101–107.
Jahoor, F., 2012. Effects of decreased availability of sulfur amino acids in severe childhood undernutrition. Nutr. Rev. 70 (3), 176–187.
Jahoor, F., Wykes, L.J., Reeds, P.J., Henry, J.F., del Rosario, M.P., Frazer, M.E., 1995. Protein-deficient pigs cannot maintain reduced glutathione homeostasis when subjected to the stress of inflammation. J. Nutr. 125 (6), 1462–1472.
Jahoor, F., Jackson, A., Gazzard, B., et al., 1999. Erythrocyte glutathione deficiency in symptom-free HIV infection is associated with decreased synthesis rate. Am. J. Physiol. 276 (1 Pt 1), E205–211.
Kanaya, S., Nose, O., Harada, T., et al., 1984. Total parenteral nutrition with a new amino acid solution for infants. J. Pediatr. Gastroenterol. Nutr. 3 (3), 440–445.
Kim, K.I., Elliott, J.I., Bayley, H.S., 1983a. Oxidation of an indicator amino acid by young pigs receiving diets with varying levels of lysine or threonine, and an assessment of amino acid requirements. Br. J. Nutr. 50 (2), 391–399.
Kim, K.I., McMillan, I., Bayley, H.S., 1983b. Determination of amino acid requirements of young pigs using an indicator amino acid. Br. J. Nutr. 50 (2), 369–382.
Kurpad, A.V., Regan, M.M., Varalakshmi, S., et al., 2003. Daily methionine requirements of healthy Indian men, measured by a 24-h indicator amino acid oxidation and balance technique. Am. J. Clin. Nutr. 77 (5), 1198–1205.
Kurpad, A.V., Regan, M.M., Varalakshmi, S., Gnanou, J., Lingappa, A., Young, V.R., 2004a. Effect of cystine on the methionine requirement of healthy Indian men determined by using the 24-h indicator amino acid balance approach. Am. J. Clin. Nutr. 80 (6), 1526–1535.
Kurpad, A.V., Regan, M.M., Varalakshmi, S., Gnanou, J., Young, V.R., 2004b. Daily requirement for total sulfur amino acids of chronically undernourished Indian men. Am. J. Clin. Nutr. 80 (1), 95–100.
Kwon, Y.H., Stipanuk, M.H., 2001. Cysteine regulates expression of cysteine dioxygenase and gamma-glutamylcysteine synthetase in cultured rat hepatocytes. Am. J. Physiol. Endocrinol. Metab. 280 (5), E804–815.
Lee, J.I., Londono, M., Hirschberger, L.L., Stipanuk, M.H., 2004. Regulation of cysteine dioxygenase and gamma-glutamylcysteine synthetase is associated with hepatic cysteine level. J. Nutr. Biochem. 15 (2), 112–122.
Li, Q., Lancaster Jr., J.R., 2013. Chemical foundations of hydrogen sulfide biology. Nitric Oxide 35, 21–34.
Liu, H., Harrell, L.E., Shenvi, S., Hagen, T., Liu, R.M., 2005. Gender differences in glutathione metabolism in Alzheimer's disease. J. Neurosci. Res. 79 (6), 861–867.
Luo, J.L., Hammarqvist, F., Andersson, K., Wernerman, J., 1998. Surgical trauma decreases glutathione synthetic capacity in human skeletal muscle tissue. Am. J. Physiol. 275 (2 Pt 1), E359–365.
Lyons, J., Rauh-Pfeiffer, A., Yu, Y.M., et al., 2000. Blood glutathione synthesis rates in healthy adults receiving a sulfur amino acid-free diet. Proc. Natl. Acad. Sci. U. S. A. 97 (10), 5071–5076.
Lyons, J., Rauh-Pfeiffer, A., Ming-Yu, Y., et al., 2001. Cysteine metabolism and whole blood glutathione synthesis in septic pediatric patients. Crit. Care Med. 29 (4), 870–877.
Malloy, M.H., Rassin, D.K., Richardson, C.J., 1984. Total parenteral nutrition in sick preterm infants: effects of cysteine supplementation with nitrogen intakes of 240 and 400 mg/kg/day. J. Pediatr. Gastroenterol. Nutr. 3 (2), 239–244.
Malmezat, T., Breuille, D., Capitan, P., Mirand, P.P., Obled, C., 2000a. Glutathione turnover is increased during the acute phase of sepsis in rats. J. Nutr. 130 (5), 1239–1246.
Malmezat, T., Breuille, D., Pouyet, C., et al., 2000b. Methionine transsulfuration is increased during sepsis in rats. Am. J. Physiol. Endocrinol. Metab. 279 (6), E1391–1397.
Marti-Carvajal, A.J., Sola, I., Lathyris, D., 2015. Homocysteine-lowering interventions for preventing cardiovascular events. Cochrane Database Syst. Rev. 1, CD006612.
Matthews, D.E., Motil, K.J., Rohrbaugh, D.K., Burke, J.F., Young, V.R., Bier, D.M., 1980. Measurement of leucine metabolism in man from a primed, continuous infusion of L-[1-3C]leucine. Am. J. Physiol. 238 (5), E473–479.
Meister, A., Anderson, M.E., 1983. Glutathione. Annu. Rev. Biochem. 52, 711–760.
Mercier, S., Breuille, D., Buffiere, C., et al., 2006. Methionine kinetics are altered in the elderly both in the basal state and after vaccination. Am. J. Clin. Nutr. 83 (2), 291–298.
Miller, R.G., Jahoor, F., Jaksic, T., 1995a. Decreased cysteine and proline synthesis in parenterally fed, premature infants. J. Pediatr. Surg. 30 (7), 953–957 (discussion 957–958).
Miller, R.G., Jahoor, F., Reeds, P.J., Heird, W.C., Jaksic, T., 1995b. A new stable isotope tracer technique to assess human neonatal amino acid synthesis. J. Pediatr. Surg. 30 (9), 1325–1329.
Mudd, S.H., Poole, J.R., 1975. Labile methyl balances for normal humans on various dietary regimens. Metabolism 24 (6), 721–735.
Mudd, S.H., Ebert, M.H., Scriver, C.R., 1980. Labile methyl group balances in the human: the role of sarcosine. Metabolism 29 (8), 707–720.
Picou, D., Taylor-Roberts, T., 1969. The measurement of total protein synthesis and catabolism and nitrogen turnover in infants in different nutritional states and receiving different amounts of dietary protein. Clin. Sci. 36 (2), 283–296.
Pohlandt, F., 1974. Cystine: a semi-essential amino acid in the newborn infant. Acta Paediatr. Scand. 63 (6), 801–804.
Polhemus, D.J., Lefer, D.J., 2014. Emergence of hydrogen sulfide as an endogenous gaseous signaling molecule in cardiovascular disease. Circ. Res. 114 (4), 730–737.
Predmore, B.L., Lefer, D.J., Gojon, G., 2012. Hydrogen sulfide in biochemistry and medicine. Antioxid. Redox Signal. 17 (1), 119–140.
Rand, W.M., Young, V.R., 1999. Statistical analysis of nitrogen balance data with reference to the lysine requirement in adults. J. Nutr. 129 (10), 1920–1926.

Rand, W.M., Young, V.R., Scrimshaw, N.S., 1976. Change of urinary nitrogen excretion in response to low-protein diets in adults. Am. J. Clin. Nutr. 29 (6), 639–644.

Refsum, H., Ueland, P.M., Nygard, O., Vollset, S.E., 1998. Homocysteine and cardiovascular disease. Annu. Rev. Med. 49, 31–62.

Reid, M., Jahoor, F., 2000. Methods for measuring glutathione concentration and rate of synthesis. Curr. Opin. Clin. Nutr. Metab. Care 3 (5), 385–390.

Reid, M., Badaloo, A., Forrester, T., et al., 2000. In vivo rates of erythrocyte glutathione synthesis in children with severe protein-energy malnutrition. Am. J. Physiol. Endocrinol. Metab. 278 (3), E405–412.

Reid, M., Badaloo, A., Forrester, T., Jahoor, F., 2006. In vivo rates of erythrocyte glutathione synthesis in adults with sickle cell disease. Am. J. Physiol. Endocrinol. Metab. 291 (1), E73–79.

Riedijk, M.A., Stoll, B., Chacko, S., et al., 2007a. Methionine transmethylation and transsulfuration in the piglet gastrointestinal tract. Proc. Natl. Acad. Sci. U. S. A. 104 (9), 3408–3413.

Riedijk, M.A., van Beek, R.H., Voortman, G., de Bie, H.M., Dassel, A.C., van Goudoever, J.B., 2007b. Cysteine: a conditionally essential amino acid in low-birth-weight preterm infants?. Am. J. Clin. Nutr. 86 (4), 1120–1125.

Roediger, W.E., 1995. New views on the pathogenesis of kwashiorkor: methionine and other amino acids. J. Pediatr. Gastroenterol. Nutr. 21 (2), 130–136.

Rose, W.C., 1937. The nutritive significance of the amino acids and certain related compounds. Science 86 (2231), 298–300.

Rose, W.C., Wixom, R.L., 1955. The amino acid requirements of man. XIII. The sparing effect of cystine on the methionine requirement. J. Biol. Chem. 216 (2), 753–773.

Rose, W.C., Johnson, J.E., Haines, W.J., 1950. The amino acid requirements of man I. The role of valine and methionine. J. Biol. Chem. 182, 541–546.

Rose, W.C., Coon, M.J., Lockhart, H.B., Lambert, G.F., 1955. The amino acid requirements of man. XI. The threonine and methionine requirements. J. Biol. Chem. 215 (1), 101–110.

Sekhar, R.V., Patel, S.G., Guthikonda, A.P., et al., 2011. Deficient synthesis of glutathione underlies oxidative stress in aging and can be corrected by dietary cysteine and glycine supplementation. Am. J. Clin. Nutr. 94 (3), 847–853.

Selhub, J., 1999. Homocysteine metabolism. Annu. Rev. Nutr. 19, 217–246.

Selhub, J., Miller, J.W., 1992. The pathogenesis of homocysteinemia: interruption of the coordinate regulation by S-adenosylmethionine of the remethylation and transsulfuration of homocysteine. Am. J. Clin. Nutr. 55 (1), 131–138.

Shew, S.B., Keshen, T.H., Jahoor, F., Jaksic, T., 2005. Assessment of cysteine synthesis in very low-birth weight neonates using a [13C6]glucose tracer. J. Pediatr. Surg. 40 (1), 52–56.

Shoveller, A.K., Brunton, J.A., House, J.D., Pencharz, P.B., Ball, R.O., 2003a. Dietary cysteine reduces the methionine requirement by an equal proportion in both parenterally and enterally fed piglets. J. Nutr. 133 (12), 4215–4224.

Shoveller, A.K., Brunton, J.A., Pencharz, P.B., Ball, R.O., 2003b. The methionine requirement is lower in neonatal piglets fed parenterally than in those fed enterally. J. Nutr. 133 (5), 1390–1397.

Shoveller, A.K., House, J.D., Brunton, J.A., Pencharz, P.B., Ball, R.O., 2004. The balance of dietary sulfur amino acids and the route of feeding affect plasma homocysteine concentrations in neonatal piglets. J. Nutr. 134 (3), 609–612.

Smith, C.V., Jones, D.P., Guenthner, T.M., Lash, L.H., Lauterburg, B.H., 1996. Compartmentation of glutathione: implications for the study of toxicity and disease. Toxicol. Appl. Pharmacol. 140 (1), 1–12.

Snyderman, S.E., Boyer, A., Norton, P.M., Roitman, E., Holt Jr., L.E., 1964. The essential amino acid requirements of infants. X. Methionine. Am. J. Clin. Nutr. 15, 322–330.

Stegink, L.D., Baker, G.L., 1971. Infusion of protein hydrolysates in the newborn infant: plasma amino acid concentrations. J. Pediatr. 78 (4), 595–602.

Stegink, L.D., Den Besten, L., 1972. Synthesis of cysteine from methionine in normal adult subjects: effect of route of alimentation. Science 178 (60), 514–516.

Stipanuk, M.H., 1986. Metabolism of sulfur-containing amino acids. Annu. Rev. Nutr. 6, 179–209.

Stipanuk, M.H., 2004. Sulfur amino acid metabolism: pathways for production and removal of homocysteine and cysteine. Annu. Rev. Nutr. 24, 539–577.

Stipanuk, M.H., Ueki, I., 2011. Dealing with methionine/homocysteine sulfur: cysteine metabolism to taurine and inorganic sulfur. J. Inherit. Metab. Dis. 34 (1), 17–32.

Stipanuk, M.H., Coloso, R.M., Garcia, R.A., Banks, M.F., 1992. Cysteine concentration regulates cysteine metabolism to glutathione, sulfate and taurine in rat hepatocytes. J. Nutr. 122 (3), 420–427.

Stipanuk, M.H., Londono, M., Lee, J.I., Hu, M., Yu, A.F., 2002. Enzymes and metabolites of cysteine metabolism in nonhepatic tissues of rats show little response to changes in dietary protein or sulfur amino acid levels. J. Nutr. 132 (11), 3369–3378.

Stipanuk, M.H., Dominy Jr., J.E., Lee, J.I., Coloso, R.M., 2006. Mammalian cysteine metabolism: new insights into regulation of cysteine metabolism. J. Nutr. 136 (6 Suppl), 1652S–1659S.

Stoll, B., Henry, J., Reeds, P.J., Yu, H., Jahoor, F., Burrin, D.G., 1998. Catabolism dominates the first-pass intestinal metabolism of dietary essential amino acids in milk protein-fed piglets. J. Nutr. 128 (3), 606–614.

Storch, K.J., Wagner, D.A., Burke, J.F., Young, V.R., 1988. Quantitative study in vivo of methionine cycle in humans using [methyl-2H3]- and [1-13C]methionine. Am. J. Physiol. 255 (3 Pt 1), E322–331.

Storch, K.J., Wagner, D.A., Burke, J.F., Young, V.R., 1990. [1-13C; methyl-2H3]methionine kinetics in humans: methionine conservation and cystine sparing. Am. J. Physiol. 258 (5 Pt 1), E790–798.

Sturman, J.A., Gaull, G., Raiha, N.C., 1970. Absence of cystathionase in human fetal liver: is cystine essential? Science 169 (940), 74–76.

Szabo, C., Ransy, C., Modis, K., et al., 2014. Regulation of mitochondrial bioenergetic function by hydrogen sulfide. Part I. Biochemical and physiological mechanisms. Br. J. Pharmacol. 171 (8), 2099–2122.

Troen, A.M., 2005. The central nervous system in animal models of hyperhomocysteinemia. Prog. Neuropsychopharmacol. Biol. Psychiatry 29 (7), 1140–1151.

van Beynum, I.M., Smeitink, J.A., den Heijer, M., te Poele Pothoff, M.T., Blom, H.J., 1999. Hyperhomocysteinemia: a risk factor for ischemic stroke in children. Circulation 99 (16), 2070–2072.

Vidal, K., Breuille, D., Serrant, P., et al., 2014. Long-term cysteine fortification impacts cysteine/glutathione homeostasis and food intake in ageing rats. Eur. J. Nutr. 53 (3), 963–971.

Vina, J., Vento, M., Garcia-Sala, F., et al., 1995. L-cysteine and glutathione metabolism are impaired in premature infants due to cystathionase deficiency. Am. J. Clin. Nutr. 61 (5), 1067–1069.

Waterlow, J.C., Golden, M.H., Garlick, P.J., 1978. Protein turnover in man measured with 15N: comparison of end products and dose regimes. Am. J. Physiol. 235 (2), E165–174.

Wernerman, J., Hammarqvist, F., 1999. Modulation of endogenous glutathione availability. Curr. Opin. Clin. Nutr. Metab. Care 2 (6), 487–492.

Winters, R.W., Heird, W.C., Dell, R.B., et al., 1977. Plasma amino acids in infants receiving parenteral nutrition. In: Greene, H.L., Holliday, M.A., Munro, M.A. (Eds.), Clinical Nutrition Update: Amino Acids. American Medical Association, Chicago, pp. 147–157.

Womack, M., Rose, W.C., 1941. The partial replacement of dietary methionine by cystine for purposes of growth. J. Biol. Chem. 141 (2), 375–379.

Wu, G., Fang, Y.Z., Yang, S., Lupton, J.R., Turner, N.D., 2004. Glutathione metabolism and its implications for health. J. Nutr. 134 (3), 489–492.

Wu, P.Y., Edwards, N., Storm, M.C., 1986. Plasma amino acid pattern in normal term breast-fed infants. J. Pediatr. 109 (2), 347–349.

Wykes, L.J., Ball, R.O., Pencharz, P.B., 1993. Development and validation of a total parenteral nutrition model in the neonatal piglet. J. Nutr. 123 (7), 1248–1259.

Young, V.R., Bier, D.M., 1987. Amino acid requirements in the adult human: how well do we know them? J. Nutr. 117 (8), 1484–1487.

Zello, G.A., Pencharz, P.B., Ball, R.O., 1990. Phenylalanine flux, oxidation, and conversion to tyrosine in humans studied with L-[1-13C] phenylalanine. Am. J. Physiol. 259 (6 Pt 1), E835–843.

Zello, G.A., Wykes, L.J., Ball, R.O., Pencharz, P.B., 1995. Recent advances in methods of assessing dietary amino acid requirements for adult humans. J. Nutr. 125 (12), 2907–2915.

Zlotkin, S.H., Anderson, G.H., 1982a. Sulfur balances in intravenously fed infants: effects of cysteine supplementation. Am. J. Clin. Nutr. 36 (5), 862–867.

Zlotkin, S.H., Anderson, G.H., 1982b. The development of cystathionase activity during the first year of life. Pediatr. Res. 16 (1), 65–68.

Zlotkin, S.H., Bryan, M.H., Anderson, G.H., 1981. Cysteine supplementation to cysteine-free intravenous feeding regimens in newborn infants. Am. J. Clin. Nutr. 34 (5), 914–923.

SECTION IV

DIETARY AMINO ACID AND PROTEIN ON GENE EXPRESSION

CHAPTER

20

Adaptation to Amino Acid Availability: Role of GCN2 in the Regulation of Physiological Functions and in Pathological Disorders*

J. Averous, C. Jousse, A.-C. Maurin, A. Bruhat and P. Fafournoux

Unité de Nutrition Humaine, UMR 1019, INRA, Université d'Auvergne, Centre INRA de Clermont-Ferrand-Theix, Saint Genès Champanelle, France

20.1. INTRODUCTION

In contrast to other macronutrients (lipids or carbohydrates), amino acids exhibit two important characteristics. Firstly, in healthy adult mammals, nine amino acids, called essential amino acids (EAA), cannot be synthesized de novo (valine, isoleucine, leucine, lysine, methionine, phenylalanine, threonine, histidine, and tryptophan). Secondly, there is no specific storage of amino acids analogous to glycogen for glucose or triglyceride in adipose tissue for lipids. Consequently, when necessary, the organism has to hydrolyze endogenous proteins to produce free amino acids. This loss of proteins will be at the expense of tissues such as liver and muscle.

The size of the pool of each amino acid is the result of a balance between their input and removal (Fig. 20.1). The metabolic outlets for amino acids are protein synthesis and amino acid degradation, whereas the inputs are de novo synthesis (for non-EAA), protein breakdown, and dietary supply. Changes in the rates of either one of these systems lead to an adjustment in nitrogen balance. For example, the level of free amino acids in the plasma has been reported to rise after consumption of a protein-containing meal. Conversely, feeding a diet devoid or partially devoid in one or several EAA leads to a dramatic decrease of the plasmatic concentrations of the limiting amino acids. In addition to nutritional factors, various forms of stresses (trauma, thermal burn, sepsis, fevers, etc.) or several chronic illnesses (cancer, AIDS, chronic renal, cardiac, hepatic and pulmonary diseases, etc.) can affect nitrogen metabolism. In such a situation, changes in the patterns of free amino acids are observed in plasma and urine. Therefore, complex mechanisms that take into account amino acid characteristics are required for maintaining the free amino acid pools.

20.1.1 Consequences of a Dietary Amino Acid Deficiency

Protein undernutrition: Prolonged feeding on a low-protein diet causes a fall in the plasma level of most essential amino acids. One of the main consequences is a dramatic inhibition of the growth of young animals. Straus et al. (1993) demonstrated that a dramatic overexpression of IGFBP-1 was responsible for growth inhibition in response to prolonged feeding in a low-protein diet. IGFBP-1 binds the growth factors IGF1 and IGF2 and inhibits their metabolic and mitogenic effects. It has been demonstrated that a fall in the amino acid concentration was directly responsible for IGFBP-1 induction (Straus et al., 1993; Jousse et al., 1998). Therefore, amino acid

*The authors have equally participated to this work.

The Molecular Nutrition of Amino Acids and Proteins.
DOI: http://dx.doi.org/10.1016/B978-0-12-802167-5.00021-9

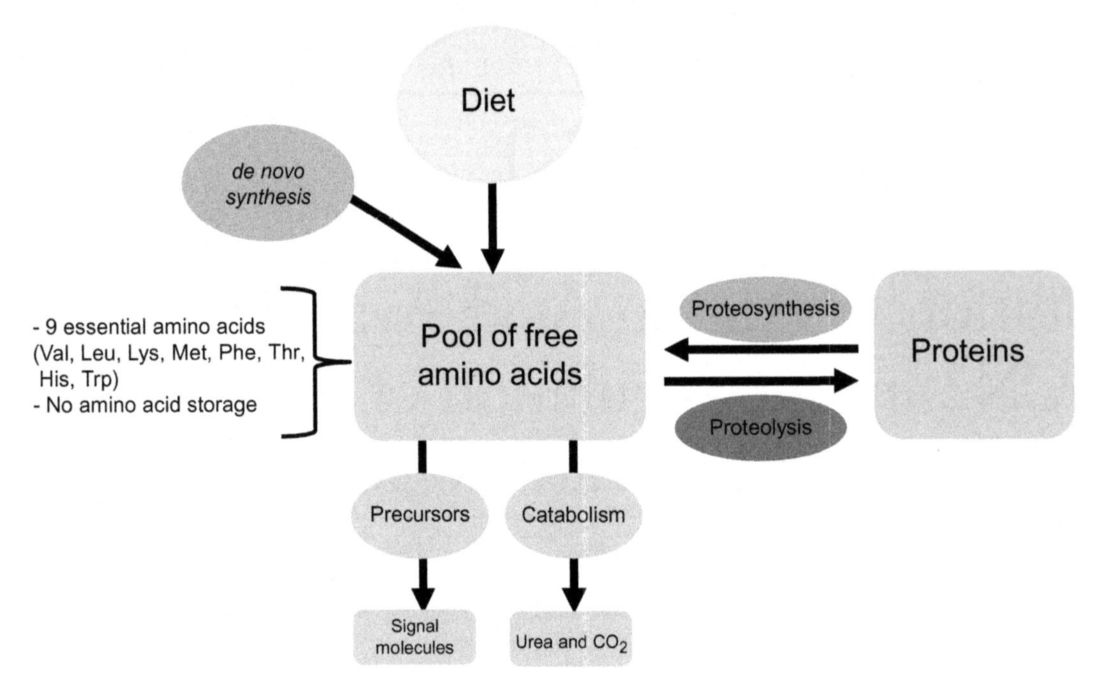

FIGURE 20.1 Biochemical systems involved in the homeostasis of proteins and amino acids.

limitation, as occurring during dietary protein deficiency, participates in the downregulation of growth through the induction of IGFBP-1.

Amino acid imbalanced diet: In the wild, an amino acid-imbalanced diet can be a frequent nutritional situation for omnivorous animals. This is particularly the case when only a single plant protein source is available, which is most likely partially deficient for one EAA. In an animal eating a diet partially lacking an EAA, the blood concentration of the missing amino acid decreases rapidly and to a great extent. The animal responds to this EAA imbalance in the blood by reducing its food intake. This observation has been made for various omnivores (rats, mice, pigs, chickens, etc.). The ability to refuse a diet devoid of one EAA prevents the animal from eating a meal potentially deleterious for its health and encourages the animal to forage for another better source of proteins (Rose, 1957; Gietzen, 1993).

These two examples suggest that a variation in blood amino acid concentrations can regulate several physiological functions including growth and appetite. In the last decade, significant progress has been made in the understanding of the signaling mechanisms involved in the control of physiological functions in response to amino acid limitations. The present review focuses mainly on the role of the GCN2-eIF2α pathway activation upon amino acid scarcity. Finally, we shall discuss how GCN2 function is relevant for the control of several biological processes.

20.2. THE GCN2-EIF2α PATHWAY

At the level of individual cells, amino acid deprivation activates a number of signal transduction pathways that lead to activation of specific transcriptional programs playing a key role in adaptation to nutrient stress or apoptosis. GCN2 is currently the only well-characterized sensor for EAA deficiency and the GCN2-eIF2α-ATF4 pathway is the predominant amino acid responsive signaling mechanism in mammalian cells.

20.2.1 Induction of the GCN2-eIF2α Pathway

The protein kinase GCN2, that is conserved in all eukaryotes, was first identified in yeast as an essential component of the adaptive response to amino acid starvation (Hinnebusch, 1984). This multidomain protein is composed of an N-terminal domain named RWD-domain—this domain being identified in three proteins: RING finger-containing proteins, WD-repeat-containing proteins, and yeast DEAD (DEXD)-like helicases—a

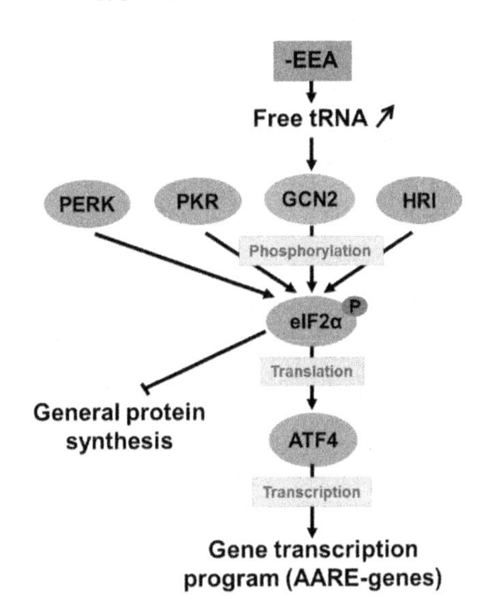

FIGURE 20.2 The mammalian GCN2-eIF2α-ATF4 signaling pathway. The signal transduction pathway triggered in response to amino acid starvation is referred to as the GCN2-eIF2α-ATF4 signaling pathway. The initial step in this pathway is activation by uncharged tRNAs of GCN2 kinase which phosphorylates the α subunit of translation initiation factor eIF2 (eIF2α) on serine 51. This phosphorylation decreases protein synthesis by inhibiting the formation of the preinitiation complex. However, eIF2α phosphorylation also triggers the translation of specific mRNAs including the transcription factor ATF4. Once induced, ATF4 binds to CARE sequence called Amino Acid Response Element (AARE) and induces a gene transcription program. In mammals, three other eIF2α kinases leading to ATF4 expression have been identified: PKR (activated by double-stranded RNA during viral infection), HRI (activated by heme deficiency), and PERK (activated by protein load in the endoplasmic reticulum).

pseudokinase domain without enzymatic activity, a kinase catalytic domain, a histidyl-tRNA synthetase-like domain, and a C-Terminal Domain (CTD) involved in dimerization and binding to the ribosome. The inherent function of this kinase is to adapt the level of protein synthesis to the amount of amino acids (Dever et al., 1992). The other function, related to the first one, is to engage the regulation of the specific gene involved in the adaptation to amino acid deprivation (Hinnebusch, 1984; Fig. 20.2). Indeed, GCN2 can sense amino acid scarcity through its histidyl-tRNA synthetase-like domain that has the ability to bind uncharged tRNA (Dong et al., 2000). This binding leads to a conformational change of GCN2 and induces the autophosphorylation of threonine residues located in the activation loop domain. In yeast, two sites have been characterized, Thr-882 and Thr-887; in mammals, it exists in two equivalent sites but so far only one has been shown to be phosphorylated, Thr-898 in human and Thr-899 in mouse (Harding et al., 2000). This autophosphorylation is required for an effective phosphorylation of the GCN2 target, the eukaryotic translation initiation factor alpha (eIF2α). The phosphorylation of eIF2α on serine 51 induces the inhibition of protein synthesis. In its phosphorylated state eIF2α binds the regulatory subunits of the eIF2 guanine exchange factor (GEF) eIF2B, this event inhibits the eIF2B GEF activity for eIF2 and blocks the formation of the ternary complex (Krishnamoorthy et al., 2001). In addition to the global protein synthesis inhibition, the phosphorylation of eIF2α derepresses the translation of specific mRNAs such as those coding for ATF4, ATF5, CHOP, IBTKα, and gadd34 (Zhou et al., 2008; Palam et al., 2011; Baird et al., 2014; Lee et al., 2009). This mechanism requires the presence of upstream open reading frames (uORFs) in the 5′ untranslated region (UTR) of these mRNAs and has been well described for the mRNA coding for the transcription factor ATF4 (GCN4 in yeast). This mRNA possesses in its 5′UTR two uORFs (GCN4 mRNA possesses 4 uORF) that bypass the translation machinery from ATF4 ORF, resulting in a low level of ATF4 protein. Nevertheless, by decreasing the amount of the ternary complex, the phosphorylation of eIF2α decreases the level of translation occurring at the uORFs and by consequence increases the probability of the ORF of ATF4 to be translated (Vattem and Wek, 2004). Thereafter, ATF4 engages a transcription program of specific genes involved in the adaptive response to amino acid deprivation (see Section 20.3). This response is completed by the induction, at the translational level, of gadd34, a subunit of a phosphatase of eIF2α, that alleviates the inhibition of protein synthesis and allows the translation of the mRNAs encoded by the ATF4 target genes (Novoa et al., 2001). It has to be mentioned that in mammals, three other eIF2α kinases leading to ATF4 expression have been identified: PKR (activated by double-stranded RNA during viral infection); HRI (activated by heme deficiency); and PERK (activated by protein load in the endoplasmic reticulum) (Donnelly et al., 2013; Fig. 20.2).

	AARE core
Trb3 (+287/+272)(+320/+305)(+338/+353)	CGGGTGATGCAAACCG
Chop (−295/−313)	GGGATGATGCAATGTT
Atf3 (−27/−12)	GGGGTGATGCAACGCT
Asns (−72/−57)	GGCATGATGAAACTTC
Snat2 (+724/+709)	AAACTGATGCAATATC
Sqstm1 (−1345/−1360)	AGCATGATGACACACA
Consensus	TGATGMMAH

FIGURE 20.3 Sequence comparison of the AARE. *p62* (−1345/−1360), *Trb3* (+287/+272, +320/+305, +338/+353), *Chop* (−295/−313), *Atf3* (−27/−12), *Asns* (−72/−57), and *Snat2* (+724/+709). The position of the minimum AARE core sequence is indicated by the gray box. The resulting minimum consensus sequence is shown at the bottom (M = A or C; H = A or C or T).

Beside amino acid starvation, other stresses have been described to stimulate GCN2 activity. In mammals, UV exposure has been shown to induce eIF2α phosphorylation in a GCN2-dependent manner (Jiang and Wek, 2005). However, the mechanisms by which UV activates GCN2 are not yet clearly understood. A study demonstrates that the effect of UVB is related to DNA damage and the activity of the DNA-PK (Powley et al., 2009). Another group has shown that UVB induces nitric oxide production, leading to an arginine depletion that in turn activates GCN2 (Bjorkoy et al., 2009). The model of UV radiation has also revealed a new target for GCN2 since under UV exposure this kinase phosphorylates the methionyl-tRNA synthetase (MRS). This event contributes notably to the inhibition of protein synthesis but it also modifies the interaction of MRS with the tumor suppressor AIMP3/P18, a factor involved in DNA repair (Kwon et al., 2011).

20.2.2 Role of the GCN2-eIF2α-ATF4 Pathway in the Transcriptional Regulation of Mammalian Genes by Amino Acid Starvation

The activation of the GCN2-eIF2α-ATF4 pathway triggers a gene transcription program of many genes involved in adaptation to stresses through the binding of ATF4 and of a number of regulatory proteins to specific promoter sequences.

20.2.2.1 Amino Acid Response Elements (AARE) Are CARE Sequences

ATF4 whose translation is induced upon amino acid deprivation, triggers an increased transcription of specific target genes by binding to C/EBP-ATF Response Element (CARE), so named because they are composed of a half-site for the C/EBP family and a half-site for the ATF family of the basic leucine zipper (bZIP) transcription factors (Wolfgang et al., 1997; Fawcett et al., 1999). In the context of amino acid starvation, the CAREs are called Amino Acid Response Elements (AARE). In cultured cell lines, several amino acid-responsive genes such as *Asparagine synthetase* (*Asns*) (Barbosa-Tessmann et al., 2000; Siu et al., 2002; Chen et al., 2004), *Chop* (Bruhat et al., 1997, 2000, 2002), or *Trb3* (Carraro et al., 2010) have been reported to contain an AARE (Fig. 20.3). The AARE sites have a 9-bp core element but the sequences can differ by one or two nucleotides between genes. Consistent with the role of ATF4 as the primary activating factor in the amino acid response pathway, the ATF half-site is well conserved, whereas the C/EBP half-site is often divergent. These AAREs are organized as a single copy of the core sequence in the *Chop*, *Atf3*, *Snat2*, or *Sqstm1* promoters, or as a repetition of three copies in the *Trb3* promoter.

20.2.2.2 ATF4, a Master Regulator of Transcription

ATF4 belongs to the ATF/CREB family of bZIP transcription factors (Ameri and Harris, 2008; Kilberg et al., 2009). Its key role in amino acid regulated transcription has been clearly established (Chen et al., 2004; Pan et al., 2007; Averous et al., 2004). This factor activates transcription by binding to AARE sequences, probably as heterodimers with members of the C/EBP family, although the identity and properties of these proposed heterodimers have not been studied extensively. The Coactivator p300/CBP-associated factor (PCAF) has also been identified as an interaction partner of ATF4 involved in the enhancement of *CHOP* transcription following amino acid starvation (Cherasse et al., 2007). All of the known AARE sites bind ATF4 whereas the binding activity and the role of the other bZIP proteins appear to vary according to the AARE sequence and chromatin structure. One major role of ATF4 is to mediate the induction of a gene expression program referred to as the Integrated Stress Response (ISR), involved in amino acid metabolism, differentiation, metastasis, angiogenesis, resistance to oxidative stress (Harding et al., 2003), and drug resistance (Rzymski et al., 2009).

20.2.2.3 CHOP, a Major Partner of ATF4 to Modulate Transcription of AARE-Containing Genes

Chop is an ATF4 target gene encoding a transcription factor that regulates the expression of a set of stress-induced target genes and modulates the signal initiated by the original stress. CHOP is a nuclear protein related to the CCAAT/enhancer-binding protein (C/EBP) family of transcription factors that dimerize with other members of its family (Ron and Habener, 1992). Several studies have identified CHOP as an interacting partner of C/EBP family members and ATF4 and demonstrated that this factor is an important member of the transcription factor network that controls the stress-induced regulation of specific genes. CHOP can negatively regulate ATF4-dependent transcription of the *ASNS* gene and therefore controls the amino acid-induced regulation of specific AARE-containing genes (Su and Kilberg, 2008). By contrast, CHOP can also be essential for transcriptional activation of several ATF4-dependent genes that can be divided into two classes (Fig. 20.4). Regarding the first class of genes, including *Trb3* and several autophagy genes such as *Sqstm1*, *Nbr1* and *Atg7*, both ATF4 and CHOP are bound to the AARE (Ohoka et al., 2005; B'Chir et al., 2013). Thus, ATF4 and CHOP need to cooperate to regulate AARE-dependent transcription but the nucleotides of the AARE involved in the binding of CHOP remain to be identified. The second class of ATF4-CHOP-dependent genes includes *Atg10*, *Gabarap*, and *Atg5* to the promoter of which CHOP was bound without interacting with ATF4 (B'Chir et al., 2013). In this situation, CHOP and C/EBPβ were bound to a CHOP-RE rather than an AARE suggesting that activation takes place by the binding

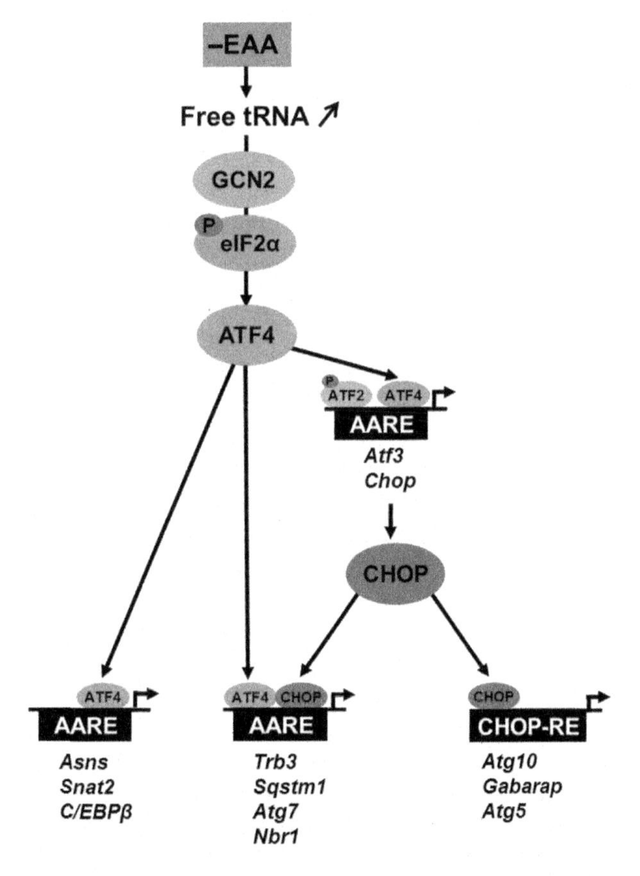

FIGURE 20.4 Role of ATF4 and CHOP in the transcriptional activation of genes in response to amino acid starvation. ATF4 and the phosphorylation of ATF2 are essential in the transcriptional activation of *Atf3* and *Chop* genes. Once expressed, CHOP is itself a transcriptional regulator of amino acid-regulated genes. Three classes of genes have been identified according to their dependence on ATF4 and CHOP and the binding of these factors to the AARE or to CHOP-RE.

of a CHOP-C/EBP heterodimer as described previously (Wang et al., 1998). Therefore, ATF4 can upregulate either directly or indirectly through the induction of CHOP activity, the transcription of a number of AARE-containing genes in response to amino acid starvation.

20.2.2.4 Other Factors Involved in the Transcription of ATF4-Regulated Genes

For some ATF4-regulated genes including *Chop* and *Atf3*, increased transcription also requires phosphorylation of ATF2, another member of the bZIP family of transcription factors (Fig. 20.4; Averous et al., 2004; Bruhat et al., 2007). ATF2 phosphorylation results from the activation of a Gα12 protein-, MEKK1-, MKK7-, and JNK2-dependent pathway in response to amino acid starvation (Chaveroux et al., 2009). ATF2 binds to the AARE in both a starved and unstarved condition and the transactivation capacity of its N-terminal domain is enhanced through phosphorylation of two N-terminal threonine residues, Thr-69 and Thr-71. Phosphorylation of ATF2 has a key role in stimulating an histone acetyl transferase (HAT) activity (Kawasaki et al., 2000; Bhoumik et al., 2005). Thus, ATF2 phosphorylation appears to be involved in promoting the modification of the chromatin structure to enhance *CHOP* and *ATF3* transcription in response to amino acid starvation.

It has also been demonstrated that several other bZIP proteins such as JDP2, ATF3, and C/EBPβ are involved in the control of some of the ATF4-regulated genes.

- The bZIP protein JDP2 (Jun Dimerization Protein 2) can bind to *CHOP* AARE in nonstarved conditions and its binding decreased following amino acid starvation (Cherasse et al., 2008). As this protein was shown to interact with ATF2 and to repress ATF2-mediated transcription by recruiting histone deacetylase HDAC3 to the promoter of target genes (Jin et al., 2002), it was suggested that JDP2 could act as a repressor of CHOP transcription. In fed cells, the ability of the AARE-bound JDP2 to recruit HDAC3 could contribute to the silencing of *CHOP* transcription via maintenance of the hypoacetylation status of histones. However, the mechanism by which amino acid starvation leads to a decrease in JDP2 binding merits further investigation.
- *ATF3* (Pan et al., 2003; Jiang et al., 2004), *C/EBPβ* (Thiaville et al., 2008), and *TRB3* (Carraro et al., 2010) are other ATF4-dependent genes whose expressions are induced in response to amino acid limitation. ATF3 and C/EBPβ proteins are also bZIP factors that act as feedback repressors of the ATF4 signaling in the context of amino acid starvation. Using the *ASNS* gene as a model, Kilberg's laboratory has characterized a self-limiting mechanism in which a prolonged amino acid limitation leads to feedback suppression due to ATF4 activation of ATF3 and C/EBPβ (Chen et al., 2004). TRB3 has a scaffold-like regulatory role for a number of signaling pathways, and in particular it can bind and inhibit ATF4 function (Ord and Ord, 2003, 2005). This protein was identified as a negative feedback regulator of the ATF4-dependent transcription and participates in the fine regulation of the eIF2α-ATF4 pathway (Jousse et al., 2007).

20.2.2.5 Binding Kinetics of ATF4 and Other Factors to AARE-Containing Genes During Amino Acid Deprivation

To illustrate the complexity of the amino acid-dependent regulation of transcription, the kinetics of events that occur at the level of the AARE sequence was investigated. Chromatin immunoprecipitation (ChIP) analysis of the *CHOP* and *ASNS* genes highlights that ATF4 binding to AARE sequences occurs 30–60 min after amino acid deprivation and elevated ATF4 binding continues for 3–4 h (Chen et al., 2004; Bruhat et al., 2007). In the case of *CHOP*, phosphorylation of ATF2 precedes ATF4 binding and the increase of CHOP mRNA. Other regulatory proteins bind AARE directly or indirectly in response to amino acid starvation to finely control gene transcription. JDP2, TRB3, PCAF, and ATF3 are involved in the fine control of *CHOP* transcription while C/EBPβ, ATF3 and CHOP antagonize ATF4 action in the *ASNS* promoter. This self-limiting mechanism of ATF4 action has also been demonstrated for a number of AARE-containing genes.

Taken together, these results demonstrate that following amino acid starvation there is a highly coordinated time-dependent program of interactions between a precise set of bZIP transcription factors and the AARE, leading to the transcriptional activation of AARE-containing genes. Although most of the amino acid responsive genes have AARE sites that are similar in sequence, the key regulator ATF4 is able to associate with various transcription factors and coactivators involved in modulating transcriptional activation. These differences in mechanism would permit flexibility among amino acid-regulated genes in the rapidity and magnitude of the transcriptional response for the same initial signal.

20.3. CONTROL OF PHYSIOLOGICAL FUNCTIONS BY GCN2

Our current understanding of GCN2 function and regulation largely originates from studies in yeast and mammalian cultured cells. From these data, it is clear that the activation of the GCN2-eIF2α-ATF4 pathway has far-reaching consequences for cell physiology in response to a number of stresses (Fig. 20.5). Moreover, an increasing number of data from animal models or clinical studies reveal an important role of the GCN2-eIF2α-ATF4 pathway in the regulation of physiological processes.

20.3.1 GCN2 and Food Intake

Besides cultural and hedonic aspects, both motivation to eat and food choices largely depend on metabolic needs (Lenard and Berthoud, 2008). Part of this homeostatic regulation arises from the capacity to sense nutrient availability and to adapt food selection accordingly (Berthoud et al., 2012). The control of food intake is highly complex in the case of omnivores that have to choose among a variety of available food sources. Notably, the selection of a balanced diet is crucial to maintain the homeostasis of essential amino acids, which cannot be synthesized de novo (Harper and Peters, 1989; Morrison et al., 2012). A remarkable example of an innate mechanism governing food choice is presented by the fact that omnivorous animals will consume substantially less of an otherwise identical meal lacking a single essential amino acid (Gietzen, 1993; Harper et al., 1970). The ability to reject amino acid-imbalanced food sources likely improves fitness by stimulating the search for healthier balanced diets (Chaveroux et al., 2010; Leung et al., 1968).

It has been established that GCN2 contributes to the aversive response to amino acid-imbalanced foods (Hao et al., 2005; Maurin et al., 2005). Following the consumption of a diet deficient in one essential amino acid, the corresponding amino acid concentration in the blood drops rapidly and dramatically, leading to GCN2 activation. Using mice models of genetic ablation of GCN2, it was shown that the onset of food intake inhibition requires the activation of GCN2 specifically in the brain (Maurin et al., 2005). Data further demonstrated that this activation takes place mainly in the mediobasal hypothalamus (Maurin et al., 2014), a major site for the integration of nutritionally relevant information originating from the periphery and mediated by circulating metabolites, hormones, and/or neural pathways (Lenard and Berthoud, 2008; Blouet and Schwartz, 2010). Knockdown experiments of GCN2 in vivo showed that GCN2 activity in this particular area controls food intake according to amino acid availability in the diet. Importantly, pharmacological experiments demonstrated that the level of eIF2α phosphorylation in the mediobasal hypothalamus is sufficient to regulate food intake (Maurin et al., 2014). Interestingly, mTORC1 activity in the same area was also shown to regulate food intake (Blouet et al., 2008; Cota et al., 2006; Harlan et al., 2013). Thus, two amino acid sensors, conserved form yeast to mammals, coexist in the hypothalamus to control food intake according to amino acid availability. While mTORC1 may sense either the body's energy status or postprandial increases in amino-acidemia resulting from protein consumption to downregulate appetite, GCN2 may rather be involved in the adaptation to a nutritional stress leading to the decrease in the concentration of one amino acid in the blood. Moreover, it was also revealed that genetic ablation of GCN2 led to alterations in the selection of macronutrients, although the mechanisms involved remain to be identified (Maurin et al., 2012).

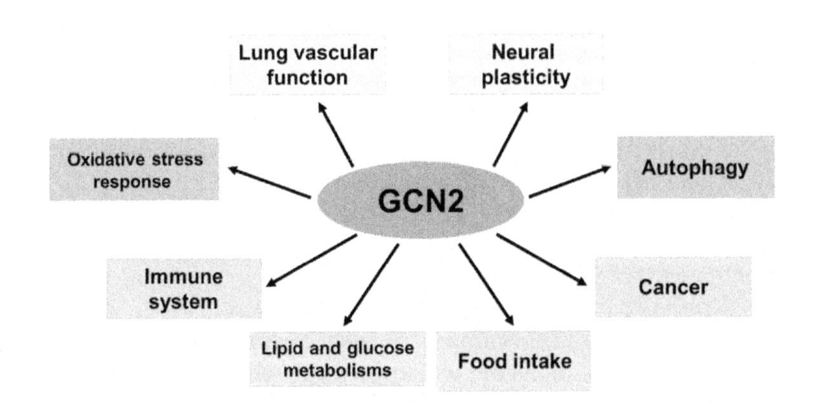

FIGURE 20.5 Implications of GCN2 in the regulation of physiological functions and the occurrence of pathophysiological disorders. In recent years, the literature describes an increasingly important role for EAA-activated GCN2 in regulating several cell (in orange (gray in print versions)) and physiological functions and pathological disorders (in green (light gray in print versions)). In addition, the basal level of GCN2 can control some other functions (in yellow (white in print versions)).

20.3.2 GCN2 and Autophagy

Mammals have the ability to adapt their metabolism to survive in a variable and sometimes hostile environment. The external stimuli to which they must be able to respond include intermittent intake of food and periods of malnutrition. As already mentioned above, adaptation to a low availability of nutrients is especially important for amino acids, and cells employ a number of mechanisms to sense and maintain their homeostatic levels. The animal responds to decreases in aminoacidemia by hydrolyzing body protein in order to produce free amino acids to maintain their homeostasis in all tissues (Goldberg and St John, 1976; Mortimore and Poso, 1987; Mortimore and Schworer, 1977; Schworer and Mortimore, 1979). Several studies revealed that the first tissue to hydrolyze resident proteins when amino acid content is limited is the liver (Mortimore and Poso, 1987; Mortimore and Schworer, 1977; Neely et al., 1977), which is a central organ in whole-body metabolism and contributes to the support of other tissues. Pioneer studies have demonstrated that, in the liver, amino acid limitation increases macroautophagy (hereafter referred to as autophagy), a degradation process involving the engulfment of cytoplasmic components within double-membrane vesicles that finally fuse with lysosomes (Mortimore and Schworer, 1977; He and Klionsky, 2009; Schworer et al., 1981).

The autophagic process involves about 35 autophagy-related genes (Atgs); these genes encode proteins involved in multiprotein complexes that act sequentially (Yang and Klionsky, 2010). Most cells have relatively high amounts of Atgs under normal circumstances. A basal level of autophagy allows a constitutive turnover of cell components, whereas restriction of essential factors, such as amino acids, can trigger an "induced autophagy" (Schworer et al., 1981; Mizushima, 2007; Sarkar, 2013). The autophagic process relies on a machinery that operates in a tightly coordinated fashion (He and Klionsky, 2009; Galluzzi et al., 2014), particularly through numerous posttranslational modifications of proteins. During the first hours of starvation, a cell should be able to generate autophagosomes with Atgs that are already in the cytosol. However, if starvation persists, the renewal of these proteins becomes rapidly vital, requiring the induction of Atgs expression at a transcriptional level (Galluzzi et al., 2014; Cuervo, 2011).

It appears that regulation of autophagy according to amino acid availability inside cells involves the two known amino acid sensors mTORC1 and GCN2 (Galluzzi et al., 2014; Blommaart et al., 1995; Carroll et al., 2014; Kroemer et al., 2010; Meijer et al., 2014; Roczniak-Ferguson et al., 2012). mTORC1 activity opposes autophagy in amino acid-rich conditions, whereas its inhibition upregulates autophagy upon amino acid deprivation, by promoting both posttranslational modifications and transcriptional induction of Atgs (Roczniak-Ferguson et al., 2012; Ganley et al., 2009; Hosokawa et al., 2009; Jung et al., 2009; Kim et al., 2011; Martina et al., 2012; Settembre et al., 2012). However, a set of data clearly shows that GCN2 also regulates autophagy depending on amino acid availability inside cells. Indeed, it has been demonstrated that eIF2α signaling can regulate autophagy in both yeast and mammalian cells (Talloczy et al., 2002; Ye et al., 2010). Whether or not GCN2 activation contributes to the early steps of initiating the autophagic process is still not known (Kroemer et al., 2010). However, during amino acid starvation, it has been clearly established that the GCN2/eIF2α/ATF4 pathway enhances the transcription of a number of *Atgs* involved in the formation, maturation, and functioning of autophagosomes (B'Chir et al., 2013), allowing the cell to maintain the level of autophagy required to cope with stress and restore amino acid homeostasis.

20.3.3 Role of GCN2 in Neural Plasticity

The late phase of long-term potentiation (L-LTP) and long-term memory (LTM) formation require long-lasting changes in synaptic function (Bailey and Kandel, 1993), a cellular mechanism that is dependent on new protein synthesis (Kandel, 2001). Recent molecular and genetic studies have provided new insights into the molecular mechanisms underlying these processes. Particularly, it has been shown that translational control by the eIF2α signaling pathway plays an important role in long-term synaptic plasticity and memory consolidation (Costa-Mattioli et al., 2007). Interestingly, GCN2 is the only eIF2α kinase that is evolutionarily conserved from yeast to mammals (Costa-Mattioli et al., 2007; Hinnebusch, 1990) and that is enriched in the brain of flies (Santoyo, 1997) and mammals (Berlanga et al., 1999; Sood et al., 2000), especially in the hippocampus (Costa-Mattioli et al., 2005). Moreover, both LTM and L-LTP were found to be enhanced in the hippocampus of mice lacking GCN2 (Costa-Mattioli et al., 2007). Conversely, hippocampal infusion with a small molecule which prevents eIF2α dephosphorylation (Sal003, a potent derivative of salubrinal; Robert et al., 2006) blocks both L-LTP and LTM formation (Costa-Mattioli et al., 2007).

20.3.3.1 eIF2α Phosphorylation Control L-LTP and LTM

Recent evidence supports the idea that eIF2α phosphorylation regulates L-LTP and LTM storage through translational control of specific mRNAs, such as ATF4 mRNA. As previously described, eIF2α phosphorylation causes both the upregulation of ATF4 mRNA translation and inhibition of protein synthesis. Importantly, ATF4 and its homologs are repressors of cAMP response element binding protein (CREB)-mediated gene expression, which is widely considered to be required for the expression of long-lasting synaptic plasticity genes and thus memory formation.

Thus, eIF2α phosphorylation regulates two fundamental processes that are crucial for the storage of new memories: new protein synthesis and CREB-mediated gene expression via translational control of ATF4 mRNA.

20.3.3.2 Developmental Role of the Activation of the Pathway and Role of Impact

IMPACT is an inhibitor of GCN2 that is highly abundant in the brain. Neurons expressing high levels of IMPACT are found in most areas of the brain (Bittencourt et al., 2008). Given the physiological relevance of GCN2 in food intake regulation, altered synaptic plasticity and memory, the available data thus indicate that proper regulation of GCN2 activity is crucial in the central nervous system (CNS).

Neuronal IMPACT is developmentally upregulated, promoting protein synthesis and neuritogenesis (a fundamental event in brain development), opposing GCN2 activity (Roffe et al., 2013). The increased abundance of IMPACT may promote translation by maintaining low levels of active GCN2 in a timely manner to support neurite outgrowth. Regions that display high levels of IMPACT exhibit an inhibition of basal GCN2 activity and therefore of ATF4 expression, which might contribute to the differences in the long-term synaptic plasticity with regions where IMPACT is not expressed (Roffe et al., 2013).

20.3.4 Role of GCN2 in Lipid and Glucose Metabolism During Leucine Deprivation

There is a body of evidence that GCN2 functions as a regulator of metabolic adaptation to long-term deprivation of essential amino acid. Guo and Cavener (2007) highlighted the role of GCN2 in regulating lipid metabolism in the liver during leucine deprivation. These authors showed that lipid synthesis was repressed in the livers of GCN2$^{+/+}$ mice during prolonged leucine starvation, whereas lipid synthesis continued unabated in GCN2$^{-/-}$ mice, resulting in severe steatosis. Failure to downregulate lipid synthesis was found to be due to persistent expression of sterol regulatory element-binding protein 1c (Srebp-1c) protein and its downstream transcriptional targets involved in fatty-acid and triglyceride synthesis. Interestingly, this phenomenon was shown not to be dependent on ATF4 as ATF4$^{-/-}$ mice did not develop fatty liver and were able to repress expression of the fatty acid synthase mRNA (Guo and Cavener, 2007). Therefore, the signaling pathway linking GCN2 activity to the regulation of Srebp-1c expression remains to be identified.

Recent studies reveal that GCN2 is also involved in regulating insulin sensitivity and glucose metabolism during individual branched-chain amino acids (BCAAs) deprivation (Schneider et al., 2011; Xiao et al., 2011). BCAAs-deficient diets improve insulin signaling in the liver by activation of GCN2 as measured by increased phosphorylation of the insulin receptor, and whole-body insulin sensitivity as measured by an insulin tolerance test. In GCN2-knockout mice fed on a BCAAs-deficient diet, mTOR signaling in the liver is increased and the improvement in insulin sensitivity is lost indicating that GCN2 functions as an upstream inhibitor of mTOR under BCAAs deprivation.

For a long time, controls of lipid and protein metabolisms were considered to be relatively independent. These data give an overview of the role of GCN2 in integrated regulation mechanisms taking account of variations in the availability of diverse types of nutrients.

20.3.5 Role of GCN2 in the Immune System

As uncontrolled immune activation can be lethal, the immune cell function has to be finely controlled. Metabolic inputs such as insulin, oxygen, and amino acids indirectly influence T-cell growth and function, notably by modulating both molecular pathways and cytokine signaling. The signaling pathway controlled by GCN2 has been involved in the regulation of the immune system at different levels. First, it has been involved in the management of the fate of pathogens in infected epithelial cells. Indeed, infection of epithelial cells with *Shigella* and *Salmonella* triggers an acute intracellular amino acid starvation due to host membrane damage. This pathogen-induced amino acid starvation activates the cellular GCN2-eIF2α-ATF4 pathway, triggering a protective innate immune response against bacteria (Lemaitre and Girardin, 2013; Tattoli et al., 2012). This response involves a net increase of autophagic activity in infected cells, together with an increased production of inflammatory cytokines through AFT3 and ATF4 induction, and potentiation of the NF-κB pathway. Secondly, the GCN2-eIF2α pathway is used in the dialog between dendritic cells and T lymphocytes in order to promote anergy, by moderating T-cell response. This process is notably used in the context of body immune tolerance regarding its own components. Indeed, activated dendritic cells express high amounts of the amino acid-consuming enzymes (AACE) Indoleamine 2,3 dioxygenase (IDO) and arginase, the activity of which results in the local depletion of tryptophan and arginine, respectively (Mellor and Munn, 2008; Munn et al., 2005; Pierre, 2009). The rise in the levels of uncharged transfer RNAs (uncharged $tRNA^{Trp}$ or $tRNA^{Arg}$) in neighboring $CD4^+$ T cells activates GCN2, which in turn promotes cell cycle arrest, as well as differentiation in regulatory T cells. Consequently, GCN2-knock-out T cells are refractory to IDO-induced anergy (Munn et al., 2005). Thus, GCN2 has been proposed to act as a molecular sensor in T cells, allowing them to detect and respond to the IDO-dependent immunoregulatory signal generated by dendrictic cells. Thirdly, GCN2 activation in T helper 17 (T_H17) cells, a subset of $CD4^+$ effectors that have recently emerged as important and broad mediators of immunity, inhibits cell differentiation and function (Carlson et al., 2014; Sundrud et al., 2009), thereby emerging as an important regulator of processes involved in autoimmunity and cancer (Sundrud and Trivigno, 2013). Thus, GCN2-eIF2α activation may protect against pathophysiologic inflammation by enforcing the tolerogenic effects of IDO-expressing dendritic cells and concomitantly blunting T_H17 differentiation. Finally, a recently published study highlights the role of GCN2 activation in dendritic cells in modulating the adaptive immune response (Ravindran et al., 2014). These data demonstrate a key role for virus-induced GCN2 activation in programming dendritic cells to initiate autophagy and enhanced antigen presentation to both $CD4^+$ and $CD8^+$ T cells.

20.4. INVOLVEMENT OF GCN2 IN PATHOLOGY

20.4.1 GCN2 and Cancer

Several studies have established that GCN2 is necessary for the adaption of tumor cells to the hostile condition that they generate (Ye et al., 2010; Wang et al., 2013). Due to their uncontrolled proliferation rate tumor cells are rapidly exposed to an environment that is deprived in oxygen but also in nutrients. It has been shown by Wang et al. that GCN2 is required for the expression of VEGF (vascular endothelial growth factor), a major angiogenic factor, in tumor context. This regulation is coherent with the need for the tumor to increase the supply in nutrients, notably amino acids, when its environment is deprived. Interestingly, it appears that some tumors express a higher level of GCN2 (Wang et al., 2013). The study of Ye et al. has demonstrated that GCN2 and ATF4 were required for tumor growth and survival (Ye et al., 2010). They provide the evidence that the transcription of the ATF4 target gene *ASNS* is necessary for tumor cell survival. ASNS is an enzyme that participates to the synthesis of asparagine. It is known that certain types of tumors present a low level of ASNS, rendering tumor cells sensitive to asparaginase treatment. This is the case of childhood acute lymphoblastic leukemia (ALL) primary cells (Balasubramanian et al., 2013). Contrariwise, ALL cell lines selected for their resistance to asparaginase treatment present a higher expression of ASNS (Aslanian and Kilberg, 2001). Moreover, it has been demonstrated that GCN2 is required for the adaptation to the toxic effect of asparaginase treatment (Wilson et al., 2013). This suggests that the inhibition of GCN2 could represent a suitable strategy to improve the efficiency of asparaginase treatment.

The ability of GCN2 to control amino acid synthesis is all the more important because amino acids exert important effects on energetic metabolism, that is, by providing metabolic intermediates. This aspect may be crucial in the singular case of tumors. The example of serine, an allosteric activator of PKM2 (pyruvate kinase muscle 2) should be mentioned. This enzyme of the TCA cycle is found to be preferentially expressed in cancer cells. In a serine-starved condition the inhibition of PKM2 activity contributes to the accumulation of glycolytic intermediates that feed the synthesis of serine by enzymes that are dependent on GCN2 for their expression. This defines a key role for GCN2 in a mechanism that links serine synthesis to glycolytic flux and allows sustaining cancer cell proliferation in a serine-deprived condition. In addition to control of the supply of amino acids through the regulation of angiogenesis or amino acid synthesis, GCN2 can also impact more directly on the metabolic programming of cancer cell. Notably, GCN2 activation has been shown to reduce the translation level of the β-F1-ATPase, a subunit of the mitochondrial H^+-ATP synthase (Martinez-Reyes et al., 2012). A low level of expression of this enzyme contributes to the decrease of oxidative phosphorylation and is considered as a marker of many cancer cells (Cuezva et al., 2002). In addition, GCN2 activation induces PEPCK-M (mitochondrial phosphoenolpyruvate carboxykinase) expression, an enzyme that has been shown to be involved in tumor cell survival under nutrient deprivation (Méndez-Lucas et al., 2014). The precise contribution of PEPCK-M in tumor metabolism is not clearly defined, but it has been proposed that it could contribute to the synthesis of serine. Thus, by promoting glycolysis and by limiting oxidative metabolism GCN2 could contribute to the Warburg effect, a hallmark of cancer cell (Warburg, 1956).

Another important feature of tumors is their capacity to deal with oxidative stress. There is a growing interest in the role of the GCN2 substrate eIF2α, in this capacity to manage oxidative stress (Rajesh et al., 2015). The study of Rajesh et al. demonstrates that the phosphorylation of eIF2α protects cells from oxidative stress induced by antitumor treatments (Rajesh et al., 2013). So, targeting eIF2α phosphorylation could represent an efficient way to improve the effect of pro-oxidant drugs. Interestingly, in specific situations, GCN2 seems to contribute to maintaining the redox status. Indeed, we established, in mice, that in the absence of GCN2 the consumption of a leucine-devoid diet provokes an oxidative stress (Chaveroux et al., 2011). It is certain that several other functions regulated by GCN2 might represent interesting targets to tackle cancer cells. GCN2, by its role in immune cell programming, could be targeted in order to increase the antitumor efficiency of the immune system (Platten et al., 2014). In addition, targeting GCN2 could be also a way to modulate the level of autophagy in tumor cells, as this process appears to have a major role in tumor growth and survival (Galluzzi et al., 2015).

It has also to be mentioned, that if eIF2α phosphorylation can protect cells from stresses, it can also promote apoptosis (Srivastava et al., 1998). It has been notably proposed that inducing eIF2α phosphorylation could increase the efficiency of antitumor treatments (Schewe and Aguirre-Ghiso, 2009). That illustrates that the comprehension of the role of GCN2 in tumor formation and its survival capacity is complex and cannot be restricted to one model.

20.4.2 Role in Lung Vascular Function

Pulmonary capillary hemangiomatosis (PCH) and pulmonary veno-occlusive disease (PVOD) are causes of pulmonary hypertension, which is a relatively uncommon disorder affecting the lung and the heart. The clinical features of these disorders are progressive dyspnea, cough, occasional hemoptysis, and profound reductions of carbon monoxide diffusion (Montani et al., 2010). Recently, two research groups independently identified mutations in the gene encoding GCN2 as the cause of PCH and PVOD (Best et al., 2014b; Eyries et al., 2014). The link between a dysfunction of the GCN2 pathway and a failure in vascular cell proliferation and/or lung vessels remodeling remain difficult to understand for the moment (Eyries et al., 2014). GCN2 could play a protective role of blood vessels against oxidative stress and protein carbonylation, as suggested by previous data (Chaveroux et al., 2011). Studies are currently ongoing to better characterize the role of GCN2 loss-of-function in pathogenesis of these disorders (Best et al., 2014a).

20.5. CONCLUSION

The last few years have seen a growing amount of experimental evidence implicating the GCN2-eIF2α pathway in multiple unsuspected physiological pathways and diseases (Fig. 20.5). However, as highlighted throughout this review, there are still several important gaps that need to be filled in the molecular mechanisms involved in the regulation of the basal and activated level of the GCN2-eIF2α pathway. Thus, it is fundamental to gain a detailed understanding of the function and regulation of all the steps of this pathway in order to provide new drug targets for correcting and preventing diseases/disorders associated with GCN2/eIF2α deregulation.

References

Ameri, K., Harris, A.L., 2008. Activating transcription factor 4. Int. J. Biochem. Cell. Biol. 40, 14–21.

Aslanian, A.M., Kilberg, M.S., 2001. Multiple adaptive mechanisms affect asparagine synthetase substrate availability in asparaginase-resistant MOLT-4 human leukaemia cells. Biochem. J. 358, 59–67.

Averous, J., Bruhat, A., Jousse, C., Carraro, V., Thiel, G., Fafournoux, P., 2004. Induction of CHOP expression by amino acid limitation requires both ATF4 expression and ATF2 phosphorylation. J. Biol. Chem. 279, 5288–5297.

Bailey, C.H., Kandel, E.R., 1993. Structural changes accompanying memory storage. Annu. Rev. Physiol. 55, 397–426.

Baird, T.D., Palam, L.R., Fusakio, M.E., Willy, J.A., Davis, C.M., McClintick, J.N., et al., 2014. Selective mRNA translation during eIF2 phosphorylation induces expression of IBTKα. Mol. Biol. Cell 25, 1686–1697.

Balasubramanian, M.N., Butterworth, E.A., Kilberg, M.S., 2013. Asparagine synthetase: regulation by cell stress and involvement in tumor biology. Am. J. Physiol. Endocrinol. Metab. 304, E789–E799.

Barbosa-Tessmann, I.P., Chen, C., Zhong, C., Siu, F., Schuster, S.M., Nick, H.S., et al., 2000. Activation of the human asparagine synthetase gene by the amino acid response and the endoplasmic reticulum stress response pathways occurs by common genomic elements. J. Biol. Chem. 275, 26976–26985.

B'Chir, W., Maurin, A.C., Carraro, V., Averous, J., Jousse, C., Muranishi, Y., et al., 2013. The eIF2α/ATF4 pathway is essential for stress-induced autophagy gene expression. Nucleic Acids Res. 41, 7683–7699.

Berlanga, J.J., Santoyo, J., de Haro, C., 1999. Characterization of a mammalian homolog of the GCN2 eukaryotic initiation factor 2α kinase. Eur. J. Biochem. 265, 754–762.

Berthoud, H.R., Munzberg, H., Richards, B.K., Morrison, C.D., 2012. Neural and metabolic regulation of macronutrient intake and selection. Proc. Nutr. Soc. 71, 390–400.

Best, D.H., Austin, E.D., Chung, W.K., Elliott, C.G., 2014a. Genetics of pulmonary hypertension. Curr. Opin. Cardiol. 29, 520–527.

Best, D.H., Sumner, K.L., Austin, E.D., Chung, W.K., Brown, L.M., Borczuk, A.C., et al., 2014b. EIF2AK4 mutations in pulmonary capillary hemangiomatosis. Chest 145, 231–236.

Bhoumik, A., Takahashi, S., Breitweiser, W., Shiloh, Y., Jones, N., Ronai, Z., 2005. ATM-dependent phosphorylation of ATF2 is required for the DNA damage response. Mol. Cell 18, 577–587.

Bittencourt, S., Pereira, C.M., Avedissian, M., Delamano, A., Mello, L.E., Castilho, B.A., 2008. Distribution of the protein IMPACT, an inhibitor of GCN2, in the mouse, rat, and marmoset brain. J. Comp. Neurol. 507, 1811–1830.

Bjorkoy, G., Lamark, T., Pankiv, S., Overvatn, A., Brech, A., Johansen, T., 2009. Monitoring autophagic degradation of p62/SQSTM1. Methods Enzymol. 452, 181–197.

Blommaart, E.F., Luiken, J.J., Blommaart, P.J., van Woerkom, G.M., Meijer, A.J., 1995. Phosphorylation of ribosomal protein S6 is inhibitory for autophagy in isolated rat hepatocytes. J. Biol. Chem. 270, 2320–2326.

Blouet, C., Schwartz, G.J., 2010. Hypothalamic nutrient sensing in the control of energy homeostasis. Behav. Brain Res. 209, 1–12.

Blouet, C., Ono, H., Schwartz, G.J., 2008. Mediobasal hypothalamic p70 S6 kinase 1 modulates the control of energy homeostasis. Cell Metab. 8, 459–467.

Bruhat, A., Jousse, C., Wang, X.Z., Ron, D., Ferrara, M., Fafournoux, P., 1997. Amino acid limitation induces expression of CHOP, a CCAAT/enhancer binding protein-related gene, at both transcriptional and post- transcriptional levels. J. Biol. Chem. 272, 17588–17593.

Bruhat, A., Jousse, C., Carraro, V., Reimold, A.M., Ferrara, M., Fafournoux, P., 2000. Amino acids control mammalian gene transcription: activating transcription factor 2 is essential for the amino acid responsiveness of the CHOP promoter. Mol. Cell. Biol. 20, 7192–7204.

Bruhat, A., Averous, J., Carraro, V., Zhong, C., Reimold, A.M., Kilberg, M.S., et al., 2002. Differences in the molecular mechanisms involved in the transcriptional activation of the CHOP and asparagine synthetase genes in response to amino acid deprivation or activation of the unfolded protein response. J. Biol. Chem. 277, 48107–48114.

Bruhat, A., Cherasse, Y., Maurin, A.C., Breitwieser, W., Parry, L., Deval, C., et al., 2007. ATF2 is required for amino acid-regulated transcription by orchestrating specific histone acetylation. Nucleic Acids Res. 35, 1312–1321.

Carlson, T.J., Pellerin, A., Djuretic, I.M., Trivigno, C., Koralov, S.B., Rao, A., et al., 2014. Halofuginone-induced amino acid starvation regulates Stat3-dependent Th17 effector function and reduces established autoimmune inflammation. J. Immunol. 192, 2167–2176.

Carraro, V., Maurin, A.C., Lambert-Langlais, S., Averous, J., Chaveroux, C., Parry, L., et al., 2010. Amino acid availability controls TRB3 transcription in liver through the GCN2/eIF2α/ATF4 pathway. PLoS One 5 (12), e15716.

Carroll, B., Korolchuk, V.I., Sarkar, S., 2014. Amino acids and autophagy: cross-talk and co-operation to control cellular homeostasis. Amino Acids 47 (10), 2065–2088.

Chaveroux, C., Jousse, C., Cherasse, Y., Maurin, A.C., Parry, L., Carraro, V., et al., 2009. Identification of a novel amino acid response pathway triggering ATF2 phosphorylation in mammals. Mol. Cell. Biol. 29, 6515–6526.

Chaveroux, C., Lambert-Langlais, S., Cherasse, Y., Averous, J., Parry, L., Carraro, V., et al., 2010. Molecular mechanisms involved in the adaptation to amino acid limitation in mammals. Biochimie 92, 736–745.

Chaveroux, C., Lambert-Langlais, S., Parry, L., Carraro, V., Jousse, C., Maurin, A.C., et al., 2011. Identification of GCN2 as new redox regulator for oxidative stress prevention in vivo. Biochem. Biophys. Res. Commun. 415, 120–124.

Chen, H., Pan, Y.X., Dudenhausen, E.E., Kilberg, M.S., 2004. Amino acid deprivation induces the transcription rate of the human asparagine synthetase gene through a timed program of expression and promoter binding of nutrient-responsive basic region/leucine zipper transcription factors as well as localized histone acetylation. J. Biol. Chem. 279, 50829–50839.

Cherasse, Y., Maurin, A.C., Chaveroux, C., Jousse, C., Carraro, V., Parry, L., et al., 2007. The p300/CBP-associated factor (PCAF) is a cofactor of ATF4 for amino acid-regulated transcription of CHOP. Nucleic Acids Res. 35, 5954–5965.

Cherasse, Y., Chaveroux, C., Jousse, C., Maurin, A.C., Carraro, V., Parry, L., et al., 2008. Role of the repressor JDP2 in the amino acid-regulated transcription of CHOP. FEBS Lett. 582, 1537–1541.

Costa-Mattioli, M., Gobert, D., Harding, H., Herdy, B., Azzi, M., Bruno, M., et al., 2005. Translational control of hippocampal synaptic plasticity and memory by the eIF2α kinase GCN2. Nature 436, 1166–1173.

Costa-Mattioli, M., Gobert, D., Stern, E., Gamache, K., Colina, R., Cuello, C., et al., 2007. eIF2α phosphorylation bidirectionally regulates the switch from short- to long-term synaptic plasticity and memory. Cell 129, 195–206.

Cota, D., Proulx, K., Smith, K.A., Kozma, S.C., Thomas, G., Woods, S.C., et al., 2006. Hypothalamic mTOR signaling regulates food intake. Science 312, 927–930.

Cuervo, A.M., 2011. Cell biology. Autophagy's top chef. Science 332, 1392–1393.

Cuezva, J.M., Krajewska, M., de Heredia, M.L., Krajewski, S., Santamaria, G., Kim, H., et al., 2002. The bioenergetic signature of cancer: a marker of tumor progression. Cancer Res. 62, 6674–6681.

Dever, T.E., Feng, L., Wek, R.C., Cigan, A.M., Donahue, T.F., Hinnebusch, A.G., 1992. Phosphorylation of initiation factor 2 alpha by protein kinase GCN2 mediates gene-specific translational control of GCN4 in yeast. Cell 68, 585–596.

Dong, J., Qiu, H., Garcia-Barrio, M., Anderson, J., Hinnebusch, A.G., 2000. Uncharged tRNA activates GCN2 by displacing the protein kinase moiety from a bipartite tRNA-binding domain. Mol. Cell 6, 269–279.

Donnelly, N., Gorman, A.M., Gupta, S., Samali, A., 2013. The eIF2α kinases: their structures and functions. Cell. Mol. Life. Sci. 70, 3493–3511.

Eyries, M., Montani, D., Girerd, B., Perret, C., Leroy, A., Lonjou, C., et al., 2014. EIF2AK4 mutations cause pulmonary veno-occlusive disease, a recessive form of pulmonary hypertension. Nat. Genet. 46, 65–69.

Fawcett, T.W., Martindale, J.L., Guyton, K.Z., Hai, T., Holbrook, N.J., 1999. Complexes containing activating transcription factor (ATF)/cAMP-responsive-element-binding protein (CREB) interact with the CCAAT/enhancer-binding protein (C/EBP)-ATF composite site to regulate Gadd153 expression during the stress response. Biochem. J. 339, 135–141.

Galluzzi, L., Pietrocola, F., Levine, B., Kroemer, G., 2014. Metabolic control of autophagy. Cell 159, 1263–1276.

Galluzzi, L., Pietrocola, F., Bravo-San Pedro, J.M., Amaravadi, R.K., Baehrecke, E.H., Cecconi, F., et al., 2015. Autophagy in malignant transformation and cancer progression. EMBO J. 34, 856–880.

Ganley, I.G., Lam du, H., Wang, J., Ding, X., Chen, S., Jiang, X., 2009. ULK1.ATG13.FIP200 complex mediates mTOR signaling and is essential for autophagy. J. Biol. Chem. 284, 12297–12305.

Gietzen, D.W., 1993. Neural mechanisms in the responses to amino acid deficiency. J. Nutr. 123, 610–625.

Goldberg, A.L., St John, A.C., 1976. Intracellular protein degradation in mammalian and bacterial cells: Part 2. Annu. Rev. Biochem. 45, 747–803.

Guo, F., Cavener, D.R., 2007. The GCN2 eIF2α kinase regulates fatty-acid homeostasis in the liver during deprivation of an essential amino acid. Cell Metab. 5, 103–114.

Hao, S., Sharp, J.W., Ross-Inta, C.M., McDaniel, B.J., Anthony, T.G., Wek, R.C., et al., 2005. Uncharged tRNA and sensing of amino acid deficiency in mammalian piriform cortex. Science 307, 1776–1778.

Harding, H.P., Novoa, I.I., Zhang, Y., Zeng, H., Wek, R., Schapira, M., et al., 2000. Regulated translation initiation controls stress-induced gene expression in mammalian cells. Mol. Cell 6, 1099–1108.

Harding, H.P., Zhang, Y., Zeng, H., Novoa, I., Lu, P.D., Calfon, M., et al., 2003. An integrated stress response regulates amino acid metabolism and resistance to oxidative stress. Mol. Cell 11, 619–633.

Harlan, S.M., Guo, D.F., Morgan, D.A., Fernandes-Santos, C., Rahmouni, K., 2013. Hypothalamic mTORC1 signaling controls sympathetic nerve activity and arterial pressure and mediates leptin effects. Cell Metab. 17, 599–606.

Harper, A.E., Peters, J.C., 1989. Protein intake, brain amino acid and serotonin concentrations and protein self-selection. J. Nutr. 119, 677–689.

Harper, A.E., Benevenga, N.J., Wohlhueter, R.M., 1970. Effects of ingestion of disproportionate amounts of amino acids. Physiol. Rev. 50, 428–558.

He, C., Klionsky, D.J., 2009. Regulation mechanisms and signaling pathways of autophagy. Annu. Rev. Genet. 43, 67–93.

Hinnebusch, A.G., 1984. Evidence for translational regulation of the activator of general amino acid control in yeast. Proc. Natl. Acad. Sci. U.S. A. 81, 6442–6446.

Hinnebusch, A.G., 1990. Involvement of an initiation factor and protein phosphorylation in translational control of GCN4 mRNA. Trends Biochem. Sci. 15, 148–152.

Hosokawa, N., Hara, T., Kaizuka, T., Kishi, C., Takamura, A., Miura, Y., et al., 2009. Nutrient-dependent mTORC1 association with the ULK1-Atg13-FIP200 complex required for autophagy. Mol. Biol. Cell 20, 1981–1991.

Jiang, H.Y., Wek, R.C., 2005. Phosphorylation of the alpha-subunit of the eukaryotic initiation factor-2 (eIF2α) reduces protein synthesis and enhances apoptosis in response to proteasome inhibition. J. Biol. Chem. 280, 14189–14202.

Jiang, H.Y., Wek, S.A., McGrath, B.C., Lu, D., Hai, T., Harding, H.P., et al., 2004. Activating transcription factor 3 is integral to the eukaryotic initiation factor 2 kinase stress response. Mol. Cell. Biol. 24, 1365–1377.

Jin, C., Li, H., Murata, T., Sun, K., Horikoshi, M., Chiu, R., et al., 2002. JDP2, a repressor of AP-1, recruits a histone deacetylase 3 complex to inhibit the retinoic acid-induced differentiation of F9 cells. Mol. Cell. Biol. 22, 4815–4826.

Jousse, C., Bruhat, A., Ferrara, M., Fafournoux, P., 1998. Physiological concentration of amino acids regulates insulin-like- growth-factor-binding protein 1 expression. Biochem. J. 334, 147–153.

Jousse, C., Deval, C., Maurin, A.C., Parry, L., Cherasse, Y., Chaveroux, C., et al., 2007. TRB3 inhibits the transcriptional activation of stress-regulated genes by a negative feedback on the ATF4 pathway. J. Biol. Chem. 282, 15851–15861.

Jung, C.H., Jun, C.B., Ro, S.H., Kim, Y.M., Otto, N.M., Cao, J., et al., 2009. ULK-Atg13-FIP200 complexes mediate mTOR signaling to the autophagy machinery. Mol. Biol. Cell 20, 1992–2003.

Kandel, E.R., 2001. The molecular biology of memory storage: a dialogue between genes and synapses. Science (New York, NY) 294, 1030–1038.

Kawasaki, H., Schiltz, L., Chiu, R., Itakura, K., Taira, K., Nakatani, Y., et al., 2000. ATF-2 has intrinsic histone acetyltransferase activity which is modulated by phosphorylation. Nature 405, 195–200.

Kilberg, M.S., Shan, J., Su, N., 2009. ATF4-dependent transcription mediates signaling of amino acid limitation. Trends Endocrinol. Metab. 20, 436–443.

Kim, J., Kundu, M., Viollet, B., Guan, K.L., 2011. AMPK and mTOR regulate autophagy through direct phosphorylation of Ulk1. Nat. Cell. Biol. 13, 132–141.

Krishnamoorthy, T., Pavitt, G.D., Zhang, F., Dever, T.E., Hinnebusch, A.G., 2001. Tight binding of the phosphorylated alpha subunit of initiation factor 2 (eIF2α) to the regulatory subunits of guanine nucleotide exchange factor eIF2B is required for inhibition of translation initiation. Mol. Cell. Biol. 21, 5018–5030.

Kroemer, G., Marino, G., Levine, B., 2010. Autophagy and the integrated stress response. Mol. Cell. 40, 280–293.

Kwon, N.H., Kang, T., Lee, J.Y., Kim, H.H., Kim, H.R., Hong, J., et al., 2011. Dual role of methionyl-tRNA synthetase in the regulation of translation and tumor suppressor activity of aminoacyl-tRNA synthetase-interacting multifunctional protein-3. Proc. Natl. Acad. Sci. U.S.A. 108, 19635–19640.

Lee, Y.Y., Cevallos, R.C., Jan, E., 2009. An upstream open reading frame regulates translation of GADD34 during cellular stresses that induce eIF2α phosphorylation. J. Biol. Chem. 284, 6661–6673.

Lemaitre, B., Girardin, S.E., 2013. Translation inhibition and metabolic stress pathways in the host response to bacterial pathogens. Nat. Rev. Microbiol. 11, 365–369.

Lenard, N.R., Berthoud, H.R., 2008. Central and peripheral regulation of food intake and physical activity: pathways and genes. Obesity (Silver Spring) 16 (Suppl. 3), S11–22.

Leung, P.M., Rogers, Q.R., Harper, A.E., 1968. Effect of amino acid imbalance on dietary choice in the rat. J. Nutr. 95, 483–492.

Martina, J.A., Chen, Y., Gucek, M., Puertollano, R., 2012. MTORC1 functions as a transcriptional regulator of autophagy by preventing nuclear transport of TFEB. Autophagy 8, 903–914.

Martinez-Reyes, I., Sanchez-Arago, M., Cuezva, J.M., 2012. AMPK and GCN2-ATF4 signal the repression of mitochondria in colon cancer cells. Biochem. J. 444, 249–259.

Maurin, A.C., Jousse, C., Averous, J., Parry, L., Bruhat, A., Cherasse, Y., et al., 2005. The GCN2 kinase biases feeding behavior to maintain amino acid homeostasis in omnivores. Cell Metab. 1, 273–277.

Maurin, A.C., Chaveroux, C., Lambert-Langlais, S., Carraro, V., Jousse, C., Bruhat, A., et al., 2012. The amino acid sensor GCN2 biases macronutrient selection during aging. Eur. J. Nutr. 51, 119–126.

Maurin, A.C., Benani, A., Lorsignol, A., Brenachot, X., Parry, L., Carraro, V., et al., 2014. Hypothalamic eIF2α signaling regulates food intake. Cell Rep. 6, 438–444.

Meijer, A.J., Lorin, S., Blommaart, E.F., Codogno, P., 2014. Regulation of autophagy by amino acids and MTOR-dependent signal transduction. Amino Acids 47 (10), 2037–2063.

Mellor, A.L., Munn, D.H., 2008. Creating immune privilege: active local suppression that benefits friends, but protects foes. Nat. Rev. Immunol. 8, 74–80.

Méndez-Lucas, A., Hyroššová, P., Novellasdemunt, L., Viñals, F., Perales, J.C., 2014. Mitochondrial PEPCK is a pro-survival, ER-stress response gene involved in tumor cell adaptation to nutrient availability. J. Biol. Chem. 289, 22090–22102. Available from: http://dx.doi.org/10.1074/jbc.M114.566927.

Mizushima, N., 2007. Autophagy: process and function. Genes Dev. 28, 2861–2873.

Montani, D., Dorfmuller, P., Maitre, S., Jais, X., Sitbon, O., Simonneau, G., et al., 2010. Pulmonary veno-occlusive disease and pulmonary capillary hemangiomatosis. Presse Med. 39, 134–143.

Morrison, C.D., Reed, S.D., Henagan, T.M., 2012. Homeostatic regulation of protein intake: in search of a mechanism. Am. J. Physiol. Regul. Integr. Comp. Physiol. 302, R917–928.

Mortimore, G.E., Poso, A.R., 1987. Intracellular protein catabolism and its control during nutrient deprivation and supply. Annu. Rev. Nutr. 7, 539–564.

Mortimore, G.E., Schworer, C.M., 1977. Induction of autophagy by amino-acid deprivation in perfused rat liver. Nature 270, 174–176.

Munn, D.H., Sharma, M.D., Baban, B., Harding, H.P., Zhang, Y., Ron, D., et al., 2005. GCN2 kinase in T cells mediates proliferative arrest and anergy induction in response to indoleamine 2,3-dioxygenase. Immunity 22, 633–642.

Neely, A.N., Cox, J.R., Fortney, J.A., Schworer, C.M., Mortimore, G.E., 1977. Alterations of lysosomal size and density during rat liver perfusion. Suppression by insulin and amino acids. J. Biol. Chem. 252, 6948–6954.

Novoa, I., Zeng, H., Harding, H.P., Ron, D., 2001. Feedback inhibition of the unfolded protein response by GADD34-mediated dephosphorylation of eIF2α. J. Cell. Biol. 153, 1011–1022.

Ohoka, N., Yoshii, S., Hattori, T., Onozaki, K., Hayashi, H., 2005. TRB3, a novel ER stress-inducible gene, is induced via ATF4-CHOP pathway and is involved in cell death. EMBO J. 24, 1243–1255.

Ord, D., Ord, T., 2003. Mouse NIPK interacts with ATF4 and affects its transcriptional activity. Exp. Cell. Res. 286, 308–320.

Ord, D., Ord, T., 2005. Characterization of human NIPK (TRB3, SKIP3) gene activation in stressful conditions. Biochem. Biophys. Res. Commun. 330, 210–218.

Palam, L.R., Baird, T.D., Wek, R.C., 2011. Phosphorylation of eIF2 facilitates ribosomal bypass of an inhibitory upstream ORF to enhance CHOP translation. J. Biol. Chem. 286, 10939–10949.

Pan, Y., Chen, H., Siu, F., Kilberg, M.S., 2003. Amino acid deprivation and endoplasmic reticulum stress induce expression of multiple activating transcription factor-3 mRNA species that, when overexpressed in HepG2 cells, modulate transcription by the human asparagine synthetase promoter. J. Biol. Chem. 278, 38402–38412.

Pan, Y.X., Chen, H., Thiaville, M.M., Kilberg, M.S., 2007. Activation of the ATF3 gene through a co-ordinated amino acid-sensing response programme that controls transcriptional regulation of responsive genes following amino acid limitation. Biochem. J. 401, 299–307.

Pierre, P., 2009. Immunity and the regulation of protein synthesis: surprising connections. Curr. Opin. Immunol. 21, 70–77.

Platten, M., von Knebel Doeberitz, N., Oezen, I., Wick, W., Ochs, K., 2014. Cancer immunotherapy by targeting IDO1/TDO and their downstream effectors. Front. Immunol. 5, 673.

Powley, I.R., Kondrashov, A., Young, L.A., Dobbyn, H.C., Hill, K., Cannell, I.G., et al., 2009. Translational reprogramming following UVB irradiation is mediated by DNA-PKcs and allows selective recruitment to the polysomes of mRNAs encoding DNA repair enzymes. Genes Dev. 23, 1207–1220.

Rajesh, K., Papadakis, A.I., Kazimierczak, U., Peidis, P., Wang, S., Ferbeyre, G., et al., 2013. eIF2α phosphorylation bypasses premature senescence caused by oxidative stress and pro-oxidant antitumor therapies. Aging 5, 884–901.

Rajesh, K., Krishnamoorthy, J., Kazimierczak, U., Tenkerian, C., Papadakis, A.I., Wang, S., et al., 2015. Phosphorylation of the translation initiation factor eIF2α at serine 51 determines the cell fate decisions of Akt in response to oxidative stress. Cell Death Dis. 6, e1591.

Ravindran, R., Khan, N., Nakaya, H.I., Li, S., Loebbermann, J., Maddur, M.S., et al., 2014. Vaccine activation of the nutrient sensor GCN2 in dendritic cells enhances antigen presentation. Science 343, 313–317.

Robert, F., Kapp, L.D., Khan, S.N., Acker, M.G., Kolitz, S., Kazemi, S., et al., 2006. Initiation of protein synthesis by hepatitis C virus is refractory to reduced eIF2.GTP.Met-tRNA(i)(Met) ternary complex availability. Mol. Biol. Cell 17, 4632–4644.

Roczniak-Ferguson, A., Petit, C.S., Froehlich, F., Qian, S., Ky, J., Angarola, B., et al., 2012. The transcription factor TFEB links mTORC1 signaling to transcriptional control of lysosome homeostasis. Sci. Signal. 5, ra42.

Roffe, M., Hajj, G.N.M., Azevedo, H.F., Alves, V.S., Castilho, B.A., 2013. IMPACT is a developmentally regulated protein in neurons that opposes the eukaryotic initiation factor 2α kinase GCN2 in the modulation of neurite outgrowth. J. Biol. Chem. 288, 10860–10869.

Ron, D., Habener, J.F., 1992. CHOP, a novel developmentally regulated nuclear protein that dimerizes with transcription factors C/EBP and LAP and functions as a dominant-negative inhibitor of gene transcription. Genes Dev. 6, 439–453.

Rose, W.C., 1957. The amino acid requirement of adult man. Nutr. Abstr. Rev. 27, 489–497.

Rzymski, T., Milani, M., Singleton, D.C., Harris, A.L., 2009. Role of ATF4 in regulation of autophagy and resistance to drugs and hypoxia. Cell Cycle 8, 3838–3847.

Santoyo, J., 1997. Cloning and characterization of a cDNA encoding a protein synthesis initiation factor-2alpha (eIF-2α) kinase from drosophila melanogaster. Homology to yeast GCN2 protein kinase. J. Biol. Chem. 272, 12544–12550.

Sarkar, S., 2013. Regulation of autophagy by mTOR-dependent and mTOR-independent pathways: autophagy dysfunction in neurodegenerative diseases and therapeutic application of autophagy enhancers. Biochem. Soc. Trans. 41, 1103–1130.

Schewe, D.M., Aguirre-Ghiso, J.A., 2009. Inhibition of eIF2α dephosphorylation maximizes bortezomib efficiency and eliminates quiescent multiple myeloma cells surviving proteasome inhibitor therapy. Cancer Res. 69, 1545–1552.

Schneider, L., Giordano, S., Zelickson, B.R., Johnson, M.S., Benavides, G.A., Ouyang, X., et al., 2011. Differentiation of SH-SY5Y cells to a neuronal phenotype changes cellular bioenergetics and the response to oxidative stress. Free Radic. Biol. Med. 51, 2007–2017.

Schworer, C.M., Mortimore, G.E., 1979. Glucagon-induced autophagy and proteolysis in rat liver: mediation by selective deprivation of intracellular amino acids. Proc. Natl. Acad. Sci. U.S.A. 76, 3169–3173.

Schworer, C.M., Shiffer, K.A., Mortimore, G.E., 1981. Quantitative relationship between autophagy and proteolysis during graded amino acid deprivation in perfused rat liver. J. Biol. Chem. 256, 7652–7658.

Settembre, C., Zoncu, R., Medina, D.L., Vetrini, F., Erdin, S., Huynh, T., et al., 2012. A lysosome-to-nucleus signalling mechanism senses and regulates the lysosome via mTOR and TFEB. EMBO J. 31, 1095–1108.

Siu, F., Bain, P.J., LeBlanc-Chaffin, R., Chen, H., Kilberg, M.S., 2002. ATF4 is a mediator of the nutrient-sensing response pathway that activates the human asparagine synthetase gene. J. Biol. Chem. 277, 24120–24127.

Sood, R., Porter, A.C., Olsen, D.A., Cavener, D.R., Wek, R.C., 2000. A mammalian homologue of GCN2 protein kinase important for translational control by phosphorylation of eukaryotic initiation factor-2alpha. Genetics 154, 787–801.

Srivastava, S.P., Kumar, K.U., Kaufman, R.J., 1998. Phosphorylation of eukaryotic translation initiation factor 2 mediates apoptosis in response to activation of the double-stranded RNA-dependent protein kinase. J. Biol. Chem. 273, 2416–2423.

Straus, D.S., Burke, E.J., Marten, N.W., 1993. Induction of insulin-like growth factor binding protein-1 gene expression in liver of protein-restricted rats and in rat hepatoma cells limited for a single amino acid. Endocrinology 132, 1090–1100.

Su, N., Kilberg, M.S., 2008. C/EBP homology protein (CHOP) interacts with activating transcription factor 4 (ATF4) and negatively regulates the stress-dependent induction of the asparagine synthetase gene. J. Biol. Chem. 283, 35106–35117.

Sundrud, M.S., Trivigno, C., 2013. Identity crisis of Th17 cells: many forms, many functions, many questions. Semin. Immunol. 25, 263–272.

Sundrud, M.S., Koralov, S.B., Feuerer, M., Calado, D.P., Kozhaya, A.E., Rhule-Smith, A., et al., 2009. Halofuginone inhibits TH17 cell differentiation by activating the amino acid starvation response. Science 324, 1334–1338.

Talloczy, Z., Jiang, W., Virgin, H.Wt, Leib, D.A., Scheuner, D., Kaufman, R.J., et al., 2002. Regulation of starvation- and virus-induced autophagy by the eIF2α kinase signaling pathway. Proc. Natl. Acad. Sci. U.S.A. 99, 190–195.

Tattoli, I., Sorbara, M.T., Vuckovic, D., Ling, A., Soares, F., Carneiro, L.A., et al., 2012. Amino acid starvation induced by invasive bacterial pathogens triggers an innate host defense program. Cell Host Microbe 11, 563–575.

Thiaville, M.M., Dudenhausen, E.E., Zhong, C., Pan, Y.X., Kilberg, M.S., 2008. Deprivation of protein or amino acid induces C/EBPbeta synthesis and binding to amino acid response elements, but its action is not an absolute requirement for enhanced transcription. Biochem. J. 410, 473–484.

Vattem, K.M., Wek, R.C., 2004. Reinitiation involving upstream ORFs regulates ATF4 mRNA translation in mammalian cells. Proc. Natl. Acad. Sci. U.S.A. 101, 11269–11274.

Wang, X.Z., Kuroda, M., Sok, J., Batchvarova, N., Kimmel, R., Chung, P., et al., 1998. Identification of novel stress-induced genes downstream of chop. EMBO J. 17, 3619–3630.

Wang, Y., Ning, Y., Alam, G.N., Jankowski, B.M., Dong, Z., Nör, J.E., et al., 2013. Amino acid deprivation promotes tumor angiogenesis through the GCN2/ATF4 pathway. Neoplasia (New York, NY) 15, 989.

Warburg, O., 1956. On the origin of cancer cells. Science 123, 309–314.

Wilson, G.J., Bunpo, P., Cundiff, J.K., Wek, R.C., Anthony, T.G., 2013. The eukaryotic initiation factor 2 kinase GCN2 protects against hepatotoxicity during asparaginase treatment. Am. J. Physiol. Endocrinol. Metab. 305, E1124–1133.

Wolfgang, C.D., Chen, B.P., Martindale, J.L., Holbrook, N.J., Hai, T., 1997. gadd153/Chop10, a potential target gene of the transcriptional repressor ATF3. Mol. Cell. Biol. 17, 6700–6707.

Xiao, F., Huang, Z., Li, H., Yu, J., Wang, C., Chen, S., et al., 2011. Leucine Deprivation increases hepatic insulin sensitivity via GCN2/mTOR/S6K1 and AMPK pathways. Diabetes 60, 746–756.

Yang, Z., Klionsky, D.J., 2010. Mammalian autophagy: core molecular machinery and signaling regulation. Curr. Opin. Cell. Biol. 22, 124–131.

Ye, J., Kumanova, M., Hart, L.S., Sloane, K., Zhang, H., De Panis, D.N., et al., 2010. The GCN2-ATF4 pathway is critical for tumour cell survival and proliferation in response to nutrient deprivation. EMBO J. 29, 2082–2096.

Zhou, D., Palam, L.R., Jiang, L., Narasimhan, J., Staschke, K.A., Wek, R.C., 2008. Phosphorylation of eIF2 directs ATF5 translational control in response to diverse stress conditions. J. Biol. Chem. 283, 7064–7073.

CHAPTER

21

Amino Acid-Related Diseases

I. Knerr

National Centre for Inherited Metabolic Disorders, Temple Street Children's University Hospital, Dublin, Ireland

21.1. INTRODUCTION

Aminoacidopathies are a heterogeneous group of inborn metabolic disorders which are usually genetic. In this chapter, the main focus is on Phenylketonuria, Tyrosinemia Type 1, Urea Cycle Disorders, and organic acidurias, for example, branched-chain organic acidurias and Homocystinuria (Table 21.1).

In principle, the term "amino aciduria/acidemia" refers to a condition in which one or more amino acids accumulate and may exert direct toxic effects; they are subsequently excreted in larger quantities in urine. Conversely, the term "organic acidemia/aciduria" refers to a metabolic defect in amino acid catabolism that results in a buildup of toxic organic acids which are usually not present and are often toxic.

Aminoacidopathies are caused by inborn defects in enzymatic steps in the metabolic pathway of one or more amino acids or, in some cases, a transport abnormality. A deficiency of an enzymatic step involved in the breakdown of a single amino acid, or a group of amino acids, can result in an accumulation of (toxic) metabolites and also endogenous product deficiencies. Clinical symptoms result from the toxicity of accumulating metabolites and also deficiencies in downstream intermediates. The clinical picture depends on the severity of the underlying defect and also the amount of metabolic stress, such as protein catabolism or high protein intake. Among the organs that are most frequently affected are the brain and the liver. Many aminoacidopathies can present with acute symptoms due to acute metabolic decompensation. Typical presenting features comprise acute deterioration and neurological symptoms, along with biochemical abnormalities such as hyperammonemia (particularly in Urea Cycle Disorders (UCD)) or metabolic acidosis (in organic acidurias). Milder variants, usually due to milder mutations with higher residual enzyme activity compared with severe forms of the disease, may be episodic and not become symptomatic until late childhood or even adult life. Some disorders can lead to chronic neurological impairment without acute crisis, such as Phenylketonuria. In patients, particularly in individuals not screened as newborns, nonspecific symptoms such as failure to thrive or developmental delay should trigger a metabolic evaluation that may identify a diagnostic metabolite.

Diagnosis of these disorders is best accomplished by measurement of (1) quantitative amino acids in plasma (urine), (2) urinary organic acids by gas chromatography-mass spectrometry (GC/MS), and (3) identification of elevated/abnormal acylcarnitines by tandem mass spectrometry (MS/MS) in patients with organic acidurias. Urgent diagnostic tests include blood biochemistry including acid–base status, ammonia, glucose, lactate, ketones, urea/electrolytes, and liver function tests for the majority of these conditions, particularly when the patient is acutely unwell. High-risk newborns with a positive family history require an early newborn screening test, for example, at day 1 (depending on condition) along with preventative (dietary) intervention.

Treatment consists of long-term management and emergency treatment for conditions that can lead to acute metabolic decompensation. The primary aim of therapeutic intervention is to lower the levels of the accumulated toxic metabolite(s) by reducing flux through the affected metabolic pathway. This can be achieved by regulating intake of the offending amino acid(s) to the minimum necessary for de novo protein synthesis and by controlling catabolic breakdown.

The Molecular Nutrition of Amino Acids and Proteins.
DOI: http://dx.doi.org/10.1016/B978-0-12-802167-5.00022-0

As a general rule, treatment includes: (1) dietary restriction of the precursor amino acids(s) along with optimal nutritional supply; (2) adjunct therapy (eg, with appropriate cofactors, conjugating compounds); and (3) rapid intervention for metabolic decompensation.

Management of acute metabolic decompensation comprises: (1) elimination of intake of the offending amino acid(s) in the short term; (2) provision of adequate/increased caloric intake, including carbohydrates/intravenous glucose; (3) cautious surveillance and correction of fluid, electrolyte, pH, glucose, and plasma osmolality abnormalities; (4) provision of vitamin cofactors and selective chemical detoxicants and other measures as indicated.

Late complications must be anticipated in some conditions, such as neurological symptoms, liver and renal involvement.

TABLE 21.1 Nomenclature

Disorder (abbreviation)	Alternative name	Affected gene(s)	Chromosomal localization	Affected enzyme(s)	OMIM[a] No.
Phenylketonuria (PKU)	Phenylalanine hydroxylase deficiency, different subtypes from mild hyperphenylalaninemia to severe classical form	*PAH*	12q22-24.1	Phenylalanine hydroxylase	261600
Tyrosinemia Type 1 (HT1)	Fumarylacetoacetase deficiency, hepatorenal tyrosinemia	*FAH*	15q25.1	Fumarylacetoacetase	276700
N-Acetyl-glutamate synthetase (NAGS) deficiency	NAGS deficiency	*NAGS*	17q21.31	N-Acetylglutmate synthase	237310
Carbamoyl phosphate synthetase 1 (CPS1) deficiency	CPS1 deficiency	*CPS1*	2q35	Carbamoyl phosphate synthetase 1	237300
Ornithine transcarbamylase (OTC) deficiency	OTC deficiency	*OTC*	Xp11.4	Ornithine transcarbamylase	311250
Citrullinemia Type 1	Argininosuccinate synthase deficiency	*ASS1*	9q34.11	Argininosuccinate synthetase	215700
Argininosuccinate lyase (ASL) deficiency	ASL deficiency	*ASL*	7q11.21	Argininosuccinate lyase	207900
Argininemia	Arginase deficiency	*ARG1*	6q23.2	Arginase 1	207800
Maple syrup urine disease (MSUD)	Branched-chain alpha-keto acid dehydrogenase deficiency	*BCKDHA, BCKDHB, DBT, DLD*	19q13.2, 6q14.1, 1p21.2, 7q31.1	Branched-chain alpha-keto acid dehydrogenase complex	248600
Isovaleric acidemia (IVA)	Isovaleryl-CoA dehydrogenase deficiency	*IVD*	15q15.1	Isovaleryl-CoA dehydrogenase	243500
Propionic acidemia (PA)	Propionyl-CoA-carboxylase deficiency	*PCCA, PCCB*	13q32.3, 3q22.3	Propionyl-CoA-Carboxylase deficiency	232000, 232050
Methylmalonic acidemia (MMA)	Methylmalonyl-CoA mutase deficiency	*MUT*	6p12.3	Methylmalonyl-CoA mutase	251000
Classical Homocystinuria (HCU)	Cystathionine beta-synthase deficiency	*CBS*	21q22.3	Cystathionine beta-synthase	613381
Glutaric Aciduria Type 1 (GA1)	Glutaryl- CoA dehydrogenase deficiency	*GCDH*	19p13.2	Glutaryl- CoA dehydrogenase	231670
Nonketotic hyperglycinemia (NKH)	Glycine cleavage system deficiency	*GLDC, AMT, GCSH*	9p24.1, 3p21.31, 16q23.2	Glycine cleavage system (P-, T-, H-protein)	605899

[a] *Online Mendelian Inheritance in Man, http://www.omim.org/.*

21.2. DISORDER OF PHENYLALANINE AND TYROSINE METABOLISM (PHENYLKETONURIA, HYPERPHENYLALANINEMIA, TYROSINEMIA TYPE 1)

21.2.1 Phenylketonuria (PKU) and Hyperphenylalaninemia (HPA)

Phenylketonuria (PKU) is the most prevalent metabolic disorder caused by an inborn error in amino acid metabolism (Blau et al., 2010). The prevalence is approximately 1 in 10,000 with a considerable variation between different populations.

21.2.1.1 Metabolic Derangement

Phenylalanine (Phe) hydroxylase (PAH) is the key enzyme in the metabolism of the essential amino acid Phe, converting Phe into tyrosine whereby using tetrahydrobiopterin (BH4) as its cofactor.

"Classical" Phenylketonuria (PKU) is caused by a defect in the PAH apoenzyme along with a profound deficiency of PAH enzyme activity as distinct from "atypical" or "malignant" PKU which is caused by an inborn defect in BH4 metabolism. A disruption of the PAH pathway leads to an excessive accumulation of Phe in the bloodstream and subsequent neurotoxicity. The estimated frequency of this autosomal recessive disorder in the general population is approximately 1/10,000 whereas "atypical PKU" is an ultra-rare condition. Untreated, PKU causes severe mental disability, microcephaly, epilepsy, and other medical and mental health issues.

21.2.1.2 Diagnostic Principles

PKU is usually detected in newborns with raised Phe levels as part of the national newborn blood spot screening. Further work-up for PKU includes measurements of plasma amino acids Phe and tyrosine, along with pterins and dihydropterin reductase activity in blood to check for "atypical PKU." The accumulation of Phe in blood stimulates alternative metabolic pathways, resulting in the excretion of phenylketones, such as phenylpyruvate and phenylacetate, in the urine of PKU patients. Molecular genetic analysis of the PAH gene is useful for confirming the diagnosis and identification of disease-causing mutations. Large-scale studies have demonstrated a good correlation between the underlying PAH mutation and clinical phenotypes in most affected individuals; however, exceptions do occur.

On the basis of plasma Phe concentrations, for example, before initiation of treatment, PAH deficiency can be classified into classical PKU (Phe >1200 μmol/L or 20 mg/dL), mild/moderate PKU (Phe 600 to 1200 μmol/L or 10–20 mg/dL), and hyperphenylalaninemia (HPA) with Phe concentrations of 120–600 μmol/L (2–10 mg/dL). In clinical practice, however, PKU/HPA is a spectrum of PAH deficiencies which give rise to a continuum of conditions from very mild HPA, requiring no intervention, to severe classical PKU, requiring immediate attention and long-term treatment.

21.2.1.3 Therapeutic Principles

A Phe-restricted diet is the mainstay of treatment for patients with classical PKU. In PAH deficiency, treatment is usually started when blood Phe concentrations are above 360–600 μmol/L but there is no international consensus as yet (Blau and van Spronsen, 2014). Strict dietary restriction of natural protein intake, and hence Phe ingestion, requires supplementation of amino acids other than Phe. These manufactured amino acid supplements are usually enriched with vitamins and other micronutrients. However, the diet is difficult to maintain and compliance issues may be encountered, particularly in adolescents. For the treatment of PKU/HPA, the least restrictive dietary approach should be taken along with close amino acids/Phe monitoring in order to avoid overtreatment and nutritional deficiencies.

Patients with a rare cofactor defect (BH4 deficiency) require additional medications life-long, including BH4 or supplementation of the neurotransmitter precursors L-dopa and 5-hydroxy tryptophan to prevent/reduce severe neurological symptoms.

There is also a subgroup of PKU patients with mutations in the PAH gene that are "BH4-responsive" (ie, reduction in blood Phe concentration of 30% or greater). For this subgroup (estimated 20% of PKU patients), a supplementation of BH4, a requisite cofactor of PAH, can be beneficial, for example, adjunct to a low-Phe diet (Blau et al., 2010).

In pregnant women with PKU, an additional complication is the effect of maternal PKU on fetal development. Therefore, maternal Phe levels must be strictly controlled before conception and throughout pregnancy to prevent fetal anomalies such as microcephaly, mental retardation, and congenital heart defects.

The therapeutic range of Phe concentration varies considerably depending, for example, on the age of the patient and across countries (Blau and van Spronsen, 2014). In children under two years of age, the target concentration is usually in the range of 120–360 (<400) μmol/L, in adolescents and young adults <600 to <700 μmol/L, and in the range of 120–250 (<360) μmol/L during pregnancy.

The frequency of blood tests is adjusted to age and actual demands, that is, weekly or twice weekly in newborns and pregnant women, weekly during infancy, fortnightly at age 4–10 years, monthly at >10 years of age, including finger prick tests.

21.2.2 Tyrosinemia Type 1

Hereditary tyrosinemia Type 1 (HT1) is a severe inborn error of tyrosine catabolism. The clinical manifestations of this condition are variable and comprise acute or chronic liver disease, hypoglycemia, kidney disease, neurological manifestations, including acute neurological crisis with painful paresthesia and other manifestations.

21.2.2.1 Metabolic Derangement

The metabolic pathway of tyrosine degradation includes five major enzymatic steps. In HT1 or hepatorenal tyrosinemia (or fumarylacetoacetase deficiency) the underlying defect occurs in the last enzymatic step of the tyrosine degradation pathway. HT1 can cause severe hepatic, renal, and peripheral nerve damage. In younger patients, the most common presentation is severe liver disease or liver failure. Long-term complications include neuropathies, kidney dysfunction, and hepatocellular carcinoma.

21.2.2.2 Diagnostic Principles

Tyrosine levels show variable elevations in patients with HT1, for example, in the range of 150–1300 μmol/L (reference range 30–130 μmol/L). Typical biochemical findings are abnormal coagulation tests, moderate increase in transaminases and bilirubin, and characteristic tyrosine metabolites detected by urine organic acid analysis. Succinylacetone is one of the major toxic compounds that accumulate in affected individuals and the presence of succinylacetone in blood/urine is a diagnostic hallmark of HT1. Patients can also present with renal tubular dysfunction which can lead to growth failure or hypophosphatemic rickets. In some countries, affected individuals are detected on newborn screening.

21.2.2.3 Therapeutic Principles

Treatment of patients with tyrosinemia type 1 comprises a special diet that contains limited amounts of phenylalanine and tyrosine. As diet alone does not prevent disease progression, it is usually combined with the orphan drug nitisinone to suppress the endogenous production of toxic compounds. This drug should be administered by an experienced metabolic physician as the dosage must be tailored to each patient. Nitisinone induces proximal blockade of the tyrosine pathway and, therefore, increases tyrosine concentrations in plasma. It must be used in conjunction with a diet restricted in the amino acids tyrosine and phenylalanine. Biochemical monitoring includes plasma amino acids, succinylacetone levels, kidney and liver function tests, including coagulation screen, and alpha-fetoprotein as a screening test. Imaging (ultrasound and magnetic resonance imaging) should be performed regularly (eg, every 6 and 12 months, respectively) and patients who develop findings suggestive of liver cancer should be urgently considered for a possible liver transplantation (Holme and Mitchell, 2014).

21.3. UREA CYCLE DISORDERS/HYPERAMMONEMIAS

Patients with an inherited defect in one of the enzymes of the urea cycle can present at any age with a rapidly progressive intoxication-like resemblance and an overwhelming illness. Toxic plasma ammonia concentrations may also occur in many different conditions with hyperammonemia as a surrogate marker (Häberle, 2013). The analysis of blood ammonia should, therefore, be part of the baseline investigations in patients with unclear encephalopathy.

21.3.1 Metabolic Derangement

The urea cycle is the main ammonia/nitrogen-disposal pathway in humans. It requires the coordinated function of both cytosolic and mitochondrial enzymes and also transporters/carriers to catalyze the conversion of a molecule of ammonia, the alpha-nitrogen of aspartate, and bicarbonate into urea. Ammonia is toxic, particularly neurotoxic, as opposed to urea which is nontoxic, water-soluble, and readily excreted by the kidneys. Normal plasma ammonia levels in adults are <50 and <100 μmol/L in newborns.

There are six major disease entities characterized by a specific enzyme deficiency of the urea cycle:

- N-Acetylglutamate synthetase (NAGS) deficiency
- Carbamoyl phosphate synthetase 1 (CPS1) deficiency
- Ornithine transcarbamylase (OTC) deficiency
- Argininosuccinate synthase (ASS) deficiency (=Citrullinemia Type 1)
- Argininosuccinate lyase (ASL) deficiency (=Argininosuccinic aciduria)
- Arginase deficiency (Argininemia).

OTC deficiency is an X-linked disorder whereas the other UCD are autosomal-recessive conditions.

21.3.2 Diagnostic Principles

Urea cycle disorders (UCD) may present at any age from the neonatal period to adulthood, however, the more severely affected patients usually present early in life.

The most important clinical presentation of an inborn UCD is a neurological phenotype. Accumulation of ammonia or other toxic intermediates of the urea cycle can lead to progressive neurological symptoms, including acute encephalopathy, ataxia, seizures and coma, and potentially to death. Other symptoms may comprise tachypnea with respiratory alkalosis, vomiting, hepatic involvement, and neurological/psychiatric symptoms. Patients are at risk for metabolic crisis with hyperammonemia throughout life, often triggered by illness, fasting with catabolic stress, or increased protein intake.

Diagnosis of UCD is best achieved by (1) analysis of blood ammonia (emergency analysis in patients with encephalopathy), (2) quantitative plasma amino acids, and (3) urinary organic acids by GC/MS, including orotic acid. Elevated urinary orotic acid levels are found in UCD distal to the formation of carbamoylphosphate, including OTC, ASS, ASL, and arginase deficiency. Molecular genetic confirmation is available.

21.3.3 Therapeutic Principles

Early diagnosis and treatment are essential for successful patient outcome. Acute symptomatic hyperammonemia is an acute emergency. In principle, treatment aims at reducing ammonia levels by increasing ammonia removal and reducing protein breakdown.

Emergency treatment includes promotion of anabolism using a special high-calorie diet, transient stop of natural protein intake, carefully adjusted IV treatment including adequate amounts of IV dextrose, detoxification (ie, "nitrogen scavenger" drugs such as sodium benzoate and sodium phenylbutyrate and also supplements to increase residual urea cycle function, such as arginine when levels are low). Prevention of brain edema is crucial and additional extracorporal detoxification by hemodialysis or hemodiafiltration should be performed if plasma ammonia exceeds 400–500 μmol/L in classic forms of the disorders or in an encephalopathic patient.

Patients require lifelong treatment, and in some cases hepatocyte/liver transplantation might be considered.

In patients with severe UCD, outcomes remain poor. This might be related to the severity of the underlying defect, and potentially delayed diagnosis due to the nonspecific clinical presentation of these rare diseases (Häberle et al., 2012).

21.4. DISORDERS OF BRANCHED-CHAIN AMINO ACID METABOLISM (MAPLE SYRUP URINE DISEASE, ISOVALERIC ACIDEMIA, PROPIONIC ACIDEMIA, METHYLMALONIC ACIDEMIA)

Disorders of branched-chain amino acid (BCAA)/keto acid metabolism encompass diverse diseases entities (Knerr et al., 2012). Severe forms of these disorders usually present as acute, overwhelming illness in the neonatal period if not detected through newborn screening. Ongoing clinical assessments of patients in conjunction with monitoring of disorder-specific biochemical parameters are essential.

21.4.1 Metabolic Derangement

Among the main defects of branched-chain amino acids metabolism are Maple Syrup Urine Disease (MSUD), which is both amino aciduria and organic aciduria, and the organic acidurias isovaleric acidemia (IVA), propionic acidemia (PA), and methylmalonic acidemia (MMA). They are rare autosomal recessively-inherited inborn metabolic disorders.

MSUD is caused by a deficiency of the branched-chain 2-ketoacid dehydrogenase complex of the valine, leucine, and isoleucine catabolic pathways. IVA is caused by isovaleryl-CoA dehydrogenase deficiency in the leucine metabolic pathway.

PA and MMA are disorders of propionic acid degradation, mainly but not exclusively derived from the catabolism of isoleucine and valine, methionine, and threonine. Most patients with the severe forms of these disorders present in the neonatal period or during infancy with clinical deterioration, severe metabolic acidosis, and moderate hyperammonemia.

The pathophysiology of severe branched-chain amino/organic acidopathies is multifactorial. It includes accumulation of the substrate prior to the enzymatic block with intracellular toxicity due to specific metabolites and secondary effects such as metabolic imbalances, transport competition, and inhibition of other enzymes.

21.4.2 Diagnostic Principles

Diagnosis of these disorders is best accomplished by measurement of (1) quantitative plasma amino acids, (2) urinary organic acids by GC/MS, and (3) identification of elevated/abnormal acylcarnitines by MS/MS in patients with organic acidurias.

Although the characteristic metabolites are often extremely elevated in acute episodes, at other times these increases may be modest and sometimes difficult to distinguish from normal concentrations. A diagnosis rarely depends on the level of a single compound, but rather on a pattern of elevated amino acids, organic acids (as in MSUD, IVA, PA, and MMA) and/or acylcarnitines (as in IVA, PA, and MMA). MS/MS of dried blood spots is now used in many countries as a newborn screening test to diagnose severe disorders of the branched chain amino acid metabolism in neonates before they become symptomatic und critically unwell.

Biochemical monitoring includes measurements of blood glucose, electrolytes, pH, blood gases, ammonia, lactate, osmolality, plasma amino acids, urinary organic acids, and dipstick for ketones as indicated by the underlying condition, clinical history, and examination.

PA and MMA can lead to severe ketoacidotic episodes in patients which are accompanied by profound biochemical alterations, including metabolic acidosis, raised lactate and ammonia levels, and marked elevations in plasma and urinary glycine concentrations. Because of these observations, the name "ketotic hyperglycinemia" was used in the past (as opposed to nonketotic hyperglycinemia, see Section 21.6).

21.4.3 Therapeutic Principles

As a general rule, treatment includes:

1. Long-term dietary restriction of the precursor amino acid along with optimal nutritional supply.
2. Adjunct therapy, for example, with carnitine and cofactors as required. Carnitine is used in organic acidurias to support excretion of organic acids as carnitine adducts, thus sparing coenzyme-A, and preserving the function of the Krebs cycle.
3. Rapid intervention for metabolic decompensation.

Regarding (1) The mainstay of treatment is to limit intake of the offending amino acid(s) while preventing catabolism. With severe forms of the disorders, special medical foods (devoid of all BCAA, Leucine, or Isoleucine/Methionine/Threonine/Valine, as required for the different conditions) are needed to allow for adequate caloric, protein, and other nutrient intake. Milder forms may only require a moderately reduced natural protein intake. The amount of natural whole protein tolerated is determined by monitoring parameters, such as excretion of abnormal metabolites, blood amino acid levels, testing for body protein stores, control of acidosis, and development/growth. The least restrictive dietary approach should be taken in order to avoid nutritional deficiencies (Knerr et al., 2014).

The primary paradigm of treatment is to lower the levels of the accumulated toxic metabolites by reducing flux through the affected pathway. As the branched chain amino acids are essential and cannot be synthesized in humans, increased flux through a catabolic pathway can result only via exogenous protein intake or degradation of endogenous protein. Therefore, reducing flux through the affected catabolic pathway is achieved by reducing intake of the offending amino acid(s) to the minimum required for anabolic protein synthesis (ie, maintenance protein turnover and in children, that necessary for growth), and by minimizing catabolic breakdown induced by, for example, fasting, malnutrition, or intercurrent illnesses. Sufficient metabolic control cannot be achieved solely through restriction of natural protein, and the use of specialized medical formula and foods selectively deficient in a given amino acid(s) permits limitation of the offending amino acids(s) to that necessary for anabolic functions while providing sufficient quantities of the other amino acids to promote growth and sustain anabolic functions. Avoidance of a chronic catabolic state and quick intervention during acute catabolic crises is essential.

Regarding (2) Carnitine is used in many metabolic disorders to support excretion of organic acids as carnitine adducts and to prevent carnitine deficiency. In those conditions where an enzyme defect is known to be responsive to cofactor supplementation, for example, cobalamin-responsive MMA, treatment with cofactor should be intensified during acute illness and maintained over the long term. Glycine therapy has largely been limited to patients with IVA in which large amounts of urinary isovalerylglycine conjugates are excreted.

Regarding (3) Rapid intervention for metabolic decompensation, during an acute illness, for instance, is critical for resolution of a metabolic crisis. The primary focus should be on restoring an anabolic state. In brief, emergency treatment in MSUD or the organic acidurias requires: (1) transient elimination of intake of the offending amino acid(s)/natural protein for a very short period of time; (2) provision of adequate caloric intake in the range of 120–150% maintenance, including carefully adjusted IV treatment with glucose, and also special high-calorie tube feeding, quickly adding back small amounts of essential amino acids to prevent further catabolism; (3) appropriate fluid/electrolyte management including cautious surveillance and correction of blood glucose, electrolytes, pH, and plasma osmolality abnormalities; (4) provision of vitamin cofactors as indicated (eg, vitamin B12 injections in MMA); (5) the use of selective chemical detoxicants such as carnitine; (6) maintenance of cerebral function with adequate perfusion and oxygenation; (7) avoidance of cerebral edema; and (8) emergency hemodialysis in case of coma/encephalopathy with severe hyperammonemia or profound metabolic acidosis in severe forms of the diseases. Moderate hyperammonemia may occur during a metabolic crisis, particularly in patients with PA and MMA, but nitrogen-conjugating medications are typically not indicated for therapy, as plasma ammonia usually normalizes when the metabolic decompensation is reversed. Furthermore, other mitochondrial alterations associated with disturbed energy production and increased oxidative stress can be seen.

The long-term treatment of severe forms of the disorders, particularly of PA and MMA, is also challenging (Baumgartner et al., 2014). Late complications must be anticipated, including neurological complications, osteoporosis, progressive kidney failure (particularly in patients with MMA), or cardiomyopathy. Liver transplantation and combined liver/kidney transplantation have been applied in a number of patients and there is evidence of benefit. However, concerns arise because of documented cases of neurological complications (eg, in MMA patients).

21.5. CLASSICAL HOMOCYSTINURIA (HCU)

The complex catabolic pathway of methionine to sulfate involves several enzymatic steps. It leads to the formation of homocysteine which is normally either further metabolized by transsulfuration or recycled.

21.5.1 Metabolic Derangement

Classical homocystinuria (HCU) due to cystathionine beta-synthase (CBS) deficiency is an autosomal recessively inherited disorder of methionine metabolism. Methionine is a sulfur-containing amino acid. Of note, CBS is a pyridoxine (vitamin B6)-dependent enzyme.

21.5.2 Diagnostic Principles

HCU affects four organ systems, the brain, the eye, the skeleton, and the vascular system. Clinical manifestations include mental retardation, dislocation of the lens, vascular lesions, skeletal abnormalities/osteoporosis, arterial and venous thromboembolism, and cerebrovascular accidents. Patients are usually asymptomatic at birth and, if left undiagnosed and untreated, develop the typical clinical features early in life. However, the spectrum of clinical findings is wide.

The initial diagnosis can be achieved by quantitative amino acid analysis, including total homocysteine or free homocystine. Autooxidation of homocysteine leads to an accumulation of homocystine. Normal plasma homocysteine levels are usually less than 15 μmol/L and free homocystine is usually undetectable in healthy individuals. The main biochemical features of HCU are considerably increased concentrations of plasma homocyst(e)ine and methionine and also increased concentration of urine homocystine. Conversely, plasma cystine concentrations can be low. Enzyme activity of CBS in cultured fibroblasts is deficient in HCU patients. Molecular genetic analysis of the CBS gene is available. Neonates can be identified by newborn screening.

21.5.3 Therapeutic Principles

The aim of treatment is to lower homocyst(e)ine levels to as close to normal as possible, and to reduce risk of arterial and venous thromboembolism.

Response to pyridoxine (vitamin B6) needs to be tested in all newly diagnosed patients with HCU. Pyridoxine-responsive HCU is usually milder than the nonresponsive form. An approach to assessing pyridoxine responsiveness could be to begin with a moderate dose (eg, 50 mg three times a day) and to increase the dose if required (eg, to 100 mg three times a day). Those patients who are responsive to vitamin B6 receive pyridoxine therapy long-term. However, the dose of pyridoxine should be kept at the lowest dose possible in the long term to avoid possible side effects, particularly neuropathy.

A strict protein-restricted and, therefore, methionine-restricted diet is necessary for achieving biochemical control in patients who are less pyridoxine-responsive or nonresponsive. Synthetic methionine-free amino acid mixtures are important supplements in achieving metabolic control, anabolism and maintaining normal growth rate and body protein stores. Some patients may require cystine supplements for low plasma cystine concentrations (Adam et al., 2013). Additional vitamin supplements, including folate and vitamin B12, are usually required.

Betaine, given orally at a dose of approximately 50–100 mg/kg/day (6–9 g maximum in adults) can be used as adjunctive therapy, particularly in adolescents and adults. Betaine serves as a methyl donor for the methylation of homocysteine and, therefore, betaine intake does raise the methionine concentrations. Plasma methionine as well as homocysteine levels must be monitored in patients receiving betaine. Cerebral edema has been described in patients with excessive plasma methionine levels (above 1000 μmol/L, for instance) and betaine treatment must be immediately stopped if symptoms of cerebral edema occur along with appropriate medical care.

In addition to this, complications of HCU should be managed appropriately and in a timely manner, including, for example, eye surgery for ectopia lentis.

21.6. MISCELLANEOUS

21.6.1 Glutaric Aciduria Type 1 (GA1)

GA1 is an autosomal recessive metabolic disorder caused by deficiency of the enzyme glutaryl-CoA dehydrogenase in the metabolic pathways of lysine, hydroxylysin, and tryptophan. The clinical picture is characterized by macrocephaly, acute encephalopathy, and dystonic movements due to neuronal loss, particularly in the basal ganglia. Patients can be identified by Newborn Screening, using glutarylcarnitine in blood as a diagnostic parameter, and urinary organic acid analysis, where elevated excretion of glutaric acid and 3-hydroxyglutaric acid can be identified. Treatment consists of a low lysine diet with lysine-free, tryptophan-reduced special amino acid supplements, carnitine supplementation, and a strict emergence regimen, particularly during intermittent illnesses with fever, to prevent catabolism and acute encephalopathic crises along with neurological sequelae (Kölker et al., 2011).

21.6.2 Nonketotic Hyperglycinemia (NKH)

Nonketotic hyperglycinemia (NKH, "Glycine Encephalopathy") is an autosomal recessively-inherited disorder of glycine metabolism. The underlying defect is a deficiency in the mitochondrial glycine cleavage system, leading to accumulation of large quantities of glycine in the body, particularly in the brain. Clinical symptoms comprise hypotonia, seizures/myoclonic jerks, respiratory failure, hiccoughs, movement disorders, and developmental delay. Patients with severe NKH may succumb to an early death secondary to respiratory problems and neurological deterioration. Diagnosis can be achieved by amino acid analysis in plasma and cerebrospinal fluid (paired samples) followed by molecular genetic confirmation (*GLDC*, *AMT*, and *GCSH* genes). A well-recognized diagnostic hallmark is an elevated cerebrospinal fluid/plasma glycine ratio (normal <0.02) (Van Hove and Thomas, 2014). Patients may present soon after birth (neonatal, classical NKH) or later in life (attenuated, mild NKH). A different cohort of patients may have a transient elevation of glycine in cerebrospinal fluid in the neonatal period with no mutation in the underlying genes. Treatment options are limited, including antiepileptic drugs, sodium benzoate (to reduce plasma concentrations of glycine), and N-methyl D-aspartate receptor antagonists, as well as symptomatic treatment.

21.6.3 Disorders of Amino Acid Transport

Amino acid transporters are essential for the absorption/uptake of amino acids in the body and also the transport of amino acids between different compartments within the cell, including the mitochondria. Malfunction of an amino acid transporter can affect, for example, the function of the brain, kidneys, liver, and intestine. As the main focus in this chapter is on inborn metabolic disorders due to enzyme deficiencies, as distinct from cellular transporter defects, these conditions will not be discussed here in detail. However, the clinical relevance and range of these conditions will be demonstrated here by listing typical examples:

- An inborn defect in a mitochondrial glutamate carrier (GC1) can lead to neuronal excitability along with early-onset epilepsy and progressive encephalopathy.
- A malfunction of the heteromeric amino acid transporter SLC3A1/SLC7A9 is the underlying defect in cystinuria. This affects renal reabsorption of cystine, lysine, arginine, and ornithine, which are excreted in urine in large quantities, causing urolithiasis/cystine stones.
- An inborn defect in the amino acid transporter system SLC7A7 leads to a rare condition called lysinuric protein intolerance (LPI); this disorder is characterized by a high excretion of lysine in urine, altered plasma amino acid levels, and dysfunction of the urea cycle along with hyperammonemia and orotic acid excretion. Patients may present with a wide range of symptoms including vomiting and diarrhea, altered cognitive state, and growth faltering.

References

Adam, S., Almeida, M.F., Carbasius Weber, E., Champion, H., Chan, H., Daly, A., et al., 2013. Dietary practices in pyridoxine non-responsive homocystinuria: a European survey. Mol. Genet. Metab. 110 (4), 454–459.

Baumgartner, M.R., Hörster, F., Dionisi-Vici, C., Haliloglu, G., Karall, D., Chapman, K.A., et al., 2014. Proposed guidelines for the diagnosis and management of methylmalonic and propionic acidemia. Orphanet J. Rare Dis. 9, 130.

Blau, N., van Spronsen, F.J., 2014. Disorders of phenylalanine and tetrahydrobiopterin metabolism. In: Blau, N., Duran, M., Gibson, K.M., Dionisi-Vici, C. (Eds.), Physician's Guide to the Diagnosis, Treatment and Follow-Up of Inherited Metabolic Diseases. Springer, Heidelberg, pp. 3–21.

Blau, N., van Spronsen, F.J., Levy, H.L., 2010. Phenylketonuria. Lancet 376, 1417–1427.

Häberle, J., 2013. Clinical and biochemical aspects of primary and secondary hyperammonemic disorders. Arch. Biochem. Biophys. 536 (2), 101–108.

Häberle, J., Boddaert, N., Burlina, A., Chakrapani, A., Dixon, M., Huemer, M., et al., 2012. Suggested guidelines for the diagnosis and management of urea cycle disorders. Orphanet J. Rare Dis. 7, 32.

Holme, E., Mitchell, G.A., 2014. Tyrosine metabolism. In: Blau, N., Duran, M., Gibson, K.M., Dionisi-Vici, C. (Eds.), Physician's Guide to the Diagnosis, Treatment and Follow-Up of Inherited Metabolic Diseases. Springer, Heidelberg, pp. 23–31.

Kölker, S., Christensen, E., Leonard, J.V., Greenberg, C.R., Boneh, A., Burlina, A.B., et al., 2011. Diagnosis and management of glutaric aciduria type 1 – revised recommendations. J. Inherit. Metab. Dis. 34 (3), 677–694.

Knerr, I., Weinhold, N., Vockley, J., Gibson, K.M., 2012. Advances and challenges in the treatment of branched-chain amino/keto acid metabolic defects. J. Inherit. Metab. Dis. 35 (1), 29–40.

Knerr, I., Vockley, J., Gibson, K.M., 2014. Disorders of leucine, isoleucine and valine metabolism. In: Blau, N., Duran, M., Gibson, K.M., Dionisi-Vici, C. (Eds.), Physician's Guide to the Diagnosis, Treatment and Follow-Up of Inherited Metabolic Diseases. Springer, Heidelberg, pp. 103–141.

Van Hove, J., Thomas, J.A., 2014. Disorders of glycine, serine, GABA and proline metabolism. In: Blau, N., Duran, M., Gibson, K.M., Dionisi-Vici, C. (Eds.), Physician's Guide to the Diagnosis, Treatment and Follow-Up of Inherited Metabolic Diseases. Springer, Heidelberg, pp. 63–83.

CHAPTER

22

Genes in Skeletal Muscle Remodeling and Impact of Feeding: Molecular and Cellular Aspects

Y.-W. Chen[1,2], *M.D. Barberio*[2] *and M.J. Hubal*[2,3]

[1]Department of Integrative Systems Biology, George Washington University, Washington DC, USA
[2]Center for Genetic Medicine Research, Children's National Healthy System, Washington DC, USA
[3]Department of Exercise and Nutrition Sciences, George Washington University, Washington DC, USA

Skeletal muscle is highly plastic and exhibits a wide spectrum of adaptations in response to different environmental and physiological stimuli, including physical activity or different types of exercise, muscle disuse, immobilization, denervation, and microgravity. After the initial acute responses to the stimuli, skeletal muscle remodeling can take place to achieve sustainable signaling and structural adaptations in order to accommodate specific physiological requirements. Positive skeletal remodeling typically occurs following increased physical activity or exercise. Negative remodeling can occur with decreased physical activity, as well as in many disease conditions, such as various muscular dystrophies, myositis, cancer cachexia, and sarcopenia. Adaptive responses of skeletal muscle are also significantly affected by dietary nutrient intake. A diet rich in healthy macro- and micronutrients enables positive muscle adaptation, while starvation or malnutrition negatively affects skeletal muscle health. Certain dietary practices, such as calorie restriction, may have varied effects on skeletal muscle remodeling in the short and long term. In this chapter, we will review the major cellular and molecular mechanisms involved in muscle remodeling and then discuss current knowledge of dietary effects on muscle remodeling under physiological or disease conditions.

22.1. CELLULAR EVENTS INVOLVED IN SKELETAL MUSCLE REMODELING

Skeletal muscle is the largest organ of the adult human body. In addition to movement, it is critical to our overall health by assisting in thermoregulation and serving as a major site of energy storage and metabolism. Skeletal muscle is mainly composed of multinucleated myofibers which are responsible for muscle contraction. Myofibers can increase or decrease in size in response to chronic environmental conditions or stimuli, in addition to changing their metabolic characteristics. In addition to myofibers, other tissues and cell types in the skeletal muscle also play essential roles in muscle remodeling, including changes in numbers and activities of satellite cells, immune cells, fibroblasts, smooth muscle cells, and endothelial cells. Fig. 22.1 depicts an overview of cellular mechanisms involved in skeletal muscle remodeling. For the purpose of this review, we have subdivided muscle adaptation into three overall adaptive responses: fatigability, hypertrophy, and atrophy. The following section will briefly summarize the cellular events that occur during each of these three adaptive processes.

22.1.1 Fatigability

Muscle fatigue is an inability of a muscle to sustain force production over time. Many factors can impact the ability of muscle to produce force, including both central and peripheral (ie, within the muscle) variables. Central

The Molecular Nutrition of Amino Acids and Proteins.
DOI: http://dx.doi.org/10.1016/B978-0-12-802167-5.00023-2

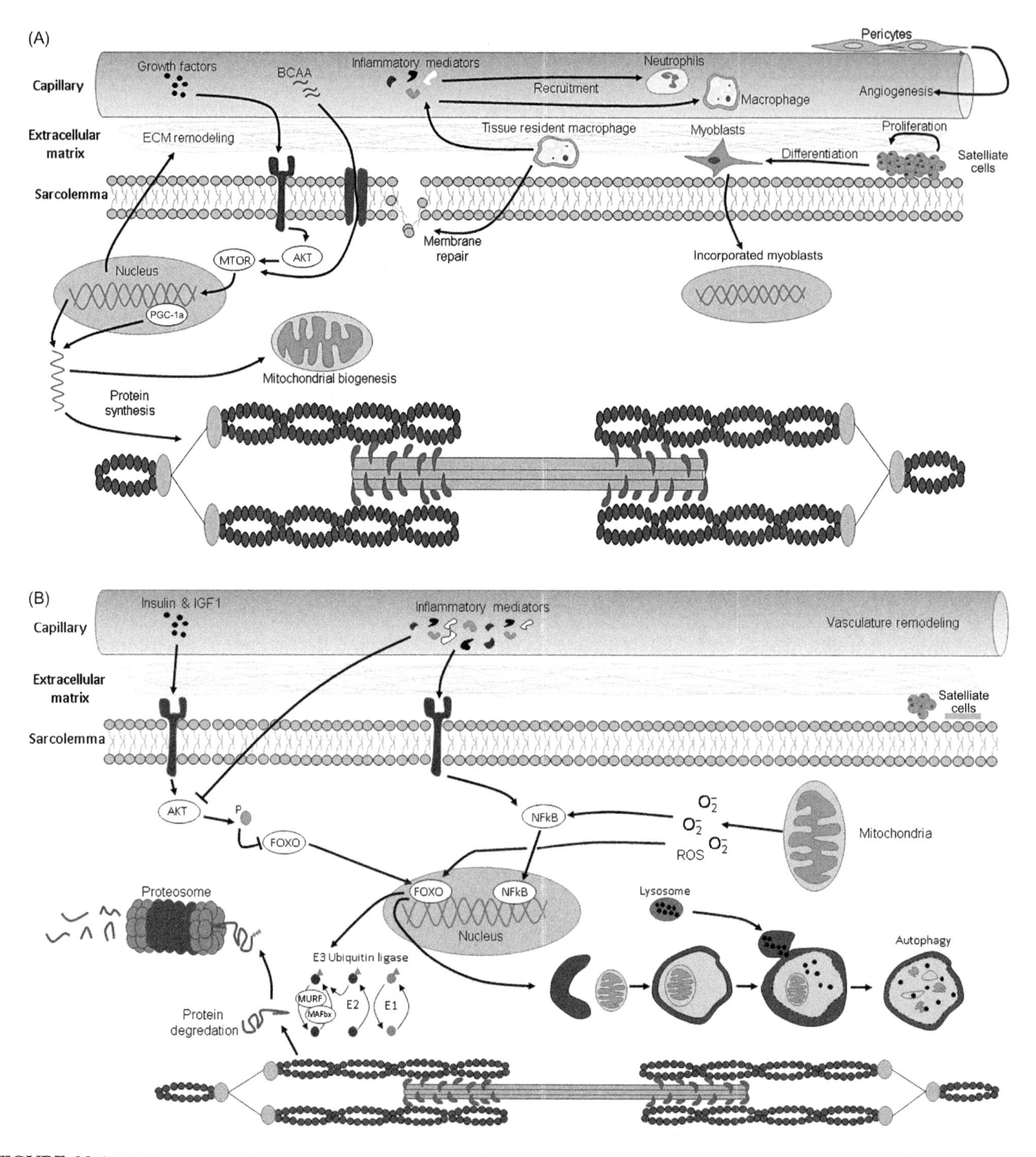

FIGURE 22.1 Cellular mechanisms of (A) Hypertrophy and (B) Atrophy. (A) *AKT*, protein kinase B; *BCAA*, branched-chain amino acid; *ECM*, extracellular matrix; *MTOR*, mechanistic target of rapamycin; *PGC-1α*, peroxisome proliferator-activated receptor gamma coactivator 1-alpha. (B) *AKT*, protein kinase B; *E1*, ubiquitin-activating enzyme; *E2*, ubiquitin-conjugating enzyme; *E3*, ubiquitin ligase; *FOXO*, Forkhead box proteins; *IGF1*, insulin-like growth factor 1; *MAFbx*, F-box only protein 32; *MURF*, muscle RING-finger protein; *NFkB*, nuclear factor kappa-light chain-enhancer of activated B cells; O_2^-, superoxide.

factors include motor neuron firing rates and motor unit innervation patterns, and there is evidence that acute neuroendocrine changes in serotonin, dopamine, or acetylcholine can affect central fatigue, especially during prolonged endurance exercise (Davis and Bailey, 1997). However, there is little data to support central fatigue as a significant factor in muscle adaptation, outside of certain disease states. Several excellent papers provide more comprehensive reviews of central (ie, neural) and peripheral muscle fatigue (Hunter et al., 2004; Ortenblad et al.,

2013; Westerblad and Allen, 2003; Westerblad et al., 2010; Fitts, 1994). The purpose of this section is to discuss fatigability specifically in relation to muscle adaptation.

Peripheral fatigability in muscle can adapt over time, with a reshaping of the metabolic capacity of the muscle driven by training, disuse or disease. Factors that can modify peripheral fatigue include changes in the intracellular metabolic environment and the number and activity of mitochondria in a muscle or functional muscle group. Some aspects of the metabolic environment include the availability and utilization of muscle glycogen, blood glucose, and triglycerides/fatty acids as energy substrates, the accumulation of metabolic byproducts such as lactic acid and their effects on the pH within the cell, and changes in phosphate compounds such as increases in inorganic phosphate (Pi) and ADP/AMP/IMP or decreases in phosphocreatine or ATP. Energy capacity in the cell is largely determined by the amount and activity of the mitochondria, which are modified by variables such as age, sex, fiber type distribution, and training history. Increased physical activity stimulates mitochondrial biogenesis, defined as the growth and division of resident mitochondria, as well as increases in expression and production of various metabolic enzymes.

Several diseases cause maladaptations that increase fatigability of muscle, including various metabolic disorders and mitochondrial myopathies. Many of these diseases are inherited conditions in which optimal use of energy for muscle contraction is impaired, such as phosphorylase deficiency (McArdle's Disease), carnitine deficiency, or phosphofructokinase deficiency (Tarui's Disease) (Sharp and Haller, 2014). Mitochondrial myopathies are typically inherited conditions in which mitochondria do not function properly, often leading to muscle weakness and exercise intolerance. The use of some drugs can also alter fatigability by affecting the metabolic status of muscle or by impacting mitochondrial function. For example, subclinical differences in mitochondrial function in certain individuals may be responsible for muscle pain and exercise intolerance when coupled with statin use (Hubal et al., 2011).

Positive adaptation in fatigability occurs most often with increased physical activity and aerobic exercise training, which increases enzymes associated with energy utilization, as well as the function and number of mitochondria. Muscle can experience a shift in fiber type distribution with aerobic training, such that more Type I (slow, oxidative) and Type IIA (fast, oxidative) fibers are present, as compared to Type IIB (fast, glycolytic) fibers (Fitts and Widrick, 1996). Specific molecular pathways associated with improved mitochondrial function and biogenesis will be addressed later in this chapter.

22.1.2 Hypertrophy

Hypertrophy is a gain in muscle size, typically through increases in protein synthesis that outpace protein breakdown. Size gains are often associated with gains in strength, though strength is dependent on both muscle size and neuromuscular signaling, such that the strength to size relationship is only partially dependent. Adaptations in muscle quality (strength per unit size) can vary depending on the source of stimulation, with greater muscle quality gains with progressive resistance training as compared to size gains evoked by anabolic steroid use (Schroeder et al., 2003). Prolonged resistance training can result in fiber type shifting of some hybrid fibers from Type IIA to IIB.

Satellite cells are muscle stem cells that play essential roles in homeostasis of skeletal muscles and muscle damage repair. The cells are located between the sarcolemmal membrane of myofibers and the basal lamina of extracellular matrix (ECM). When satellite cells are activated, the cells proliferate to maintain the stem cell population and generate myoblasts which proliferate, differentiate, and fuse into myofibers. Myoblasts can also be incorporated into existing myofibers, adding to the myonuclear pool. The complex behavior of satellite cells during skeletal muscle regeneration is tightly regulated through the dynamic interplay between intrinsic factors within satellite cells and extrinsic factors constituting the muscle stem cell niche/microenvironment (Aziz et al., 2012; Collins et al., 2005). While hypertrophy can occur without the activation of resident satellite cells, long-term maintenance of hypertrophy depends on the cycle of satellite cell activation, proliferation, and differentiation (Fry et al., 2014). Previous studies have shown that exercise evokes this cycle of events, providing the stimulus for adaptation (Macaluso and Myburgh, 2012; Pallafacchina et al., 2013).

In addition to muscle and satellite cells, various other cell types can be involved in the hypertrophic response to stimuli (Fig. 22.1A). Inflammatory cells are required for successful muscle remodeling by cleaning up debris and providing cytokines to facilitate repair processes. Cells in the ECM and fibroblasts also participate in adaptive processes. Vasculature remodeling can occur, involving smooth muscle and endothelial cells lining the vasculature. Exercise known to evoke hypertrophy, such as high-force eccentric exercise, is also known to cause micro damage in muscle, and it is during the process of repairing that damage that remodeling typically occurs (Hyldahl and Hubal, 2014). During the first 24–48 h after damaging exercise, activated resident inflammatory cells and those infiltrating from the circulation create a largely pro-inflammatory environment. After that, the balance shifts to more anti-inflammatory signaling, such that the acute proinflammatory signaling is attenuated and that repairs to structures can occur. There is abundant evidence that the cell types involved in the damage and repair cycle communicate with one another, including satellite cells, macrophages, fibroblasts, and endothelial cells, which signal to one another using various chemokine signals like monocyte chemoattractant protein 1 (MCP1).

22.1.3 Atrophy

Skeletal muscle atrophy at the cellular level is not simply the reverse of hypertrophy (Fig. 22.1B). Much research has been done to study muscle atrophy at the cellular level due to the high prevalence of muscle atrophy in muscle disuse conditions, diseases, and in normal aging. Muscle atrophy can involve reduction of myofiber size, number of myonuclei, and active satellite cells (Kudryashova et al., 2012; Hauerslev et al., 2014; Darr and Schultz, 1989; Allen et al., 1995; Day et al., 1995; Hikida et al., 1997; Schmalbruch et al., 1991). Various myofiber types are affected differently, depending on the conditions that evoke the muscle atrophy. In addition, remodeling of ECM and vasculature can also take place, depending on the atrophic stimuli (Giannelli et al., 2005; Vandervoort, 2002; Wang et al., 2014).

While various tissues are involved in muscle atrophy, active removal of muscle specific proteins via protein degradation is believed to play the most critical role. The proteasome (ubiquitin-proteasome) and autophagy (autophagy-lysosome) systems are two major protein degradation systems in eukaryotic cells. Both systems have been shown to be involved in the muscle atrophy process. For the proteasome system, targeted proteins are tagged by polyubiquitin and then sequestered to proteasomes for degradation. Ubiquitin is an 8-kDa peptide which is covalently attached to proteins targeted for degradation by the proteasome or autophagy pathways (Chau et al., 1989; Marmor and Yarden, 2004). The ubiquitination of proteins involves several steps: (1) activation of ubiquitin by E1 ubiquitin-activating enzyme; (2) the activated ubiquitin is transferred to an E2 ubiquitin-conjugating enzyme; and (3) the ubiquitin-E2 binds to an E3 ubiquitin ligase which carries targeted substrates and transfers the ubiquitin to the target proteins (Bartke et al., 2004; Berleth and Pickart, 1996). The most studied muscle-specific E3 ubiquitin ligases are MuRF1 and MAFbx/Atrogin-1 (Bodine et al., 2001; Gomes et al., 2001). MuRF1 targets the myofibrillar protein troponin I, titin, myosin-binding protein C, myosin light chain, and myosin heavy chain for degradation (McElhinny et al., 2002; Clarke et al., 2007; Kedar et al., 2004; Cohen et al., 2009). F-box only protein 32 (MAFbx) targets MyoD and eIF3f, which are a master myogenic factor and a translation initiation factor, respectively (Tintignac et al., 2005; Lagirand-Cantaloube et al., 2008).

Autophagy is a critical process for the degradation of cytoplasmic components, including damaged and aging organelles. Upon induction of autophagy, cytoplasmic cargos are engulfed by a double membraned phagophore to form an autophagosome. The autophagosome then fuses with a lysosome to form an autophagolysosome. The targeted organelles or proteins are then degraded by lysosomal hydrolases in the autophagolysosome (Xie and Klionsky, 2007). The autophagy system plays critical roles in muscle atrophy. Autophagy pathways are activated during muscle atrophy including starvation and denervation (Zhao et al., 2007; Mammucari et al., 2007). In $SOD1^{G93A}$ transgenic mice, muscle atrophy was reported to be caused by increased autophagy activities. Knocking down a key autophagy gene, microtubule-associated protein 1a/1b-light chain 3 (LC3B), attenuated the muscle atrophic phenotype (Dobrowolny et al., 2008). Muscle-specific deletion of an autophagy gene, *ATg7*, caused profound muscle atrophy and accumulation of abnormal organelles in muscles (Masiero et al., 2009). On the other hand, reduction of the basal autophagy activities, which is necessary for maintaining cellular homeostasis, contributes to collagen VI myopathy. In the study, induction of autophagy was due to muscle pathology (Grumati et al., 2010, 2011). The studies support that basal autophagy plays an essential role in maintaining muscle mass. Too much or too little autophagy can contribute to muscle atrophy and pathology.

In addition to altered protein degradation systems that contribute to the size reduction of myofibers, additional cellular changes are involved in muscle atrophy. In disuse atrophy, myofiber type shifts from slow Type I to fast Type II myofibers (Wittwer et al., 2002). In sarcopenia and cancer cachexia, instead of the loss of Type I fiber, Type II fast fibers are predominantly affected, causing a higher proportion of slow versus fast fibers in muscles (Vandervoort, 2002; Dedkov et al., 2003). Remodeling of ECM and neuromuscular junction (NMJ) also takes place in different types of muscle atrophy (Fahim and Robbins, 1986; Herbison et al., 1978). In disuse atrophy, remodeling of NMJ includes retraction of the nerve terminal and decentralization of nicotinic acetylcholine receptors (nAChR). This decentralized localization of nAChR away from the NMJ also takes place during denervation, which induces rapid muscle atrophy. NMJ function was also reported to be affected in sarcopenia (Gonzalez-Freire et al., 2014). In addition to cellular changes in myofibers, defects in angiogenesis have been linked to aging associated muscle wasting (sarcopenia). Apoptosis in both muscle cells and capillary endothelial cells has been reported in muscle wasting associated with aging (Wang et al., 2014). Differences among muscle atrophy with different causes suggest different pathophysiological mechanisms involved in each of these muscle atrophy conditions.

22.2. MOLECULAR PATHWAYS INVOLVED IN SKELETAL MUSCLE REMODELING

To understand the underlying mechanisms driving adaptations in fatigability or muscle size, cellular processes need to be understood in a more detailed manner by examining specific molecular responses in tissues or cell types. With the completion of the Human Genome Project, researchers were afforded an unprecedented opportunity to understand how genes and their interactions (ie, molecular pathways) with one another are involved in important biological processes like skeletal muscle adaptation. Investigating molecular pathways during muscle remodeling using genome-wide expression profiling has provided an overview of the transcriptomic changes involved in adaptation. While specific molecular pathways may be involved in responses to a specific stimulus, these profiling studies have identified key molecular pathways involved in muscle remodeling triggered by various types of stimuli. Several major molecular pathways have been identified as important in muscle remodeling, including signaling pathways that affect mitochondrial function and biogenesis (peroxisome-proliferator-activated receptor γ co-activator-1α (PPARGC1A) and AMP-activated protein kinase (AMPK)), protein synthesis and cell growth (IGF1, PI3K/AKT, mTOR, p70S6K, and 4E-BP1), and protein degradation and cell death (FOXO, NFκB, MAPK, MuRF1, and MAFbx). Each of these pathways will be explored in more detail in the following sections.

22.2.1 Fatigability

As mentioned previously, many factors can modify fatigability, from the whole muscle level (ie, innervation and motor unit firing patterns) to the cellular environment. Each of these factors has molecular drivers that respond to a variety of environmental cues or to the presence and severity of disease. This review will briefly cover some of the interrelated pathways involved in shifting fatigability of muscle with exercise. These pathways cluster around the important "master regulator" protein peroxisome proliferator-activated receptor-gamma coactivator 1 alpha (PGC-1α) (encoded by the PPARGC1A gene), which regulates mitochondrial biogenesis, angiogenesis, and oxidative metabolism (Fig. 22.2) (Handschin and Spiegelman, 2008; Lanza and Sreekumaran Nair, 2010; Ventura-Clapier et al., 2008; Baar, 2014). Exercise or feeding can alter important upstream regulators of PGC-1α, such as AMPK, calcineurin, calcium/calmodulin-dependent protein kinase type IV, and p38 mitogen-activated protein kinase (MAPK). We will explore the feeding effects in more detail in a later section of this review.

AMPK is a key sensor of energy availability via various stress signals, including AMP, ADP, calcium, or reactive oxygen species (Hardie et al., 2012). AMPK regulation can modify glucose uptake, fatty acid beta oxidation, biogenesis of GLUT4 glucose transporter proteins, and overall mitochondrial biogenesis. Each of these effects has specific molecular drivers, many of which are interrelated. For example, to regulate fatty acid oxidation, AMPK phosphorylates and inactivates acetyl-CoA carboxylase, which converts acetyl-CoA to malonyl-CoA. Lower malonyl-CoA would increase carnitine palmitoyltransferase 1 transport of fatty acids into mitochondria, promoting fatty acid oxidation.

PGC-1α, when activated, binds to various nuclear receptors to affect downstream events (Fig. 22.2). These receptors include nuclear respiratory factors 1 and 2 (NRF-1 and NRF-2), which interact with mitochondrial transcription factor A (Tfam or mtTFA), a mitochondrial-encoded transcription factor that drives mitochondrial DNA transcription and replication. PGC-1α binding to the NRFs is a key pathway in mitochondrial biogenesis, as the replication of mitochondria is driven by Tfam activity. PGC-1α also binds to estrogen-related receptors, which are orphan nuclear receptors that also play roles in mitochondrial biogenesis, as well as energy homeostasis.

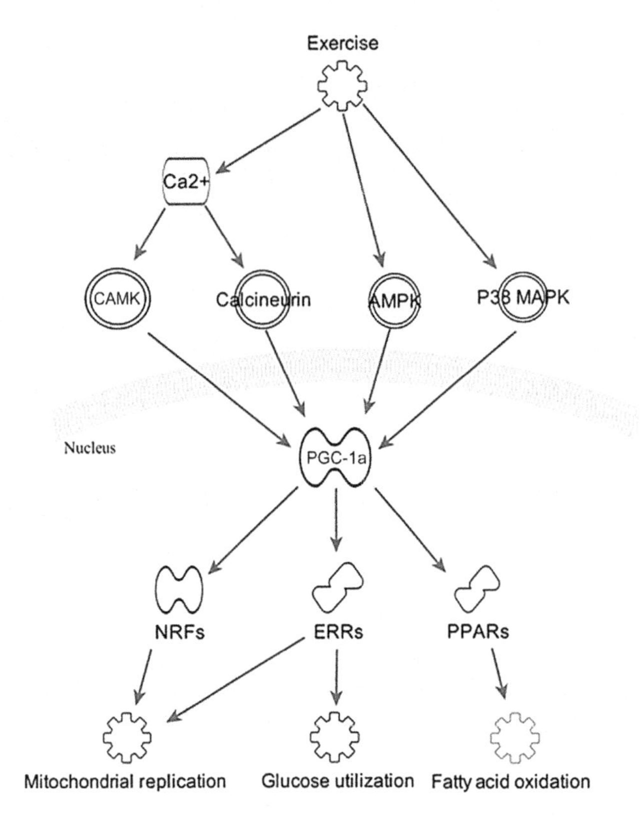

FIGURE 22.2 Interrelated pathways of exercise-induced adaptation. *AMPK*, 5′ AMP-activated protein kinase; *Ca^{2+}*, calcium; *CAMK*, Ca^{2+}/calmodulin-dependent protein kinase; *ERRs*, estrogen-related receptors; *NRFs*, nuclear respiratory factors; *P38 MAPK*, P38 mitogen-activated protein kinases; *PGC-1α*, peroxisome proliferator-activated receptor gamma coactivator 1-alpha; *PPARs*, peroxisome proliferator-activated receptors.

As its name implies, PGC-1α also binds to peroxisome proliferator-activated receptors, with downstream effects on fatty acid oxidation, among other effects.

PGC-1α has feedback effects on its upstream regulators, including calcineurin, and can also activate AKT1 levels in muscle. Interestingly, while PGC-1α is most linked to the effects of endurance exercise, there is evidence that progressive resistance training activates an alternate promoter site within the PPARGC1A gene to increase expression of the PGC-1A4 isoform, which can inhibit myostatin and increase insulin-like growth factor (IGF1) signaling, leading to muscle hypertrophy (Ruas et al., 2012), possibly via G protein-coupled receptor 56 (White et al., 2014). More traditional molecular modifiers of muscle hypertrophy (and atrophy) are summarized in the next section of this review.

22.2.2 Hypertrophy

Muscle hypertrophy generally occurs in healthy muscle with progressive mechanical overload, as with resistance training. Anabolic signaling cascades that promote muscle protein synthesis (MPS) are stimulated by tension and load, as well as the presence of amino acids. A key pathway in muscle hypertrophy is the phosphatidylinositol-3-kinase (PI3K)/protein kinase B (AKT)/mammalian target of rapamycin (mTOR)/$p70^{s6k}$ pathway (Fig. 22.3). Contributing pathways also include the MAPK pathway, growth hormone (GH) signaling, and various inflammatory pathways. Evolving evidence has also pointed to growth factor-independent pathways stimulated by load that provide an alternate activation route for mTOR (Goodman et al., 2011). Additionally, muscle hypertrophy is often dependent on satellite cell activation, proliferation, and differentiation. Myogenic regulatory factors (MRFs), such as MyoD, myogenin, myogenic factor 5 (Myf5), and myogenic factor 6 (Myf6), also help regulate hypertrophy (Blaauw and Reggiani, 2014; Schiaffino et al., 2013).

Overall protein synthesis can be influenced by many factors. IGF1 is an important growth factor that stimulates protein synthesis. IGF1 is produced primarily in the liver and most resistance training studies indicate that IGF levels are responsive to increased load, as well as to increased protein intake (Schiaffino et al., 2013; Glass,

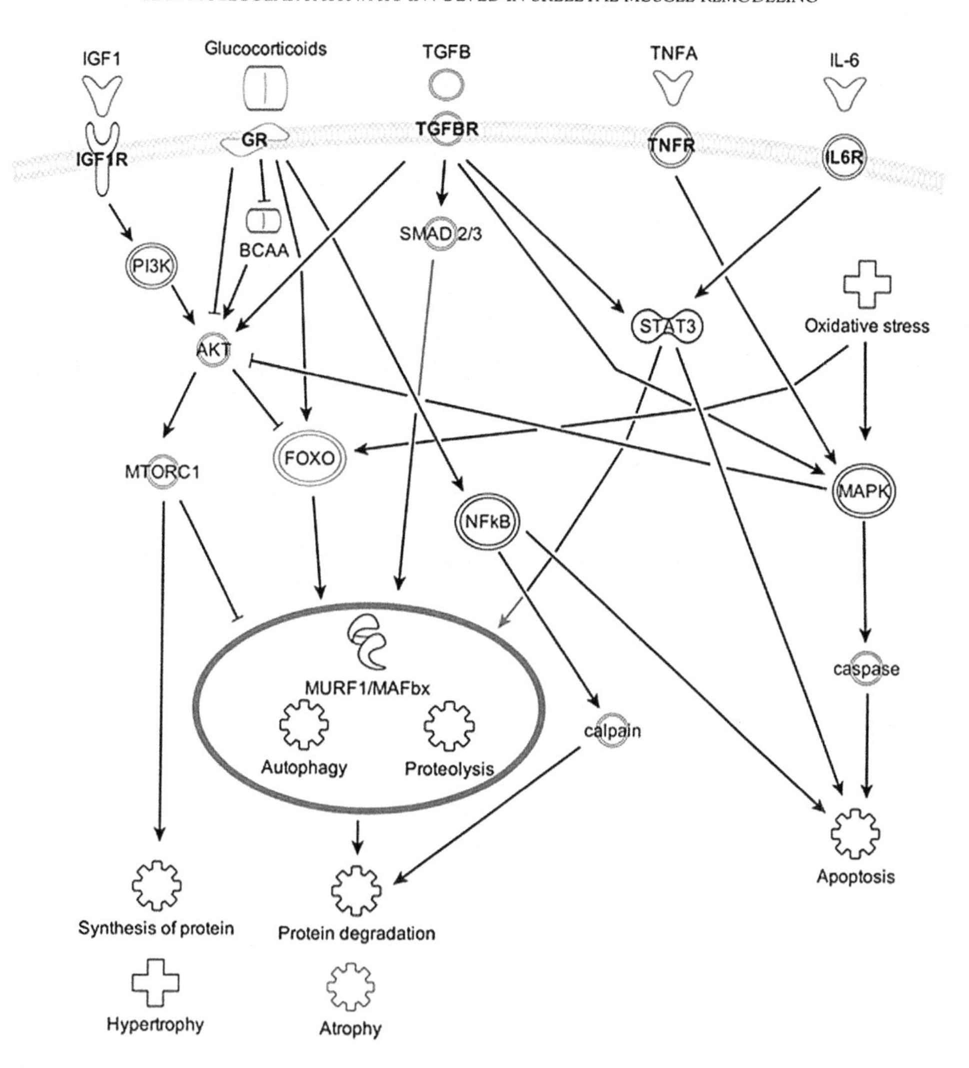

FIGURE 22.3 Molecular pathways of hypertrophy and atrophy. *AKT*, protein kinase B; *BCAA*, branched-chain amino acid; *FOXO*, Forkhead box proteins; *GR*, glucocorticoid receptor; *IGF1*, insulin-like growth factor 1; *IFG1R*, insulin-like growth factor receptor; *IL-6*, interleukin 6; *IL6R*, interleukin 6 receptor; *MAPK*, mitogen-activated protein kinases; *MAFbx*, F-box only protein 32; *MTORC1*, mammalian target of rapamycin complex 1; *MURF*, muscle RING-finger protein; *NFkB*, nuclear factor kappa-light chain-enhancer of activated B cells; *P13K*, phosphatidylinositol-4,5-bisphosphate 3-kinase; *SMAD2/3*, mothers against decapentaplegic homolog 2/3; *STAT3*, signal transduce and activator of transcription 3; *TGFB*, transforming growth factor beta; *TGFBR*, transforming growth factor beta receptor; *TNFA*, tumor necrosis factor alpha; *TNFR*, tumor necrosis factor receptor superfamily.

2010). IGF1 is stimulated by the presence of GH, a peptide hormone that stimulates growth and regeneration. GH is released by the pituitary gland under the control of the hypothalamus, and stimulates IGF1 production via the JAK-STAT pathway. Binding of IGF1 to its receptor results in the phosphorylation of insulin receptor substrate 1 (IRS1), which in turn activates PI3K and results in the phosphorylation of AKT1. AKT1 can activate a variety of downstream targets, the most important of which to muscle hypertrophy is mTOR, which in turn activates $p70^{s6k}$.

The importance of the IGF1/PI3K/AKT/mTOR/$p70^{s6k}$ pathway in muscle hypertrophy was demonstrated in various animal knockout models for PI3K, AKT, mTOR, or $p70^{s6k}$, resulting in smaller cell or muscle sizes. Variations in IGF-related genes have long been associated with growth and body composition differences across animal breeding stocks (Bass et al., 1999). In addition to being affected by GH, IGF1 can be regulated by activity and levels of calcineurin, IGF2, and IGF-binding proteins.

While mTOR is activated by upstream signaling via IGF1/PI3K/AKT, evolving evidence suggests that mTOR/p70s6k can be stimulated by loads during resistance training independently of upstream IGF1 signaling (Goodman et al., 2011; Philp et al., 2011). Data suggest an alternative mechanosensing pathway that moves tuberous sclerosis complex 2 (TSC2) away from the mTOR activator Rheb. Amino acids (particularly leucine) taken up by the muscle cell then help activate mTOR via binding to Rheb (Marcotte et al., 2015; Wolfson et al., 2015). While these studies provide interesting new insights into load-activated hypertrophy, more work is needed to fully elucidate how each of the pathways contribute to overall muscle growth.

Myostatin (MSTN; also known as growth differentiation factor 8), which inhibits muscle growth, is another important mediator of hypertrophy, as evidenced by the double-muscled phenotype displayed by animals with mutations in the MSTN or related genes like Belgian Blue cattle or the bully whippet (Allen et al., 2011; Rodriguez et al., 2014). MSTN is also a key regulator of adiposity. MSTN, a member of TFGβ family, binds via activin receptors ACVR2A and ACVR2B, which can be inhibited by the glycoprotein follistatin (FST). Genetic variants in MSTN-related genes such as MSTN, FST and ACVR2B have been studied in relation to muscle traits such as strength and size gains with resistance training, with varied results.

In addition to growth factors and the main hypertrophy-related molecular mediators, additional regulation of muscle size can be found in various inflammatory factors. Muscle hypertrophy often happens in conjunction with muscle repair, especially following exercise involving lengthening actions, which are potent stimulators of the growth pathways but also produce exercise induced muscle damage. As such, inflammatory cytokines, such as tumor necrosis factor alpha (TNFα), interleukin-6 (IL6), and MCP1, are closely linked to muscle growth. Some of these molecules directly play roles in proteolysis (including TNFα), while others could mediate communications between the myofibers, resident satellite cells, and infiltrating inflammatory cells. Optimizing the satellite cell response to damage is important to training-induced size gains. Therefore, the activity of cytokines like MCP1, in addition to regulation of MRFs, such as MyoD and myogenin, is vital to muscle growth. Interestingly, emerging evidence suggests cross-regulation between the MRFs and members of the IGF family, highlighting the complexity of the system.

22.2.3 Atrophy

Muscle atrophy can be caused by both physiological and disease conditions. Studies showed that activations of the E3 ubiquitin ligases MuRF1 and MAFbx play major roles in muscle atrophy (Bodine et al., 2001; Gomes et al., 2001). E3 ubiquitin ligases are an important part of the proteasome pathway, which is responsible for protein degradation in cells. In cells, proteins destined for proteasome degradation are tagged by ubiquitin ligases. Substrate selection for protein degradation depends on ubiquitin ligases E3. MuRF1 and MAFbx have been shown to target muscle specific proteins for degradation in various conditions including muscle disuse, immobilization, denervation, and steroid treatment (McElhinny et al., 2002; Clarke et al., 2007; Tintignac et al., 2005; Lagirand-Cantaloube et al., 2008). MuRF1 and MAFbx are regulated by FOXO transcription factors (Sandri et al., 2004, 2006; Stitt et al., 2004). The activities of FOXO are regulated by insulin and growth factor signaling. Studies showed that insulin and IGF activated the PI3K/AKT signaling, which lead to phosphorylation of FOXO (Biggs et al., 1999; Brunet et al., 1999, 2001; Kops et al., 1999). When FOXO is not phosphorylated, it is localized in the nuclei and activates downstream transcription targets such as MuRF1, MAFbx, and other genes that contribute to muscle atrophy. When it is phosphorylated, it is excluded from the nucleus, therefore FOXO transcriptional activity is repressed, reducing expression of MuRF1 and MAFbx. In addition to FOXO, activation of NFκB and p38 MAPK signaling activated by inflammatory signals and oxidative stresses have also been shown to activate transcription of MuRF1 and/or MAFbx, promoting muscles atrophy (Brunet et al., 2004; Cai et al., 2004; Mastrocola et al., 2008; Powers et al., 2007).

In addition to proteasome-mediated protein degradation, dysregulated autophagy activities and activation of caspase-3 and calpains were also reported to play major roles in skeletal muscle atrophy (Salazar et al., 2010; Enns and Belcastro, 2006; Milan et al., 2015; Nascimbeni et al., 2012). Autophagy removes damaged or unwanted organelles and proteins in cells. The process involves coordination of a group of autophagy-related genes, which encode proteins that selectively interact with the targets, formation of autophagosomes, fusion with the lysosome, and degradation of targeted organelles and proteins. Imbalanced autophagy activities in cells have been associated with muscle wasting and diseases. While basal level of autophagy activity is required to keep muscles healthy, excessive autophagy causes muscle wasting. FOXO3 induces the expression of a number of autophagy genes involved in various stages of the process, including LC3b, Gabarapl1, Pi3kIII, Ulk2, Atg12l, Beclin1, Atg4b, and Bnip3 (Zhao et al., 2007; Mammucari et al., 2007). In addition, previous studies showed that FOXO activates autophagy in addition to proteasome degradation, which includes mitophagy, a specialized form of autophagy (Milan et al., 2015; Zhao et al., 2008). Autophagy in disused skeletal muscles can also be activated by p38 MAPK signaling (McClung et al., 2010).

An increase in certain circulating signaling molecules, such as inflammatory cytokines (ie, TNFα, IL1β and IL6), TGFβ and steroids can activate molecular pathways that induce muscle atrophy (Bruunsgaard and

Pedersen, 2003; Schakman et al., 2012; Spate and Schulze, 2004; Watson et al., 2012; Narola et al., 2013). Oxidative stress has also have been shown contribute to muscle atrophy. Both have been shown to activate NFκB and/or MAPK signaling which reduce myoblast differentiation, induce apoptosis, and increase protein degradation (Powers et al., 2011; Archuleta et al., 2009; Langen et al., 2012; Hunter et al., 2002; Lu et al., 2012). In addition, IL6 has been shown to activate STAT3 and contribute to cancer cachexia and sarcopenia (Budui et al., 2015; Bonetto et al., 2012; Gilabert et al., 2014). In addition to inflammatory cytokine, TGFβ signaling has been shown to contribute to muscle atrophy (Narola et al., 2013; Mendias et al., 2012). Both TGFβ1 and myostatin have been shown to activate smad2/3 and lead to muscle wasting. We also recently reported a novel interaction between the TGFβ1 signaling and STAT3, which contribute to more severe muscle wasting in a conditional muscle-specific TGFβ1 mouse model (Guadagnin et al., 2015).

22.3. EFFECTS OF FEEDING ON SKELETAL MUSCLE REMODELING

22.3.1 Fatigability

Skeletal muscle fatigability can adapt over time in response to chronic changes in diet and/or physical activity, among other environmental stimuli. Much work has been done to examine the effects of physical activity or exercise on factors that impact muscle fatigability, from single fiber work to various whole body performance models, in healthy people and across various disease states. Various animal models have also been used to study fatigability adaptations to feeding (with the benefit of being able to better control environment and diet), though the reproducibility of findings from animal studies in human physiology can vary widely. Acute or chronic exercise effects on overall performance fatigue are beyond the scope of this chapter, which will address primarily the effects of feeding on muscle adaptation related to fatigability. Throughout the section, citations note review papers in which each of these complex issues is addressed in more detail.

As noted earlier, the primary biological changes associated with muscle fatigability include those associated with mitochondrial biogenesis and function, angiogenesis, and shifts in fiber type distribution toward more Type I oxidative fibers, as well as enhancing or maintaining the ability to utilize a variety of different energy sources, termed "metabolic flexibility" (Galgani and Ravussin, 2008). In cases of obesity or various metabolic diseases, muscle can lose its ability to shift from carbohydrate to fat utilization during fasting periods, making it overreliant on carbohydrate sources. On a positive note, metabolic flexibility can be regained via diet or exercise intervention, especially during early disease development.

During exercise, higher ratios of AMP:ATP or other stress signals sensed by AMPK inhibit downstream anabolic processes, like glycogen or MPS, and stimulate catabolism, such as the liberation and oxidation of fatty acids. Following feeding, the AMP: ATP ratio and other stress signals drop, reversing these effects and favoring anabolism. AMPK is also expressed in various other tissues, such as the brain, liver, and adipose tissue, with its levels affected by various hormones such as ghrelin, adiponectin, and leptin (Hardie et al., 2012).

Many factors related to diet affect the nutrient sensing pathways, including macronutrient composition, timing of feeding, and modification by physical activity demands/effects, among others. Caloric restriction can affect PGC-1α via parallel AMPK and NAD/sirtuin 1 (SIRT1) pathways (Rodgers et al., 2005). Some have argued that caloric restriction can stimulate mitochondrial biogenesis downstream of PGC-1α via this mechanism (Civitarese et al., 2007), while others argue against the idea (Miller et al., 2012). Ketogenic diets can also activate AMPK activity (Paoli et al., 2015), while diets with low protein/carbohydrate ratios can inhibit mTOR, therefore activating AMPK and PGC-1α. Dietary modifications can also alter metabolic flexibility. For example, diets high in certain protein components like branched chain amino acids (BCAAs; particularly leucine) can significantly affect nutrient signaling pathways. Chronic excess of BCAAs have been reported to be associated with diabetes, heart disease, and obesity (Newgard, 2012). However, some studies (using both animal and human models) have reported increases in glucose excursion in insulin resistant subjects, indicating that BCAA intake could have varied effects on metabolism depending on the metabolic health of subjects (Jitomir and Willoughby, 2008; Layman and Walker, 2006; She et al., 2007; Zhang et al., 2007).

22.3.2 Hypertrophy

Muscle hypertrophy is a coordinated adaptive effect reliant upon several cell types within the muscle, including myofibers, satellite cells, inflammatory cells, and endothelial cells. Anabolic signaling among these cell types, as well as sufficiently high MPS rates above muscle protein degradation rates, will drive hypertrophy at the cellular level over time, though the exact nature of these relationships are complex and remain somewhat unclear. Many molecular pathways are likely involved in hypertrophy, including those that act in parallel to one another, in addition to sharing upstream effectors or downstream effects.

MPS is dependent on two major sources of stimulation: feeding and exercise. Malnutrition, caloric restriction, or other dietary practices that limit amino acid availability for MPS (subsequently tipping the proteostasis balance toward breakdown) will be discussed in the following section on muscle atrophy. The current section will focus on the effects on skeletal muscle remodeling pathways of feeding and feeding in combination with exercise, as these two sources of increased MPS are interdependent. While a full review of muscle protein turnover and the various mechanisms that modify it is beyond the scope of this chapter, several excellent review papers are available on this topic (Philp et al., 2011; Drummond et al., 2009; Phillips, 2014).

Ingestion of dietary protein can elevate MPS acutely in the hours after a meal (Moore et al., 2009a). After a short initial time lag, MPS rises and peaks ~90 min postmeal, returning to baseline by ~2 h after feeding (Atherton et al., 2010b). The amount of MPS induced with feeding is largely dependent on the essential amino acid content of the meal (Smith et al., 1992), and is dose-dependent, with a saturation point occurring at ~20 g of protein without the concurrent effect of exercise (Moore et al., 2009b).

Exercise can further stimulate MPS for up to ~48 h postexercise (Phillips et al., 1997). Furthermore, the relationship between feeding- and exercise-induced MPS stimulation is interdependent (Phillips, 2014), as dietary protein ingested following an exercise bout (up to ~24 h afterwards) increases MPS more than intake at rest (Burd et al., 2011), and some evidence suggests that protein ingested during or prior to exercise also boosts basal MPS stimulation (van Loon, 2014). Many factors can modify the overall MPS response to feeding, ranging from age to acute and chronic history of physical activity, as well as intrinsic genetic variations. In addition, the type of protein and the timing of ingestion are key factors that can modify size gains.

Most of the MPS effects described above are modulated through the mTOR pathway. Activation of mTOR stimulates multiple elements of the 48S pre-initiation complex (4E-BP1, p70s6k and EIF4), promoting protein synthesis. As mentioned previously, evolving evidence suggests that mTOR can be regulated via multiple pathways. In addition to the traditional IRS1/PI3K/AKT pathway, recent evidence suggests alternate pathways for amino acids (especially leucine) and mechanical load signaling, including regulation of Rag (Rag guanosine triphosphatases) proteins via molecules like sestrin2 for leucine sensing (Goodman et al., 2011; Philp et al., 2011; Wolfson et al., 2015). The rapid evolution of this research field precludes definitive description here, as more work on these pathways needs to be done to understand them fully.

22.3.3 Atrophy

While many different physiological and environmental stimuli can cause muscle atrophy, decrease in protein synthesis and increase in protein turnover in the muscle are shared among most conditions. At cellular level, suppression of satellite activation increases cell apoptosis contributing to muscle atrophy in addition to cellular and molecular changes in myofibers, which actively reduce the volume of the myofibers. To date PI3K/AKT signaling is believed to play a central role in the protein synthesis and degradation pathways in skeletal muscle remodeling. MPS can be activated by amino acids and insulin via PI3K/AKT/mTOR signaling, which results in cellular mass growth. Suppression of PI3K/AKT/FOXO signaling leads to activation of protein degradation pathways. The activities of these pathways are affected by calories and nutrients in food, which affect skeletal muscle remodeling with or without other factors (eg, exercise, disuse, and diseases). Studies showed that proper calorie and protein intake is critical in maintaining healthy muscle mass. Excessive food intake, including overload of calorie, fat, and protein, have been shown to cause insulin resistance in peripheral insulin responsive tissues, which lead to skeletal muscle atrophy. A lack of nutrients, such as during fasting, has been reported to activate AMPK and NAD/SIRT1 pathways, which suppresses the mTOR and other substrates, including S6K, 4EBP-1, and EIF2, which resulted in reduced protein synthesis and reduced cell size and growth rates. In addition, it was shown that starvation increased catabolic activities, including protein degradation via both autophagy and proteasome pathways. The change of theses pathways contribute to muscle wasting caused by starvation.

While adequate energy is critical for protein homeostasis, mild calorie restriction has been shown to be beneficial to preserve muscle function and mass against sarcopenia during aging. Sarcopenia is a loss of skeletal muscle mass and function associated with aging, which is very important for individuals' health. Sarcopenia is associated with higher incidence of falls, disability, and mortality in seniors. The protective effect of calorie restriction against sarcopenia involves both cellular and molecular events. Studies showed that calorie restriction protected muscle cells against TNFα-dependent apoptosis (Phillips and Leeuwenburgh, 2005; Marzetti et al., 2008). In addition, calorie restriction was found to lower protein degradation and oxidative stress. Both proteasome and autophagy pathways were affected by the calorie restriction. In a human study, chronic calorie restriction (30% of recommended daily intake) for 4–20 years resulted in reduced IGF1 level and a threefold reduction in AKT activity. In addition, the FOXO3a and FOXO4 expression was increased (Mercken et al., 2013). The study showed that superoxide dismutase 2, a transcriptional target of FOXOs, was increased by calorie restriction. These studies suggest that chronic calorie restriction reprograms transcriptional network, which shift cellular homeostasis from growth to maintenance, including reduction of inflammatory activities and oxidative stress. The findings suggest a mild calorie restriction may be beneficial to maintain muscle mass during aging. In addition to calorie restriction, a study showed that supplementation of different types of fat during calorie restriction had different effects on the antiapoptotic effect of calorie restriction. A study showed that fish oil supplementation augmented the protective effect of calorie restriction, which was evident by reduction of proapototic genes and plasma membrane neutral sphingomyelins. In addition, levels of the antioxidant coenzyme Q at the plasma membrane were increased by the fish oil supplement. On the other hand, the changes were not found when lard was supplemented. Effects of antioxidants have been studied because oxidative stress increases with age and is postulated to be one of the major factors contributing to sarcopenia. Some of the studies showed positive findings of the protective effects of antioxidants to sarcopenia.

Muscle remodeling is significantly affected by protein intake, which enhances MPS in a dose-responsive manner in young and old adults (Moore et al., 2009b; Cuthbertson et al., 2005). It is not surprising that high dietary protein promotes muscle hypertrophy in response to exercise and helps maintain muscle mass during muscle disuse (Cermak et al., 2012; Wall and van Loon, 2013). When combined with calorie restriction, higher protein content in combination with exercise promotes maintenance of muscle mass (Josse et al., 2011; Mojtahedi et al., 2011). Studies showed that supplementation with BCAAs such as leucine, isoleucine, valine or metabolites of leucine such as β-hydroxy-β-methylbutyrate activate mTOR and protein synthesis in skeletal muscle to a greater extent compared with other amino acids (Atherton et al., 2010a; Churchward-Venne et al., 2012; Pimentel et al., 2011; Salles et al., 2013). Supplementation of mixed BCAAs in combination with calorie restriction in rats results in a reduction in total body mass and fat mass, while muscle mass was maintained (Mourier et al., 1997). The results suggest a potential role for BCAAs in maintaining skeletal muscle mass under calorie restricted conditions. Additional studies suggested that reduction of ubiquitin ligase, MAFbx, and protein degradation, as well as activation of AKT and mTOR signaling may contribute to the muscle maintenance.

Prolonged periods of skeletal muscle disuse due to physiological or disease conditions (eg, bed rest, limb immobilization, or microgravity) can lead to muscle atrophy and impair muscle functions. During muscle disuse, the pathways involved in muscle growth and protein synthesis were suppressed (PI3K/AKT, mTOR, p30S6K, and 4E-BP1), while pathways involved in regulating protein degradation (FOXO, NFκB, MuRF1, and MAFbx) are activated. While ROS produced by mitochondria increases during exercises, it is also increased in inactive myofibers (Kondo et al., 1991; Powers et al., 2012; Radak et al., 2013). Studies showed that muscle disuse lead to changes in redox state in muscle cells by increasing production of ROS and decreasing antioxidative capacity (Powers et al., 2007, 2012; Li et al., 2003; Salo et al., 1991). The changes contribute to increase protein degradation and reduced protein synthesis in skeletal muscle. In addition to oxidative stress, several inflammatory cytokines, such as TNFα and IL6, were shown to activate NFκB and p38 STAT3 signaling, which contribute to muscle atrophy by increasing cell apoptosis and muscle protein degradation. Supplements that modulate oxidative stress responses and inflammation have been shown to affect skeletal muscle remodeling by modulating the related pathways (Magne et al., 2013).

22.3.4 Summary

Skeletal muscle remodeling involves changes in structural and functional properties, which can be modulated by feeding. The skeletal muscle fatigability has been shown to be affected by caloric restriction and BCAAs via the PGC-1α and AMPK/NAD/SIRT1 pathways. Cellular and molecular pathways that affect MPS (PI3K/AKT and mTOR) and protein degradation (proteasome and autophagy systems) regulate changes of myofiber size during remodeling (hypertrophy and atrophy). These pathways can be affected by GHs, inflammatory cytokines, and oxidative stress. Calorie restriction, protein/amino acid intake and antioxidants have all been shown to modulate the main pathways involved in the process. Many have synergistic effects with exercise interventions.

In addition, the effects of the dietary intervention (with or without exercise) often depend on the intensity and time period of the treatment. Additional studies are needed to further dissect the pathways involved and to understand long-term effects of different dietary strategies on muscle remodeling.

References

Allen, D.L., et al., 1995. Plasticity of myonuclear number in hypertrophied and atrophied mammalian skeletal muscle fibers. J. Appl. Physiol. (1985) 78 (5), 1969–1976.

Allen, D.L., Hittel, D.S., McPherron, A.C., 2011. Expression and function of myostatin in obesity, diabetes, and exercise adaptation. Med. Sci. Sports Exerc. 43 (10), 1828–1835.

Archuleta, T.L., et al., 2009. Oxidant stress-induced loss of IRS-1 and IRS-2 proteins in rat skeletal muscle: role of p38 MAPK. Free Radic. Biol. Med. 47 (10), 1486–1493.

Atherton, P.J., et al., 2010a. Distinct anabolic signalling responses to amino acids in C2C12 skeletal muscle cells. Amino Acids 38 (5), 1533–1539.

Atherton, P.J., et al., 2010b. Muscle full effect after oral protein: time-dependent concordance and discordance between human muscle protein synthesis and mTORC1 signaling. Am. J. Clin. Nutr. 92 (5), 1080–1088.

Aziz, A., Sebastian, S., Dilworth, F.J., 2012. The origin and fate of muscle satellite cells. Stem Cell Rev. 8 (2), 609–622.

Baar, K., 2014. Nutrition and the adaptation to endurance training. Sports Med. 44 (Suppl. 1), S5–12.

Bartke, T., et al., 2004. Dual role of BRUCE as an antiapoptotic IAP and a chimeric E2/E3 ubiquitin ligase. Mol. Cell 14 (6), 801–811.

Bass, J., et al., 1999. Growth factors controlling muscle development. Domest. Anim. Endocrinol. 17 (2–3), 191–197.

Berleth, E.S., Pickart, C.M., 1996. Mechanism of ubiquitin conjugating enzyme E2-230K: catalysis involving a thiol relay? Biochemistry 35 (5), 1664–1671.

Biggs III, W.H., et al., 1999. Protein kinase B/Akt-mediated phosphorylation promotes nuclear exclusion of the winged helix transcription factor FKHR1. Proc. Natl. Acad. Sci. U.S.A. 96 (13), 7421–7426.

Blaauw, B., Reggiani, C., 2014. The role of satellite cells in muscle hypertrophy. J. Muscle Res. Cell. Motil. 35 (1), 3–10.

Bodine, S.C., et al., 2001. Identification of ubiquitin ligases required for skeletal muscle atrophy. Science 294 (5547), 1704–1708.

Bonetto, A., et al., 2012. JAK/STAT3 pathway inhibition blocks skeletal muscle wasting downstream of IL-6 and in experimental cancer cachexia. Am. J. Physiol. Endocrinol. Metab. 303 (3), E410–E421.

Brunet, A., et al., 1999. Akt promotes cell survival by phosphorylating and inhibiting a Forkhead transcription factor. Cell 96 (6), 857–868.

Brunet, A., et al., 2001. Protein kinase SGK mediates survival signals by phosphorylating the forkhead transcription factor FKHRL1 (FOXO3a). Mol. Cell Biol. 21 (3), 952–965.

Brunet, A., et al., 2004. Stress-dependent regulation of FOXO transcription factors by the SIRT1 deacetylase. Science 303 (5666), 2011–2015.

Bruunsgaard, H., Pedersen, B.K., 2003. Age-related inflammatory cytokines and disease. Immunol. Allergy Clin. North Am. 23 (1), 15–39.

Budui, S.L., Rossi, A.P., Zamboni, M., 2015. The pathogenetic bases of sarcopenia. Clin. Cases Miner. Bone Metab. 12 (1), 22–26.

Burd, N.A., et al., 2011. Enhanced amino acid sensitivity of myofibrillar protein synthesis persists for up to 24 h after resistance exercise in young men. J. Nutr. 141 (4), 568–573.

Cai, D., et al., 2004. IKKbeta/NF-kappaB activation causes severe muscle wasting in mice. Cell 119 (2), 285–298.

Cermak, N.M., et al., 2012. Protein supplementation augments the adaptive response of skeletal muscle to resistance-type exercise training: a meta-analysis. Am. J. Clin. Nutr. 96 (6), 1454–1464.

Chau, V., et al., 1989. A multiubiquitin chain is confined to specific lysine in a targeted short-lived protein. Science 243 (4898), 1576–1583.

Churchward-Venne, T.A., et al., 2012. Supplementation of a suboptimal protein dose with leucine or essential amino acids: effects on myofibrillar protein synthesis at rest and following resistance exercise in men. J. Physiol. 590 (Pt 11), 2751–2765.

Civitarese, A.E., et al., 2007. Calorie restriction increases muscle mitochondrial biogenesis in healthy humans. PLoS Med. 4 (3), e76.

Clarke, B.A., et al., 2007. The E3 ligase MuRF1 degrades myosin heavy chain protein in dexamethasone-treated skeletal muscle. Cell Metab. 6 (5), 376–385.

Cohen, S., et al., 2009. During muscle atrophy, thick, but not thin, filament components are degraded by MuRF1-dependent ubiquitylation. J. Cell Biol. 185 (6), 1083–1095.

Collins, C.A., et al., 2005. Stem cell function, self-renewal, and behavioral heterogeneity of cells from the adult muscle satellite cell niche. Cell 122 (2), 289–301.

Cuthbertson, D., et al., 2005. Anabolic signaling deficits underlie amino acid resistance of wasting, aging muscle. FASEB J. 19 (3), 422–424.

Darr, K.C., Schultz, E., 1989. Hindlimb suspension suppresses muscle growth and satellite cell proliferation. J. Appl. Physiol. (1985) 67 (5), 1827–1834.

Davis, J.M., Bailey, S.P., 1997. Possible mechanisms of central nervous system fatigue during exercise. Med. Sci. Sports Exer. 29 (1), 45–57.

Day, M.K., et al., 1995. Adaptations of human skeletal muscle fibers to spaceflight. J. Gravit. Physiol. 2 (1), P47–P50.

Dedkov, E.I., Borisov, A.B., Carlson, B.M., 2003. Dynamics of postdenervation atrophy of young and old skeletal muscles: differential responses of fiber types and muscle types. J. Gerontol. A Biol. Sci. Med. Sci. 58 (11), 984–991.

Dobrowolny, G., et al., 2008. Skeletal muscle is a primary target of SOD1G93A-mediated toxicity. Cell Metab. 8 (5), 425–436.

Drummond, M.J., et al., 2009. Nutritional and contractile regulation of human skeletal muscle protein synthesis and mTORC1 signaling. J. Appl. Physiol. (1985) 106 (4), 1374–1384.

Enns, D.L., Belcastro, A.N., 2006. Early activation and redistribution of calpain activity in skeletal muscle during hindlimb unweighting and reweighting. Can. J. Physiol. Pharmacol. 84 (6), 601–609.

Fahim, M.A., Robbins, N., 1986. Remodelling of the neuromuscular junction after subtotal disuse. Brain Res. 383 (1–2), 353–356.

Fitts, R.H., 1994. Cellular mechanisms of muscle fatigue. Physiol. Rev. 74 (1), 49–94.

Fitts, R.H., Widrick, J.J., 1996. Muscle mechanics: adaptations with exercise-training. Exerc. Sport Sci. Rev. 24, 427–473.

Fry, C.S., et al., 2014. Regulation of the muscle fiber microenvironment by activated satellite cells during hypertrophy. FASEB J. 28 (4), 1654–1665.

Galgani, J., Ravussin, E., 2008. Energy metabolism, fuel selection and body weight regulation. Int. J. Obes. (Lond.) 32 (Suppl. 7), S109–S119.

Giannelli, G., et al., 2005. Matrix metalloproteinase imbalance in muscle disuse atrophy. Histol. Histopathol. 20 (1), 99–106.

Gilabert, M., et al., 2014. Pancreatic cancer-induced cachexia is Jak2-dependent in mice. J. Cell. Physiol. 229 (10), 1437–1443.

Glass, D.J., 2010. PI3 kinase regulation of skeletal muscle hypertrophy and atrophy. Curr. Top. Microbiol. Immunol. 346, 267–278.

Gomes, M.D., et al., 2001. Atrogin-1, a muscle-specific F-box protein highly expressed during muscle atrophy. Proc. Natl. Acad. Sci. U.S.A. 98 (25), 14440–14445.

Gonzalez-Freire, M., et al., 2014. The neuromuscular junction: aging at the crossroad between nerves and muscle. Front. Aging Neurosci. 6, 208.

Goodman, C.A., Mayhew, D.L., Hornberger, T.A., 2011. Recent progress toward understanding the molecular mechanisms that regulate skeletal muscle mass. Cell Signal. 23 (12), 1896–1906.

Grumati, P., et al., 2010. Autophagy is defective in collagen VI muscular dystrophies, and its reactivation rescues myofiber degeneration. Nat. Med. 16 (11), 1313–1320.

Grumati, P., et al., 2011. Autophagy induction rescues muscular dystrophy. Autophagy 7 (4), 426–428.

Guadagnin, E., et al., 2015. Tyrosine 705 Phosphorylation of STAT3 is associated with phenotype severity in TGFbeta1 transgenic mice. Biomed. Res. Int. 2015, 843743.

Handschin, C., Spiegelman, B.M., 2008. The role of exercise and PGC1alpha in inflammation and chronic disease. Nature 454 (7203), 463–469.

Hardie, D.G., Ross, F.A., Hawley, S.A., 2012. AMPK: a nutrient and energy sensor that maintains energy homeostasis. Nat. Rev. Mol. Cell Biol. 13 (4), 251–262.

Hauerslev, S., Vissing, J., Krag, T.O., 2014. Muscle atrophy reversed by growth factor activation of satellite cells in a mouse muscle atrophy model. PLoS ONE 9 (6), e100594.

Herbison, G.J., Jaweed, M.M., Ditunno, J.F., 1978. Muscle fiber atrophy after cast immobilization in the rat. Arch. Phys. Med. Rehabil. 59 (7), 301–305.

Hikida, R.S., et al., 1997. Myonuclear loss in atrophied soleus muscle fibers. Anat. Rec. 247 (3), 350–354.

Hubal, M.J., et al., 2011. Transcriptional deficits in oxidative phosphorylation with statin myopathy. Muscle Nerve 44 (3), 393–401.

Hunter, R.B., et al., 2002. Activation of an alternative NF-kappaB pathway in skeletal muscle during disuse atrophy. FASEB J. 16 (6), 529–538.

Hunter, S.K., Duchateau, J., Enoka, R.M., 2004. Muscle fatigue and the mechanisms of task failure. Exerc. Sport Sci. Rev. 32 (2), 44–49.

Hyldahl, R.D., Hubal, M.J., 2014. Lengthening our perspective: morphological, cellular, and molecular responses to eccentric exercise. Muscle Nerve 49 (2), 155–170.

Jitomir, J., Willoughby, D.S., 2008. Leucine for retention of lean mass on a hypocaloric diet. J. Med. Food 11 (4), 606–609.

Josse, A.R., et al., 2011. Increased consumption of dairy foods and protein during diet- and exercise-induced weight loss promotes fat mass loss and lean mass gain in overweight and obese premenopausal women. J. Nutr. 141 (9), 1626–1634.

Kedar, V., et al., 2004. Muscle-specific RING finger 1 is a bona fide ubiquitin ligase that degrades cardiac troponin I. Proc. Natl. Acad. Sci. U.S.A. 101 (52), 18135–18140.

Kondo, H., Miura, M., Itokawa, Y., 1991. Oxidative stress in skeletal muscle atrophied by immobilization. Acta Physiol. Scand. 142 (4), 527–528.

Kops, G.J., et al., 1999. Direct control of the forkhead transcription factor AFX by protein kinase B. Nature 398 (6728), 630–634.

Kudryashova, E., Kramerova, I., Spencer, M.J., 2012. Satellite cell senescence underlies myopathy in a mouse model of limb-girdle muscular dystrophy 2H. J. Clin. Invest. 122 (5), 1764–1776.

Lagirand-Cantaloube, J., et al., 2008. The initiation factor eIF3-f is a major target for atrogin1/MAFbx function in skeletal muscle atrophy. EMBO J. 27 (8), 1266–1276.

Langen, R.C., et al., 2012. NF-kappaB activation is required for the transition of pulmonary inflammation to muscle atrophy. Am. J. Respir. Cell Mol. Biol. 47 (3), 288–297.

Lanza, I.R., Sreekumaran Nair, K., 2010. Regulation of skeletal muscle mitochondrial function: genes to proteins. Acta Physiol. (Oxf) 199 (4), 529–547.

Layman, D.K., Walker, D.A., 2006. Potential importance of leucine in treatment of obesity and the metabolic syndrome. J. Nutr. 136 (1 Suppl), 319S–323SS.

Li, Y.P., et al., 2003. Hydrogen peroxide stimulates ubiquitin-conjugating activity and expression of genes for specific E2 and E3 proteins in skeletal muscle myotubes. Am. J. Physiol. Cell Physiol. 285 (4), C806–C812.

Lu, A., et al., 2012. NF-kappaB negatively impacts the myogenic potential of muscle-derived stem cells. Mol. Ther. 20 (3), 661–668.

Macaluso, F., Myburgh, K.H., 2012. Current evidence that exercise can increase the number of adult stem cells. J. Muscle Res. Cell. Motil. 33 (3–4), 187–198.

Magne, H., et al., 2013. Nutritional strategies to counteract muscle atrophy caused by disuse and to improve recovery. Nutr. Res. Rev. 26 (2), 149–165.

Mammucari, C., et al., 2007. FoxO3 controls autophagy in skeletal muscle in vivo. Cell Metab. 6 (6), 458–471.

Marcotte, G.R., West, D.W., Baar, K., 2015. The molecular basis for load-induced skeletal muscle hypertrophy. Calcif. Tissue Int. 96 (3), 196–210.

Marmor, M.D., Yarden, Y., 2004. Role of protein ubiquitylation in regulating endocytosis of receptor tyrosine kinases. Oncogene 23 (11), 2057–2070.

Marzetti, E., et al., 2008. Modulation of age-induced apoptotic signaling and cellular remodeling by exercise and calorie restriction in skeletal muscle. Free Radic. Biol. Med. 44 (2), 160–168.

Masiero, E., et al., 2009. Autophagy is required to maintain muscle mass. Cell Metab. 10 (6), 507–515.

Mastrocola, R., et al., 2008. Muscle wasting in diabetic and in tumor-bearing rats: role of oxidative stress. Free Radic. Biol. Med. 44 (4), 584–593.

McClung, J.M., et al., 2010. p38 MAPK links oxidative stress to autophagy-related gene expression in cachectic muscle wasting. Am. J. Physiol. Cell Physiol. 298 (3), C542–C549.
McElhinny, A.S., et al., 2002. Muscle-specific RING finger-1 interacts with titin to regulate sarcomeric M-line and thick filament structure and may have nuclear functions via its interaction with glucocorticoid modulatory element binding protein-1. J. Cell Biol. 157 (1), 125–136.
Mendias, C.L., et al., 2012. Transforming growth factor-beta induces skeletal muscle atrophy and fibrosis through the induction of atrogin-1 and scleraxis. Muscle Nerve 45 (1), 55–59.
Mercken, E.M., et al., 2013. Calorie restriction in humans inhibits the PI3K/AKT pathway and induces a younger transcription profile. Aging Cell 12 (4), 645–651.
Milan, G., et al., 2015. Regulation of autophagy and the ubiquitin-proteasome system by the FoxO transcriptional network during muscle atrophy. Nat. Commun. 6, 6670.
Miller, B.F., et al., 2012. A comprehensive assessment of mitochondrial protein synthesis and cellular proliferation with age and caloric restriction. Aging Cell 11 (1), 150–161.
Mojtahedi, M.C., et al., 2011. The effects of a higher protein intake during energy restriction on changes in body composition and physical function in older women. J. Gerontol. A Biol. Sci. Med. Sci. 66 (11), 1218–1225.
Moore, D.R., et al., 2009a. Differential stimulation of myofibrillar and sarcoplasmic protein synthesis with protein ingestion at rest and after resistance exercise. J. Physiol. 587 (Pt 4), 897–904.
Moore, D.R., et al., 2009b. Ingested protein dose response of muscle and albumin protein synthesis after resistance exercise in young men. Am. J. Clin. Nutr. 89 (1), 161–168.
Mourier, A., et al., 1997. Combined effects of caloric restriction and branched-chain amino acid supplementation on body composition and exercise performance in elite wrestlers. Int. J. Sports Med. 18 (1), 47–55.
Narola, J., et al., 2013. Conditional expression of TGF-beta1 in skeletal muscles causes endomysial fibrosis and myofibers atrophy. PLoS ONE 8 (11), e79356.
Nascimbeni, A.C., et al., 2012. Impaired autophagy contributes to muscle atrophy in glycogen storage disease type II patients. Autophagy 8 (11), 1697–1700.
Newgard, C.B., 2012. Interplay between lipids and branched-chain amino acids in development of insulin resistance. Cell Metab. 15 (5), 606–614.
Ortenblad, N., Westerblad, H., Nielsen, J., 2013. Muscle glycogen stores and fatigue. J. Physiol. 591 (Pt 18), 4405–4413.
Pallafacchina, G., Blaauw, B., Schiaffino, S., 2013. Role of satellite cells in muscle growth and maintenance of muscle mass. Nutr. Metab. Cardiovasc. Dis. 23 (Suppl. 1), S12–S18.
Paoli, A., et al., 2015. Ketosis, ketogenic diet and food intake control: a complex relationship. Front. Psychol. 6, 27.
Phillips, S.M., 2014. A brief review of higher dietary protein diets in weight loss: a focus on athletes. Sports Med. 44 (Suppl. 2), S149–S153.
Phillips, S.M., et al., 1997. Mixed muscle protein synthesis and breakdown after resistance exercise in humans. Am. J. Physiol. 273 (1 Pt 1), E99–E107.
Phillips, T., Leeuwenburgh, C., 2005. Muscle fiber specific apoptosis and TNF-alpha signaling in sarcopenia are attenuated by life-long calorie restriction. FASEB J. 19 (6), 668–670.
Philp, A., Hamilton, D.L., Baar, K., 2011. Signals mediating skeletal muscle remodeling by resistance exercise: PI3-kinase independent activation of mTORC1. J. Appl. Physiol. (1985) 110 (2), 561–568.
Pimentel, G.D., et al., 2011. beta-Hydroxy-beta-methylbutyrate (HMbeta) supplementation stimulates skeletal muscle hypertrophy in rats via the mTOR pathway. Nutr. Metab. (Lond.) 8 (1), 11.
Powers, S.K., Kavazis, A.N., McClung, J.M., 2007. Oxidative stress and disuse muscle atrophy. J. Appl. Physiol. (1985) 102 (6), 2389–2397.
Powers, S.K., et al., 2011. Reactive oxygen species: impact on skeletal muscle. Compr. Physiol. 1 (2), 941–969.
Powers, S.K., Smuder, A.J., Judge, A.R., 2012. Oxidative stress and disuse muscle atrophy: cause or consequence? Curr. Opin. Clin. Nutr. Metab. Care 15 (3), 240–245.
Radak, Z., et al., 2013. Oxygen consumption and usage during physical exercise: the balance between oxidative stress and ROS-dependent adaptive signaling. Antioxid. Redox Signal. 18 (10), 1208–1246.
Rodgers, J.T., et al., 2005. Nutrient control of glucose homeostasis through a complex of PGC-1alpha and SIRT1. Nature 434 (7029), 113–118.
Rodriguez, J., et al., 2014. Myostatin and the skeletal muscle atrophy and hypertrophy signaling pathways. Cell. Mol. Life Sci. 71 (22), 4361–4371.
Ruas, J.L., et al., 2012. A PGC-1alpha isoform induced by resistance training regulates skeletal muscle hypertrophy. Cell 151 (6), 1319–1331.
Salazar, J.J., Michele, D.E., Brooks, S.V., 2010. Inhibition of calpain prevents muscle weakness and disruption of sarcomere structure during hindlimb suspension. J. Appl. Physiol. (1985) 108 (1), 120–127.
Salles, J., et al., 2013. 1,25(OH)2-vitamin D3 enhances the stimulating effect of leucine and insulin on protein synthesis rate through Akt/PKB and mTOR mediated pathways in murine C2C12 skeletal myotubes. Mol. Nutr. Food Res. 57 (12), 2137–2146.
Salo, D.C., Donovan, C.M., Davies, K.J., 1991. HSP70 and other possible heat shock or oxidative stress proteins are induced in skeletal muscle, heart, and liver during exercise. Free Radic. Biol. Med. 11 (3), 239–246.
Sandri, M., et al., 2004. Foxo transcription factors induce the atrophy-related ubiquitin ligase atrogin-1 and cause skeletal muscle atrophy. Cell 117 (3), 399–412.
Sandri, M., et al., 2006. PGC-1alpha protects skeletal muscle from atrophy by suppressing FoxO3 action and atrophy-specific gene transcription. Proc. Natl. Acad. Sci. U.S.A. 103 (44), 16260–16265.
Schakman, O., et al., 2012. Role of IGF-I and the TNFalpha/NF-kappaB pathway in the induction of muscle atrogenes by acute inflammation. Am. J. Physiol. Endocrinol. Metab. 303 (6), E729–E739.
Schiaffino, S., et al., 2013. Mechanisms regulating skeletal muscle growth and atrophy. FEBS J. 280 (17), 4294–4314.
Schmalbruch, H., al-Amood, W.S., Lewis, D.M., 1991. Morphology of long-term denervated rat soleus muscle and the effect of chronic electrical stimulation. J. Physiol. 441, 233–241.

Schroeder, E.T., Terk, M., Sattler, F.R., 2003. Androgen therapy improves muscle mass and strength but not muscle quality: results from two studies. Am. J. Physiol. Endocrinol. Metab. 285 (1), E16–E24.
Sharp, L.J., Haller, R.G., 2014. Metabolic and mitochondrial myopathies. Neurol. Clin. 32 (3), 777–799, ix.
She, P., et al., 2007. Obesity-related elevations in plasma leucine are associated with alterations in enzymes involved in branched-chain amino acid metabolism. Am. J. Physiol. Endocrinol. Metab. 293 (6), E1552–E1563.
Smith, K., et al., 1992. Flooding with L-[1-13C]leucine stimulates human muscle protein incorporation of continuously infused L-[1-13C]valine. Am. J. Physiol. 262 (3 Pt 1), E372–E376.
Spate, U., Schulze, P.C., 2004. Proinflammatory cytokines and skeletal muscle. Curr. Opin. Clin. Nutr. Metab. Care 7 (3), 265–269.
Stitt, T.N., et al., 2004. The IGF-1/PI3K/Akt pathway prevents expression of muscle atrophy-induced ubiquitin ligases by inhibiting FOXO transcription factors. Mol. Cell 14 (3), 395–403.
Tintignac, L.A., et al., 2005. Degradation of MyoD mediated by the SCF (MAFbx) ubiquitin ligase. J. Biol. Chem. 280 (4), 2847–2856.
Vandervoort, A.A., 2002. Aging of the human neuromuscular system. Muscle Nerve 25 (1), 17–25.
van Loon, L.J., 2014. Is there a need for protein ingestion during exercise? Sports Med. 44 (Suppl. 1), S105–S111.
Ventura-Clapier, R., Garnier, A., Veksler, V., 2008. Transcriptional control of mitochondrial biogenesis: the central role of PGC-1alpha. Cardiovasc. Res. 79 (2), 208–217.
Wall, B.T., van Loon, L.J., 2013. Nutritional strategies to attenuate muscle disuse atrophy. Nutr. Rev. 71 (4), 195–208.
Wang, H., et al., 2014. Apoptosis in capillary endothelial cells in ageing skeletal muscle. Aging Cell 13 (2), 254–262.
Watson, M.L., et al., 2012. A cell-autonomous role for the glucocorticoid receptor in skeletal muscle atrophy induced by systemic glucocorticoid exposure. Am. J. Physiol. Endocrinol. Metab. 302 (10), E1210–E1220.
Westerblad, H., Allen, D.G., 2003. Cellular mechanisms of skeletal muscle fatigue. Adv. Exp. Med. Biol. 538, 563–570 (discussion 571).
Westerblad, H., Bruton, J.D., Katz, A., 2010. Skeletal muscle: energy metabolism, fiber types, fatigue and adaptability. Exp. Cell Res. 316 (18), 3093–3099.
White, J.P., et al., 2014. G protein-coupled receptor 56 regulates mechanical overload-induced muscle hypertrophy. Proc. Natl. Acad. Sci. U.S. A. 111 (44), 15756–15761.
Wittwer, M., et al., 2002. Prolonged unloading of rat soleus muscle causes distinct adaptations of the gene profile. FASEB J. 16 (8), 884–886.
Wolfson, R.L., et al., 2015. Sestrin2 is a leucine sensor for the mTORC1 pathway. Science.
Xie, Z., Klionsky, D.J., 2007. Autophagosome formation: core machinery and adaptations. Nat. Cell Biol. 9 (10), 1102–1109.
Zhang, Y., et al., 2007. Increasing dietary leucine intake reduces diet-induced obesity and improves glucose and cholesterol metabolism in mice via multimechanisms. Diabetes 56 (6), 1647–1654.
Zhao, J., et al., 2007. FoxO3 coordinately activates protein degradation by the autophagic/lysosomal and proteasomal pathways in atrophying muscle cells. Cell Metab. 6 (6), 472–483.
Zhao, J., et al., 2008. Coordinate activation of autophagy and the proteasome pathway by FoxO transcription factor. Autophagy 4 (3), 378–380.

CHAPTER

23

Brain Amino Acid Sensing: The Use of a Rodent Model of Protein-Malnutrition, Lysine Deficiency

K. Torii[1] *and T. Tsurugizawa*[2]

[1]Torii Nutrient-Stasis Institute, Inc., Tokyo, Japan [2]Neurospin, Commissariat à l'Energie Atomique et aux Energies Alternatives, Gif-sur-Yvette, France

23.1. INTRODUCTION

It is well known that gustatory and anticipatory cephalic phases during a meal provide nutritional information and promote efficient digestion, absorption, and subsequent metabolic control for many nutrients (Torii et al., 2013). Foods having a familiar or pleasant taste could be swallowed without caution and they will be ingested until satiation (Torii et al., 1987). Each L-amino acid (herein called amino acid (AA)) has a different taste profile. Essential AA, with the exception of threonine (Thr, sweet), are generally bitter. In contrast, nonessential AAs are generally sweet or have an umami taste. Both nonessential AAs L-glutamate (Glu) and L-aspartate (Asp) make up a large component of dietary protein. Foods having an umami taste are generally pleasant and palatable much like sweeteners and they serve as a beneficial biomarker in food for animal and/or plant tissues containing dietary protein (Torii et al., 1987).

Animals, including humans, can detect information on both the amount and quality of dietary protein ingested during meal. They often use umami taste information, transmitted via cephalic relays, to initiate digestion processes. Subsequently, initial digestion produces free AAs and oligopeptides with Glu and Asp moieties in the stomach and throughout the alimentary tract. These provide visceral information via vagal afferent signals to the brain. This results in efficient metabolic processes that maintain AA homeostasis within normal limits (Torii et al., 2013).

We have shown previously that insufficient protein intake in growing rats leads to a preference for Thr and glycine (Gly) in drinking water in a choice paradigm. Both AAs are known to induce anorexia or to spare protein degradation when rodents are protein malnourished. Such rats also display a preference for the umami-tasting AAs, Glu and L-arginine (Arg). Intake of these AAs leads to a decrease in plasma ammonia arising from AA metabolism particularly during catch-up growth following recovery from protein-malnutrition. Animals consuming adequate protein exhibit a preference for umami-tasting substances relative to salt (sodium chloride, NaCl). But once they become protein malnourished, striking preferences for sodium chloride and Gly appear.

When rats develop a deficit of a particular essential amino acid, they are able to easily detect it, when is placed in the drinking water. When they have the option to intake a drinking solution that contains the lacked essential amino acids, they increase the amount of their intake of this solution to relieve the deficiency. Once they are no longer protein deficient, their preference for this AA is reduced to that of the normal level.

Like other basic tastes (sweet, sour, salty, bitter), umami taste information is received by receptors taste buds located in the oral cavity. What is unique for umami taste perception is that the intensity of Glu taste is enhanced synergistically by 5′-ribonucleotides, like inosinate and guanylate which are both DNA and RNA derivatives.

The Molecular Nutrition of Amino Acids and Proteins.
DOI: http://dx.doi.org/10.1016/B978-0-12-802167-5.00024-4

This unique synergism has been demonstrated by means of electrophysiology in rats and mice as well as biochemistry in the cow tongue (Torii and Cagan, 1980). We hypothesized that umami taste may represent a basic nutritional signal to identify and motivate intake of dietary protein. This would be similar to saltiness as a signal for electrolytes and sweetness as a signal for an energy source.

The current review is organized as follows: (1) we describe the central mechanism of AA homeostasis using the model of an essential AA, L-lysine (Lys) deficiency. We discuss how neuronal plasticity can guide recovery from the deficiency by motivating selective ingestion of Lys solution; (2) we discuss the physiological and nutritional significance of Glu signaling. Here we address both gustatory (oral—conscious) and visceral (via the gut–brain axis—unconscious) Glu signaling and sensing. We particularly focus on diet-induced thermogenesis and the prevention obesity when energy intake is high.

23.2. BRAIN ESSENTIAL AA SENSING: THE CASE OF THE RODENT MODEL OF LYSINE DEFICIENCY

We must ingest animal and plant origin dietary protein to fulfill our body's needs. Generally plants contain about 10% protein which is sufficient for life if the AAs in the protein are balanced as they generally are for animal protein. However, it is often the case that plants lack one or more essential AA, such as Lys, Thr, and L-tryptophan (Trp). Importantly, Lys deficiency in human is often observed in developing countries where people consume plants as their major dietary protein and energy sources. Lys deficiency causes children and young adults to be stunted and to suffer from higher infant mortality associated with immunological dysfunction. For such children Lys supplementation is quite beneficial, aiding them to catch-up in their growth. However, the economics of supplementation make this difficult (Torii et al., 1987). To better understand the underlying factors in Lys deficiency, we investigated Lys deficiency in an animal model, the rat.

Our goal was to identify the primary nucleus in the brain that recognizes ingested Lys in Lys deficiency. To accomplish this, we used a functional magnetic resonance imaging (fMRI) technique in a 4.7 tesla MRI system in Lys-deficient rats receiving Lys (Tsurugizawa et al., 2012b). Awake rats were trained to be habituated to MRI experiment in a specifically designed apparatus with a dedicated volume coil in the super conductance magnet bore. After acclimatization to the functional MRI procedure, rats were offered a Lys-deficient diet for six consecutive days for induction of Lys deficiency. These Lys-deficient rats received an intragastric infusion of Lys solution through a catheter (3 mmol/10 min/kg body weight in distilled water) during fMRI acquisition. Intragastric load was continued for 10 min. Then, BOLD signal increase at 5–10 min after the stop of the intragastric infusion was compared with the basal period (before the 5–0 min of intragastric load). Rats that were fed a Lys-normal diet served as controls. Brain functional responses to intragastric Lys infusion in the Lys-deficient group were clearly different from those in control group (Fig. 23.1). In both groups, intragastric infusion of Lys solution induced the responses in the hippocampus (HIP—important for memory function) and the hypothalamus, particularly the lateral hypothalamic area (LHA—a potent feeding center). The increase of fMRI signals occurred in a wider area of the brain in the Lys-deficient rats compared to the control group. In addition, Lys-rescued rats, who were fed Lys-normal diet for 6 days after fMRI scanning in Lys-deficient condition, elicited similar brain functional responses to intragastric Lys as control group.

Notably, the nucleus accumbens (NAc) responded to intragastric Lys solution only in Lys-deficient rats. This nucleus is connected with the ventral tegmental area, a source of the dopaminergic system, which is involved with addiction and craving behavior in response to certain nutrients like lipids and sugars and pharmacological agents like ethanol and nicotine. Lys solution (400 mM) is strongly bitter and generally animals refuse to ingest it. But Lys-deficient rats, which crave Lys, prefer its taste until they have consumed sufficient quantities to reduce the deficiency, indicating a kind of reward system without addiction (Fig. 23.2; Mori et al., 1992). In summary, intragastric infusion of Lys solutions caused neuronal excitation in the NAc that supports Lys preference and intake helping the animal to recover from its deficiency.

Neurons in the LHA, a feeding center, efficiently respond to the Lys concentration in blood as well as in the brain (Tabuchi et al., 1991). When Lys levels were normalized, rats exhibited a normal appetite for the Lys-deficient diet (Fig. 23.2; Mori et al., 1991, 1992). Therefore, neurons in the NAc modulate responses to nutrients, stimulating ingestion beyond the normal requirements in unpleasant foods containing the deficient nutrient, or stimulating overeating of foods providing energy sources that allow the animal to store energy as body fat in preparation for upcoming cold weather or seasonal migration.

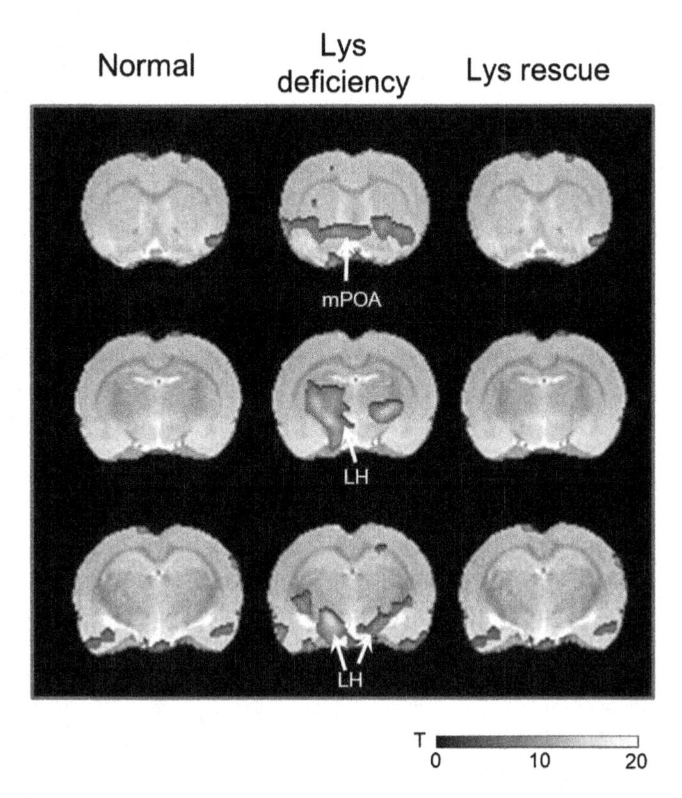

FIGURE 23.1 BOLD signal changes following an intragastric load of Lys (3 mmol/10 min/kg body weight in distilled water) in normal rats, Lys-deficient rats and a Lys-rescue group of rats following recovery from Lys deficiency. Intragastric load was continued for 10 min. The bar, *t*-values of regression analysis in comparison of BOLD signal increase (5–10 min after the stop of the intragastric infusion) with the basal period (before 5 min of intragastric load). *Modified with permission from Turugizawa, T., Uematsu, A., Uneyama, H., Torii K., 2012b. Reversible brain response to an intragastric load of L-Lysine under L-Lysine depletion in conscious rats. Br. J. Nutr. 24:1–7.*

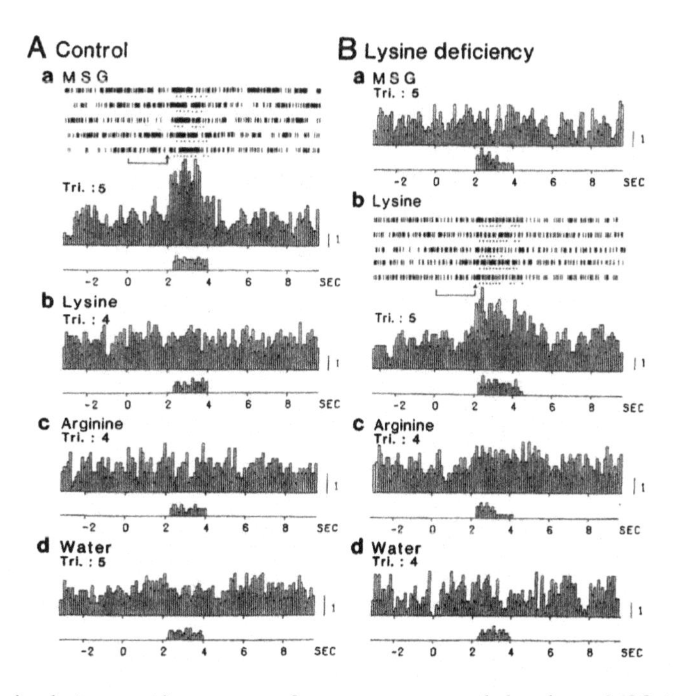

FIGURE 23.2 Two examples of solution-specific neurons. One neuron responded only to MSG in control conditions and another responded only to Lys when the diet was Lys-deficient. The neuron in (A) was excited during licking for MSG, but not for any other solution. In (B), the neuron was excited during licking for Lys, but not for MSG, arginine, or water. Horizontal brackets indicate the cue tone period. Arrow heads indicate presentation of the tube for licking (Mori et al., 1992).

In other studies, multibarreled glass electrodes were inserted into the LHA of awake rats that had been well-trained and surgically provided with a stainless steel-made clamp with a microguage. We recorded single neuronal extracellular responses to orally presented AA solutions in rats fed either a Lys-deficient or normal diet to identify the neurons responding preferentially to 150 mM orally presented monosodium Glu (MSG) and other AA solutions. The oral sensitivity to each AA was monitored when the Lys-deficient diet was offered to the rats for 1 week. The specific umami (MSG) taste-responding neurons changed to be responsive to oral 400 mM Lys solutions instead of MSG solutions. Thus, these neurons, responding to oral Glu signaling (umami taste), could change their sensitivity to promote detection of the deficient nutrient taste sensation (Fig. 23.2; Mori et al., 1992).

This neuronal plasticity was likely to be controlled by neurotrophic factors. To test this, we identified changes of activin A activity in the cerebrospinal fluid of rats with or without Lys deficiency using a *Hydra japonica* behavioral bioassay that is sensitive to neurotrophic factors in varied concentration as a behavior of the tentacle formation or not (Torii et al., 1993). AA sequence of activin A in all mammals is exactly the same but that in avians is slightly different. To test activin A activity, we prepared activin A using a gene engineering technique and antisera against activin A extracted from chicken yolk (IgY). Purification and concentration of activin A from egg yolk containing activin A antisera were performed using affinity chromatography (Murata et al., 1996). Also, either purified porcine inhibin, a competitive inhibitor of activin A receptor binding, or follistatin, which renders activin A-binding protein inert, thereby inhibiting its physiological activity, were also employed. The localization of activin A, its receptor, and inhibin were histologically identified in the brain. We found results that were remarkably similar to those obtained from our functional MRI observations previously described (Funaba et al., 1997; Fujimura et al., 1999).

Operant behaviors of rats with a Lys deficiency were also investigated. Each rat was surgically implanted with an intracranial catheter into the LHA bilaterally. They received a continuous microinjection of isotonic concentration of AA solutions including Lys, in combination with Activin A, inhibin, and follistatin, or anti-activin A sera. Rats were trained to receive a small pellet containing 1 mg Lys as a reward following 30 bar presses. Data from these trials suggests that Lys and activin A infusions regulated bar-pressing behavior, depending upon hunger and the degree of Lys deficiency (Fig. 23.3; Mori et al., 1992; Hawkins et al., 1995). Both supplementation of Lys into the LHA and suppression of activin A activity are effective to suppress the bar-pressing behavior (Mori et al., 1992; Hawkins et al., 1995). Lys-deficient diets with infusion of inhibin or anti-activin A antisera suppressed the behavioral changes. This indicates that activin A is a key molecule for sensing Lys deficiency centrally, as long as Lys in the blood and in the brain remains at deficient levels.

This kind of neuronal plasticity could occur in other examples of essential AA deficiency in rats. Once the rats recovered from a specific essential AA deficiency during growth, the memory of the deficiency and behavioral adaptation is retained for a considerable period of time. For example, Trp deficient rats specifically select and ingest Trp when they experienced Lys deficiency previously.

In our studies we have not discovered any cases of heightened Glu preference during recovery from an essential AA deficiency (Torii et al., 1987; Mori et al., 1992). When there is a nutritional deficiency in an essential AA such as Lys, there are no ways to detect protein bound Lys without the bitter taste sensation that is obtained in response to

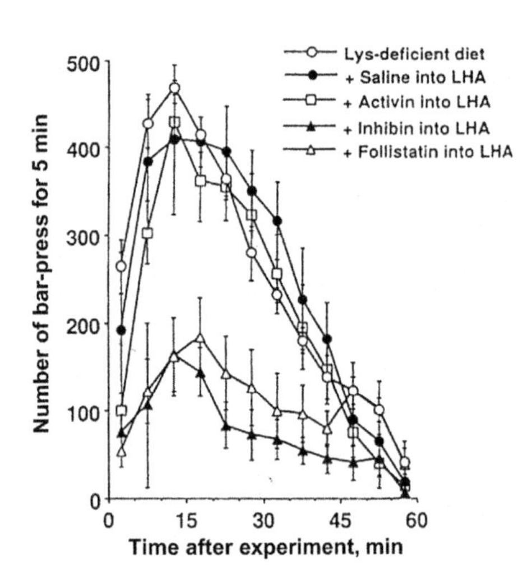

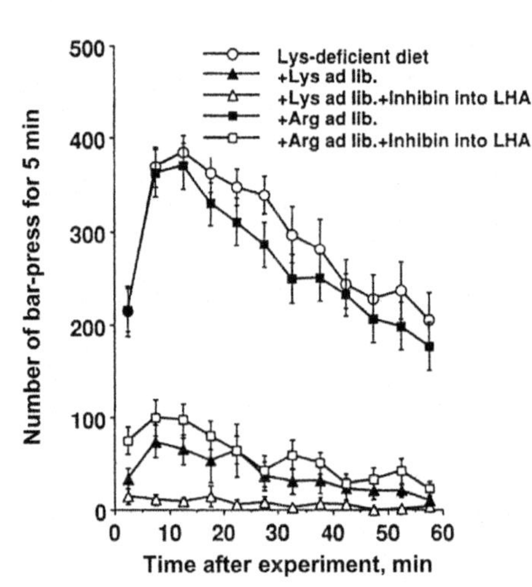

FIGURE 23.3 Bar-pressing behavior in rats receiving amino acids and an inhibitor of activin activity in the lateral hypothalamic area (LHA) during lysine (Lys) deficiency. Adult Wistar male rats were surgically implanted with a catheter into the LHA. After they recovered from this treatment, bar-pressing in order to obtain a 30 mg pellet containing 1 mg Lys was evaluated. Continuous microinjection of Lys alone suppressed bar-pressing rate. Both antiserum against activin A and a specific inhibitor of activin A binding were administered into the LHA along with Lys (Mori et al., 1992; Hawkins et al., 1995).

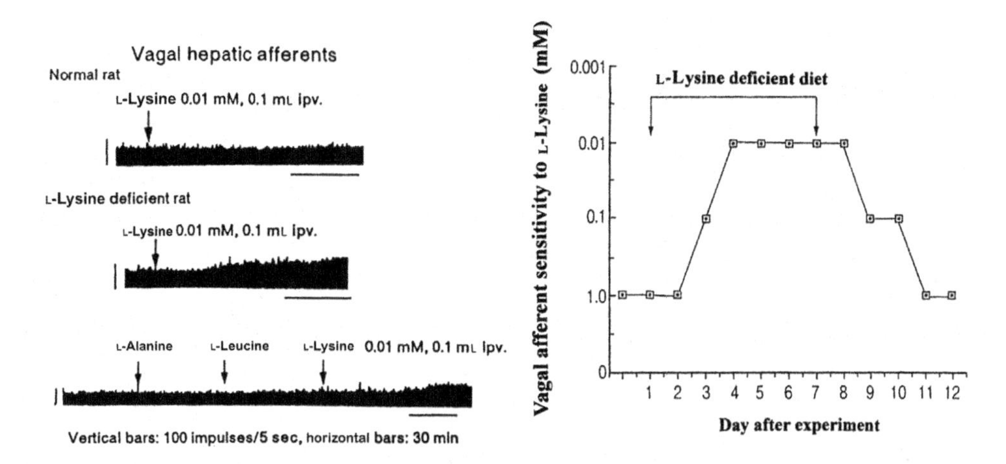

FIGURE 23.4 Vagal afferent sensitivity to Lys changes during Lys deficiency. Adult rats were fed a lysine (Lys)-deficient diet for 1 week. Their vagal afferent hepatic branch responses were recorded under anesthesia. The sensitivity to the L-form of Lys alone was enhanced during deficiency. The threshold for neural Lys sensitivity was reduced from 10 to 0.1 mM following 4 days on the Lys-deficient diet. This decrease in sensitivity disappeared gradually during the 4-day recovery period (Torii and Niijima, 2001).

the free AA (Ninomiya et al., 1994). After dietary protein is digested, the free form of Lys is available in the small intestine and in blood of the hepatic portal vein (Torii et al., 1987; Mori et al., 1992).

We recorded electrical activity from the hepatic branch of vagal afferent in young adult rats after they had been fed a Lys-deficient diet for 1 week. The goal was to determine sensitivity to a range of AAs. For each rat the vagal afferent fiber was exposed and a platinum electrode was used to record neural responses of afferent fibers to injection of various AAs intraperitoneally. Non-Lys-deficient rat vagal afferent fibers never responded to Lys stimulation. In contrast, responses were evident in rats with Lys deficiency. Over time, the threshold concentration for activation by Lys solution decreased from 10 mM to 0.1 mM (100-fold), suggesting that neural plasticity in the vagal afferent occurred under Lys deficiency (Torii and Niijima, 2001). These Lys-deficient rats exhibited essentially the same sensitivity to other AAs including D-lysine in comparison with Lys-sufficient rats (Fig. 23.4).

These data suggest that when an AA deficiency is detected in the hepatic portal-brain circulation, foods containing Lys are detected by characteristics such as color, shape, smell, taste, and texture, and animals with successful memory are motivated to consume the diet with Lys. The sensory characteristics of these diets are remembered and thus a future deficiency is rapidly alleviated. Thus, deficient animals can access previous memories to select the food that contains the required nutrient. Overall, nutrient information transmitted via the vagal nerve from the gastrointestinal tract to the brain areas including the nucleus tractus solitarius, the hypothalamus—particularly the LHA in the hypothalamus—could be involved in nutrient selection and recognition. This is what we term the gut–brain axis (Torii et al., 1987; Mori et al., 1992; Tsurugizawa et al., 2009).

In summary, once an animal is exposed to a deficiency of an essential AA like Lys, neural plasticity plays the very important role of allowing the animal to select food containing the deficient AA. This mechanism operates through the gut–brain axis and both central and peripheral plasticity are essential for the smooth function of AA homeostasis. These phenomena do not merely happen in the case of essential AA deficiency. This adaptation mechanism to any essential nutrient deficiency should operate to survive in the living body.

23.3. BRAIN FUNCTIONAL CHANGES ELICITED BY INTRAGASTRIC STIMULATION BY NUTRIENTS, GLUCOSE, GLUTAMATE, AND SODIUM CHLORIDE

Which regions in the forebrain are involved in the process of the visceral nutrient sensing pathway? To answer this question, functional MRI studies to intragastric infusion of nutrient were conducted using awake or slightly anesthetized experimental animals, mainly Wistar strain male young adult rats. Rats were habituated to the circumstance in the MRI bore, as previously described. Awake functional MRI is required to investigate the brain activation related to palatable response because anesthetic agents disturb the hemodynamic response to the neuronal activation, which is essential to the blood oxygenation level dependent response in fMRI (Tsurugizawa

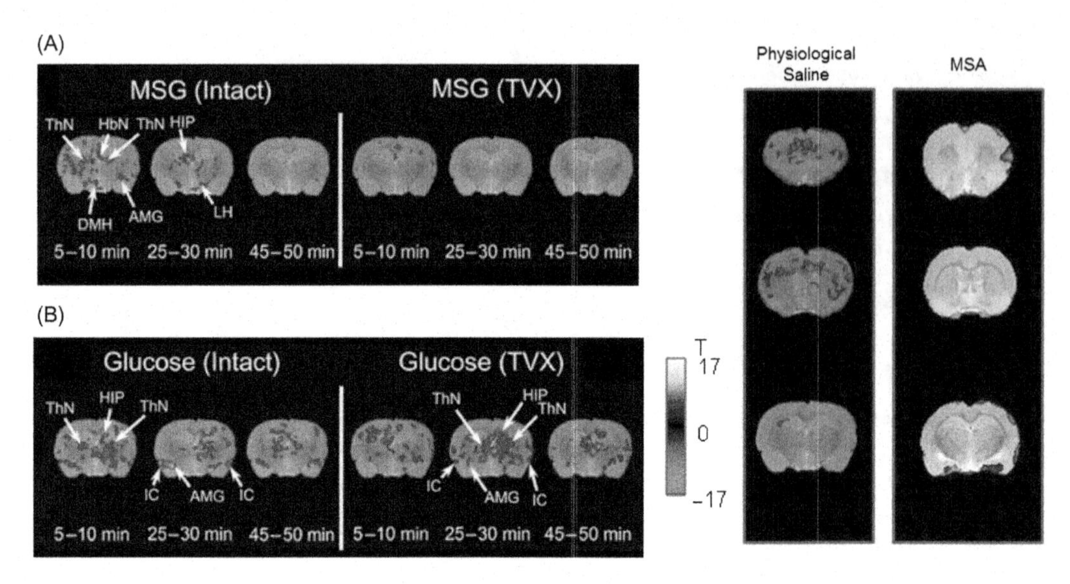

FIGURE 23.5 Left panel, BOLD signal changes induced by an intragastric load of 60 mM monosodium Glu (MSG) and glucose in intact rats and totally vagotomized (TVX) rats. Right panel, BOLD signal changes by intragastric load of 60 mM NaCl (physiological saline) and monosodium aspartate (MSA) during the infusion. The bar, *t*-values of BOLD signal increase compared with the basal period before the 5 min intragastric load. BOLD responses to intragastric load of MSG and glucose are from Tsurugizawa et al. (2009). *Data of MSA and NaCl are from Tsurugizawa, T., Uneyama, H., Torii, K., 2014. Brain amino acid sensing. Diabetes Obes. Metab. 16(Suppl. 1): 41–48.*

et al., 2008, 2013). Furthermore, as the rats are conscious during experiments, it is possible to observe the brain activation during learning (Uematsu et al., 2015) or motivation (Tsurugizawa et al., 2012a) during/after eating.

Experimental animals were given intragastric administrations of various solutions at 60 mM. The nutrients of interest included glucose, MSG, monosodium Asp (components of protein and similar in character to Glu), and NaCl (a major electrolyte and a sodium control for administration of MSG). Administration of these nutrients was performed through a catheter that was surgically installed from the back into the stomach one week before the acclimation training. The brain responses to both glucose and MSG were evident in the HIP and the amygdala, structures involved in learning and memory and in feeding motivation respectively. Only gut stimulation of glucose increased the activity in the NAc; this is recognized as a brain center for reward and addiction of caloric sweetener. Remarkably, there were no responses to MSG or NaCl in the Nac. In contrast, MSG induced responses in the dorsomedial hypothalamus as well as the medial preoptic area (mPOA), which are brain areas involved in primitive metabolic regulation and body temperature control respectively, while NaCl did not activate these areas. These results from the functional MRI observations suggest that diet-induced thermogenesis might be triggered by Glu signaling in the gut. There was no particular response in the brain following gastric infusion of either NaCl or Asp (Fig. 23.5; Tsurugizawa et al., 2013, 2014). Next, effects of subdiaphragmatic total vagotomy treatment on the brain functional responses to glucose, MSG, and NaCl were compared with the intact group. Brain functional changes in total vagotomized rats were almost absent when 60 mM MSG was infused intragastrically. But in the case of treatment with 60 mM glucose, there were essentially the same brain functional changes between the groups with and without total vagotomy (Fig. 23.5; Tsurugizawa et al., 2009). These brain functional changes observed using the functional MRI were confirmed by studies evaluating c-fos expression as an indicator of neural activation (Otsubo et al., 2011).

In addition, we studied responsiveness of gastric branch vagal afferent fiber to AAs, sugars, and NaCl in anesthetized Wistar strain male adult rats. Remarkably, free Glu (MSG) alone among AAs elicited neural responses in the vagal gastric afferents. This gastric vagal response to MSG was abolished by pretreatment intravenously with a nonselective inhibitor, (N^{ω}-nitro-L-arganic methyl ester; L-NAME) or nitrate monoxide (NO) synthase and 95% of the response to Glu was similarly inhibited by the monoamine deplete, parachlorol L-phenylalanine (Uneyama et al., 2006, 2009). These phenomena suggested that there might be Glu specific receptors, possibly metabotropic Glu receptor type I (mGluR1), localized in the luminal layer of the stomach, particularly in the mucus cells. These cells express NO synthase and thereby release NO on the basal side. This released NO stimulates the release of serotonin (5-HT) from the enterochromaffin (EC) cells also localized in the luminal layer. EC cells also express another umami taste receptor, the heterodimer (taste receptor type 1, member 1/taste receptor type 1, member 3 TIRI/TIR3).

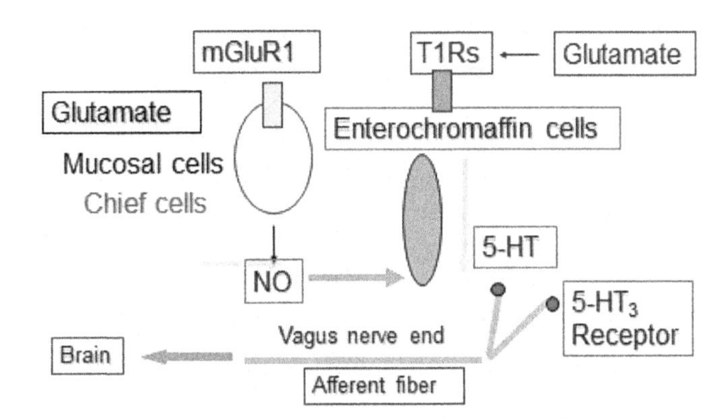

FIGURE 23.6 Mechanism of Glu signaling perception in the stomach. mGluR1 is found in the apical membrane of mucosal and chief cells. The transduction mechanism involves NO and 5-HT. Glu alone can be perceived in the stomach via vagal afferents (Nakamura et al., 2013).

Therefore, we proposed that ingested food contains free Glu that binds to mGluR1 in the luminal layer of the stomach to release NO from the basal side of mucus cells and then NO stimulates the EC cells to release 5-HT. Finally, released 5-HT interacts with the 5-HT_3 receptor at the nerve end of the vagus nerve. This Glu signaling in the stomach provides the brain with information on food intake, particularly dietary protein intake (Fig. 23.6; Nakamura et al., 2013).

It is well known that dietary protein intake elicits the diet-induced thermogenesis. These findings strongly suggest that this diet-induced thermogenesis may be triggered by Glu in the stomach via the gut–brain axis. The brain functional changes previously mentioned in describing work using functional MRI support this conclusion.

23.4. GLUTAMATE SIGNALING IN THE GUT TRIGGERS DIET-INDUCED THERMOGENESIS AND AIDS IN THE PREVENTION OF OBESITY

The ingested Glu signaling orally induces excitation of vagal efferents in the gastric, celiac, and pancreatic branches in overnight fasting rat. It also activates the sympathetic nerve innervating adipose tissue to sustain lipolysis during fasting periods and maintain body temperature. But the vagal afferents of the hepatic branch never respond to oral Glu signaling because there is no incorporation of nutrients into the liver from the small intestine (Torii et al., 2013).

Vagal efferent information evoked by oral Glu signaling can provide anticipatory activation of the digestive system, for example by stimulating digestive juice secretion and production of gastric and pancreatic juices. But there is no uptake of nutrients into the portal vein during oral stimulation of Glu signaling; both the vagal efferent hepatic branch and the sympathetic nerve innervating the adipose tissue respond as if the animals are still in the fasting state (Tsurugizawa et al., 2012b).

In addition, we investigated weaned rats fed high-fat high-sugar diets ad libitum during growth. These animals were allowed to ingest 1% (w/v) MSG solution and water in a choice paradigm or water alone as the control. Both groups of rats consumed the same amount of this high-fat high-sugar experimental diet, but their body shapes, body weights, and subcutaneous and abdominal fat deposits were substantially different. Animals permitted to ingest 1% MSG solution did not become obese. In contrast, the control group that had only water to consume developed a typical obesity phenotype (Fig. 23.7; Kondoh and Torii, 2008).

Dietary free Glu is almost completely catabolized into CO_2 in digestion and absorption processes in the alimentary tracts when MSG is added to foods in concentrations in the preferable range (0.5–0.6%, w/v). This suggests that there is a mechanism that prevents abundant free Glu as well as Asp, both from proteins and free form in foods, and from direct absorption and incorporation into the body fluid that maintains strict Glu homeostasis to be low. Given that both Glu and Asp are neurotransmitters in the peripheral tissues as well as in the brain, Glu concentration in the blood should be maintained. Thus, ingested Glu should be only effective in the luminal side of the alimentary tract and Glu information is conveyed to the brain via the vagus nerves, not in the blood circulation (Uneyama et al., 2006, 2009). Umami taste materials such as free dietary Glu and partially digested glutamyl oligopeptides as ligands to Glu receptors remain in the stomach until gastric emptying is completed. It is likely that Glu signaling in the stomach and subsequently in the small intestine and probably large

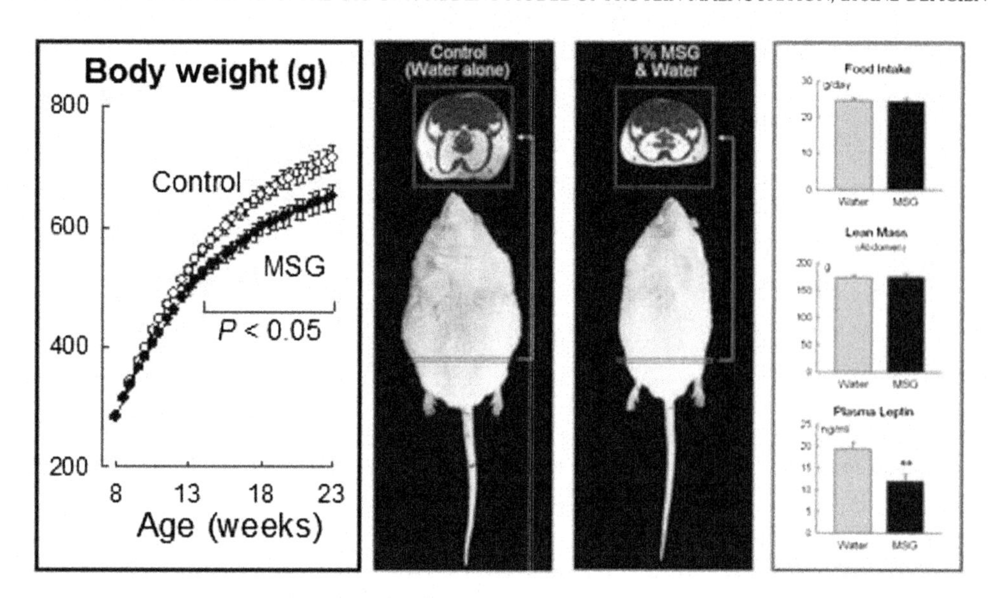

FIGURE 23.7 Spontaneous drinking of monosodium Glu (MSG) solution suppresses the development of obesity in a rat model of diet-induced obesity. Weaned Sprague-Dawley strain male rats, fed a high-sugar high-fat diet, were given free access to a 1% (w/v) MSG solution and water in a choice paradigm. Food intake in both the MSG and control groups was comparable. Body weight and fat deposition in the MSG group were significantly lower than that in the control group. Plasma leptin levels differed in the same direction (Kondoh and Torii, 2008).

intestine is sensed by vagal afferents to efficiently control digestive processes. In contrast to Glu catabolization of Glu into CO_2 at more than 90% efficiency after digestion and absorption, dietary glucose is absorbed from the small intestine without being catabolized and thus glucose reaches the circulating blood without being altered and ultimately can reach the brain directly (Tsurugizawa et al., 2013). Therefore, the brain is able to directly recognize the absolute amount of glucose uptake after a meal.

In contrast to the benefits of added Glu to the diet, people who consume high-sugar and high-fat foods may become obese because of their relative shortage of energy expenditure. In the diet-induced obese rat model, using male Sprague-Dawley rats fed high-sugar and high-fat diets chronically, these animals typically became obese. However, when similar rats were provided access to 1% (w/v) MSG solution and drinking water in a choice paradigm, they consumed considerable MSG and failed to develop typical obesity. Their body weight and subcutaneous and abdominal fat deposits were significantly lower than that in controls which were provided with drinking water alone. Also, plasma leptin levels in controls were significantly reduced. This resembled the prevention from leptin resistance induced by consumption of MSG solutions. Together, these data suggest that Glu signaling induces thermogenesis via gut–brain axis and ultimately this additional energy expenditure prevents body fat deposition and hence obesity (Fig. 23.7; Kondoh and Torii, 2008).

It is a well-known effect that totally vagotomized rats do not grow normally because of both anorexia and maldigestion in the alimentary tract. Consequently, we hypothesized that partially vagotomized human patients (following the removal of stomach tissue with cancer tumors), as well as some elderly individuals with nerve response retardation, would benefit from consuming added MSG in food and/or soup in a preferable concentration. To test this hypothesis, intervention clinical studies were performed using rice gruel supplemented with 0.5% MSG (w/v) for 3 months in a group of elderly people ($N = 90$, average age approximately 90 years old) in an attempt to improve their quality of life (Akiba and Kannitz, 2009; Yamamoto et al., 2009). Positive effects were observed as follows: enhanced appetite; improved eating behavior; increased body weight and albumin level in the blood; and skin temperature elevation with improvement of the peripheral circulation. Glu signaling during and after a meal thus provides very important physiological stimulation in the whole body, triggering diet-induced thermogenesis and thereby maintaining optimal protein nutritional status. The brain functional changes after Glu signaling intragastrically described above support these findings from the human intervention studies.

Recently, the incidence of juvenile obesity has increased in parallel with the use of infant milk formula. These formulas are often chosen because of their convenience for young parents. However there is a lower frequency of juvenile obesity in breast-fed infants compared to bottle-fed infants. Perhaps significantly, there is also substantially freer Glu in the breast milk than in infant milk formula. We suggest that this indicates that Glu signaling in breast milk is beneficial, preventing juvenile obesity development. Recently it was reported that infants fed regular milk formulas with low Glu levels grew faster than infants fed formulas that were made up of hydrolyzed casein and therefore had high levels of Glu (Yamamoto et al., 2009). The infants fed the hydrolyzed casein formula grew at the same rate as breast-fed infants whereas those fed the cow milk formulas grew at a faster rate. In a subsequent short-term study it was shown that when infants were given a test with regular milk formula with or without 0.2% (w/v) MSG added, they consumed less of the MSG-fortified formula to reach satiety (Mennella et al., 2011; Ventura et al., 2012). These human experimental data are consistent with our hypothesis that Glu signaling in early development is effective and beneficial for maintenance of normal growth and helps prevent obesity. Epidemiological surveys of obesity incidence in industrialized countries that have similar food cultures with relatively high levels of consumption of umami taste substances, specifically Japan and Korea, have among the lowest incidences of obesity (less than 3% of the population) and longevity.

Recently MSG consumption globally exceeded 3 million tons per year and was spread across many cultures around the world. These data reflect the findings that umami taste is a basic taste that enhances the palatability of food and diet-induced thermogenesis but does not induce overeating. Consequently, the addition of umami taste materials to foods at appropriate levels helps prevent the development of obesity even in the face of high energy intake by means of Glu-induced thermogenesis during and after meals.

23.5. CONCLUSION

The present review describes the mechanisms of gut–brain axis to regulate the food eating behavior depending on the condition of the nutrition, using the examples of the animal deficiency model of the essential AA, Lys. Glu, which is a nonessential AA and elicits umami taste, acts as representative signals for the protein contained in the foods. The umami-taste signaling pathway in the gastrointestinal tract, as well as the oral cavity, are important for smooth digestion and postingestive regulation of feeding behavior.

Acknowledgments

This research was primarily performed in Japan Science and Technology Agency ERATO Torii nutrient state project (1990–96) and the Institute for Innovation, Ajinomoto Co. Inc., Kawasaki, Japan.

The authors appreciate both the research assistance of their colleagues over many years and the constructive suggestions and collaborations of Monell Chemical Senses Center, Philadelphia, PA, USA, particularly the former director and current Distinguished Member, Dr G.K. Beauchamp.

References

Akiba, Y., Kannitz, J.D., 2009. Luminal chemosensing and upper gastrointestinal mucosal defenses. Am. J. Clin. Nutr. 90 (3), 826S–831S.

Fujimura, H., Ohsawa, K., Funaba, M., Murata, T., et al., 1999. Immunological localization and ontogenetic development inhibin α submit in rat brain. J. Neuroendocrinol. 11 (3), 157–163.

Funaba, M., Murata, T., Fujimura, H., Murata, E., et al., 1997. Immunolocalization of type I or type II activin receptors in the rat brain. J. Neuroendocrinol. 9 (2), 105–111.

Hawkins, R.L., Inoue, M., Mori, M., Torii, K., 1995. Effect of inhibin, follistatin, or activin infusion into the lateralhypothalamus on operant behavior of rats fed lysine deficient diet. Brain Res. 704 (1), 1–9.

Kondoh, T., Torii, K., 2008. MSG intake suppresses weight gain, fat deposition, and plasma leptin levels in male Sprague-Dawley rats. Physiol. Behav. 95, 135–144.

Mennella, J.A., Ventura, A.K., Beauchamp, G.K., 2011. Differential growth patterns among healthy infants fed protein hydrolysate or cow-milk formulas. Pediatrics 127 (1), 110–118.

Mori, M., Kawada, T., Ono, T., Torii, K., 1991. Taste preference and protein nutrition and L-amino acid homeostasis in male Sprague–Dawley rats. Physiol. Behav. 49 (5), 987–995.

Mori, M., Tabuchi, E., Ono, T., Torii, K., 1992. Sensitivity changes to L-amino acid by central and ingestive application in the lateral hypothalamus of rats under L-lysine deficiency. In: Takai, K. (Ed.), Frontiers and New Horizons in Amino Acid Research. Elsevier, Amsterdam, pp. 283–291.

Murata, T., Saito, S., Shiozaki, M., Lu, R.Z., et al., 1996. Anti-activin A antibody(IgY) specifically neutralizes various activin A activities. Proc. Soc. Exp. Biol. Med. 211 (1), 100–107.

Nakamura, H., Kawamata, Y., Kuwahara, T., Torii, K., et al., 2013. Nitrogen in dietary glutamate is utilized exclusively for the synthesis of amino acids in the rat intestine. Am. J. Physiol. Endocrinol. Metab. 304, 100–108.

Ninomiya, Y., Kajiura, H., Naito, Y., Mochizuki, K., et al., 1994. Glossopharyngeal denervation alters responses to nutrients and toxic substances. Physiol. Behav. 56 (6), 1179–1184.

Otsubo, H., Kondoh, T., Shibata, M., Torii, K., et al., 2011. Introduction of Fos expression in the rat forebrain after intragastric administration of monosodium L-Glutamate, glucose and NaCl. Neuroscience 196, 97–103.

Tabuchi, E., Ono, T., Nishijo, H., Torii, K., 1991. Amino acid and NaCl appetite, and LHA neuron responses of lysine-deficient rat. Physiol. Behav. 49 (5), 951–964.

Torii, K., Cagan, R.H., 1980. Biochemical studies of taste sensation. IX. Enhancement of 1-[^{3}H]glutamate binding to bovine taste papillae by 5′rebonucleotides. Biochim. Biophys. Acta 627 (3), 313–323.

Torii, K., Niijima, A., 2001. Effect of lysine on afferent activity of the hepatic branch of the vagus nerve in normal and L-lysine deficient rats. Physiol. Behav. 72 (5).

Torii, K., Mimura, T., Yugari, Y., 1987. Biochemical mechanism of umami taste perception and effect of dietary protein on the taste preference for amino acids and sodium chloride in rats. In: Kawamura, Y., Kare, M.R. (Eds.), Umami; A Basic Taste. Marcel Dekker Inc., New York, NY, pp. 513–563.

Torii, K., Hanai, K., Oosawa, K., Funaba, M., et al., 1993. Activin A: serum levels and immunohistochemical brain localization in rats give diets deficient in l-lysine or protein. Physiol. Behav. 54 (3), 459–466.

Torii, K., Uneyama, H., Nakamura, E., 2013. Physiological roles of dietary glutamate signaling via gut-brain axis due to efficient digestion and absorption. J. Gastroenterol. 48, 442–451.

Tsurugizawa, T., Kondoh, T., Torii, K., 2008. Forebrain activation induced by postoral nutritive substances in rats. Neuroreport 19 (11), 1111–1115.

Tsurugizawa, T., Uematsu, A., Nakamura, E., Uneyama, H., et al., 2009. Mechanisms of neural response to gastrointestinal nutritive stimuli: the gut-brain axis. Gastroenterology 137 (1), 262–273.

Tsurugizawa, T., Uematsu, A., Uneyama, H., Torii, K., 2012a. Functional brain mapping of conscious rats during reward anticipation. J. Neurosci. Methods 206 (2), 132–137.

Tsurugizawa, T., Uematsu, A., Uneyama, H., Torii, K., 2012b. Reversible brain response to an intragastric load of L-lysine under L-lysine depletion in conscious rats. Br. J. Nutr. 24, 1–7.

Tsurugizawa, T., Ciobanu, L., Le Bihan, D., 2013. Water diffusion in brain cortex closely tracks underlying neuronal activity. Proc. Natl. Acad. Sci. U.S.A. 10 (28), 11636–11641.

Tsurugizawa, T., Uneyama, H., Torii, K., 2014. Brain amino acid sensing. Diabates Obes. Metab. 16 (Suppl. 1), 41–48.

Uematsu, A., Kitamura, A., Iwatsuki, K., Uneyama, H., et al., 2015. Correlation between activation of prelimbic cortex, the basolateral amygdala, and agranular insular cortex during taste memory formation. Cereb. Cortex 25 (9), 2719–2728.

Uneyama, H., Niijima, A., San Gabriel, Torii, K., et al., 2006. Luminal amino acid sensing in the rat gastric mucosa. Am. J. Physiol. Gastrointest. Liver Physiol. 291, G1163–G1170.

Uneyama, H., Niijima, A., Kitamura, A., Torii, K., 2009. Existence of NO-triggered vagal afferent activation in the rat gastric mucosa. Life Sci. 85, 782–787.

Ventura, A.K., Beauchamp, G.K., Mennella, J.A., 2012. Infant regulation of intake: the effect of free glutamate content in infant formulas. Am. J. Clin. Nutr. 95 (4), 875–881.

Yamamoto, S., Tomoe, M., Toyama, K., Kawai, M., et al., 2009. Can dietary supplementation of monosodium glutamate improve the health of the elderly? Am. J. Clin. Nutr. 90 (3), 844S–849S.

Index

Note: Page numbers followed by "*f*" and "*t*" refer to figures and tables, respectively.

B

D

E

F

G

H

R

S

Printed in the United States
By Bookmasters